FRAME RELAY NETWORKING

Covers all aspects of Frame Relay Networking, from the evolution and rationale to architecture and equipment.

1999 0 471 98578 3

INTERNETWORKING LANs AND WANs
Concepts, Techniques and Methods
Second Edition

Internetworking is one of the fastest growing markets in the field of computer communications. However, the interconnection of LANs and WANs tends to cause significant technological and administrative difficulties. This updated version provides valuable guidance, enabling the reader to avoid the pitfalls and achieve successful connection.

1998 0 471 97514 1

THE MULTIPLEXER REFERENCE MANUAL

Designed to provide the reader with a detailed insight into the operation, utilization and networking of six distinct types of multiplexers, this book will appeal to practising electrical, electronic and communications engineers, students in electronics, network analysts and designers.

1993 0 471 93484 4

PRACTICAL NETWORK DESIGN TECHNIQUES

Many network design problems are addressed and solved in this informative volume. Gil Held confronts a range of issues including through-put problems, line facilities, economic trade-offs and multiplexers. Readers are also shown how to determine the numbers of ports, dial-in lines and channels to install on communications equipment in order to provide a defined level of service.

1991 0 471 92938 7 (Set)

Please refer to the back of the book for further details

DATA COMMUNICATIONS NETWORKING DEVICES: OPERATION, UTILIZATION AND LAN AND WAN INTERNETWORKING

Fourth Edition

DATA COMMUNICATIONS NETWORKING DEVICES: OPERATION, UTILIZATION AND LAN AND WAN INTERNETWORKING

Fourth Edition

Gilbert Held
4-Degree Consulting
Macon, Georgia
USA

JOHN WILEY & SONS
Chichester · New York · Weinheim · Brisbane · Singapore · Toronto

Baffins Lane, Chichester,
West Sussex PO19 1UD, England

National 01243 779777
International (+44) 1243 779777
e-mail (for orders and customer service enquiries): cs-books@wiley.co.uk
Visit our Home Page on http://www.wiley.co.uk or http://www.wiley.com

Other Wiley Editorial Offices

John Wiley & Sons, Inc., 605 Third Avenue,
New York, NY 10158-0012, USA

WILEY-VCH Verlag GmbH, Pappelallee 3,
D-69469 Weinheim, Germany

Jacaranda Wiley Ltd, 33 Park Road, Milton,
Queensland 4064, Australia

John Wiley & Sons (Canada) Ltd, 22 Worcester Road,
Rexdale, Ontario M9W 1L1, Canada

John Wiley & Sons (Asia) Pte Ltd, Clementi Loop #02-01,
Jin Xing Distripark, Singapore 129809

Libaray of Congress Cataloging-in-Publication Data

Held, Gilbert, 1943–
Data communications networking devices : operation, utilization, and LAN and WAN internetworking / Gilbert Held. — 4th ed.
p. cm.
Includes index.
ISBN 0-471-97515-X (alk. paper)
1. Computer networks. 2. Computer networks—Equipment and supplies. 3. Data transmission systems. I. Title.
TK5105.5.H44 1998
004.6—dc21 98-27200
CIP

British Library Cataloguing in Publication Data

A catalogue record for this book is available from the British Library

ISBN 0 471 97515-X

Typeset in 10/12pt Imprint by Thomson Press (India) Ltd, New Delhi, India
Printed and bound in Great Britain by Bookcraft (Bath) Ltd
This book is printed on acid-free paper responsibly manufactured from sustainable forestry, in which at least two trees are planted for each one used for paper production.

To Beverly, Jonathan and Jessica
for their patience understanding and support—
I love you all

To Dr Alexander Ioffe and family of Moscow—
congratulations on next year in Jerusalem being each year!

CONTENTS

PREFACE

Over fifteen years ago I introduced the first edition of this book with the statement 'data communications networking devices are the building blocks upon which networks are constructed.' Although networking technology has made significant advances, that statement retains its validity. Today you can use devices such as bridges and routers that were non-existent in the late 1970s to link local and wide area networks together, while boosting LAN productivity and access through the use of switches and remote access servers that represent products of the 1990s. Thus, the basic rationale and goal of this fourth edition, which is to provide readers with an intimate awareness of the operation and utilization of important networking products that can be used in the design, modification, or optimization of a data communications network, has not changed from the rationale and goal of the first edition. What has changed is the scope and depth of the material included in this book.

In developing this new edition I have taken into consideration and acted upon comments received from both individuals and professors who used the book for a college course on networking. Major changes include an expansion and subdivision of the Fundamental Concepts chapter, which now covers both WANs and LANs in a series of separate chapters focused upon fundamental concepts and advanced networking topics. Other significant changes in this new edition include a chapter covering Wide Area Networks as a separate entity and another covering LAN internetworking devices. In addition, a significant amount of material was revised and updated to provide detailed information covering the operation and utilization of additional networking devices and the updating of information concerning the operating characteristics of other devices. To facilitate the use of this book as a text as well as for reader review purposes, the questions at the end of each chapter reference the sections in each chapter. Through the use of a numbering scheme, students can easily reference an appropriate section in the book for assistance in answering a question while instructors can easily reference the assignment of questions to reading assignments based upon specific sections within chapters.

The expansion of the Fundamental Concepts chapter followed by the addition of two new chapters covering wide area networks and local area networks provides readers new to the field of data communications with the ability to use these chapters as a detailed introduction to this field. For more experienced readers the information in these chapters can be used as a reference to the many facets of data communications.

The new chapter covering wide area networks first explains the different types of networks and then examines network architecture and the flow of data in several popular networks. Similarly, the new chapter on LANs provides a solid foundation concerning the topology, access methods, and operation of several types of popular local area networks, laying the groundwork for detailed information concerning the operation of WAN and LAN internetworking devices presented in later chapters in this book.

Similar to prior editions of this book, this edition was structured for a two semester course at a high level undergraduate or first-year graduate course level. In addition, this book can be used as a comprehensive reference to the operation and utilization of different networking devices and as a self-study guide for individuals who wish to pace themselves at their leisure.

As I once again rewrote this book, I again focused attention upon explaining communications concepts which required an expansion of an already comprehensive introductory chapter into a series of three chapters in order to cover the fundamental concepts common to all phases of data communications. All three chapters should be read first by those new to this field and can be used as a review mechanism for readers with a background in communications concepts. Thereafter, each chapter is written to cover a group of devices based upon a common function.

Through the use of numerous illustrations and schematic diagrams, I believe readers will easily be able to see how different devices can be integrated into networks, and some examples should stimulate new ideas for even the most experienced person. At the end of each chapter I have included a comprehensive series of questions that cover many of the important concepts covered in the chapter. These questions can be used by the reader as a review mechanism prior to going forward in the book.

For those readers actually involved in the sizing of network devices I have include several appendices in this book that cover this area. Since the mathematics involved in the sizing process can result in a considerable effort to obtain the required data, I have enclosed computer program listings that readers can use to generate a series of sizing tables. Then after reading the appendices and executing the computer programs, you can reduce many sizing problems to a table lookup procedure. As always I look forward to receiving reader comments, either though my publisher whose address is on the back cover of this book or via email to 235-8068@meimail.com

Gilbert Held
Macon, Georgia

ACKNOWLEDGEMENTS

The preparation of a manuscript that gives birth to a book requires the cooperation and assistance of many persons.

First and foremost, I must thank my family for enduring those long nights and missing weekends while I drafted and redrafted the manuscript to correspond to each of the editions of this book. The preparation of the first edition was truly a family affair, since both my wife and my son typed significant portions of the manuscript on our mobile Macintosh, with both my family and the Macintosh traveling a considerable distance during the preparation of the manuscript. For the preparation of the second and third editions of this book I am grateful for the efforts of Mrs Carol Ferrell who turned my handwritten inserts and drawings into a legible manuscript. As a frequent traveler I write the old-fashioned way—with pen and paper—to avoid battery drain and electrical outlet incompatibilities affecting my productivity. However, it still requires a talent to decipher my handwriting, especially since aircraft turbulence periodically affects my writing effort. Thus, I am most appreciative of Mrs Linda Hayes's efforts in turning my latest manuscript into typed pages that resulted in the book you are reading. In addition, I would also like to thank Auerbach Publishers, Inc., for permitting me to use portions of articles I previously wrote for their *Data Communications Management* publication. Excerpts from these articles were used for developing the section covering integrated services digital network (ISDN) presented in Chapter 1, for expanding the statistical and T1 multiplexing in Chapter 5, and for the voice digitization, data compression and fiber optic transmission systems presented in Chapter 7.

Last but not least, one's publishing editor, editorial supervisor and desk editor are the critical link in converting the author's manuscript into the book you are now reading. To Ian Shelley, who enthusiastically backed the first edition of this book, I would like to take the opportunity to thank you again for your efforts. To Ann-Marie Halligan and Ian McIntosh who provided me with the opportunity to produce the third and fourth editions, I would again like to acknowledge your efforts in a multinational way. Cheers! To Stuart Gale, Robert Hambrook, and Sarah Lock who moved my manuscripts through proofs and into each edition of this book, many thanks for your fine effort.

1

FUNDAMENTAL WIDE AREA NETWORKING CONCEPTS

The main purpose of this chapter is to provide readers with a common level of knowledge concerning wide area networking (WAN) communications concepts. To achieve this goal we will examine the fundamental concepts associated with wide area network communications. Commencing with a description of the three components necessary to establish communications, we will expand our base of knowledge by discussing the types of line connections available for use, different types of transmission services and transmission devices, carrier offerings, transmission modes and techniques, and other key concepts. In doing so we will obtain a base of knowledge that will allow readers to better understand how devices and transmission facilities are interconnected to establish networks and interconnect geographically separated local area networks which is the focus of Chapters 2 and 3. In addition, the material presented in this chapter will enable readers to better understand the operation and utilization of devices explained in subsequent chapters.

While the transmission of data may appear to be a simple process, many factors govern the success or failure of a communications session. In addition, the exponential increase in the utilization of personal computers and a corresponding increase in communications between personal computers and other personal computers and large-scale computers had enlarged the number of hardware and software parameters you must consider. Although frequently we will use the terms 'terminal' and 'personal computers' interchangeably and refer to them collectively as 'terminals' in this book, in certain instances we will focus our attention upon personal computers in order to denote certain hardware and software characteristics unique to such devices. In these instances we will use the term 'personal computer' to explicitly reference this terminal device. In other instances we will use the term 'workstation' to refer to any computational device from a personal computer to a mainframe that is connected to a local area network. Such general use of this term should not be confused with its usage to represent a specialized powerful computer designed to facilitate the mathematical operations that are required to generate 3-D graphics, perform computer-aided design or similar compute-intensive operations, a topic beyond the scope of this book.

1.1 COMMUNICATIONS SYSTEM COMPONENTS

To transmit information between two locations it is necessary to have a transmitter, a receiver, and a transmission medium which provides a path or link between the transmitter and the receiver. In addition to transmitting signals, a transmitter must be capable of translating information from a form created by humans or machines into a signal suitable for transmission over the transmission medium. The transmission medium provides a path to convey the information to the receiver without introducing a prohibitive amount of signal distortion that could change the meaning of the transmitted signal. The receiver then converts the signal from its transmitted form into a form intelligible to humans or machines.

1.2 LINE CONNECTIONS

Three basic types of line connections are available to connect terminal devices to computers or to other terminals via a wide area network: dedicated, switched, and leased lines.

Dedicated line

A dedicated line is similar to a leased line in that the terminal is always connected to the device on the distant end, transmission always occurs on the same path, and, if required, the line may be able to be tuned to increase transmission performance.

The key difference between a dedicated and a leased line is that a dedicated line refers to a transmission medium internal to a user's facility, where the customer has the right of way for cable laying, whereas a leased line provides an interconnection between separate facilities. The term facility is usually employed to denote a building, office, or industrial plant. Dedicated lines are also referred to as direct connect lines and normally link a terminal or business machine on a direct path through the facility to another terminal or computer located at that facility. The dedicated line can be a wire conductor installed by the employees of a company or by the computer manufacturer's personnel, or it can be a local line installed by the telephone company.

Normally, the only cost associated with a dedicated line in addition to its installation cost is the cost of the cable required to connect the devices that are to communicate with one another.

Leased line

A leased line is commonly called a private line and is obtained from a communications company to provide a transmission medium between two facilities which could be in separate buildings in one city or in distant cities. In addition to a one-time installation charge, the communications carrier will normally bill the user on a monthly basis for the leased line, with the cost of the line usually based upon the distance between the locations to be connected.

Switched line

A switched line, often referred to as a dial-up line, permits contact with all parties having access to the analog public switched telephone network (PSTN) or the digital switched network. If the operator of a terminal device wants access to a computer, he or she dials the telephone number of a telephone which is connected to the computer. In using switched or dial-up transmission, telephone company switching centers establish a connection between the dialing party and the dialed party. After the connection is set up, the terminal and the computer conduct their communications. When communications are completed, the switching centers disconnect the path that was established for the connection and restore all paths used so they become available for other connections.

The cost of a call on the PSTN is based upon many factors which include the time of day when the call was made, the distance between called and calling parties, the duration of the call and whether or not operator assistance was required in placing the call. Direct dial calls made from a residence or business telephone without operator assistance are billed at a lower rate than calls requiring operator assistance. In addition, most telephone companies have three categories of rates: 'weekday', 'evening' and 'night and weekend'. Typically, calls made between 8 a.m. and 5 p.m. Monday to Friday are normally billed at a 'weekday' rate, while calls between 5 p.m. and 10 p.m. on weekdays are usually billed at an 'evening' rate, which reflects a discount of approximately 25% over the 'weekday' rate. The last rate category, 'night and weekend', is applicable to calls made between 10 p.m. and 8 a.m. on weekdays as well as anytime on weekends and holidays. Calls during this rate period are usually discounted 50% from the 'weekday' rate.

Table 1.1 contains a sample PSTN rate table which is included for illustrative purposes but which should not be used by readers for determining the actual cost of a PSTN call as the cost of intrastate calls by state and interstate calls varies. In addition, the cost of using different communications carriers to place a call between similar locations will typically vary from vendor to vendor and readers should obtain a current interstate and/or state schedule from the vendor they plan to use in order to determine or project the cost of using PSTN facilities.

Table 1.1 Sample PSTN rate table (cost per minute in cents)

	Rate category					
	Weekend		Evening		Night and weekend	
Mileage between location	First minute	Each additional minute	First minute	Each additional minute	First minute	Each additional minute
1–100	0.31	0.19	0.23	0.15	0.15	0.10
101–200	0.35	0.23	0.26	0.18	0.17	0.12
201–400	0.48	0.30	0.36	0.23	0.24	0.15

Cost trends

Although many vendors continue to maintain a rate table similar to the one shown in Table 1.1, other vendors have established a variety of flat-rate billing schemes in which calls made anywhere within a country are billed at a uniform cost per minute regardless of distance. During 1996 Sprint introduced a 10 cents per minute long-distance charge for calls made between 7 p.m. and 7 a.m. Monday through Friday and all day at weekends. During 1997 AT&T introduced a flat 15 cents per minute charge for calls made anywhere in the United States at any time. Both offerings require the selection of one communications carrier as your primary long-distance carrier and the selection of an appropriate calling plan to obtain flat-rate billing.

Factors to consider

Cost, speed of transmission, and degradation of transmission are the primary factors used in the selection process between leased and switched lines.

As an example of the economics associated with comparing the cost of PSTN and leased line usage, assume that a personal computer user located 50 miles from a mainframe needs to communicate between 8 a.m. and 5 p.m. with the mainframe once each business day for a period of 30 minutes. Using the data in Table 1.1, each call would cost $0.31 \times 1 + 0.19 \times 29$ or $5.82. Assuming there are 22 working days each month, the monthly PSTN cost for communications between the PC and the mainframe would be $5.82 × 22 or $128.04. If the monthly cost of a leased line between the two locations was $250, it is obviously less expensive to use the PSTN for communications. Suppose the communications application lengthened in duration to 2 hours per day. Then, from Table 1.1, the cost per call would become $0.31 \times 1 + 0.19 \times 119$ or $22.92. Again assuming 22 working days per month, the monthly PSTN charge would increase to $504.24, making the leased line more economical.

Thus, if data communications requirements involve occasional random contact from a number of terminals at different locations and each call is of short duration, dial-up service is normally employed. If a large amount of transmission occurs between a computer and a few terminals, leased lines are usually installed between the terminal and the computer.

Since a leased line is fixed as to its routing, it can be conditioned to reduce errors in transmission as well as permit ease in determining the location of error conditions since its routing is known. Normally, analog switched circuits are used for transmission at speeds up to 33 600 bits per second (bps); however, in certain situations data rates as high as 56 000 bps are achievable when transmission on the PSTN occurs through telephone company offices equipped with modern electronic switches.

Some of the limiting factors involved in determining the type of line to use for transmission between terminal devices and computers are listed in Table 1.2. Information in this table is applicable to both analog and digital transmission facilities and as such was generalized. For more specific information concerning the speed of transmission obtainable on analog and digital transmission facilities, readers are referred to the analog facilities and digital facilities subsections in Section 1.3.

Table 1.2 General line selection guide

Line type	Distance between transmission points	Speed of transmission	Use of transmission
Dedicated (direct connect)	Local	Limited by conductor	Short or long duration
Switched (dial-up)	Limited by telephone access availability	Normally up to 33 600 bps (analog), 1.544 Mbps (digital)	Short-duration transmission
Leased (private)	Limited by telephone company availability	Limited by type of facility	Long duration or numerous short duration calls

1.3 TYPES OF SERVICES AND TRANSMISSION DEVICES

Digital devices which include terminals, mainframe computers, and personal computers transmit data as unipolar digital signals, as indicated in Figure 1.1(a). When the distance between a terminal device and a computer is relatively short, the transmission of digital information between the two devices may be obtained by cabling the devices together. As the distance between the two devices increases, the pulses of the digital signals become distorted because of the resistance, inductance, and capacitance of the cable used as a transmission medium. At a certain distance between the two devices the pulses of the digital data will distort, such that they are unrecognizable by the receiver, as illustrated in Figure 1.1(b). To extend the transmission distance between devices, specialized equipment must be employed, with the type of equipment used dependent upon the type of transmission medium employed.

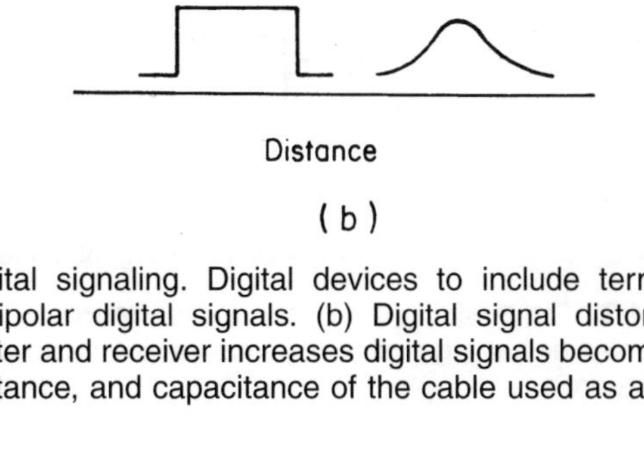

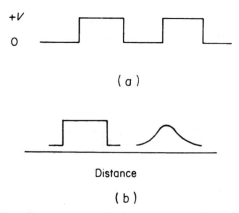

Figure 1.1 (a) Digital signaling. Digital devices to include terminals and computers transmit data as unipolar digital signals. (b) Digital signal distortion. As the distance between the transmitter and receiver increases digital signals become distorted because of the resistance, inductance, and capacitance of the cable used as a transmission medium

Digital repeaters

You can transmit data in a digital or analog form. To transmit data long distances in digital form requires repeaters to be placed on the line at selected intervals to reconstruct the digital signals. The repeater is a device that essentially scans the line looking for the occurrence of a pulse and then regenerates the pulse into its original form. Thus, another name for the repeater is a data regenerator. As illustrated in Figure 1.2, a repeater extends the communications distance between terminal devices to include personal computers and mainframe computers.

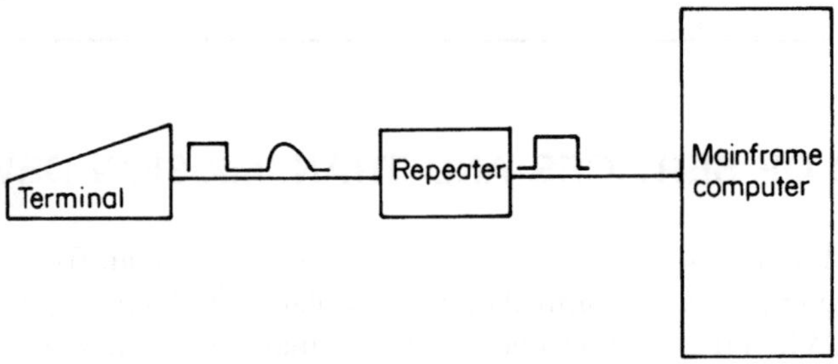

Figure 1.2 Transmitting data in digital format. To transmit data long distances in digital format requires repeaters to be placed on the line to reconstruct the digital signals

Unipolar and bipolar signaling

Since unipolar signaling results in a dc voltage buildup when transmitting over long distance, digital networks require unipolar signals to be converted into a modified bipolar format for transmission on this type of network. This requires the installation at each end of the circuit of a device known as a data service unit (DSU) in the United States and a network terminating unit (NTU) in the United Kingdom. The utilization of DSUs for transmission of data on a digital network is illustrated in Figure 1.3. Although not shown, readers should note that repeaters are placed on the path between DSUs to regenerate the bipolar signals. Later in this chapter we will examine digital facilities in more detail.

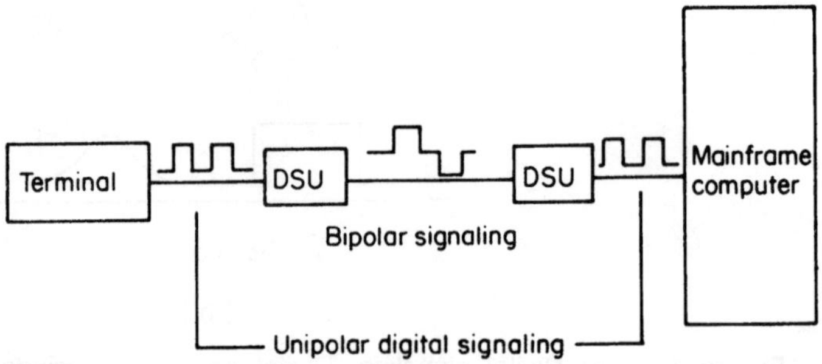

Figure 1.3 Transmitting data on a digital network. To transmit data on a digital network, the unipolar digital signals of terminal devices and computers must be converted into a bipolar signal

Repeaters are primarily used on wide area network digital transmission facilities at distances of approximately 6000 feet from one another on lines connecting subscribers to telephone company offices serving those subscribers. From local telephone company offices data will travel either by microwave or via fiber optic cable to a higher level telephone company office for routing through the telephone network hierarchy. By the late 1990s, over 99.9% of long-distance transmission was being carried in digital form via fiber optic cable. A vast majority of connections between telephone company subscribers and the local office serving those subscribers were, however, over twisted-pair copper cables that have amplifiers inserted to boost the strength of analog signals. Such connections require the conversion of digital signals into an analog form to enable the signal to be carried over the analog transmission facility.

Other digital signaling methods

In a LAN environment the full bandwidth of the cable is usually available for use. In comparison, the communications carrier commonly uses filters to limit the bandwidth usable on the local loop between a telephone company office and a subscriber's premises to 4 kHz or less. Although the absence of filters enables LAN designers to obtain a much higher data rate than that obtainable on a local loop, other operational considerations, to include the potential buildup of dc voltage and the cost of constructing equipment to operate at a high signaling rate to provide a high data transmission rate, resulted in the development of several digital signaling techniques used on LANs. Two of the more popular techniques are Manchester and Differential Manchester which are used on Ethernet and Token–Ring networks, respectively. In Chapter 3, when we focus our attention on local area networks, we will also turn our attention to the digital signaling methods used by different types of LANs.

Modems

Since telephone lines were originally designed to carry analog or voice signals, the digital signals transmitted from a terminal to another digital device must be converted into a signal that is acceptable for transmission by the telephone line. To effect transmission between distant points, a data set or modem is used. A modem is a contraction of the compound term modulator–demodulator and is an electronic device used to convert digital signals generated by computers and terminal devices into analog tones for transmission over telephone network analog facilities. At the receiving end, a similar device accepts the transmitted tones, reconverts them to digital signals, and delivers these signals to the connected device.

Signal conversion

Signal conversion performed by modems is illustrated in Figure 1.4. This illustration shows the interrelationship of terminals, mainframe computers, and

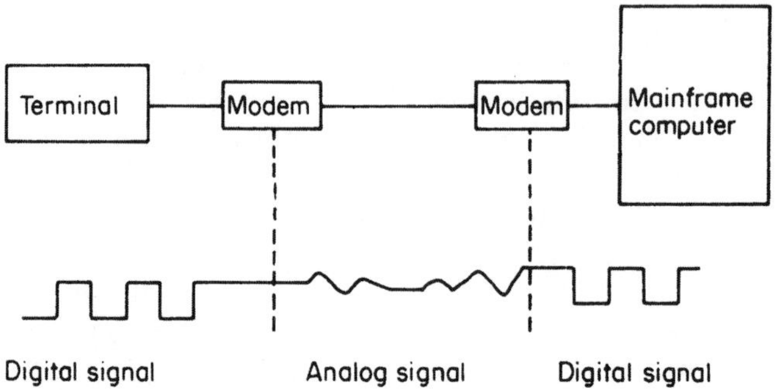

Figure 1.4 Signal conversion performed by modems. A modem converts (modulates) the digital signal produced by a terminal into an analog tone for transmission over an analog facility

transmission lines when an analog transmission service is used. Analog transmission facilities include both leased lines and switched lines; therefore, modems can be used for transmission of data over both types of analog line connections. Although an analog transmission medium used to provide a transmission path between modems can be a direct connect, a leased, or a switched line, modems are connected (hard-wired) to direct connect and leased lines, whereas they are interfaced to a switched facility; thus, a terminal user can communicate only with one distant location on a leased line, but with many devices when there is access to a switched line.

Acoustic couplers

Although popular with data terminal users in the 1970s, today only a very small percentage of persons use acoustic couplers for communications. The acoustic coupler is a modem whose connection to the telephone line is obtained by acoustically coupling the telephone headset to the coupler. The primary advantage of the acoustic coupler was the fact that it required no hard-wired connection to the switched telephone network, enabling terminals and personal computers to be portable with respect to their data transmission capability. Owing to the growth in modular telephone jacks, modems that interface the switched telephone network via a plug, in effect, are portable devices. Since many hotels and older office buildings still have hard-wired telephones, the acoustic coupler permits terminal and personal computer users to communicate regardless of the method used to connect a telephone set to the telephone network.

Signal conversion

The acoustic coupler converts the signals generated by a terminal device into a series of audible tones, which are then passed to the mouthpiece or transmitter of the telephone and in turn onto the switched telephone network. Information transmitted from the device at the other end of the data link is converted into audible

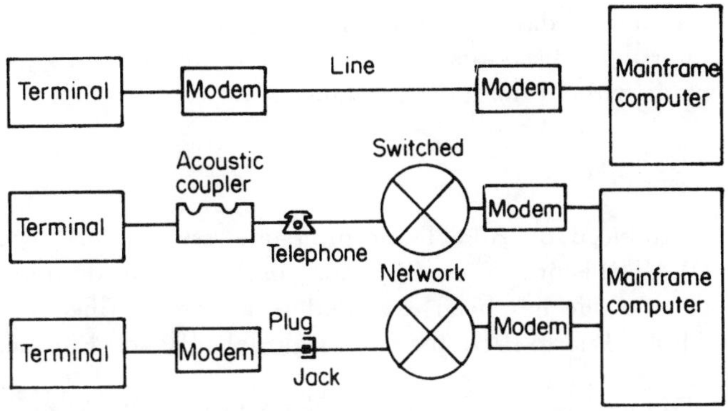

Figure 1.5 Interrelationship of terminals, modems, acoustic couplers, computers and analog transmission mediums. When using modems on an analog transmission medium, the line can be a dedicated, leased, or switched facility. Terminal devices can use modems or acoustic couplers to transmit via the switched network

tones at the earpiece of the telephone connected to the terminal's acoustic coupler. The coupler then converts those tones into the appropriate electrical signals recognized by the attached terminal. The interrelationship of terminals, acoustic couplers, modems and analog transmission media is illustrated in Figure 1.5.

In examining Figure 1.5, you will note that a circle subdivided into four equal parts by two intersecting lines is used as the symbol to denote the public switched telephone network or PSTN. This symbol will be used in the remainder of the book to illustrate communications occurring over this type of line connection.

Analog facilities

Several types of analog switched and leased line facilities are offered by communications carriers. Each type of facility has its own set of characteristics and rate structure. Normally, for extensive communications requirements, an analytic study is conducted to determine which type or types of service should be utilized to provide an optimum cost-effective service for the user. Common types of analog switched facilities are direct distance dialing, wide area telephone service, and foreign exchange service. The most common type of analog line is a voice grade private line.

DDD

Direct distance dialing (DDD) permits a person to dial directly any telephone connected to the public switched telephone network. The dialed telephone may be connected to another terminal device or mainframe computer. The charge for this service, in addition to installation costs, may be a fixed monthly fee if no long-distance calls are made, a message unit rate based upon the number and duration of local calls, or a fixed fee plus any long-distance charges incurred. Depending upon the time of day a long-distance call is initiated and its destination (intrastate or interstate), discounts from normal long-distance tolls are available for selected calls made without operator assistance.

WATS

Introduced by AT&T for interstate use in 1961, wide area telephone service (WATS) is now offered by most long-distance communications carriers. Its scope of coverage has been extended from the continental United States to Hawaii, Alaska, Puerto Rico, the US Virgin Islands, and Europe, as well as selected Pacific and Asian countries.

Wide area telephone service (WATS) may be obtained in two different forms, each designed for a particular type of communications requirement. Outward WATS is used when a specific location requires placing a large number of outgoing calls to geographically distributed locations. Inward WATS service provides the reverse capability, permitting a number of geographically distributed locations to communicate with a common facility. Calls on WATS are initiated in the same manner as a call placed on the public switched telephone network. However, instead of being charged on an individual call basis, the user of WATS facilities pays a flat rate per block of communications hours per month occurring during weekday, evening, and night and weekend time periods.

A voice-band trunk called an access line is provided to the WATS users. This line links the facility to a telephone company central office. Other than cost considerations and certain geographical calling restrictions which are a function of the service area of the WATS line, the user may place as many calls as desired on this trunk if the service is outward WATS or receive as many calls as desired if the service is inward. Inward WATS, the well-known '800' area code which was extended to the '888' area code during 1996, permits remotely located personnel to call your facility toll-free from the service area provided by the particular inward WATS-type of service selected. The charge for WATS is a function of the service area. This can be intrastate WATS, a group of states bordering the user's state where the user's main facility is located, a grouping of distant states, or International WATs which extends inbound 800 service to the United States from selected overseas locations. Another service very similar to WATS is AT&T's 800 READYLINE^{SM} service. This service is essentially similar to WATS; however, calls can originate or be directed to an existing telephone in place of the access line required for WATS service.

Figure 1.6 illustrates the AT&T WATS service area one for the state of Georgia. If this service area is selected and a user in Georgia requires inward WATS service, he or she will pay for toll-free calls originating in the states surrounding Georgia–Florida, Alabama, Mississippi, Tennessee, Kentucky, South Carolina, and North Carolina. Similarly, if outward WATS service is selected for service area one, a person in Georgia connected to the WATS access line will be able to dial all telephones in the states previously mentioned. The states comprising a service area vary based upon the state in which the WATS access line is installed. Thus, the states in service area one when an access line is in New York would obviously differ from the states in a WATS service area one when the access line is in Georgia. Fortunately, AT&T publishes a comprehensive book which includes 50 maps of

Figure 1.6 AT&T WATS service area one for an access line located in Georgia

the United States, illustrating the composition of the service areas for each state. Similarly, a time-of-day rate schedule for each state based upon state service areas is also published by AT&T.

In general, since WATS is a service based upon volume usage its cost per hour is less than the cost associated with the use of the PSTN for long-distance calls. Thus, one common application for the use of WATS facilities is to install one or more inward WATS access lines at a data processing center. Then, terminal and personal computer users distributed over a wide geographical area can use the inward WATS facilities to access the computers at the data processing center.

Since International 800 service enables employees and customers of US companies to call them toll-free from foreign locations, this service may experience a considerable amount of data communications usage. This usage can be expected to include applications requiring access to such databases as hotel and travel reservation information as well as order entry and catalog sales data updating. Persons traveling overseas with portable personal computers as well as office personnel using terminals and personal computers in foreign countries who desire access to computational facilities and information utilities located in the United States represent common International 800 service users. Due to the business advantages of WATS its concept has been implemented in several foreign countries, with inward WATS in the United Kingdom marketed under the term Freefone.

FX

Foreign exchange (FX) service may provide a method of transmission from a group of terminal devices remotely located from a central computer facility at less

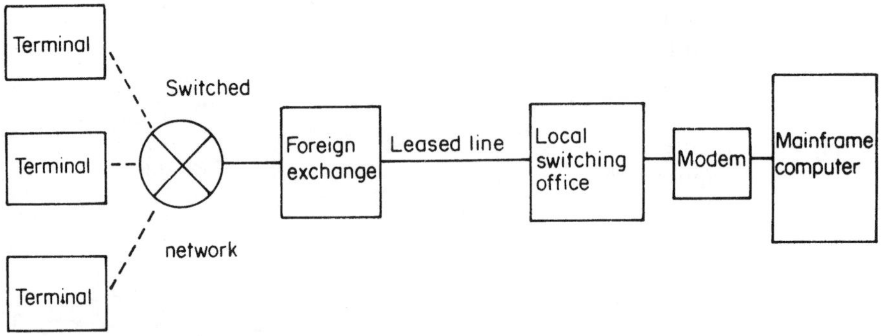

Figure 1.7 Foreign exchange (FX) service. A foreign exchange line permits many terminal devices to use the facility on a scheduled or on a contention basis

than the cost of direct distance dialing. An FX line can be viewed as a mixed analog switched and leased line. To use an FX line, a user dials a local number which is answered if the FX line is not in use. From the FX, the information is transmitted via a dedicated voice line to a permanent connection in the switching office of a communications carrier near the facility with which communication is desired. A line from the local switching office which terminates at the user's home office is included in the basic FX service. This is illustrated in Figure 1.7.

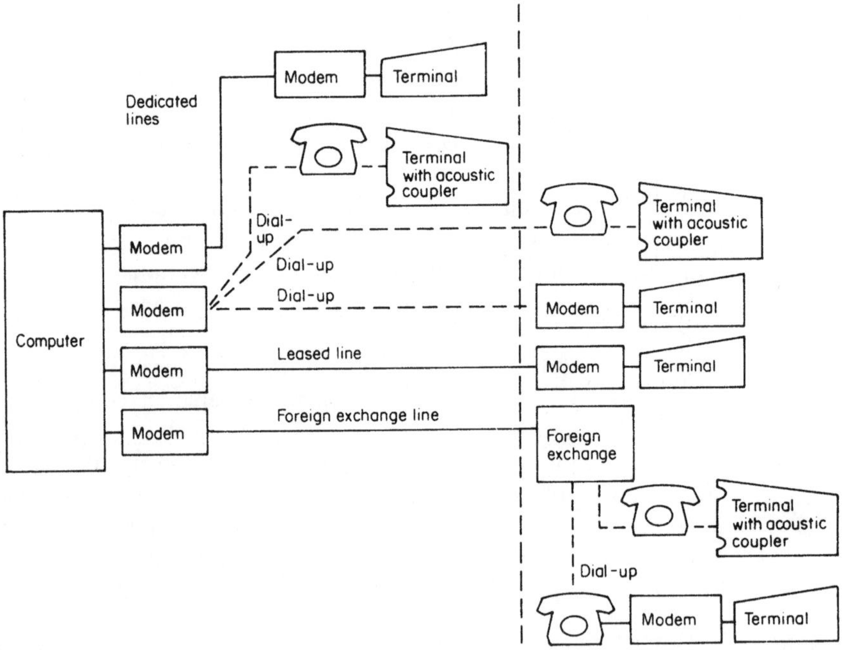

Figure 1.8 Terminal-to-computer connections via analog mediums. A mixture of dedicated, dialup, leased and foreign exchange lines can be exployed to connect local and remote terminals to a central computer facility

The use of an FX line permits the elimination of long-distance charges that would be incurred by users directly dialing a distant computer facility. Since only one person at a time may use the FX line, normally only groups of users whose usage can be scheduled are suitable for FX utilization. Figure 1.8 illustrates the possible connections between remotely located terminal devices and a central computer where transmission occurs over an analog facility.

The major difference between an FX line and a leased line is that any terminal dialing the FX line provides the second modem required for the transmission of data over the line; whereas a leased line used for data transmission normally has a fixed modem attached at both ends of the circuit.

Leased lines

The most common type of analog leased line is a voice grade private line. This line obtains its name from its ability to permit one voice conversation with frequencies between 300 and 3300 Hz to be carried on the line. In actuality, the bandwidth or range of frequencies that can be transmitted over a twisted-pair analog switched or analog leased line extends from 0 to approximately 1 MHz. However, to economize on the transmission of multiple voice conversations routed between telephone company offices, the initial design of the telephone company cable infrastructure resulted in the use of filters to remove frequencies below 300 Hz and above 3300 Hz, resulting in a 3000 Hz bandwidth for voice conversations. At telephone company offices voice conversations destined to another office are multiplexed onto a trunk or high speed line by frequency, requiring only 300 Hz of bandwidth per conversation, enabling one trunk to transport a large number of voice conversations shifted in frequency from one another. At the distant office other frequency division multiplexing equipment shifts each conversation back into its original frequency range as well as routing the call to its destination. Although the use of filters has considerably economized on the cost of routing multiple calls on trunks connecting telephone company offices, they have resulted in a bandwidth limit of 3000 Hz which makes high speed transmission on an analog loop most difficult to obtain. As we turn our attention to the operation of different types of modems later in this book, we will also obtain an appreciation of how the 3000 Hz bandwidth of analog lines limits the communications rate to most homes and many offices.

Figure 1.9 illustrates the typical routing of a leased line in the United States. The routing from each subscriber location to a telephone company central office serving the subscriber is known as a local loop. Normally the local loop is a two-wire or four-wire copper single or dual twisted-pair cable with amplifiers inserted

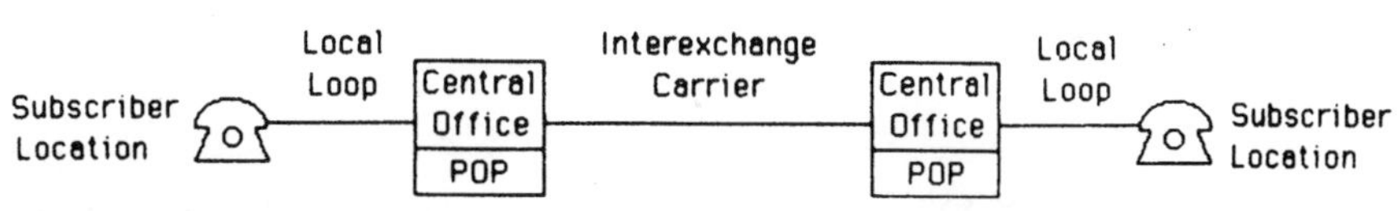

Figure 1.9 Leased line routing. Leased lines are routed from a local telephone company serving a subscriber to an interexchange carrier at the point of presence (POP)

on the local loop to boost the strength of the signal. Both the local loop and the central office are operated by the telephone company serving each subscriber location. If the leased line is routed outside the local telephone company's serving area it must be connected to an interexchange carrier (IXC), such as AT&T, MCI, or Sprint. The location where this interconnection takes place is called the point of presence (POP), which is normally located in the central office of the local telephone company. Although data on an analog leased line flows in an analog format, by the early 1990s most interexchange carriers digitized analog signals at the POP. Thus, between POPs most analog data is actually carried in digital form. Since the local loop is still an analog medium, it is the local loop which limits the data transmission rate obtainable through the use of an analog leased line. By 1998 modems permitting a 33.6 kbps data transmission rate on leased lines were commonly available, and some vendors had introduced products that allow data rates of up to 56 kbps in one direction and up to 33.6 kbps in the opposite direction, a technique referred to as asymmetrical transmission.

Digital facilities

In addition to the analog service, numerous digital service offerings have been implemented by communications carriers over the last decade. Using a digital service, data is transmitted from source to destination in digital form without the necessity of converting the signal into an analog form for transmission over analog facilities as is the case when modems or acoustic couplers are interfaced to analog facilities.

To understand the ability of digital transmission facilities to transport data requires an understanding of digital signaling techniques. Those techniques provide a mechanism to transport data end-to-end in modified digital form on LANs as well as on wide area networks that can connect locations hundreds or thousands of miles apart.

Digital signaling

Digital signaling techniques have evolved from use in early telegraph systems to provide communications for different types of modern technology, ranging in scope from the data transfer between a terminal and a modem to the flow of data on a LAN and the transport of information on high speed wide area network digital communications lines. Instead of one signaling technique numerous techniques are used, each technique having been developed to satisfy different communications requirements. In this section we will focus our attention upon digital signaling used on wide area network transmission facilities, deferring a discussion of LAN signaling until later in this book.

Unipolar non-return to zero

Unipolar non-return to zero (NRZ) is a simple type of digital signaling which was originally used in telegraphy. Today, unipolar non-return to zero signaling is used

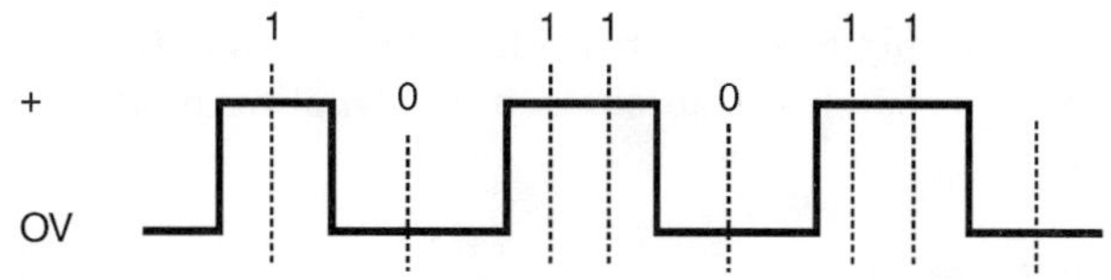

Figure 1.10 Unipolar non-return to zero signaling

in computers as well as by the common RS–232/V.24 interface between data terminal equipment and data communications equipment.

Figure 1.10 illustrates an example of unipolar non-return to zero signaling. In this signaling scheme, a dc current or voltage represents a mark, while the absence of current or voltage represents a space. Since voltage or current does not return to zero between adjacent set bits, this signaling technique is called non-return to zero. Since voltage or current is only varied from 0 to some positive level the pulses are unipolar, hence the name unipolar non-return to zero.

There are several problems associated with unipolar non-return to zero signaling which make it unsuitable for use as a signaling mechanism on wide area network digital transmission facilities. These problems include the need to sample the signal and the fact that it provides residual dc voltage buildup.

Since two or more repeated marks or spaces can stay at the same voltage or current level, sampling is required to distinguish one bit value from another. The ability to sample requires clocking circuitry which drives up the cost of communications. A second problem related to the fact that a sequence of marks or set bits can occur is that this condition results in the presence of residual dc levels. Residual dc requires the direct attachment of transmission components, while the absence of residual dc permits ac coupling via the use of a transformer. When communications carriers engineered their early digital networks they were based upon the use of copper conductors, as fiber optics did not exist. Communications carriers attempt to do things in an economical manner. Rather than install a separate line to power repeaters, they examined the possibility of carrying both power and data on a common line, removing the data from the power at the distant end as illustrated in Figure 1.11. This required transformer coupling at the distant end,

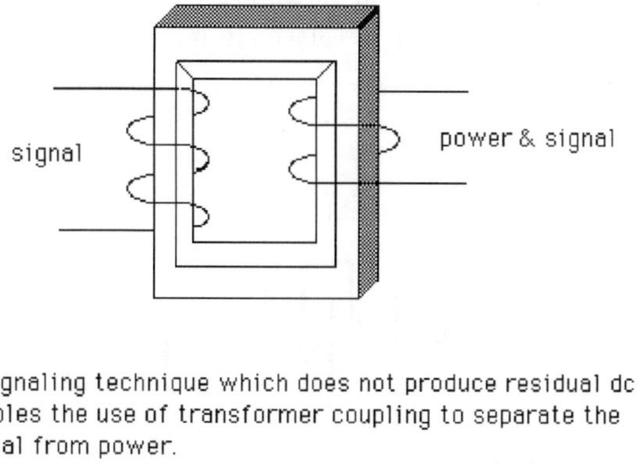

signal
power & signal
A signaling technique which does not produce residual dc
enables the use of transformer coupling to separate the
signal from power.

Figure 1.11 Transformer coupling

which is only possible if residual dc is eliminated. Thus, communications carriers began to search for an alternative signaling method.

Unipolar return to zero

One of the first alternative signaling methods examined was unipolar return to zero (RTZ). Under this signaling technique, which is illustrated in Figure 1.12, the current or voltage always returns to zero after every '1' bit. While this signal is easier to sample since each mark has a pulse rise, it still results in residual dc buildup and was unsuitable for use as the signaling mechanism on communications carrier digital transmission facilities.

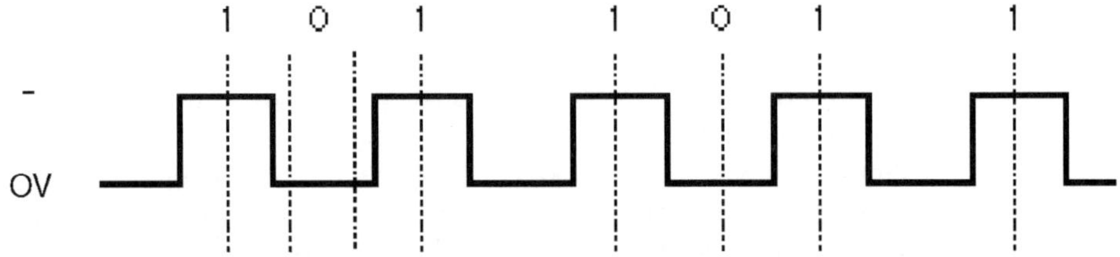

Figure 1.12 Unipolar return to zero signaling

Bipolar return to zero

After examining a variety of signaling methods communications carriers focused their attention upon a technique known as bipolar return to zero. Under bipolar signaling alternating polarity pulses are used to represent marks, while a zero level pulse is used to represent a space. In the bipolar return to zero signaling method the bipolar signal returns to zero after each mark. Figure 1.13 illustrates an example of bipolar return to zero signaling.

The key advantage of bipolar return to zero signaling is the fact that it precludes dc voltage buildup on the line. This enables both power and data to be carried on the same line, enabling repeaters to be powered by a common line. In addition, repeaters can be placed relatively far apart in comparison to other signaling techniques, which reduces the cost of developing the digital transmission infrastructure.

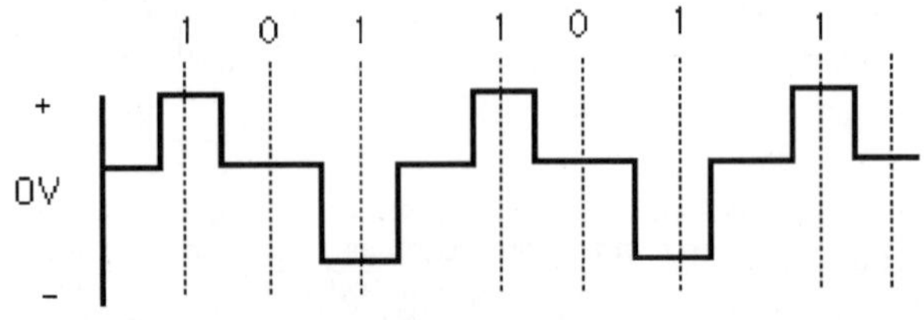

Figure 1.13 Bipolar return to zero signaling

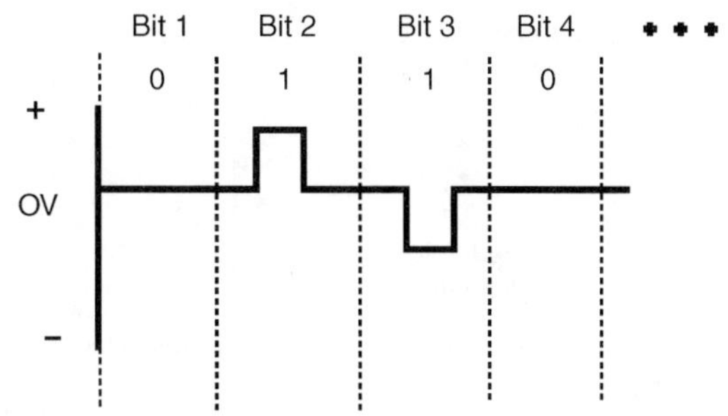

Figure 1.14 Bipolar (AMI) RTZ 50 percent duty cycle

On modern wide area network digital transmission facilities a modified form of bipolar return to zero signaling is employed. That modification involves the placement of pulses in their bit interval so that they occupy 50% of the interval, with the pulse centered at the center of the interval. This positioning eliminates high frequency components of the signal that can interfere with other transmissions and results in a bipolar pulse known as a 50% duty cycle alternate mark inversion (AMI). An example of this pulse is illustrated in Figure 1.14.

Now that we have a general appreciation for the type of digital signaling used on digital transmission facilities and the rationale for the use of that type of signaling, we will discuss some of the types of digital transmission facilities available for use. In doing so we will first consider the manner by which the digitization of voice conversations was performed, as the voice digitization effort had a significant effect upon the evolution of digital service offerings.

Evolution of service offerings

The evolution of digital service offerings can be traced to the manner by which telephone company equipment was developed to digitize voice and the initial development of digital multiplexing equipment to combine multiple voice conversations for transmission between telephone company offices. The digitization of voice was based upon the use of a technique referred to as Pulse Code Modulation (PCM) which requires an analog voice conversation to be sampled 8000 times per second. Each sample is encoded using an 8-bit byte, resulting in a digital data stream of 64 kbps to carry one digitized voice conversation.

The first high speed digital transmission circuit used in North America was designed to transport 24 digitized voice conversations. This circuit, which is referred to as a T1 line, transports a Digital Signal Level 1 (DS1) signal. That signal represents 24 digitized voice samples plus one framing bit repeated 8000 times per second. Thus, the operating rate of a T1 circuit becomes $(24 \times 8 + 1)$ bits/sample $\times$ 8000 samples/second, or 1.544 Mbps.

Each individual time slot within a DS1 signal is referred to as DS0 or Digital Signal Level 0 and represents a single digitized voice conversation transported at 64 kbps. Figure 1.15 illustrates the relationship between DS0s and a DS1. In examining Figure 1.15 note that the DS1 frame is repeated 8000 times per second,

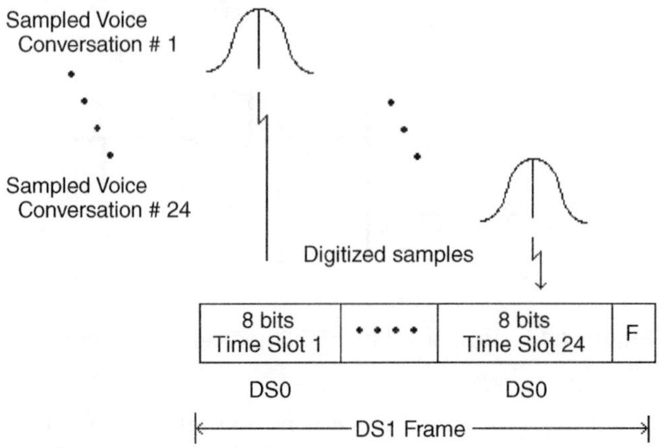

Figure 1.15 Relationship between DS0s and a DS1 signal (F = frame bit)

resulting in 8000 bps of framing information transmitted between telephone company central offices. Initially the framing bits were used for synchronization and the transmission of certain types of alarm indications. Later the sequence of framing bits was altered to enable control information to be transmitted as part of the framing. Later in this book when we investigate the operation and utilization of T1 multiplexers, we will examine the use of framing bits in detail.

Due to the necessity of transporting certain types of telephone information such as on-hook and off-hook information with each DS0, a technique referred to as bit-robbing was developed. Under bit-robbing one of the bits used to represent the height of a digitized sample was periodically 'stolen' for use to convey telephone set information. This bit-robbing process only occurred in the 6th and 12th frames in each continuous sequence of 12 frames, resulting in the inability of the human ear to recognize that a few digitized samples were encoded in seven bits instead of eight. However, when early digital transmission facilities were developed, the bit-robbing process limited data transmission to seven bits per eight-bit byte, which explains why switched 56 service operates at 56 kbps instead of 64 kbps. Later, the development of a separate network by telephone companies to convey signaling information enabled the full transmission capacity of DS0s to be used for data transmission. Today many communications carriers offer switched 56 and switched 64 kbps digital transmission as well as digital fractional T1 leased lines that operate in increments of 56 or 64 kbps.

AT&T offerings

In the United States, AT&T offers several digital transmission facilities under the ACCUNET[SM] Digital Service mark. Dataphone® Digital Service was the charter member of the ACCUNET family and is deployed in over 100 major metropolitan cities in the United States as well as having an interconnection to Canada's digital network. Dataphone Digital Service operates at synchronous data transfer rates of 2.4, 4.8, 9.6, 19.2 and 56 kbps, providing users of this service with dedicated, two-way simultaneous transmission capability.

Originally all AT&T digital offerings were leased line services where a digital leased line is similar to a leased analog line in that it is dedicated for full-time use to one customer. In the late 1980s, AT&T introduced its Accunet Switched 56 service, a dial-up 56 kbps digital data transmission service. This service enables users to maintain a dial-up backup for previously installed 56 kbps AT&T Dataphone Digital Services leased lines or a partial backup for ACCUNET T1.5 service which is described later in this section. In addition, this service can be used to supplement existing services during peak transmission periods or for applications that only require a minimal amount of transmission time per day since the service is billed on a per minute basis.

Access to Switched 56 service is obtained by dialing area code 700 numbers available in approximately 100 cities in the United States. All numbers are 10-digit, of the form 700-56X-XXXX.

Transmission on both leased line and switched Dataphone Digital Service requires the use of a Data Service Unit and Channel Service Unit (DSU/CSU) in comparison to the use of modems when transmission occurs on an analog transmission facility. Originally separate devices, most vendors now market combined DSU/CSU products that are commonly and collectively referred to as DSUs. The operation of DSUs is described later in this section and in significantly more detail in Chapter 5.

Although DDS was a very popular digital transmission service during the 1980s, the expansion of communications carriers' digital infrastructure based upon the installation of tens of thousands of miles of fiber cable during the late 1980s and early 1990s resulted in a range of new digital offerings. Those offerings are based upon the use of portions of, or entire, T1 circuits, with the former referred to as fractional T1, and are considerably more cost-effective than DDS. Thus, although DDS was still in use during 1998 its future days of use are probably limited.

Another offering from AT&T, ACCUNET T1.5 Service is a high capacity 1.544 Mbps terrestrial digital service which permits 24 voice-grade channels or a mixture of voice and data to be transmitted in digital form. This service was originally only obtainable as a leased line and is more commonly known as a T1 channel or circuit. Transmission on a T1 circuit also requires the use of a DSU. However, the DSU portion of the DSU/CSU is commonly built into terminal equipment connected to T1 lines. Separate CSUs are required, therefore, to interface T1 circuits. Channel Service Units manufactured for use on T1 lines are described in detail in Chapter 5. Readers should note that field trials of switched 384 kbps and 1.544 Mbps services during the mid-1990s resulted in their availability for commercial usage.

Until 1989 there was a significant gap in the transmission rates obtainable on digital lines. DDS users could transmit at data rates up to 56 kbps, while the use of a T1 line resulted in the transmission of data at 1.544 Mbps. Recognizing the requirements of many organizations for the transmission of data at rates above that obtainable on DDS but below the T1 rate, several communications carriers introduced fractional T1 digital service. AT&T's fractional T1 service is called ACCUNET Spectrum of Digital Service (ASDS). Under ASDS digital transmission is furnished via leased lines at data rates ranging from 9.6, 56, or 64 kbps up to 1.544 Mbps, in increments of 64 kbps from 64 kbps to 1.544 Mbps. In actuality, 9.6 and 56 kbps ASDS services are not fractional T1 as they represent special digital services that allow DDS to be carried on a fraction of a T1 circuit at a considerable reduction in cost.

Data rates of 64, 128, 256, 384, 512 and 768 kbps available under ASDS can be considered as true fractional T1 as they represent distinct fractions of a T1 circuit. A 64 kbps data rate represents 1/24th of a T1 circuit since 64 kbps is the data rate of one digitized voice channel on a T1 circuit and that circuit carries 24 digitized voice channels.

The majority of access to a fractional T1 line requires the use of a T1 local loop. Although the data transmission rate on the local loop is 1.544 Mbps, the fractional T1 subscriber in actuality uses one or more 64 kbps channels on the local loop which is routed to the telephone company central office serving the subscriber. At that location the fractions of the T1 local loop used by the subscriber are removed from the T1 line and input into an interexchange carrier's equipment at the point of presence. The interexchange carrier combines the fractions of T1 circuits used by many subscribers into a full T1 circuit operating at 1.544 Mbps which is then routed through the carrier's transmission facilities to the point of presence serving the distant location. At that point of presence the 64 kbps channels representing the fractional T1 channel used by the subscriber are passed to another telephone company which, more than likely, routes the transmission via a T1 line to the subscriber.

In addition to T1 lines, AT&T and other communications carriers offer T3 digital circuits operating at 44.736 Mbps. A T3 circuit transports a DS3 signal which is formed via the multiplexing of 28 DS1 signals. Similar to the recognition that many organizations cannot use a full T1, AT&T and other communications carriers recognized that only a limited number of organizations can use the capacity of a full T3 digital circuit. As you might expect, this realization resulted in the development of fractional T3 (FT3) offerings.

European offerings

In Europe, a number of countries have established digital transmission facilities. One example of such offerings is British Telecom's KiloStream service. KiloStream provides synchronous data transmission at 2.4, 4.8, 9.6, 48 and 64 kbps and is very similar to AT&T's Dataphone Digital Service. Each KiloStream circuit is terminated by British Telecom with a network terminating unit (NTU), which is the digital equivalent of the modem required on an analog circuit. In comparison with the T1 circuit used in North America which was based upon a design for carrying 24 digitized voice conversations, European countries use E1 circuits. Such circuits were constructed based upon the placement of 32 channels on one circuit and operate at 2.048 Mbps. In the United Kingdom British Telecom's E1 service is marketed as MegaStream.

DSUs

A data service unit (DSU) provides a standard interface to a digital transmission service and handles such functions as signal translation, regeneration, reformatting, and timing. Most DSUs are designed to operate at one of four speeds—2.4, 4.8, 9.6, and 56 kbps—while some DSUs also support 19.2 kbps operations. The transmitting portion of the DSU processes the customer's signal into bipolar pulses suitable for transmission over the digital facility. The receiving portion of the DSU is used both to extract timing information and to regenerate mark and space information from the received bipolar signal. The second interface arrangement originally developed for AT&T's Dataphone Digital Service is called a channel service unit (CSU) and was provided by the communication carrier to those customers who wish to perform the signal processing to and from the bipolar line, as well as to retime and regenerate the incoming line signals through the utilization of their own equipment. Originally marketed as separate devices, almost all DSUs and CSUs designed for use on AT&T Dataphone Digital Service and equivalent facilities from other carriers are now manufactured as one integrated device which is commonly referred to as a DSU or a DSU/CSU. Since most terminal devices connected to T1 lines contain a built-in data service unit, a separate channel service unit is required for transmission on that type of digital transmission facility.

As data is transmitted over digital facilities, the signal is regenerated by the communications carrier numerous times prior to its arrival at its destination. In general, digital service gives data communications users improved performance and reliability when compared to analog service, owing to the nature of digital transmission and the design of digital networks. This improved performance and reliability is due to the fact that digital signals are regenerated whereas, when analog signals are amplified, any distortion to the analog signal is also being amplified.

Although a digital service is offered in many locations, for those locations outside the serving area of a digital facility the user will have to employ analog devices as an extension in order to interface to the digital facility. The utilization of digital service via an analog extension is illustrated in Figure 1.16. As depicted in Figure 1.16, if the closest city to the terminal located in city 2 that offers digital

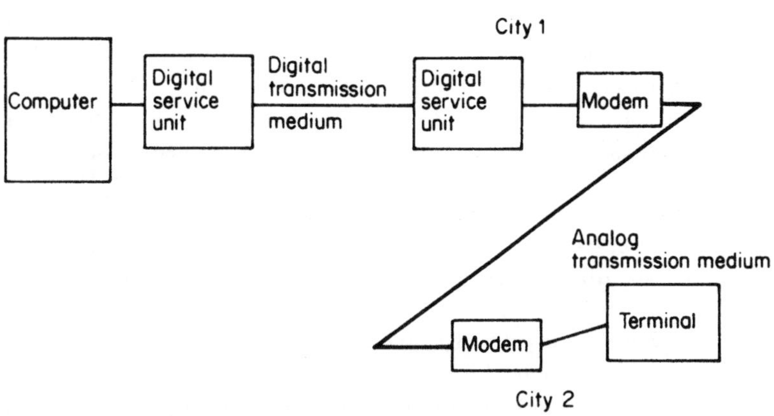

Figure 1.16 Analog extension to digital service. Although data is transmitted in digital form from the computer to city 1, it must be modulated by the modern at that location for transmission over the analog extension

service is city 1, then to use digital service to communicate with the computer an analog extension must be installed between the terminal location in city 2 and city 1. In such cases, the performance, reliability, and possible cost advantages of using digital service may be completely dissipated.

1.4 TRANSMISSION MODE

One method of characterizing lines, terminal devices, mainframe computers, and modems is by their transmission or communications mode. The three classes of transmission modes are simplex, half-duplex, and full-duplex.

Simplex transmission

Simplex transmission occurs in one direction only, disallowing the receiver of information a means of responding to the transmission. A home AM radio which receives a signal transmitted from a radio station is an example of a simplex communications mode. In a data transmission environment, simplex transmission might be used to turn on or off specific devices at a certain time of the day or when a certain event occurs. An example of this would be a computer-controlled environmental system where a furnace is turned on or off depending upon the thermostat setting and the current temperature in various parts of a building. Normally, simplex transmission is not utilized where human-machine interaction is required, owing to the inability to turn the transmitter around so that the receiver can reply to the originator.

Half-duplex transmission

Half-duplex transmission permits transmission in either direction; however, transmission can occur in only one direction at a time. Half-duplex transmission is used in citizen band (CB) radio transmission where the operator can either transmit or receive but cannot perform both functions at the same time on the same channel. When the operator has completed a transmission, the other party must be advised that they have finished transmitting and is ready to receive by saying the term 'over'. Then the other operator can begin transmission.

When data is transmitted over the telephone network, the transmitter and the receiver of the modem or acoustic coupler must be appropriately turned on and off as the direction of the transmission varies. Both simplex and half-duplex transmission require two wires to complete an electrical circuit. The top of Figure 1.17 illustrates a half-duplex modem interconnection while the lower portion of that illustration shows a typical sequence of events in the terminal's sign-on process to access a computer. In the sign-on process, the user first transmits the word NEWUSER to inform the computer that a new user wishes a connection to the computer. The computer responds by asking for the user's password, which is then furnished by the user.

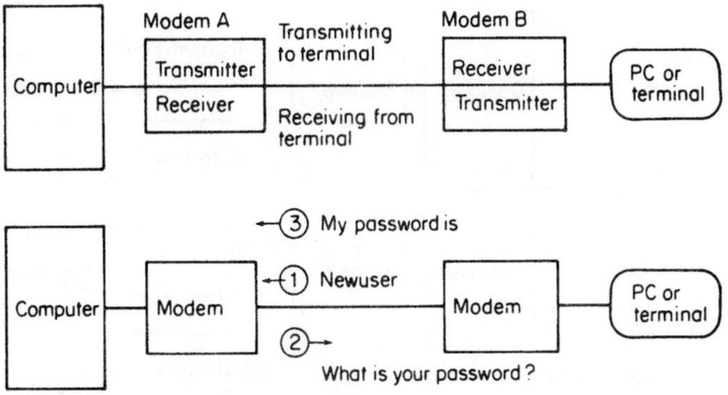

Figure 1.17 Half-duplex transmission. Top: control signals from the mainframe computer and terminal operate the transmitter and receiver sections of the attached modems. When the transmitter of modem A is operating, the receiver of modem B operates; when the transmitter of modem B operates; the reveiver of modem A operates. However, only one transmitter operates at any one time in the half-duplex mode of transmission. Bottom: during the sign-on sequence, transmission is turned around several times

In the top portion of Figure 1.17, when data is transmitted from a computer to a terminal, control signals are sent from the computer to modem A which turns on the modem A transmitter and causes the modem B receiver to respond. When data is transmitted from the terminal to the computer, the modem B receiver is disabled and its transmitter is turned on while the modem A transmitter is disabled and its receiver becomes active. The time necessary to effect these changes is called a transmission turnaround time, and during this interval transmission is temporarily halted. Half-duplex transmission can occur on either a two-wire or four-wire circuit. The switched network is a two-wire circuit, whereas leased lines can be obtained as either two-wire or four-wire links. A four-wire circuit is essentially a pair of two-wire links which can be used for transmission in both directions simultaneously. This type of transmission is called full-duplex.

Full-duplex transmission

Although you would normally expect full-duplex transmission to be accomplished over a four-wire connection that provides two two-wire paths, full-duplex transmission can also occur on a two-wire connection. This is accomplished by the use of modems that subdivide the frequency bandwidth of the two-wire connection into two distinct channels, permitting simultaneous data flow in both directions on a two-wire circuit. This technique will be examined and explained in more detail later in this book, when the operating characteristics of modems are examined in detail.

Full-duplex transmission is often used when large amounts of alternate traffic must be transmitted and received within a fixed time period. If two channels were used in our CB example, one for transmission and another for reception, two simultaneous transmissions could be effected. While full-duplex transmission

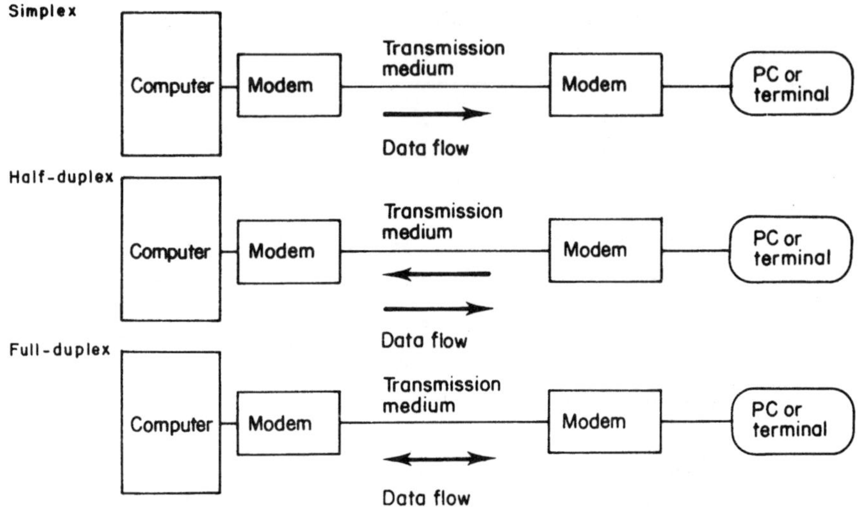

Figure 1.18 Transmission modes. Top: simplex transmission is in one direction only; transmission cannot reverse direction. Center: half-duplex transmission permits transmission in both directions but only one way at a time. Bottom: full-duplex transmission permits transmission in both directions simultaneously

provides more efficient throughput, this efficiency was originally negated by the cost of two-way lines and more complex equipment required by this mode of transmission. The development of low cost digital signal processor chips enabled high speed modems to operate in a full-duplex transmission mode on a two-wire circuit through a technique referred to as echo cancellation. This technique will be described when we discuss the operation of modems in detail in Chapter 5.

In Figure 1.18, the three types of transmission modes are illustrated, while Table 1.3 summarizes the three transmission modes previously discussed.

In Table 1.3, the column headed ITU refers to the International Telecommunications Union, a United Nations agency headquartered in Geneva, Switzerland. Previously, the standardization body of the ITU was known as the Consultative Committee on International Telephone and Telegraph (CCITT) and, even though the renaming occurred several years ago, many people still use the term CCITT when referencing certain standards.

Table 1.3 Transmission mode comparison

Symbol	ANSI	US telecommunications industry	ITU	Historical physical line requirement
←	One-way only	Simplex		Two-wire
← →	Two-way alternate	Half-duplex (HDX)	Simplex	Two-wire
⇆	Two-way simultaneous	Full-duplex (FDX)	Duplex	Four-wire

Until the 1980s most ITU standards were primarily followed in Europe. Since then, ITU modem standards have achieved a worldwide audience of followers, and enable true global communications compatibility. Although most modern modems are compatible with ITU standards, a large base of modems are still being used that were designed to Bell System standards that were popular in North America through the 1980s. In Chapter 5 we will examine some of the more common ITU modem standards as well as a few Bell System standards that will provide us with an understanding of some of the compatibility problems that can occur when communications are attempted between Bell System and ITU compatible modems.

Terminal and mainframe computer operations

When referring solely to terminal operations, the terms half-duplex and full-duplex operation take on meanings different from the communications mode of the transmission medium. Vendors commonly use the term half-duplex to denote that the terminal device is in a local copy mode of operation. This means that each time a character is pressed on the keyboard it is printed or displayed on the local terminal as well as transmitted. Thus, a terminal device operated in a half-duplex mode would have each character printed or displayed on its monitor as it is transmitted.

When one says a terminal is in a full-duplex mode of operation this means that each character pressed on the keyboard is transmitted but not immediately displayed or printed. Here the device on the distant end of the transmission path must 'echo' the character back to the originator, which, upon receipt, displays or prints the character. Thus, a terminal in a full-duplex mode of operation would only print or display the characters pressed on the keyboard after the character is echoed back by the device at the other end of the line. Figure 1.19 illustrates the terms full- and half-duplex as they apply to terminal devices. Note that although most conventional terminals have a switch to control the duplex setting of the device, personal computer users normally obtain their duplex setting via the

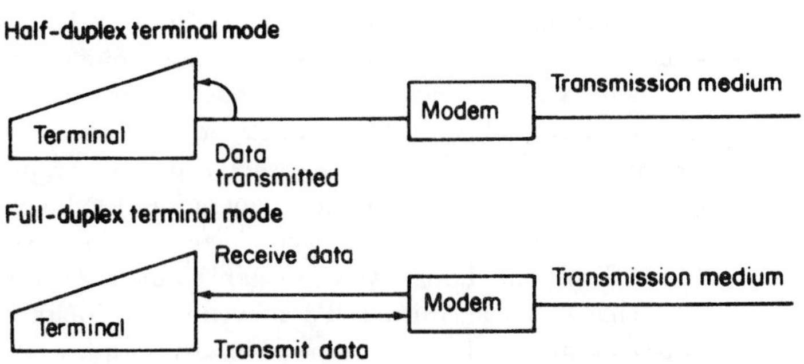

Figure 1.19 Terminal operation modes. Top: the term half-duplex terminal mode implies that data transmitted is also printed on the local terminal. This is known as local copy. Bottom: the term full-duplex terminal mode implies that no local copy is provided

software program they are using. Thus, the term 'echo on' during the initialization of a communications software program would refer to the process of displaying each character on the user's screen as it is transmitted.

When we refer to half- and full-duplex with respect to mainframe computer systems we are normally specifying whether or not they echo received characters back to the originator. A half-duplex computer system does not echo characters back, while a full-duplex computer system echoes each character it receives.

Different character displays

When considering the operating mode of the terminal device, the transmission medium, and the operating mode of the mainframe computer on the distant end of the transmission path as an entity, three things could occur in response to each character you press on a keyboard. Assuming a transmission medium is employed that can be used for either half- or full-duplex communications, your terminal device could print or display no character for each character transmitted, one character for each character transmitted, or two characters for each character transmitted. Here the resulting character printed or displayed would be dependent upon the operating mode of the terminal device and the host computer you are connected to as indicated in Table 1.4.

Table 1.4 Operating mode and character display

Operating mode		
Terminal device	Host computer	Character display
Half-duplex	Half-duplex	1 character
Half-duplex	Full-duplex	2 characters
Full-duplex	Half-duplex	No characters
Full-duplex	Full-duplex	1 character

To understand the character display column in Table 1.4, let us examine the two-character display result caused by the terminal device operating in a half-duplex mode while the host computer operates in a full-duplex mode.

When the terminal is in a half-duplex mode it echoes each transmitted character onto its printer or display. At the other end of the communications path, if the computer is in a full-duplex mode of operation it will echo the received character back to the terminal, causing a second copy of the transmitted character to be printed or displayed. Thus, two characters would appear on your printer or display for each character transmitted. To alleviate this situation, you would change the transmission mode of your terminal to full-duplex. This would normally be accomplished by turning 'echo' off during the initialization of a communications software program, if using a personal computer; or you would turn a switch to half-duplex if operating a conventional terminal.

1.5 TRANSMISSION TECHNIQUES

Data can be transmitted either synchronously or asynchronously. Asynchronous transmission is commonly referred to as a start–stop transmission where one character at a time is transmitted or received. Start and stop bits are used to separate characters and synchronize the receiver with the transmitter, thus providing a method of reducing the possibility that data becomes garbled.

Most devices designed for human–machine interaction that are teletype compatible transmit data asynchronously. By teletype compatible, we refer to terminals and personal computers that operate similarly to the Teletype® terminal manufactured by Western Electric, originally a subsidiary of AT&T and now known as Lucent Technology after this equipment manufacturing arm of AT&T was spun off during 1996. Various versions of this popular terminal have been manufactured for over 30 years and an installed base of approximately one million such terminals at one time during the late 1970s were in operation worldwide. As characters are depressed on the device's keyboard they are transmitted, with idle time occurring between the transmission of characters. This is illustrated in Figure 1.20(b). Although many teletype terminals have been replaced by personal computers, asynchronous transmission pioneered by those terminals remains very popular in use.

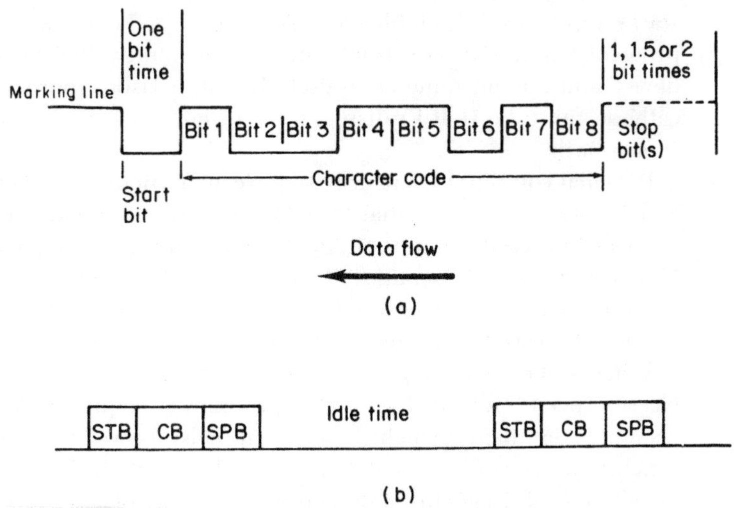

Figure 1.20 Asynchronous (start-stop) transmission. (a) Transmission of one 8-bit character. (b) Transmission of many characters. STB = start bit; CB = character bits; SPB = stop bit(s); idle time is time between character transmission

Asynchronous transmission

In asynchronous transmission, each character to be transmitted is encoded into a series of pulses. The transmission of the character is started by a start pulse equal in length to a code pulse. The encoded character (series of pulses) is followed by a stop pulse that may be equal to or longer than the code pulse, depending upon the transmission code used.

The start bit represents a transition from a mark to a space. Since in an idle condition when no data are transmitted the line is held in a marking condition, the start bit serves as an indicator to the receiving device that a character of data follows. Similarly, the stop bit causes the line to be placed back into its previous 'marking' condition, signifying to the receiver that the data character is completed.

As illustrated in Figure 1.20(a), the transmission of an 8-bit character requires either 10 or 11 bits, depending upon the length of the stop bit. In actuality the eighth bit may be used as a parity bit for error detection and correction purposes. The use of the parity bit is described in detail in Section 1.12.

In the start–stop mode of transmission, transmission starts anew on each character and stops after each character. This is indicated in Figure 1.20(b). Since synchronization starts anew with each character, any timing discrepancy is cleared at the end of each character, and synchronization is maintained on a character-by-character basis. Asynchronous transmission normally is used for transmission at speeds at or under 33 600 bps over the switched telephone network or on leased lines, while data rates up to 115 200 bps are possible over a direct connect cable whose distance is normally limited to approximately 50 feet.

The term asynchronous TTY, or TTY compatible, refers to the asynchronous start–stop protocol employed originally by Teletype® terminals and is the protocol where data are transmitted on a line-by-line basis between a terminal device and a mainframe computer. In comparison, more modern terminals with cathode ray tube (CRT) displays are usually designed to transfer data on a full screen basis.

Personal computer users only require an asynchronous communications adapter and a software program that transmits and receives data on a line-by-line basis to connect to a mainframe that supports asynchronous TTY compatible terminals. Here the software program that transmits and receives data on a line-by-line basis is normally referred to as a TTY emulator program and is the most common type of communications program written for use with personal computers.

When a personal computer is used to transmit and receive data on a full screen basis a specific terminal emulator program is required. Most terminal emulator programs emulate asynchronous terminals; however, some programs emulate synchronous terminals. Concerning the latter, such programs usually require the installation of a synchronous communications adapter in the personal computer and the use of a synchronous modem. Since a personal computer includes a video display onto which characters and graphics can be positioned, the PC can be used to emulate a full-screen addressable terminal. Thus, with appropriate software or a combination of hardware and software you can use a personal computer as a replacement for proprietary terminals manufactured to operate with a specific type of mainframe computer, as well as to perform such local processing as spreadsheet analysis and word processing functions. In fact, during the early 1990s a large majority of conventional terminal devices in business use were replaced by PC-operating terminal emulation software.

Synchronous transmission

A second type of transmission involves sending a grouping of characters in a continuous bit stream. This type of transmission is referred to as synchronous or bit-stream synchronization. In the synchronous mode of transmission, modems located at each end of the transmission medium normally provide a timing signal or clock that is used to establish the data transmission rate and enable the devices attached to the modems to identify the appropriate characters as they are being transmitted or received. In some instances, timing may be provided by the terminal device itself or a communication component, such as a multiplexer or front-end processor channel. No matter what timing source is used, prior to beginning the transmission of data the transmitting and receiving devices must establish synchronization among themselves. In order to keep the receiving clock in step with the transmitting clock for the duration of a stream of bits that may represent a large number of consecutive characters, the transmission of the data is preceded by the transmission of one or more special characters. These special synchronization or 'syn' characters are at the same code level (number of bits per character) as the coded information to be transmitted. They have, however, a unique configuration of zero and one bits which are interpreted as the syn character. Once a group of syn characters is transmitted, the receiver recognizes and synchronizes itself onto a stream of those syn characters.

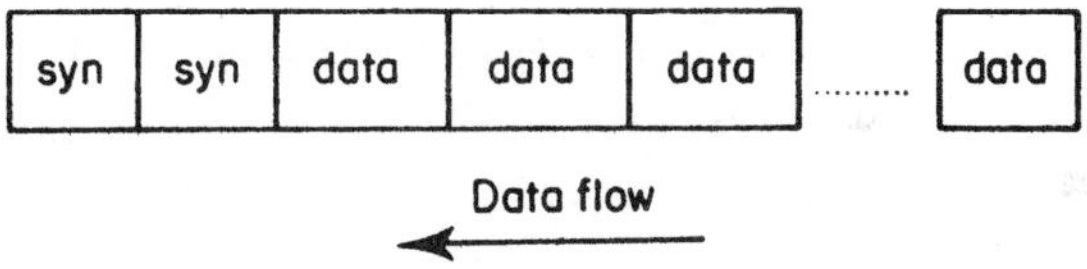

Figure 1.21 Synchronous transmission. In synchronous transmission, one or more syn characters are transmitted to establish clocking prior to the transmission of data

After synchronization is achieved, the actual data transmission can proceed. Synchronous transmission is illustrated in Figure 1.21. In synchronous transmission, characters are grouped or blocked into groups of characters, requiring a buffer or memory area so characters can be grouped together. In addition to having a buffer area, more complex circuitry is required for synchronous transmission since the receiving device must remain in phase with the transmitter for the duration of the transmitted block of information. Synchronous transmission is normally used for data transmission rates in excess of 2000 bps. The major characteristics of asynchronous and synchronous transmission are denoted in Table 1.5.

In examining the entries in Table 1.5 a word of explanation is in order concerning the fifth entry for asynchronous transmission. The ability to transmit at 56 000 bps is based upon a relatively new modem technology that allows one analog to digital conversion to occur. This means that one end of a dialed circuit path must directly connect to digital equipment. Normally this can only be accomplished in one direction when a call is originated over the switched telephone network.

Table 1.5 Transmission technique characteristics

Asynchronous

1. Each character is prefixed by a start bit and followed by one or more stop bits.
2. Idle time (period of inactivity) can exist between transmitted characters.
3. Bits within a character are transmitted at prescribed time intervals.
4. Timing is established independtly in the computer and terminal.
5. Transmission speeds normally do not exceed 33 600 bps bidirectional (or 56 000 in one direction and 33 600 in the opposite direction) over switched facilities or leased lines and 56 000 bps over analog dedicated links and leased lines.

Synchronous

1. Syn characters prefix transmitted data.
2. Syn characters are transmitted between blocks of data to maintain line synchronization.
3. No gaps exist between characters.
4. Timing is established and maintained by the transmitting and receving modems, the terminal, or other devices.
5. Terminals must have buffers.
6. Transmission speeds normally are in excess of 2000 bps.

However, when dedicated or leased lines are used it becomes theoretically possible to obtain a 56 kbps transmission capability in both directions. Although modem equipment marketed during 1998 only enabled 56 kbps transmission in one direction via the switched network, it is quite possible that by the time you read this book a bidirectional 56 kbps transmission capability will be available for dedicated and leased analog lines.

1.6 TYPES OF TRANSMISSION

The two types of data transmission one can consider are serial and parallel. For serial transmission the bits which comprise a character are transmitted in sequence over one line, whereas in parallel transmission characters are transmitted serially but the bits that represent the character are transmitted in parallel. If a character consists of eight bits, then parallel transmission would require a minimum of eight lines. Additional lines may be necessary for control signals and for the transmission of a parity bit. Although parallel transmission is used extensively in computer-to-peripheral unit transmission, it is not normally employed other than in dedicated data transmission usage over relatively short distances owing to the cost of the extra circuits required.

A typical use of parallel transmission is the in-plant connection of badge readers and similar devices to a computer in that facility. Parallel transmission can reduce the cost of terminal circuitry since the terminal does not have to convert the internal character representation to a serial data stream for transmission. The cost of the transmission medium and interface will, however, increase because of the additional number of conductors required. Since the total character can be transmitted at the same moment in time using parallel transmission, higher data transfer rates can be obtained than are possible with serial transmission facilities.

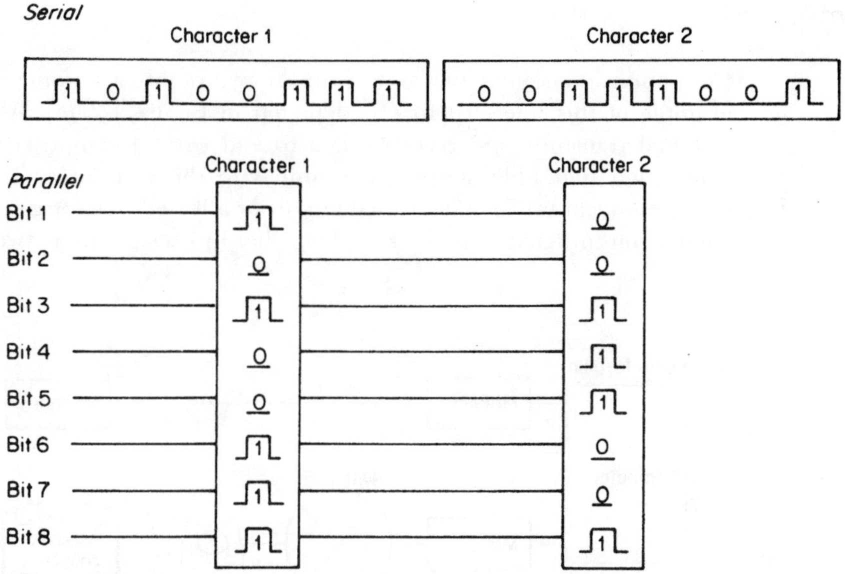

Figure 1.22 Types of data transmission. In serial transmission, the bits that comprise the character to be transmitted are sent in sequence over one line. In parallel transmission, the characters are transmitted serially but the bits that represent the character are transmitted in parallel

For this reason, most local facility communications between computers and their peripheral devices are accomplished using parallel transmission. In comparison, communications between terminal devices and computers normally occur serially, since this requires only one line to interconnect the two devices that need to communicate with one another. Figure 1.22 illustrates serial and parallel transmission.

1.7 LINE STRUCTURE

The geographical distribution of terminal devices and the distance between each device and the device it transmits to are important parameters that must be considered in developing a wide area network configuration. The method used to interconnect personal computers and terminals to mainframe computers or to other devices is known as line structure and results in a computer's network configuration.

Types of line structure

The two types of line structure used in networks are point-to-point and multipoint, the latter also commonly referred to as multidrop lines.

Point-to-point

Communications lines that only connect two points are point-to-point lines. An example of this line structure is depicted in Figure 1.23(a). As illustrated, each terminal transmits and receives data to and from a computer via an individual connection that links a specific terminal to the computer. The point-to-point connection can utilize a dedicated circuit or a leased line, or can be obtained via a connection initiated over the switched (dial-up) telephone network.

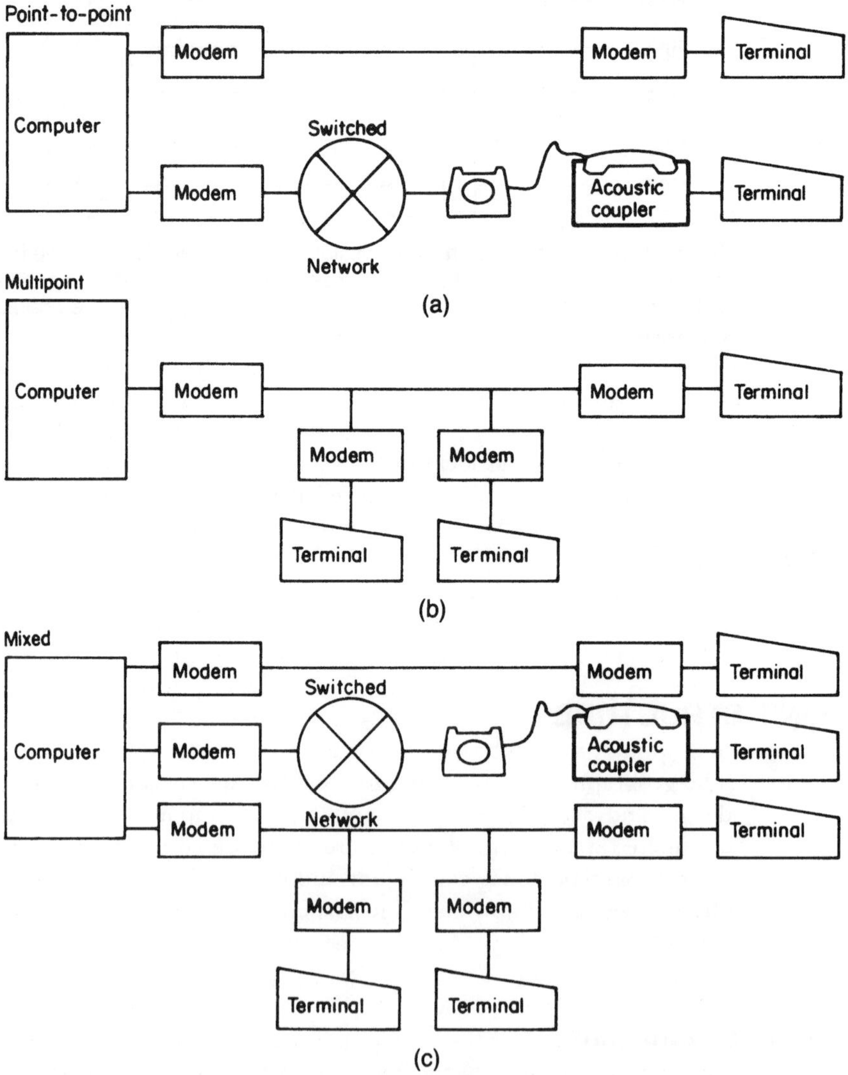

Figure 1.23 Line structures in networks. Top: point-to-point line structure. Center: multipoint (multidrop) line structure. Bottom: mixed network line structure

Multipoint

When two or more terminal locations share portions of a common line, the line is a multipoint or multidrop line. Although no two devices on such a line can transmit data at the same time, two or more devices may receive a message at the same time. The number of devices receiving such a message is dependent upon the addresses assigned to the message recipients. In some systems a 'broadcast' address permits all devices connected to the same multidrop line to receive a message at the same time. When multidrop lines are employed, overall line costs may be reduced since common portions of the line are shared for use by all devices connected to that line. To prevent data transmitted from one device from interfering with data transmitted from another device, a line discipline or control must be established for such a link. This discipline controls transmission so no two devices transmit data at the same time. A multidrop line structure is depicted in Figure 1.23(b).

Although modems are shown on point-to-point and multipoint line structures illustrated in Figure 1.23, both line structures are also applicable to digital transmission services. Dataphone Digital Service facilities support both point-to-point and multipoint line structures, with the use of modems and acoustic couplers shown in Figure 1.23 replaced by the use of DSUs. T1 and fractional T1 lines, however, are only available as a point-to-point line structure and require the use of DSUs/CSUs in place of the modems shown at the top of Figure 1.23.

Depending upon the type of transmission facilities used, it may be possible to intermix both point-to-point and multipoint lines in developing a network. For example, analog facilities and the use of DDS support both line structures which enable the development of a mixed network line structure as shown in Figure 1.23(c).

1.8 LINE DISCIPLINE

When several devices share the use of a common, multipoint communications line, only one device may transmit at any one time, although one or more devices may receive information simultaneously. To prevent two or more devices from transmitting at the same time, a technique known as 'poll and select' is utilized as the method of line discipline for multidrop lines. To utilize poll and select, each device on the line must have a unique address of one or more characters as well as circuitry to recognize a message sent from the computer to that address. When the computer polls a line, in effect it asks each device in a predefined sequence if it has data to transmit. If the device has no data to transmit, it informs the computer of this fact and the computer continues its polling sequence until it encounters a device on the line that has data to send. Then the computer acts on that data transfer.

As the computer polls each device, the other devices in the line must wait until they are polled before they can be serviced. Conversely, transmission of data from the computer to each device on a multidrop line is accomplished by the computer selecting the device address to which those data are to be transferred, informing the device that data are to be transferred to it, and then transmitting data to the

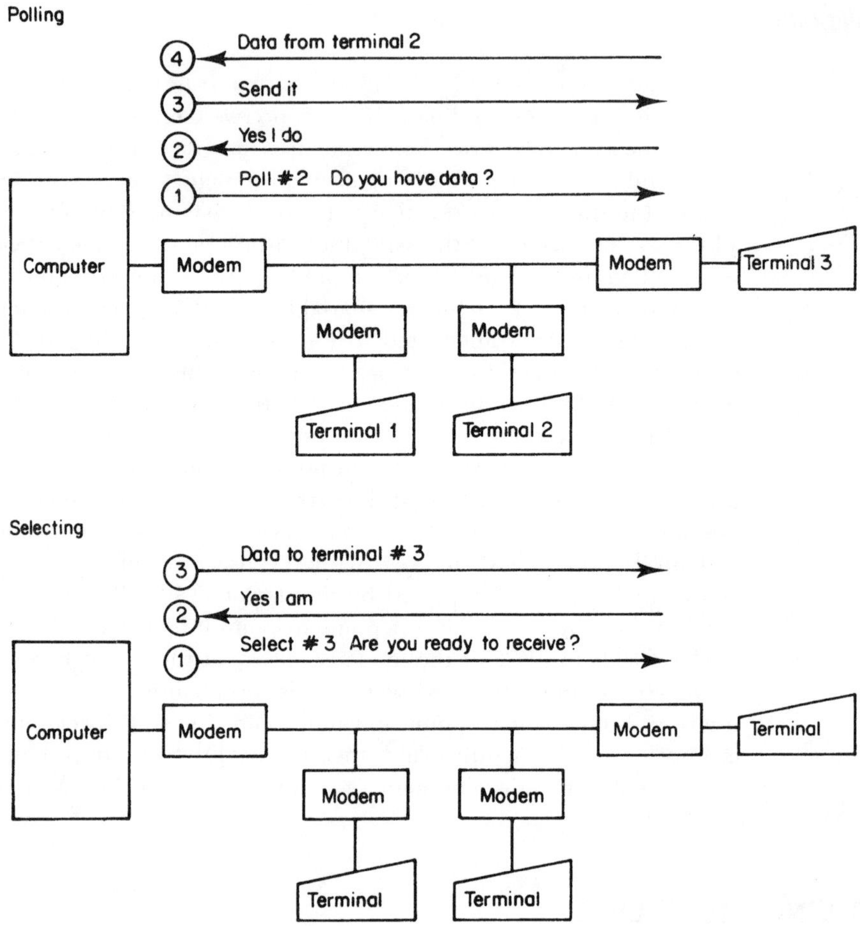

Figure 1.24 Poll and select line discipline. Poll and select is a line discipline which permits several devices to use a common line facility in an orderly manner

selected device. Polling and selecting can be used to service both asynchronous or synchronous operating terminal devices connected to independent multidrop lines. Owing to the control overhead of polling and selecting, synchronous devices are normally serviced in this type of environment. By the use of signals and procedures, polling and selecting line control ensures the orderly and efficient utilization of multidrop lines. An example of a computer polling the second terminal on a multipoint line and then receiving data from that device is shown at the top of Figure 1.24. At the bottom of that illustration, the computer first selects the third terminal on the line and then transfers a block of data to that device.

When terminals transmit data on a point-to-point line to a computer or another terminal, the transmission of that data occurs at the discretion of the terminal operator. This method of line control is known as 'non-poll-and-select' or 'free-wheeling' transmission.

1.9 NETWORK TOPOLOGY

The topology or structure of a wide area network is based upon the interconnection of point-to-point and/or multipoint lines used to develop the network. Although an infinite number of network topologies can be developed by connecting lines to one another, there are several basic types of topologies that are commonly used to construct wide area networks or parts of those networks. These topologies include the point-to-point, 'v', star, and mesh.

Figure 1.25(a) illustrates the four basic WAN topology structures, with dots used to reference end points or network nodes. These structures can be interconnected to one another to link organizational locations that represent different types of clusters of locations. Figure 1.25(b) illustrates the creation of two examples of different network topologies based upon connecting two 'v' structures and joining a 'v' to a star structure.

The design of a wide area network topology is commonly based upon economics and reliability. From an economic perspective the WAN should use one or more network structures that minimize the distance of circuits used to connect geographically separated locations, since the monthly cost of a circuit is in general proportional to its distance. From a reliability perspective, the ability to have multiple paths between nodes enables an alternative path to be used in the event of a transmission impairment making the primary path inoperative. Since redundancy requires additional communications paths, a trade-off occurs between the need to obtain an economically efficient network and a highly reliable network. For example, the use of a mesh structure permits multiple paths between nodes but requires more circuits than a star topology.

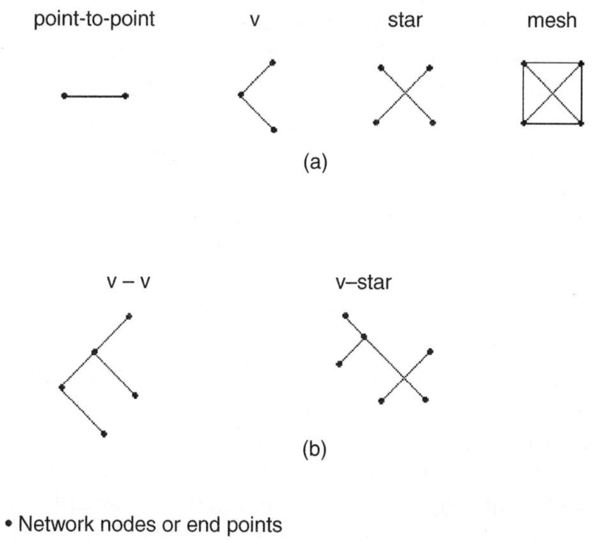

Figure 1.25 Wide Area Network Topologies: (a) basic topology structures; (b) connecting different structures

1.10 TRANSMISSION RATE

Many factors can affect the transmission rate at which data is transferred on a wide area network. The types of modems and acoustic couplers used on analog facilities or the types of DSUs or CSUs used on digital facilities, as well as the line discipline and the type of computer channel to which a terminal is connected via a transmission medium, play governing roles that affect transmission rates. However, the transmission medium itself is a most important factor in determining transmission rates.

Data transmission services offered by communications carriers such as AT&T, MCI, Sprint, and British Telecom are based on their available plant facilities. Depending upon terminal and computer locations, two types of transmission services may be available. The first type of service, analog transmission, is readily available in the form of switched and leased telephone lines. Digital transmission is available in most large cities; however, the type of digital service offered can vary from city to city as well as between different locations within a city. For example, AT&T's DDS service, while available in over 100 cities in North America, is only supported by certain telephone company offices in each city. Thus, an analog extension is required to connect to this service from non-digital service locations as previously illustrated in Figure 1.16. Within each type of service several grades of transmission are available for consideration.

Analog service

In general, analog service offers the user three grades of transmission: narrowband, voice-band and wideband. The data transmission rates capable on each of these grades of service depend upon the bandwidth and electrical properties of each type of circuit offered within each grade of service. Basically, transmission speed is a function of the bandwidth of the communications line: the greater the bandwidth, the higher the possible speed of transmission.

Narrowband facilities are obtained by the carrier subdividing a voice-band circuit or by grouping a number of transmissions from different users onto a single portion of a circuit by time. Transmission rates obtained on narrowband facilities range between 45 and 300 bps. Teletype® terminals that connect to message switching systems are the primary example of the use of narrowband facilities.

The primary use of analog narrowband transmission facilities is by information distribution organizations, such as news services and weather bureaux. Although the utilization of analog narrowband transmission facilities was popular through the 1970s, since that time many organizations that used that medium have migrated to transmission facilities that permit higher data transmission rates.

While narrowband facilities have a bandwidth in the range of 200 to 400 Hz, voice-band facilities have a bandwidth in the range of 3000 Hz. Data transmission speeds obtainable on voice-band facilities are differentiated by the type of voice-band facility utilized—switched dial-up transmission or transmission via a leased line. For transmission over the switched telephone network data rates up to 56 kbps may be obtainable in one direction and up to 33.6 kbps in the opposite direction when communicating with an information utility or corporate remote access device that is directly connected to the switched network, resulting in only one analog–digital conversion in the downstream direction. Although the fact that leased lines can be conditioned enabled higher speed transmission to occur on analog leased lines than the switched telephone network through the 1980s, since then most modems have been designed to operate on both transmission facilities at equal operating rates. When operating on the switched network a high speed modem must use echo cancellation to obtain a full-duplex transmission capability on a two-wire circuit. In comparison, when used on a four-wire leased line a modem can use each two-wire pair for transmission in different directions.

Although low data speeds can be transmitted on both narrowband and voice-band circuits, one should not confuse the two, since a low data speed on a voice circuit is transmission at a rate far less than the maximum permitted by that type of circuit, whereas a low rate on a narrowband facility is at or near the maximum transmission rate permitted by that type of circuit.

Analog facilities which have a higher bandwidth than voice-band are termed wideband or group-band facilities since they provide a wider bandwidth through the grouping of a number of voice-band circuits. Wideband facilities are available only on leased lines and permit transmission rates at or in excess of 40 800 bps. Transmission rates on wideband facilities vary with the offerings of communications carriers. Speeds normally available include 40.8, 50 and 230.4 kbps. The growth in the availability of high speed digital transmission facilities during the late 1980s resulted in a corresponding decrease in the use of high speed analog transmission facilities. By the late 1990s, most wideband analog transmission facilities were replaced by the use of high speed digital transmission facilities.

For direct connect circuits, transmission rates are a function of the distance between the terminal and the computer as well as the gauge of the conductor used.

Digital service

In the area of digital service, several offerings are currently available for user consideration, that can be categorized by the data transmission rates they support—low, medium, and high data rates.

Low speed digital transmission services include AT&T's Dataphone Digital Service (DDS) and equivalent offerings from other communications carriers. DDS provides full-duplex, point-to-point, and multipoint synchronous transmission via leased lines at speeds of 2.4, 4.8, 9.6, 19.2 and 56 kbps as well as switched 56 kbps service.

Medium speed digital transmission service is represented by AT&T's Accunet Spectrum of Digital Services (ASDS) and equivalent offerings from other communication carriers. As discussed earlier in this chapter, ASDS provides a fractional T1 transmission capability from 64 kbps to 1.544 Mbps in increments of 64 kbps. Common data transmission rates supported by ASDS and equivalent offerings from other carriers include 64, 128, 256, 512 and 768 kbps.

High speed digital transmission service is represented by T1 facilities operating at 1.544 Mbps through T3 facilities that operate at 44.736 Mbps. A relatively new

Table 1.6 Common wide area network transmission facilities

Facility	Transmission speed	Typical use
Analog		
Narrowband	45–300 pbs	Message switching
Voice-band		
Switched	Up to 56 000 bps	Time sharing; remote job entry; information utility access; file transfer
Leased	Up to 56 000 bps	Computer-to-computer; remote job entry; tape-to-tape transmission
Wideband	At or over 40 800 bps	High-speed terminal to high-speed terminal
Digital		
Switched	56 kbps, 384 kbps, 1.544 Mbps	LAN internetworking, Videoconferencing, computer-to-computer
Leased line	2.4, 4.8, 9.6, 19.2 kbps	Remote job entry; computer-to-computer
	56 kbps, $n \times 56/64$ kbps	High speed fax; LAN internetworking
	1.544 Mbps	Integrated voice and data
	44 Mbps	Integrated voice and data

digital offering known as a synchronous optical network (SONET) in North America, and as the Synchronous Digital Hierarchy (SDH) in Europe, provides a mechanism for interconnecting high speed networks via optical media. SONET speeds start at 51.84 Mbps and increase in multiples of that data rate to 2.488 Gbps.

Table 1.6 lists the main analog and digital facilities, the range of transmission speeds over those facilities, and the general use of such facilities. The entry $n \times 56/64$ represents fractional service, with n varying from 1 to 12 on some carrier offerings while other carriers support n varying up to 24, the latter providing a full T1 service. When we discuss the T1 carrier later in this book we will also discuss in additional detail the reason why the fractional T1 operating rate increments at either 56 or 64 kbps intervals.

1.11 TRANSMISSION CODES

Data within a computer or terminal device is structured according to the architecture of the device. The internal representation of data in either device is seldom suitable for transmission other than to peripheral units attached to terminals or computers. In most cases, to effect data transmission, internal formatted data must be redesigned or translated into a suitable transmission code. This transmission code creates a correspondence between the bit encoding of data for transmission or internal device representation and printed symbols. The end result of the translation is usually dictated by the character code that the devices communicating with one another mutually support. Frequently available codes include Baudot, which is a 5-level (5 bits per character) code; binary-coded decimal (BCD), which is a 6-level code; American Standard Code for Information Interchange (ASCII), which is normally a 7-level code; and the extended binary-coded decimal interchange code (EBCDIC), which is an 8-level code.

In addition to information being encoded into a certain number of bits based upon the transmission code used, the unique configuration of those bits to represent certain control characters can be considered as a code that can be used to effect line discipline. These control characters may be used to indicate the acknowledgement of the receipt of a block of data without errors (ACK), the start of a message (SOH), or the end of a message (ETX), with the number of permissible control characters standardized according to the code employed. With the growth of computer-to-computer data transmission, a large amount of processing can be avoided by transferring data in the format used by the computer for internal processing. Such transmission is known as binary mode transmission, transparent data transfer, code-independent transmission, or native mode transmission.

Morse code

One of the most commonly known codes, the Morse code, is not practical for utilization in a computer communications environment. This code consists of a series of dots and dashes, which, while easy for the human ear to decode, are of unequal length and not practical for data transmission implementation. In addition, since each character in the Morse code is not prefixed with a start bit and terminated with a stop bit, it was initially not possible to construct a machine to automatically translate received Morse transmissions into their appropriate characters.

Baudot code

The Baudot code, which is a 5-level (5 bits per character) code, was the first code to provide a mechanism for encoding characters by an equal number of bits, in this case five. The 5-level Baudot code was devised by Emil Baudot to permit teletypewriters to operator faster and more accurately than relays used to transmit information via telegraph.

Since the number of different characters which can be derived from a code having two different (binary) states is 2^m, where m is the number of positions in the code, the 5-level Baudot code permits 32 unique character bit combinations. Although 32 characters could be represented normally with such a code, the necessity of transmitting digits, letters of the alphabet, and punctuation marks made it necessary to devise a mechanism to extend the capacity of the code to include additional character representations. The extension mechanism was accomplished by the use of two 'shift' characters: 'letters shift' and 'figures shift'. The transmission of a shift character informs the receiver that the characters which follow the shift character should be interpreted as characters from a symbol and numeric set or from the alphabetic set of characters.

Table 1.7 Five-level Baudot code

		Bit Selection				
Letters	Figures	1	2	3	4	5
Characters						
A	-	1	1			
B	?	1			1	1
C	:		1	1	1	
D	$	1			1	
E	3	1				
F	!	1		1	1	
G	&		1		1	1
H				1		
I	8		1	1		
J	'	1	1		1	
K	(	1	1	1	1	
L	)		1			1
M	.			1	1	1
N	,			1	1	
O	9					1
P	Ø		1	1		1
Q	1	1	1	1		1
R	4		1		1	
S		1		1		
T	5				1	
U	7	1	1	1		
V	;		1	1	1	1
W	2	1	1			1
X	/	1		1	1	1
Y	6	1		1		1
Z	"	1				1
Functions						
Carriage return					1	
Line feed			1			
Space				1		
Lettershift	<	1	1	1	1	1
Figures shift	=	1	1		1	1

The 5-level Baudot code is illustrated in Table 1.7 for one particular terminal pallet arrangement. A transmission of all ones in bit positions 1 to 5 indicates a letter shift, and the characters following the transmission of that character are interpreted as letters. Similarly, the transmission of ones in bit positions 1, 2, 4 and 5 would indicate a figures shift, and the following characters would be interpreted as numerals or symbols based upon their code structure. Although the Baudot code is quite old in comparison to the age of personal computers, it is the transmission code used by the Telex network which at one time was extensively employed in the business community to send messages throughout the world.

BCD code

The development of computer systems required the implementation of coding systems to convert alphanumeric characters into binary notation and the binary notation of computers into alphanumeric characters. The BCD system was one of the earliest codes used to convert data to a computer-acceptable form. This coding technique permits decimal numeric information to be represented by four binary bits and permits an alphanumeric character set to be represented through the use of six bits of information. This code is illustrated in Table 1.8. An advantage of this code is that two decimal digits can be stored in an 8-bit computer word and

Table 1.8 Binary-coded decimal (BCD) system

Bit position						
b_6	b_5	b_4	b_3	b_2	b_1	Character
0	0	0	0	0	1	A
0	0	0	0	1	0	B
0	0	0	0	1	1	C
0	0	0	1	0	0	D
0	0	0	1	0	1	E
0	0	0	1	1	0	F
0	0	0	1	1	1	G
0	0	1	0	0	0	H
0	0	1	0	0	1	I
0	1	0	0	0	1	J
0	1	0	0	1	0	K
0	1	0	0	1	1	L
0	1	0	1	0	0	M
0	1	0	1	0	1	N
0	1	0	1	1	0	O
0	1	0	1	1	1	P
0	1	1	0	0	0	Q
1	1	1	0	0	1	R
1	0	0	0	1	0	S
1	0	0	0	1	1	T
1	0	0	1	0	0	U
1	0	0	1	0	1	V
1	0	0	1	1	0	W
1	0	0	1	1	1	X
1	0	1	0	0	0	Y
1	0	1	0	0	1	Z
1	1	0	0	0	0	0
1	1	0	0	0	1	1
1	1	0	0	1	0	2
1	1	0	0	1	1	3
1	1	0	1	0	0	4
1	1	0	1	0	1	5
1	1	0	1	1	0	6
1	1	0	1	1	1	7
1	1	1	0	0	0	8
1	1	1	0	0	1	9

manipulated with appropriate computer instructions. Although only 36 characters are shown for illustrative purposes, a BCD code is capable of containing a set of 2^6 or 64 different characters.

EBCDIC code

In addition to transmitting letters, numerals, and punctuation marks, a considerable number of control characters may be required to promote line discipline. These control characters may be used to switch on and off devices which are connected to the communications line, control the actual transmission of data, manipulate message formats, and perform additional functions. Thus, an extended character set is usually required for data communications. One such character set is EBCDIC code. The extended binary coded decimal interchange code (EBCDIC) is an extension of the BCD system and uses 8 bits for character representation. This code permits 2^8 or 256 unique characters to be represented, although currently a lesser number is assigned meanings. This code is primarily used for transmission by byte-oriented computers, where a byte is normally a grouping of eight consecutive binary digits operated on as a unit by the computer. The use of this code by computers may alleviate the necessity of the computer performing code conversion if the connected terminals operate with the same character set.

One of the more interesting examples of computer terminology is the use of the terms bytes and octets. In the early days of computer development machines were constructed using different groupings of bits that were operated upon as an entity and referred to as a byte. At various times during the 1960s groupings of 4, 6, 8, 12 and 16 bits operated upon as an entity were referred to as bytes. Owing to this ambiguity, the term octet became commonly used in communications to refer to an 8-bit byte. However, since virtually all modern computers use 8-bit bytes, this author will use the term byte to refer to a grouping of 8 bits.

Several subsets of EBCDIC exist that have been tailored for use with certain devices. As an example, IBM 3270 type terminal products would not use a paper feed and its character representation is omitted in the EBCDIC character subset used to operate that type of device, as indicated in Table 1.9.

ASCII code

As a result of the proliferation of data transmission codes, several attempts to develop standardized codes for data transmission were made. One such code is the American Standard Code for Information Interchange (ASCII). This 7-level code is based upon a 7-bit code developed by the International Standards Organization (ISO) and permits 128 possible combinations or character assignments, to include 96 graphic characters that are printable or displayable and 32 control characters to include device control and information transfer control characters. Table 1.10 lists the ASCII character set while Table 1.11 lists the ASCII control characters by position and their meaning. A more detailed explanation of these control characters is contained in the section covering protocols in this chapter.

Table 1.9 EBCIDIC code implemented for the IBM 3270 information display system

Bits	Hex1	00				01				10				11				← Bits 0, 1
		00	01	10	11	00	01	10	11	00	01	10	11	00	01	10	11	← 2, 3
4567		0	1	2	3	4	5	6	7	8	9	A	B	C	D	E	F	← Hex 0
0000	0	NUL	DLE			SP	&	-									0	
0001	1	SOH	SBA					/		a	j			A	J		1	
0010	2	STX	EUA		SYN					b	k	s		B	K	S	2	
0011	3	ETX	IC							c	l	t		C	L	T	3	
0100	4									d	m	u		D	M	U	4	
0101	5	PT	NL							e	n	v		E	N	V	5	
0110	6			ETB						f	o	w		F	O	W	6	
0111	7			ESC	EOT					g	p	x		G	P	X	7	
1000	8									h	q	y		H	Q	Y	8	
1001	9		EM							i	r	z		I	R	Z	9	
1010	A					c	!	1p	:									
1011	B					.	$	,	#									
1100	C		DUP		RA	<	*	%	@									
1101	D		SF	ENQ	NAK	(	)	—	'									
1110	E		FM			+	;	>	=									
1111	F		ITB		SUB	|	—ar	?	"									

The primary difference between the ASCII character set listed in Table 1.10 and other versions of the ITU International Alphabet Number 5 is the currency symbol. Although the bit sequence 0 1 0 0 0 1 0 is used to generate the dollar ($) currency symbol in the United States, in the United Kingdom that bit sequence results in the generation of the pound sign (£). Similarly, this bit sequence generates other currency symbols when the ITU International Alphabet Number 5 is used in other countries.

Extended ASCII

Members of the IBM PC series and compatible computers use an extended ASCII character set which is represented as an 8-level code. The first 128 characters in the character set, ASCII values 0 through 127, correspond to the ASCII character set listed in Table 1.10 while the next 128 characters can be viewed as an extension of that character set since they require an 8-bit representation.

Caution is advised when transferring IBM PC files since characters with ASCII values greater than 127 will be received in error when they are transmitted using 7

Table 1.10 The ASCII character set. This coded character set is to be used for the general interchange of information among information processing systems, communications systems, and associated equipment

Bits												
b_7					0	0	0	0	1	1	1	1
b_6					0	0	1	1	0	0	1	1
b_5					0	1	0	1	0	1	0	1
b_4	b_3	b_2	b_1	COLUMN / ROW	0	1	2	3	4	5	6	7
0	0	0	0	0	NUL	DLE	SP	0	@	P	\	p
0	0	0	1	1	SOH	DC1	!	1	A	Q	a	q
0	0	1	0	2	STX	DC2	"	2	B	R	b	r
0	0	1	1	3	ETX	DC3	#	3	C	S	c	s
0	1	0	0	4	EOT	DC4	$	4	D	T	d	t
0	1	0	1	5	ENQ	NAK	%	5	E	U	e	u
0	1	1	0	6	ACK	SYN	&	6	F	V	f	v
0	1	1	1	7	BEL	ETB	/	7	G	W	g	w
1	0	0	0	8	BS	CAN	(	8	H	X	h	x
1	0	0	1	9	HT	EM	)	9	I	Y	i	y
1	0	1	0	10	LF	SUB	★	:	J	Z	j	z
1	0	1	1	11	VT	ESC	+	;	K	[	k	{
1	1	0	0	12	FF	FS	,	<	L	\	l	∫
1	1	0	1	13	CR	GS	–	=	M	j	m	}
1	1	1	0	14	SO	RS	.	>	N	Λ	n	~
1	1	1	1	15	SI	US	/	?	O	—	o	DEL

Note that b_7 is the higher order bit and b_1 is the low order bit as indicated by the following example for coding the letter C.

b_7	b_6	b_5	b_4	b_3	b_2	b_1
1	0	0	0	0	1	1

data bits. This is because the ASCII values of these characters will be truncated to values in the range 0 to 127 when transmitted with 7 bits from their actual range of 0 to 255. To alleviate this problem from occurring you can initialize your communications software for 8-bit data transfer; however, the receiving device must also be capable of supporting 8-bits ASCII data.

Table 1.11 ASCII control characters

Column/row	Control character	Mnemonic and meaning	
0/0	^@	NUL	Null (CC)
0/1	^A	SOH	Start of heading (CC)
0/2	^B	STX	Start of text (CC)
0/3	^C	ETX	End of text (CC)
0/4	^D	EOT	End of transmission (CC)
0/5	^E	ENQ	Enquiry (CC)
0/6	^F	ACK	Acknowledgement (CC)
0/7	^G	BEL	Bell
0/8	^H	BS	Backspace (FE)
0/9	^I	HT	Horizonatal tabulation (FE)
0/10	^J	LF	Line feed (FE)
0/11	^K	VT	Vertical tabulation (FE)
0/12	^L	FF	Form feed (FE)
0/13	^M	CR	Carriage return (FE)
0/14	^N	SO	Shift out
0/15	^O	SI	Shift in
1/0	^P	DLE	Date link escape (CC)
1/1	^Q	DC1	Device control 1
1/2	^R	DC2	Device control 2
1/3	^S	DC3	Device control 3
1/4	^T	DC4	Device control 4
1/5	^U	NAK	Negative acknowledge (CC)
1/6	^V	SYN	Synchronous idle (CC)
1/7	^W	ETB	End of transmission block (CC)
1/8	^X	CAN	Cancel
1/9	^Y	EM	End of medium
1/10	^Z	SUB	Substitute
1/11	^[	ESC	Escape
1/12	^/	FS	File separator (IS)
1/13	^]	GS	Group separator (IS)
1/14	~	RS	Record separator (IS)
1/15	,	US	Unit separator (IS)
7/15	^-	DEL	Delete

(CC) communications control; (FE) format effector; (IS) information separator.

Although conventional ASCII files can be transmitted in a 7-bit format, many word processing and computer programs contain text graphics represented by ASCII characters whose values exceed 127. In addition, EXE and COM files which are produced by assemblers and compilers contain binary data that must also be transmitted in 8-bit ASCII to be accurately received. While most communications programs can transmit 7- or 8-bit ASCII data, some programs, especially those developed in the early 1980s and handed down with antiquated computers, may not be able to transmit binary files accurately. This is due to the fact that communications programs that use the control Z character (ASCII SUB) to identify the end of a file transfer will misinterpret a group of 8 bits in an EXE or COM file being transmitted when they have the same 8-bit format as a control Z, and upon detection prematurely close the file. To avoid this situation you should obtain a communications software program that transfers files by blocks of bits or converts the data into a hexadecimal or octal ASCII equivalent prior to transmission if this type of data transfer will be required.

Another communications problem you can encounter occurs when attempting to transmit files using some electronic mail services that are limited to transferring 7-bit ASCII. If you created a file using a word processor which employs 8-bit ASCII codes to indicate special character settings, such as bold and underlined text, attempting to transmit the file via a 7-bit ASCII electronic mail service will result in the proverbial gobbledygook being received at the distant end. Instead, you should first save the file as a 'text' file prior to transmission. Although you will lose any embedded 8-bit ASCII control codes, you will be able to transmit the document over an electronic mail system limited to the transfer of 7-bit ASCII.

Code conversion

A frequent problem in data communications is that of code conversion. Consider what must be done to enable a computer with an EBCDIC character set to transmit and receive information from a terminal with an ASCII character set. When that terminal transmits a character, that character is encoded according to the ASCII character code. Upon receipt of that character, the computer must convert the bits of information of the ASCII character into an equivalent EBCDIC character. Conversely, when data is to be transmitted to the terminal, it must be converted from EBCDIC to ASCII so the terminal will be able to decode and act according to the information in the character that the terminal is built to interpret.

One of the most frequent applications of code conversion occurs when personal computers are used to communicate with IBM mainframe computers.

Normally, ASCII to EBCDIC code conversion is implemented when an IBM PC or compatible personal computer is required to operate as a 3270 type terminal. This type of terminal is typically connected to an IBM or IBM compatible mainframe computer and the terminal's replacement by an IBM PC requires the PC's ASCII coded data to be translated into EBCDIC. There are many ways to obtain this conversion, including emulation boards that are inserted into the system unit of a PC and protocol converters that are connected between the PC and the mainframe computer. Later in this book, we will explore these and other methods that enable the PC to communicate with mainframe computers that transmit data coded in EBCDIC.

Table 1.12 lists the ASCII and EBCDIC code character value for the 10 digits for comparison purposes. In examining the difference between ASCII and EBCDIC coded digits you will note that each EBCDIC coded digit has a value precisely Hex C0 (decimal 192) higher than its ASCII equivalent. Although this might appear to make code conversion a simple process of adding or subtracting a fixed quantity depending upon which way the code conversion takes place, in reality many of the same ASCII and EBCDIC coded characters differ by varying quantities. As an example, the slash (/) character is Hex 2F in ASCII and Hex 61 in EBCDIC, a difference of Hex 92 (decimal 146). In comparison, other characters such as the carriage return and form feed have the same coded value in ASCII and EBCDIC, while other characters are displaced by different amounts in these two

Table 1.12 ASCII and EBCDIC digits comparison

ASCII				
Dec	Oct	Hex	EBCDIC	Digit
048	060	30	F0	0
049	061	31	F1	1
050	062	32	F2	2
051	063	33	F3	3
052	064	34	F4	4
053	065	35	F5	5
054	066	36	F6	6
055	067	37	F7	7
056	070	38	F8	8
057	071	39	F9	9

codes. Due to this, code conversion is typically performed as a table lookup process, with two buffer areas used to convert between codes in each of the two conversion directions. Thus, one buffer area might have the ASCII character set in Hex order in one field of a two-field buffer area, with the equivalent EBCDIC Hex values in a second field in the buffer area. Then, upon receipt of an ASCII character its Hex value is obtained and matched to the equivalent value in the first field of the buffer area, with the value of the second field containing the equivalent EBCDIC Hex value which is then extracted to perform the code conversion.

1.12 ERROR DETECTION AND CORRECTION

As a signal propagates down a transmission medium several factors can cause it to be received in error: the transmission medium employed and impairments caused by nature and machinery.

The transmission medium will have a certain level of resistance to current flow that will cause signals to attenuate. In addition, inductance and capacitance will distort the transmitted signals and there will be a degree of leakage which is the loss in a transmission line due to current flowing across, though insulators, or changes in the magnetic field.

Transmission impairments result from numerous sources. First, Gaussian or white noise is always present as it is the noise level that exists due to the thermal motions of electrons in a circuit. Next, impulse can occur from line hits due to atmospheric static or poor contacts in a telephone system.

Regardless of the cause of a transmission disturbance, its duration and the operating rate of your communications session are the governing factors that determine the effect of the disturbance. For example, consider a noise burst of 0.01 s which is not uncommon and which sounds like a short click during a voice conversation. At a 1200 bps transmission rate the noise burst will result in 12 bits having the potential to be received in error. At 2400 bps the number of bits having the potential to be received in error is increased to 24. Similarly, increasing the duration of the noise burst increases the number of bits that have the potential to be received in error. For example, a noise burst of 0.1 s would result in 120 bits having the potential to be received in error when the data rate is 1200 bps, while 240 bits would have the potential to be received in error when the data rate is 2400 bps.

Asynchronous transmission

In asynchronous transmission the most common form of error control is the use of a single bit, known as a parity bit, for the detection of errors. Owing to the proliferation of personal computer communications, more sophisticated error detection methods have been developed which resemble the methods employed with synchronous transmission.

Parity checking

Character parity checking, also known as vertical redundancy checking (VRC), requires an extra bit to be added to each character in order to make the total quantity of 1 s in the character either odd or even, depending upon whether you are employing odd parity checking or even parity checking. When odd parity checking is employed, the parity bit is set to 1 if the number of 1 s in the character's data bits is even; or it is set at 0 if the number of 1 s in the character's data bits is odd. When even parity checking is used, the parity bit is set to 0 if the number of 1 s in the character's data bits is even; or if it is set to 1 if the number of 1 s in the character's data bits is odd.

Two additional terms used to reference parity settings are 'mark' and 'space'. When the parity bit is set to a mark condition the parity bit is always 1 while space parity results in the parity bit always set to 0. Although not actually a parity setting, parity can be set to none, in which case no parity checking will occur. When transmitting binary data asynchronously, such as between personal computers via a wide area network transmission facility, parity checking must be set to none or off. This enables all 8 bits to be used to represent a character. Table 1.13 summarizes the effect of five types of parity checking upon the eighth data bit in asynchronous transmission.

Table 1.13 Parity effect upon eighth data bit

Parity type	Parity effect
Odd	Eighth bit is logical zero if the total number of logical 1s in the first seven data bits is odd
Even	Eighth bit is logical zero if the total number of logical 1s in the first seven data bits is even.
Mark	Eighth bit is always logical 1.
Space	Eighth bit is always logical zero.
None/Off	Eighth bit is ignored.

For an example of parity checking, let us examine the ASCII character R whose bit composition is 1 0 1 0 0 1 0. Since there are three 1 bits in the character R, a 0 bit would be added if odd parity checking is used or a 1 bit would be added as the parity bit if even parity checking is employed. Thus, the ASCII character R would appear as follows:

data bits	parity bit	
1 0 1 0 0 1 0	0	odd parity check
1 0 1 0 0 1 0	1	even parity check
1 0 1 0 0 1 0	1	mark parity check
1 0 1 0 0 1 0	0	space parity check

Since there are three bits set in the character R, a 0 bit is added if odd parity checking is employed while a 1 bit is added if even parity checking is used. Similarly, mark parity results in the parity bit being set to 1 regardless of the composition of the data bits in the character, while space parity results in the parity bit always being set to 0.

Undetected errors

Although parity checking is a simple mechanism to investigate if a single bit error has occurred, it can fail when multiple bit errors occur. This can be visualized by returning to the ASCII R character example and examining the effect of additional bits erroneously being transformed as indicated in Table 1.14. Here the ASCII R character has three set bits and a one-bit error could transform the number of set bits to four. If parity checking is employed, the received set parity bit would result in the character containing five set bits, which is obviously an error since even parity checking is employed. Now suppose two bits are transformed in error as indicated in Table 1.14. This would result in the reception of a character containing six set bits, which would appear to be correct under even parity checking. Thus, two bit errors in this situation would not be detected by a parity error detection technique.

Table 1.14 Character parity cannot detect an even number of bit errors

ASCII character R	1 0 1 0 0 1 0
Adding an even parity bit	1 0 1 0 0 1 0 1
	1
1 bit in error	1 ∅ 1 0 0 1 0 1
	1 1
2 bits in error	1 ∅ 1 ∅ 0 1 0 1
	1 0 1
3 bits in error	1 ∅ ~~1~~ ∅ 0 1 0 1
	0 1 0 1
4 bits in error	~~1~~ ∅ ~~1~~ ∅ 0 1 0 1

Now consider the effect of three bit errors shown in Table 1.14. In this example, the number of set bits was charged to five which is a detectable error when even parity checking is employed. Let us, however, examine the effect of four bits being in error. In this example at the bottom of Table 1.14 the number of set bits was changed to four, which does not result in the detection of errors when even parity is employed. From the preceding we should recognize that parity can detect the occurrence of an odd number of bits errors within a character. Errors that effect an even number of bits will not be detected. This problem exists regardless of the method of parity checking used.

In addition to the potential of undetected errors, parity checking has several additional limitations. First, the response to parity errors will vary based upon the type of computer with which you are communicating. Certain mainframes will issue a 'Retransmit' message upon detection of a parity error. Some mainframes will transmit a character that will appear as a 'fuzzy box' on your screen in response to detecting a parity error, while other computers will completely ignore parity errors.

When transmitting data asynchronously on a personal computer, most communications programs permit the user to set parity to odd, even, off, space, or mark. Off or no parity would be used if the system with which your are communicating does not check the parity bit for transmission errors. No parity would be used when you are transmitting 8-bit EBCDIC or an extended 8-bit ASCII coded data, such as that available on the IBM PC and similar personal computers. Mark parity means that the parity bit is set to 1, while space parity means that the parity bit is set to 0.

In the asynchronous communications world, two common sets of parameters are used by most bulletin boards, information utilities and supported by mainframe computers. The first set consists of seven data bits and one stop bit with even parity checking employed, while the second set consists of eight data bits and one stop bit using no parity checking. Table 1.15 compares the communications parameter settings of two popular information utilities.

Table 1.15 Communication parameter settings

Parameter	Information utility	
	CompuServe	Dow Jones
Data bits	7/8	8
Parity	even/none	none
Stop bits	1	1
Duplex	full	full

File transfer problems

Although visual identification of parity errors in an interactive environment is possible, what happens if you transfer a large file over the switched telephone network? During the 1970s a typical call over the switched telephone network resulted in the probability of a random bit error occurring of approximately 1 in 100 000 bits at a data transmission rate of 1200 bps. If you desired to upload or download a 1000-line program containing an average of 40 characters per line, a total of 320 000 data bits would have to be transmitted. During the 4.4 minutes required to transfer this file you could expect 3.2 bit errors to occur, probably resulting in several program lines being received incorrectly if the errors occur randomly. In such situations you would prefer an alternative to visual inspection. Thus, a more efficient error detection and correction method was needed for large data transfers.

Block checking

In this method, data is grouped into blocks for transmission. A checksum character is generated and appended to the transmitted block and the checksum is also calculated at the receiver, using the same algorithm. If the checksums match, the data block is considered to be received correctly. If the checksums do not match, the block is considered to be in error and the receiving station will request the transmitting station to retransmit the block.

One of the most popular asynchronous block checking methods is included in the XMODEM protocol, which was the first method developed to facilitate file transfer and is still extensively used in personal computer communications. Although the operation of this protocol is covered in detail later in this chapter, we will focus our attention at his time on its error detection and correction capability. For information concerning the actual operation of data transfers under the XMODEM protocol the reader is referred to Section 1.15.

Under the XMODEM protocol groups of asynchronous characters are blocked together for transmission and a checksum is computed and appended to the end of the block. The checksum is obtained by first summing the ASCII value of each data character in the block and dividing that sum by 255. Then, the quotient is discarded and the remainder is appended to the block as the checksum. Thus, mathematically the XMODEM checksum can be represented as

$$\text{Checksum} = R\left[\frac{\sum_{1}^{128} \text{ASCII value of characters}}{255}\right]$$

where R is the remainder of the division process.

When data is transmitted using the XMODEM protocol, the receiving device at the other end of the link performs the same operation upon the block being received. This 'internally' generated checksum is compared to the transmitted checksum. If the two checksums match, the block is considered to have been received error-free. If the two checksums do not match, the block is considered to be in error and the receiving device will then request the transmitting device to resend the block.

Start of header	Block number	One's complement block number	128 data characters	Checksum

Figure 1.26 XMODEM protocol block format. The start of header is the ASCII SOH character whose bit composition is 00000001, while the one's complement of the block number is obtained by subtracting the block number from 255. The checksum is formed by first adding the ASCII values of each of the characters in the 128 character block, dividing the sum by 255 and using the remainder

Figure 1.26 illustrates the XMODEM protocol block format. The start of header is the ASCII SOH character whose bit composition is 0 0 0 0 0 0 0 1, while the one's complement of the block number is obtained by subtracting the block number from 255. The block number and its complement are contained at the beginning of each block to reduce the possibility of a line hit at the beginning of the transmission of a block causing the block number to be corrupted but not detected.

The construction of the XMODEM protocol format permits errors to be detected in one of three ways. First, if the start of header is damaged, it will be detected by the receiver and the data block will be negatively acknowledged. Next, if either the block number or the one's complement field is damaged, they will not be the one's complement of each other, resulting in the receiver negatively acknowledging the data block. Finally, if the checksum generated by the receiver does not match the transmitted checksum, the receiver will transmit a negative acknowledgement. For all three situations the negative acknowledgement will serve as a request to the transmitting station to retransmit the previously transmitted block.

Data transparency

Since the XMODEM protocol supports an 8-bit, no parity data format it is transparent to the data content of each byte. This enables the protocol to support ASCII, binary and extended ASCII data transmission, where extended ASCII is the additional 128 graphic characters used by the IBM PC and compatible computers through the employment of an 8-bit ASCII code.

Error detection efficiency

While the employment of a checksum reduces the probability of undetected errors in comparison to parity checking, it is still possible for undetected errors to occur under the XMODEM protocol. This can be visualized by examining the construction of the checksum character and the occurrence of multiple errors when a data block is transmitted.

Assume a 128-character data block of all 1 s is to be transmitted and each data character has the format 0 0 1 1 0 0 0 1, which is an ASCII 49. When the checksum

```
          0                    10
..100110001|00110001|001100 01|...
```

(In the figure, the last bit of the first character, and the last two bits of the third character, are struck through; the corrected values 0 and 10 are printed above them.)

Figure 1.27 Multiple errors on an XMODEM data block

is computed the ASCII value of each data character is first added, resulting in a sum of 6272 (128 × 49). Next, the sum is divided by 255, with the remainder used as the checksum, which in this example is 152.

Suppose two transmission impairments occur during the transmission of a data block under the XMODEM protocol, affecting two data characters as illustrated in Figure 1.27.

Here the first transmission impairment converted the ASCII value of the character from 49 to 48, while the second impairment converted the ASCII value of the character from 49 to 50. Assuming no other errors occurred, the receiving device would add the ASCII value of each of the 128 data characters and obtain a sum of 6272. When the receiver divides the sum by 255, it obtains a checksum of 152, which matches the transmitted checksum and the errors remain undetected. Although the preceding illustration was contrived, it illustrates the potential for undetected errors to occur under the XMODEM protocol.

To make the protocol more efficient with respect to undetected errors, several derivatives of the XMODEM protocol have gained popularity and are now commonly available in most communications programs designed for operation on personal computers as well as supported by information utilities and bulletin board systems. These derivatives of the XMODEM protocol include XMODEM/CRC and YMODEM, each of which uses a cyclic redundancy check (CRC) in place of the checksum for error detection. The use of CRC error detection reduces the probability of undetected errors to less than one in a million blocks and is the preferred method for ensuring data integrity. The concept of CRC error detection is explained later in this section under synchronous transmission, as it was first employed with this type of transmission. The operation of the XMODEM, XMODEM/CRC and YMODEM protocols are examined in more detail in the protocol section of this chapter.

Synchronous transmission

The majority of error-detection schemes employed in synchronous transmission involve geometric codes or cyclic code. However, several modifications to the original XMODEM protocol, such as XMODEM-CRC, use a cyclic code to protect asynchronously transmitted data.

Geometric codes attack the deficiency of parity by extending it to two dimensions. This involves forming a parity bit on each individual character as well as on all the characters in the block. Figure 1.28 illustrates the use of block parity checking for a block of 10 data characters. As indicated, this block parity character is also known as the 'longitudinal redundancy check' (LRC) character.

Geometric codes are similar to the XMODEM error-detection technique in the fact that they are also far from foolproof. As an example of this, suppose a 2-bit

			Character parity bit
Character	1	1 0 1 1 0 1 1	0
Character	2	0 1 0 0 1 0 1	0
	3	0 1 1 0 1 0 0	0
	4	1 0 0 1 0 0 1	0
	5	0 1 1 1 1 0 1	0
	6	1 0 1 0 0 0 0	1
	7	0 1 0 1 1 1 0	1
	8	0 1 1 1 0 0 1	1
	9	1 0 0 0 1 1 0	0
	10	0 1 1 0 1 0 1	1
Block parity character (LRC)	. .	1 1 1 0 1 0 1	1

Figure 1.28 VRC/LRC geometric code (odd parity checking)

duration transmission impairment occurred at bit positions 3 and 4 when characters 7 and 9 in Figure 1.28 were transmitted. Here the two 1 s in those bit positions might be replaced by two 0 s. In this situation each character parity bit as well as the block parity character would fail to detect the errors.

A transmission system using a geometric code for error detection has a slightly better capability to detect errors than the method used in the XMODEM protocol and is hundreds of times better than simple parity checking. While block parity checking substantially reduces the probability of an undetected error in comparison to simple parity checking on a character by character basis, other techniques can be used to further decrease the possibility of undetected errors. Among these techniques is the use of cyclic or polynomial code.

Cyclic codes

When a cyclic or polynomial code error-detection scheme is employed the message block is treated as a data polynomial $D(x)$, which is divided by a predefined generating polynomial $G(x)$, resulting in a quotient polynomial $Q(x)$ and a remainder polynomial $R(x)$, such that

$$D(x)/G(x) = Q(x) + R(x)$$

The remainder of the division process is known as the cyclic redundancy check (CRC) and is normally 16 bits in length or two 8-bit bytes. The CRC checking method is used in synchronous transmission and in asynchronous transmission using derivatives of the XMODEM protocol similar to the manner in which the checksum is employed in the XMODEM protocol previously discussed. That is, the CRC is appended to the block of data to be transmitted. The receiving devices uses the same predefined generating polynomial to generate its own CRC based upon the received message block and then compares the 'internally' generated CRC with the transmitted CRC. If the two match, the receiver transmits a positive acknowledgement (ACK) communications control character to the transmitting device which not only informs the distant device that the data was received

Table 1.16 Common generating polynomials

Standard	Polynomial
CRC-16 (ANSI)	$X^{16}+X^{15}+X^5+1$
CRC (ITU)	$X^{16}+X^{12}+X^5+1$
CRC-12	$X^{12}+X^{11}+X^3+1$
CRC-32	$X^{32}+X^{26}+X^{23}+X^{22}+X^{16}+X^{12}+X^{11}+X^{10}+X^8$ $+X^7+X^5+X^4+X^2+X+1$

correctly but also serves to inform the device that if additional blocks of data remain to be transmitted the next block can be sent. If an error has occurred, the internally generated CRC will not match the transmitted CRC and the receiver will transmit a negative acknowledgement (NAK) communications control character which informs the transmitting device to retransmit the block previously sent.

Table 1.16 lists four generating polynomials in common use today. The CRC-16 is based upon the American National Standards Institute and is commonly used in the United States. The ITU CRC is commonly used in transmissions in Europe while the CRC-12 is used with 6-level transmission codes and has been basically superseded by the 16-bit polynomials. The 32-bit CRC is defined for use in local networks by the Institute of Electrical and Electronic Engineers (IEEE) and the American National Standards Institute (ANSI). For further information concerning the use of the CRC-32 polynomial the reader is referred to the IEEE/ANSI 802 standards publications.

The column labelled polynomial in Table 1.16 actually indicates the set bits of the 32-bit, 16-bit or 12-bit polynomial. Thus, the CRC-16 polynomial has a bit composition of 1 1 0 0 0 0 0 0 0 0 0 0 1 0 0 0 1.

Hardware computations

Originally CRC computations were associated with synchronous transmission in which the computations were performed using special hardware known as shift registers. Figure 1.29 illustrates the basic format of a multisection shift register used for the computation of the ITU CRC. The contents of this register are initially set to zero.

As each bit in a data block is output onto the communications line it is also applied to the location marked X in Figure 1.29. This results in that bit and subsequent bits being placed into bit positions 15, 10, and 3. Note that those bit positions reference the positions within the register. Since the least significant bit (LSB) is output from the register first, that bit position corresponds to the X^{16} term of the CRC.

As bits are shifted from left to right through the segments of the shift register they eventually reach exclusive-OR gates located between each section of the register. The input to each of these exclusive-OR gates consists of bits shifted through sections of the registers and new bits output onto the communications line.

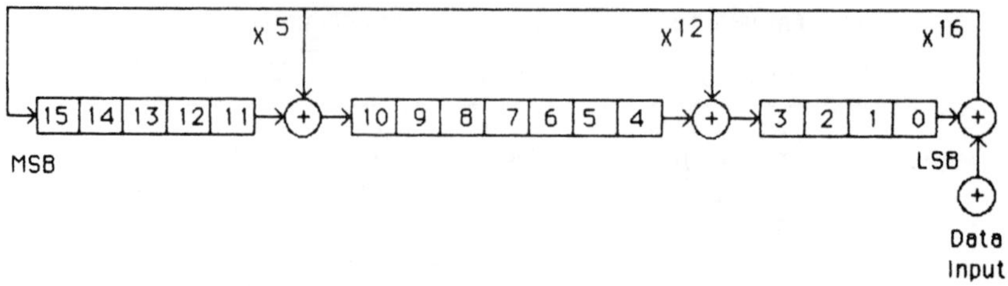

Figure 1.29 Multisection shift used for ITU CRC computation + exclusive-OR gate.

Input	Output
00	0
01	1
10	1
11	0

The operation of an exclusive-OR gate results in the output of a 0 if the inputs are both 0 or both 1. If the inputs differ, the output of the exclusive-OR gate is a 1. Once all of the bits in a data block entered the shift register the transmitter retrieves the contents of the shift register and transmits the 16 bits as a two-byte CRC. This two-byte CRC is commonly referred to as the block check characters (BCC) in communications literature. Even though a CRC-32 uses a four-byte CRC, that CRC is also commonly referred to as a BCC. After transmitting the CRC the transmitter initializes the shift register positions to zero prior to sending bits from the next block into the register. At the opposite end of the transmission link the receiving device applies the incoming bit stream to point X of a similar multisection shift register. Once the receiver's CRC is computed it is then extracted from the shift register and compared to the CRC appended to the transmitted data block. If the two CRCs match the data block is considered to be received without error.

Although hardware-based CRC computations are still normally used with synchronous transmission, almost all such computations are performed by software using the microprocessor of personal computers. To calculate the 16-bit CRC the message bits are considered to be the coefficients of a polynomial. This message polynomial is first multiplied by X^{16} and then divided by the generator polynomial $(X^{16}+X^{12}+X^{5}+1)$ using modulo 2 arithmetic. The remainder after the division is the desired CRC.

International transmission

Due to the growth in international communications, one frequently encountered transmission problem is the employment of dissimilar CRC generating polynomials. This typically occurs when an organization in the United States attempts to communicate with a computer system in Europe or a European organization attempts to transmit to a computer system located in the United States.

When dissimilar CRC generating polynomials are employed, the two-byte block-check character appended to the transmitted data block will never equal the block-check character computed at the receiver. This will result in each transmitted data block being negatively acknowledged, eventually resulting in a threshold of negative acknowledgements being reached. When this threshold is reached the protocol aborts the transmission session, causing the terminal operator to reinitiate the communications procedure required to access the computer system they wish to connect to. Although the solution to this problem requires either the terminal or a port on the computer system to be changed to use the appropriate generating polynomial, the lack of publication of the fact that there are different generating polynomials has caused many organizations to expend a considerable amount of needless effort. One bank which the author is familiar with monitored transmission attempts for almost 3 weeks. During this period, they observed each block being negatively acknowledged and blamed the communications carrier, insisting that the quality of the circuit was the culprit. Only after a consultant was called and spent approximately a week examining the situation was the problem traced to the utilization of dissimilar generating polynomials.

Forward error correcting

During the 1950s and 1960s when mainframe computers used core memory circuits, designers spent a considerable amount of effort developing codes that carried information which enabled errors to be detected and corrected. Such codes are collectively called forward error correction (FEC) and have been employed in Trellis Coded Modulation modems under the ITU-T V.32, V.32 bis, V.33, and V.34 recommendation, and will be described later in this book.

One popular example of a forward error correcting code is the Hamming code. This code can be used to detect one or more bits in error at a receiver as well as to determine which bits are in error. Since a bit can have only one of two values, knowledge that a bit is in error allows the receiver to reverse or reset the bit, correcting its erroneous condition.

The Hamming code uses m parity bits with a message length of n bits, where $n = 2^m - 1$. This permits k information bits where $k = n - m$. The parity bits are then inserted into the message at bit positions 2^{j-1} where $j = 1, 2, \ldots, m$. Table 1.17 illustrates the use of a Hamming code error correction for $m = 3$, $k = 4$, and $n = 7$.

In the Hamming code encoding process the data and parity bits are exclusive-ORed with all possible data values to determine the value of each parity bit. This is illustrated in Figure 1.30 which results in P_1, P_2, and P_3 having values of 0, 1, and 1, respectively.

Table 1.17 Hamming code error correction example

Message length	= 7 bits
Information bits	= 4 bits = 1100
Parity	= 3 bits = P_1, P_2, P_3

$$[P_1 P_2 1 P_3 1 0 0] \begin{bmatrix} 0&0&1 \\ 0&1&0 \\ 0&1&1 \\ 1&0&0 \\ 1&0&1 \\ 1&1&0 \\ 1&1&1 \end{bmatrix} = \begin{cases} P_1 + 0 + 1 + 0 + 1 + 0 + 0 = 0 \\ 0 + P_2 + 1 + 0 + 0 + 0 + 0 = 0 \\ 0 + 0 + 0 + P_3 + 1 + 0 + 0 = 0 \end{cases}$$

Figure 1.30 The Hamming code encoding process. Using exclusive-OR arithmetic the three equations yield $P_1 = 0$, $P_2 = 1$, $P_3 = 1$

$$[0\ 1\ 1\ 1\ 0\ 0\ 0] \begin{bmatrix} 0&0&1 \\ 0&1&0 \\ 0&1&1 \\ 1&0&0 \\ 1&0&1 \\ 1&1&0 \\ 1&1&1 \end{bmatrix} = \begin{cases} 0\ 0\ 1\ 0\ 0\ 0\ 0 = 1 \\ 0\ 1\ 1\ 0\ 0\ 0\ 0 = 0 \\ 0\ 0\ 0\ 1\ 0\ 0\ 0 = 1 \end{cases}$$

Figure 1.31 The Hamming code decoding process

Using the values obtained for each parity bit results in the transmitted message becoming 0111100. Now assume an error occurs in bit 5, resulting in the received message becoming 0111000. Figure 1.31 illustrates the Hamming code decoding process in which the received message is exclusive-ORed against a matrix of all possible values of the three parity bits, yielding the sum 101_2, which is the binary value for 5. Thus, bit 5 is in error and is corrected by reversing or resetting its value.

1.13 STANDARDS ORGANIZATIONS, ACTIVITIES AND THE OSI REFERENCE MODEL

The importance of standards and the work of standards organizations have proved essential for the growth of worldwide communications. In the United States and many other countries, national standards organizations have defined physical and operational characteristics that enable vendors to manufacture equipment compatible with line facilities provided by communications carriers as well as equipment produced by other vendors. At the international level, standards organizations have promulgated several series of communications related recommendations. These recommendations, while not mandatory, have become highly influential on a worldwide basis for the development of equipment and facilities and have been adopted by hundreds of public companies and communications carriers.

In addition to national and international standards a series of *de facto* standards has evolved through the licensing of technology among companies. Such *de facto* standards, as an example, have facilitated the development of communications software for use on personal computers. Today, consumers can purchase communications software that can control modems manufactured by hundreds of vendors since most modems are now constructed to respond to a uniform set of control codes.

In this section we will first focus our attention upon national and international standards bodies as well as discuss the importance of *de facto* standards. Due to the importance of the Open Systems Interconnection (OSI) Reference Model in data communications we will conclude this section with an examination of that model. Although this examination will only provide the reader with an overview of the seven layers of the OSI model, it will provide a foundation for a detailed investigation of several layers of that model presented in subsequent chapters in this book.

National standards organizations

Table 1.18 lists six representative national standards organizations—four whose activities are primarily focused within the United States and two whose activities are directed within specific foreign countries. Although each of the organizations listed in Table 1.18 is an independent entity, their level of coordination and cooperation with one another and international organizations is very high. In fact, many standards adopted by one organization have been adopted either 'as is' or in modified form by other organizations.

Table 1.18 Representative national standards organizations

United States
American National Standards Institute
Electronic Industries Association
Federal Information Processing Standards
Institute of Electrical and Electronic Engineers
Foreign
British Standards Institution
Canadian Standards Association

ANSI

The American National Standards Institute (ANSI) is the principal standards forming body in the United States. This non-profit, non-governmental organization is supported by approximately 1000 professional societies, companies and trade organizations. ANSI represents the United States in the International Organization for Standardization (ISO) and develops voluntary standards that are designed to benefit both consumers and manufacturers.

Perhaps the most well known part of ANSI is its X3 standards committee. Established in 1960 to investigate computer industry related standards, the X3

Table 1.19 ANSI X3S3 data communications technical committee task groups

Task group	Responsibility
X3S31	Planning
X3S32	Glossary
X3S33	Transmission format
X3S34	Control procedures
X3S35	System performance
X3S36	Signaling speeds
X3S37	Public data networks

standards committee has grown to 25 technical committees to include the X3S3 Data Communications Technical Committee which is composed of seven task groups. Table 1.19 lists the current task groups of the ANSI X3S3 Data Communications Technical Committee and their responsibilities.

Foremost among ANSI standards are X3.4 which defines the ASCII code and X3.1 which defines data signaling rates for synchronous data transmission on the switched telephone network.

Table 1.20 lists representative ANSI data communications standards publications by number and title. Many of these standards have been adopted by the US government as Federal Information Processing Standards (FIPS). The reader is referred to Table 1.23 at the end of this section for a list of the addresses of standards organizations mentioned in this section.

EIA

The Electronic Industries Association (EIA) is a trade organization headquartered in Washington, DC, which represents most major US electronics industry manufacturers. Since its founding in 1924, the EIA Engineering Department has published over 500 documents related to standards.

In the area of communications, the EIA established its Technical Committee TR-30 in 1962. The primary emphasis of this committee is upon the development and maintenance of interface standards governing the attachment of data terminal equipment (DTE), such as terminals and computer ports, to data communications equipment (DCE), such as modems and digital service units.

TR-30 committee standards activities include the development of the ubiquitous RS-232 interface standard which describes the operation of a 25-pin conductor which is the most commonly used physical interface for connecting DTE to DCE.

Two other commonly known EIA standards and one emerging standard are RS-366A, RS-449, and RS-530. RS-366 describes the interface used to connect terminal devices to automatic calling units; RS-449 was originally intended to replace the RS-232 interface due to its ability to extend the cabling distance between devices, while RS-530 may eventually evolve as a replacement for both RS-232 and RS-449 as it eliminates many objections to RS-449 that inhibited its

Table 1.20 Representative ANSI publications

Publication number	Publication title
X3.1	Synchronous Signaling Rates for Data Transmission
X3.4	Code for Information Interchange
X3.15	Bit Sequencing for the American National Standard Code for Information Interchange in Serial-by-Bit Data Transmission
X3.16	Character Structure and Character Parity Sense for Serial-by-Bit Data Communications Information Interchange
X3.24	Signal Quality at Interface Between Data Processing Technical Equipment for Synchronous Data Transmission
X3.25	Character Structure and Character Parity Sense for Parallel-by-Bit Communications in American National Standard Code for Information Interchange
X3.28	Procedures for the Use of the Communications Control Characters for American National Standard Code for Information Interchange interchange in Specific Data Communications Links
X3.36	Synchronous High-Speed Signaling Rates Between Data Terminal Equipment and Data Communication Equipment
X3.41	Code Extension Techniques for Use with 7-Bit Coded Character Set of American National Standard Code for Information Interchange
X3.44	Determination of the Performance of Data Communications Systems
X3.57	Structure for Formatting Message Headings for Information Interchange for Data Communication System Control
X3.66	American National Standard for Advanced Data Communication Control Procedures (ADCCP)
X3.79	Determination of Performance of Data Communications System that Use Bit-Oriented Control Procedures
X3.39	Data Encryption Algorithm

adoption. The reader is referred to Section 1.14 for detailed information covering the previously mentioned EIA standards.

The TR-30 committee works closely with both ANSI Technical Committee X3S3 and with groups within the International Telecommunications Union (ITU) Standardizations Sector (ITU-T) which was formerly known as the Consultative Committee for International Telephone and Telegraph (CCITT). In fact, the ITU V.24 standard is basically identical to the EIA RS-232 standard, resulting in hundreds of communications vendors designing RS-232/V.24 compatible equipment.

As a result of the widespread acceptance of the RS-232/V.24 interface standard, a cable containing up to 25 conductors with a predefined set of connectors can be used to cable most DTEs to DCEs. Even though there are exceptions to this interface standard, this standard has greatly facilitated the manufacture of communications products, such as terminals, computer ports, modems, and digital service units that are physically compatible with one another and which can be easily cabled to one another.

Another important EIA standard resulted from the joint efforts of the EIA and the Telecommunications Industry Association (TIA). Known as EIA/TIA-568, this standard defines structured wiring within a building and is primarily used for supporting local area networking. This standard will be covered in Chapter 3 when we focus our attention upon LANs.

FIPS

As a result of US Public Law 89-306 (the Brooks Act) which directed the Secretary of the Department of Commerce to make recommendations to the President concerning uniform data processing standards, that agency developed a computer standardization program. Since Public Law 89-306 did not cover telecommunications, the enactment of Public Law 99-500, known as the Brooks Act Amendment, expanded the definition of automatic data processing (ADP) to include certain aspects related to telecommunications. FIPS, an acronym for Federal Information Processing Standards, are the indirect result of Public Law 89-306 and is the term applied to standards developed under the US government's computer standardization program.

FIPS specifications are drafted by the National Institute of Standards and Technology (NIST), formerly known as the National Bureau of Standards (NBS). Approximately 100 FIPS have been adopted, ranging in scope from the ASCII code to the Hollerith punched card code and such computer languages as COBOL and FORTRAN. Most FIPS have an ANSI national counterpart. The key difference between FIPS and their ANSI counterparts is that applicable FIPS must be met for the procurement, management and operation of ADP and telecommunications equipment by federal agencies, whereas commercial organizations in the private sector can choose whether or not to obtain equipment that complies with appropriate ANSI standards.

IEEE

The Institute of Electrical and Electronic Engineers (IEEE) is a US-based engineering society that is very active in the development of data communications standards. In fact, the most prominent developer of local area networking standards is the IEEE, whose subcommittee 802 began its work in 1980 prior to the establishment of a viable market for the technology.

The IEEE Project 802 efforts are concentrated upon the physical interface of equipment and the procedures and functions required to establish, maintain and release connections among network devices to include defining data formats, error control procedures and other control activities governing the flow of information. This focus of the IEEE actually represents the lowest two layers of the ISO model, physical and data link, which are discussed later in this section and in more detail in Chapter 3.

BSI

The British Standards Institution (BSI) is the national standards body of the United Kingdom. In addition to drafting and promulgating British National Standards, BSI is responsible for representing the United Kingdom at ISO and other international bodies. BSI responsibilities include ensuring that British standards are in technical agreement with relevant international standards, resulting in, as an example, the widespread use of the V.24/RS-232 physical interface in the United Kingdom.

CSA

The Canadian Standards Association (CSA) is a private, non-profit organization which produces standards and certifies products for compliance with their standards. CSA functions similar to a combined US ANSI and Underwriters Laboratory, the latter also a private organization which is well known for testing electrical equipment ranging from ovens and toasters to modems and computers.

In many instances, CSA standards are the same as international standards, with many ISO and ITU standards adopted as CSA standards. In other instances, CSA standards represent modified international standards.

International standards organizations

Two important international standards organizations are the International Telecommunications Union (ITU) and the International Organization for Standardization (ISO). The ITU can be considered as a governmental body as it functions under the auspices of an agency of the United Nations. Although the ISO is a non-governmental agency, its work in the field of data communications is well recognized.

ITU

The International Telecommunications Union (ITU) is a specialized agency of the United Nations headquartered in Geneva, Switzerland. The ITU is tasked with direct responsibility for developing data communications standards and consists of 15 Study Groups, each tasked with a specific area of responsibility.

The work of the ITU is performed on a four-year cycle which is known as a study period. At the conclusion of each study period, a plenary session occurs. During the plenary session, the work of the ITU during the previous four years is reviewed, proposed recommendations are considered for adoption, and items to be investigated during the next four-year cycle are considered. The ITU's Tenth Plenary Session met in 1992 and its 11th session occurred during 1996. Although approval of recommended standards is not intended to be mandatory, ITU recommendations have the effect of law in some Western European countries and many of its recommendations have been adopted by both communications carriers and vendors in the United States.

Recommendations

Recommendations promulgated by the ITU are designed to serve as a guide for technical, operating and tariff questions related to data and telecommunications.

ITU recommendations are designated according to the letters of the alphabet, from Series A to Series Z, with technical standards included in Series G to Z. In the field of data communications, the most well known ITU recommendations include Series I which pertains to Integrated Services Digital Network (ISDN) transmission, Series Q which describes ISDN switching and signaling systems, Series V which covers facilities and transmission systems used for data transmission over the PSTN and leased telephone circuits, the DTE–DCE interface and modem operations, and Series X which covers data communications networks to include Open Systems Interconnection (OSI).

The ITU V-series

To provide readers with a general indication of the scope of ITU recommendations, Table 1.21 lists those promulgated for the V-series at the time this book was prepared. In examining the entries in Table 1.21, note that the ITU Recommendation V.3 is actually a slightly modified ANSI X3.4 standard, the ASCII code. For international use, the V.3 recommendation specifies national currency symbols in place of the dollar sign ($) as well as a few other minor differences. You should also note that certain ITU recommendations, such as V.21, V.22 and V.23, among others, while similar to AT&T Bell modems, may or may not provide operational compatibility with modems manufactured to Bell specifications. Chapter 5 provides detailed information concerning modem operations and compatibility issues.

ISO

The International Organization for Standardization (ISO) is a non-governmental entity that has consultative status within the UN Economic and Social Council. The goal of the ISO is to 'promote the development of standards in the world with a view to facilitating international exchange of goods and services'.

The membership of the ISO consists of the national standards organizations of most countries, with approximately 100 countries currently participating in its work.

Perhaps the most notable achievement of the ISO in the field of communications is its development of the seven-layer Open Systems Interconnection (OSI) Reference Model. This model is discussed in detail at the end of this section.

De facto standards

Prior to the breakup of AT&T, a process referred to as divestiture, US telephone interface definitions were the exclusive domain of AT&T and its research subsidiary, Bell Laboratories. Other vendors which developed equipment for use on the AT&T network had to construct their equipment to those interface definitions. In addition, since AT&T originally had a monopoly on equipment that could be connected to the switched telephone network, upon liberalization of that

Table 1.21 CCITT V-series recommendations

General

V.1 Equivalence between binary notation symbols and the significant conditions of a two-condition code.

V.2 Power levels for data transmission over telephone lines.

V.3 International Alphabet No. 5.

V.4 General structure of signals of International Alphabet No. 5 code of data transmission over public telephone networks.

V.5 Standardization of data signaling rates for synchronous data transmission in the general switched telephone network.

V.6 Standardization of data signaling rates for synchronous data transmission on leased telephone-type circuits.

V.7 Definitions of terms concerning data communication over the telephone network.

Interface and voice-band modems

V.10 Electrical characteristic for unbalanced double-current interchange circuits for general use with integrated circuit equipment in the field of data communications. Electrically similar to RS-423.

V.11 Electrical characteristics of balanced double-current interchange circuits for general use with integrated circuit equipment in the field of data communications. Electrically similar to RS-422.

V.15 Use of acoustic coupling for data transmission.

V.16 Medical analogue data transmission modems.

V.19 Modems for parallel data transmission using telephone signaling frequencies.

V.20 Parallel data transmission modems standardized for universal use in the general switched telephone network.

V.21 300 bps duplex modem standardized for use in the general switched telephone network. Similar to the Bell 103.

V.22 1200 bps duplex modem standardized for use on the general switched telephone network and on leased circuits. similar to the Bell 212.

V.22 bis 2400 bps full-duplex two-wire modem standard.

V.23 600/1200 baud modem standardized for use in the general switched telephone network. Similar to the Bell 202.

V.24 List of definitions for interchange circuits between data terminal equipment and data circuit-terminating equipment. Similar to and operationally compatible with RS-232.

V.25 Automatic calling and/or answering on the general switched telephone network, including disabling of echo suppressors on manually established calls. RS-366 parallel interface.

V.25 bis Serial RS-232 interface.

V.26 2400 bps modem standardized for use on four-wire leased telephone-type circuits. Similar to the Bell 201 B.

V.26 bis 2400/1200 pbs modem standardized for use in the general switched telephone network. Similar to the Bell 201 C.

V.26 ter 2400 pbs modem that uses echo cancellation techniques suitable for application in the general switched telephone network.

V.27 4800 bps modem with manual equalizer standardized for use on leased telephone-type circuits. Similar to the Bell 208A.

(*continued*)

Table 1.21 (*cont.*)

V.27 bis 4800/2400 bps modem with automatic equalizer standardized for use on leased telephone-type circuits.

V.27 ter 4800/2400 bps modem standardized for use in general switched telephone network. Similar to the Bell 208B.

V.28 Electrical characteristics for unbalanced double-current interchange circuits (defined by V.24; similar to and operational with RS-232).

V.29 9600 bps modem standardized for use on point-to-point four-wire leased telephone-type circuits. Similar to the Bell 209.

V.31 Electrical characteristics for single-current interchange circuits controlled by contact closure.

V.32 Family of 4800/9600 bps modems operating full-duplex over two-wire facilities.

V.33 14.4 kbps modem standardized for use on point-to-point four-wire leased telephone-type circuits.

V.34 28.8/33.6 kbps modem standardization for operating full duplex over two-wire facilities.

V.35 Data transmission at 48 kbps using 60–108 kHz group band circuits. CCITT balanced interface specification for data transmission at 48 kbps, using 60–108 kHz group band circuits. Usually implemented on a 34-pin M block type connector (M 34) used to interface to a high-speed digital carrier such as DDS.

V.36 Modems for synchronous data transmission using 60–108 kHz group band circuits.

V.37 Synchronous data transmission at a data signalling rate higher than 72 kbps using 60-108 kHz group band circuits.

V.40 Error indication with electromechanical equipment.

V.41 Code-independent error control system.

V.42 Error control for modems.

V.42 bis Data compression for use in switched network modems.

V.50 Standard limits for transmission quality of data transmission.

V.51 Organization of the maintenance of international telephone-type circuits used for data transmission.

V.52 Characteristic of distortion and error-rate measuring apparatus for data transmission.

V.53 Limits for the maintenance of telephone-type circuits used for data transmission.

V.54 Loop test devices for modems.

V.55 Specification for an impluse noise measuring instrument for telephone-type circuits.

V.56 Comparative tests of modems for use over telephone-type circuit.

V.90 56 kbps modem standardization for operating downstream and 33.6 kbps upstream over two-wire facilities.

policy third-party vendors had to design communications equipment, such as modems, that was compatible with the majority of equipment in use.

As some third-party vendor products gained market acceptance over other products, vendor licensing of technology resulted in the development of *de facto* standards. Another area responsible for the development of a large number of *de facto* standards is the Internet community. In this section we will examine both vendor and Internet *de facto* standards.

AT&T compatibility

Since AT&T originally had a monopoly on equipment connected to its network, when third-party vendors were allowed to manufacture products for use on AT&T facilities they designed most of their products to be compatible with AT&T equipment. This resulted in the operational characteristics of many AT&T products becoming *de facto* standards. In spite of the breakup of AT&T, this vendor still defines format and interface specifications for many facilities that third-party vendors must adhere to for their product to be successfully used with such facilities. AT&T, like standards organizations, publishes a variety of communications reference publications that define the operational characteristics of its facilities and equipment.

AT&T's catalog of technical documents is contained in two publications. AT&T's *Publication 10000* lists over 140 publications and includes a synopsis of their contents, date of publication and cost. Formally known as the *Catalog of Communications Technical Publications, Publication 10000* includes several order forms as well as a toll-free 800 number for persons who wish to call in their order. AT&T's *Publication 10000A*, which was issued as an addendum to *Publication 10000* lists new and revised technical reference releases as well as technical references deleted from *Publication 10000* and the reason for each deletion. In addition, *Publication 10000A* includes a supplementary list of select codes for publications listed in *Publication 10000*. The select code is the document's ordering code, which must be entered on the AT&T order form. Readers should obtain both documents from AT&T to simplify future orders.

Publication 10000 and 10000*A* and the publication listed therein can be obtained by writing or calling AT&T at the address or telephone numbers listed in Table 1.23.

Table 1.22 is an extract of some of the AT&T technical publications listed in *Publications 10000* and *10000A*. As can be seen, a wide diversity of publications can be ordered directly from AT&T.

Table 1.22 Selected AT&T publications

Publication number	Publication title
CB 142	The Extended Superframe Format Interface Specification
CB 143	Digital Access and System Technical Reference and Compatibility Specification
PUB 41449	Integrated Service Digital Network (ISDN) Primary Rate Interface Specification
PUB 41457	SKYNET Digital Service
PUB 52411	ACCUNET T1.5 Service Description and Interface Specification
PUB 54010	X.25 Interface Specification and Packet Switching Capabilities
PUB 54012	X.75 Interface Specification and Packet Switching Capability
PUB 54075	56 kbps Subrate Data Multiplexing

Cross-licensed technology

As the deregulation of the US telephone industry progressed, hundreds of vendors developed products for the resulting market. Many vendors cross-licensed technology, such as the command set which defines the operation of intelligent modems. Due to this, another area of *de facto* standards developed based upon consumer acceptance of commercial products. In certain cases, *de facto* standards have evolved into *de jure* standards with their adoption by one or more standard-making organizations.

Bellcore

A third area of *de facto* standards is Bellcore. Upon divestiture in 1984, AT&T formed Bell Communications Research, Inc. (Bellcore) with its seven regional Bell holding companies. Bellcore provides technical and research support to these holding companies in much the same way that Bell Laboratories supported AT&T. Bellcore maintains common standards for the nation's telephone systems, ensures a smoothly operating telephone network and coordinates telecommunications operations during national emergencies. With approximately 7000 employees, hundreds of research projects and an annual budget of approximately $1 billion, Bellcore is among the largest research and engineering organizations in the United States.

Like AT&T, Bellcore publishes a catalog of technical publications called *Catalog 10000*. The Bellcore catalog list approximately 500 publications that vary in scope from compatibility guides, which list the interface specifications that must

Table 1.23 Communications reference publication sources

ANSI 1430 Broadway New York, NY 10018, USA (212) 354-3300	ITU General Secretariat International Telecommunication Union Place des Nations 1211 Geneva 20, Switzerland
AT&T Customer Information Center Indianapolis, IN 46219, USA (800) 432-6600 (317) 352-8557	IEEE 345 East 47th Street New York, NY 10017, USA (212) 705-7900
Bell Communications Research Information Operations Center 60 New England Avenue Piscataway, NJ 08854, USA (201) 981-5600	US Department of Commerce National Technical Information Service 5285 Port Royal Rd. Springfield, VA 22161, USA (703) 487-4650
EIA Standard Sales 2001 Eye Street NW Washington DC 20006, USA (202) 457-4966	

be adhered to by manufacturers building equipment for connection to telephone company central offices, to a variety of technical references. *Catalog 10000* and the publications listed therein can be ordered directly from Bellcore or from any one of the seven regional Bell operating companies by mail or telephone. The address of Bell Communications Research is listed in Table 1.23.

Internet standards

The Internet can be considered as a collection of interconnected networks that use the Transmission Control Protocol/Internet Protocol (TCP/IP) suite. The Internet has its roots in experimental packet switching work sponsored by the US Department of Defense Advanced Research Projects Agency (ARPA) which resulted in the development of the ARPANet. That network was responsible for the development of file transfer, electronic mail and remote terminal access to computers which became applications incorporated into the TCP/IP protocol suite. The efforts of ARPA during the 1960s and 1970s were taken over by the Internet Activities Board (IAB) whose name was changed to the Internet Architecture Board in 1992. The IAB is responsible for the development of Internet protocols to include deciding if and when a protocol should become an Internet standard.

While the IAB is responsible for setting the general direction concerning the standardization of protocols, the actual effort is carried out by the Internet Engineering Task Force (IETF). The IETF is responsbile for the development of documents called Requests For Comments (RFCs) which are normally issued as a preliminary draft. After a period allowed for comments the RFC will be published as a proposed standard or, if circumstances warrant, it may be dropped from consideration. If favorable comments occur concerning the proposed standard it can be promoted to a draft standard after a minimum period of six months. After a review period of at least four months the draft standard can be recommended for adoption as a standard by the Internet Engineering Steering Group (IESG). The IESG consists of the chairperson of the IETF and other members of that group and performs an oversight and coordinating function for the IETF. Although the IESG must recommend the adoption of an RFC as a standard the IAB is responsible for the final decision concerning its adoption. Figure 1.32 illustrates

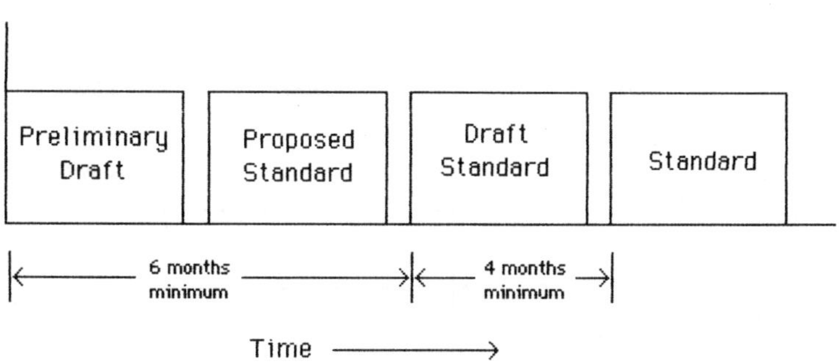

Figure 1.32 Internet standards time track

the time track for the development of an Internet standard. RFCs cover a variety of topics, ranging from TCP/IP applications to the Simple Network Management Protocol (SNMP) and the composition of databases of network management information. Over 1500 RFCs were in existence when this book was published and some of those RFCs will be described in detail later in this book.

The ISO reference model

The International Organization for Standardization (ISO) established a framework for standardizing communications systems called the Open Systems Interconnection (OSI) Reference Model. The OSI architecture defines the communications process as a set of seven layers, with specific functions isolated to and associated with each layer. Each layer, as illustrated in Figure 1.33, covers lower layer processes, effectively isolating them from higher layer functions. In this way, each layer performs a set of functions necessary to provide a set of services to the layer above it.

Layer isolation permits the characteristics of a given layer to change without impacting the remainder of the model, provided that the supporting services remain the same. One major advantage of this layered approach is that users can mix and match OSI conforming communications products to tailor their communications system to satisfy a particular networking requirement.

The OSI Reference Model, while not completely viable with current network architectures, offers the potential to directly interconnect networks based upon the use of different vendor equipment. This interconnectivity potential will be of substantial benefit to both users and vendors. For users, interconnectivity will remove the shackles that in many instances tie them to a particular vendor. For vendors, the ability to easily interconnect their products will provide them with

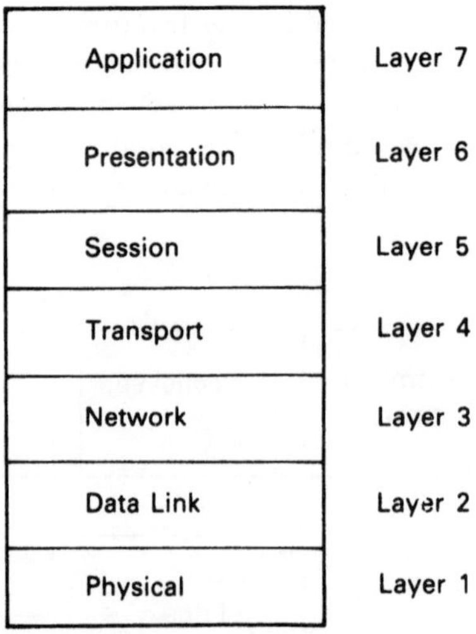

Figure 1.33 ISO reference model

access to a larger market. The importance of the OSI model is such that it has been adopted by the ITU as Recommendation X.200.

Layered architecture

As previously discussed, the OSI reference model is based upon the establishment of a layered, or partitioned, architecture. This partitioning effort can be considered as being derived from the scientific process whereby complex problems are subdivided into functional tasks that are easier to implement on an aggregate individual basis than as a whole.

As a result of the application of a partitioning approach to communications network architecture, the communications process was subdivided into seven distinct partitions, called layers. Each layer consists of a set of functions designed to provide a defined series of services which relate to the mission of that layer. For example, the functions associated with the physical connection of equipment to a network are referred to as the physical layer.

With the exception of layers 1 and 7, each layer is bounded by the layers above and below it. Layer 1, the physical layer, can be considered to be bound below by the interconnecting medium over which transmission flows, while layer 7 is the upper layer and has no upper boundary. Within each layer is a group of functions which can be viewed as providing a set of defined services to the layer which bounds it from above, resulting in layer n using the services of layer $n-1$. Thus, the design of a layered architecture enables the characteristics of a particular layer to change without affecting the rest of the system, assuming the services provided by the layer do not change.

OSI layers

An understanding of the OSI layers is best obtained by first examining a possible network structure that illustrates the components of a typical network. Figure 1.34

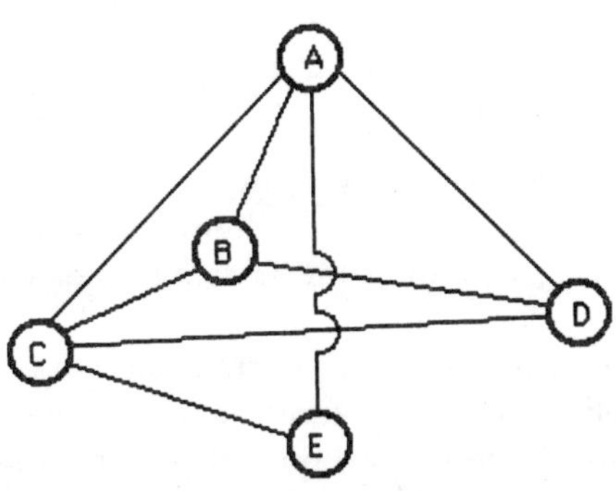

Figure 1.34 Network components. The path between a source and a destination node established on a temporary basis is called a logical connection. (O = Node. Lines represent paths)

illustrates a network structure which is only typical in the sense that it will be used for a discussion of the components upon which networks are constructed.

The circles in Figure 1.34 represents nodes which are points where data enters or exits a network or is switched between two paths. Nodes are connected to other nodes via communications paths within the network where the communications paths can be established on any type of communications media, such as cable, microwave or radio.

From a physical perspective, a node can be based upon the use of one of several types of computers to include a personal computer, minicomputer or mainframe computer or specialized computer, such as front-end processor. Connections to network nodes into a network can occur via the use of terminals directly connected to computers, terminals connected to a node via the use of one or more intermediate communications devices or via paths linking one network to another network.

The routes between two nodes, such as C–E–A, C–D–A, C–A and C–B–A which could be used to route data between nodes A and C are information paths. Due to the variability in the flow of information through a network, the shortest path between nodes may not be available for use or may represent a non-efficient path with respect to other paths constructed through intermediate nodes between a source and destination node. A temporary connection established to link two nodes whose route is based upon such parameters as current network activity is known as a logical connection. This logical connection represents the use of physical facilities to include paths and node switching capability on a temporary basis.

The major functions of each of the seven OSI layers are described in the following seven subsections.

Layer 1—the physical layer

At the lowest or most basic level, the physical layer (level 1) is a set of rules that specifies the electrical and physical connection between devices. This level specifies the cable connections and the electrical rules necessary to transfer data between devices. Typically, the physical link corresponds to established interface standards, such as RS-232. The reader is referred to Section 1.14 for detailed information concerning several physical layer interface standards.

Layer 2—the data link layer

The next layer, which is known as the data link layer (level 2), denotes how a device gains access to the medium specified in the physical layer; it also defines data formats, including the framing of data within transmitted messages, error control procedures, and other link control activities. From defining data formats to include procedures to correct transmission errors, this layer becomes responsible for the reliable delivery of information. Data link control protocols such as binary synchronous communications (BSC) and high-level data link control (HDLC) reside in this layer. The reader is referred to Section 1.15 for detailed information concerning data link control protocols and to Chapter 3 for information concerning the subdivision of that layer for local area network communications.

Layer 3—the network layer

The network layer (level 3) is responsible for arranging a logical connection between a source and destination on the network to include the selection and management of a route for the flow of information between source and destination based upon the available data paths in the network. Service provided by this layer are associated with the movement of data through a network, to include addressing, routing, switching, sequencing and flow control procedures. In a complex network the source and destination may not be directly connected by a single path, but instead require a path to be established that consists of many subpaths. Thus, routing data through the network onto the correct paths is an important feature of this layer.

Several protocols have been defined for layer 3, including the ITU X.25 packet switching protocol and the ITU X.75 gateway protocol. X.25 governs the flow of information through a packet network while X.75 governs the flow of information between packet networks. In the TCP/IP protocol suite the Internet Protocol (IP) represents a network layer protocol. Packet switching networks are described in Chapter 2.

Layer 4—the transport layer

The transport layer (level 4) is responsible for guaranteeing that the transfer of information occurs correctly after a route has been established through the network by the network level protocol. Thus, the primary function of this layer is to control the communications session between network nodes once a path has been established by the network control layer. Error control, sequence checking, and other end-to-end data reliability factors are the primary concern of this layer. Examples of transport layer protocols include the Transmission Control Protocol (TCP) and the User Datagram Protocol (UDP).

Layer 5—the session layer

The session layer (level 5) provide a set of rules for establishing and terminating data streams between nodes in a network. The services that this session layer can provide include establishing and terminating node connections, message flow control, dialog control, and end-to-end data control.

Layer 6—the presentation layer

The presentation layer (level 6) services are concerned with data transformation, formatting and syntax. One of the primary functions performed by the presentation layer is the conversion of transmitted data into a display format appropriate for a receiving device. This can include any necessary conversion between different data codes. Data encryption/decryption and data compression and decompression are examples of the data transformation that could be handled by this layer.

Layer 7—the application layer

Finally, the application layer (level 7) acts as a window through which the application gains access to all of the services provided by the model. Examples of functions performed at this level include file transfers, resource sharing and database access. While the first four layers are fairly well defined, the top three layers may vary considerably, depending upon the network used. Figure 1.35 illustrates the OSI model in schematic format, showing the various levels of the model with respect to a terminal accessing an application on a host computer system.

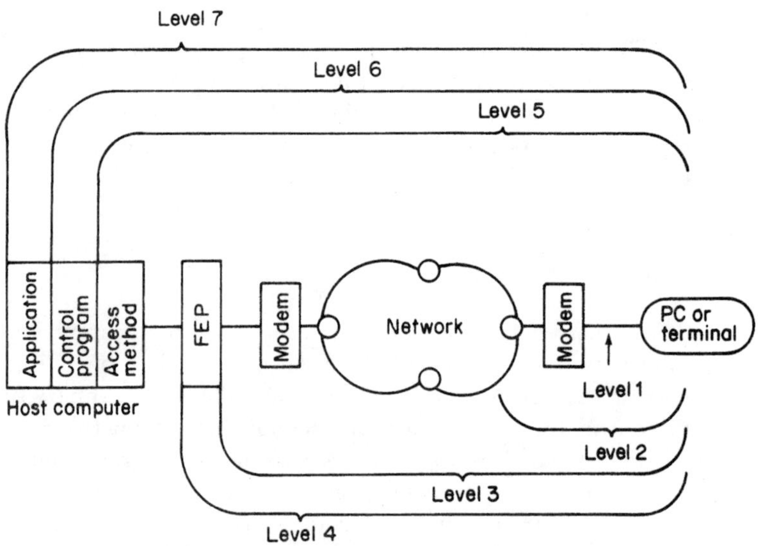

Figure 1.35 OSI model schematic

Data flow

As data flows within an ISO network each layer appends appropriate heading information to frames of information flowing within the network while removing the heading information added by a lower layer. In this manner, layer (n) interacts with layer ($n - 1$) as data flows through an ISO network.

Figure 1.36 illustrates the appending and removal of frame header information as data flows through a network constructed according to the ISO reference model. Since each higher level removes the header appended by a lower level, the frame traversing the network arrives in its original form at its destination.

As the reader will surmise from the previous illustrations, the ISO reference model is designed to simplify the construction of data networks. This simplification is due to the eventual standardization of methods and procedures to append appropriate heading information to frames flowing through a network, permitting data to be routed to its appropriate destination following a uniform procedure.

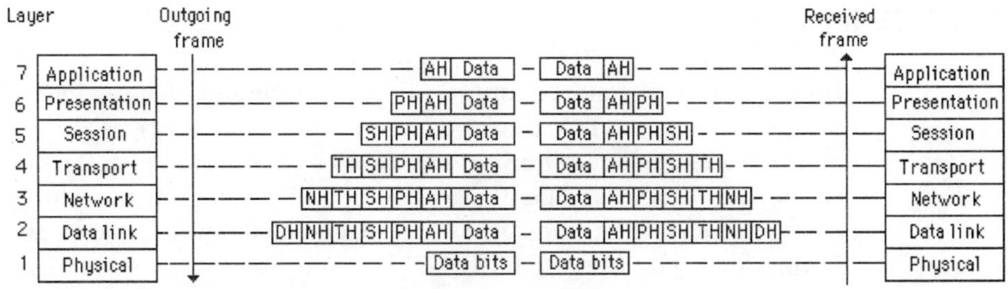

Figure 1.36 Data flow within an ISO reference model network DH, NH, TH, SH, PH and AH are appropriate headers Data Link, Network Header, Transport Header, Session Header, Presentation Header and Application Header added to data as the data flows through an ISO reference model network

1.14 THE PHYSICAL LAYER: CABLES, CONNECTORS, PLUGS AND JACKS

As discussed in Section 1.13, the physical layer is the lowest layer of the ISO reference model. This layer can be considered to represent the specifications required to satisfy the electrical and mechanical interface necessary to establish a communications path. Standards for the physical layer are concerned with connector types, connector pin-outs and electrical signaling, to include bit synchronization and the identification of each signal element as a binary one or binary zero. This results in the physical layer providing those service necessary for establishing, maintaining and disconnecting the physical circuits that form a communications path.

Since one part of communications equipment utilization involves connecting data terminal equipment (DTE) to data communications equipment (DCE), the physical interface is commonly thought of as involving such standards as RS-232, RS-449, ITU V.24 and ITU X.21. Another less recognized aspect of the physical layer is the method whereby communications equipment is attached to communications carrier facilities.

In this section we will first focus attention upon the DTE/DCE interface, examining several popular standards and emerging standards. This examination will include the signal characteristics of several interface standards to include the interchange circuits defined by the standard and their operation and utilization. Since the RS-232/V.24 interface is by far the most popularly employed physical interface, we will examine the cable used for this interface in the second part of this section. This examination will provide the foundation for illustrating the fabrication of several types of null modem cables as well as the presentation of other cabling tricks.

Since communications, in most instances, depend upon the use of facilities provided by a common carrier, in the last of this section we will discuss the interface between customer equipment and carrier facilities. In doing so we will examine the use of plugs and connectors to include the purpose of different types of jacks.

DTE/DCE interfaces

In the world of data communications, equipment that includes terminals and computer ports is referred to as data terminal equipment or DTEs. In comparison, modems and other communications devices are referred to as data communications equipment or DCEs. The physical, electrical and logical rules for the exchange of data between DTEs and DCEs are specified by an interface standard; the most commonly used are the EIA RS-232-C and RS-232-D standards which are very similar to the ITU V.24 standard used in Europe and other locations outside North America.

The EIA refers to the Electronic Industries Association which is a national body that represents a large percentage of the manufacturers in the US electronics industry. The EIA's work in the area of standards has become widely recognized and many of its standards have been adopted by other standards bodies. RS-232-C is a recommended standard (RS) published by the EIA in 1969, with the number 232 referring to the identification number of one particular communications standard and the suffix C designating the revision to that standard.

In the late 1970s it was intended that the RS-232-C standard would be gradually replaced by a set of three standards—RS-449, RS-422 and RS-423. These standards were designed to permit higher data rates than are obtainable under RS-232-C as well as to provide users with added functionality. Although the EIA and several government agencies heavily promoted the RS-449 standard, its adoption by vendors has been limited. Recognizing the fact that the universal adoption of RS-449 and its associated standards was basically impossible, the EIA issued RS-232-D (Revision D) in January 1987 and RS-232-E (Revision E) in July 1991, as well as a new standard known as RS-530.

Four other DTE/DCE interfaces that warrant attention are the EIA RS-366-A and the ITU X.20, X.21, and V.35 standards. The RS-366-A interface governs the attachment of DTEs to a special type of DCE called an automatic calling unit. The ITU X.20 and X.21 standards govern the attachment of DTE to DCE for asynchronous and synchronous operation on public networks, respectively. The ITU V.35 standard governs high-speed data transmission, typically at 48 kbps and above, with a limit occurring at approximately 6 Mbps.

Two emerging standards we will also examine are the High Speed Serial Interface (HSSI) and the High Performance Parallel Interface (HIPPI). HSSI provides support for serial operating rates up to 52 Mbps, while for extremely high bandwidth requirements, such as extending the channel on a supercomputer, HIPPI supports maximum data rates of either 800 Mbps or 1.6 Gbps on a parallel interface.

Figure 1.37 compares the maximum operating rate of the six interfaces we will discuss in this section. Although RS-232 is 'officially' limited to approximately 20 kbps, that limit is for a maximum cable length of 50 feet, which explains why that interface can still be used to connect data terminal equipment devices to include personal computers to modems operating at data rates up to 28.8, 33.6, or even 56 kbps. That is, since pulse distortion is proportional to the cable distance between two devices, shortening the cable enables a higher speed data transfer capability to be obtained between a PC and a modem. However, when the

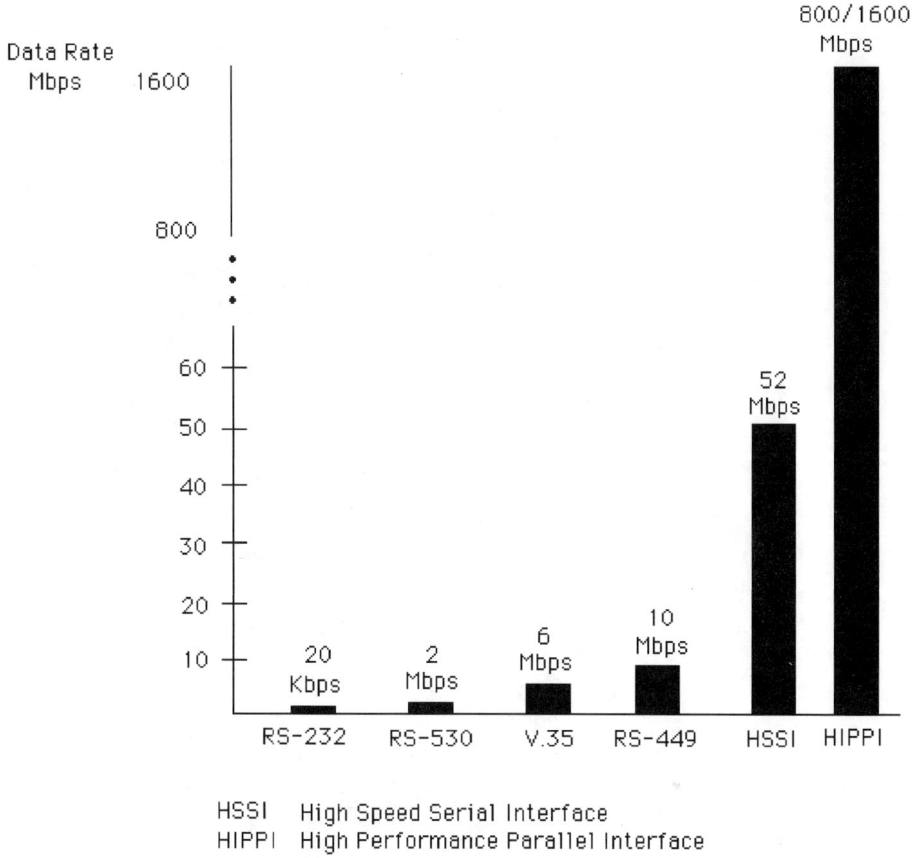

Figure 1.37 Maximum interface data rates

compression feature built into modems is enabled, most vendors recommend you set the interface speed between the DTE and DCE to four times the modem operating rate to permit the modem to perform compression effectively. Unfortunately, even at very short cable lengths the RS-232 interface will lose data at an operating rate above 115.2 kbps. Rather than incorporate a V.35 or another less commonly available interface into their modem, many modem manufacturers currently include a Centronics parallel interface in their products to support a data transfer rate above 115.2 kbps. Since just about every personal computer has a parallel interface, it is readily available or an additional port can be installed at a nominal cost in comparison to the cost associated with obtaining a different interface.

Connector overview

RS-232-D and the ITU V.24 standard as well as RS-530 formally specify the use of a D-shaped 25-pin interface connector similar to the connector illustrated in

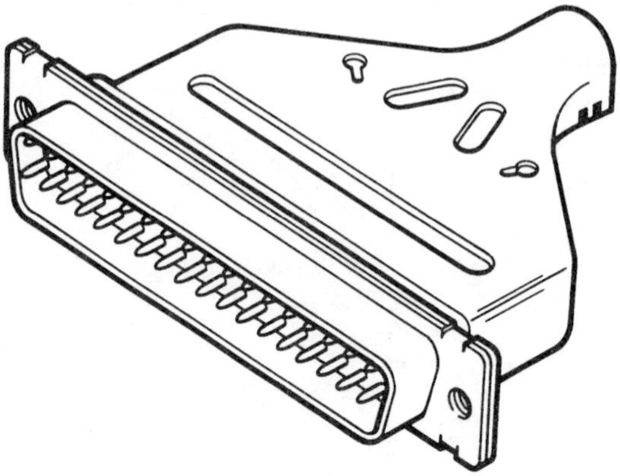

Figure 1.38 The D connector

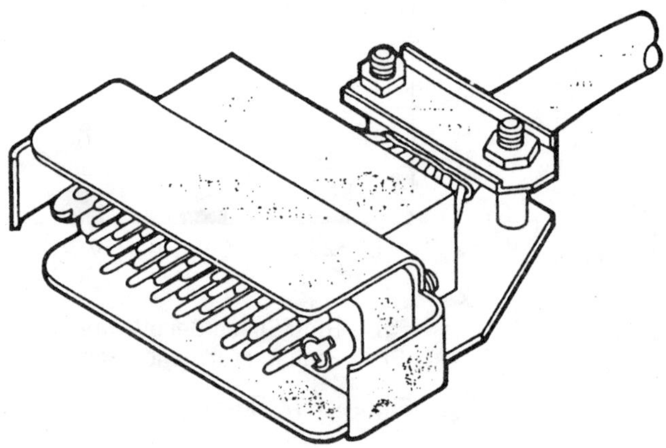

Figure 1.39 The V.35 M series connector

Figure 1.38. A cable containing up to 25 individual conductors is fastened to the narrow part of the connector, while the individual conductors are soldered to predefined pin connections inside the connector.

ITU X.20 and X.21 compatible equipment use a 15-position subminiature D connector that is both smaller and has 10 fewer predefined pins than the 25-position D connector illustrated in Figure 1.38. ITU V.35 compatible equipment uses a 34-position 'M' series connector similar to the connector illustrated in Figure 1.39. RS-449 compatible equipment uses a 37-position D connector and may optionally use a 9-position D connector, while RS-366 compatible equipment uses a 25-position D connector similar to the connector illustrated in Figure 1.38. Although the HSSI connector has 50 pins, it is actually smaller than the 32-position V.35 connector. Another interesting feature of HSSI connectors is their 'genders'. The cable connectors are specified as male, while both DTE and DCE connectors are specified as receptacles. This specification minimizes the need for male–male and female–female adapters commonly required to mate equipment and cables based upon the use of other interface standards.

In comparison to RS-232-D and RS-530, the RS-232-C standard only referenced the connector in an appendix and stated that it was not part of the standard. In spite of this omission, the use of a 25-pin D-shaped connector with RS-232-C is considered as a *de facto* standard. Although a *de facto* standard, many RS-232-C devices, in fact, use other types of connectors. Perhaps the most common exception to the 25-pin connector resulted from the manufacture of a serial/parallel adapter card for use in the IBM PC AT and compatible personal computers, such as the Compaq Deskpro and AST Bravo. The serial RS-232-C port on that card uses a 9-pin connector, which resulted in the development of a viable market for 9-pin to 25-pin converters consisting of a 9-pin and 25-pin connector on opposite ends of a short cable whose interchange circuits are routed in a specific manner to provide a required level of compatibility.

The major differences between RS-232-D and RS-232-C are that the new revision supports the testing of both local and remote communications equipment by the addition of signals to support this function and modified the use of the protective ground conductor to provide a shielding capability.

A more recent revision, RS-232-E, added a specification for a smaller alternative 26-pin connector and slightly modified the functionality of a few of the interface pins. Since RS-232-C and RS-232-D are by far the most commonly supported serial interfaces, we will first focus our attention upon those interfaces. Once this is accomplished we will discuss the newest revision to this popular interface, RS-232-E.

RS-232-C/D

Since the use of RS-232-C is basically universal since its publication by the EIA in 1969, we will examine both revisions C and D in this section, denoting the differences between the revisions when appropriate. When both revisions are similar, we will refer to them as RS-232. In general, devices built to either standard as well as the equivalent ITU V.24 recommendation are compatible with one another. There are, however, some slight differences that can occur due to the addition of signals to support modem testing under RS-232-D.

Since the RS-232-C/D standards define the most popular method of interfacing between DTEs and DCEs in the United States, they govern, as an example, the interconnection of most terminal devices to stand-alone modems. The RS-232-C/D standards apply to serial data transfers between a DTE and DCE in the range from 0 to 19 200 bps. Although the standards also limit the cable length between the DTE and DCE to 50 feet, since the pulse width of digital data is inversely proportional to the data rate, you can normally exceed this 50-foot limitation at lower data rates as wider pulses are less susceptible to distortion than narrower pulses. When a cable length in excess of 50 feet is required, it is highly recommended that low capacitance shielded cable be used and tested prior to going on-line, to ensure that the signal quality is acceptable. This type of cable is discussed later in this section.

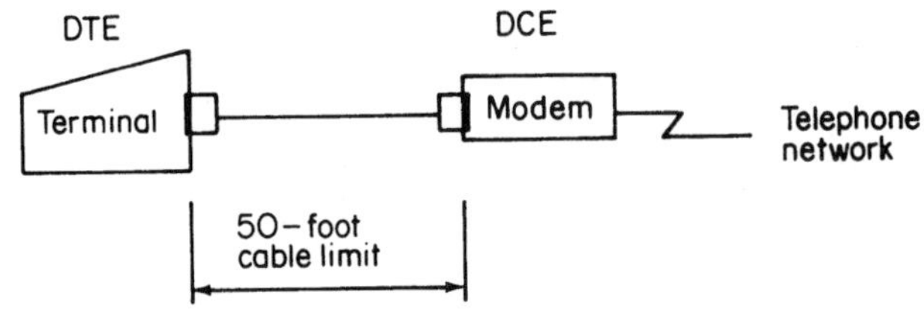

Figure 1.40 The RS-232 physical interface standard cables are typically 6, 10, or 12 feet in length with 'male' connectors on each end

Another part of the RS-232-D standard specifies the cable heads that serve as connectors to the DTEs and DCEs. Here the connector is known as a DB-25 connector and each end of the cable is equipped with this 'male' connector that is designed to be inserted into the DB-25 'female' connectors normally built into modems. Figure 1.40 illustrates the RS-232 interface between a terminal and a stand-alone modem.

Signal characteristics

The RS-232 interface specifies 25 interchange circuits or conductors that govern the data flow between the DTE and DCE. Although you can purchase a 25-conductor cable, normally fewer conductors are required. For asynchronous transmission, normally 9 to 12 conductors are required, while synchronous transmission typically requires 12 to 16 conductors, with the number of conductors required a function of the operational characteristics of the devices to be connected to one another.

The signal on each of the interchange circuits is based upon a predefined voltage transition occurring as illustrated in Figure 1.41. A signal is considered to be ON when the voltage (V) on the interchange circuit is between $+3$ V and $+15$ V. In comparison, a voltage between -3 V and -15 V causes the interchange circuit to be placed in the OFF condition. The voltage range from $+3$ V to -3 V is a transition region that has no effect upon the condition of the circuit.

Although the RS-232 and V.24 standards are similar to one another, the latter differs with respect to the actual electrical specification of the interface. The ITU V.24 recommendation is primarily concerned with how the interchange circuits operate. Thus, another recommendation, known as V.28, actually defines the electrical specifications of the interface that can be used with the V.24 standard.

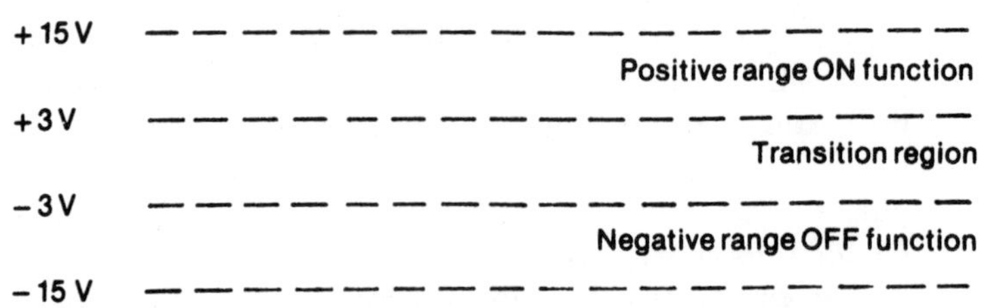

Figure 1.41 Interchange circuit voltage ranges

Table 1.24 Interchange circuit comparison

	Interchange circuit voltage	
	Negative	Positive
Binary state	1	0
Signal condition	Mark	Space
Function	OFF	ON

Table 1.24 provides a comparison between the interchange circuit voltage, its binary state, signal condition and function. As a binary 1 is normally represented by a positive voltage with a terminal device, this means that data signals are effectively inverted for transmission.

Since the physical implementation of the RS-232 standard is based upon the conductors used to interface a DTE to a DCE, we will examine the functions of each of the interchange circuits. Prior to discussing these circuits, an explanation of RS-232 terminology is warranted since there are three ways you can refer to the circuits in this interface.

Circuit/conductor reference

The most commonly used method to refer to the RS-232 circuits is by specifying the number of the pin in the connector which the circuit uses. A second method used to refer to the RS-232 circuits is by the two- or three-letter designation used by the standards to label the circuits. The first letter in the designator is used to group the circuits into one of six circuit categories as indicated by the second column labeled 'interchange circuit' in Figure 1.42. As an example of the use of this method, the two ground circuits have the letter A as the first letter in the circuit designator the signal ground circuit is called 'AB', since it is the second circuit in the 'A' ground category. Since these designators are rather cryptic, they are not commonly used.

A third method used is to describe the circuits by their functions. Thus, pin 2, which is the transmit data circuit, can be easily referenced as transmit data. Many persons have created acronyms for the descriptions which are easier to remember than the RS-232 pin number or interchange circuit designator. For example, transmit data is referred to as 'TD', which is easier to remember than any of the RS-232 designators previously discussed.

Although the list of circuits in Figure 1.42 may appear overwhelming at first glance, in most instances only a subset of the 25 conductors are employed. To better understand this interface standard, we will first examine those interchange circuits required to connect an asynchronously operated terminal device to an asynchronous modem. Then we can expand upon our knowledge of these interchange circuits by examining the functions of the remaining circuits, to include those additional circuits that would be used to connect a synchronously operated terminal to a synchronous modem.

PIN Number	Interchange circuit	ITU equivalent	Description	Gnd	Data		Control		Timing		Testing	
					From DCE	To DCE	From DCE	To DCE	From DCE	To DCE	From DCE	To DCE
1	AA	101	Protective Ground (Shield)	X								
7	AB	102	Signal Ground/Common Return	X								
2	BA	103	Transmitted Data			X						
3	BB	104	Received Data		X							
4	CA	105	Request to Send					X				
5	CB	106	Clear to Send				X					
6	CC	107	Data Set Ready (DCE Ready)				X					
20	CD	108.2	Data Terminal Ready (DTE Ready)					X				
22	CE	125	Ring Indicator				X					
8	CF	109	Received Line Signal Detector				X					
21	(RL)/ CG	110	(Remote Loopback)/Signal Quality Detector				X					
23	CH	111	Data Signal Rate Selector (DTE)					X				
23	CI	112	Data Signal Rate Selector (DCE)				X					
24	DA	113	Transmitter Signal Element Timing (DTE)							X		
15	DB	114	Transmitter Signal Element Timing (DCE)						X			
17	DD	115	Receiver Signal Element Timing (DCE)						X			
14	SBA	118	Secondary Transmitted Data			X						
16	SBB	119	Secondary Received Data		X							
19	SCA	120	Secondary Request to Send					X				
13	SCB	121	Secondary Clear to Send				X					
12	SCF	122	Secondary Received Line Signal Detector				X					
8	—	—	Reserved for Testing									X
9	—	—	Reserved for Testing								X	
18	(LL)		(Local Loopback)									X
25	(TM)		(Test Mode)								X	

Figure 1.42 RS-232-C/D and ITU V.24 interchange circuit by category, RS-232-D additions/changes to RS-232-C are indicated in parentheses

Asynchronous operations

Figure 1.43 illustrates the general signals that are required to connect an asynchronous terminal device to a asynchronous modem. Note that although a 25-conductor cable can be used to cable the terminal to the modem, only 10 conductors are actually required.

By reading the modem vendor's specification sheet you can easily determine the number of conductors required to cable DTEs to DCEs. Although most cables have straightthrough conductors, in certain instances the conductor pins at one end of a cable may require reversal or two conductors may be connected onto a common pin, a process called strapping. In fact, many times only one conductor will be used for both protective ground and signal ground, with the common conductor cabled to pins 1 and 7 at both ends of the cable. In such instances a 9-conductor cable could be employed to satisfy the cabling requirement illustrated in Figure 1.43. With this in mind, let us review the functions of the 10 circuits illustrated in Figure 1.43.

Protective ground (GND, Pin 1)

This interchange circuit is normally electrically bonded to the equipment's frame. In some instances, it can be further connected to external grounds as required by

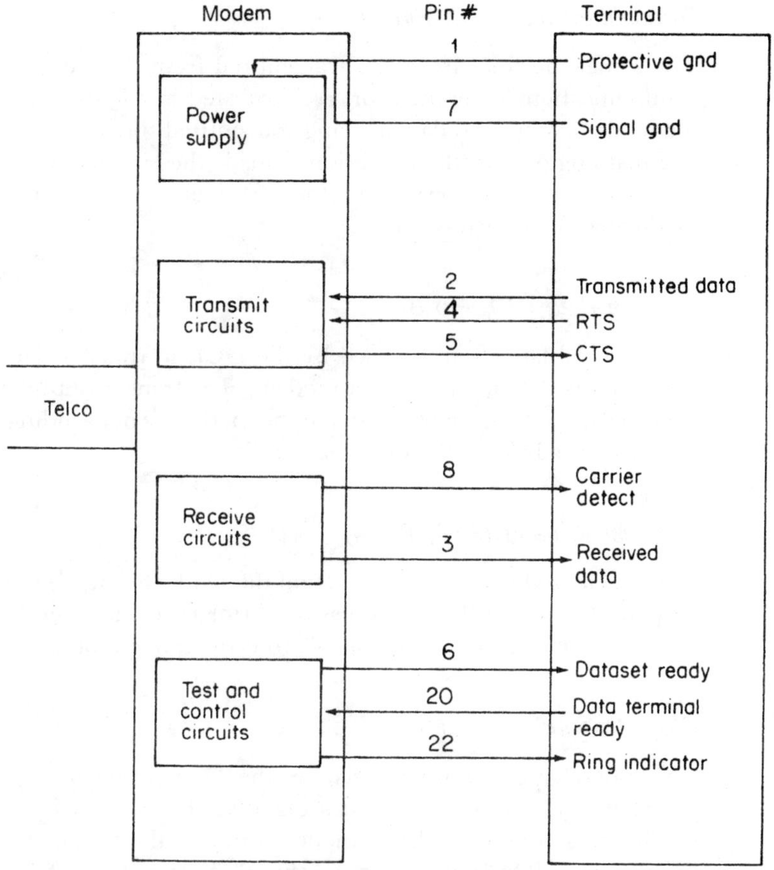

Figure 1.43 DTE-DCE interface example. In this example, pin 7 can be tied to pin 1, resulting in the use of a common 9-conductor cable

applicable regulations. Under RS-232-D this conductor use is modified to provide shielding for protection against electromagnetic or other interference occurring in high-noise environments.

Signal ground (SG, Pin 7)

This circuit must be included in all RS-232 interfaces as it establishes a ground reference for all other lines. The voltage on this circuit is set to zero to provide a reference for all other signals. Although the conductors for pins 1 and 7 can be independent of one another, typical practice is to 'strap' pin 7 to pin 1 at the modem. This is known as tying signal ground to frame ground. Since RS-232 uses a single ground reference circuit the standard results in what is known as an electrically unbalanced interface. In comparison, RS-422 uses differential signaling in which information is conveyed by the relative voltage levels in two conductors, enhancing the data rate and distance for that standard.

Transmitted data (TD, Pin 2)

The signals on this circuit are transmitted from data terminal equipment to data communications equipment or, as illustrated in Figure 1.43, a terminal device to the modem. When no data is being transmitted the terminal maintains this circuit in a marking or logical 1 condition. This is the circuit over which the actual serial bit stream of data flows from the terminal device to the modem where it is modulated for transmission.

Receive data (RD, Pin 3)

The receive data circuit is used by the DCE to transfer data to the DTE. Thus, after data is demodulated by a modem, it is transferred to the attached terminal over this interchange circuit. When the modem is not sending data to the terminal, this circuit is held in the marking condition.

Request to send (RTS, Pin 4)

The signal on this circuit is sent from the DTE (terminal) to the DCE (modem) to prepare the modem for transmission. Prior to actually sending data, the terminal must receive a clear to send signal from the modem on pin 5.

Clear to send (CTS, Pin 5)

This interchange circuit is used by the DCE (modem) to send a signal to the attached DTE (terminal); indicating that the modem is ready to transmit. By turning this circuit OFF, the modem informs the terminal that it is not ready to receive data. The modem raises the CTS signal after the terminal initiates a request to send (RTS) signal.

Carrier detect (CD, Pin 8)

Commonly referred to as received line signal detector (RLSD), a signal on this circuit is used to indicate to the DTE (terminal) that the DCE (modem) is receiving a carrier signal from a remote modem. The presence of this signal is also used to illuminate the carrier detect light-emiting diode (LED) indicator on modems equipped with that display indicator. If this light indicator should go out during a communications session, it indicates that the session has terminated owing to a loss of carrier, and software that samples for this condition will display the message ‘carrier lost’ or a similar message to indicate this condition has occurred.

Data set ready (DSR, Pin 6)

Signals on this interchange circuit flow from the DCE to the DTE and are used to indicate the status of the data set connected to the terminal. When this circuit is in the ON (logic 0 as in 1, 2, 3) condition, it serves as a signal to the terminal that the modem is connected to the telephone line and is ready to transmit data. Since the RS-232 standard specifies that the DSR signal is ON when the modem is connected to the communications channel and not in any test condition, a modem using a self-testing feature or automatic dialing capability would pass this signal to the terminal after the self-test is completed or after the telephone number of a remote location was successfully dialed. Under RS-232-D this signal was renamed DCE ready.

Data terminal ready (DTR, Pin 20)

The signal on this circuit flow from the DTE to the DCE and is used to control the modem's connection to the telephone line. An ON condition on this circuit prepares the modem to be connected to the telephone line, after which the connection can be established by manual or automatic dialing. If the signal on this circuit is placed in an OFF condition, it causes the modem to drop any telephone connection in progress, providing a mechanism for the terminal device to control the line connection. Under RS-232-D this signal was renamed DTE ready.

Ring indicator (RI, Pin 22)

The signal on this interchange circuit flows from the DCE to the DCE and indicates to the terminal device that a ringing signal is being received on the communications channel. This circuit is used by an auto-answer modem to 'wake-up' the attached terminal device. Since a telephone rings for 1 s and then pauses for 4 s prior to ringing again, this line becomes active for 1 s every 5 s when an incoming call occurs.

The ring indicator circuit is turned ON by the DCE when it detects the ON phase of a ring cycle. Depending upon how the DCE is optioned, it may either keep the RI signal high until the DTE turns DTR low or the DCE may turn the RI signal ON and OFF to correspond to the telephone ring sequence.

If the DTE is ready to accept the call its DTR lead will either by high, which is known as a hot-DTR state, or be placed into on ON condition in response to the RI signal turning ON. Once the RI and DTR circuits are both ON, the DCE will actually answer the incoming call and place a carrier tone onto the line.

If a computer port connected to a modem is not in a hot-DTR state the first ring causes the modem to turn ON its RI circuit momentarily, alerting the computer port to the incoming call. As the computer port turns on its DTR circuit the modem's RI circuit will be turned off as it cycles in tandem with the telephone company ringing signal. Thus, the DCE must wait for the next ON phase of the ring cycle to answer the call, explaining why many modems may require two rings to answer a call.

Control signal timing relationship

The actual relationship of RS-232 control signals varies by time based upon the operational characteristics of devices connected as well as the strapping option

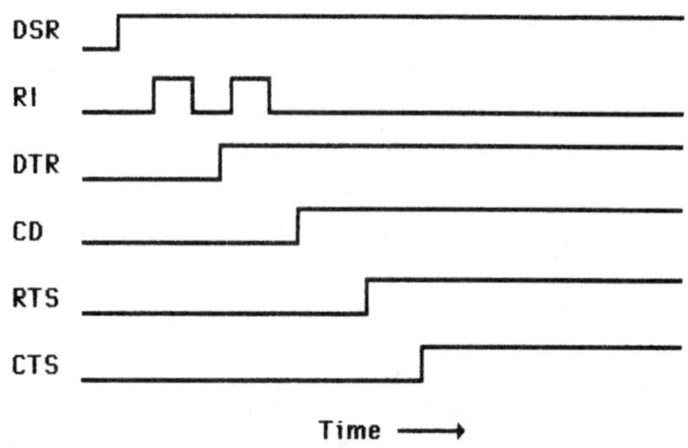

Figure 1.44 Control signal timing relationship. The state of the control signals varies by time based upon the operational characteristics of devices connected as well as the strapping option settings of those devices

settings of those devices. Figure 1.44 illustrates a common timing relationship of control signals between a computer port and a modem.

At the top of Figure 1.44 it is assumed that the data set is powered on, resulting in the data set ready (DSR) control signal being high or in the ON state. Next, two ring indicator (RI) signals are passed to the computer port, resulting in the computer responding by raising its data terminal ready (DTR) signal. The DTR signal in conjunction with the second ring indicator (RI) signal results in the modem answering the call, presenting the carrier detect (CD) signal to the computer port. Assuming the computer is programmed to transmit a sign-on message, it will raise its request to send (RTS) signal. The modem will respond by raising its clear to send (CTS) signal if it is ready to transmit, which enables the computer port to begin the actual transmission of data.

Synchronous operations

One major difference between asynchronous and synchronous modems is the timing signals required for synchronous transmission.

Timing signals

When a synchronous modem is used, it puts out a square wave on pin 15 at a frequency equal to the modem's bit rate. This timing signal serves as a clock from which the terminal would synchronize its transmission of data onto pin 2 to the modem. Thus, pin 15 is referred to as transmit clock as well as its formal designator of transmission signal element timing (DCE), with DCE referencing the fact that the communications device supplies the timing.

Whenever a synchronous modem receives a signal from the telephone line it puts out a square wave on pin 17 to the terminal at a frequency equal to the modem's bit rate, while the actual data is passed to the terminal on pin 3. Since pin 17 provides receiver clocking, it is known as 'receive clock' as well as its more formal designator of receiver signal element timing.

In certain cases a terminal device such as a computer port can provide timing signals to the DCE. In such situations the DTE will provide a clocking signal to the DCE on pin 24 while the formal designator of transmitter signal element timing (DTE) is used to reference this signal.

The process whereby a synchronous modem or any other type of synchronous device generates timing is known as internal timing. If a synchronous modem or any other type of synchronous DCE is configured to receive timing from an attached DTE, such as a computer port or terminal, the DCE must be set to external timing when the DTE is set to internal timing.

Intelligent operations

There are three interchange circuits that can be employed to change the operation of the attached communications device. One circuit can be used to first determine that a deterioration in the quality of a circuit has occurred, while the other two circuits can be employed to change the transmission rate to reflect the circuit quality.

Signal quality detector (CG, Pin 21)

Signals on this circuit are transmitted from the DCE (modem) to the attached DTE (terminal) whenever there is a high probability of an error in the received data due to the quality of the circuit falling below a predefined level. This circuit is maintained in an ON condition when the signal quality is acceptable and turned to an OFF condition when there is a high probability of an error. Under RS-232-D this circuit can also be used to indicate that a remote loopback is in effect.

Data signal rate selector (CH/C1, Pin 23)

When an intelligent terminal device such as a computer port receives an OFF condition on pin 21 for a predefined period of time, it may be programmed to change the data rate of the attached modem, assuming that the modem is capable of operating at dual data rates. This can be accomplished by the terminal device providing an ON condition on pin 23 to select the higher data signaling rate or range of rates while an OFF condition would select the lower data signaling rate or range of rates. When the data terminal equipment selects the operating rate the signal on pin 23 flows from the DTE to the DCE and the circuit is known as circuit CH. If the data communications equipment is used to select the data rate of the terminal device, the signal on pin 23 flows from the DCE to the DTE and the circuit is known as circuit CI.

Secondary circuits

In certain instances a synchronous modem will be designed with the capability to transmit data on a secondary channel simultaneously with transmission occurring on the primary channel. In such cases the data rate of the secondary channel is normally a fraction of the data rate of the primary channel.

To control the data flow on the secondary channel RS-232 standards employ five interchange circuits. Pins 14 and 16 are equivalent to the circuits on pins 2 and 3, except that they are used to transmit and receive data on the secondary channel. Similarly, pins 19, 13 and 12 perform the same functions as pins 4, 5 and 8 used for controlling the flow of information on the primary data channel.

In comparing the interchange circuits previously described to the connector illustrated in Figure 1.38, the reader should note that the location of each interchange circuit is explicitly defined by the pin number assigned to the circuit. In fact, the RS-232-D connector is designed with two rows of pins, with the top row containing 13 while the bottom row contains 12. Each pin has an explicit signal designation that corresponds to a numbering assignment that goes from left to right across the top row and then left to right across the bottom row of the connector. For ease of illustration the assignment of the interchange circuits to each of the pins in the D connector is presented in Figure 1.45 by rotating the connector 90° clockwise. In this illustration, RS-232-D conductor changes from RS-232-C are denoted in parentheses.

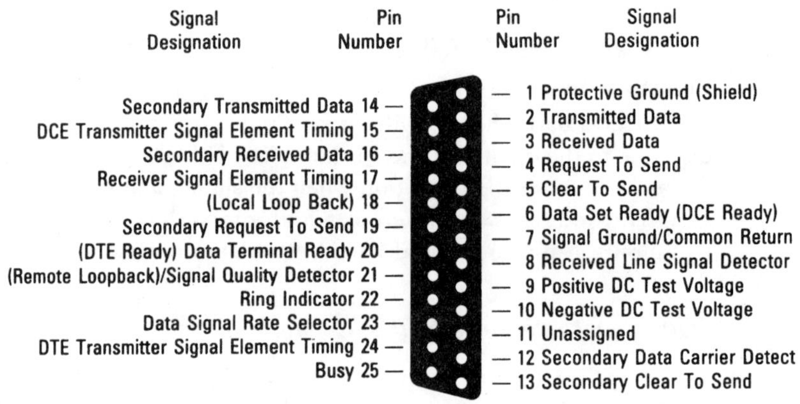

Figure 1.45 RS-232 interface on D connector

Test circuits

RS-232-D adds three circuits for testing that were not part of the earlier RS-232-C standard. The DTE can request the DCE to enter remote loopback (RL, pin 21) or local loopback (LL, pin 18) mode by placing either circuit in the ON condition. The DCE, if built to comply with RS-232-D, will respond by turning the test mode (TM, pin 25) circuit ON and performing the appropriate test.

Connector conversion

Table 1.25 contains a list of the corresponding pins between a DB-9 connector used on an IBM PC AT Compaq Deskpro and other personal computer serial ports and a standard RS-232 DB-25 connector. Data in this table can be used to develop an appropriate DB-9 to DB-25 converter. As an example of the use of

Table 1.25 DB-9 to DB-25 pin correspondence

DB-9		DB-25
1	Carrier detect	8
2	Receive data	3
3	Transmitted data	2
4	Data terminal ready	20
5	Signal ground	7
6	Data set ready	6
7	Request to send	4
8	Clear to send	5
9	Ring indicator	22

Table 1.25, the conductor for carrier detect would be wired to connect pin 1 at the DB-9 connector to pin 8 at the DB-25 connector.

RS-232-E

A more recent revision to RS-232, RS-232-E (Revision E), resulted in several minor changes to the operation of some interface circuits and the specification of an alternate interface connector (Alt A). Although none of the changes were designed to create compatibility problems with prior versions of RS-232, the use of the alternative physical interface can only be accomplished if a Revision E device is cabled via an adapter to an earlier version of that interface or if a dual connector cable is used. The 26-pin Alt A connector is about half the size of the 25-pin version and was designed to support hardware where connector space is at a premium, such as laptop and notebook computers. Pin 26 is only contained on the Alt A connector and presently is functionless.

In addition to specifying an alternative interface connector RS-232-E slightly modified the functionality of certain pins or interchange circuits. First, pin 4 (Request to Send) is defined as Ready for Receiving when the DTE enables that circuit. Next, pin 18 which was used for Local Loopback will now generate a 'Busy Out' when enabled. A third modification is the use of Clear to Send for hardware flow control, a function used by countless vendors over the past 10 years but only now formally recognized by the standard. The term 'flow control' represents the orderly control of the flow of data. By toggling the state of the Clear to Send signal, DCE equipment can regulate the flow of information from DTE equipment, a topic we will discuss in detail later in this book.

RS-232/V.24 limitations

There are several limitations associated with the RS-232 standard and the V.28 recommendation which defines the electrical specification of the interface that can be used with the V.24 standard. Foremost among these limitations are data rate and cabling distance.

RS-232/V.24 is limited to a maximum data rate of 19.2 kbps at a cabling distance of 50 feet. In actuality, speeds below 19.2 kbps allow greater transmission distances while for cable lengths of only a few feet a data rate over 100 kbps becomes possible.

Differential signaling

Over long cabling distances the cumulative cable capacitance and resistance combine to cause a significant amount of signal distortion. At some cabling distance, which decreases in an inverse relationship to the data rate, the signal cannot be recognized. Thus, to overcome the cabling distance and speed limitations associated with RS-232 a different method of signaling was devised. This signaling technique, known as differential signaling, results in information being conveyed by the relative voltage levels in two wires. Instead of using one driver to produce a large voltage swing as RS-232 does, differential signaling uses two drivers to split a signal into two parts.

Figure 1.46 illustrates differential signaling as specified by the RS-422 interface standard. To transmit a logical '1' the A driver output is driven more positive while the B driver output is more negative. Similarly, to transmit a logical '0' the A driver output is driven more negative while the B driver output is driven more positive. At the receiver a comparator is used to examine the relative voltage levels on the two signal wires.

With the use of two wires, RS-449 specifies a mark or space by the difference between the voltages on the two wires. This difference is only 0.4 V under RS-422, whereas it is 6 V (+3 V and −3 V) under RS-232/V.24. Thus, if the difference signal between the two wires is positive and greater than 0.2 V, the receiver will read a mark. Similarly, if the difference signal is negative and more negative than −0.2 V, the receiver will read a space. In addition to requiring a lower voltage

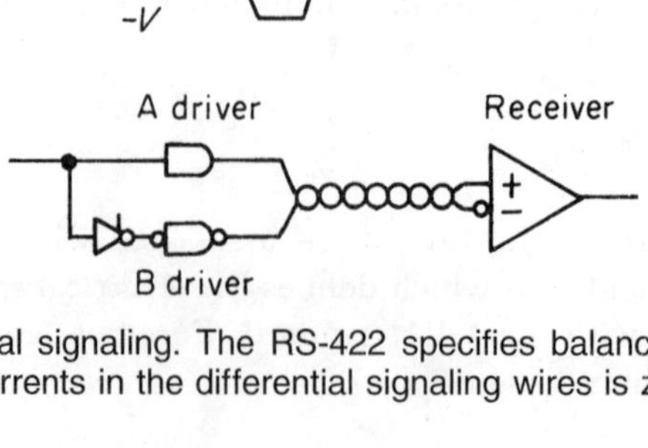

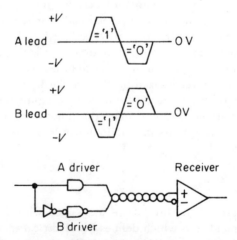

Figure 1.46 Differential signaling. The RS-422 specifies balanced differential signaling since the sum of the currents in the differential signaling wires is zero

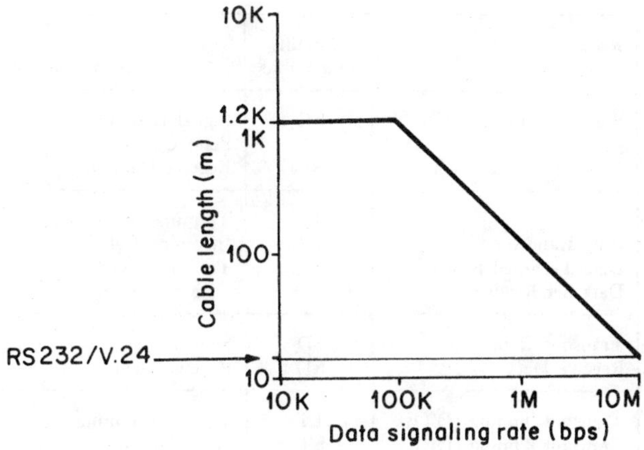

Figure 1.47 RS-422 cable distance plotted against signaling rate

shift, source and load impedance of RS-232, which is approximately 5 kΩ, is reduced to 100 Ω by the use of differential signaling. Another benefit of this signaling method is the effect of noise on signal distortion. Since the presence of noise results in the same voltage being imposed on both wires, there is no change in the voltage difference between the signal wires. Thus, the combination of lower signaling levels and impedances coupled with voltage difference immunity to noise permits differential signaling to drive longer cable distances at higher speeds. In Figure 1.47 you will find a plot of cable distances against signaling rate for the RS-422 standard.

For comparison purposes, the RS-232/V.24 cable distance is plotted against the signaling rate in Figure 1.47. As indicated, RS-422 offers significant advantages over RS-232 with respect to both cabling distance and data signaling rate.

RS-449

RS-449 was introduced in 1977 as an eventual replacement for RS-232-C. This interface specification calls for the use of a 37-pin connector as well as an optional 9-pin connector for devices using a secondary channel. Unlike RS-232, RS-449 does not specify voltage levels. Two additional specifications known as RS-442-A and RS-423 cover voltage levels for a specific range of data speeds. RS-442-A and its counterpart, ITU X.27 (V.11), define the voltage levels for data rates from 20 kbps to 10 Mbps, while RS-423-A and its ITU counterpart, X.26 (V.10), define the voltage levels for data rates between 0 and 20 kbps.

As previously mentioned, RS-442 (as well as its ITU counterpart) defines the use of differential balanced signaling. RS-422 is designed for twisted-pair telephone wire transmission ranging from 10 Mbps at distances up to 40 feet to 100 kbps at distances up to 4000 feet. RS-423 defines the use of unbalanced signaling similar to RS-232. This standard supports data rates ranging from 100 kbps at distances up to 40 feet to 10 kbps at distances up to 200 feet.

RS-232 Designation	Circuit mnemonic	Circuit name	Circuit direction	Circuit type	
Signal Ground	SG	Signal Ground	—	Common	
—	SC	Send Common	to DCE		
—	RC	Receive Common	from DCE		
—	IS	Terminal in Service	to DCE	Control	
Ring Indicator	IC	Incoming Call	from DCE		
Data Terminal Ready	TR	Terminal Ready	to DCE		
Data Set Ready	DM	Data Mode	from DCE		
Transmit Data	SD	Send Data	to DCE	Data	Primary channel
Receive Data	RD	Receive Data	from DCE		
Transmit Timing (DTE)	TT	Terminal Timing	to DCE	Timing	
Transmit Timing (DCE)	ST	Send Timing	from DCE		
Receive Timing	RT	Receive Timing	from DCE		
Request to Send	RS	Request to Send	to DCE	Control	
Clear to Send	CS	Clear to Send	from DCE		
Receive Signal Detector	RR	Receiver Ready	from DCE		
Signal Quality Detector	SO	Signal Quality	from DCE		
—	NS	New Signal	to DCE		
—	SF	Select Frequency	to DCE		
Data Rate Selector (DTE)	SR	Signaling Rate Selector	to DCE		
Data Rate Selector (DCE)	SI	Signaling Rate Indicator	from DCE		
Secondary Transmit Data	SSD	Secondary Send Data	to DCE	Data	Secondary channel
Secondary Receive Data	SRD	Secondary Receive Data	from DCE		
Secondary Request to Send	SRS	Secondary Request to Send	to DCE	Control	
Secondary Clear to Send	SCS	Secondary Clear to Send	from DCE		
Secondary Receiver Signal Detector	SRR	Secondary Receiver Ready	from DCE		
Local Loopback (D/E)	LL	Local Loopback	to DCE	Control	
—	RL	Remote Loopback	to DCE		
—	TM	Test Mode	from DCE		
—	SS	Select Standby	to DCE	Control	
—	SB	Standby Indicator	from DCE		

Figure 1.48 RS-449 interchange circuits

The use of RS-422, RS-423, and RS-449 permits the cable distance between DTEs and DCEs to be extended to 4000 feet in comparison to RS-232's 50-foot limitation. Figure 1.48 indicates the RS-449 interchange circuits. In comparing RS-449 to RS-232, you will note the addition of 10 circuits which are either new control or status indicators and the deletion of three functions formerly provided by RS-232. The most significant functions added by RS-449 are local and remote loopback signals. These circuits enable the operation of diagnostic features built into communications equipment via DTE control, permitting, as an example, the loopback of the device to the DTE and its placement into a test mode operation. With the introduction of RS-232-D a local loopback function was supported. Thus, the column labeled RS-232 Designation with the row entry 'Local

Loopback' indicates that that circuit is only applicable to revisions D and E of that standard by the entries D/E in parentheses after the circuit name.

Although a considerable number of articles have been written describing the use of RS-449, its complexity has served as a constraint in implementing this standard by communications equipment vendors. Other constraints limiting its acceptance include the cost and size of the 37-pin connector arrangement and the necessity of using another connector for secondary operations. By mid-1998, less than a few percent of all communications devices were designed to operate with this interface. Due to the failure of RS-449 to obtain commercial acceptance the EIA issued RS-530 in March 1987. This new standard, which is described later in this section, is intended to gradually replace RS-449.

V.35

The V.35 standard was developed to support high-speed transmission, typically 48, 56 and 64 kbps. Originally the V.35 interface was designed into adapter boards inserted into mainframe computers to support 48 kbps transmission on analog wideband facilities. Today, the V.35 standard is the prevalent interface to 56 kbps common carrier digital transmission facilities in the United States and 48 kbps common carrier digital transmission facilities in the UK. In addition, the V.35 standard can support data transfer operations at operating rates up to approximately 6 Mbps. This has resulted in the V.35 interface being commonly employed in videoconferencing equipment, routers and other popularly used communications devices.

The V.35 electrical signaling characteristics are a combination of an unbalanced voltage and a balanced current. Although control signals are electrically unbalanced and compatible with RS-232 and ITU V.28, data and clock interchange circuits are driven by balanced drivers using differential signaling similar to RS-422 and ITU V.11. V.35 uses a 34-pin connector specified in ISO 2593 similar to the connector illustrated in Figure 1.39.

Figure 1.49 illustrates the correspondence between RS-232 and V.35. Note that the V.35 pin pairs tied together by a brace are differential signaling circuits that use a wire pair.

RS-366-A

The RS-366-A interface is employed to connect terminal devices to automatic calling units. This interface standard uses the same type 25-pin connector as RS-232; however, the pin assignments are different. A similar interface to RS-366-A is the ITU V.25 recommendation, which is also designed for use with automatic calling units. Figures 1.50 and 1.51 illustrate the RS-366-A and ITU V.25 interfaces. Note that for both interfaces each actual digit to be dialed is transmitted as parallel binary information over circuits 14 to 17. The pulse on pin 14 represents the value 2^0 while the pulses on pins 15 to 17 represent the values 2^1, 2^2 and 2^3, respectively. Thus, to indicate to the automatic calling unit that it should dial the digit 9, circuits 14 and 17 would become active.

V.35 Pin	V.35 Name	Direction DCE DTE	Function	RS–232 Pin	RS–232 Name
A	FG		Frame GND	1	AA
B	SG		Signal GND	7	AB
C	RTS	⟶	Request to Send	4	CA
D	CTS	⟵	Clear to Send	5	CB
E	DSR	⟶	Data Set Ready	6	CC
F	RLSD	⟶	Received Line Signal	8	CF
H	DTR	⟵	Data Terminal Ready	20	CD
J	RI	⟶	Ring Indicator	22	CE
R } T	RD	⟶	Receive Data	3	BB
U } X	SGR	⟶	Receive Clock	17	DD
P } S	SD	⟵	Send Data	2	BA
U } W	SCTE	⟵	Send Clock (EXT)	24	DA
Y } a	SCT	⟶	Send Clock	15	DB
m	TST	⟶	Reserved for Test (D/E)	25	TM

Figure 1.49 V.35/RS-232 signal correspondence. Illustrates the correspondence between RS-232 and V.35. Note that the V.35 pin pairs tied together by a brace are differential signaling circuits that use a wire pair

Originally, automatic calling units provided the only mechanism to automate communications dialing over the PSTN. This enabled the use of RS-366 automatic dialing equipment under computer control to re-establish communications via the PSTN if a leased line became inoperative, a process called dial-backup. Another common use of automatic calling units is to poll remote terminals from a centrally located computer in the evening when rates are lower. Under software control the central computer would dial each remote terminal and request the transmission of the day's transactions for processing. Due to the development and wide acceptance of the use of intelligent modems with automatic dialing capability, the use of automatic calling units has greatly diminished.

Until the mid-1980s only intelligent asynchronous modems had an automatic dialing capability, restricting the use of automatic calling units to mainframe computers that required a method to originate synchronous data transfers over the PSTN. In such situations a special adapter needed to be installed in the communications controller of the mainframe, which controlled the operation of the automatic calling unit. The introduction of synchronous modems with automatic dialing capability significantly diminished the requirement for automatic calling units since their use eliminates the requirement to install an expensive adapter in the communications controller as well as the cost of the automatic calling unit. The operation and utilization of intelligent modems is discussed in Chapter 5.

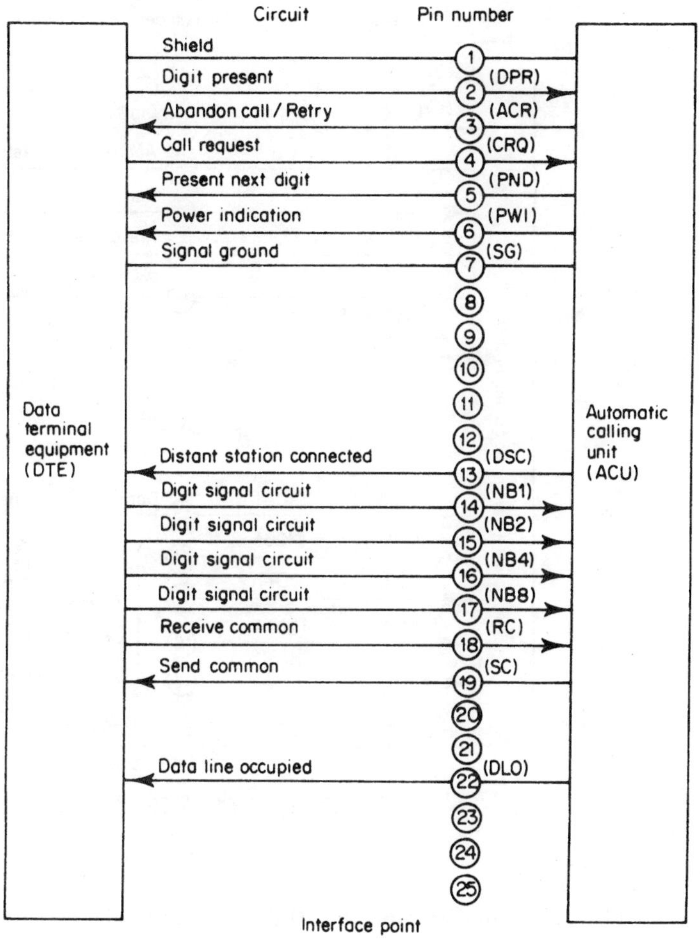

Figure 1.50 RS-336-A interface

X.21 and X.20

Interface standards X.21 and X.20 were developed to accommodate the growing use of public data networks. X.21 is designed to allow synchronous devices to access a public data network while X.20 provides a similar capability for asynchronous devices.

Instead of assigning functions to specific pins on a connector like RS-232, RS-449 and V.35, X.21 assigns coded character strings to each function for establishing connections through a public data network. For example, a dial tone is presented to a computer as a continuous sequence of ASCII '+' characters on the X.21 receive circuit. The computer can then dial a stored number by transmitting it as a series of ASCII characters on the X.21 transmit circuit. Once the call dialing process is completed, the computer will receive call progress signals from the modem on the receive circuit indicating such conditions as number busy and call in progress.

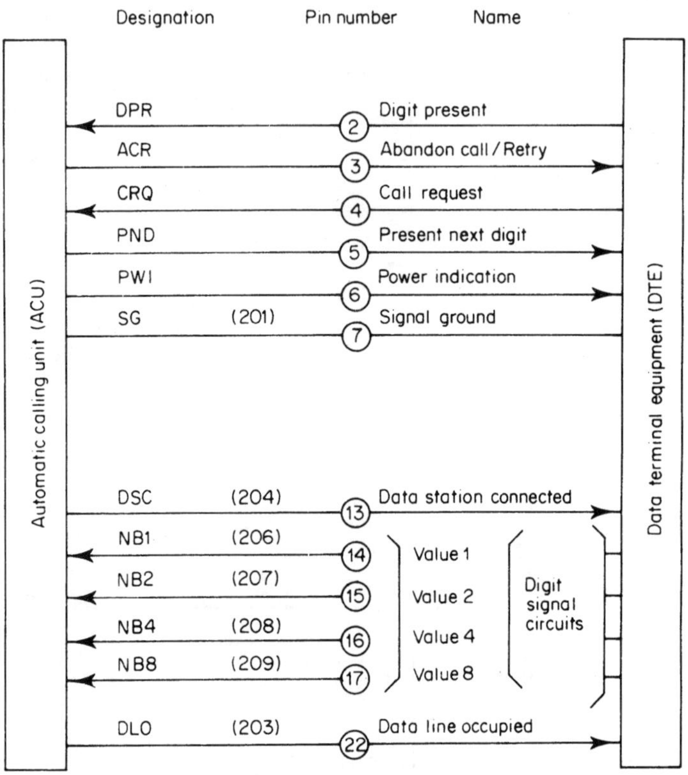

Figure 1.51 V.25 interface

The X.21 interface specifies the use of the balanced signaling characteristics of ITU X.27 (RS-422) for the network side of the interface and either X.27 or X.26 (RS-423) for the terminal equipment side. This specification enables terminal equipment to be designed for several applications. Unlike RS-232 and V.24, the X.21 standard specifies the use of a 15-pin connector. The X.20 interface uses the same 15-pin connector as X.21; however, since it supports asynchronous transmission it only needs to transmit data, and to receive data and ground signals.

Figure 1.52 illustrates the X.21 interchange circuits by circuit type. As indicated, X.21 specifies four categories of interchange circuits—ground, data transfer, control and timing. The operation of the circuits in each category is described below.

Ground signals

Circuit G, signal ground or common return, is used to connect the zero volt reference of the transmitter and receiver ends of the circuit. If X.26 differential signaling is used the G circuit is split into two. The Ga circuit is used as the DTE common return and is connected to ground at the DTE. The G circuit becomes Gb and is used as the DCE common return and is connected to ground at the DCE.

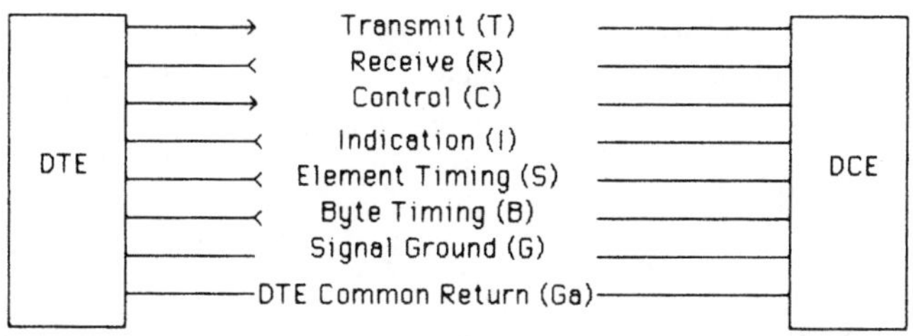

Figure 1.52 CCITT X.21 interchange circuits. Illustrates the X.21 interchange circuits by circuit type

Data transfer circuits

Circuit T is the transmit circuit used by the DTE to transmit data to the DCE. Circuit R is the receive circuit which is used by the DCE to transmit data to the DTE.

Control circuits

The X.21 specification has two control circuits—control and indication. Circuit C is the control circuit used by the DTE to indicate to the DCE the state of the interface. During the data transfer phase in which coding flows over the transmit circuit, circuit C remains in the ON condition.

Circuit I, which is the indication circuit, is used by the DCE to indicate the state of the interface to the DTE. When circuit I is ON the representation of the signal occurs in coded form over the receive circuit. Thus, during the data transfer phase circuit I is always ON.

Timing circuits

There are two timing circuits specified by X.21—signal element timing and byte timing.

Circuit S, which is signal element timing, is generated by the DCE and controls the time of data on the transmit and receive circuits. In providing a clocking signal, circuit S turns ON and OFF for nominally equal periods of time. The second timing circuit, circuit B, which is byte timing, provides the DTE with 8-bit timing information for synchronous transmission generated by the DCE. Circuit B turns OFF whenever circuit S is ON, indicating the last bit of an 8-bit byte. At other times within the period of the 8-bit byte circuit B remains ON. This circuit is not mandatory in X.21 and is only used occasionally.

Limitations

The X.21 standard has not gained wide acceptance for several reasons to include the popularity of the RS-232/V.24 standard and the cost of implementing X.21. With respect to cost, X.21 transmits and interprets coded character strings. This requires more intelligence to be built into the interface, adding to the cost of the interface. Due to the preceding limitations, the ITU defined an alternate interface for public data network access known as X.21 bis, where bis is the Latin term for secondary.

X.21 bis

The X.21 bis recommendation is both physically and functionally equivalent to the V.24 standard, which is compatible with RS-232. The X.21 bis recommendation is designed as an interim interface for X.25 network access and will gradually be replaced by the X.21 standard as more equipment is manufactured to meet the X.21 specification. The X.21 bis connector is the common DB25 connector used by RS-232 and V.24. The connector pins for X.21 bis are defined in an ISO specification called DIS 2110.

RS-530

Like RS-232, RS-530 uses the near-universal 25-pin D-shaped interface connector. Although this standard is intended to replace RS-449, both RS-422 and RS-423 standards specify the electrical characteristics of the interface and will continue in existence. These standards are referenced by the RS-530 standard.

Similar to RS-449, RS-530 provides equipment meeting this specification with the ability to transmit at data rates above the RS-232 limit of 19.2 kbps. This is accomplished by the standard originally specifying the utilization of balanced signals in place of several secondary signals and the Ring Indicator signal included in RS-232. As previously mentioned in our discussion of differential signaling, this balanced signaling technique is accomplished by using two wires with opposite polarities for each signal to minimize distortion.

The RS-530 specification was first outlined in March 1987 and was officially released in April 1988. A revision known as RS-530-A was approved in May 1992. By supporting data rates up to 2 Mbps and using the standard 'D' type 25-pin connector, RS-530 offers the potential to achieve a high level of adoption during the 1990s. One significant change resulting from Revision A is the specification of an alternative 26-position interface connector (Alt A) which is the same optional connector as specified in Revision E to RS-232. Another significant change resulting from Revision A was the addition of support for Ring Indicator which enables the interface to support switched network applications.

Figure 1.53 summarizes the RS-530 interchange circuits and compares those circuits to both RS-232 and RS-449. Note that RS-530 has maintained the standard RS-232 circuit structure for data, clock and control, all of which are balanced signals based upon the RS-442 standard. RS-530 has also adopted the three test circuits specified in RS-232-D: local loop, remote loop and test mode. Like RS-232-D, these three circuits are single ended. RS-530 has maintained pin 1 as frame ground or shield and pin 7 as signal ground.

The original RS-530 interface specified balanced circuits for DCE Ready and DTE ready, using pins 22 and 23 for one pair of each signal, respectively. Under

Designation	RS-530	RS-232	RS-449
Shield	1	1	1
Transmitted data	BA (A) 2 BA (B) 14	2 —	4 Send data 22
Received data	BB (A) 3 BB 16	3 —	6 Received data 24
Request to send	CA (A) 4 CA (B) 19	4 —	7 Request to send 25
Clear to send	CB (A) 5 CB (B) 13	5 —	9 Clear to send 27
DCE ready	CC (A) 6 CC (B) 22	6 —	11 Data mode 29
DTE ready	CD (A) 20 CD (B) 23	20 —	12 Terminal ready 30
Signal ground	AB 7	7	19 Signal ground
Received line signal detector	CF (A) 8 CF (B) 10	8 —	13 Receiver ready 31
Transmit signal element timing (DCE source)	DB (A) 15 DB (B) 12	15 —	5 Send timing 23
Receive signal element timing (DCE source)	DD (A) 17 DD (B) 9	17 —	8 Receive timing 26
Local loopback	LL 18	—	10 Local loopback
Remote loopback	RL 21	—	14
Transmit signal element timing (DTE source)	DA (A) 24 DA (B) 11	24 —	17 Terminal timing 35
Test mode	TM 25	—	18 Test mode
Ring indicator (Revision A)	CD 22	22	15 Incoming call
Signal common (Revision A)	AC 23	—	20 Receive common

Figure 1.53 Pin comparison—RS-530 to RS-232 and RS-449

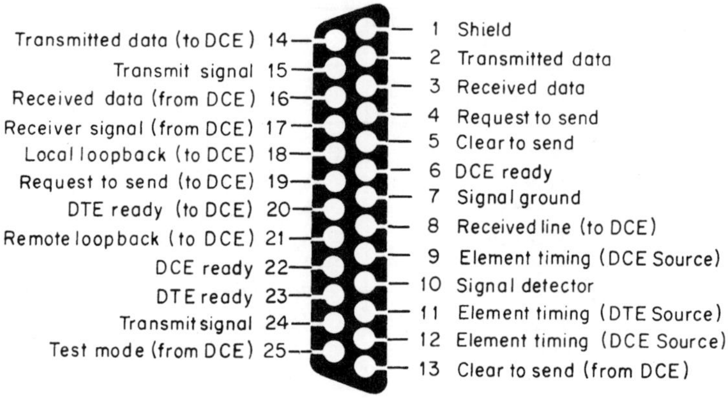

Figure 1.54 RS-530 interface on D connector

Revision A, Ring Indicator support was added through the use of pin 22 while Signal Common support was added through the use of pin 23.

Based upon the RS-530 pin assignments contained in Figure 1.53, the interchange circuits for the D connector specified by the standard are illustrated in Figure 1.54. You can compare Figure 1.54 with Figure 1.45 to see the similarity between RS-530 and RS-232 D-connector interfaces.

High Speed Serial Interface

The High Speed Serial Interface (HSSI) was jointly developed by T3Plus Networking, Inc. and Cisco Systems, Inc. as a mechanism to satisfy the growing demands of high-speed data transmission applications. Although the development of HSSI dates from 1989, its developers were forward looking, recognizing that the practical use of T3 and Synchronous Optical Network (SONET) terminating products would require equipment to transfer information well beyond the capability of the V.35 and RS 422/449 interfaces. The result of their efforts was HSSI, which is a full-duplex synchronous serial interface capable of transmitting and receiving information at data rates up to 52 Mbps between a DTE and a DCE. This standard was ratified by the American National Standards Institute (ANSI) in July 1992 and was being reviewed by the International Organization for Standardization (ISO) and the ITU-T for standardization when this book was revised. ANSI document SP2795 defines the electrical specifications for the interface at data rates up to 52 Mbps while document SP2796 specifies the operation of the DTE–DCE interface circuits.

Rationale for development

The rationale for the development of HSSI was not only the operating limit of 6 Mbps for V.35 and 10 Mbps for RS-422/449 but, in addition, the problems that occur when those standards are extended to higher operating rates. Several manufacturers developed proprietary methods to increase the data transfer rate of those standards; however, doing so resulted in an increase in radio frequency interference (RFI) which in some instances resulted in the disruption of the operation of other nearby equipment and cable connections.

HSSI eliminates potential RFI problems while obtaining a high speed data transfer capability through the use of emitter-coupled logic (ECL). ECL is faster than complementary metal-oxide semiconductor (CMOS) or transistor-to-transistor logic (TTL) commonly used in other interfaces, while generating a lower amount of noise. To accomplish this, ECL has a voltage swing of 0.8 between defining 0 and 1 bits, which is considerably smaller than the voltage swings in CMOS and TTL logic.

Signal definitions

HSSI can be viewed as a simple V.35 type interface based upon the use of emitter-coupled logic for transmission levels, with 12 signals currently defined. Figure 1.55 illustrates the 12 HSSI currently defined interchange circuits, with the normal dataflow indicated by arrowheads on each circuit.

Under HSSI signaling the DCE manages clocking, similar to the V.35 and RS-499 standards. The DCE generates Receive Timing (RT) and Send Timing (ST) signals from the network clock. In comparisons, the DTE returns the clocking signal as Terminal Timing (TT) with data on circuit SD (Send Data) to ensure data is in phase with timing.

HSSI signaling was designed to support continuous as well as gapped, or discontinuous clocking. The latter is associated with the DS3 signal used for T3 transmission at 44.736 Mbps. Under DS3 signaling every 85th frame is a control frame, requiring the DCE clock to run for 84 contiguous pulses and then miss one

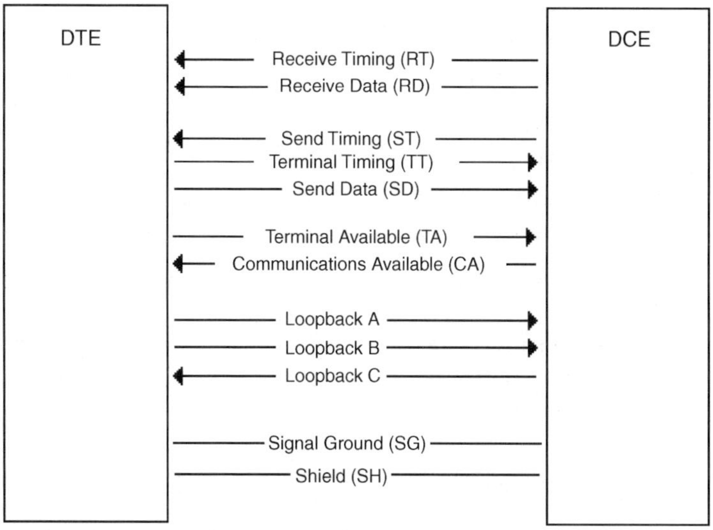

Figure 1.55 HSSI signaling between DTE and DCE

pulse or gap over it to correctly achieve the DS3 frequency. We can obtain an appreciation of the operation of HSSI by examining the operation and functionality of each of the 12 signals currently supported by the interface.

Receive Timing (RT)

The Receive Timing circuit presents the DCE clocking obtained from the network to the attached terminal device. As previously discussed, RT is a gapped clock and has a maximum bit rate of 52 Mbps. The clocking signal on RT provides the timing information necessary for the DTE to receive data on circuit RD.

Receive Data (RD)

Data received by the DCE from an attached communications circuit are transferred on the RD circuit to the DTE.

Send Timing (ST)

The Send Timing circuit transports a gapped clocking signal with a maximum bit rate of 52 Mbps from the DCE to the DTE. This clock provides transmit signal element timing information to the DTE which is returned via the Terminal Timing (TT) circuit.

Terminal Timing (TT)

The Terminal Timing circuit provides the path for the echo of the Send Timing clocking signal from the DTE to the DCE. The clocking signal on this circuit provides transmit signal element timing information to the DCE which is used for sampling data forwarded to the DCE on circuit SD.

Send Data (SD)

The flow of data from the DTE to the DCE occurs on circuit SD. As previously mentioned, clocking on circuit TT provides the DCE with the timing signals to correctly sample the SD circuit.

Terminal Available (TA)

Terminal Available can be considered as the functional equivalent of Request to Send (RTS); however, unlike TRS, TA is asserted by the DTE independently of DCE when the DTE is ready to both send and receive data. Actual data transmission will not occur until the DCE has asserted a Communications Available (CA) signal.

Communications Available (CA)

The Communications Available (CA) signal can be considered as functionally similar to the Clear to Send (CTS) signal. However, the CA signal is asserted by the DCE independently of the TA signal whenever the DCE is prepared to both transmit and receive data to and from the DCE. The assertion of voltage on circuit CA indicates that the DCE has a functional data communications channel; however, transmission will not occur until the TA signal is asserted by the DTE.

Through the elimination of Data Set Ready (DSR) and Data Terminal Ready (DTR) signals commonly used in other interfaces, the HSSI interface becomes relatively simple to implement. This in turn simplifies the DTE-to-DCE data exchange by eliminating the complex handshaking procedures required when using other interfaces.

Loopback circuits

Through the use of three loopback circuits HSSI provides expanded diagnostic testing capability that can be extremely valuable when attempting to isolate transmission problems. Circuits LA and LB are asserted by the DTE to inform the near or far end DCE to initiate one of three diagnostic loopback modes—loopback at the remote DCE line, loopback at the local DCE line, or loopback at the remote DTE. The third loopback circuit, LC, is optional and is used to request the local DTE to provide a loopback path to the DCE. When the LC circuit is asserted the DTE would set TT = RT and SD = RD, enabling testing of the DCE to DTE interface independent of the DTE. The ST circuit would not be used as it cannot be relied upon as a valid clocking source.

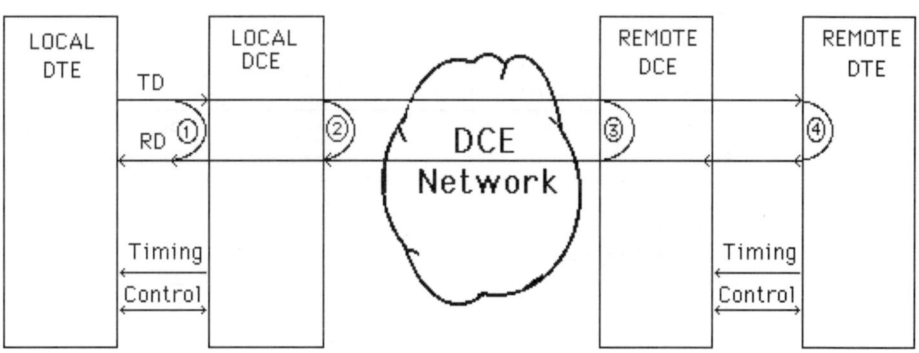

1. Local digital loopback
2. Local 'analog' loopback
3. Remote digital loopback
4. Remote 'analog' loopback

Figure 1.56 HSSI supports four loopbacks

Figure 1.56 illustrates the four possible HSSI loopbacks with respect to the local DTE. Although data flows end-to-end in digital form, the term 'analog' used to reference two loopbacks is a carry-over from modem loopback terminology and indicates that data is converted from unipolar to bipolar in the same manner as certain modem loopbacks convert digital to analog to test the modulator of a modem.

Signal Ground (SG)

Signal Ground is used to ensure that transmit signal levels remain within the common input range of receivers. The SG circuit is grounded at both ends.

Shield (SH)

The shield is used to limit electromagnetic interference. To accomplish this the shield encapsulates the HSSI cable.

Pin Assignments

HSSI employs a 50-pin plug connector and receptacle. The connectors are mated to a cable consisting of 25 twisted pairs of 28 AWG cable and are limited to a 50 foot length. The 25 twisted pairs are encapsulated in a polyvinyl chloride (PVC) jacket. Table 1.26 lists the pin assignments. Note that the signal direction is indicated with respect to the DCE.

Table 1.26 HSSI pin assignments

Signal name	Signal direction	Pin no. (+side)	Pin no. (−side)
SG Signal Ground	N/A	1	26
RT Receive Timing	←	2	27
CA DCE Available	←	3	28
RD Receive Data	←	4	29
LC Loopback circuit C (optional)	←	5	30
ST Send Timing	←	6	31
SG Signal Ground	N/A	7	32
TA DTE Available	→	8	33
TT Terminal Timing	→	9	34
LA Loopback circuit A	→	10	35
SD Send Data	→	11	36
LB Loopback circuit B	→	12	37
SG Signal Ground	N/A	13	38
Reserved for future use	→	14–18	39–43
SG Signal Ground	N/A	19	44
Reserved for future use	←	20–24	45–49
SG Signal Ground	N/A	25	50

Applications

Since its initial development in 1989 HSSI has been incorporated into a large number of products designed to support high speed serial communications. In addition, its relatively simple interface has resulted in its use at data rates that would normally require the use of a V.35 or RS-499 interface. For example, many routers, multiplexers, inverse multiplexers and Channel Service Units operating at 1.544 Mbps can now be obtained with HSSI as well as products designed to operate at T3 (44.736 Mbps) and SONET Synchronous Transmission Service Level 1 (STS-1) (51.84 Mbps).

High Performance Parallel Interface

The High Performance Parallel Interface (HIPPI) represents an ANSI switched network standard which was originally developed to support direct communications between mainframes, supercomputers and directly attached storage devices. A series of ANSI standards currently define the physical layer operation of HIPPI as well as data framing, disk and tape connections and link encapsulation. Table 1.27 lists ANSI HIPPI related standards and their areas of standardization.

Table 1.27 ANSI HIPPI-related standards

ANSI standard	Area covered
X3.183-1991	Physical layer
X3.222-1993	Switch control
X3.218-1993	Link encapsulation
X3.210-1992	Framing protocol
ANSI/ISO 9318-3	Disk connections
ANSI/ISO 9318-4	Tape connection

Transmission distance

Although its name implies the use of a parallel interface, a number of extensions to HIPPI resulted in its ability to support a 300 meter serial interface over multimode copper as well as parallel transmission via a 50-pair shielded twisted-pair wiring group for relatively short distances. In its basic mode of operation a HIPPI based network consists of two computers connected via a pair of 50-pair copper cables to HIPPI channels on each device. Each 50-pair cable supports transmission in one direction, resulting in the pair of 50-pair cables providing a full-duplex transmission facility. That transmission facility can extend up to 25 meters and operate at either 800 Mbps or 1.6 Gbps, the latter accomplished by doubling the data path.

Through the use of one or more HIPPI switches you can develop an extended HIPPI network. That network can use copper cable between switches which permits cabling runs up to 200 meters in length, or a fiber extender can be used to support extending the distance between switches up to 10 km. The fiber extender

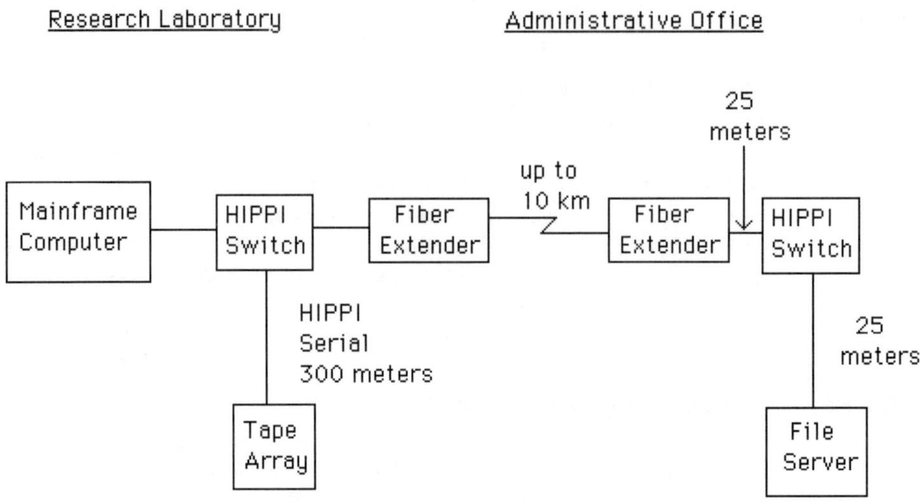

Figure 1.57 Creating a HIPPI-based network

functions as a parallel to serial converter as well as an electrical to optical converter to support serial light transmission between switches.

Figure 1.57 illustrates the creation of HIPPI based network on a college campus to connect a research laboratory to an administrative file server. HIPPI interfaces are now available for a wide range of products to include personal computers, routers and gateways as well as mainframes and supercomputers.

Operation

HIPPI operates by framing data to be transmitted as well as by using messages to control data transfer operations. A HIPPI connection is set up through the use of three messages. A Request message is used by the data originator to request the establishment of a connection. A Connect message is returned by the desired destination to inform the requestor that a connection was established. The third message is Ready, which is issued by the destination to inform the originator that it is ready to accept data.

Cables and connectors

Numerous types of cables and connectors can be employed in data transmission systems. To familiarize readers with cabling options that can be considered, we will now focus our attention upon several types of cables and their connectors, as well as several cabling tricks based upon our previous examination of the operation of RS-232/V.24 interchange circuits.

Twisted-pair cable

The most commonly employed data communications cable is the twisted-pair cable. This cable can usually be obtained with 4, 7, 9, 12, 16 or 25 conductors, where each conductor is insulated from another by a PVC shield.

For EIA RS-232 and ITU V.24 applications, those standards specify a maximum cabling distance of 50 feet between DTE and DCE equipment for data rates ranging from 0 to 19 200 bps; and, normal industry practice is to use male connectors at the cable ends which mate with female connectors normally built into such devices as terminals and modems. Figure 1.58 illustrates the typical cabling practice employed to connect a DTE to a DCE.

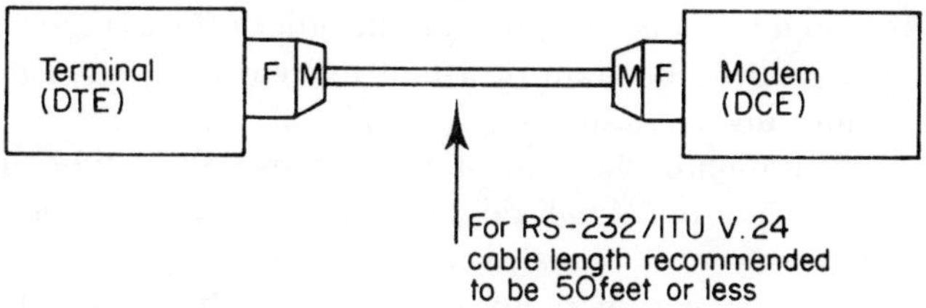

Figure 1.58 DTE to DCE cabling

Low-capacitance shielded cable

In certain environments where electromagnetic interference and radio frequency emissions could be harmful to data transmission, you should consider the utilization of low-capacitance shielded cable in place of conventional twisted-pair cable. Low-capacitance shielded cable includes a thin wrapper of lead foil that is wrapped around the twisted-pair conductors contained in the cable, thereby providing a degree of immunity to electrical interference that can be caused by machinery, fluorescent ballasts and other devices.

Ribbon cable

Since an outer layer of PVC houses the individual conductors in a twisted-pair cable, the cable is rigid with respect to its ability to be easily bent. Ribbon or flat cable consists of individually insulated conductors that are insulated and positioned in a precise geometric arrangement that results in a rectangular rather than a round cross-section. Since ribbon cable can be easily bent and folded, it is practical for those situations where you must install a cable that must follow the contour of a particular surface.

The RS-232 null modem

No discussion of cabling would be complete without a description of a null modem, which is also referred to as a modem eliminator. A null modem is special cable that is designed to eliminate the requirement for modems when interconnecting two collocated data terminal equipment devices. One example of this would be a requirement to transfer data between two collocated personal computers that do not have modems and use different types of diskettes, such as an IBM PC which uses a $5\frac{1}{4}$-inch diskette and an IBM PS/2 which uses a $3\frac{1}{2}$-inch diskette. In this situation, the interconnection of the two computers via a null modem cable would permit programs and data to be transferred between each personal computer in spite of the media incompatibility of the two computers. Since DTEs transmit data on pin 2 and receive data on pin 3, you could never connect two such devices together with a conventional cable as the data transmitted from one device would never be received by the other. In order for two DTEs to communicate with one another, a connector on pin 2 of one device must be wired to connector pin 3 on the other device. Figure 1.59 illustrates an example of the wiring diagram of a null modem cable used to connect two DB-25 connectors together, showing how pins 2 and 3 are cross-connected as well as the configuration of the control circuit pins on this type of cable.

Since a terminal will raise or apply a positive voltage in the 9 V to 12 V range to turn on a control signal, you can safely divide this voltage to provide up to three different signals without going below the signal threshold of 3 V previously illustrated in Figure 1.41. In examining Figure 1.59, we should note the following control signal interactions are caused by the pin cabling:

(1) Data terminal ready (DTR, pin 20) raises data set ready (DSR, pin 6) at the other end of the cable. This makes the remote DTE think a modem is connected to the other end and powered ON.
(2) Request to send (RTS, pin 4) raises data carrier detect (CD, pin 8) on the other end and signals clear to send (CTS, pin 5) at the original end of the cable. This makes the DTE believe that an attached modem received a carrier signal and is ready to modulate data.
(3) Once the handshaking of control signals is completed, we can transmit data onto one end of the cable (TD, pin 2) which becomes receive data (RD, pin 3) at the other end.

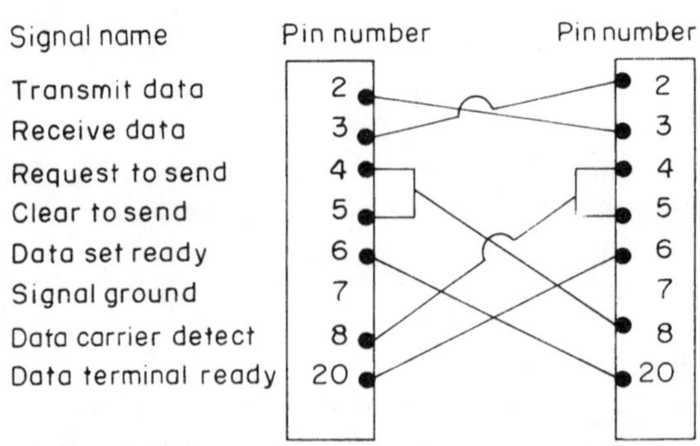

Figure 1.59 DB-25 to DB-25 null modem cable

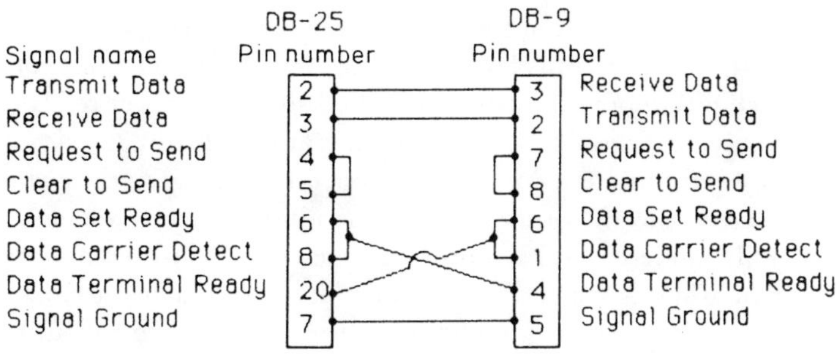

Figure 1.60 DB-25 to DB-9 null modem cable

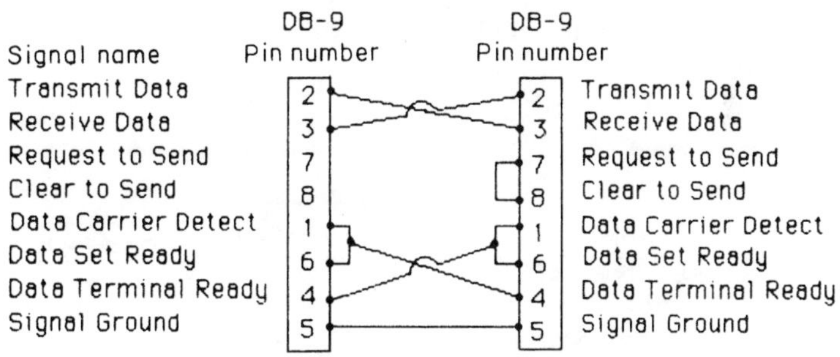

Figure 1.61 DB-9 to DB-9 null modem cable

You can also use a null modem cable to connect a DB-25 connector port to a DB-9 connector port as well as two DB-9 connector ports to each other. Figure 1.60 illustrates the connector wiring for a DB-25 to DB-9 null modem cable, while Figure 1.61 illustrates the conductor wiring for a null modem cable used to connect two DB-9 connectors.

Since a large number of personal computers use DB-9 connectors, the null modem cables illustrated in Figures 1.60 and 1.61 are popularly employed to directly cable two personal computers to one another as well as PCs to other types of terminal devices, including mainframe computer ports and the ports of a data PBX.

In comparing the wiring of the conductors in Figure 1.59 to the wiring of conductors illustrated in Figures 1.60 and 1.61, you will note the similarity between each type of null modem cable. That is, transmit data is always routed to receive data at the opposite end of the cable, RTS and CTS control signals are always tied together, and the tying of DSR to DCD is routed to the DTR signal at the other end of the cable. You will also note that, although the routing of conductors is consistent for all three types of null modem cables, the actual routing of conductors to specific pins will vary due to the difference in the assignment of conductors to pins on the DB-25 and DB-9 conductors.

The cable configurations illustrated in Figures 1.59 and 1.61 will work for most data terminal equipment interconnections; however, there are a few exceptions. The most common exception is when a terminal device is to be cabled to a port on a mainframe computer that operates as a 'ring-start' port. This means that the computer port must obtain a ring indicator (RI, pin 22) signal. In this situation, each null modem must be modified so that data set ready (DSR, pin 6) on a DB-25 connector is jumpered to ring indicator (RI, pin 22) at the other end of the DB-25 cable to initiate a connect sequence to a 'ring-start' system.

Owing to the omission of transmit and receive clocks, the previously described null modem cables can only be used for asynchronous transmission. For synchronous transmission you must either drive a clocking device at one end of the cable or employ another technique. Here you would use a modem eliminator which differs from a null modem by providing a clocking signal to the interface. If a clocking source is to be used, DTE timing (pin 24) on a DB-25 connector is normally selected to develop a synchronous null modem cable. In developing this cable, pin 24 is strapped to pins 15 and 17 at each end of the cable as illustrated in Figure 1.62. Then, DTE timing provides transmit and receive clocking signals at both ends of the cable.

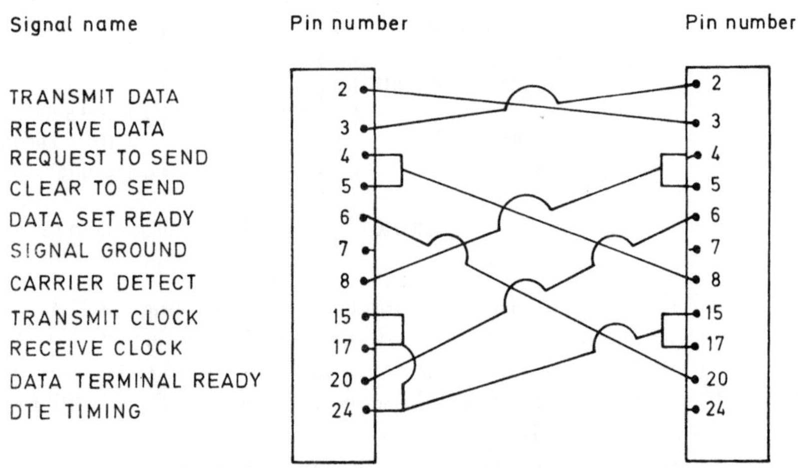

Figure 1.62 Synchronous null modem cable

RS-232 cabling tricks

A general purpose 3-conductor cable can be used when there is no requirement for hardware flow control and a modem will not be controlled. Here the term flow control refers to the process that causes a delay in the flow of data between DTE and DCE, DCE and DTE or two DCEs or two DTEs resulting from the changing of control circuit states. Figure 1.63 illustrates the use of a 3-conductor cable for DTE–DCE and DTE–DTE or DCE–DCE connections. When this situation occurs it becomes possible to use a 9-conductor cable with three D-shaped connectors at each end, with each connected to three conductors on the cable connector which eliminates the necessity of installing three separate cables.

Figure 1.64 illustrates a 5-conductor cable that can be installed between a DTE and DCE (modem) when asynchronous control signals are required. Similar to the

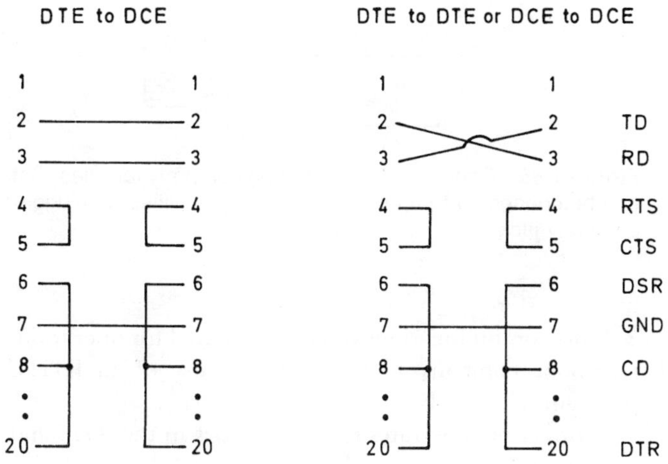

Figure 1.63 General purpose 3-conductor cable

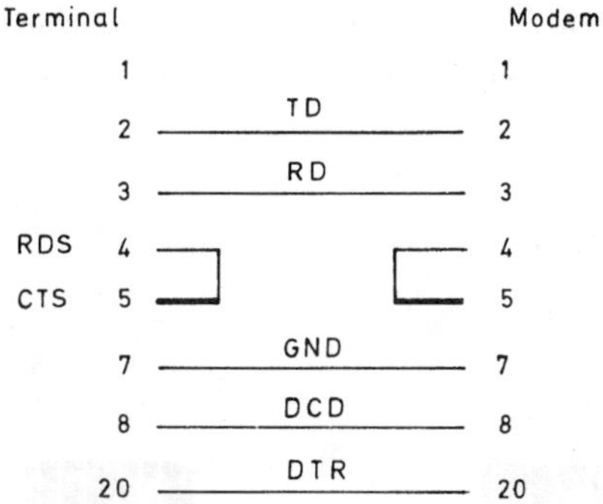

Figure 1.64 General purpose 5-conductor cable

use of 9-conductor cable to derive three 3-conductor connections, standard 12-conductor cable can be used to derive two 5-conductor connections.

Plugs and jacks

Modern data communications equipment is connected to telephone company facilities by a plug and jack arrangement as illustrated in Figure 1.65. Although the connection appears to be, and in fact is, simplistic, the number of connection arrangements and differences in the types of jacks offered by telephone companies usually ensures that the specification of an appropriate jack can be a complex task. Fortunately, most modems and other communications devices include explicit instructions covering the type of jack the equipment must be connected to as well as providing the purchaser with information that must be furnished to the telephone company in order to legally connect the device to the telephone company line.

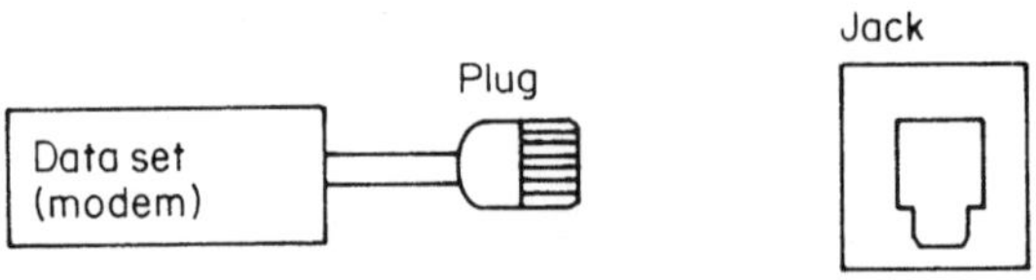

Figure 1.65 Connection to telephone company facilities. Data communications equipment can be connected to telephone company facilities by plugging the device into a telephone company jack

Most communications devices designed for operation on the PSTN interface the telephone company network via the use of an RJ11C permissive or an RJ45S programmable jack.

Figure 1.66 illustrates the conductors in the RJ11 and RJ45 modular plugs. The RJ11 plug is primarily used on two-wire dial lines. This plug is used in both the home and office for connecting a single instrument telephone to the PSTN. In addition, the RJ11 also serves as an optional connector for four-wire private lines. Although the RJ11 connector is fastened to a cable containing four or six stranded-copper conductors, only two wires in the cable are used for switched network applications. When connected to a four-wire leased line, four conductors are used.

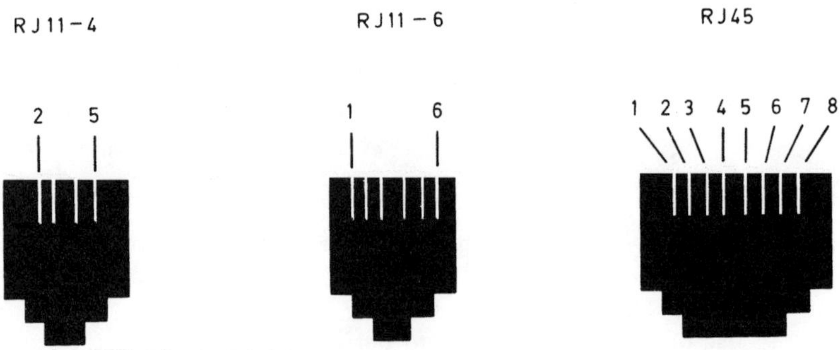

Figure 1.66 RJ11 and RJ45 modular plugs

The development of the RJ11 connector can be traced to the evolution of the switchboard. The plugs used with switchboards had a point known as the 'tip' which was colored red, while the adjacent sleeve known as the ring was colored green.

The original color coding used with switchboard plugs was carried over to telephone wiring. If you examine a four-wire (two-pair) telephone cable, you will note that the wires are coloured yellow, green, red and black. The green wire is the tip of the circuit while the red wire is the ring. The yellow and black wires can be used to supply power to the light in a telephone or used to control a secondary telephone using the same four-wire conductor cable.

Table 1.28 Color identification of telephone cables

Four-wire		Six-wire	
Pair	Color	Pair	Color
1	Yellow Green	1	Blue Yellow
2	Red Black	2	Green Red
		3	Black White

The most common types of telephone cable used for telephone installation are four-wire and six-wire conductors. Normally, a four-wire conductor is used in a residence that requires one telephone line. A six-wire conductor is used in either a residential or business location that requires two telephone lines and can also be used to provide three telephone lines from one jack. Table 1.28 compares the color identification of the conductors in four-wire and six-wire telephone cable.

During the late 1970s, telephone companies replaced the use of multiprong plugs by the introduction of modular plugs which in turn are connected to modular jacks.

The RJ11C plug was designed for use with any type of telephone equipment that requires a single telephone line. Thus, regardless of the use of either four-wire or six-wire cable only two wires in the cable need be connected to an RJ11C jacks. The RJ11 plug can also be used to service an instrument that supports two or three telephone lines; however, RJ14C and RJ25C jacks must then be used to provide that service. These two jacks are only used for voice. For data transmission both four- and six-conductor plugs are available for use, with conductors 1, 2, 5 and 6 in the jack normally reserved for use by the telephone company. Then, conductor 4 functions as the ring circuit while conductor 5 functions as the tip to the telephone company network.

The RJ45 plug is also designed to support a single line although it contains eight positions. In this plug, positions 4 and 5 are used for ring and tip and a programmable resistor on position 8 in the jack is used to control the transmit level of the device connected to the switched network.

The RJ45 plug and jack connectors are also used in some communications products to provide an RS-232 DTE–DCE interface via twisted-pair telephone wire. In certain cases an RJ45 to DB25 adapter may be needed. This adapter will, as an example, permit the cabling of a cable terminated with an RJ45 plug to DB25 connector or a DB25 connector cable end to a RJ45 socket. RJ45 connectors typically support the transmitted data, received data, data terminal ready (DTE ready), data set ready (DCE ready), data detect (received line signal detector), request to send, clear to send and signal ground circuits.

The physical size of the plugs used to wire equipment to each jack as well as the size of each of the previously discussed jacks are the same. The only difference between jacks is in the number of wires cabled to the jack and the number of contacts in the jack which are used to pass telephone wire signals.

Connecting arrangements

There are three connecting arrangements that can be used to connect data communications equipment to telephone facilities. The object of these arrangements is to ensure that the signal received at the telephone company central office in the United States does not exceed −12 dBm.

Permissive arrangement

The permissive arrangement is used when you desire to connect a modem to your organization's switchboard, such as a private branch exchange (PBX). When a permissive arrangement is employed, the output signal from the modem is fixed at a maximum of −9 dBm and the plug that is attached to the data set cable can be connected to three types of telephone company jacks as illustrated in Figure 1.67. The RJ11 jack can be obtained as a surface mounting (RJ11C) for desk sets or as a wall-mounted (RJ11W) unit; however the RJ41S and RJ45S are available only for surface mounting.

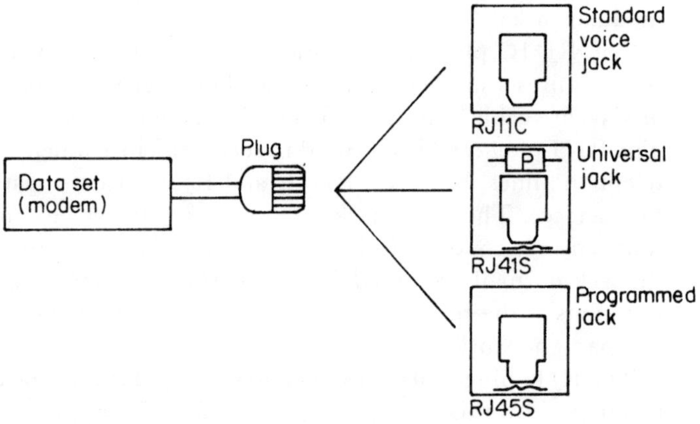

Figure 1.67 Permissive arrangement jack options

Since permissive jacks use the same six-pin capacity miniature jack used for standard voice telephone installations, this arrangement provides for good mobility of terminals and modems.

Fixed loss loop arrangement

Under the fixed loss loop arrangement the output signal from the modem is fixed at a maximum of −4 dBm and the line between the subscriber's location and the telephone company central office is set to 8 dBm of attenuation by a pad located within the telephone company provided jack. As illustrated in Figure 1.68, the

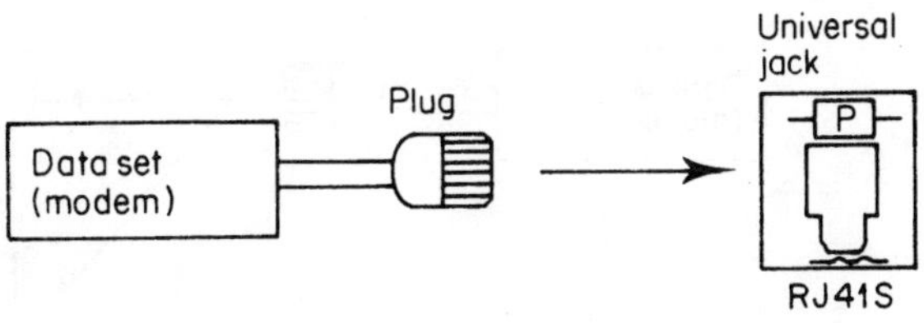

Figure 1.68 Fixed loss loop arrangement

only jack that can be used under the fixed loss loop arrangement is the RJ41S. This jack has a switch FLL-PROG, which must be placed in the FLL position under this arrangement. Since the modem output is limited to −4 dBm, the 8 dB attenuation of the pad ensures that the transmitted signal reaches the telephone company office at −12 dBm. As the pad in the jack reduces the receiver signal-to-noise ratio by 8 dB, this type of arrangement is more susceptible to impulse noise and should only be used if one cannot use either of the two other arrangements.

Programmable arrangement

Under the programmable arrangement configuration a level setting resistor inside the standard jack provided by the telephone company is used to set the transmit level within a range between 0 and −12 dBm. Since the line from the user is directly routed to the local telephone company central office at installation time, the telephone company will measure the loop loss and set the value of the resistor based upon the loss measurement. As the resistor automatically adjusts the transmitted output of the modem so the signal reaches the telephone company office at −12 dBm, the modem will always transmit at its maximum allowable level. As this is a different line interface in comparison to permissive or fixed loss data sets, the data set must be designed to operate with the programmability feature of the jack.

Either the RJ41S universal jack or the RJ45S programmed jack can be used with the programmed arrangement as illustrated in Figure 1.69. The RJ41S jack is installed by the telephone company with both the resistor and pad for programmed and fixed loss loop arrangements. By setting the switch to PROG, the programmed arrangement will be set. Since the RJ45S jack can operate in either the permissive or programmed arrangement without a switch, it is usually preferred as it eliminates the possibility of an inadvertent switch reset.

Telephone options

Prior to the use of modular jacks, telephones were hardwired to the switched telephone network. Even with the growth in the use of modular connecting arrangements, there are still a few locations where telephones are connected the 'old-fashioned way'. Those telephone sets require the selection of specific options to be used with communications equipment. As part of the ordering procedure you must specify a series of specific options that are listed in Table 1.29.

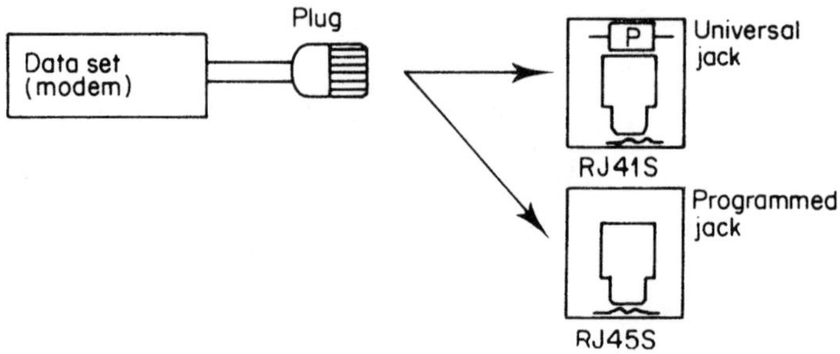

Figure 1.69 Programmed arrangement jack

Table 1.29 Telephone ordering options

Decision	Description	
A	1	Telephone set controls line
	2	Data set controls line
B	3	No aural monitoring
	4	Aural monitoring provided
C	5	Touchtone dialing
	6	Rotary dialing
D	7	Switchhook indicator
	8	Mode indicator

When the telephone set is optioned for telephone set controls the line, calls are originated or answered with the telephone by lifting the handset off-hook. To enable control of the line to be passed to a modem or data set an 'exclusion key' is required.

The exclusion key telephone permits calls to be manually answered and then transferred to the modem using the exclusion key. The exclusion key telephone is wired for either 'telephone set controls line' or 'data set controls line'. Data set control is normally selected if you have an automatic call or automatic answer modem since this permits calls to be originated or answered without taking the telephone handset off-hook. To use the telephone for voice communications the handset must be raised and the exclusion key placed in an upward location.

The telephone set control of the line option is used with manual answer or manual originate modems or automatic answer or originate modems that will be operated manually. To connect the modem to the line the telephone must be off-hook and the exclusion key placed in an upward position. To use telephone for voice communications the telephone must be off-hook while the exclusion key is placed in the downward position.

When the dat set controls the line option is selected, calls can be automatically originated or answered by the data equipment without lifting the telephone handset.

Aural monitoring enables the telephone set to monitor call progress tones as well as voice answer back messages without requiring the user to switch from data to voice.

You can select option B3 if aural monitoring is not required, while option B4 should be selected if it is required. Option C5 should be selected if touchtone dialing is to be used, while option C6 should be specified for rotary dial telephones. Under option D7, the exclusion key will be bypassed, resulting in the lifting of the telephone handset causing the closure of the switchhook contact in the telephone. In comparison, option D8 results in the exclusion key contacts being wired in series with the switchhook contacts, indicating to the user whether he or she is in a voice or data mode.

Ordering the business line

Ordering a business line to transmit data over the switched telephone network currently requires you to provide the telephone company with four items of information. First, you must supply the telephone company with the Federal Communications Commission (FCC) registration number of the device to be connected to the switched telephone network. This 14-character number can be obtained from the vendor who must first register their device for operation on the switched network prior to making it available for use on that network.

Next, you must provide the ringer equivalence number of the data set to be connected to the switched network. This is a three-character number, such as 0.4A, and represents a unitless quotient formed in accordance with certain circuit parameters. Finally, you must provide the jack numbers and arrangement to be used as well as the telephone options if you intend to use a handset.

1.15 THE DATA LINK LAYER

In the ISO model, the data link layer is responsible for the establishment, control, and termination of connections among network devices. To accomplish these tasks the data link layer assumes responsibility for the flow of user data as well as for detecting and providing a mechanism for recovery from errors and other abnormal conditions, such as a station failing to receive a response during a predefined time interval.

In this section, we will first examine the key element that defines the data link layer—its protocol. In this examination, we will differentiate between terminal protocols and data link protocols to eliminate this terminology as a potential area of confusion. Next, we will focus our attention upon several specific wide area networking protocols, starting with simple asynchronous line-by-line protocols. Protocols examined in the second portion of this section include an asynchronous teletype protocol, several popular asynchronous file transfer protocols used to transfer data to and from personal computers, IBM's character-oriented binary synchronous communication (commonly referred to as BSC or bisync), Digital Equipment Corporation's Digital Data Communications Message Protocol (DDCMP), and the bit-oriented Higher Level Data Link Control (HDLC) as well as its IBM near-equivalent, Synchronous Data Link Control (SDLC).

Terminal and data link protocols

Two types of protocols should be considered in a data communications environment: terminal protocols and data link protocols.

The data link protocol defines the control characteristics of the network and is a set of conventions that are followed which govern the transmission of data and control information. A terminal or a personal computer can have a predefined control character or set of control characters which are unique to the terminal and are not interpreted by the line protocol. This internal protocol can include such control characters as the bell, line feed and carriage return for conventional teletype terminals, blink and cursor positioning characters for a display terminal and form control characters for a line printer.

For experimenting with members of the IBM PC series and compatible computers, you can execute the one line BASIC program PRINT CHR$(X) "DEMO", substituting different ASCII values for the value of *X* to see the effect of different PC terminal control characters. As an example, using the value 7 for X, the IBM PC will beep prior to displaying the message DEMO, since ASCII 7 is interpreted by the PC as a request to beep the speaker. Using the value 9 for X will cause the message DEMO to be printed commencing in position 9, since ASCII 9 is a tab character which causes the cursor to move on the screen 8 character positions to the right. Another example of a terminal control character is ASCII 11, which is the home character. Using the value 11 for X will cause the message DEMO to be printed in the upper left-hand corner of the screen since the cursor is first placed at that location by the home character.

Although poll and select is normally thought of as a type of line discipline or control, it is also a data link protocol. In general, the data link protocol enables the exchange of information according to an order or sequence by establishing a series of rules for the interpretation of control signals which will govern the exchange of information. The control signals govern the execution of a number of tasks which are essential in controlling the exchange of information via a communications facility. Some of these information control tasks are listed in Table 1.30.

Table 1.30 Information control tasks

Connection establishment	Transmission sequences
Connection verification	Data sequence
Connection disengagement	Error control procedures

Connection establishment and verification

Although all of the tasks listed in Table 1.30 are important, not all are required for the transmission of data, since the series of tasks required is a function of the total data communications environment. As an example, a single terminal or personal computer connected directly to a mainframe or another terminal device by a leased line may not require the establishment and verification of the connection. Several devices connected to a mainframe computer on a multidrop or multipoint line would, however, require the verification of the identification of each terminal device on the line to insure that data transmitted from the computer would be received by the proper device. Similarly, when a device's session is completed, this fact must be recognized so that the mainframe computer's resources can be made available to other users. Thus, connection disengagement on devices other than those connected on a point-to-point leased line permits a port on the front-end processor to become available to service other users.

Transmission sequence

Another important task is the transmission sequence which is used to establish the precedence and order of transmission, to include both data and control information. As an example, this task defines the rules for when devices on a multipoint circuit may transmit and receive information. In addition to the transmission of information following a sequence, the data itself may be sequenced. Data sequencing is normally employed in synchronous transmission and in asynchronous file transfer operations where a long block is broken into smaller blocks for transmission, with the size of the blocks being a function of the personal computer's or terminal's buffer area and the error control procedure employed. By dividing a block into smaller blocks for transmission, the amount of data that must be retransmitted, in the event that an error in transmission is detected, is reduced.

Although many error-checking techniques are more efficient when short blocks of information are transmitted, the efficiency of transmission correspondingly decreases since an acknowledgement (negative or positive) is returned to the device transmitting after each block has been received and checked. For communications between remote job entry terminals and computers, blocks of up to several thousand characters are typically used. Block lengths from 80 to 1024 characters are, however, the most common sizes. Although some protocols specify block length, most protocols permit the user to set the size of the block, while other protocols automatically vary the block size based upon the error rate experienced by the transmission progress.

Error control

The simplest method of error control does not actually ensure errors are corrected. This method of error control is known as echoplex and results in each character transmitted to a receiving device being sent back or echoed from the receiver to the transmitting device, hence, the term 'echoplex'.

Under the echoplex method of error correction, the transmitting device examines the echoed data. If the echoed data differs from the transmitted data an error is assumed to have occurred and the data must then be retransmitted. Since a transmission error can occur in either direction, it is possible for a character corrupted to another character during transmission in one direction to be corrupted back into its original bit form when echoed to the transmitting device. In addition, a character received correctly may be corrupted during its echo, resulting in the false impression that an error occurred.

Echoplex was a popular method for detecting transmission errors that was used with teletype terminals. This method of error detection was also used in such message switching systems as TWX and is currently used with many types of asynchronous transmission, including personal computers. Concerning the latter, a PC communicating in an asynchronous full-duplex mode to another full-duplex computer will have each character transmitted echoed back. Since the detection of erroneous characters, however, depends upon the visual accuity of the operator more modern methods of error detection and correction have replaced the use of echoplex in applications where we cannot rely upon an operator to correct errors. Thus, a large number of file transfer protocols that group characters into data blocks and append a checking mechanism were developed to automatically detect and correct transmission errors.

Today, the most commonly employed method to correct transmitted errors is to inform the transmitting device simply to retransmit a block. This procedure requires coordination between the sending and receiving devices, with the receiving device either continuously informing the sending device of the status of each previously transmitted block or transmitting a negative acknowledgement only when a block is received in error.

If the protocol used requires a response to each block and the block previously transmitted contained no detected errors, the receiver will transmit a positive acknowledgement and the sender will transmit the next block. If the receiver detects an error, it will transmit a negative acknowledgement and discard the block containing an error. The transmitting station will then retransmit the previously sent block. Depending upon the protocol employed, a number of retransmissions may be attempted. However, if a default limit is reached owing to a bad circuit or other problems, then the computer or terminal device acting as the master station may terminate the session, and the operator will have to re-establish the connection.

If the protocol supports transmission of a negative acknowledgement only when a block is received in error, additional rules are required to govern transmission. As an example, the sending device could transmit several blocks and, in fact, could be transmitting block $n+4$ prior to receiving a negative acknowledgement concerning block n. Depending upon the protocol's rules, the transmitting device could retransmit block n and all blocks after that block or finish transmitting block $n+4$, then transmit block n and resume transmission with block $n+5$.

Types of protocols

Now that we have examined protocol tasks, let us focus our attention upon the characteristics, operation and utilization of several types of protocols that provide a predefined agreement for the orderly exchange of information. To facilitate this examination we will start with an overview of one of the simplest protocols in use and structure our overview of protocols with respect to their complexity.

Teletype protocols

Teletype and teletype compatible terminals support relatively simple protocols used for conveying information. In general, a teletype protocol is a line-by-line protocol that requires no acknowledgement of line receipt. Thus, the key elements of this protocol define how characters are displayed and when a line is terminated and the next line is to be displayed. Some additional elements included in line by line teletype protocols actually are part of the terminal protocol, since they define how the terminal should respond to specific control characters.

Teletype Model 33

One commonly used teletype protocol is the Teletype® Model 33 data terminal. This terminal transmits and receives data asynchronously on a line by line basis using a modified ASCII code in which lower-case character received by a Model 33 are actually printed as their upper-case equivalent, a term known as 'fold-over' printing. Although the ASCII code defines the operation of 32 control characters, only 11 control characters can be used for communications control purposes. Prior to examining the use of communications control characters in the teletype protocol, let us first review the operational function and typical use of each control character. These characters were previously listed in Table 1.11 with the two-character designator CC following their meaning and will be reviewed in the order of their appearance in the referenced table.

Communications control characters

NUL

The null (NUL) character is a non-printable time delay or filler character. This character is primarily used for communicating with printing devices that require a defined period of time after each carriage return in which to reposition the printhead to the beginning of the next line. In the early days of PC communications many mainframe computers would be programmed to prompt users to 'Enter the number of nulls'; this is a mechanism to permit electromechanical terminal devices that require a delay to return the print head to the first position on the next line without obtaining garbled output.

SOH

The start of heading (SOH) is a communications control character used in several character-oriented protocols to define the beginning of a message heading data block. In synchronous transmission on a multipoint or multidrop line structure, the SOH is followed by an address which is checked by all devices on the common line to ascertain if they are the recipient of the data. In asynchronous transmission, the SOH character can be used to signal the beginning of a filename during multiple file transfers, permitting the transfer to occur without treating each file transfer as a separate communications session. Since asynchronous communications typically involve point-to-point communications, no address is required after the SOH character; however, both devices must have the same communications software program that permits multiple file transfers in this manner.

STX

The start of text (STX) character signifies the end of heading data and the beginning of the actual information contained within the block. This communications control character is used in the bisynchronous protocol that will be examined later in this section.

ETX

The end of text (ETX) character is used to inform the receiver that all the information within the block has been transmitted and normally terminates a block of data started with an STX or SOH. This character is also used to denote the beginning of the block check characters appended to a transmission block as an error detection mechanism. This communications control character is primarily used in the bisynchronous protocol and its receipt requires a status acknowledgement, such as an ACK or NAK.

EOT

The end of transmission (EOT) character defines the end of transmission of all data associated with a message transmitted to a device. If transmission occurs on a multidrop circuit the EOT also informs other devices on the line to check later transmissions for the occurrence of messages that could be addressed to them. The EOT is also used as a response to a poll when the polled station has nothing to send and as an abort signal when the sender cannot continue transmission. In the XMODEM protocol the EOT is used to indicate the end of a file transfer operation.

ENQ

The enquiry (ENQ) communications control character is used in the bisynchronous protocol to request a response or status from the other station on a point to point line or to a specifically addressed station on a multidrop line. In response to the ENQ character, the receiving station may respond with the number of the last block of data it successfully received. In a multidrop environment, the mainframe computer would poll each device on the line by addressing the ENQ to one particular station at a time. Each station would respond to the poll positively or negatively, depending upon whether or not they had information to send to the mainframe computer at that point in time.

ACK and DLE

The acknowledgement (ACK) character is used to verify that a block of data was received correctly. After the receiver computes its own 'internal' checksum or cyclic code and compares it to the one appended to the transmitted block, it will transmit the ACK character if the two checksums match. In the XMODEM protocol the ACK character is used to inform the transmitter that the next block of data can be transmitted. In the bisynchronous protocol the data link escape (DLE) character is normally used in conjunction with the 0 and 1 characters in place of the ACK character. Alternating DLE0 and DLE1 as positive acknowledgement to each correctly received block of data eliminates the potential of a lost or garbled acknowledgement resulting in the loss of data. In some literature, DLE0 and DLE1 are referred to as ACK0 and ACK1.

NAK

The negative acknowledgement (NAK) communications control character is transmitted by a receiving device to request the transmitting device to retransmit the previously sent data block. This character is transmitted when the receiver's internally generated checksum or cyclic code does not match the one transmitted, indicating that a transmission error has occurred. In the XMODEM protocol this character is used to inform the transmitting device that the receiver is ready to commence a file transfer operation as well as to inform the transmitter of any blocks of data received in error. In the bisynchronous protocol, the NAK is also used as a station-not-ready reply to an ENQ line bid or a station selection.

SYN

The synchronous idle (SYN) character is employed in the bisynchronous protocol to establish and then maintain line synchronization between the transmitter and receiver during periods when no data is transmitted on the line. When a series of SYN characters is interrupted, this indicates to the receiver that a block of data is being transmitted.

ETB

The end of transmission block (ETB) character is used in the bisynchronous protocol in place of an ETX character when data is transmitted in multiple blocks. This character then indicates the end of a particular block of transmitted data that commenced with an SOH or STX character. A block check character (BCC) is sent after an ETB. The receipt of an ETB is followed by an acknowledgement by the receiving device, such as an ACK or NAK.

Information flow

Figure 1.70 illustrates in a time chart format the possible flow of information between a teletype compatible terminal and a computer system employing a basic

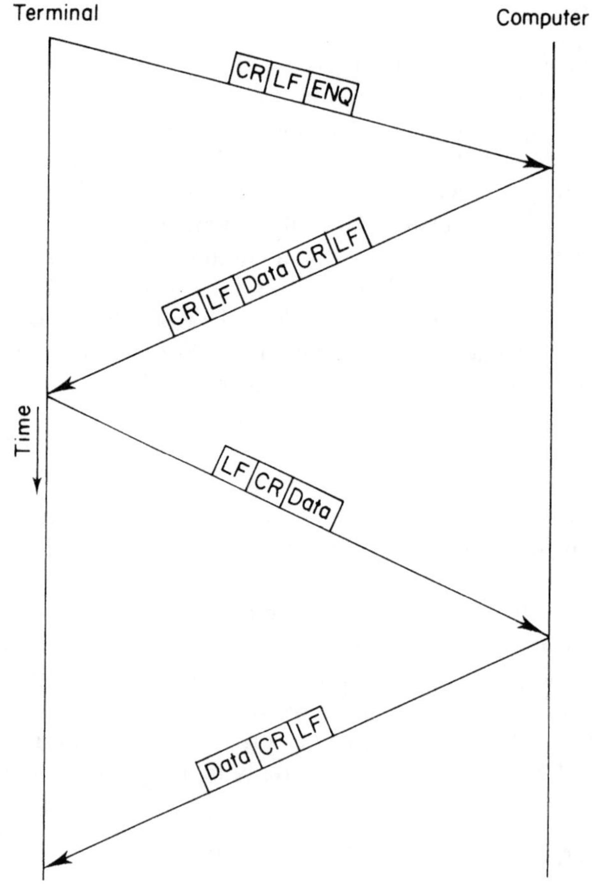

Figure 1.70 Basic teletype protocol

teletype protocol. In this protocol the terminal operator might first transmit the ENQ character, which is formed by pressing the shift and E keys simultaneously. If the call originated over the PSTN, the ENQ character, in effect, tells the computer to respond with its status. Since the computer is beginning its servicing of a new connection request, it normally responds with a log-on message. This log-on message can contain one or more lines of data.

The first line of the log-on message shown in Figure 1.70 is prefixed with a carriage return (CR) line feed (LF) sequence, which positions the printhead to the first column on a new line prior to printing the data in the received log-on message line. The log-on message line, as well as all following lines transmitted by the computer, will have a CR LF suffix, in effect, preparing the terminal for the next line of data. Upon receipt of the log-on message the terminal operator keys in his or her log-on code, which is transmitted to the computer as data followed by the CR LF suffix which terminates the line entry.

Variations

There are numerous variations to the previously discussed teletypewriter protocol, of which space permits mentioning only two.

Some computers will not recognize an ENQ character on an asynchronous ASCII port. Those computers are normally programmed to respond to a sequence of two or more carriage returns. Thus, the sequence ENQ LF CR would be replaced by CR CR or CR CR CR.

With the growth in popularity and use of personal computers as terminals, it was found that the time delay transmitted by computers in the form of null characters to separate multiple lines of output from one another by time was not necessary. Originally, the transmission of one line was separated through the use of NUL characters by several character intervals from the next line. This separation was required to provide the electromechanical printer used on teletype terminals with a sufficient amount of time to reposition its printhead from the end of one line to the beginning of the next line prior to receiving the first character to be printed on the next line. Since a cursor on a video display can be repositioned almost instantly, the growth in the use of personal computers and video display terminals resulted in the removal of time delays between computer transmitted lines.

Today, some computer software designed to service asynchronous terminals, as well as personal computers, will prompt the terminal operator with a message similar to: "ENTER NULLS (0 TO 5)-". This message provides the terminal operator with the ability to inform the computer whether he or she is using an electromechanical terminal. If 0 is entered, the computer assumes the terminal has a CRT display and does not separate multiple lines transmitted from the computer by anything more than the standard CR LF sequence. If a number greater than zero is entered, the computer separates multiple lines by the use of the indicated number of null characters. The NUL character, also called a PAD character, is considered to be a blank character which is discarded by the receiver. Thus, transmitting one or more NUL characters between lines only serves to provide time for the terminal's printhead to be repositioned to column 1 and has no effect upon the received data. If you are accessing a computer system that assumes all users have CRT terminals or personal computers, more than likely the null message will not be displayed. Such systems assume all users do not require a timing delay between transmitted lines and do not insert NUL characters between lines.

Error control

What happens if a line hit occurs during the transmission of data when a teletype protocol is used? Unfortunately, the only error detection mechanisms employed by teletype terminals and computer ports that supports this protocol are parity checking and echoplex.

If parity checking is supported by the terminal, it may simply substitute and display a special error character received with a parity error. This places the responsibility for error detection and correction upon the terminal operator, who must first visually observe the error and then request the computer to retransmit the line containing the parity error. Similarly, echoplex requires the operator to visually note that the echoed character differs from the character key just pressed.

As previously discussed in Section 1.11, the response of a computer to a parity error can range from no action to the generation of a special symbol to denote the occurrence of a parity error. In fact, most asynchronous line by line protocols do not check for parity errors. These protocols use echoplex, which as previously explained can result in a false indication of a transmission error or the appearance that all is well even though a transmitted character was received in error.

XMODEM protocol

The XMODEM protocol which was originally developed by Ward Christensen has been implemented into many asynchronous personal computer communications software programs and is supported by a large number of bulletin boards. Figure 1.71 illustrates in a time chart format the use of the XMODEM protocol for a file transfer consisting of two blocks of data. As illustrated, under the XMODEM protocol the receiving device transmits a negative acknowledgement (NAK) character to signal the transmitter that it is ready to receive data.

The XMODEM protocol is a 'receiver-driven' protocol in that the receiver transmits a character as a signal for the transmitter to start its data transfer operation. Under the XMODEM protocol the receiver has a 10 s timeout. It transmits a NAK each time it times out; hence, if the software on the personal computer that is to transmit a file is not set up to do so a period of 10 s can transpire until transmission actually starts. In response to the NAK the transmitter sends a start of header (SOH) communications control character followed by two characters that represent the block number and the one's complement of the block number.

The block number used in the XMODEM protocol starts at 01, increments by 1, and wraps from a maximum value of 0FFH to 00H and not to 01H. The one's complement is obtained by subtracting the block number from 255. Next a 128-character data block is transmitted which in turn is followed by the checksum character. As previously discussed in Section 1.11, the checksum is computed by first adding the ASCII values of each of the characters in the 128-character block and dividing the sum by 255. Next, the quotient is discarded and the remainder is retained as the checksum.

If the data blocks are damaged during transmission, the receiver can detect the occurrence of an error in one of three ways. If the start of header is damaged, it will be detected by the receiver and the data block will be negatively acknowledged. If either the block count or the one's complement field are damaged, they will not be the one's complement of each other. Finally, the receiver will compute its own checksum and compare it to the transmitted checksum. If the checksums do not match this is also an indicator that the transmitted block was received in error.

Each of the preceding situations results in the block being considered to have been received in error. Then the receiving station will transmit a NAK character which serves as a request to the transmitting station to retransmit the previously transmitted block. As illustrated in Figure 1.71, a line hit occurring during the transmission of the second block resulted in the receiver transmitting a NAK and

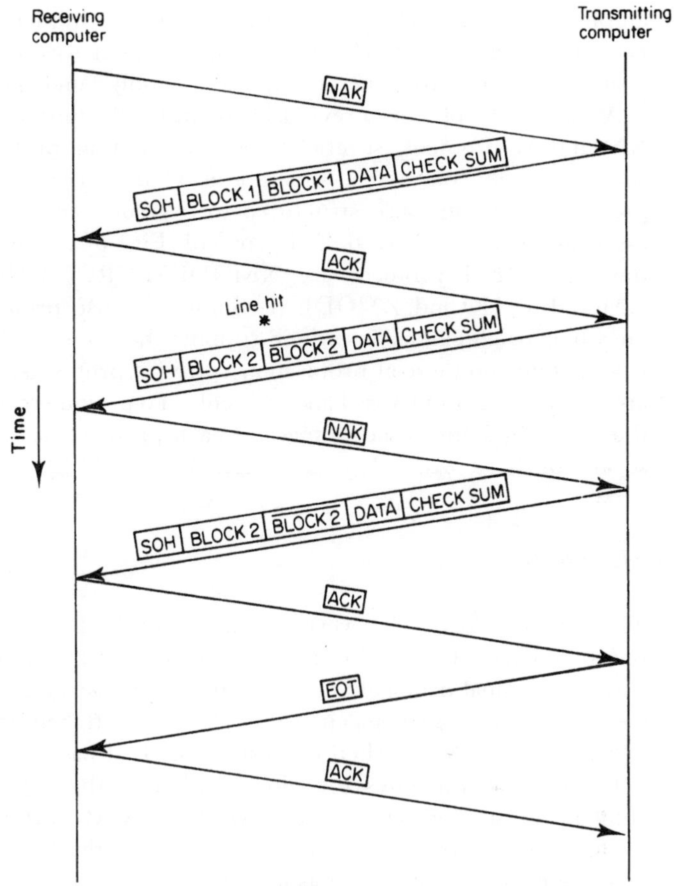

Figure 1.71 XMODEM protocol file transfer operation

the transmitting device resending the second block. Suppose more line hits occur which affects the retransmission of the second block. Under the XMODEM protocol the retransmissions process will be repeated until the block is correctly received or until ten retransmission attempts occur. If, owing to a thunderstorm or other disturbance, line noise is a problem, after 10 attempts to retransmit a block the file transfer process will be aborted. This will require a manual operator intervention to restart the file transfer at the beginning and is one of many deficiencies of the XMODEM protocol. Other deficiencies of the XMODEM protocol include its relatively small block size, its half-duplex transmission scheme, and its use of the checksum that provides a less reliable error detection capability in comparison to the use of a CRC.

In spite of the limitations of the XMODEM protocol, it is one of the most popular protocols employed by personal computer users for asynchronous data transfer because of several factors. First, the XMODEM protocol is in the public domain which means it is readily available at no cost for software developers to incorporate into their communications programs. Secondly, the algorithm employed to generate the checksum is easy to implement using a higher level language such as BASIC or Pascal. In comparison, a CRC-16 block-check character is normally generated using assembly language.

As a result of the previously mentioned limitations associated with the XMODEM protocol, several extensions to that protocol were developed. In addition, many commercial software developers designed proprietary file transfer protocols that were also structured to overcome one or more of the limitations associated with the XMODEM protocol. Five of the more popular extensions of the XMODEM protocol are XMODEM/CRC, YMODEM, YMODEM-G, XMODEM-1K and ZMODEM. Some of those protocols also have what are known as batch extensions which support the transfer of multiple (batched) files. Concerning commercial proprietary software protocols, two of the more popular are the BLAST and CrossTalk protocols. To provide readers with an indication of the advantages and disadvantages of each protocol we will compare and contrast several of those protocols to the original XMODEM protocol.

XMODEM/CRC protocol

The XMODEM/CRC is very similar to the XMODEM protocol, except that a two-byte CRC-16 is used in place of the one-character arithmetic checksum used with the original protocol. To differentiate the use of a CRC-16 from the use of a checksum, the receiver specifies the CRC-16 by transmitting the character C (Hex 43) instead of a NAK when requesting the first packet.

Figure 1.72 illustrates the block format of the XMODEM/CR protocol. In comparing the format of the XMODEM/CRC protocol to the XMODEM protocol, you will note the similarity between the two protocols, since only the error detection mechanism has changed.

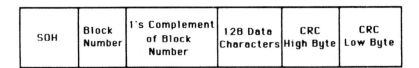

SOH	Block Number	1's Complement of Block Number	128 Data Characters	CRC High Byte	CRC Low Byte

Figure 1.72 XMODEM/CRC block format

XMODEM/CRC

Through the use of a CRC, the probability of an undetected error is significantly reduced in comparison to the use of the XMODEM checksum. The CRC will detect all single- and double-bit errors, all errors with an odd number of bits, all bursts of errors up to 16 bits in length, 99.997% of 17-bit error bursts, and 99.998% of 18-bit and longer bursts.

Although the XMODEM/CRC protocol significantly reduces the probability of an undetected error, it is a half-duplex protocol similar to XMODEM and uses the same size data block. Thus, it removes only one of the three constraints associated with the XMODEM protocol.

YMODEM and YMODEM batch protocols

The YMODEM protocol was developed as an extension to XMODEM to overcome several constraints of the latter as well as to provide additional capabilities beyond those provided by both the XMODEM and XMODEM/CRC protocols. Under the YMODEM protocol, a header block was added to relay the filename and other information and multiple file transfers are supported in a batch mode. In addition, data is normally transferred in 1024-byte blocks, which results in more time being spent actually transferring data and less time spent computing checksums or CRCs and sending acknowledgements.

The original development of the YMODEM protocol was limited to transferring one file at a time using 1024-byte (1 K) blocks. Although many communications software programmes implemented YMODEM correctly as it was designed—as a single file protocol, other programs implemented it as a multiple file protocol. In actuality, the multiple file protocol version of YMODEM is normally and correctly referenced as YMODEM BATCH. Since YMODEM BATCH is the same as YMODEM except that the former allows multiple file (batch) transfers, we will examine both protocols and collectively refer to them as YMODEM, although this is not absolutely correct.

The format of the YMODEM protocol is illustrated in Figure 1.73. Under this protocol, the start of text (STX) character whose ASCII value is 02H replaces the SOH character used by the XMODEM and XMODEM/CRC protocols. The use of the STX character informs the receiver that the block contains 1024 data characters; however, the receiver can also accept 128 data character blocks. When 128 data character blocks are sent, the SOH character replaces the STX character.

STX	Block Number	1's Complement of Block Number	1024 Data Characters	CRC High Byte	CRC Low Byte

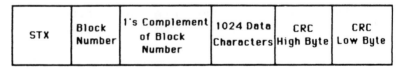

Figure 1.73 YMODEM block format

Under the YMODEM protocol multiple files can be transmitted through the use of a single command which, in some implementations, accept global characters. For example, specifying *.DAT would result in an attempt to transmit all files with the extension DAT. If no files with that extension are found, then zero files are transmitted.

When a data transfer is initiated, the receiver transmits the ASCII C character to the sender to synchronize transmission startup as well as to indicate that CRC checking is to be employed. The sender then opens the first file and transmits block number 0 instead of block number 1 used with the XMODEM and XMODEM/CRC protocols. Block number 0 will contain the filename of the file being transmitted and may optionally contain the file length and file creation date. Due to the manner in which most personal computer operating systems work, the creation or modification date of a file being downloaded will be modified to the current date when the file is received. For example, if the file being downloaded was created on JUN 15 1998 and today's date is JLY 19 1998, the file data would be changed to JLY 19 1998 when the file is downloaded onto your computer. The remaining data characters in block 0 are then set to nulls.

Once block 0 is correctly received, it will be ACKed if the receiver can perform a 'write open' operation. Otherwise, the receiver will transmit the CAN character to cancel the file transfer operation. After block 0 is acknowledged, the sender will commence transferring the contents of the file similar to the manner in which data is transmitted using the XMODEM/CRC protocol. During actual data transfer, the sender can switch between 128 and 1024 data character blocks by prefixing 128 character blocks with the SOH character and 1024 data character blocks with the STX character. After the contents of a file are successfully transmitted, the receiver will transmit an ASCII C which serves as a request for the next file. If no additional files are to be transmitted, the sender will transmit a 128 character data block, with the value of each character set to an ASCII 00H or null character.

Figure 1.74 illustrates the transmission of the file named STOCK.DAT which was last modified on JLY 19 1998 at 20:30 hours and which contains 2276 characters of information.

To initiate the file transfer, the receiver transmits an ASCII C to the sender. Upon its receipt the sender transmits a 128 character data block numbered as block 0. This block is prefixed with the SOH character to differentiate it from a 1024 data character block. The 'file info' field in block number 0 contains the filename (STOCK.DAT) followed by the time the file was created or last modified (20:30), the date the file was created or last modified (JLY 19 1998), and the file size in bytes (2276). A single space is used to separate the date from the file size, resulting in a total of 30 characters used to convey file information. Since the smallest data block contains 128 characters, 98 nulls are added to complete this block. After this block is acknowledged, the sender then transmits the first 1024 data characters through the use of a 1024 character data block, prefixing the block with the STX character.

In examining Figure 1.74, note that blocks 01 and 02 are 1024 characters in length, while blocks 03 and 04 are 128 characters in length. Since the file size was 2276 characters, the YMODEM protocol attempts to use as many 1024 data character blocks as possible and then transmits 128 character data blocks to complete the transmission. In doing so, the last 28 characters in block 4 are set to NULs.

Once the last block is successfully transferred, the sender transmits the EOT character to denote the completion of the file transfer. The receiver then transmits the ASII C as an indicator to the sender to initiate the transfer of the next file. Since only one file was to be transmitted, the sender transmits a new block number 00H that contains 128 NUL characters, signifying that no more files remain to be transmitted. Once this block is acknowledged, the transmission session is completed.

In addition to providing an increase in throughput over XMODEM and XMODEM/CRC, when transmission occurs over relatively noise-free lines the header information carried by YMODEM enables communications programs to compute the expected duration of the file transfer operation. This explains why

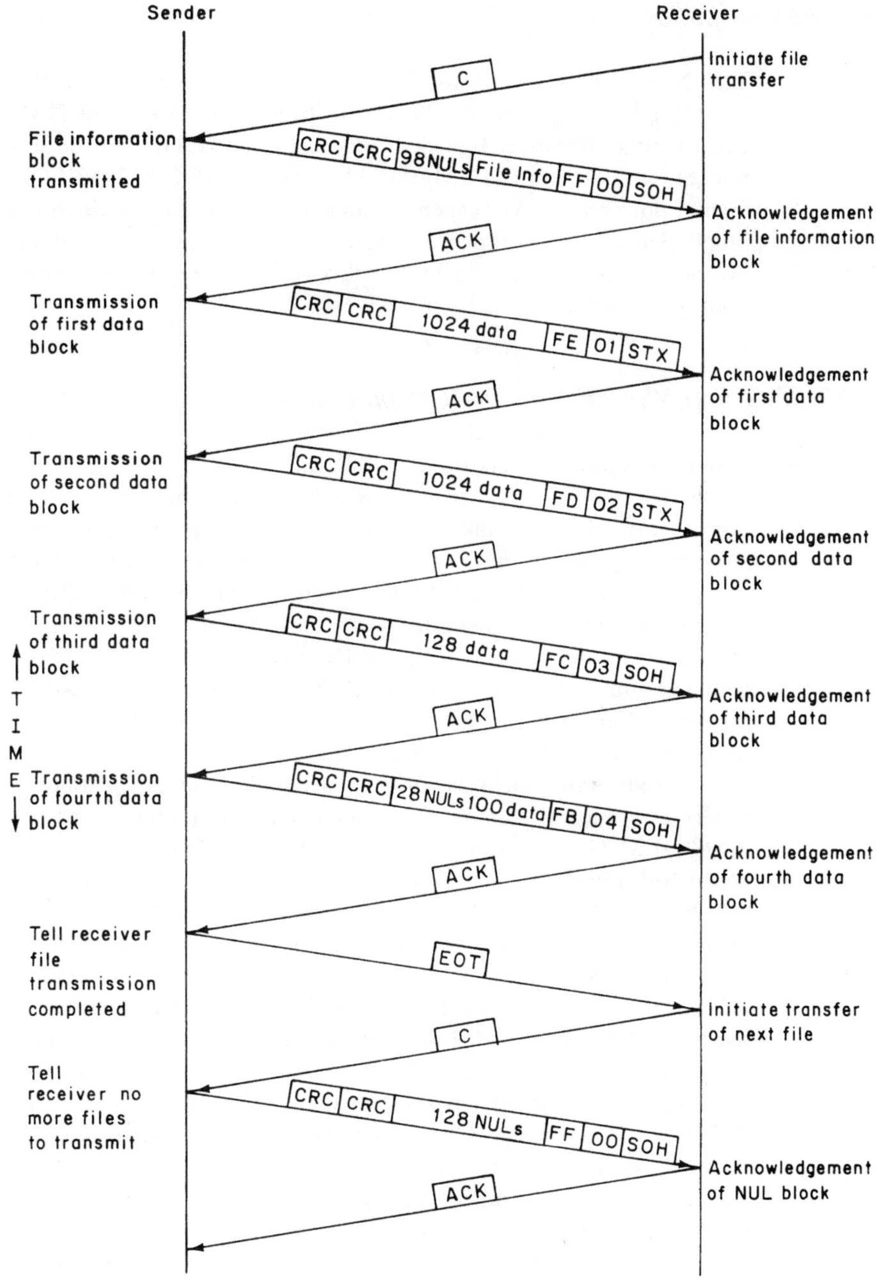

Figure 1.74 YMODEM protocol file transfer example

most communications programs will visually display the file transfer time and some programs will provide an updated bar chart of the progress of a YMODEM file transfer, while they cannot do the same when the XMODEM or XMODEM/CRC protocols are used.

XMODEM-1K protocol

The XMODEM-1K protocol is a derivative of the XMODEM standard. The XMODEM-1K protocol follows the previously described XMODEM protocol, substituting 1024-byte blocks in place of byte data blocks. The XMODEM-1K is not compatible with the YMODEM nor the YMODEM BATCH protocols, as the former does not send or accept a block 0, which contains file information. Since the block size of this protocol is significantly longer than that of the XMODEM protocol, you can expect a higher level of throughput when transmitting on good quality circuits using XMODEM-1K.

YMODEM-G and YMODEM-G BATCH protocols

The development of error correction and detection modems essentially made the use of CRC checking within a protocol redundant. In recognition of this, a 'G' option was originally added to the YMODEM protocol which changed it into a 'streaming' protocol in which all data blocks are transmitted one after another, with the receiver then acknowledging the entire transmission. This acknowledgement simply acknowledges the entire transmission without the use of error detection and correction. In fact, the two-byte CRC field is set to zero during a YMODEM-G transmission. Thus, this protocol should only be used with error correcting modems that provide data integrity. The use of error correcting modems is described in detail in Chapter 5.

Although some software programs enable users to initiate YMODEM-G by entering the character G as an optional parameter, most programs consider YMODEM-G as a separate protocol selected from a pull-down menu or via a command line entry.

Like the YMODEM BATCH protocol, YMODEM-G BATCH protocol permits multiple files to be transmitted and sends the first 128 data character block with file information in the same manner as carried by the YMODEM BATCH protocol. Typically, the multiple file transfer capability is selected by the use of a YMODEM-G BATCH option available with many communications programs. To differentiate YMODEM-G BATCH from YMODEM-G, the receiver will initiate the batch transfer by sending the ASCII G instead of the ASCII C. When the sender recognizes the ASCII G, it bypasses the wait for an ACK to each transmitted block and sends succeeding blocks one after another, subject to any flow control signals issued by an attached modem or by a packet network if that network is used to obtain a transmission path. When the transmission is completed, the sender transmits an EOT character and the receiver returns an ACK which serves to acknowledge the entire file transmission. The ACK is then followed by the receiver transmitting another ASCII G to initiate the transmission of the next file. If no additional files are to be transmitted, the sender then transmits a block of 128 characters with each character set to an ASCII 00H or NUL character.

Figure 1.75 illustrates the transmission of the previously described STOCK.-DAT file using the YMODEM-G protocol. In comparing Figure 1.75 with Figure

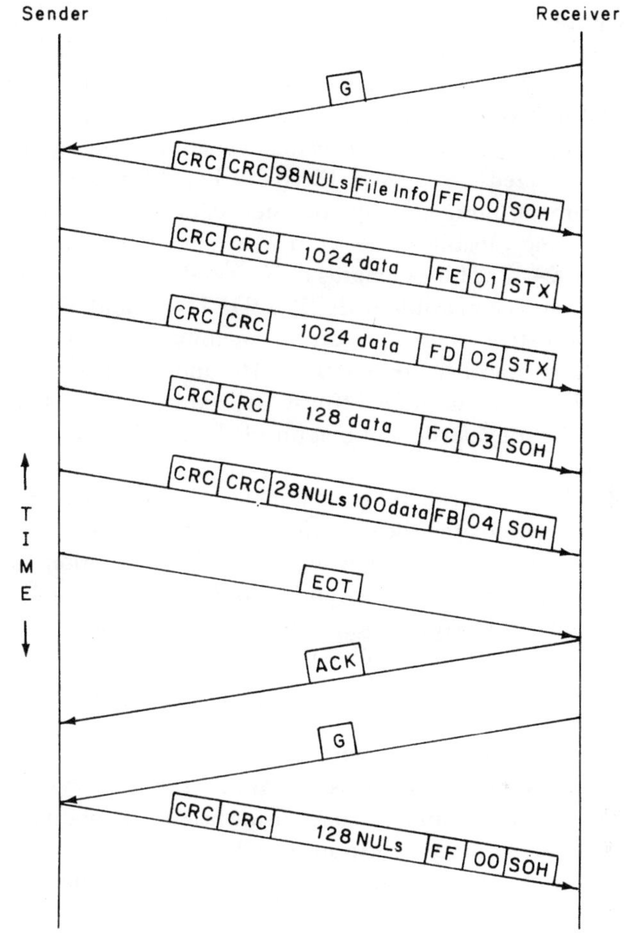

Figure 1.75 YMODEM-G protocol file transfer example

1.74, it becomes obvious that the streaming nature of YMODEM-G and YMODEM-G BATCH increases transmission throughout, resulting in a decrease in the time required to transmit a file or group of files.

ZMODEM

The development of the ZMODEM protocol was funded by Telenet which is now operated by Sprint as SprintNet. This packet switching vendor turned to Mr Chuck Forsberg, the author of the original YMODEM protocol, to develop a file transfer protocol that would provide a more suitable mechanism for transferring information via packet networks. The resulting file transfer protocol, which was called ZMODEM, corrected many of the previously described constraints associated with the use of the XMODEM and YMODEM protocols. Significant features of the ZMODEM protocol include its streaming file transfer operation, an extended error detection capability, automatic file transfer capability, the use of data compression, and downward compatibility with the XMODEM-1K and YMODEM protocols.

The streaming file transfer capability of ZMODEM is similar to that incorporated into YMODEM-G—that is, the sender will not receive an acknowledgement until the file transfer operation is completed. In addition to the streaming capability, ZMODEM supports the transmission of conventional 128- and 1024-byte block lengths of XMODEM-based protocols. In fact, ZMODEM is backward-compatible with XMODEM-1K and YMODEM.

The extended error detection capability of ZMODEM is based upon the ability of the protocol to support both 16- and 32-bit CRCs. According to the protocol developer, the use of a 32-bit CRC reduces the probability of an undetected error by at least five orders of magnitude below that obtainable from the use of a 16-bit CRC. In fact, 32-bit CRCs are commonly used with local area network protocols to reduce the probability of undetected errors occurring on LANs.

The automatic file transfer capability of ZMODEM enables a sending or receiving computer to trigger file transfer operations. In comparison, XMODEM and YMODEM protocols and their derivatives are receiver driven. Concerning the file transfer startup process, a file transfer begins immediately under ZMODEM while XMODEM and YMODEM protocols and their derivatives have a 10-second delay as the receiver transmits NAKs or another character during protocol startup operations.

An additional significant feature associated with the ZMODEM protocol is its support of data compression. When transmitting data between Unix systems, ZMODEM compresses data using a 12-bit modified Lempel–Ziv compression technique, similar to the modified Lempel–Ziv technique incorporated into the ITU-T V.42 bis modem recommendation. When ZMODEM is used between non-Unix systems, compression occurs through the use of Run Length Encoding similar to MNP Class 5.

Kermit

Kermit was developed at Columbia University in New York City primarily as a mechanism for downloading files from mainframes to microcomputers. Since its original development this protocol has evolved into a comprehensive communications system which can be employed for transferring data between most types of intelligent devices. Although the name might imply some type of acronym, in actuality, this protocol was named after Kermit the Frog, the star of the well-known Muppet television show.

Kermit is a half-duplex communications protocol which transfers data in variable sized packets, with a maximum packet size of 96 characters. Packets are transmitted in alternate directions since each packet must be acknowledged in a manner similar to the XMODEM protocol.

In comparison to the XMODEM protocol and its derivatives which permit 7- and 8-level ASCII as well as binary data transfers in their original data composition, all Kermit transmissions occur in 7-level ASCII. The reason for this restriction is the fact that Kermit was originally designed to support file transfers to 7-level ASCII mainframes. Binary file transfers are supported by the protocol prefixing each byte whose eighth bit is set by the ampersand (&) character. In addition, all characters transmitted to include 7-level ASCII must be printable, resulting in Kermit transforming each ASCII control character with the pound (£) character. This transformation is accomplished through the complementation of the seventh bit of the control character. Thus, 64 modulo 64 is added or subtracted from each control character encountered in the input data stream. When an 8-bit byte is encountered whose low order 7 bits represent a control character, Kermit appends a double prefix to the character. Thus, the byte 100000001 would be transmitted as &£A.

Although character prefixing adds a considerable amount of overhead to the protocol, Kermit includes a run length compression facility which may partially reduce the extra overhead associated with control character and binary data transmission. Here, the tilde (~) character is used as a prefix character to indicate run length compression. The character following the tilde is a repeat count, while the third character in the sequence is the character to be repeated. Thus, the sequence XA is used to indicate a series of 88 As, since the value of X is 1011000 binary or decimal 88. Through the use of run length compression the requirement to transmit printable characters results in an approximate 25% overhead increase in comparison to the XMODEM protocol for users transmitting binary files. If ASCII data is transmitted, Kermit's efficiency can range from more efficient to less efficient in comparison to the XMODEM protocol, with the number of control characters in the file to be transferred and the susceptibility of the data to run length compression the governing factors in comparing the two protocols.

Figure 1.76 illustrates the format of a Kermit packet. The header field is the ASCII start of header (SOH) character. The length field is a single character whose value ranges between 0 and 94. This one-character field defines the packet length in characters less two, since it indicates the number of characters to include the checksum that follow this field.

Header	Length	Sequence	Type	Data	Check

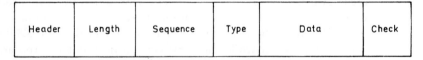

Figure 1.76 The Kermit packet format. The first three fields in the Kermit packet are one character in length and the maximum total packet length is 96 or fewer characters

The sequence field is another one-character field whose value varies between 0 and 63. The value of this field wraps around to 0 after each group of 64 packets is transmitted.

The type field is a single printable character which defines the activity the packet initiates. Packet types include D (data), Y (acknowledgement), N (negative acknowledgement), B (end of transmission or break), F (file header), Z (end of file) and E (error).

The information contents of the packet are included in the data field. As previously mentioned, control characters and binary data are prefixed prior to their placement in this field.

The check field can be one, two or three characters in length depending upon which error detection method is used since the protocol supports three options. A single character is used when a checksum method is used for error detection. When this occurs, the checksum is formed by the addition of the ASCII values of all characters after the Header character through the last data character and the low order 7 bits are then used as the checksum. The other two error detection methods supported by Kermit include a two-character checksum and a three-character 16-bit CRC. The two-character checksum is formed similar to the one-character checksum; however, the low order 12 bits of the arithmetic sums are used and broken into two 7-bit printable characters. The 16-bit CRC is formed using the CCITT standard polynomial, with the high order 4 bits going into the first character while the middle 6 and low order 6 bits are placed into the second and third characters, respectively.

By providing the capability to transfer both the filename and contents of files, Kermit provides a more comprehensive capability for file transfers than XMODEM. In addition, Kermit permits multiple files to be transferred in comparison to XMODEM, which requires the user to initiate file transfers on an individual basis.

Bisynchronous protocols

During the 1970s IBM's BISYNC (binary synchronous communications) protocol was one of the most frequently used for synchronous transmission. This particular protocol is actually a set of very similar protocols that provides a set of rules which effect the synchronous transmission of binary-coded data.

Although there are numerous versions of the bisynchronous protocol in existence, three versions account for the vast majority of devices operating in a bisynchronous environment. These three versions of the bisynchronous protocol are known as 2780, 3780 and 3270. The 2780 and 3780 bisynchronous protocols are used for remote job entry communications into a mainframe computer, with the major difference between these versions the fact that the 3780 version performs space compression while the 2780 version does not incorporate this feature. In comparison to the 2780 and 3780 protocols that are designed for point to point communications, the 3270 protocol is designed for operation with devices connected to a mainframe on a multidrop circuit or devices connected to a cluster controller which, in turn, is connected to the mainframe. Thus, 3270 is a poll and select software protocol.

Originally, 2780 and 3780 workstations were large devices that controlled such peripherals as card readers and line printers. Today, an IBM PC or compatible computer can obtain a bisynchronous communications capability through the installation of a bisynchronous communications adapter card into the PC's system unit. This card is designed to operate in conjunction with a bisynchronous communications software program which with the adapter card enables the PC to operate as an IBM 2780 or 3780 workstation or as an IBM 3270 type of interactive terminal.

The bisynchronous transmission protocol can be used in a variety of transmission codes on a large number of medium- to high-speed equipment. Some of the constraints of this protocol are that it is limited to half-duplex transmission and that it requires the acknowledgement of the receipt of every block of data transmitted. A large number of protocols have been developed owing to the success of the BISYNC protocol. Some of these protocols are bit-oriented, whereas BISYNC is a character-oriented protocol, and some permit full-duplex transmission, whereas BISYNC is limited to half-duplex transmission.

Data code use

Most bisynchronous protocols support several data codes including the 6-bit transcode (SBT), 7-bit ASCII and 8-bit EBCDIC. Normally, error control is obtained by using a two-dimensional parity check (LRC/VRC) when transmission is in ASCII. When transmission is in EBCDIC the CRC-16 polynomial is used to generate a block-check character, while the use of the SBT code is accompanied by the use of a CRC-12 polynomial.

Figure 1.77 illustrates the generalized bisynchronous block structure. For synchronization, most BISYNC protocols require the transmission and detection of two successive synchronization (SYN) characters. The start of message control code is normally the STX communications control character. The end of message control code can be either the end of text (ETX), end of transmission block (ETB), or the end of transmission (EOT) character; the actual character, however, depends upon whether the block is one of many blocks, the end of the transmission block, or the end of the transmission session.

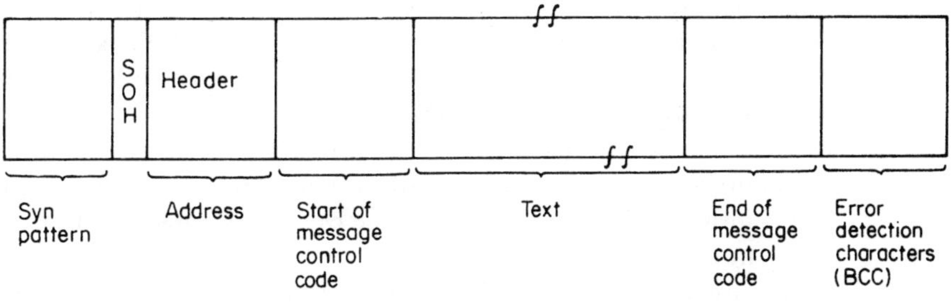

Figure 1.77 Generalized BSC block structure

The ETX character is used to terminate a block of data started with a SOH or STX character which was transmitted as an entity. SOH identifies the beginning of a block of control information, such as a destination address, priority and message sequence number. The STX character denotes both the end of the message header and the beginning of the actual content of the message. A BCC character always follows an ETX character. Since the ETX only signifies the end of a message, it

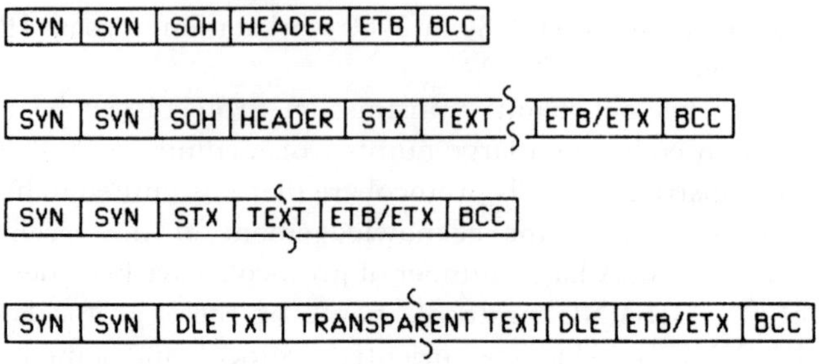

Figure 1.78 Common BISYNC data block formats

requires a status reply from the receiving station prior to subsequent communications occurring. A status reply can be an DLE0, DLE1, NAK, WACK or RVI character, with the meaning of the last three characters discussed later in this section.

The ETB character identifies the end of a block that was started with a SOH or STX. Similar to ETX, a BCC is sent immediately after the ETB and the receiving station is required to furnish a status reply.

The EOT code defines the end of message transmission for a single or multiple block message. The effect of the EOT is to reset all receiving stations. In a multidrop environment, the EOT is used as a response to a poll when an addressed station has no data to transmit.

The generalized block structure illustrated in Figure 1.77 can vary considerably based upon many factors, including the link topology (point-to-point or multipoint), operational mode (contention or polled), and the type of data to be transported by the protocol (alphanumeric or binary). Figure 1.78 illustrates four specific BISYNC data block formats commonly used.

The use of a header field is optional and the format and contents of that field are specified by the user. Typically, the header field is used for device selection purposes or for other routing information. The first block format shown in Figure 1.78 will normally precede the third block format, while the second block format can be viewed as a combination of the first and third formats. The fourth block format allows any bit pattern to be carried into the text field of the block while avoiding the possibility of the pattern being misinterpreted as a control character. This format, which permits text to be treated as transparent data, will be described later in this section.

Figure 1.79 illustrates the error control mechanism employed in a bisynchronous protocol to handle the situation where a line hit occurs during transmission or if an acknowledgement to a previously transmitted data block becomes lost or garbled.

In the example on the left portion of Figure 1.79, a line hit occurs during the transmission of the second block of data from the mainframe computer to a terminal or a personal computer. Note that, although Figure 1.79 is an abbreviated illustration of the actual bisynchronous block structure and does not show the actual block-check characters in each block, in actuality they are contained in each block. Thus, the line hit which occurs during the transmission of the second block

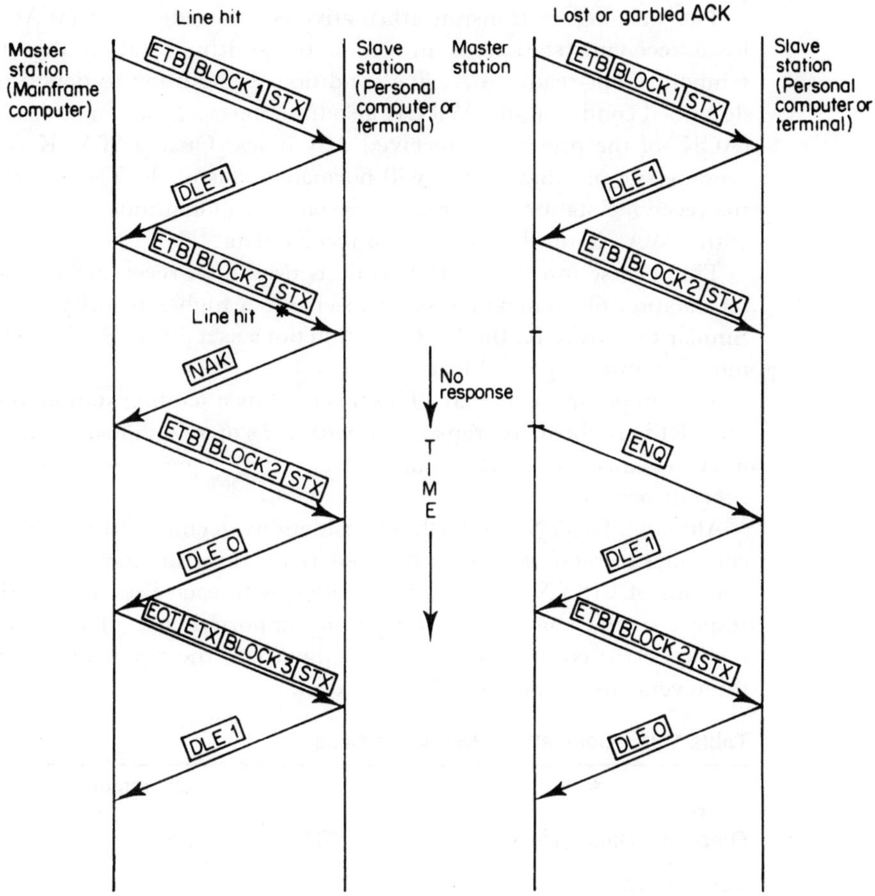

Figure 1.79 BSC, error control methods

results in the 'internally' generated BCC being different from the BCC that was transmitted with the second block. This causes the terminal device to transmit a NAK to the mainframe, which results in the retransmissions of the second block.

In the example on the right-hand part of Figure 1.79, let us assume that the terminal received block 2 and sent an acknowledgement which was lost or garbled. After a predefined timeout period occurs, the master station transmits an ENQ communications control character to check the status of the terminal. Upon receipt of the ENQ, the terminal will transmit the alternating acknowledgement, currently DLE1; however, the mainframe was expecting DLEO. Thus, the mainframe is informed by this that block 2 was never acknowledged and as a result retransmits that block.

Other control codes

Three additional transmission codes commonly used in a bisynchronous protocol are the WACK, RVI and TTD characters.

The wait-before-transmit affirmative acknowledgement (WACK) code is used by a receiving station to inform a transmitting station that the former is a temporary not-ready-to-receive condition. In addition to denoting the previously described condition, the WACK also functions as an affirmative acknowledgement (ACK) of the previously received data block. Once a WACK is received by the sending station, that station will normally transmit ENQs at periodic intervals to the receiving station. The receiving station will continue to respond to each ENQ with a WACK until it is ready to receive data.

The reverse interrupt (RVI) code is used by a receiving station to request the termination of a current session to enable a higher priority message to be sent. Similar to a WACK, the RVI also functions as a positive acknowledgement to the most recently received block.

The temporary text delay (TTD) is used by a sending station to keep control of a line. TTD is normally transmitted within 2 s of a previously transmitted block and indicates that the sender cannot transmit the next block within a predefined timeout period.

Although BISYNC usage has considerably declined during the past few years, it continues to retain a degree of popularity. Unfortunately, there are actually three versions of BISYNC you must consider, with each version slightly different with respect to the character composition supported, use of control codes, and the method error control. Table 1.31 summarizes the major differences between the three versions of the BISYNC protocol.

Table 1.31 Comparing BISYNC versions

	Character codes		
Function/control code;	SBT	ASCII	EBCDIC
Composition			
Number of bits per character	6	7	8
Number of defined characters	64	128	144 (256 possible)
Control code			
DLEO code	DLE-	DLEO	DLE70
DLEI code	DLET	DLE1	DLE/
WACK	DLEW	DLE;	DLE,
RVI	DLE2	DLE	DLE@
TTD	STX ENQ	STX ENQ	STX ENQ
Error control			
No transparency	CRC-12	VRC/LRC	CRC-16
Tranparent mode installed and operating	CRC-12	CRC-16	CRC-16
Transparent mode installed but not operating	CRC-12	VRC/CRC-16	CRC-16

Timeouts

Timeouts are incorporated into most communications protocols to preclude the infinite seizure of a facility due to an undetected or detected but not corrected error condition. The bisynchronous protocol defines four types of timeouts–transmit, receive, disconnect and continue.

The transmit timeout defines the rate of insertion of synchronous idle character sequences used to maintain synchronization between a transmitting and receiving station. Normally, the transmitting station will insert SYN SYN or DLE SYN sequences between blocks to maintain synchronization. Transmit timeout is normally set for 1 s.

The receive timeout can be used to limit the time a transmitting station will wait for a reply, signal a receiving station to check the line for synchronous idle characters or to set a limit on the time a station on a multidrop line can control the line. The typical default setting of the receive timeout is 3 s.

The disconnect timeout causes a station communicating on the PSTN to disconnect from the circuit after a predefined period of inactivity. The default setting for a disconnect timeout is normally 20 s of inactivity.

The fourth timeout supported by bisynchronous protocols is the continue timeout. This timeout causes a sending station transmitting a TTD to send another TTD character if it is unable to send text. A receiving station must transmit a WACK within two seconds of receiving the TTD if it is unable to receive.

Although the default timeout values are sufficient for most applications, there is one area where they almost always result in unnecessary problems—the situation where satellite communications facilities are used. Satellite communications add at least a 52 000-mile round trip delay to signal propagation, resulting in a built-in round trip delay of approximately 0.5 s. Due to this, you may always experience transmit and continue timeouts and can even experience many receive timeouts that are unwarranted if default timeout values are used. To eliminate the occurrence of unwarranted timeouts you should add 1 s to the default timeout values for each satellite 'hop' in a communications path, where a 'hop' can be defined as the transmission from one earth station to another earth station via the use of a satellite.

To illustrate the deterioration in a bisynchronous protocol when transmission occurs on a satellite circuit, assume you wish to transmit 80-character data blocks at 9.6 kbps and use modems whose internal delay time is 5 ms. Let us further assume there is a single satellite hop transmission will flow over, resulting in a one-way propagation delay of 250 ms. Since each message block must be acknowledged prior to the transmission of the next block, let us assume there are eight characters in each acknowledgement message. Based upon those assumptions, Table 1.32 lists each of the delay times associated with the transmission of one message block until an acknowledgement is received as well as the computation of the protocol efficiency.

There are three methods you can consider to improve throughput efficiency. You can increase the size of the message block, use high-speed modems or employ a full-duplex protocol. The first two methods have distinct limitations. As the size of the data block increases a point will be reached where the error rate on the data link results in the retransmission of the larger size message every so often, negating the efficiency increase from an increased block size. Since the data rate obtainable is a function of the bandwidth of a channel, it may not be practical to increase the data transmission rate, resulting in a switch to a full-duplex protocol being the

Table 1.32 Bisynchronous protocol efficiency example

		Time (ms)
Message transmission time	$\frac{80 \text{ characters} \times 8 \text{ bits/character}}{9600 \text{ bps}}$	67
Propagation delay		250
Modem delay time		10
Acknowledgement delay time	$\frac{8 \text{ characters} \times 8 \text{ bits/character}}{9600 \text{ bps}}$	7
Propagation delay		250
Modem dalay		10
		594

$$\text{Efficiency} = \frac{\text{Time spent transmitting data}}{\text{Total time to transmit and acknowledge}} = \frac{67}{594} = 11.3\%$$

method used by most organizations to increase efficiency when transmitting via a satellite link.

Data transparency

In transmitting data between two devices there is always a probability that the composition of an 8-bit byte will have the same bit pattern as a bisynchronous control character. This probability significantly increases if, as an example, you are transmitting the binary representation of a compiled computer program.

Since 8-bit groupings are examined to determine if a specific control character has occurred, a bisynchronous protocol would normally be excluded from use if you wished to transmit binary data. To overcome this limitation, protocols have what is known as a transparent mode of operation.

The control character pair DLE STX is employed to initiate transparent mode operations while the control character pairs DLE ETB or DLE ETX are used to terminate this mode of operation. Any control characters formed by data when the transparent mode is in operation are ignored. In fact, if a DLE character should occur in the data during transparent mode operations, a second DLE character will be inserted into the data by the transmitter. Similarly, if a receiver recognizes two DLE character in sequence, it will delete one and treat the second one as data, eliminating the potential of the composition of the bit patterns of the data causing a false ending to the transparent mode of operation.

DDCMP

Digital Equipment Corporation's Digital Data Communications Message Protocol (DDCMP) is a character-oriented data link protocol similar to IBM's bisynchronous protocol. Unlike IBM's protocol that is restricted to synchronous transmission, DDCMP can operate either asynchronously or synchronously over switched or non-switched facilities in a full- or half-duplex transmission mode.

Figure 1.80 illustrates the DDCMP protocol format, in which the header contains 56 bits partitioned into six distinct fields.

SYN	SYN	Type	Header						Information (any number of 8-bit characters)	CRC-16 (16 bits)
			Count (14 bits)	Flags (2 bits)	Response (8 bits)	Sequence (8 bits)	Address (8 bits)	CRC-16 (16 bits)		

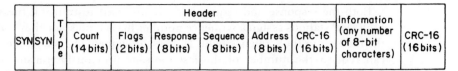

Figure 1.80 Digital data communications message protocol (DDCMP) format

Structure

Like IBM's bisynchronous protocol, DDCMP uses two SYNC characters for synchronization. The type field is a one-character field which defines the type of message being transmitted. Data messages are indicated by a SOH character, while control messages which in DDCMP are either an ACK or NAK, are indicated by the ENQ character. A third type of message, maintenance, is denoted by the use of the DLE character in the type field.

When data is transferred the count field defines the number of bytes in the information field to include CRC bytes. One advantage of this structure is its inherent transparency, since the count field defines the number of bytes in the information field, the composition of the bytes will not be misinterpreted as they are not examined as part of the protocol. If the message is a control message, the count field is used to clarify the type of NAK indicated in the type field. Although there is only one type of ACK, DDCMP supports several types of NAKs, such as buffer overrun or the occurrence of a block-check error on a preceding message.

Table 1.33 lists the composition of the rightmost six bits of the count field which are used to define the reason for a NAK. Both ACK and NAK are denoted by an ENQ in the type field (00000101), so the count field is also used to distinguish between an ACK and NAK. If the first eight bits in the count field are binary 00000001, an ACK is defined, whereas, if the first eight bits in the count field are binary 00000010, this bit composition defines a NAK. Thus, the bit compositions listed in Table 1.33 are always prefixed by binary 00000010 which defines a NAK.

The high order bit of the flag field denotes the occurrence of a SYNC character at the end of the current message. This allows the receiver to reinitialize its synchronization detection logic. The low order bit of the FLAG field indicates the current message to be the last of a series the transmitter intends to send. This allows the addressed station to begin transmission at the end of the current message.

Table 1.33 Count field NAK definitions

Count field value (rightmost 6 bits)	NAK definition
000001	CRC header error
000010	CRC data error
000011	Reply response
001000	Buffer unavailable
001001	Receiver overrun
010000	Message too long
010001	Header format error

Both the response and sequence fields are used to transmit message numbers. DDCMP stations assign a sequence number to each message they transmit, placing the number in the sequence field. If message sequencing is lost, the control station can request the number of the last message previously transmitted by another station. When this request is received, the answering station will place the last accepted sequence number in the response field of the message it transmits back to the control station.

The address field is used in a multipoint line configuration to denote stations destined to receive a specific message. The following CRC1 field provides a mechanism for the detection of errors in the header portion of the message. This CRC field is required since error-free transmission depends upon the count field being detected correctly. The actual data is placed in the information field and, as previously mentioned, can include special control characters. Finally the CRC2 field provides an error detection and correction mechanism for the data in the information field.

Operation

Unlike IBM's bisynchronous protocol, DDCMP does not require the transmission of an acknowledgement to each received message. Only when a transmission occurs or if traffic is light in the opposite direction, a condition where no data messages are to be sent, is it necessary to transmit a special NAK or ACK.

The number in the response field of a normal header or in either a special NAK or ACK message is used to specify the sequence number of the last good message received. To illustrate this, assume messages 3, 4, 5 and 6 were received since the last time an acknowledgement was sent and message 7 contains an error. Then, the header in the NAK message would have a response field value of 6, indicating that messages 3, 4, 5 and 6 were received correctly and message 7 was received incorrectly. Under the DDCMP protocol up to 255 messages can be outstanding due to the use of an 8-bit response field.

Another advantage of DDCMP over IBM's bisynchronous protocols is the ability of DDCMP to operate in a full-duplex mode. This eliminates the necessity of line turnarounds and results in an improved level of throughput. Another function of the response field is to inform a transmitting station of the occurrence of a sequence error. This is accomplished by the transmitting station examining the contents of the response field. For example, if the next message the receiver expects is 4 and it receives 5, it will not change of the response field of its data messages which contains a 3. In effect, this tells the transmitting station that the receiving station has accepted all messages up through message 3 and is still awaiting message 4.

Bit-oriented protocols

A number of bit-oriented line control procedures were implemented by computer vendors that are based upon the International Organization for Standardization (ISO) procedure known as high-level data link control (HDLC). Various names for line control procedures similar to HDLC include IBM's synchronous data link control (SDLC) and Unisys' data link control (UDLC). We refer to each of these protocols as being bit oriented, as a receiver continuously monitors data bit by bit.

The advantages of bit-oriented protocols are threefold. First, their full-duplex capability supports the simultaneous transmission of data in two directions, resulting in a higher throughput than is obtainable in BISYNC. Secondly, bit-oriented protocols are naturally transparent to data, enabling the transmission of pure binary data without requiring special sequences of control characters to enable and disable a transparency transmission mode of operation as required with BISYNC. Lastly, most bit-oriented protocols permit multiple blocks of data to be transmitted one after another prior to requiring an acknowledgement. Then, if an error affects a particular block, only that block has to be retransmitted.

HDLC link structure

Under the HDLC transmission protocol one station on the line is given the primary status to control the data link and supervise the flow of data on the link. All other stations on the link are secondary stations and respond to commands issued by the primary station.

The vehicle for transporting messages on an HDLC link is called a frame and is illustrated in Figure 1.81. The frame provides a common format for all supervisory and information transfers. In addition, it provides a structure which contains fields that have a predefined general interpretation.

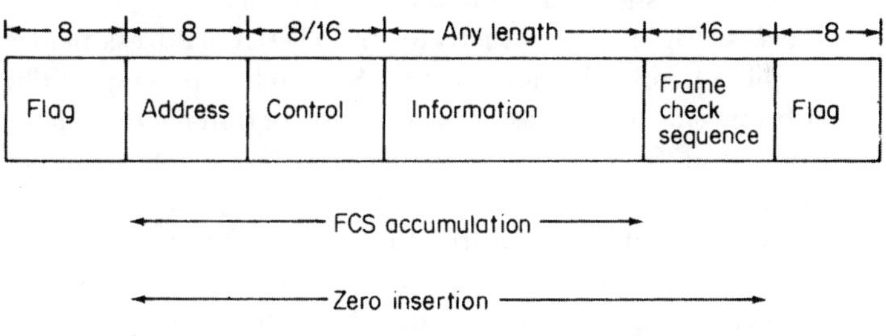

Figure 1.81 HDLC frame format. HDLC flag is 01111110 which is used to delimit an HDLC frame. To protect the flag and assure transparency the transmitter will insert a zero bit after a fifth 1 bit to prevent data from being mistaken as a flag. The receiver always deletes a zero after receiving five 1s

The HDLC frame contains six fields, wherein two fields serve as frame delimiters and are known as the HDLC flag. The HDLC flag has the unique bit combination of 01111110 (7EH), which defines the beginning and end of the frame. To protect the flag and assure transparency the transmission device will always insert a zero bit after a sequence of five 1-bits occurs to prevent data from being mistaken as a flag. This technique is known as zero insertion. The receiver will always delete a zero after receiving five ones to ensure data integrity.

The zero bit insertion technique insures that any string of more than five 1-bits will be interpreted as either a flag, a transmission error, or a deliberately

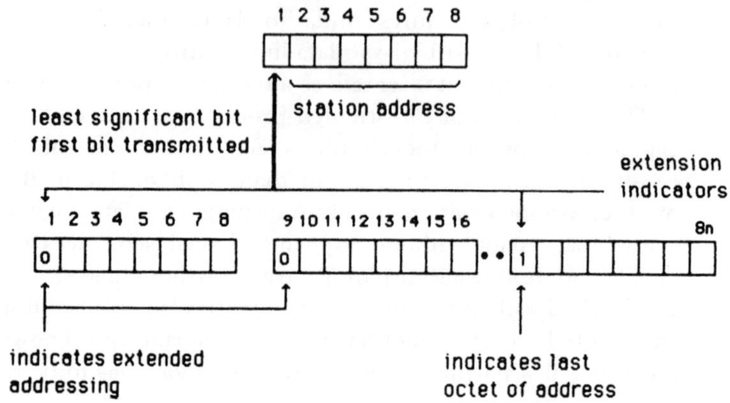

Figure 1.82 HDLC address field formats

transmitted fill pattern. Once a frame delimiter is recognized, the receiving device knows that the beginning of a frame has occurred when it receives an 8-bit non-flag address field. Upon detecting a subsequent frame delimiter, the receiving device knows that the frame has ended.

The address field is normally an 8-bit pattern that identifies the secondary station involved in the data transfer. If the least significant bit position of the address field is a zero, this indicates that extended addressing is used. The resulting extended address can contain any number of bytes and is terminated by a binary 1 in the least significant bit position of the last byte that constitutes the extended address. Figure 1.82 illustrates the HDLC address field formats.

The control field can be either 8 or 16 bits in length. This field identifies the type of frame transmitted as either an information frame or a command/response frame. Command frames are those transmitted by a primary station, while response frames are those transmitted by secondary stations. In the command frame, the address identifies the destination station for the command issued by the primary station. Similarly, in a response frame, the address field identifies the station transmitting the response. The information field can be any length and is treated as pure binary information, while the frame check sequence (FCS) contains a 16-bit value generated using a cyclic redundancy check (CRC) algorithm.

Control field formats

The 8-bit control field formats are illustrated in Figure 1.83. N(S) and N(R) are the send and receive sequence counts. They are maintained by each station for Information (I-frames) defined by the least significant bit having a value of 0 that are sent and received by that station. Each station increments its N(S) count by one each time it sends a new frame. The N(R) count indicates the expected number of the next frame to be received.

Using an 8-bit control field, the N(S)/N(R) count ranges from 0 to 7. Using a 16-bit control field the count can range from 0 to 127. There P/F bit is a poll/final bit. It is used as a poll by the primary (set to 1) to obtain a response from a

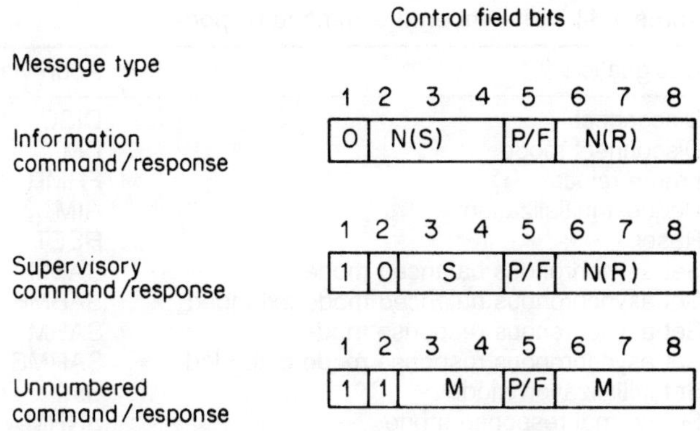

Figure 1.83 HDLC control field formats. N(S) = send sequence cont; N(R) = receive sequence count; S = supervisory function bits; M = modifier function bits; P/F = poll/final bit

secondary station. It is set to 1 as a final bit by a secondary station to indicate the last frame of a sequence of frames.

The supervisory command/response frame is used in HDLC to control the flow of data on the line. Figure 1.84 illustrates the composition of the supervisory control field: supervisory frames (S-frames) contain an N(R) count and are used to acknowledge I-frames, request retransmission of I-frames, request temporary suspension of I-frames, and perform similar functions.

As indicated in Figure 1.84, a supervisory frame is identified when the two least significant bits in the control field have a value of 01. Then, the value of the two following supervisory function bits indicate which of four functions are being invoked. If the first two bits in the control field are set to 11, a management frame referred to as an unnumbered command/response occurs. The latter reference results from the absence of N(S) and N(R) numbering fields. Five modifier (M) bits are used in the control field to define up to 32 general link control functions. Table 1.34 lists the unnumbered management functions presently defined for HDLC.

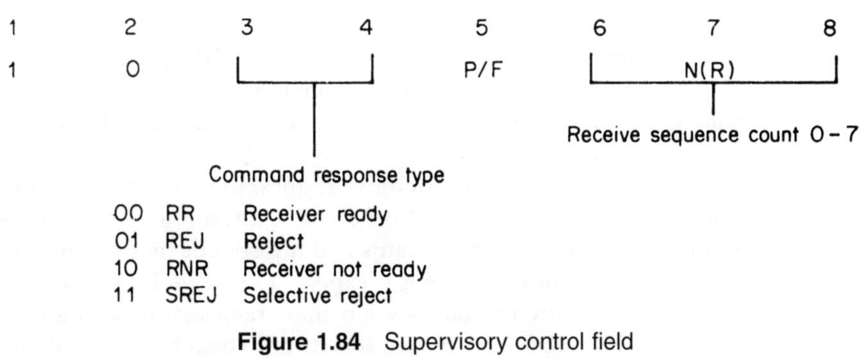

Figure 1.84 Supervisory control field

Table 1.34 Unnumbered command/response

Designation	Abbreviation	Command	Response
Disconnect	DISC	×	
Disconnect mode	DM		
Frame reject	FRMR		
Request initialization	RIM		×
Reset	RSET	×	×
Set asynchronous balanced mode	SABM	×	×
Set asynchronous balanced mode extended	SABME	×	
Set asynchronous response mode	SARM	×	
Set asynchronous response mode extended	SARME	×	
Set initialization mode	SIM		
Set normal response mode	SNRM	×	
Set normal response mode extended	SNRME	×	
Test	TEST	×	×
Unnumbered acknowledgement	UA		×
Unnumbered information	UI	×	×
Unnumbered poll	UP	×	
Exchange identification	XID	×	×

Operational modes

HDLC support three operational modes—normal response, asynchronous response, and asynchronous response balanced. In a normal response mode, a secondary station can only initiate transmission after receiving explicit permission from a controlling primary station. This operational mode is best suited for multipoint operations.

The asynchronous response mode is only applicable when there is one secondary station under primary control. In this operational mode, the secondary station can initiate transmission without having to receive explicit permission from a primary station.

The third operational mode supported by HDLC is asynchronous balanced. This operational mode enables the symmetrical transfer of data between two 'combined' stations on a point-to-point circuit. Here, each station has the ability to initialize and disconnect the circuit and is responsible for both controlling its own data flow and for recovering from error conditions. This operational mode is commonly used in packet switching. Figure 1.85 illustrates the difference between balanced and unbalanced operational modes.

To illustrate the advantages of HDLC over BISYNC transmission, consider the full-duplex data transfer illustrated in Figure 1.86. For each frame transmitted, this figure shows the type of frame, $N(S)$, $N(R)$ and poll/final (P/F) bit status.

In the transmission sequence illustrated in the left part of Figure 1.86, the primary station has transmitted five frames, numbered zero through four, when its poll bit is set in frame four. This poll bit is interpreted by the secondary station as a request for it to transmit its status and it responds by transmitting a receiver ready (RR) response, indicating that it expects to receive frame five next. This serves as an indicator to the primary station that frames zero through four were received correctly. The secondary station sets its poll/final bit as a final bit to indicate to the primary station that its transmission is completed.

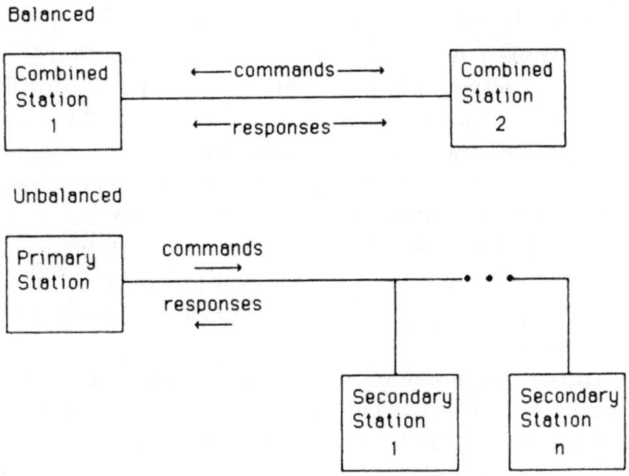

Figure 1.85 Balanced as opposed to unbalanced operations

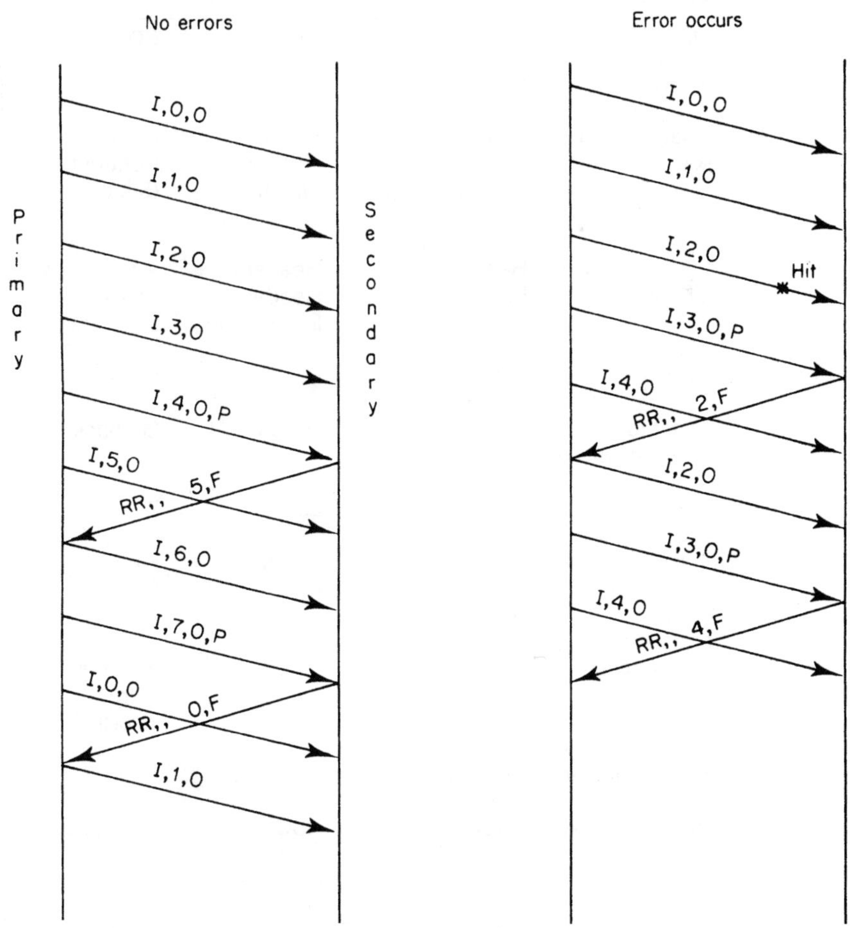

Figure 1.86 HDLC full-duplex data transfer. Format: Type, *N*(S), *N*(R), p/f

Note that since full-duplex transmission is permissible under HDLC, the primary station continues to transmit information (I) frames while the secondary station is responding to the primary's polls. If an 8-bit control field is used, the maximum frame number that can be outstanding is limited to seven since 3 bit positions are used for $N(S)$ frame numbering. Thus, after frame number seven has been transmitted, the primary station then begins frame numbering again at $N(S)$ equal to zero. Notice that when the primary station sets its poll bit when transmitting frame seven the secondary station responds, indicating that it expects to receive frame zero. This indicates to the primary station that frames five to seven were received correctly, since the previous secondary response acknowledge frames zero to four.

In the transmission sequence indicated on the right-hand side of Figure 1.86, assume a line hit occurs during the transmission of frame two. Note that in comparison to BISYNC, under HDLC the transmitting station does not have to wait for an acknowledgement of the previously transmitted data block; and it can

Table 1.35 Protocol characteristics comparison

Feature	BISYNC	DDCMP	SDLC	HDLC
Full duplex	No	Yes	Yes	Yes
Half duplex	Yes	Yes	Yes	Yes
Message format	Variable	Fixed	Fixed	Fixed
Link control	Control character, character sequences, optional header	Header (fixed)	Control field (8 bits)	Control field (8/16 bits)
Station addressing	Header	Header	Address field	Address field
Error checking	Information field only	Header information field	Entire frame	Entire frame
Error detection	VRC/LRC-8 VRC/CRC-16	CRC-16	CRC-ITU	CRC-ITU
Request for retransmission	Stop and wait	Go back N	Go back N	Go back N, selected reject
Maximum frames outstanding	1	255	7	127
Framing—start	2 SYNs	2SYNs	Flag	Flag
—end	Terminating characters	Count	Flag	Flag
Information transparency	Transparent mode	Inherent (count)	Inherent (zero insertion/ deletion)	Inherent (zero insertion/deletion
Control character	Numerous	SOH, DLE, ENQ	None	None
Character codes	ASCII EBCDIC Transcode (SBT)	ASCII (control character only)	Any	Any

continue to transmit frames until the maximum number of frames outstanding is reached; or, it can issue a poll to the secondary station to query the status of its previously transmitted frames while it continues to transmit frames up until the maximum number of outstanding frames is reached.

The primary station polled the secondary in frame three and then sent frame four while it waited for the secondary's response. When the secondary's response was received, it indicated that the next frame the secondary expected to receive $N(\mathrm{R})$ was two. This informed the primary station that all frames after frame one would have to be retransmitted. Thus, after transmitting frame four the primary station then retransmitted frames two and three prior to retransmitting frame four.

It should be noted that if selective rejection is implemented, the secondary could have issued a selective reject (SREJ) of frame two. Then, upon its receipt, the primary station would retransmit frame two and have then continued its transmission with frame five. Although selective rejection can considerably increase the throughput of HDLC, even without its use this protocol will provide the user with a considerable throughput increase in comparison to BISYNC.

For comparison purposes Table 1.35 compares the major features of BISYNC, DDCMP, IBM's SDLC and the ITU HDLC protocols.

Other protocols

Most of the previously mentioned protocols are restricted to use on wide area networks. Other protocols that are considerably more popular than those previously discussed operate on both LANs and WANs and will be covered in detail in Chapter 2 and 3. These protocols include Novell's NetWare IPX/SPX and TCP/IP, the latter being the only protocol that can be used on the Internet.

1.16 INTEGRATED SERVICES DIGITAL NETWORK

In this section we will examine both data and voice communications in the form of the Integrated Services Digital Network (ISDN), which at one time was expected to replace most, if not all, existing analog networks. Although ISDN has not lived up to the hype that surrounded its introduction, its service availability has considerably expanded since its introduction during the 1980s. One of the main limitations of ISDN availability was the requirement for local telephone companies to upgrade their switching infrastructure to support the technology. Many telephone companies postponed the upgrade of their switches through the mid-1990s as an economy measure while waiting for consumer demand for the service to develop. Since 1995 the upgrade of telephone switches has proceeded at a very high rate, resulting in ISDN service becoming available to approximately 80% of all telephone users in the United States by 1998. Thus, the limited availability of ISDN which acted as a constraint on its usage has considerably diminished over the past few years.

ISDN offers the potential for the development of a universal international digital network, with a series of standard interfaces that will facilitate the connection of a wide variety of telecommunications equipment to the network. Although the full transition to ISDN may require several decades and some ISDN functions may never be offered in certain locations, its potential cannot be overlooked. Since many ISDN features offer a radical departure from existing services and current methods of communications, we will review the concept behind ISDN, its projected features and services that can result from its implementation in this section.

Concept behind ISDN

The original requirement to transmit human speech over long distances resulted in the development of telephone systems designed for the transmission of analog data. Although such systems satisfied the basic requirement to transport human speech, the development of computer systems and the introduction of remote processing required a conversion of digital signals into an analog format. This conversion was required to enable computers and business machines to use existing telephone company facilities for the transmission of digital data. Not only was this conversion awkward and expensive due to the requirement to design high speed modems by developing and incorporating advanced encoding and error correcting techniques, but, in addition, analog facilities of telephone systems limit the data transmission rate obtainable when such facilities are used.

The evolution of digital processing and the rapid decrease in the cost of semiconductors resulted in the application of digital technology to telephone systems. By the late 1960s, telephone companies began to replace their electromechanical switches in their central offices with digital switches, while by the early 1970s, several communications carriers were offering end-to-end digital transmission services. By the mid-1980s, a significant portion of the transmission facilities of most telephone systems were digital, with over 99.9% of long distance transmission converted to digital by 1998. On such systems, human speech is encoded into digital format for transmission over the backbone network of the telephone system. At the local loop of the network, digitized speech is reconverted into its original analog format and then transmitted to the subscriber's telephone.

Based upon the preceding, ISDN can be viewed as an evolutionary progression in the conversion of analog telephone systems into an eventual all-digital network, with both voice and data to be carried end-to-end in digital form.

ISDN architecture

Under ISDN, network access functions that govern the methods by which user data flows into the network were separated from actual network functions, such as the manner by which signaling information is conveyed. In fact, ISDN's architecture resulted in a separate signaling network referred to as Common Channel Signaling Number 7 (CCS7) being used to convey signaling information. Although separate data and signaling networks govern the operation of ISDN, a single user–network interface provides a ubiquitous interface to ISDN network users.

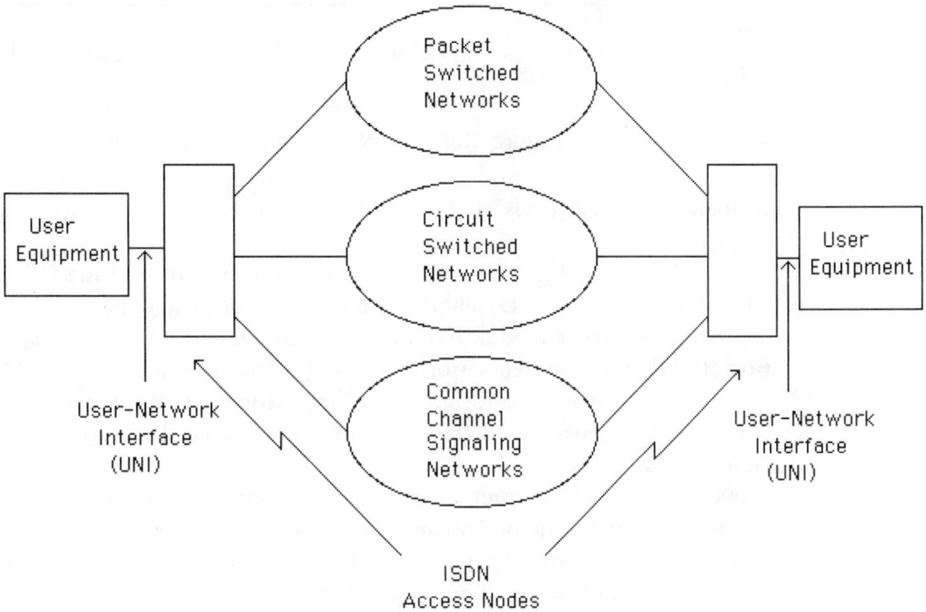

Figure 1.87 Basic ISDN architecture

Figure 1.87 illustrates the basic ISDN architecture to and between user–network interfaces. Note that under ISDN the user–network consists of both circuit switched and packet switched services. The circuit switched service provides for the routing of a call between access nodes to obtain an end-to-end connection similar to the manner by which calls are routed via the PSTN. The packet switched service enables packetized X.25 data transfer between access nodes as well as between access nodes and a device located on an X.25 packet network.

Types of service

Two types of ISDN are now standardized—Narrowband ISDN (N-ISDN) and broadband ISDN (B-ISDN). Broadband ISDN involves the logical grouping of N-ISDN facilities into a higher operating rate facility to obtain a high speed data transmission capability. Asynchronous Transfer Mode (ATM) evolved from the development work of B-ISDN. Since ATM is covered as a separate section in Chapter 2, we will focus our attention primarily upon N-ISDN in this section. In doing so we will examine the two major narrowband ISDN connection methods: basic access and primary access.

Basic access

Basic access defines a multiple channel connection derived by multiplexing data on twisted-pair wiring. This multiple channel connection is between an end-user terminal device and a telephone company office or a local Private Automated

F	B1 channel	D	B2 channel

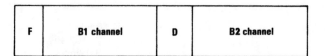

Figure 1.88 ISDN basic access channel format

Branch Exchange (PABX). The ISDN basic access channel format is illustrated in Figure 1.88.

As indicated in Figure 1.88, basic access consists of framing (F), two bearer (B) channels and a data (D) channel that are multiplexed by time onto a common twisted-pair wiring media. Each bearer channel can carry one pulse code modulation (PCM) voice conversation or data at a transmission rate of 64 kbps.

PCM is a voice digitization technique which results in a data rate of 64 kbps being used to represent a voice conversation. Since there are two B channels this enables basic access to provide the end-user with the capability to simultaneously transmit data and conduct a voice conversation on one telephone line or to be in conversation with one person and receive a second telephone call. In the case of the latter situation, assuming the end-user has an appropriate telephone instrument, he or she could place one person on hold and answer the second call.

The basic access frame

The actual manner by which basic access framing is accomplished is significantly more complex than previously illustrated in Figure 1.88. In actuality, a variety of framing, echo, activate and DC balancing bits are spread out within a 48-bit basic rate interface frame along with B and D channel bits that transport data. The 48-bit frame actually carries only 36 data bits, with the remaining 12 bits representing line overhead. Since the frame is repeated 4000 times per second this results in a line operating rate of 48 bits/frame $\times$ 4000 frames/s, or 192 kbps, with the actual data transfer rate becoming 36 bits/frame $\times$ 4000 frames/s, or 144 kbps.

There are two frame formats used for ISDN basic access. One format is used when frames are transmitted from the network to ISDN terminal equipment. Since the frame flows from the Network Termination (NT) to the terminal equipment (TE) it is referred to as a basic access NT frame. The second type of ISDN basic access frame flows from the TE to the NT and, as you might expect, is referred to as a basic access TE frame. Figure 1.89 illustrates the format of the two basic-access frames.

In examining Figure 1.89 note that both NT and TE frames commence with a framing bit for frame alignment or synchronization. In an NT frame the L bits are used to electrically balance the entire frame. In comparison, in the TE frame the L bits are used to balance each octet of B channel information and each individual D channel bit. Through electrical balancing a situation where an excess number of binary zeros could occur that would prevent receiver synchronization with data is avoided. In addition, frame balancing limits DC voltage buildup and enables devices to communicate at greater distances from one another.

The A bit in the NT frame is used to activate or deactivate terminal equipment (TE), enabling the TE to be placed in a low power consumption mode when there

Basic Access NT Frame

F	L	Eight B1 bits	E	D	A	Fa	N	Eight B2 bits	E	D	S	Eight B1 bits	E	D	S	Eight B2 bits	E	D	L

Basic Access TE Frame

F	L	Eight B1 bits	L	D	L	Fa	L	Eight B2 bits	L	D	L	Eight B1 bits	L	D	L	Eight B2 bits	L	D	L

where: A = Activate/deactivate bit
B1 = B1 channel bits
B2 = B2 channel bits
D = D channel bit
E = D channel echo bit
F = Framing bit
S = Reserved for future standardization
Fa = Auxiliary framing bit
L = DC balancing bit
NT = Network termination
TE = Terminal equipment

Figure 1.89 ISDN basic-access frame formats

is no activity or to come on-line. The S bits are reserved for future standardization, while the E bits represent the echoes of previously transmitted D channel bits in TE frames. That is, when the NT receives a D channel bit from a TE the NT echoes the bit in the next E bit position in the basic access NT frame flowing to the TE.

The D channel

The D channel was designed for both controlling the B channels through the sharing of network signaling functions on this channel as well as for the transmission of packet switched data. Concerning the transmission of packet switched data, the D channel provides the capability for a number of applications to include monitoring home alarm systems and the reading of utility meters upon demand. Since these types of applications have minimum data transmission requirements, the D channel can be expected to be used for a variety of applications in addition to providing the signaling required to set up calls on the B channels.

When terminal equipment has D channel information to transmit it monitors the flow of E bits in the NT frame for a specified number of set bits that indicates the D channel is not in use. When that number is reached the TE will transmit its

Country Code	National Destination Code	Subscriber Number	ISDN Sub-Address
≤17 digits			≤40 digits

Figure 1.90 Basic ISDN address structure

D channel data to the NT. The number of set E bits the TE must receive depends upon the type of information to be transmitted. Signaling information has a higher priority than non-signaling information, resulting in eight set E bits that must be received for the TE to transmit signaling information, and 10 set E bits for transmitting non-signaling information.

Data carried by an ISDN D channel is encoded using the link access protocol D-channel (LAPD) format. LAPD is a layer 2 protocol defined by the ITU-T Q.921 recommendation whose frame format is identical to HDLC. The composition of the address field of Q.921, however, significantly differs from HDLC. ISDN addressing is standardized by the ITU I.331 Recommendation. That recommendation specifies a primary address up to 17 digits in length and an optional sub-address up to 40 digits in length. Figure 1.90 illustrates the basic structure of an ISDN address.

The country code, national destination code, and subscriber number uniquely identify each ISDN subscriber. This variable length number can be up to 17 digits in length, with the country code standardized in the ITU E.163 Recommendation. The optional sub-address field shown in Figure 1.90 enables organizations to extend their addressing within a private network which is accessible via ISDN. To provide interoperability between ISDN and other networks, such as the PSTN and public packet networks, the ITU developed a series of additional recommendations. Its I.330 series of recommendations define internetworking between ISDN and the North American Numbering Plan (NANP) for switched network telephone calls as well as for D channel transmission to and from X.25 public packet networks which use the ITU X.121 Recommendation for network addresses.

One of the more publicized features of ISDN is its calling line identification (CLID) capability. Essentially, CLID results in the transmission of the caller's telephone number via the D channel where it will be displayed on a liquid crystal display (LCD) built into most ISDN telephones, which are commonly referred to as digital telephones. Since the calling line identification is carried as a series of binary numbers, it becomes possible to integrate incoming numbers into a computer system. This makes it possible for a business to use the incoming number as a database search element. For example, an insurance company could route the calling number to their mainframe computer. As the telephone operator answers the call, the computer could simultaneously search a database and retrieve and display policy information on the operator's terminal. This capability not only enhances customer service, but, in addition, increases the productivity of organization employees.

Although calling line information can be received via a PSTN connection, that information is transferred between rings as a sequence of modem modulated symbols. This normally requires the use of a separate caller ID box to display the called number, although a few analog telephone sets now have a built-in caller ID display. While the displayed digits can be used in a manner similar to the previously described ISDN CLID capability, a specialized device would be required to frame the digits in a manner suitable for recognition by the computer. In comparison, the X.25 frame format used by the ISDN D channel for transporting CLID information represents a standardized method for conveying information.

Primary access

Primary access can be considered as a multiplexing arrangement whereby a grouping of basic access users shares a common line facility. Typically, primary access will be employed to directly connect a Private Automated Branch Exchange (PABX) to the ISDN network. This access method is designed to eliminate the necessity of providing individual basic access lines when a group of terminal devices shares a common PABX which could be directly connected to an ISDN network via a single high-speed line. Due to the different types of T1 network facilities in North America and Europe, two primary access standards have been developed.

In North America, primary access consists of a grouping of 23 B channels and one D channel to produce a 1.544 Mbps composite data rate, which is the standard T1 carrier data rate. In Europe, primary access consists of a grouping of 30 B channels plus one D channel to produce a 2.048 Mbps data rate, which is the T1 carrier transmission rate in Europe.

Other channels

In addition to the previously mentioned B and D channels, ISDN standards define a number of additional channels. These channels include the A, C and H series of channels.

The A channel is a 56 kbps wideband analog channel. The C channel is a digital channel that is used with the A channel during the transition to ISDN. The C channel operates at 8 or 16 kbps and carries signaling information similar to the manner in which the D channel controls B channels.

H channels are also known as broadband ISDN (B-ISDN) and are formed out of multiple B channels. For example, the H0 channel operates at 384 kbps and represents 6 B channels. Other H channels include the H11 channel which operates at 1.536 Mbps, the H12 channel which operates at 1.92 Mbps, and the H4 channel which operates at approximately 135 Mbps. The use of ISDN H channels provides a mechanism to interconnect local and wide area networks as well as to support such applications as full motion video teleconferencing and high-speed packet switching, the latter commonly referred to as frame relay. Both packet switching and frame relay are discussed in Chapter 2.

Network characteristics

Four of the major characteristics of an ISDN network are listed in Table 1.36. These characteristics can also be considered as driving forces for the implementation of the network by communications carriers.

Table 1.36 ISDN characteristics

Integrates voice, data and video services
Digital end-to-end connection resulting in high transmission quality
Improved and expanded services due to B and D channel data rates
Greater efficiency and productivity resulting from the ability to have several simultaneous calls occur on one line

Due to the digital nature of ISDN, voice, data, and video services can be integrated, alleviating the necessity of end-users obtaining separate facilities for each service. Since the network is designed to provide end-to-end digital transmission, pulses can be easily regenerated throughout the network, resulting in the generation of new pulses to replace distorted pulses. In comparison, analog transmission facilities employ amplifiers to boost the strength of transmission signals, which also increases any impairments in the signal. As a result of regeneration being superior to amplification, digital transmission has a lower error rate and provides a higher transmission signal quality than an equivalent analog transmission facility.

Due to basic access in effect providing three signal paths on a common line, ISDN offers the possibilities of both improvements to existing services and an expansion of services to the end-user. Concerning existing services, current analog telephone line bandwidth limitations normally preclude bidirectional data transmission rates over 33.6 kbps occurring on the switched telephone network. In comparison, under ISDN each B channel can support a 64 kbps transmission rate while the D channel will operate at 16 kbps. In fact, if both B channels and the D channel were in simultaneous operation a data rate of 144 kbps would be obtainable on a basic access ISDN circuit, which would extend current analog circuit data rates by a factor of 4.

Since each basic access channel in effect consists of three multiplexed channels, different operations can occur simultaneously without requiring an end-user to acquire separate multiplexing equipment. Thus, an end-user could receive a call from one person, transmit data to a computer and have a utility company read their electric meter at a particular point in time. Here, the ability to conduct simultaneous operations on one ISDN line should result in both greater efficiency and productivity. Efficiency should increase since one line can now support several simultaneous operations, while the productivity of the end-user can increase due to the ability to receive telephone calls and then conduct a conversation while transmitting data.

Terminal equipment and network interfaces

One of the key elements of ISDN is a small set of compatible multipurpose user–network interfaces that were developed to support a wide range of applications. These network interfaces are based upon the concept of a series of reference points for different user terminal arrangements which is then used to define these interfaces. Figure 1.91 illustrates the relationship between ISDN reference points and network interfaces.

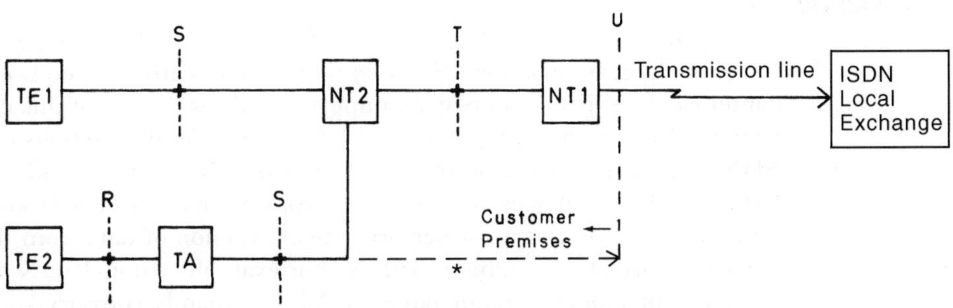

Figure 1.91 ISDN reference points and network interfaces. TE1 (Terminal Equipment 1) type devices comply with the ISDN network interface. TE2 (Terminal Equipment 2) type devices do not have an ISDN interface and must be connected through a TA (Terminal Adapter) functional grouping. NT2 (Network Termination 2) includes switching and concentration equipment which performs functions equivalent to layers 1 through 3 of the OSI Reference Model. NT1 (Network Termination 1) includes functions equivalent to layer 1 of the OSI Reference Model. A terminal adapter with a built-in NT1 can be directly connected to the U interface, eliminating the need for a separate NT1. (Reprinted with permission from *Data Communications Management,* © 1987 Auerbach Publishers, New York, NY.)

The ISDN reference configuration consists of functional groupings and reference points at which physical interfaces may exist. The functional groupings are sets of functions that may be required at an interface, while reference points are employed to divide the functional groups into distinct entities.

The TE (terminal equipment) functional grouping is comprised of TE1 and TE2 type equipment. Examples of TE equipment include digital telephones, conventional data terminals, and integrated voice/data workstations.

TE1

TE1 type equipment complies with the ISDN user–network interface and permits such equipment to be directly connected to an ISDN 'S' type interface which supports multiple B and D channels. TE1 equipment connects to ISDN via a twisted-pair four-wire circuit. Transmission is full-duplex and occurs at 192 kbps for basic access and at 1.544 or 2.048 Mbps for primary access.

TE2

TE2 type equipment are devices with non-ISDN interfaces, such as RS-232 or the ITU X or V-series interfaces. This type of equipment must be connected through a TA (terminal adapter) functional grouping, which in effect converts a non-ISDN interface (R) into an ISDN Sending interface (S), performing both a physical interface conversion and protocol conversion to permit a TE2 terminal to operate on ISDN.

Terminal adapters

Due to the large base of non-ISDN equipment currently in operation, the terminal adapter can be expected to play an important role as the use of this digital network expands. The terminal adapter (TA) performs a series of functions to convert non-ISDN equipment for use on the ISDN network. First, it must adapt the data rate of the non-ISDN device to either a 64 kbps B channel or a 16 kbps D channel operating rate. Next, it must perform the conversion of data from the non-ISDN device to a format acceptable to ISDN. For example, a non-ISDN device, such as an intelligent modem, might have its AT commands converted into ISDN D-channel signaling information. Other functions performed by TAs include the conversion of electrical, mechanical, functional, and procedural characteristics of non-ISDN equipment interfaces to those required by ISDN and the mapping of network layer data to enable a signaling terminal to be 'understood' by ISDN equipment.

Since a basic access channel operates at a multiple of most non-ISDN equipment rates, most terminal adapters include a multiple number of R interface ports. This allows, for example, an asynchronous modem connected to a personal computer, a facsimile machine, and a telephone to be connected to a basic access line via the R interface. In fact, most commercially available terminal adapters have three or four R interface ports.

Rate adaption

Rate adaption is the process during which the data rate of slow-speed devices is increased to the 64 kbps synchronous data rate of an ISDN B channel. During the rate adaption process, the data stream produced by a non-ISDN device is padded with dummy bits by the terminal adapter and clocked at a 64 kbps data rate.

In 1984, the CCITT approved its rate adaption standard known as the V.110 recommendation. Originally, this recommendation was strictly for synchronously operated devices and was modified in 1988 to support asynchronous devices. The framing specified by the V.110 recommendation is complex, with each 80-bit frame containing a 17-bit frame alignment pattern, while the actual rate adaption process can involve between one and three steps, with the actual number of steps dependent upon the operating rate of the terminal and its operating mode. Asynchronous devices require a three-stage process, while synchronous terminals operating below 64 kbps require a two-stage process as indicated in Figure 1.92.

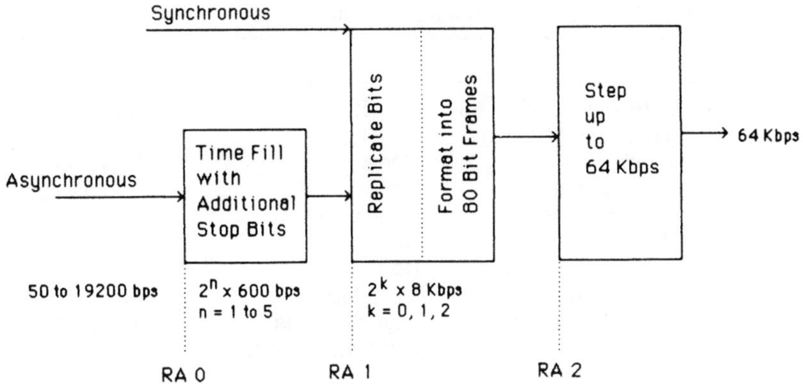

Figure 1.92 V.110 rate adaption

As indicated in Figure 1.92, asynchronous devices require a three-stage rate adaption process. In the first stage, extra stop bits are appended to each character to make the operating rate a multiple of 600 bps. The second stage of the V.110 rate adaption process services both asynchronous and synchronous data. In this stage, bits are replicated to create an 80-bit frame operating at an intermediate data rate of 8, 16 or 32 kbps. In the third stage, a process called bit positioning occurs in which one, two or four bits are added for each bit to bring the data rate up to 64 kbps. In addition to defining a complex rate adaption process, V.110 does not completely define flow control, fails to define a mechanism for error detection on its 80-bit frame, and is often difficult for the procedure defined by the recommendation to detect and adjust to a change in the data rate of a non-ISDN device. Due to these problems, a second rate adaption standard, which was originated in the United States by the ANSI T1E1 standards committee, has gained widespread acceptance. Known as V.120, this standard is based upon HDLC which makes flow control, error detection and correction, and other functions very easy to perform although they are either not included in or difficult to perform under the V.110 recommendation. The V.120 rate adaption is based upon the use of the LAPD protocol, with flag stuffing used to adapt the data rate to 64 kbps. Here the term 'flag stuffing' refers to the addition of a sufficient number of flag (01111110) bytes to bring the operating rate to 64 kbps. Today, V.120 is primarily used in the United States, while V.110 is primarily used in Europe and Japan. Both procedures provide support for V.24 and V.35 R interface.

Three additional rate adaption schemes that warrant mention include AT&T's Digital Multiplexed Interface (DMI), the ITU X.32 Recommendation, and Northern Telecom's T-Link. DMI represents AT&T's computer to PBX interface and is supported by their large 5ESS central office switches. In actuality there are three types of DMI rate adaption methods. DMI-1 supports 56 kbps data service and is compatible with the V.110 Recommendation at that data rate. DMI-2 supports rate adaption from devices operating below 20 kbps, such as regular RS-232 interfaces. DMI-3 is based on the ITU LAP-D protocol and is similar to the V.120 rate adaption scheme.

Table 1.37 ITU terminal adapter standards

Feature	V.110	V.120	X.31
ISDN Beaver service	Circuit	Circuit	Circuit/packet
B-channel multiplexing	Q.931	LinkID	Channel number
Error detection	None	CRC and V.41	CRC and V. 41
Error correction	None	Retransmission	Retransmission
Flow control	Unidirectional	Yes	Yes
HDLC-based	No	Yes	Yes
Multiple destinations	No	No	Yes
Rate adaption method	Multistep	Brit stuffing	Bit stuffing
Type of DTE/DCE at R-interface	Asynchronous and synchronous	Asynchronous, HDLC, transparent	X.25 synchronous

The X.31 Recommendation governs the rate adaption of packet mode X.25 equipment to an ISDN channel. Under the X.31 Recommendation call control procedures between X.25 and Q.931 are defined, enabling X.25 signaling to be converted to ISDN's and vice versa. The Northern Telecom T-Link rate adaption scheme is a circuit-mode terminal adapter protocol used in that vendor's ISDN terminal products. T-Link adapts the user data rate to an ISDN 64 kbps channel for transmission through Northern Telecom DMS-100 central office switches.

Until 1991 differences in rate adaption methods and ISDN line provisioning made the installation and configuration of an ISDN circuit a most challenging process. In 1991 Bellcore defined a National ISDN-1 standard which defines the method by which ISDN capable devices signal their status, such as busy, available, or no answer to carrier switches. This was a significant step in enabling different rate adaption methods to correctly pass required signaling information, for example to AT&T 5ESS and Northern Telecom DMS-100 switches.

A second significant event occurred a few years later with the development of a standard set of ISDN ordering codes. This enables telephone company installers to easily configure their ISDN line to work with user equipment manufactured to operate in a predefined manner. Table 1.37 provides a comparison of the nine features between the V.110, V.120, and X.31 standardized rate adaption Recommendations.

NT1

The NT1 (network termination 1) functional group is the ISDN digital interface point and is equivalent to layer 1 of the OSI reference model. Functions of NT1 include the physical and electrical termination of the loop, line monitoring, timing, and bit multiplexing. In Europe, where most communications carriers are government owned monopolies, NT1 and NT2 functions may be combined into a common device, such as a PABX. In such situations, the equipment serves as an NT12 functional group. In comparison, in the United States the communications carrier may provide only the NT1, while third-party equipment would connect to the communications carrier equipment at the T interface.

NT2

The NT2 (network termination 2) functional group includes devices that perform switching and data concentration functions equivalent to the first three layers of the OSI Reference Model. Typical NT2 equipment can include PABXs, terminal controllers, concentrators, and multiplexers.

Interfaces

As previously explained, the R interface is the point of connection between non-ISDN equipment and a terminal adapter. Although the R interface can consist of any common DTE interface, most terminal adapters support RS-232 and V.35.

The S interface is the standard interface between the TE and NT1 or between the TA and NT1. This is the interface to a 192 kbps, 2B + D, four-wire circuit. The S/T interface can operate up to distances of 1000 m using a pseudo-ternary coding technique. In this coding technique, a binary one is encoded by the transmission of an electrical zero, while binary zeros are encoded by transmitting alternating positive and negative pulses. Since there are three signal states (+ , 0, − voltage) used to encode two symbols, the coding method is called pseudo-ternary. An example of this coding technique is illustrated in Figure 1.93.

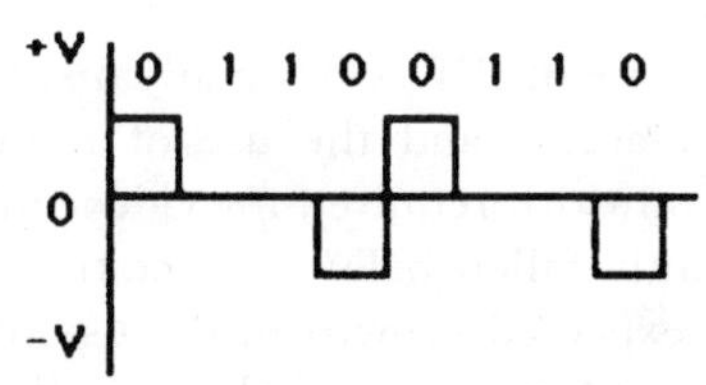

Figure 1.93 Pseudo-ternary coding example

The T interface is the customer end of an NT1 onto which you connect an NT2. For basic access, the T interface is a 192 kbps four-wire, 2B + D interface. For primary access, the T interface is a 1.544 Mbps, 23B + D, four-wire circuit or a 2.048 Mbps, 30B + D, four-wire circuit.

The U interface is the ISDN reference point that occurs between the NT1 and the network and is the first reference point at the customer premises. The coding scheme for information on the U interface is known as 2B1Q which is an acronym for 'two binary, one quaternary'. Under this coding scheme every two bits are encoded into one of four distinct states that are known as quats. The top portion of Figure 1.94 illustrates an example of 2B1Q encoding, while the lower portion of that illustration indicates the relationship between each dibit value and its quats code.

From the U interface, transmission occurs at 160 kbps to the telephone company central office. Due to the use of 2B1Q coding a maximum transmission distance of 18 000 feet is supported at a data rate of 160 kbps. This data rate represents 144 kbps used for the 2B + D channels and 16 kbps used for synchronization. In

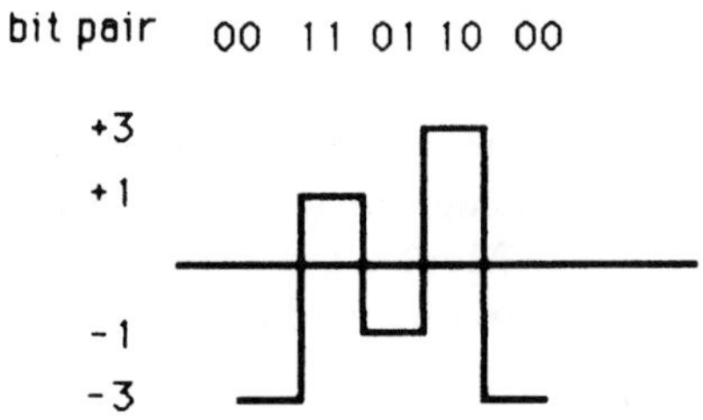

Figure 1.94 2BIQ coding example

Dibit	Quats
00	−3
01	−1
10	+3
11	+1

comparison, from the U interface to the S/T interface the data rate is 192 kbps, with 48 kbps used for synchronization and line balancing.

The future of ISDN

Although the use of ISDN has many distinct advantages, especially in the areas of dial Internet access and the use of multiple calls grouped together to obtain sufficient bandwidth required for videoconferencing, the use of this digital service has significantly fallen below expectations. Among the reasons provided for the slower than expected growth in the use of ISDN are its cost and pending competition in the form of digital subscriber lines (DSLs). Concerning its cost, installation of an ISDN line can easily exceed $300 and usage is typically billed at 5 or 10 cents/minute. When compared to the cost of a conventional analog line that can support bidirectional modem transfer at up to 33.6 kbps without incurring a $3 to $6 per hour usage surcharge, the price/performance ratio associated with the use of ISDN leaves much to be desired. Concerning DSL competition, although in its infancy, several types of digital subscriber lines have been developed that enable transmission rates up to 8 Mbps on the local analog loop between a subscriber and the serving telephone company office. Since DSL represents a modulation technology, it will be covered in detail in Chapter 5.

At the time this book revision was prepared, several field trials of DSL technology were being conducted and a few communications carriers and Internet Service Providers were offering commercial services using the technology. Although it is probably premature to predict the ultimate effect of DSL upon ISDN, it should be noted that ISDN represents a circuit switched and packet switched technology that enables calls to be routed. In comparison, DSL technology is limited to providing a high speed point-to-point access from a subscriber to a telephone company central office. Thus, the primary competition between the two will probably occur in the market for obtaining fast Internet access. This is because a subscriber requiring only fast Internet access would have previously used ISDN's basic access to group two B channels to obtain a 128 kbps data transfer capability to a single location. In those circumstances the use of DSL technology would provide a substitute that could be competitive depending upon its cost structure.

REVIEW QUESTIONS

To facilitate the reference of material in this book to review questions, each question has a three-part number. The first digit references the chapter related to the question. The second part of the number bounded by two decimal points references the section in the chapter upon which the review question is based. The third part of the question number references the question related to a specific section. For example, question 1.12.3 is the third question that references the material in Section 12 of Chapter 1.

1.1.1 Discuss the function of each of the major elements of a transmission system.

1.2.1 What is the relationship between each of the three basic types of line connections and the use of that line connection for short or long duration data transmission sessions?

1.3.1 Discuss the relationship of modems and digital service units to digital and analog transmission systems. Why are these devices required and what general functions do they perform?

1.3.2 Name four types of analog facilities offered by communications carriers and discuss the utilization of each facility for the transmission of data between terminal devices and computer systems.

1.3.3 Why is unipolar non-return to zero signaling unsuitable for use on a wide area network?

1.3.4 Why is a signaling technique that does not produce residual dc important on a wide area network?

1.3.5 What is a key advantage associated with the use of bipolar return to zero signaling?

1.3.6 What is the effect of 'bit robbing' on the ability to transmit data on a 64 kbps DS0 channel?

1.3.7 What is the function of an analog extension?

1.3.8 Name four types of digital transmission facilities offered by communications carriers and discuss the possible use of each facility by a large organization.

1.4.1 What is the difference between simplex, half-duplex, and full-duplex transmission?

1.4.2 Discuss the relationship between the modes of operation of terminals and computers with respect to the printing and display of characters on a terminal in response to pressing a key on the terminal's keyboard.

1.4.3 Assume your terminal is placed into a full-duplex mode of operation and you are accessing a similar operating computer. As you press keys on your keyboard, what do you see on your display?

1.5.1 In asynchronous transmission how does a receiving device determine the presence of a start bit?

1.5.2 What is the difference between asynchronous and synchronous transmission with respect to the timing of the data flow?

1.6.1 What is the difference between serial and parallel transmission? Why do most communications systems use serial transmission?

1.7.1 Discuss the difference in terminal requirements with respect to point-to-point and multidrop line usage.

1.8.1 What is a line discipline? Why is it required?

1.9.1 Discuss the use of three network topologies.

1.9.2 Why does a WAN designer normally attempt to structure their network to minimize the distance of circuits connecting geographically separated locations?

1.10.1 Discuss the difference between analog and digital transmission with respect to currently available operating rates.

1.11.1 Why is the Morse code basically unsuitable for transmission by terminal devices?

1.11.2 How does Baudot code, which is a 5-level code, permit the representation of more than 32 unique characters?

1.11.3 What is the bit composition of the ASCII characters A and a?

1.12.1 Assuming even parity checking is employed, what are the parity bits assigned to the ASCII characters A, E, I, O, and U? What are the parity bits if odd parity checking is employed?

1.12.2 What are the major limitations of parity checking?

1.12.3 Assume a file on your personal computer contains 3000 lines of data, with an average of 60 characters per line. If you transmit the file using 8-bit character transmission and the probability of an error occurring is 1.5 per 100 000 bits; how many characters can be expected to be received in error if the bit errors occur randomly and are singular in occurrence per transmitted character?

1.12.4 Under the XMODEM protocol, what would be the value of the checksum if the data contained in a block consisted of all ASCII X characters?

1.12.5 Discuss the relationship between a transmitted cyclic redundancy check character and an internally generated cyclic redundancy check character with respect to the data integrity of the block containing the transmitted cyclic redundancy check character.

1.13.1 Discuss the importance of having standards.

1.13.2 Discuss the difference between national, international, and *de facto* standards. Cite an example of each.

1.13.3 Why would it be in the best interest of a manufacturer to build a product compatible with appropriate standards, such as the RS-232/V.24 standard?

1.13.4 Discuss the applicability of FIPS and ANSI standards with respect to federal agencies and private sector firms.

1.13.5 Why can it take up to four years or more for the ITU to adopt a recommendation?

1.13.6 Name two sources of *de facto* communications standards.

1.13.7 What is a Request For Comment (RFC)?

1.13.8 What is the purpose of layer isolation in the OSI reference model?

1.13.9 What are the functions of nodes and paths in a network?

1.13.10 Discuss the seven OSI layers and the functions performed by each layer.

1.14.1 What is the primary difference between RS-232-C and RS-232-D with respect to interchange circuits and connectors?

1.14.2 What is the difference in connector requirements between RS-232, ITU V.24, V.35 and X.20 standards?

1.14.3 Discuss the relationship between the voltage to represent a binary one in a terminal and the RS-232 signal characteristics that represent a binary one.

1.14.4 What are three methods commonly used to refer to RS-232 circuits? Which method do you feel is most popular in industry? Why?

1.14.5 What is the purpose of the ring indicator signal? Why do some modems require two rings prior to answering a call?

1.14.6 What is the difference between internal and external timing?

1.14.7 What are two key limitations associated with RS-232? Describe how differential signaling associated with RS-449 and RS-530 alleviate a considerable portion of those limitations.

1.14.8 What is balanced signaling?

1.14.9 What is the primary application for using a V.35 interface?

1.14.10 What is the purpose of the RS-366-A interface?

1.14.11 Why is the X.21 interface more costly than an RS-232/V.24 interface?

1.14.12 What is the purpose of the X.21 bis interface?

1.14.13 What are the operating rate differences between the V.35 and HSSI interfaces?

1.14.14 What are the key differences between the use of the HSSI and HIPPI interfaces with respect to transmission distance?

1.14.15 What is a null modem? Why are pins 2 and 3 reversed on that cable?

1.15.1 What is the difference between a terminal protocol and a data link protocol?

1.15.2 What is the purpose of data sequencing in which a large block is broken into smaller blocks for transmission?

1.15.3 Define the characteristics of a teletype protocol.

1.15.4 What is the purpose in using one or more null character after a carriage return line feed sequence?

1.15.5 What error detection method is used in the teletype protocol? How are errors corrected when they are detected?

1.15.6 What is echoplex? When echoplex is used who is responsible for examining locally printed characters?

1.15.7 How can a receiver detect the occurrence of an error when the XMODEM protocol is used?

1.15.8 Discuss two limitations of the XMODEM protocol.

1.15.9 What are two advantages of the ZMODEM protocol in comparison to the XMODEM protocol?

1.15.10 What are the major differences between the 2780, 3780 and 3270 protocols?

1.15.11 What is the purpose of bisynchronous protocol transmitting alternating acknowledgements (DLE1 and DLE0)?

1.15.12 Why are alternating DLE0 and DLE1 characters transmitted as positive acknowledgements in bisynchronous transmission?

1.15.13 What procedure is used to prevent a stream of binary date from being misinterpreted as an HDLC flag? Explain the operation of this procedure.

1.15.14 What are the advantages of a bit-oriented protocol in comparison to a character-oriented protocol?

1.15.15 If a secondary station responds to the poll of a primary station by settin $N(\mathrm{R})$ equal to five in its response, what does this signify to the primary station?

1.15.16 How does the DDCMP protocol provide data transparency?

1.15.17 Assume the response field of a NAK message in a DDCMP protocol has a value of 14 and messages 12, 13, 14, 15 and 16 are outstanding. What does this indicate?

1.15.18 What is the advantage of a selective reject command?

1.16.1 Why can you expect transmission quality on ISDN facilities to be superior to existing analog facilities?

1.16.2 Discuss the data transmission rate differences between a basic access ISDN circuit and that obtainable on the switched telephone network.

1.16.3 What is the actual data transfer rate obtainable on an ISDN basic access line?

1.16.4 What is the purpose of the A bit in an NT frame?

1.16.5 What function does a terminal adapter perform?

1.16.6 What is pseudo-ternary coding? How would the bit sequence 1010 be encoded using this coding technique?

1.16.7 What is rate adaption?

1.16.8 Discuss the difference between the V.110 and V.120 rate adaption standards.

1.16.9 Discuss the use of the R, S, T and U interfaces.

1.16.10 How would the bit sequence 00101001 be encoded using 2B1Q coding?

1.16.11 In what application would digital subscriber line technology appear to compete with ISDN?

2

WIDE AREA NETWORKS

Building upon our examination of fundamental wide area networking concepts presented in Chapter 1, we will now turn our attention to the operation and utilization of several types of wide area networks. In doing so we will examine circuit switching networks, packet switching networks, the collection of interconnected networks, referred to as the Internet, and two special types of networks. The first special type of network actually represents two proprietary IBM networks—System Network Architecture (SNA) and Advanced Peer-to-Peer Networking (APPN). Although the direction of networking is towards open systems, IBM's domination of the mainframe market makes it highly probable that a large base of SNA and APPN networks will continue to be used for the foreseeable future. In fact, at the time this book was prepared there were over 50 000 SNA and APPN networks in operation, some connecting over 10 000 terminal devices. The second special type of network we will examine was developed to provide an integrated capability which combines some of the features of circuit switching and packet switching to enable voice, data, video, and images to be transported via a common network infrastructure. That network technology is referred to as Asynchronous Transfer Mode, or ATM.

Although we will briefly discuss the role of several types of networking devices in this chapter, we will defer a detailed explanation of their operation and utilization to succeeding chapters. Instead, we will primarily focus our attention upon the operation of different types of wide area networks, which will provide us with the ability to better appreciate the role of different networking devices in providing a transmission capability onto a WAN or between LANs and WANs, the latter commonly referred to as internetworking.

2.1 OVERVIEW

As its name implies, a wide area network is a network structure which interconnects locations that are geographically dispersed. While a wide area network spans a distance beyond that covered by a local area network, in actuality, it can considerably range in its geographical area of coverage. Some wide area networks can simply interconnect terminal devices and computers located in a few cities within close proximity of one another. Other wide area networks can interconnect terminal devices and computers located in cities on different continents. What each of these wide area networks has in common is the use of transmission facilities obtained from communications carriers and the purchase or lease of networking devices to develop a network structure designed to meet the communications requirements of an organization.

Transmission facilities

Today, the wide area network designer has a large variety of transmission facilities to use to construct this type of network. You can use analog leased lines, DDS leased lines, fractional T1, and T1 or T3 lines to interconnect sites on a permanent basis. For sites that require transmission periodically or to back up leased lines, you can use the PSTN or a variety of switched digital facilities. By selecting transmission facilities to match the capabilities of different types of communications devices you obtain the ability to design a wide area network to meet the communications requirements of your organization. Thus, in the remainder of this chapter we will turn our attention to obtaining an appreciation for the operation and utilization of different types of wide area networks as well as an overview of different types of equipment used on such networks.

2.2 CIRCUIT SWITCHED NETWORKS

The most popular type of network and the one almost all readers use on a daily basis is a circuit switched network. The public switched telephone network in which the number you dial results in telephone company offices switching the call between offices to establish a connection to the dialed party represents the most commonly used circuit switched network. Circuit switching is not, however, limited to the telephone company. By purchasing appropriate switching equipment, any organization can construct their own internal circuit switched network and, if desired, provide one or more interfaces to the public switched network to allow voice and data transmission to flow between the public network and their private internal network.

In a circuit switching system, switching equipment is used to establish a physical path for the duration of a call. The path that is established is temporary and the facilities used to establish the call become available for another call after the conclusion of the first call. This type of connection in which a call is established by switching equipment over a temporary path is known as a switched virtual call. Figure 2.1 illustrates the operation of a generic switch used to establish a circuit switched call. In this illustration, any incoming line can be cross connected to any outgoing line. This switching mechanism results in the establishment of a path between the caller/originator and the called party/destination.

Both the incoming and outgoing lines and the switching equipment illustrated in Figure 2.1 are shown in a generic representation. In actuality, both the type of line facilities and the switching equipment used to establish a circuit switched call can vary considerably.

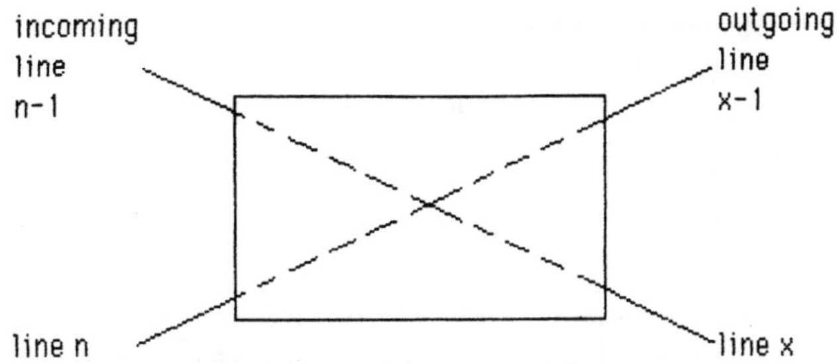

Figure 2.1 Generic switch operation. In a circuit switching system, an incoming line is cross connected to an outgoing line

The concept of circuit switching was recently applied to local area networks through the development of LAN switches. Through the use of LAN switches, described in detail later in this book, data is switched on a frame-by-frame basis between source and destination, with the cross-connection established on a temporary basis for each frame and then torn down, even when succeeding frames have the same destination. In comparison, in a circuit switching system the cross-connection remains in place until one or both parties terminates the call.

The earliest circuit switched network used a twisted-pair wire conductor to route calls from a telephone subscriber to a serving office, the latter commonly referred to as a local or end office. Depending upon the destination of the call it will be routed directly to another subscriber served by that office or into a trunk for transmission to another telephone company office. The first type of routing involves an intraoffice call, while the second type of routing involves an interoffice call.

In the early days of circuit switching, groups of lines were installed in telephone company offices, with each line used to carry one conversation at a time that was switched onto the line. As the number of telephone subscribers proliferated, telephone companies quickly realized that this method of switching individual calls onto individual circuits was not economical.

To reduce the number of physical lines required to connect telephone company offices to one another, communications carriers implemented a technique called multiplexing. Originally, frequency division multiplexing (FDM) was used exclusively by communications carriers. FDM was gradually replaced in most areas of the world by time division multiplexing that utilizes the T-carrier as a digital transport mechanism. This evolution to TDM equipment and T-carrier facilities was based upon the advantages associated with digital signaling in comparison to analog signaling.

Frequency division multiplexing

Employing frequency division multiplexing between carrier central offices requires the use of a communications circuit that has a relatively wide bandwidth. This bandwidth is then divided into subchannels by frequency.

When a communications carrier uses FDM for the multiplexing of voice conversations onto a common circuit, the 3 kHz passband of each conversation is shifted upward in frequency by a fixed amount of frequency. This frequency shifting places the voice conversation into a predefined channel of the FDM multiplexed circuit. At the opposite end of the circuit, another FDM demultiplexes the voice conversations by shifting the frequency spectrum of each conversation downward in frequency by the same amount of frequency in which it was previously shifted upward.

As previously mentioned, the primary use of FDM equipment by communications carriers was to enable those carriers to carry a large number of simultaneous voice conversations on a common circuit routed between two carrier offices. The actual process for allocating the bands of frequencies to each voice conversation has been standardized by the ITU. ITU FDM recommendations govern the channel assignments of voice multiplexed conversations based upon the use of 12, 60 and 300 derived voice channels.

ITU FDM recommendations

The standard group as defined by ITU recommendation G.232 occupies the frequency band from 60 to 108 kHz. This group can be considered as the first level of frequency division multiplexing and contains 12 voice channels, with each channel occupying the 300 to 3400 Hz spectrum shifted in frequency.

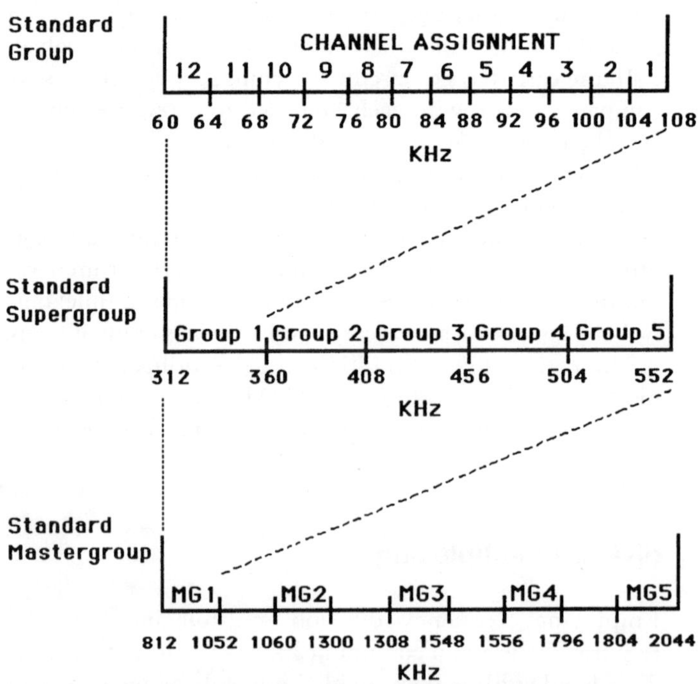

Figure 2.2 Standard ITU FDM groups. ITU FDM recommendations govern the assignment of 12, 60, and 300 voice channels on wideband analog circuits

The standard supergroup, as defined by ITU recommendation G.241 contains five standard groups, equivalent to 60 voice channels. The standard supergroup can be considered as the second level of frequency division multiplexing and occupies the frequency band from 312 to 552 kHz.

The third ITU FDM recommendation, known as the standard mastergroup, can be considered as the top of the FDM hierarchy. The standard mastergroup contains five supergroups. Since each supergroup consists of 60 voice channels, the mastergroup contains a total of 300 voice channels. The standard mastergroup occupies the frequency band from 812 to 2044 kHz. Figure 2.2 illustrates the three standard ITU FDM groups, as well as the relationship between groups.

By using one of the three types of groups illustrated in Figure 2.2, telephone companies were able to place up to 12, 60 or 300 simultaneous calls on one circuit routed between offices. Although originally used to carry voice conversations, this mechanism for sharing wideband circuits also supports the transmission of data. As FDM, however, is an analog system, the transmission of data required conversion into a modulated signal through the use of modems.

Time division multiplexing

In comparison to FDM in which a circuit is subdivided into derived channels by frequency, time division multiplexing (TDM) results in the use of a circuit being shared by time. Since TDM operates upon digital data, its utilization by communications carriers required voice conversations to be digitized prior to being transported on digital circuits routed between telephone company offices.

Voice digitization was accomplished using a technique known as pulse code modulation (PCM) in which an analog voice conversation is encoded into a 64 kbps digital data stream for transmission on telephone company digital transmission facilities. Once a call was digitized and routed to a distant telephone company office serving the destination or called party, the 64 kbps digital data stream was converted back into an analog voice signal and passed to the called party. Similarly, the voice conversation generated by the called party flowed in an analog form via the local loop to the telephone company office serving the subscriber. At that location, the conversation was digitized and was multiplexed using TDM equipment and placed onto a digital trunk linking that office to the office serving the call originator. At that office the call was removed from the trunk by the process known as demultiplexing, converted back into its analog format, and passed to the local subscriber.

The key to the successful use of time division multiplexing by telephone companies was the rapid growth in the use of T-carrier transmission facilities.

T-carrier evolution

T-carrier facilities were originally developed by telephone companies as a mechanism to relieve heavy loading on interexchange circuits. First employed in the 1960s for intra-carrier communications, T-carrier facilities only became available to the general public within the last 20 years as a commercial offering.

The first T-carrier was placed into service by American Telephone & Telegraph in 1962 to ease cable congestion problems in urban areas. Known as T1 in North America, this wideband digital carrier facility operates at a 1.544 Mbps signaling rate.

The term T1 was originally defined by AT&T and referred to 24 64 kbps PCM voice channels carried in a 1.544 Mbps wideband signal. Under the T1 framing format, each group of 24 8-bit bytes representing 24 voice samples has a framing bit added for synchronization. Since PCM sampling occurs 8000 times per second, this resulted in the use of 8000 frame bits. Thus, the T1 operating rate can be expressed as 24 channels $\times$ 64 kbps/channels + 8000 frame bits/second, resulting in an operating rate of 1.544 Mbps.

When AT&T initiated use of its T1 carrier, the company employed digital channel banks which were used to interface the analog telephone network to the T1 digital transmission facility.

Channel banks

Channel banks used by telephone companies were originally analog devices. They were designed to provide the first step required in the handling of telephone calls that originated in one central office, but whose termination point was a different central office. The analog channel bank included frequency division multiplexing equipment, permitting it to multiplex, by frequency, a group of voice channels routed to a common intermediate or final destination over a common circuit. This method of multiplexing was previously illustrated in Figure 2.2.

The development of pulse code modulation resulted in analog channel banks becoming unsuitable for use with digitized voice. AT&T then developed the D-type channel bank which actually performs several functions in addition to the time division multiplexing of digital data.

The first digital channel bank, known as D1, contained three key elements as illustrated in Figure 2.3. The codec, an abbreviation for coder-decoder, converted analog voice into a 64 kbps PCM encoded digital data stream. The TDM multiplexes 24 PCM encoded voice channels and inserts framing information to permit the TDM in a distant channel bank to be able to synchronize itself to the resulting multiplexed data stream that is transmitted on the T1 span line. The line

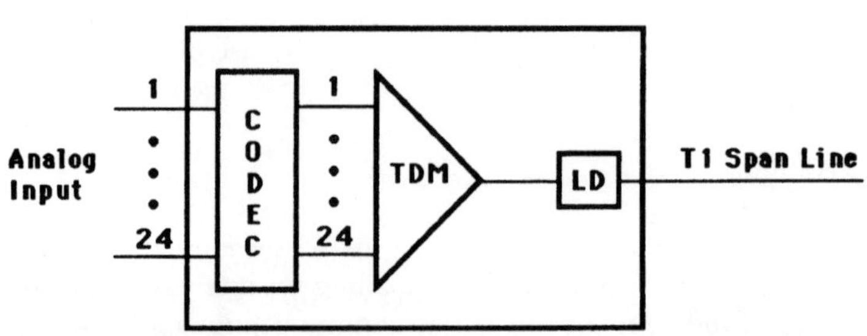

Figure 2.3 The D1 channel bank (TDM time division multiplexer, LD line driver)

driver conditions the transmitted bit stream to the electrical characteristics of the T1 span line, ensuring that the pulse width, pulse height and pulse voltages are correct. In addition, the line driver converts the unipolar digital signal transmitted by the multiplexer into a bipolar signal suitable for transmission on the T1 span line.

Due to the operation of the digital channel bank, this equipment can be viewed as a bridge from the analog world to the digital word. To ensure the quality of the resulting multiplexed digital signal, AT&T installed repeaters at intervals of 6000 feet on span lines constructed between central offices. Although repeaters are still required on local loops to a subscriber's premises and on copper wire span lines, the introduction of digital radio and fiber optic transmission has added significant flexibility to the construction and routing of T-carrier facilities, since repeaters are not required on those facilities.

Improvements made to the encoding method used by the D1 channel bank resulted in the development and installation of the D1D and D2 channel banks. Channel banks currently used include D3, D4 and D5, whose simultaneous voice channel carrying capacity varies from 24 to 96 channels.

In addition to serving as the basic networking building block for the PSTN, T1 lines are available for use by organizations from a variety of communications carriers, including AT&T, MCI, Sprint, and others. In Europe, the equivalent T1 carrier, which is known as E1 and CEPT PCM-30, is available in most countries under different names. As an example, in the United Kingdom E1 service is marketed under the name MegaStream.

T1 multiplexer

The key to the commercial use of T1 lines was the development and widespread use of T1 multiplexers. The T1 multiplexer can be viewed as a sophisticated channel bank which allows voice, data and video to be carried on a common path between organizational locations. Through the use of T1 multiplexers, organizations have been able to construct their own sophisticated circuit switching networks, with the T1 multiplexer used to switch channels from one high speed circuit to another, establishing a route through an organization's network. The operation and utilization of T1 multiplexers are described in detail in Chapter 6.

In addition to the use of high speed lines by both communications carriers and commercial organizations, many organizations constructed circuit switched data networks. Most of these networks are designed around the use of circuit switching time division multiplexers or a device known as a port selector or data PBX. Such networks operate in a manner similar to circuit switched networks that support both voice and data transmission, with the primary difference between the two typically being the use of lower capacity circuits by circuit switched networks restricted to data transmission support.

Figure 2.4 illustrates one of the many types of circuit switched network configurations you can develop using circuit switching multiplexers. In this example, it was assumed that a company has three distributed computer systems. At each location, a number of personal computer and terminal users require access to each

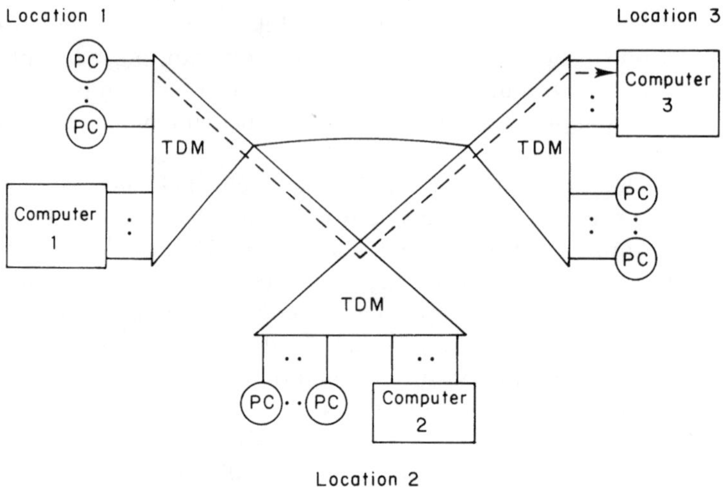

Figure 2.4 Circuit switching network using multiplexers (PC personal computer, TDM time division multiplexer)

computer as well as to the other computers in the company. As indicated by the dashed lines showing the establishment of a path from multiplexer 1 via multiplexer 2 to multiplexer 3, each of the TDMs works together to allocate channels as well as to switch channels from one circuit onto another circuit. In the example shown in Figure 2.4, a path was established which connects a personal computer at location 1 via location 2 to the computer at location 3.

Circuit switching characteristics

Once a circuit switched path is established, the resources used to establish that path cannot be used to satisfy another requirement. This is true even if transmission occurs very infrequently on the path and is a major limitation of circuit switching. However, once a physical path is set up, the transmission of data experiences a minimal delay as circuit switching is protocol independent. This means that any switches in the network used to establish the initial path do not sample nor act upon the data. Thus, the switches are transparent to the data which enables its flow through the network relatively rapidly. This is extremely important for voice as conversations are very sensitive to any delays. Data can, however, normally tolerate much greater delays than voice, which resulted in the development of packet switching networks.

2.3 LEASED LINE BASED NETWORKS

When an organization constructs a circuit switching network, it uses leased lines obtained from one or more communications carriers to construct the infrastructure that connects equipment acquired to perform circuit switching operations.

Although this represents one use of leased lines, in actuality this type of communications facility can be used to construct a number of other types of networks that can operate over a leased line network. For example, through the use of packet assemblers/disassemblers (PADs) instead of the multiplexers shown in Figure 2.4, you could construct a private packet switching network to directly link organizational locations instead of relying upon the use of a public packet switching network. Similarly, the use of Frame Relay Access Devices (FRADs) would enable your organization to construct a private Frame Relay network. Thus, the use of leased lines provides a considerable degree of flexibility that can range from directly connecting two locations that require a considerable amount of communications to developing a complex network structure for supporting a specific type of private network operating over leased lines.

Types of leased lines

As previously noted in Chapter 1, you can consider the use of several types of analog and digital leased lines. The primary types of analog leased lines include the voice grade 3000 Hz bandwidth circuit over which you can transmit data at up to 33.6 kbps, and several types of analog wideband circuits that operate at and above 40.8 kbps. Due to the near-complete digitization of the backbone communications carrier network in the United States, it is now often easier and sometimes less expensive to obtain an equivalent digital circuit.

Digital leased lines commonly available can be obtained that operate from the 2.4 kbps rate of the lowest speed of Dataphone Digital Service through the T3 carrier that represents a grouping of 28 T1 lines and provides an operating rate of approximately 45 Mbps. In between, you can select from a mixture of fractional T1 (FT1) and fractional T3 (FT3) offerings that enable you to match your organization's data transmission requirements to a specific leased line operating rate. For large organizations, some communications carriers now support Synchronous Optical Network (SONET) connections, permitting operating rates from approximately 52 Mbps to 2.4 Gbps.

Utilization examples

In addition to using leased lines to create an infrastructure for running a private packet switching or Frame Relay network, leased lines are commonly used on a point-to-point basis to connect geographically separated locations. Two of the more common uses of leased lines are for use with multiplexers and routers. Through the use of multiplexers a group of terminal devices, to include personal computers, can obtain access to a common location, such as a mainframe located at a company headquarters or regional office. Through the use of a router, two or more geographically separated LANs can be interconnected, providing workstations connected to each network with the ability to communicate with one or more devices connected to the distant network. In this section we will obtain an overview of the use of time division multiplexers and routers, while detailed information concerning their operation and utilization will be covered in subsequent chapters in this book.

Multiplexer utilization

One common use of leased lines for creating a network infrastructure is based upon different types of time division and statistical multiplexers whose operating characteristics are detailed in Chapter 6.

Suppose your organization has a group of terminal devices located throughout a geographical area, such as within different buildings in a city and its suburban area, while your mainframe computer is located in a different city. By installing a group of business lines connected to a telephone company rotary and installing automatic answering modems on each line, you can provide dial access to a multiplexer as illustrated in Figure 2.5. Then, the leased line would provide the connection facility to link the remote location to the location where the mainframe resides. In this example, the rotary serves as an automatic line searching facility. That is, if the first line in the rotary group is in use, the rotary automatically transfers the call to the first available line in the group.

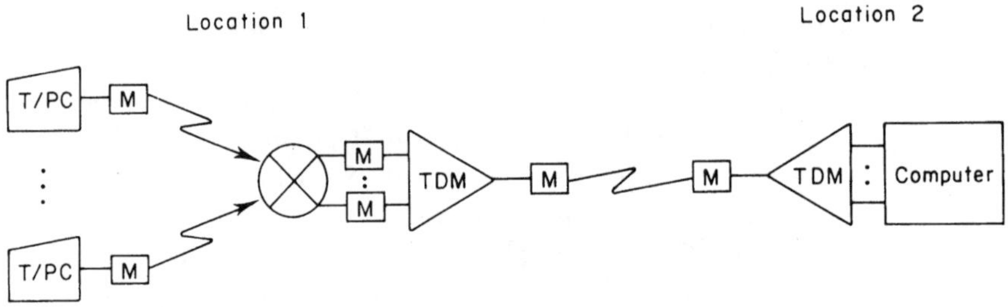

Figure 2.5 Multiple access to a common location via multiplexing (M modem, TDM time division multiplexer, T/PC terminal or personal computer)

Each terminal or personal computer at location 1 can access the computer at location 2 via a local telephone call to the rotary connected to the TDM. Since the communications facility interconnects two distant geographically separated locations, the network configuration illustrated in Figure 2.5 can be considered as a wide area network.

Now let us assume that your organization has expanded its operations to a third location. At that location your organization plans to have a number of order entry clerks that will require continuous access to a database residing on the mainframe computer. In addition, several sales personnel will be travelling to different customer locations within the suburban area and would like to access either a local minicomputer to be installed at the new location or the corporate computer located at location 2. After an analysis of this new requirement and expected transmission activity and applicable tariffs, the network designer might propose an equipment configuration for location 3 similar to that illustrated in Figure 2.6. In this example

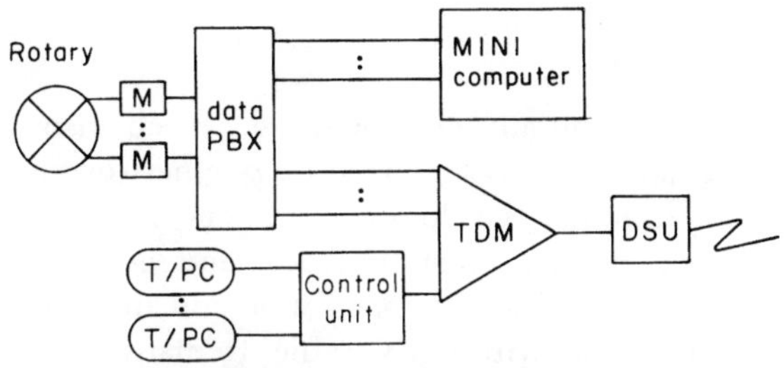

Figure 2.6 Switching at a remote location (M modem, T/PC terminal or personal computer, TDM time division multiplexer, DSU data service unit). Through the use of a data PBX or post selector a common set of communications facilities can be shared. In this example the rotary and dial-in lines can provide common access to a minicomputer or to a TDM

the data PBX, which is also known as a port selector and whose operating characteristics are described in Chapter 6, allows travelling sales personnel to dial one common rotary number. Upon access to the data PBX they obtain the ability to route their connection directly into the minicomputer at location 3 or into a port on a TDM which will multiplex data traffic from that location to the mainframe computer. To provide continuous access to the mainframe for data entry clerks, the network designer determined that a control unit whose operation is described in Chapter 6 should be installed at location 3. Through the use of a control unit at that location terminals or personal computers can be directly cabled to that device which allows a clustered group of terminal devices to share access to the mainframe on a poll and select basis. Since both the control unit and a portion of the lines from the data PBX must be routed to the mainframe, a TDM was installed to share the use of a high-speed transmission facility from the new location to the location where the mainframe is installed.

In the network configuration illustrated in Figure 2.6, it is assumed that because of the transmission requirements of the control unit as well as the maximum number of interactive dial-in users that could be routed to the TDM, the TDM would require an operating rate beyond that obtainable with an analog transmission facility. Thus, a DSU is illustrated in this example which enables transmission at 56 kbps over a DDS circuit.

Although the two prior examples represent a small selection of the many ways in which leased lines interconnecting different types of transmission devices can be used to construct wide area networks, they illustrate the basic concept of connecting devices and transmission facilities to develop networks, which is the rationale for this book. Of course, the apparent simplicity of linking equipment and facilities in actuality is a time consuming and tedious task which requires a detailed understanding of the operational characteristics of different types of communications equipment and their networking constraints and limitations. Thus, subsequent chapters in this book are devoted to providing a detailed understanding of how different types of communications products operate, why they can be used to satisfy the communications requirements of organizations, and when they should be considered.

Router utilization

Routers were initially developed as a mechanism to interconnect local area networks, enabling data originated on one network to flow onto a different network. Since their initial development routers have become much more sophisticated networking devices, enabling local and geographically separated LANs to be connected, as well as for mainframe and minicomputer network protocols to be transported along with LAN traffic. In many organizations the router has replaced the use of multiplexers, especially when mainframe-based networks were replaced by LAN-based client-server operations. In other organizations routers supplement the use of multiplexers, being either used on a stand-alone basis to interconnect geographically separated LANs, or integrated into a multiplexer-based network to provide inter-LAN communications via an existing network infrastructure. Regardless of the manner in which they are used, they are primarily used with leased lines to form a router-based network, or are integrated with multiplexers that primarily use leased lines to create a multiplexing-based infrastructure.

Figure 2.7 illustrates the use of a pair of routers to interconnect two geographically separated LANs. In this example each router has one serial and one LAN port. The LAN port enables the router to become a participant on the LAN while the serial port enables router-to-router communications to occur over leased lines. The actual routing of packets from one network to another is based upon the use of network addresses. In Chapter 3 we will examine the functions of routers with respect to their use for internetworking LANs and WANs, while in Chapter 6 we will focus our attention upon LAN concentration and transmission devices to include a detailed description of router operations and routing protocols.

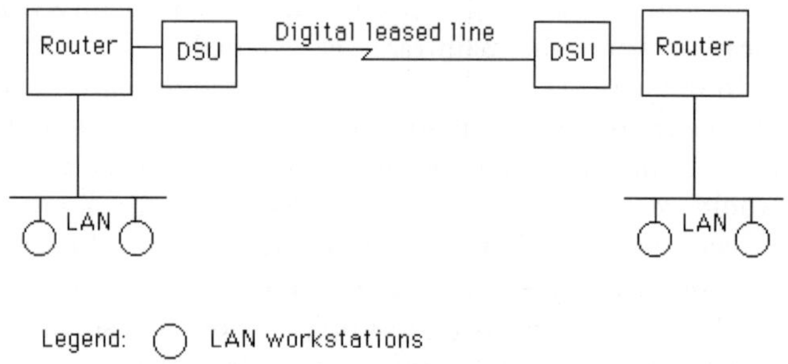

Figure 2.7 Using routers to interconnect LANs via a WAN. Routers transfer information between LANs via the use of network addresses assigned to LANs

As previously mentioned, routers use their serial ports to communicate with one another via wide area network transmission facilities. This means that the serial port represents just one more device to many high speed multiplexers, enabling a degree of LAN to LAN communications to be integrated by, for example, T1 multiplexers supporting inter-company PBX calls via the use of a pair of multiplexers. Figure 2.8 illustrates the use of a pair of T1 multiplexers to support LAN-

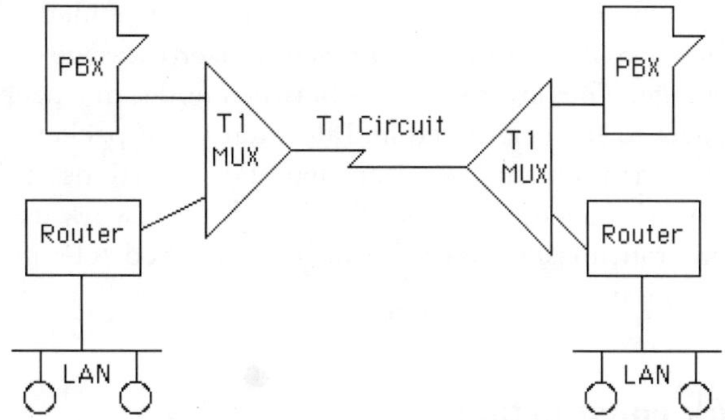

Figure 2.8 Using T1 multiplexers to transmit digitized voice and data over a T1 circuit. Although voice and data can flow over a T1 circuit, the multiplexers place digitized voice and data packets into specific time slots so they remain separated by time

to-LAN communications by means of routers interfaced to the multiplexers, and PBX to PBX communications via PBXs interfaced to the multiplexers. Although both voice and data will flow over the T1 circuit shown in Figure 2.8, the actual integration occurs by the multiplexing of time slots so that voice and data actually remain separated by time. This integration of voice and data should not be confused with the use of ATM which represents a network designed to provide an integrated transmission facility for voice, data, video, and images. ATM is described in detail in the last section in this chapter.

2.4 PACKET SWITCHING NETWORKS

One of the major limitations associated with circuit switching is the permanent assignment of a path for the exclusive use of a communications session during the duration of that session. This means that, regardless of whether a full circuit or a channel within a circuit is used to establish a path, that circuit or channel cannot be used to support other activities until the session in progress is completed. Recognizing this limitation, packet switching was developed as a technique to enable the sharing of transmission facilities among many users. Although multiplexing is also a mechanism which allows the sharing of transmission facilities, the two techniques are considerably different.

Multiplexing as opposed to packet switching

In multiplexing, bandwidth (FDM) or time (TDM) is used to establish occupancy slots into which individual data sources are assigned the use of those slots. Thereafter, information in the form of voice or data uses the reserved slot for the duration of the voice call or data transmission session. In packet switching, specialized equipment divides data into defined segments that have addressing, sequencing and error control information added. The resulting unit of data is called a packet and may represent a user message or a very small portion of a user message. The flow of packets between nodes in a packet network is intermixed with respect to the originated and desintation of packets. That is, traffic in the form of packets from many users can share large portions of the transmission facilities used to form a packet network. Thus, packet network use is normally more economical than transmission over the public switched telephone network for long distance transmission.

Packet network construction

A packet network constructed through the use of equipment that assembles and disassembles packets, equipment that routes packets, and transmission facilities used to route packets from the originator to the destination device. Some types of DTEs can create their own packets, while other types of DTEs require the conversation of their protocol into packets through the use of a packet assembler/disassembler (PAD) described later in this section and in much more detail in Chapter 6. Equipment that routes packets through the network are called packet switches. Packet switches examine the destination of packets as they flow through the network and transfer the packets onto trunks interconnecting switches based upon the packet destination and network activity.

The transmission facilities used to interconnect packet switches are leased lines since the network must be available 24 hours per day. Although a few packet networks continue to use analog leased lines to interconnect their switches, most packet networks now use high speed digital transmission facilities, such as 56 kbps DDS, fractional T1 and T1 lines, as well as T3 and fractional T3 transmission facilities, due to the growth in the use of packet networks.

ITU packet network recommendations

Most packet networks are illustrated by a drawing that resembles a cloud. A number of ITU recommendations govern the transmission of data both onto a packet network and between packet networks, as illustrated in Figure 2.9. In this illustration, two separate packet networks are shown, with a mainframe computer connected to each network.

ITU Recommendation X.25 controls the access from a packet mode DTE, such as a terminal device or computer system capable of forming packets and the DCE at a packet node. ITU Recommendation X.28 controls the interface between non-packet mode devices that cannot form packets and a PAD. ITU Recommendation X.29 specifies the interface between the PAD and the host computer. ITU Recommendation X.3 specifies the parameter settings on the PAD and X.75 specifies the interface between packet networks. Note that in Figure 2.9 there are two methods by which non-packet mode devices can be connected to an X.25 packet network. In packet network 1, the non-packet mode DTE communicates with a PAD located at the packet node, with terminal devices normally dialing the

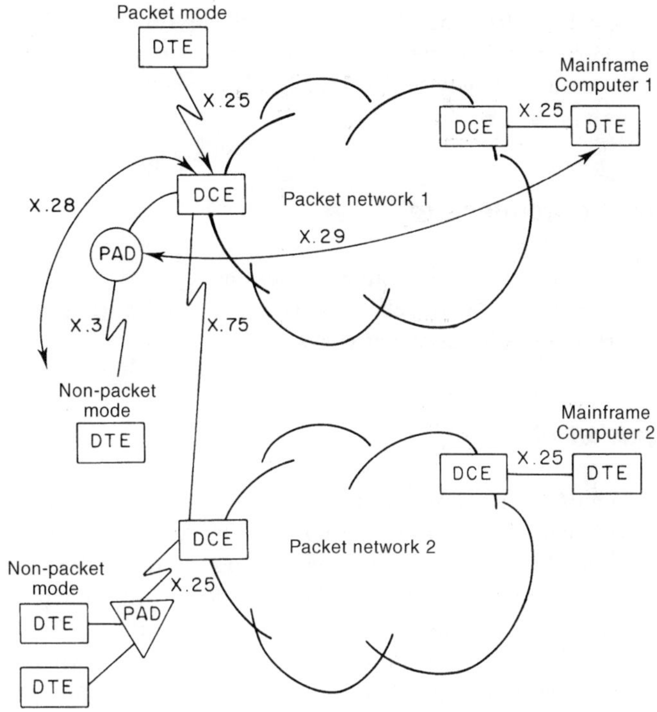

Figure 2.9 ITU packet network recommendations

number of a modem connected to a port on the PAD. In packet network 2, the PAD is located with the non-packet mode devices. This allows a permanent connection from the PAD to the packet node as well as provides the capability to service multiple terminals on one line routed to the packet node.

The PDN and value-added networks

The term 'public data network' (PDN) is used to reference a commercially available packet switching network. In the United States, British Telecom's BT Tymnet and Sprint's SprintNet (formerly known as Telenet) are examples of PDNs. Outside the United States, PDNs are usually administered by government agencies, although divestiture has occurred overseas similar to the breakup of AT&T in the United States that has resulted in British Telecom's PSS and other PDNs now belonging to private companies. In addition to public PDNs organizations can create their own private PDN. This is accomplished by purchasing appropriate packet switching equipment and leasing transmission facilities which results in the establishment of a packet switching network for the exclusive use of an organization.

Another term commonly used to reference packet switching network is 'value–added'. Networks that provide such operational facilities as reverse charging (collect calls), alternate destination routing, and closed user groups are known as value-added networks due to the extra value they provide. Since almost all packet switching networks now support one or more optional facilities, the term 'value-added network' is essentially synonymous with the term 'packet switching network'.

Packet network architecture

There are two major categories of packet networks, with each category based upon the method by which packets are routed through the network: datagram and virtual circuit based networks.

Datagram packet networks

In a datagram network, each packet is transmitted independently of other packets, with packet switches routing each packet based upon such factors as the traffic currently carried on circuits linking network switches, the error rate on those circuits, as well as whether or not specific circuits are available for use.

Figure 2.10 illustrates an example of the packet flow through a datagram packet network. In this example it was assumed that two devices creating packets have their data flow routed through a packet network consisting of four packet switches, with two computers connected to the network. Let us further assume that the packets labeled A, B, C and D are routed to computer A, while the packets labeled W, X, Y and Z are routed to computer B.

As packets are routed through a datagram packet network, each switch examines the packet destination address (computer A or B) and routes the packet based upon criteria similar to that previously discussed, which results in the selection of an optimum route at a given time. Since packets can traverse different paths, packets can be received at a destination switch out of sequence. Thus, destination switches must be capable of having sufficient memory to store packets until they can be sequenced into their appropriate order prior to their delivery to their ultimate destination. This resequencing would occur at packet switches 3 and 4 in Figure 2.10, when data flows towards the computers with the result that, although packets

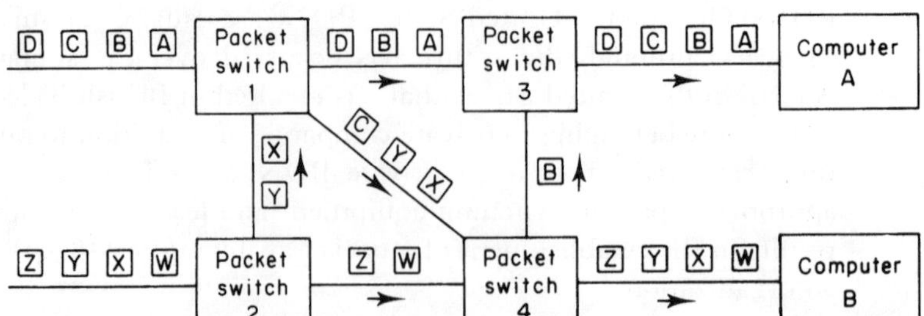

Figure 2.10 Datagram packet network data flow. In a datagram packet network packets are routed based upon the activity and availability of circuits connecting packet switches

from each source routed to a common destination took different paths, they were reassembled into their original order at their destination nodes.

Although datagram packet networks permit transmission paths to be dynamically altered to correspond to network conditions and activity, they require a substantial amount of packet switch processing overhead. This is because each switch must know the state of circuits to other switches to implement appropriate routing algorithms. While most packet switching networks were originally based upon datagram technology, a majority of those networks were converted to the use of virtual circuit packet switching.

Virtual circuit packet networks

In a virtual circuit packet network, a fixed path is established from the data originator to the recipient at the time a call is established. Thereafter, all packets flow over the same path, although the packets from two or more devices can share the use of circuits between packet switches that form a specific virtual circuit.

Figure 2.11 illustrates the data flow through a virtual circuit packet network. Note that although a route will be established based upon network activity, once established the route remains fixed for the duration of the call. Thus, packets will flow in sequence through each switch which reduces both the amount of processing required to be performed at each switch and delays associated with waiting for out of sequence packets to arrive at a destination node prior to being able to pass an ordered sequence of packets to their destination.

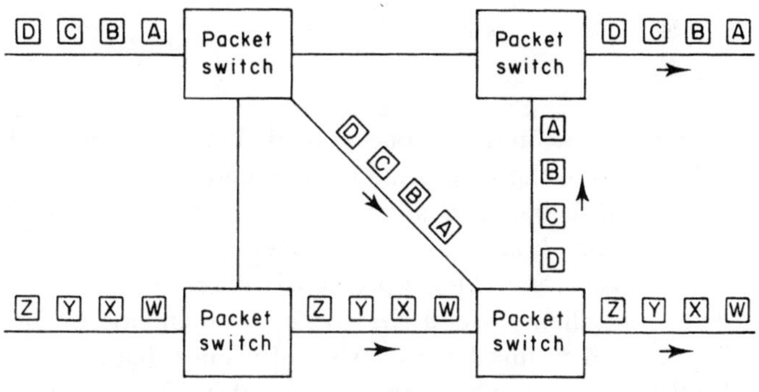

Figure 2.11 Virtual circuit packet network data flow. In a virtual circuit packet network packets are routed in sequence over a fixed path established at the time a call occurs

Packet formation

Although many devices have the ability to directly create packets, most devices rely upon the use of packet assembler/disassemblers (PADs). PADs are essentially protocol conversion devices that accept data from data terminal equipment (DTE) supporting one protocol and converting the data into packets based upon the ITU X.25 protocol.

The most common type of PAD converts an asynchronous teletype protocol into an X.25 data flow. Other common protocols supported by PADs include the IBM 2780 and 3780 bisynchronous protocols, IBM's SDLC protocol, and various protocols from other computer manufacturers that resemble those IBM protocols.

PADs are manufactured on adapter boards designed for insertion into the system unit of a personal computer as well as standalone devices. Concerning the latter, some PADs perform a one-to-one conversion and as such are single input port devices. Other PADs may contain many ports and support different protocols on each port. Figure 2.12 illustrates a multi-protocol, multiport PAD. The reader is referred to Chapter 6 for specific information concerning the operation and utilization of PADs.

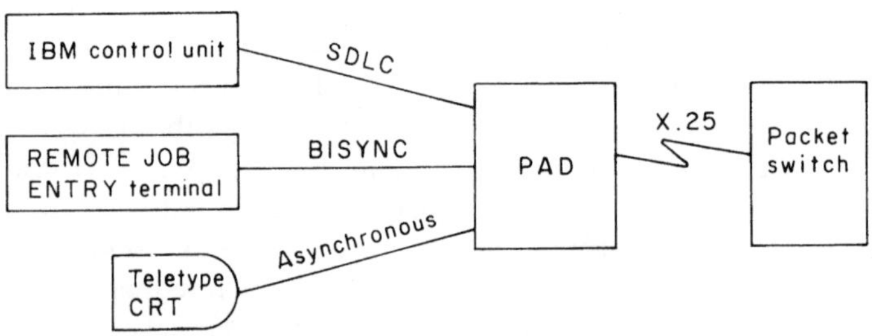

Figure 2.12 Multi-protocol multi-port PAD

X.25

The transmission from a packet mode DTE or from a PAD to a packet switch node is based upon the ITU X.25 recommendation. The X.25 recommendation defines the interface between a terminal device and a packet network, including the physical, data link, and network layers which correspond to the first three layer of the OSI Reference Model previously described in Chapter 1. The physical layer is defined by the X.21 and X.21 bis recommendations, the latter essentially RS-232. The X.25 data link layer consists of frames that are transported via the use of the HDLC protocol. The network layer of X.25 consists of packets which are carried within the information field of frames transported by the HDLC protocol. Figure 2.13 illustrates the structure of an X.25 frame.

Layer 3 which is the packet level defines the procedures for establishing and clearing calls between users, describes packet formats, and defines the procedures for such functions as flow control, packet transfer and error control.

Packet format and content

Every packet transmitted on an X.25 packet network contains a three-octet header which is illustrated in Figure 2.14.

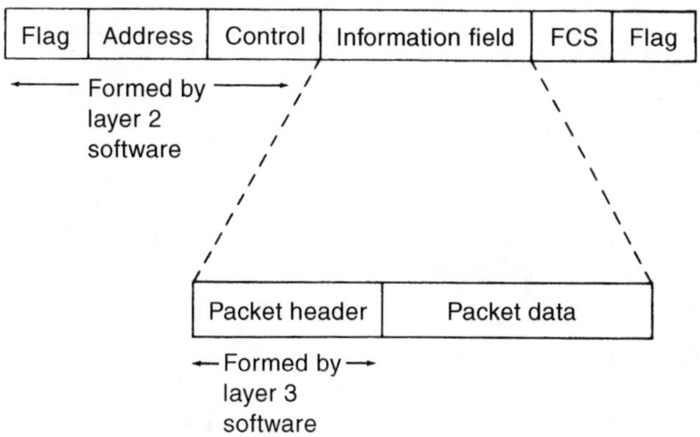

Figure 2.13 X.25 frame structure

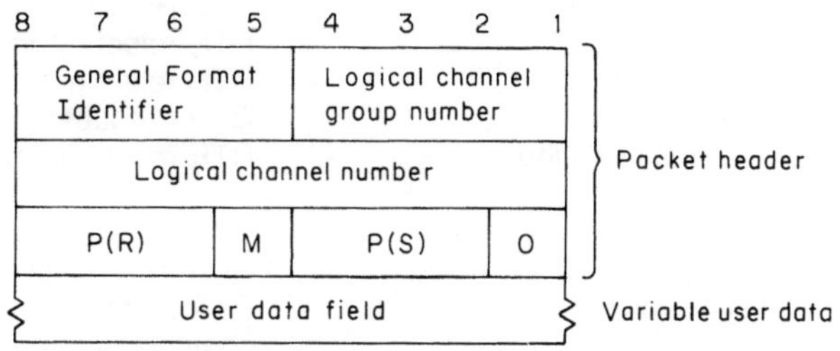

Figure 2.14 X.25 data packet format

X.25 packet switching networks use both virtual circuits and logical channels to allow many users to share the facilities of the network. As previously explained, a virtual circuit is a path established through the network for the duration of the call, similar to a direct physical circuit. The logical channel is a method which enables the multiplexing of a single subscriber access line to the packet network among several simultaneous virtual calls. For example, a PAD serving several terminal devices and connected to a packet network via a single access line would assign each terminal connected to the PAD to a different logical channel number. Thus, the logical channel number (LCN) identifies each packet as belonging to a particular logical connection.

The LCN is significant only at the interface between the subscriber and the local packet node. At the packet node, the LCNs are mapped to network paths and an independent set of LCNs are used between the destination packet node and the addressed device.

The third byte in the packet header is the packet type identifier. Currently, 20 packet types are defined in X.25. divided into six functional groups as indicated in Table 2.1.

Table 2.1 X.25 packet type identifer

Packet type third octet bit configuration From DCE to DTE	From DTE to DCE	8 7 6 5 4 3 2 1
Call setup and clearing		
Incoming call	Call request	0 0 0 0 1 0 1 1
Call connected	Call accepted	0 0 0 0 1 1 1 1
Clear indication	Clear request	0 0 0 1 0 0 1 1
DCE clear confirmation	DTE clear confirmation	0 0 0 1 0 1 1 1
Data and interrupt		
DCE data	DTE data	X X X X X X X X
DCE interrupt	DTE interrupt	0 0 1 0 0 0 1 0
DCE interrupt confirmation	DTE interrupt confirmation	0 0 1 0 0 1 1 1
Flow control and reset		
DCE RR (Mod 8)	DTE RR (Mod 8)	X X X 0 0 0 0 1
DCE RR (Mod 128)	DTE RR (Mod 128)	0 0 0 0 0 0 0 1
DCE RNR (Mod 8)	DTE RNR (Mod 8)	X X X 0 0 1 0 1
DCE RNR (Mod 128)	DTE RNR (Mod 128)	0 0 0 0 0 1 0 1
	DTE reject (Mod 8)	X X X 0 1 0 0 1
	DTE reject (Mod 128)	0 0 0 0 1 0 0 1
Reset indication	Reset request	0 0 0 1 1 0 1 1
DCE reset confirmation	DTE reset confirmation	0 0 0 1 1 1 1 1
Restart		
Restart indication	Restart request	1 1 1 1 1 0 1 1
DCE restart confirmation	DTE restart confirmation	1 1 1 1 1 1 1 1
Diagnostic		
Diagnostic		1 1 1 1 0 0 0 1
Registration		
Registration request		1 1 1 1 0 0 1 1
Registration confirmation		1 1 1 1 0 1 1 1

Call establishment

A virtual call through a packet network is initiated by the transfer of a call request packet to the network. The call request packet identifies the LCN selected by the call initiator as well as the address of the calling and called parties. This packet may also include any optional facilities desired as well as optionally contain call user data.

After the call request is routed through the packet network to the destination node, a local LCN is selected and the destination node then transmits an incoming call packet to the called party. If the called party is available to accept the call it responds to the incoming call packet by transmitting a call-accepted packet to the network using the LCN that identified the incoming call. The network then responds to the call-accepted packet by transmitting a call-connected packet to the calling party on the LCN assigned to the initial call request. This action results in the establishment of a virtual call which allows data packets to be exchanged between users.

Once a virtual call is established, the logical channel enters the data transfer phase. At this time, data packets consisting of the three-octet header field and the user data field carry the user information. The packet type identifier field (third octet) includes send (P(S)) and receiver (P(R)) sequence numbers that are used to track and acknowledge packets in a manner similar to the N(R) and N(S) field in the HDLC frame described in Chapter 1. The More-Data (M) bit when set to a 1 notifies the network that the next packet to be transmitted is a logical continuation of the data in the current packet. The actual length of the user data field depends upon the packet network. Most networks support a default maximum user data field of 128 octets. Other maxima supported by most networks include 32, 64, 256, 512, 1024, 2048 and 4096 octets.

Once a called DTE receives the first data packet, it can authorize additional transmission by returning a receiver ready (RR) packet or it can delay transmission by returning a receiver not ready (RNR) packet. Finally, when the calling DTE has completed its session, it transmits a Clear Request packet which is confirmed by the called DTE. At that time the virtual circuit is broken. Figure 2.15 illustrates the major packet flow used to establish a virtual connection, pass data, and break the virtual connection.

Flow control

Once data begins to flow through a packet network, delays can occur due to network congestion or transmission error. Either situation can result in the inability of the packet network to accept the current rate of information transfer through the network. Thus, a mechanism is required to regulate the flow of information. This mechanism is called flow control and is effected by the use of the receiver ready (RR) and receiver not ready (RNR) packets. The RR packet is used to indicate that the sending station is ready to receive packets. In addition, the RR packet carriers a receive sequence number (P(R)) which acknowledges packets previously received by the sending station. Conversely, RNR packets are transmitted by the packet network or receiving station to halt the flow of incoming data packets. RNR packets indicate that the sender should stop transmitting data as soon as possible.

Advantages of PDNs

The design of packet networks results in the allocation of capacity only when stations have packets to transmit. Almost all interactive transmission includes pauses, and because idle time on a circuit is essentially being wasted, packetizing and interleaving packets makes the most efficient use of a circuit. Thus, the more packets that can be carried on a circuit, the lower the cost per packet.

Technological advances

The growth in the use of fiber optic transmission facilities resulted in a significant lowering of the error rate experienced on long distance transmission. The advance

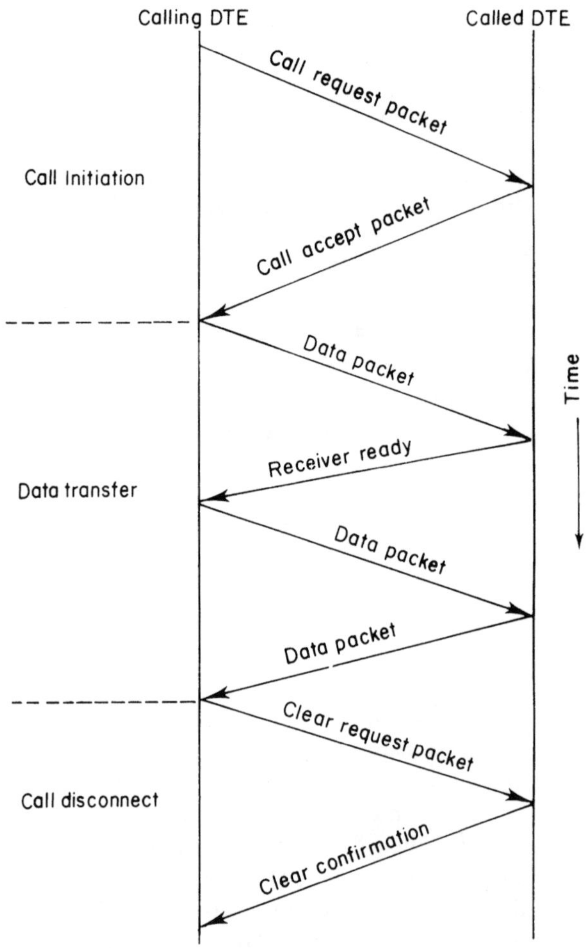

Figure 2.15 Packet network session activity

in communications transmission facilities in turn has called into question the necessity of using conventional packet switching methods that support layers 1 to 3 of the OSI reference model and resulted in the development of a series of new technologies called fast packet switching, frame relay, and cell relay with the latter providing the foundation for Asynchronous Transfer Mode (ATM). Prior to discussing these newer technologies, let us focus our attention upon the delays associated with conventional packet switching and the relevance of those delays in an era where the majority of trunks (leased lines) that interconnect packet nodes are now fiber optic cables.

Packet network delay problems

In a conventional packet switching network, each of the trunks linking the nodes in the network support layers 1 to 3 of the OSI model. This means that packets are error checked (a layer 2 function) at each node. Although this was a necessity to ensure data integrity when trunks were primarily analog leased lines with a relatively high error rate, the use of fiber optic cable for transmission between nodes reduced errors by several orders of magnitude. This means that most error checking within packet nodes only delays transmission through the network and is not actually beneficial for data integrity. In addition, the growth in intelligent terminal devices based upon microprocessors used in personal computers and workstations allows the integrity of data to be checked at each end of the link. Thus, error checking at each node can be considered as technically obsolete for many applications.

Flow control (layer 2 and layer 3) is greatly diminished when fiber optic cable is used, however, it is still applicable due to the processing delays encountered as packets are error checked at each node. Thus, the elimination of error checking at each node would substantially reduce the use of flow control in a packet network.

Fast packet switching

Recognizing the previously described packet switching problems, AT&T's Bell Laboratories began experimenting with a technique during the 1980s to reduce the processing overhead in each packet node which would allow a packet to be forwarded with lower delay. AT&T's technique was based upon the use of very large scale integration (VLSI) hardware, with the technique referred to as fast packet switching.

AT&T's fast packet switching eliminated error checking at each node so the node could begin forwarding the packet as soon as its destination was decoded. To prevent a link error which could cause a packet to be routed to an unintended destination, a CRC was placed after the routing information. If a CRC error is detected, the packet node would simply drop the packet. If a transmission error occurred that affected other data in the packet, the node would not detect the error but would simply route the packet rapidly towards its destination. In this technique, it is assumed that the recipient will ask for a retransmission of missing or errored packets.

Since all modern protocols have error detection and correction capability, as well as a timeout feature which results in a request for retransmission if a unit of data is not received within a predefined period of time, neither dropping packets nor carrying errored packets to their destination adversely affects modern equipment. In fact, by eliminating error detection processing at nodes, the flow of data through the network is substantially increased.

In general, the term 'fast packet' refers to the transmission of packets in which packet processing is streamlined to provide a much higher throughput than obtainable on an X.25-based network. From the pioneering effort of AT&T, two new technologies have emerged that were standardized in the early 1990s-frame relay and cell relay, the latter now more commonly referred to as Asynchronous Transfer Mode or ATM. Although considerable growth in the use of frame relay occurred during the mid- to late 1990s for linking LANs, which diminished the role of conventional packet switching, the latter remains a viable technology for many types of data transfer. Perhaps one of the most popular uses of X.25 networks is for the transfer of credit card verification information, since such information consists of a relatively small number of digits, and a delay of a few seconds to verify the transaction is acceptable in comparison to the time required to print a supermarket or gas station receipt. Since tens of millions of these transactions occur every day, the use of X.25 network sessions will probably continue to exceed the popular use of frame relay to interconnect LANs for remote queries of databases, file transfers and email exchanges between geographically separated LANs.

Frame relay

Frame relay is a data link protocol at the ISO layer 2 level which defines how frames of data are assembled and rapidly routed through a packet network. Frame relay is standardized by the ANSI T1S1 committee and by the ITU I.122 and Q.922 recommendations and represents the first packet mode interface to ISDN networks.

Comparison to X.25

Frame relay is very similar to X.25 packet switching; however, there are some distinct differences between the two transmission methods. Similar to X.25 packet switching, a frame relay network represents both a public transmission facility operated by communications carriers as well as a private network facility organizations can construct over a leased line infrastructure. Frame relay, like X.25, permits multiple communications sessions to share a single physical connection. To accomplish this, frame relay, like X.25 packet switching, constructs multiple virtual circuits across a common physical connection between a Frame Relay Access Device (FRAD) and a frame relay network entry node. Here the FRAD can be viewed as performing a function similar to an X.25 PAD; however, instead of translation from a protocol other than X.25 to X.25, the FRAD provides a translation into the frame relay transmission protocol. Although there are many similarities between X.25 packet switching and frame relay, there are also some distinct differences. Since those differences relate to the design goal of frame relay and its basic operation, let's turn our attention to those areas.

The design goal of frame relay assumes that the data will be transferred, in its correct order, error free. This design goal relies on the higher layers in the protocol stack to determine whether or not transmission errors occurred and, if so, to initiate a correction via retransmission. Another design goal of frame relay is to provide users with a guaranteed amount of transmission bandwidth referred to as a Committed Information Rate (CIR), while letting users obtain the ability to burst their transmission above the CIR to the maximum rate of the connection without guaranteeing that the excess above the CIR will arrive at its destination. This means that frame relay has the ability to discard frames, and the upper layers of the protocol stack become responsible for retransmission when frames are dropped.

Table 2.2 Comparing X.25 and frame relay

Performance/ operational feature	X.25	Frame relay
Performs packet sequencing	Yes	no
Performs error checking	yes	no
Performs flow control	yes	no, drops frames when congestion occurs
Network access	300 bps–64 kbps	56 kbps–2.048 Mbps, with T3 access being introduced
Switch delay	10–40 ms	2–6 ms
One-way delay	200–500 ms	40–150 ms

This also means that, by providing users with the ability to burst to high transmission rates, the physical access to a frame relay network can be much higher than access to an X.25 network.

Since a frame relay network does not have to perform packet sequencing nor error checking, the flow of data through a network is considerably faster than through an X.25 network. Similarly, a frame relay switch will have a lower delay or latency, and the ability to process more packets or frames per unit time. Table 2.2 provides a general comparison between X.25 and frame relay network performance and operation.

In examining the entries in Table 2.2, the difference in one-way delay between frame relay and X.25 networks resulted in a new application for frame relay that would be impossible to perform effectively on older X.25 networks. This application is the transmission of voice, which is very susceptible to network delays. Although we will discuss voice over frame relay later in this section, we will defer a detailed examination of how FRADs enable voice to be transported until we discuss the operation of this networking device later in this book.

Utilization

Frame relay represents a high speed internetworking technology well suited for connecting geographically separated LANs and mainframes on an any-to-any connectivity basis via the mesh structure of public frame relay providers. Figure 2.16 illustrates an example of how a public or private frame relay network could be used to interconnect three LANs and two mainframe computers. Note that a frame relay compliant router and front end processor requires only one connection to the network to obtain the ability to communicate with multiple destinations. To accomplish this, private virtual circuits (PVCs) must be established for each location that requires access to another location. For example, if the front-end processor connected to the S/390 mainframe requires the ability to communicate with each LAN, then three PVCs must be established.

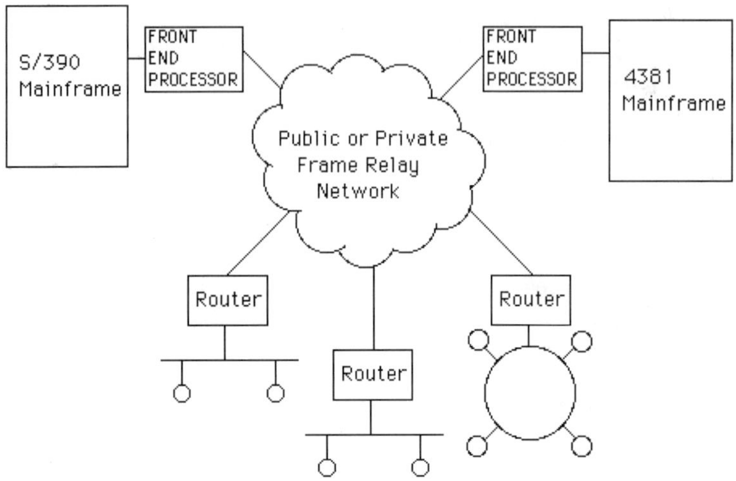

Figure 2.16 Frame relay transport application

In addition to reducing the port interface to one, frame relay can considerably reduce the cost of equipment required to interconnect geographically separated locations. This is because both router and front-end processor ports are relatively expensive. Thus, a few PVCs that a communications carrier may bill at $20 per month might eliminate a $3000 port upgrade.

Operation

Under frame relay, a 2,3 or 4 byte header is added to layer 2 information as illustrated in Figure 2.17. The data link connection identifier (DLCI) allows a packet network to route each frame through nodes along a virtual path established for a frame relay connection. The DLCI does not specify a destination address. Instead, it specifies a connection resulting from the establishment of a virtual circuit. Once the virtual circuit is cleared, the DLCI previously used can be reused on another virtual circuit established between two different subscribers.

At each node, the switch supporting frame relay monitors the utilization of its buffers. If a predefined threshold is reached which indicates that congestion could occur if the data flow is not adjusted, this situation is indicated to the end-user devices in both directions. In the forward direction, the Forward Explicit Congestion Notification (FECN) bit in the frame header is set. In the reverse direction, the switch sets the Backward Explicit Congestion Notification (BECN) bit in all frames transmitted in the opposite direction. When the devices at each end of the network receive a congestion notification, they are expected to reduce their information transfer rate until the congestion notification is cleared, in essence performing flow control. If the level of congestion should increase to a second threshold due to a slow response to FECN or BECN notifications, the frame handler at the packet node discards frames among all users in an equitable manner by the use of the discard eligibility (DE) bit. This bit when set by customer-provided equipment or the network identifies frames from users that can be

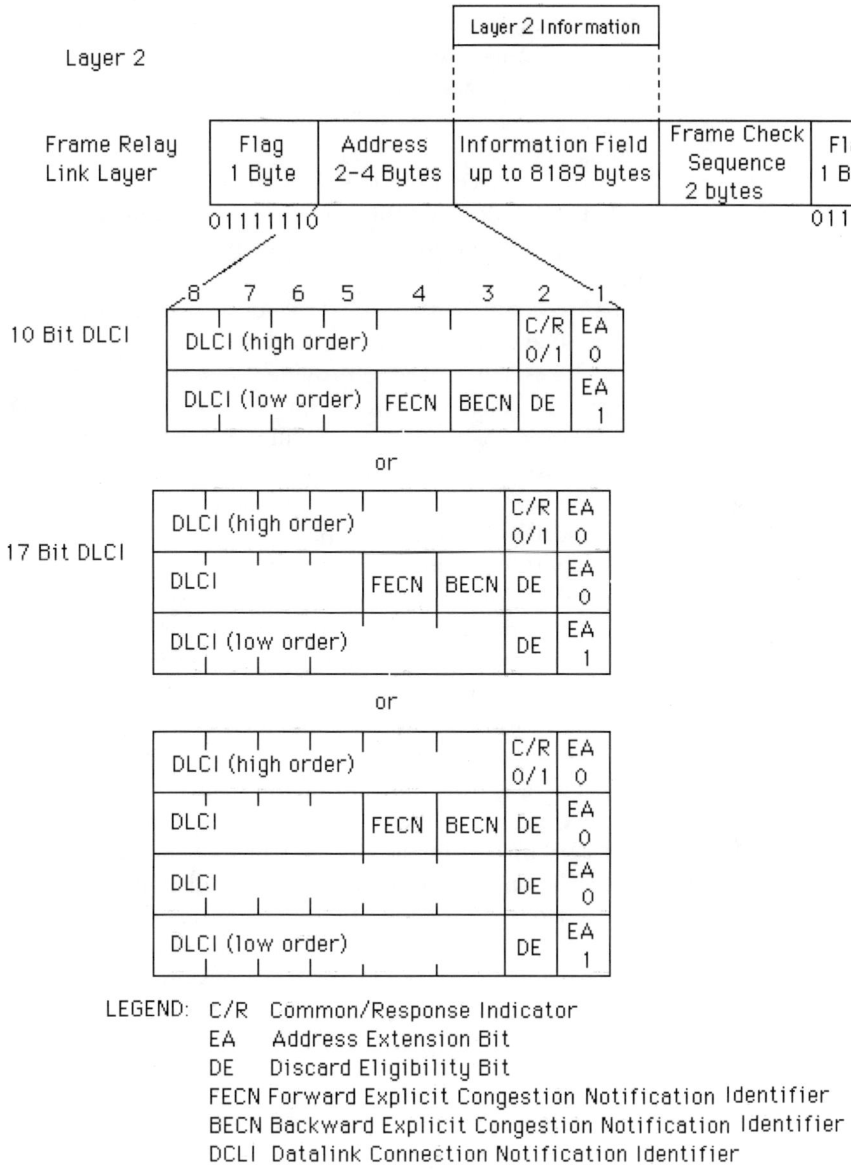

Figure 2.17 Frame relay construction and frame format

discarded if the network becomes overloaded. Thus, the DE bit enables user equipment to temporarily send more frames than it is allowed to send. The packet network will forward those frames if it has the capacity to do so; however, it will discard those frames first if the network becomes overloaded.

The CRC in the frame trailer is used to detect bit errors in the header and information fields. Unlike X.25 in which a node will delay transmission and request the originating node to retransmit the previously sent packet, a frame relaying network simply discards the errored frame. This places the burden of error recovery upon the end-user devices.

The CIR

Frame relay network service providers offer subscribers a Committed Information Rate (CIR). The CIR denotes the minimum data transfer rate a subscriber is guaranteed upon the establishment of a permanent virtual circuit routed through their network.

It is important to note that the CIR is not an instantaneous measurement of transmission but an average rate over time. For example, consider the use of a T1 line operating at 1.544 Mbps used to provide access to a frame relay service. If the CIR is 256 kbps, this does not mean that the user cannot transmit more than 256 kbps. What it means is that the user will not transmit more than 256 K bits in one second or 512 K bits in two seconds. Although the distinction may appear trivial, it is extremely important and deserves a degree of elaboration as CIRs can be assigned to each PVC and used to provide equitable access to the total bandwidth of a connection to a frame relay network.

To illustrate the use of CIRs, assume you are using a frame relay network to support inter-LAN and mainframe-to-mainframe communications. Let's further assume that you connect to a public frame relay network using a 512 kbps fractional T1 line and have the carrier provision each PVC so that the mainframe-to-mainframe communication on one PVC has a CIR of 128 kbps and LAN-to-LAN communication on another PVC has a CIR of 384 kbps. This method of provisioning assigns a greater portion of the bandwidth to LAN-to-LAN communications; however, since that type of communication is bursty in nature it could shrink to 128 kbps, 56 kbps, or even 0 bps when a mainframe-to-mainframe communications session is in progress. Thus, enabling another PVC to transmit at a rate above its CIR has a degree of merit, especially when the other PVCs are operating below their CIRs. Therefore, the mainframe-to-mainframe session can burst its transmission above its CIR when the LAN-to-LAN session falls below its CIR; however, excess transmission above the mainframe-to-mainframe CIR is not guaranteed.

Calculating the CIR

The CIR is computed based upon a measurement interval(T_c) and Committed Burst size (B_c). The measurement interval represents the period of time over which information transfer rates are computed and can vary from one frame relay service provider to another. The committed burst size (B_c) represents the maximum number of bits the network guarantees to deliver during the measurement interval (T_c). Thus, the CIR can be computed by dividing the committed burst size by the measurement interval, such that:

$$\text{CIR} = B_c/T_c$$

Carriers that use a small T_c value minimize the ability of users to burst transmission. In comparison, frame relay providers that use a relatively large value for T_c may be more suitable for users as they enable a higher degree of bandwidth aggregation over time by supporting longer bursts. Concerning transmission bursts, a third parameter known as the excess burst size (B_e) is used by frame relay service providers to specify the maximum number of bits above the CIR that the network will attempt to deliver during the measurement interval, but for which delivery is not guaranteed.

Figure 2.18 illustrates the relationship between T_c, B_c and B_e. Note that $B_c + B_e$ represents the maximum amount of information that can be transmitted during the time interval T_c. If a user transmits more than $B_c + B_e$ bits during the interval T_c, the network will immediately discard the excess frames. Many frame relay service providers set B_e as the difference between B_c and the interface access speed to the network. Thus, this setting then results in the avoidance of frame discards at the entry node to the network, and results in $(B_c + B_e)/T_c$ becoming equal to the access speed.

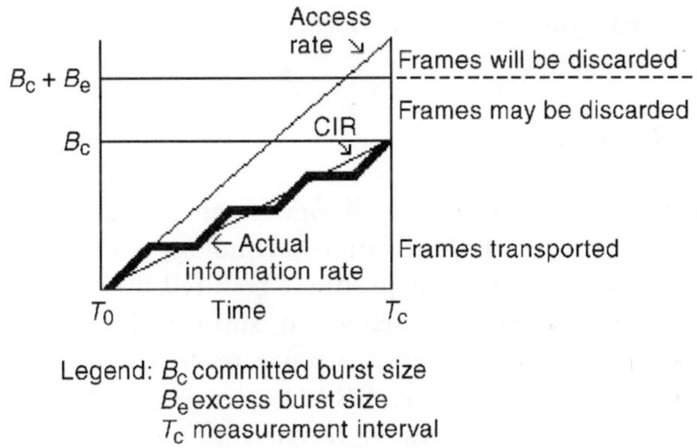

Figure 2.18 Committed Information Rate (CIR) parameters

In examining Figure 2.18, note that the actual information rate can very over measurement interval. As long as there is a sufficient amount of data to transmit, the actual information rate can never fall below the CIR and can burst above it to a maximum of $B_c + B_e$.

Cost

The cost associated with the use of a frame relay service can vary between access providers based upon the structure of their pricing for access port use, the CIR selected, and the number of PVCs. Most providers charge an increasing frame relay access port fee based upon the operating rate of the connection to the service provider. Similarly, the higher the CIR the higher the fee for providing a greater guarantee of available bandwidth. The last major frame relay provider fee is a surcharge for establishing PVCs. Typically the service provider will add a monthly charge for each PVC established through their network.

In addition to direct frame relay provider charges, you will also incur a local access charge that may be billed by the provider or a local communications carrier. The local access charge is a monthly charge for the circuit that connects your location to the nearest Point of Presence (POP) where the circuit becomes connected to the frame relay provider's network access port.

In spite of a considerable number of metrics upon which frame relay charges are based, its use may result in considerable savings in comparison to the cost associated with a private leased line based network. This potential cost saving results from the fact that similar to an X.25 network, a public frame relay network enables network resources to be shared among many users. This in turn provides greater transmission efficiencies and allows providers to set rates that can be 20–50% less than the cost associated with a private leased line network. Of course, if your organization has a substantial communications requirement between two locations and does not require the use of a mesh structure network topology, it is quite possible that a leased line will prove to be more economical. The only way to tell is to determine your transmission requirements and obtain price quotations for different transmission services.

Voice over frame relay

Although not originally designed as a digitized voice transport mechanism, advances in voice digitization technology and the incorporation of prioritization techniques into vendor products resulted in many organizations turning to frame relay as a transmission scheme to support their voice, data, and fax requirements. To do so usually requires a separate device capable of presenting a voice or fax digitized data stream to a FRAD since most FRADs are limited to supporting data protocols. This conversion device commonly attaches to a PBX to accept an analog voice or fax or a digitized voice signal and compresses and digitizes the signal, generating a data flow in a protocol the FRAD supports. Concerning voice digitization, PCM produces a 64 kbps digital data stream which is more susceptible to delays through a frame relay network when carried in a large number of time-dependent frames than a lower digitization rate carried in a lesser number of frames. Thus, many equipment developers now offer compression methods that can result in a voice conversation being carried by an 8 kbps to 16 kbps data stream.

In addition to stand-alone converters that provide an interface to FRADS, some vendors now market multifunctional FRADS that perform analog to digital conversion and compression. Included in many products is a silence detection process which further enhances voice compression and a frame sizing algorithm which limits the length of digitized voice, fax, and data frames delivered to the network. This frame limiting algorithm minimizes end-to-end delay through the frame relay network being used and ensures that voice packets which are time sensitive are not unacceptable delayed behind fax and data packets that can tolerate a relatively lengthy delay.

Figure 2.19 illustrates the use of a multifunctional FRAD to fragment and prioritize the flow of voice, data, and fax packets into a frame relay network. After a frame is created based upon a maximum length associated with the data it

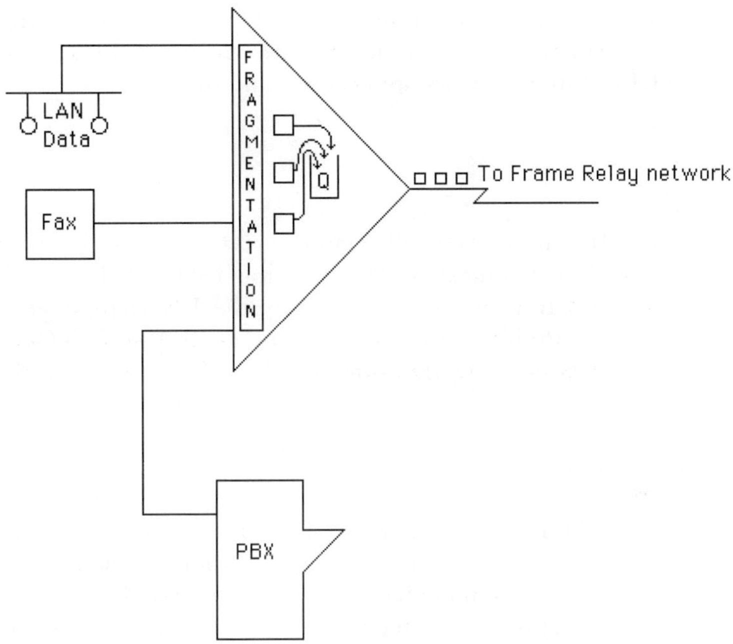

Figure 2.19 Using a multifunctional FRAD to support voice, data, and fax

transports, it is sent into a priority queue (labeled Q in Figure 2.19). Since voice has the least tolerance for delay, frames carrying digitized voice would be placed at the top of the queue while more delay-tolerant fax and data carrying frames would be placed in lower priority areas in the queue. In 1997 the Frame Relay Forum completed an Implementor's Agreement document for Voice Over Frame Relay. This document defines agreed digitization, prioritization, fragmentation, and other methods required for vendors to build equipment that will be able to inter-operate with one another, and should contribute to a further growth in the use of frame relay as a voice transport mechanism.

2.5 THE INTERNET

The Internet, with a capital 'I', is a term used to reference a network of interconnected networks. The evolution of the Internet is based upon research performed by the US Department of Defense Advanced Research Projects Agency (ARPA). During the 1960s ARPA constructed an experimental packet network which formed the basis for the network that is now globally known as the Internet.

In this section we will focus our attention upon the Internet. Since the Internet represents an interconnection of LANs and WANs operated by private and public organizations whose common thread is the use of the TCP/IP protocol suite, our primary focus in this section will be upon obtaining an appreciation of the function of protocols and applications supported by that protocol suite. Since the only protocol suite allowed to transport traffic on the Internet is TCP/IP, our examination of the Internet will be performed by focusing our attention upon TCP/IP to include its applications and protocols.

TCP/IP

TCP/IP represents a collection of network protocols that provide services at the Network and Transport layers of the ISO's OSI Reference Model. Originally developed from work performed by the US Department of Defense Advanced Research Projects Agency Network (ARPANet), TCP/IP is also commonly referred to as the DOD protocols or the Internet protocol suite.

Protocol development

In actuality, a reference to the TCP/IP protocol suite includes applications that use the TCP/IP protocol stack as a transport mechanism. Such applications range in scope from a remote terminal access program known as Telnet to a file transfer program appropriately referred to as ftp, as well as the Web browser transport mechanism referred to as the HyperText Transport Protocol. (HTTP).

The effort behind the development of the TCP/IP protocol suite has its roots in the establishment of ARPANet. The research performed by ARPANet resulted in the development of three specific protocols for the transmission of information—the Transmission Control Protocol (TCP), the Internet Protocol (IP), and the User Datagram Protocol (UDP). Both TCP and UDP represent transport layer protocols. TCP provides end-to-end reliable transmission, while UDP represents a connectionless layer-4 transport protocol. Thus, UDP operates on a best-effort basis and depends upon higher layers of the protocol stack for error detection and correction and other functions associated with end-to-end reliable transmission. TCP includes such functions as flow control, error control, and the exchange of status information, and is based upon a connection being established between source and destination prior to the exchange of information occurring. Thus, TCP provides an orderly and error-free mechanism for the exchange of information.

At the network layer, the IP protocol was developed as a mechanism to route messages between networks. To accomplish this task, IP was developed as a connectionless mode network layer protocol and includes the capability to segment or fragment and reassemble messages that must be routed between networks that support different packet sizes than the size supported by the source and/or destination networks.

The TCP/IP structure

TCP/IP represents one of the earliest developed layered communications protocols, grouping functions into defined network layers. Figure 2.20 illustrates the relationship of the TCP/IP protocol suite and the services they provide with

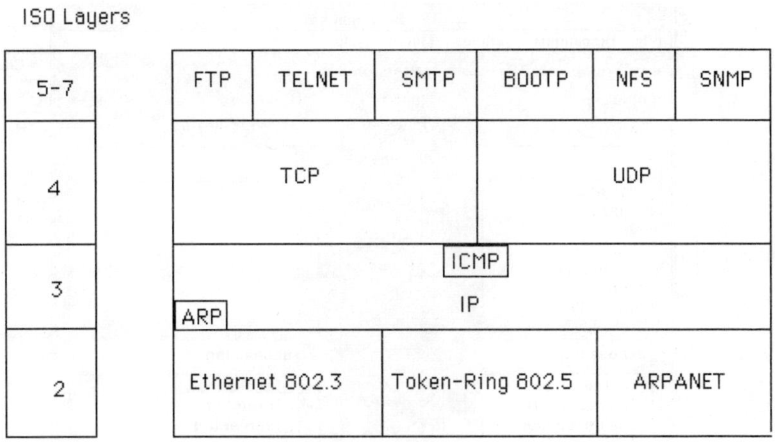

Legend

ARP	Address Resolution Protocol
BOOTP	Bootstrap Protocol
FTP	File Transfer Protocol
NFS	Network File System
SMTP	Simple Mail Transfer Protocol
SNMP	Simple Network Management Protocol

Figure 2.20 TCP/IP protocols and services

respect to the OSI Reference Model. In examining Figure 2.20 note that only seven of literally hundreds of TCP/IP application services are shown. Since TCP/IP preceded the development of the OSI Reference Model, its developers grouped what are now session, presentation, and application layers that correspond to layer 7 of the OSI Reference Model into one higher layer. Thus, TCP/IP applications, when compared to the OSI Reference Model, are normally illustrated as corresponding to the upper three layers of that model. Continuing our examination of Figure 2.20, note that the subdivision of the transport layer indicates which applications are carried via TCP and which are transported by UDP. Thus, FTP, Telnet, HTTP, and SMTP represent applications transported by TCP.

Although many people equate Web browsing with TCP/IP, that application is but one of many commonly supported by that protocol suite. In fact, many Web browsers support plug-in modules to provide file transfer, remote terminal access, and support for other applications, while other vendors market stand-alone TCP/IP applications as part of an application suite.

Figures 2.21 and 2.22 illustrate the use of the FTP and TN3270 applications from the NetManage Chameleon TCP/IP application suite. Figure 2.21 illustrates FTP access to a Windows NT FTP server with the file ACCESS. LOG being prepared for copying from the server to the local computer where it will be stored using the same name. File Transfer Protocol (FTP) is a standard method for transferring files over the Internet. Figure 2.22 illustrates the use of the NetManager

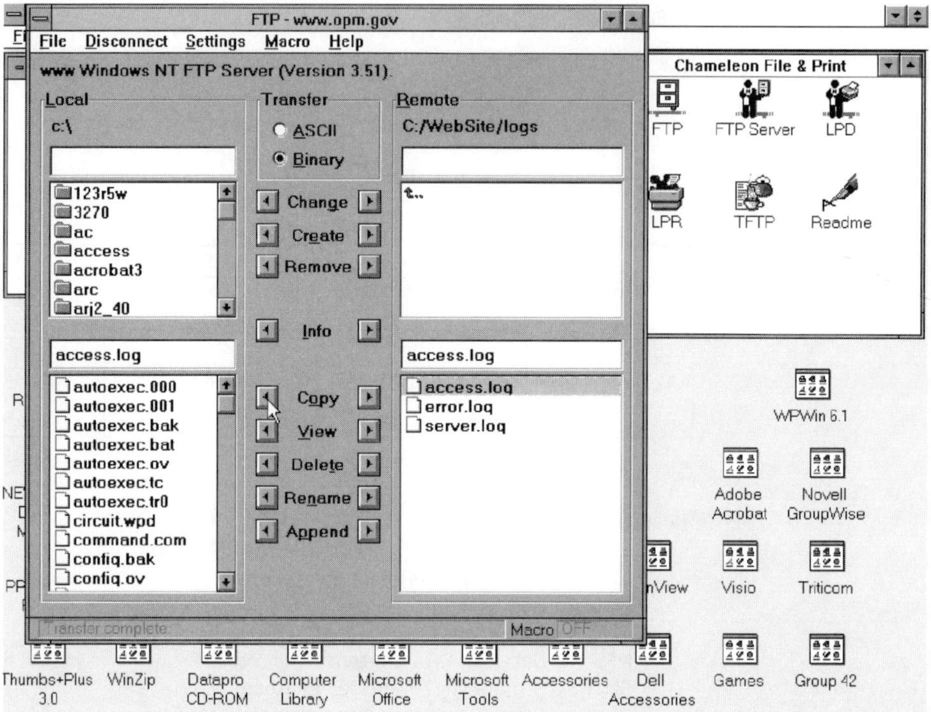

Figure 2.21 Using a client FTP program to download an access log from a Wndows NT FTP server

TN3270 program which is a Telnet derivative developed to provide remote terminal access to IBM mainframes. In this example the TN3270 program was used to enable a PC to obtain a connection as if it was a 3270 terminal device to an IBM OfficeVision application running on a S/390 mainframe which provides electronic mail, calendars, and even access to the Internet.

Returning to our examination of Figure 2.20, note that TCP/IP can be transported at the Data Link Layer by a number of popular LANs, to include Ethernet, Fast Ethernet, Token-Ring, and FDDI frames. Due to the considerable effort expended in the development of LAN adapter cards to support the bus structures used in Apple MacIntosh, IBM PCs and compatible computers, DEC Alphas and SUN Microsystem's workstations, and even IBM mainframes, the development of software based protocol stacks to facilitate the transmission of TCP/IP on LANs provides the capability to interconnect LAN-based computers to one another, whether they are on the same network and require only the transmission of frames on a common cable, or on networks separated thousands of miles from one another. Thus, TCP/IP represents both a local and a wide area network transmission capability. In the remainder of this section I will review IP and TCP packet headers as well as discuss the use of several related network and transport layer protocols and higher level protocols implemented over TCP and its related protocol suite.

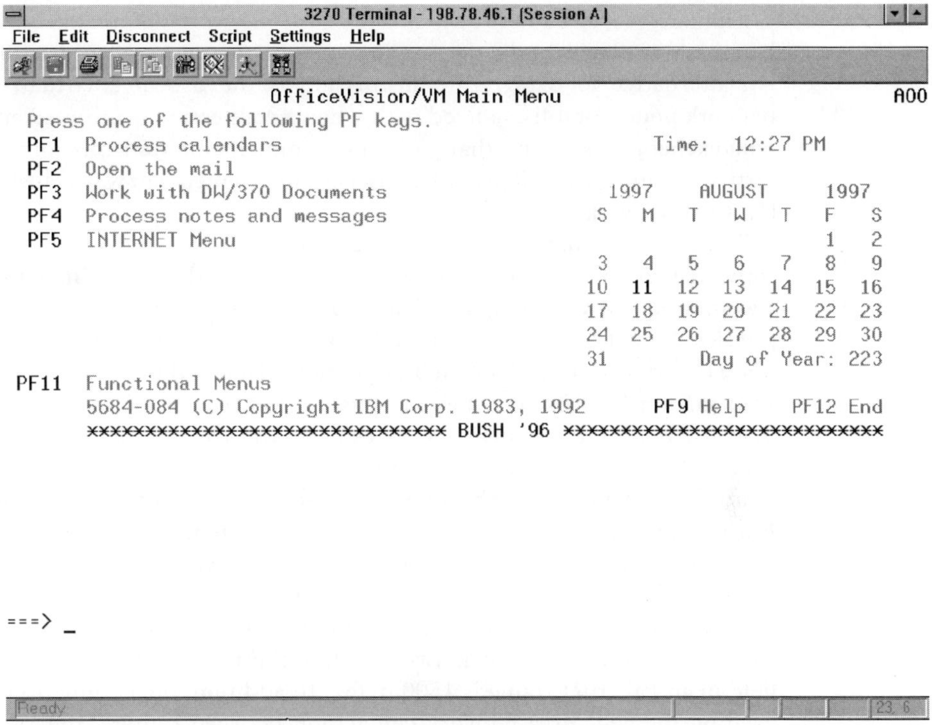

Figure 2.22 Using a TN3270 client program to enable a PC to access an IBM mainframe as if it was a 3270 terminal

Datagrams versus virtual circuits

In examining Figure 2.20 you will note that the Internet Protocol (IP) provides a common Layer 3 transport for TCP and UDP. TCP is a connection oriented protocol which requires the acknowledgment of the existence of the connection and for packets transmitted once the connection is established. In comparison, UDP, a mnemonic for User Datagram Protocol, is a connectionless mode service that provides a parallel service to TCP. Here datagram is a term used to identify the basic unit of information that represents a portion of message and which is transported across a TCP/IP network.

A datagram can be transported via either an acknowledged connection-oriented service or an unacknowledged connectionless service, where each information element is addressed to its destination and its transmission is at the mercy of network nodes. IP represents an unacknowledged connectionless service. However, although it is an unreliable transmission method you should view the term in the context that delivery is not guaranteed instead of having second thoughts concerning its use. As a non-guaranteed delivery mechanism IP is susceptible to queuing delays and other problems that can result in the loss of data. However, higher layers in the protocol suite, such as TCP, can provide error detection and correction which results in the retransmission of IP datagrams.

Similar to an X.25 network, on a TCP/IP network datagrams are routed via the best path available to the destination as the datagram is placed onto the network. An alternative to datagram transmission is the use of a virtual circuit, where network nodes establish a fixed path when a connection is initiated and subsequent data exchanges occur on that path. TCP implements transmission via the use of a virtual circuit, while IP provides a datagram oriented gateway transmission service between networks.

The routing of datagrams through a network can occur over different paths, with some datagrams arriving out of sequence from the order in which they were transmitted. In addition, as datagrams flow between networks they encounter physical limitations imposed upon the amount of data that can be transported based upon the transport mechanism used to move data on the network. For example, the Information field in an Ethernet frame is limited to 1500 bytes while a 4 Mbps Token-Ring can transport 4500 bytes in its Information field. Thus, as datagrams flow between networks, they may have to be fragmented into two or more datagrams to be transported through different networks to their ultimate destination. For example, consider the transfer of a 20 000 byte file from a file server connected to a Token-Ring network to a workstation connected to an Ethernet LAN via a pair of routers providing a connection between the two local area networks. The 4 Mbps Token-Ring network supports a maximum Information field of 4500 bytes in each frame transmitted on that network, while the maximum size of the Information field in an Ethernet frame is 1500 bytes. In addition, depending upon the protocol used on the wide area network connection between routers, the WAN protocol's Information field could be limited to 512 or 1024 bytes. Thus, the IP protocol must break up the file transfer into a series of datagrams whose size is acceptable for transmission between networks. As an alternative, IP can transmit data using a small maximum datagram size, commonly 576 bytes, to prevent fragmentation. If fragmentation is necessary, the source host can transmit using the maximum datagram size available on its network. When the datagram arrives at the router, IP operating on that communications device will then fragment each datagram into a series of smaller datagrams. Upon receipt at the destination, each datagram must then be put back into its correct sequence so that the file can be correctly reformed, a responsibility of IP residing on the destination host.

Figure 2.23 illustrates the routing of two datagrams from workstation 1 on a Token-Ring network to server 2 connected to an Ethernet LAN. As the routing of datagrams is a connectionless service, no call setup is required, which enhances transmission efficiency. In comparison, when TCP is used, it provides a connection-oriented service regardless of the lower layer delivery system (e.g. IP).

TCP requires the establishment of a virtual circuit in which a temporary path is developed between source and destination. This path is fixed and the flow of datagrams is restricted to the established path. When the User Datagram Protocol (UDP), a different layer 4 protocol in the TCP/IP protocol suite, is used in place of TCP, the flow of data at the Transport layer continues to be connectionless and results in the transport of datagrams over available paths rather than a fixed path resulting from the establishment of a virtual circuit.

The actual division of a message into datagrams is the responsibility of the layer 4 protocol, either TCP or UDP, while fragmentation is the responsibility of IP. In

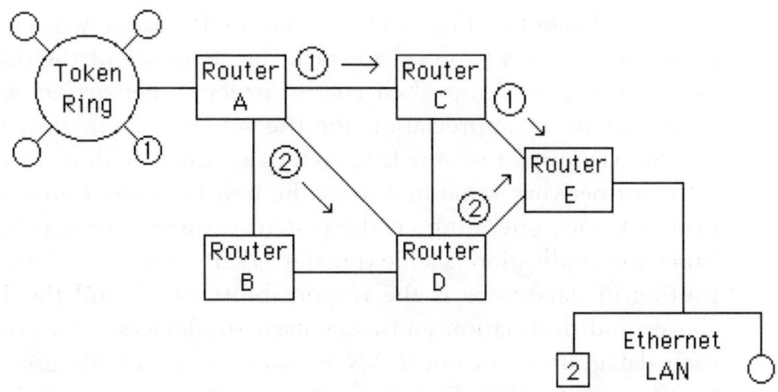

Figure 2.23 Routing of datagrams can occur over different paths

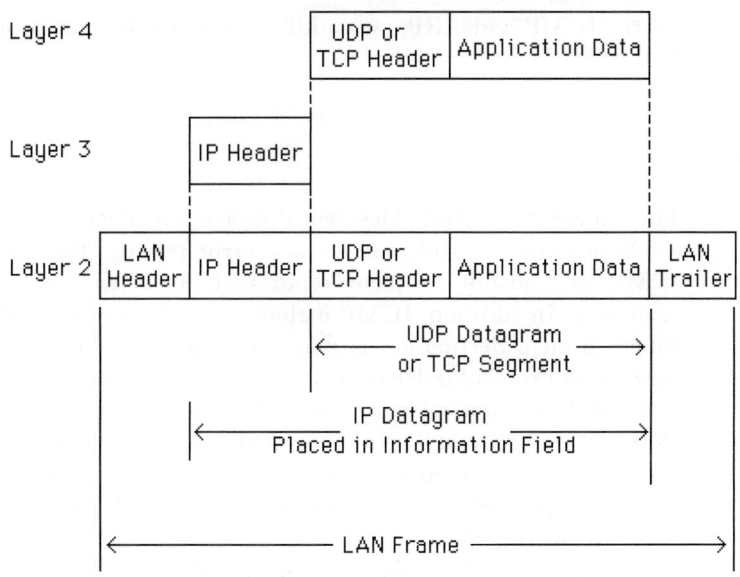

Figure 2.24 Forming a LAN frame

addition, when the TCP protocol is used, that protocol is responsible for reassembling datagrams at their destination as well as for requesting the retransmission of lost datagrams. In comparison, IP is responsible for routing of individual datagrams from source to destination. When UDP is used as the layer 4 protocol, there is no provision for the retransmission of lost or garbled datgrams. As previously noted by our discussion of IP, this is not necessarily a bad situation as applications that use UDP then become responsible for managing communications.

Figure 2.24 illustrates the relationship of an IP datagram, UDP datagram, and TCP segment to a LAN frame. The headers shown in Figure 2.24 represent a group of bytes added to the beginning of a datagram to allow a degree of control over the datagram. For example, the TCP header will contain information that allows this layer 4 protocol to track the sequence of the delivery of datagrams so they can be placed into their correct order if they arrive out of sequence.

To obtain an appreciation for the schematic illustrated in Figure 2.24, let's assume a computer on one LAN issues a request to download a file that resides on a different network separated from the first by a few hundred miles. The TCP/IP protocol stack operating on the distant computer divides the file into datagrams. Since the application is a file transfer, TCP is the layer 4 protocol used. The actual routing of datagrams is the responsibility of IP and the IP header will contain source and destination addresses used by devices referred to as routers to route each datagram from one LAN to another via a wide area network transmission facility. Since the distant computer resides on a LAN, each datagram is encapsulated in a LAN frame for transport to a router which then transmits the contents of the layer 2 frame based upon the IP source address contained in the frame. Prior to focusing our attention upon TCP and IP, let's briefly discuss the role of ICMP and ARP, two additional network layer protocols in the TCP/IP suite.

ICMP and ARP

The Internet Control Message Protocol (ICMP) provides a mechanism for communicating control message and error reports. Both gateways and hosts use ICMP to transmit problem reports about datagrams back to the datagram originator. In addition, ICMP includes an echo request/reply that can be used to determine if a destination is reachable and, if so, is responding. The Address Resolution Protocol (ARP) maps the high level IP address configured via software to a low level physical hardware address, typically the network interface card's (NIC) ROM address. The high level IP address is currently 32 bits in length and commonly represented by four decimal numbers, ranging from 0 to 255 per number, separated from one another by decimals. Thus, another term used to reference an IP address is dotted decimal address. The physical hardware address represents the MAC address. Thus, ARP provides an IP to MAC address resolution which enables an IP packet to be transported in a LAN frame to its appropriate MAC address. Later in this section we will examine IP addresses in detail.

The TCP header

The Transmission Control Protocol (TCP) represents a layer 4 connection-oriented reliable protocol. TCP provides a virtual circuit connection mode service for applications that require connection setup and error detection and automatic retransmission. In addition, TCP is structured to support multiple application programs on one host to communicate concurrently with processes on other hosts, as well as for a host to de-multiplex and service incoming traffic among different applications or processes running on the host.

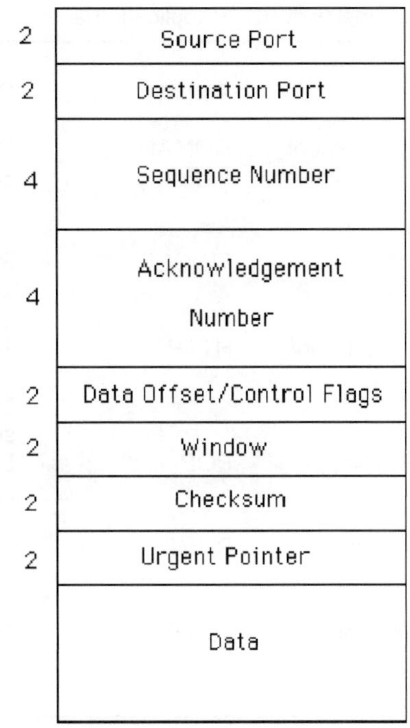

Figure 2.25 TCP protocol header

Each unit of data carried by TCP is referred to as a segment. Segments are created by TCP subdividing the stream of data passed down by application layer protocols that use its services, with each segment identified by the use of a sequence number. This segment identification process enables a receiver, if required, to reassemble data segments into their correct order.

Figure 2.25 illustrates the format of the TCP protocol header. To obtain an appreciation for the functionality and capability of TCP, let's examine the fields in its header.

Source and destination port fields

The source and destination ports are each 16 bits in length and identify a process or service at the host receiver. The source port field entry is optional and when not used is padded with zeros. Both source and destination port values up to 256 are commonly referred to as 'well-known ports' as they typically identify an application layer protocol or process. Table 2.3 lists the well-known port numbers associated with eight popular TCP/IP application layer protocols. Note that some protocols, such as FTP, use two port addresses or logical connections. In the case of FTP, one address (21) is used for the transmission of commands and responses and functions as a control path. In comparison, the second port address (20) is used for the actual file transfer.

Table 2.3 Examples of TCP/IP application layer protocol use of well-known ports

Name	Acronyn	Description	Well-known port
Domain Name Protocol	DOMAIN	Defines the DNS	53
File Transfer Protocol	FTP	Supports file transfers between hosts	20, 21
Finger Protocol	FINGER	Provides information about a specified user	79
HyperText Transmission Protocol	HTTP	Transmits information between a Web browser and a Web server	80
Post Office Protocol	POP	Enables host users to access mail from a mail server	110
Simple Mail Transfer Protocol	SNMP	Provides for the exchange of network management information	161, 162
TELENET Protocol	Telnet	Provides remote terminal access to a host	23

Sequence field

The sequence number is used to identify the data segment transported. The acknowledgement number interpretation depends upon the setting of the ACK control flag which is not directly shown in Figure 2.25. If the ACK control flag bit position is set, the acknowledgement field will contain the next sequence number the sender expects to receive. Otherwise the field is ignored.

Control field flags

There are six control field flags that are used to establish, maintain and terminate connections. Those flags include URG (urgent), SYN, ACK, RST (reset), PSH (push), and FIN (finish).

Setting URG = 1 indicates to the receiver that urgent data is arriving. The SYN flag is set to 1 as a connection request and thus serves to establish a connection. As previously discussed, the ACK flag when set indicates that the acknowledgement flag is relevant. The RST flag set means the connection should be reset, while the PSH flag tells the receiver to immediately deliver the data in the segment. Finally, the setting of the FIN flag indicates that the sender is done and the connection should be terminated.

Window field

The window field is used to convey the number of bytes the sender can accept and functions as a flow control mechanism. This 16-bit field indicates the number of octets, beginning with the one in the acknowledgement field that the originator of the segment can control. Since TCP is a full-duplex protocol, each host can use the window field to control the quantity of data that can be sent to the computer. This enables the recipient, in effect, to control its destiny. For example, if a receiving host becomes overloaded with processing or for some other reason the device is unable to receive large chunks of data, it can use the window field as a flow control mechanism to reduce the size of data chunks sent to it. At the end of our review of TCP header fields we will examine a TCP transmission sequence to note the interrelated role of the sequence, acknowledgement and window fields.

Checksum field

The checksum provides error detection for the TCP header and data carried in the segment. Thus, this field provides the mechanism for the detection of errors in each segment.

Urgent pointer field

The urgent pointer field is used in conjunction with the URG flag as a mechanism to identify the position of urgent data within a TCP segment. When the URG flag is set the value in the urgent pointer field represents the last byte of urgent data.

When an application uses TCP, TCP breaks the stream of data provided by the application into segments and adds an appropriate TCP header. Next, an IP header is prefixed to the TCP header to transport the segment via the network layer. As data arrives at its destination network it is converted into a data link layer transport mechanism. For example, on Token-Ring network TCP data would be transported within Token-Ring frames.

TCP transmission sequence example

To illustrate the interrelationship between the sequence, acknowledgement and window fields, let's examine the transmission of a sequence of TCP segments between two hosts. Figure 2.26 illustrates through the use of a time chart the transmission of a sequence of TCP segments.

At the top of Figure 2.26 it was assumed that a window size of 16 segments is in use. Although TCP supports full-duplex transmission, for simplicity of illustration we will use a half-duplex model in the time chart.

Assuming the host computer with the address ftp.fbi.gov (Internet addressing will be described later in this section) is transmitting a program or performing a similar lengthy file transfer operation, the first series of segments will have sequence

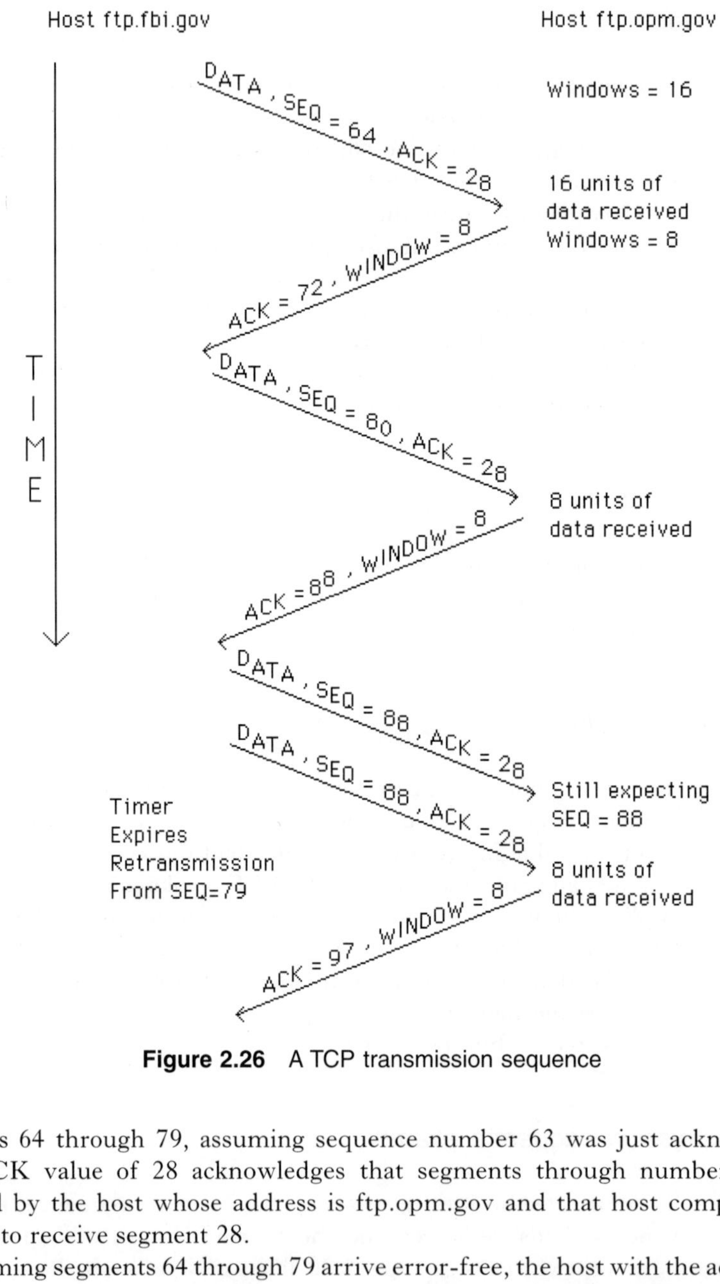

Figure 2.26 A TCP transmission sequence

numbers 64 through 79, assuming sequence number 63 was just acknowledged. The ACK value of 28 acknowledges that segments through number 27 were received by the host whose address is ftp.opm.gov and that host computer next expects to receive segment 28.

Assuming segments 64 through 79 arrive error-free, the host with the address ftp.opm.gov returns an ACK value of 80 to indicate the next segment it expects to receive. At this point in time let's assume the host with the address ftp.opm.gov is running out of buffer space and halves the window size to 8. Thus, the host with the address ftp.opm.gov sets the window field value in the TCP header it transmits to 8. Upon receipt of the TCP segment the host with the address ftp.fbi.gov reduces the number of segments it will transmit to eight and uses an initial SEQ value of 80, increasing that value by 1 each time it transmits a new segment until eight new segments are transmitted. Assuming all eight segments were received error-free, the host with the address ftp.opm.gov returns an ACK value of 88 which acknowledges the receipt of segments with sequence field numbers through 87.

Next, host ftp.fbi.gov transmits to host ftp.opm.gov another sequence of eight segments using sequence field values of 88 to 95. However, let's assume a transmission impairment occurs that results in the segments being incorrectly received or perhaps not even received at all at their intended destination. If host ftp.opm.gov does not receive anything it does not transmit anything back to host ftp. fbi.gov. Instead of waiting forever for a response, the TCP/IP protocol stack includes an internal timer which clicks down to zero while host ftp.fbi.gov waits for a response. When that vlaue is reached, the timer expires and the transmitting station retransmits its sequence of eight segments. On the second time around the sequence of eight segments are shown acknowledged at the bottom of Figure 2.26. If the impairment continued the transmitting station would attempt a predefined number of retransmissions after which it would terminate the session if no response was received.

The altering of window field values provides a 'sliding window' that can be used to control the flow of information. That is, by adjusting the value of the window field a receiving host can inform a transmitting station whether or not an adjustment in the number of segments transmitted is required. In doing so there are two special window field values that can be used to further control the flow of information. A window field value of 0 means a host shut down communications, while a value of 1 requires an acknowledgment for each unit of data transmitted, limiting transmission to a segment by segment basis.

The UDP header

The User Datagram Protocol (UDP) is the second layer 4 transport service supported by the TCP/IP protocol suite. UDP is a connectionless service which means that the higher layer application is responsible for the reliable delivery of the transported message. Figure 2.27 illustrates the composition of the UDP header.

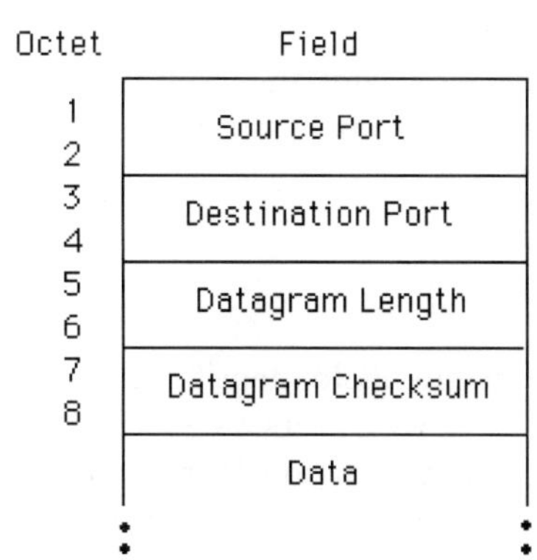

Figure 2.27 The UDP header

Source and destination port fields

The source and destination port fields are each 16 bits in length and as previously described for TCP identify the port number of the sending and receiving process, respectively. Here each port number process identifies as application running at the corresponding IP address in the IP header prefixed for the UDP header. The use of a port number provides a mechanism for identifying network services as they denote communications points where particular services can be accessed. For example, a value of 161 in a port field is used in UDP to identify SNMP.

Length field

The length field indicates the length of the UDP packets in octets to include the header and user data. The checksum, which is a one's complement arithmetic sum, is computed over a pseudo-header and the entire UDP packet. The pseudo-header is created by the conceptual prefix of 12 octets to the header previously illustrated in Figure 2.27. The first 8 octets are used by source and destination IP addresses obtained from the IP packet. This is followed by a zero-filled octet and an octet which identifies the protocol. The last two octets in the pseudo-header denote the length of the UDP packet. By computing the UDP checksum over the pseudo-header and user data a degree of additional data integrity is obtained.

The IP header

As previously mentioned, IP provides a datagram-oriented gateway service for transmission between subnetworks. This provides a mechanism for hosts to access other hosts on a best-effort basis but does not enhance reliability as it relies on upper layer protocols for error detection and correction. As a Layer 3 protocol IP is responsible for the routing and delivery of datagrams. To accomplish this task IP performs number of communications functions to include addressing, status information, management and the fragmentation and reassembly of datagrams when necessary.

Figure 2.28 illustrates the IP header format while Table 2.4 provides a brief description of the fields in the IP header.

Version field

The four-bit version field identifies the version of the IP protocol used to create the datagram. The current version of the IP protocol is 4 and is encoded as 0100 in binary. The next generation IP protocol is version 6, which is encoded as 0110 in binary. In our discussion of IP we will primarily focus on IPv4 in this section; however, we will also examine the key features of IPv6 which represent the next generation of Internet addressing.

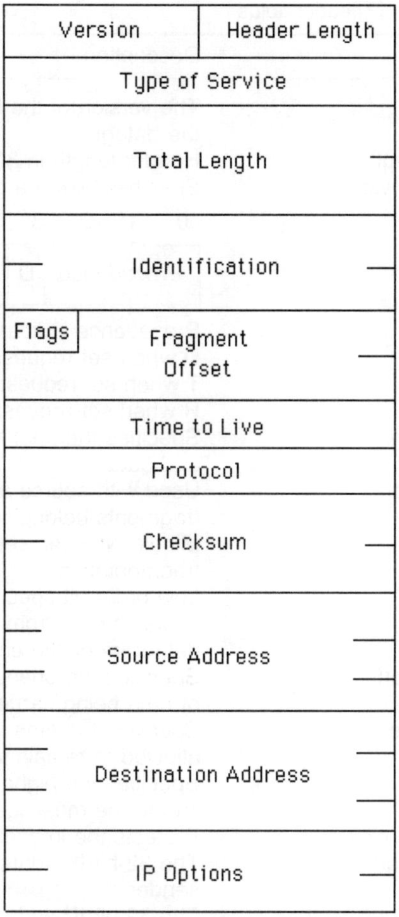

Figure 2.28 IP header format

Header length and total length fields

The header length field follows the version field and is also four bits in length. This field indicates the length of the header in 32-bit words. In comparison, the total length field indicates the total length of the datagram to include its header and higher layer information. The use of 16 bits for the total length field enables an IP datagram to be up to $2^{16}-1$ or 65 535 octets in length.

Type of service field

The type of service field identifies how the datagram is handled. Three of the eight bits in this field are used to denote the precedence or level of importance assigned by the originator. Thus, this field provides a priority mechanism for routing IP datagrams.

Table 2.4 IP header fields

Field	Description
Version	The version of the IP protocol used to create the datagram
Header length	Header length in 32-bit words
Type of service	Specifies how the datagram should be handled: 0 1 2 3 4 5 6 7 Precedence | D | T | R Unused Precedence indicates importance of the datagram D when set requests low delay T when set requests high throughput R when set requests high reliability
Total length	Specifies the total length to include header and data
Identification	Used with source address to identify fragments belonging to specific datagrams
Flags	Middle bit when set disables possible fragmentation Low-order bit specifies whether the fragment contains data from the middle of the original datagram or the end
Fragment offset	Specifies the offset in the original datagram of data being carried in a fragment
Time to live	Specifies the time in seconds a datagram is allowed to remain in the Internet
Protocol	Specifies the higher level protocol used to create the message carried in the data field
Header checksum	Protects the integrity of the header
Source IP address	The 32-bit IP address of the datagram's sender
Destination IP address	The 32-bit IP address of the datagram's intended recipient
IP options	Primarily used for network testing or debugging: 0 1 2 3 4 5 6 7 Copy | Option class | Option number Copy bit set tells gateways that the option should be copied into all fragments; when set to 0 the option is copied into the first fragment The meaning of the class field is as follows: 0 Datagram or network control 1 Reserved for future use 2 Debugging 3 Reserved for future use The option number defines a specific option within a class

Identification and fragment offset fields

The identification field enables each datagram or fragmented datagram to be identified. If a datagram was previously fragmented, the fragment offset field specifies the offset in the original datagram of the data being carried. In effect, this field indicates where the fragment belongs in the complete message. The actual value in this field is an integer which corresponds to a unit of 8 octets, providing an offset in 64-bit units.

Time to live field

The time to live (TTL) field specifies the maximum time that a datagram can live. Since an exact time is difficult to measure, almost all routers decrement this field by 1 as datagram flows between networks, with the datagram being descarded when the field value reaches zero. Thus, this field more accurately represents a hop count field. You can consider this field to represent a fail-safe mechanism as it prevents misaddressed datagrams from continuously flowing on the Internet.

Flags field

The flags field contains two bits that indicate how fragmentation occurs while a third bit is currently unassigned. The setting of one bit can be viewed as a direct fragment control mechanism, as a value of 0 indicates the datagram can be fragmented, while a value of 1 denotes don't fragment. The second bit is set to 0 to indicate that a fragment in a datagram is the last fragment and set to value of 1 to indicate more fragments follow the current protocol.

Protocol field

The protocol field specifies the higher level protocol used to create the message carried in the datagram. For example, a value of decimal 6 would indicate TCP, while a value of decimal 17 would indicate UDP.

Source and destination address fields

The source and destination address fields are both 32 bits in length. Each address represents both a network and host computer on the network.

Routing of datagrams on the Internet as well as between networks that use a TCP/IP protocol stack is presently based upon the use of 32-bit IP addresses. If you previously used ftp, Telnet, a Web browser, or another TCP/IP application you probably used near-English mnemonics, such as ftp.fbi.gov. This type of addressing is referred to as a domain name and is translated by the Internet domain name service (DNS) into an appropriate IP address. Although DNS names are much easier to remember and use, routing is based upon IP addressing with the DNS providing the mechanism required to translate DNS names into IP addresses. Since IP addresses form the basis for network addressing as well as the use of DNS, we will first focus our attention upon that topic prior to examining the role of the DNS.

IP addressing

The IP addressing scheme uses a 32-bit address which is divided into an assigned network number and host number. The latter can be further segmented into a subnet number and host number. Through subnetting you can construct multiple networks while localizing the traffic of hosts to specific subnets, a technique I will shortly illustrate.

IP addressing numbers are assigned by the InterNIC network information center and fall into one of five unique network classes, known as Class A through Class E. Figure 2.29 illustrates the IP address formats for Class A, B and C networks. Class D addresses are reserved for multicast groups, while Class E addresses are reserved for further use.

In examining Figure 2.29, note that the first bit in the IP address distinguishes a Class A address from Class B and C addresses. Thereafter, examining the composition of the second bit position enables a Class B address to be distinguished from a Class C address.

An IP 32-bit address is expressed as four decimal numbers, with each number ranging in value from 0 to 255 and separated from another number by a dot (decimal point). This explains why an IP address is commonly referred to as a dotted decimal address.

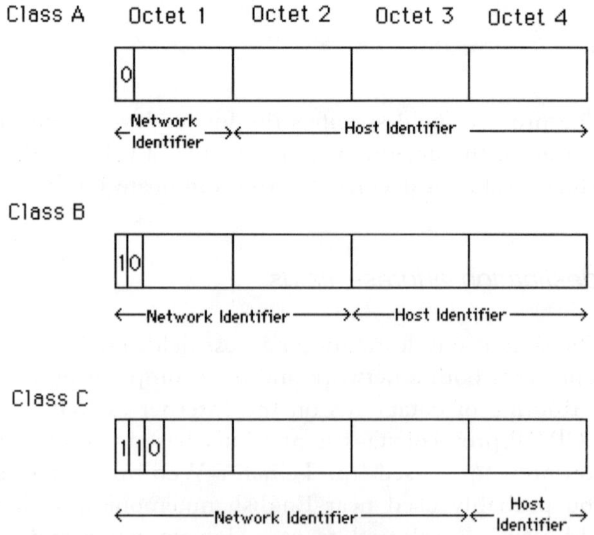

Figure 2.29 IP address formats

Class A

In examining Figure 2.29, note that a Class A address has three bytes available for identifying hosts on one network or on subnets which provides support for more hosts than other address classes. Thus, Class A addresses are only assigned to large organizations or countries. Since the first bit in a Class A address must be zero, the first byte ranges in value from 1 to 127 instead of to 255. Through the use of 7 bits for the network portion and 24 bits for the host portion of the address, 128 networks can be defined with approximately 16.78 million hosts capable of being addressed on each Class A network.

Class B

A Class B address uses two bytes for the network identifier and two for the host or subnet identifier. This permits up to 65 536 (2^{16}) hosts and/or subnets to be assigned; however, since the first two bits of the network portion of the address are used to identify a Class B address, the network portion is reduced to a width of 14 bits. Thus, up to 16 384 (2^{14}) class B networks can be assigned. Due to the manner by which Class B network addresses are subdivided into network and host portions, such addresses are normally assigned to relatively large organization with tens of thousands of employees.

Class C

In a Class C address three octets are used to identify the network, leaving one octet to identify hosts and/or subnets. The use of 21 bits for a network address enables approximately 2 million distinct networks to be supported by the Class C address class. Since one octet permits only 256 hosts or subnets to be identified, many small organizations with a requirement to provide more than 256 hosts with access to the Internet must obtain multiple Class C addresses.

Host restrictions

In actuality the host portion of an IP address has two restrictions which reduces the number of hosts that can be assigned to a network. First, the host portion cannot be set to all zero bits as an all-zeros host number is used to identify a base network or subnetwork number. Secondly, an all-ones host number represents the broadcast address for a network or subnetwork. Thus, the maximum number of hosts on a network must be reduced by two. For a Class C network a maximum of 254 hosts can then be configured for operation.

Subnetting

Through the use of subnetting you can use a single IP address as a mechanism for connecting multiple physical networks. To accomplish subnetting you logically divide the host portion of an IP address into a network address and a host address.

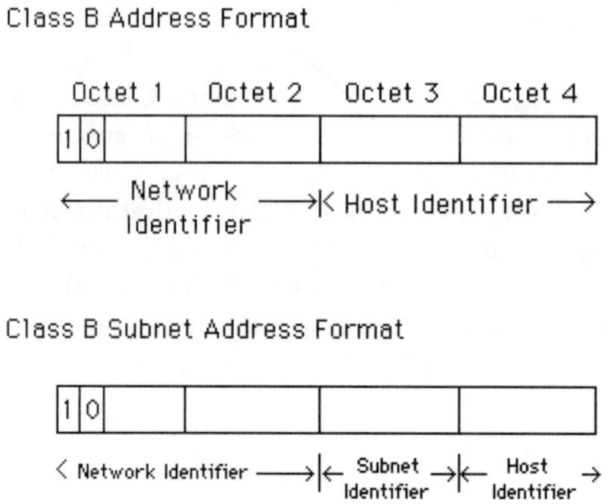

Figure 2.30 Class B subnetting

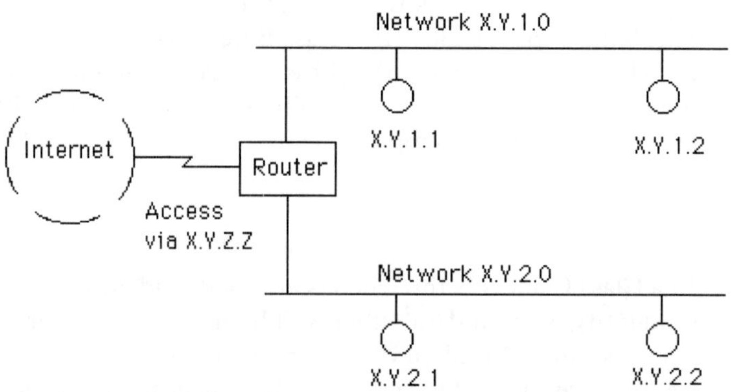

Figure 2.31 A Class B network address location with two physical networks using subnet addressing. IP datagrams with the destination address X.Y.Z.Z, where Z can be any decimal value, represent a Class B network address that can consist of 256 subnets, with 256 hosts on each subnet

Figure 2.30 illustrates an example of the IP subnet addressing format for a Class B address. In this example all traffic routed to the address XY, where X and Y represent the value of the first two Class B address octets, flows to a common location connected to the Internet, typically a router. The router in turn connects two or more Class B subnets, each with a distinct address formed by the third decimal digit which represents the subnet identifier. Figure 2.31 illustrates a Class B network address location with two physical networks using subnet addressing.

Subnet masks

The implementation of a subnet addressing scheme is accomplished by the partitioning of the host identifier portion of an IP address. To accomplish this a 32-

bit subnet mask must be created for each network, with bits set to '1' in the subnet mask to indicate the network portion of the IP address, while bits are set to '0' to indicate the host identifier portion. Thus, the Class B subnet address format illustrated in the lower portion of Figure 2.30 would require the following 32-bit subnet mask:

11111111 11111111 00000000 00000000

The prior mask would then be entered as 255.255.0.0 in dotted decimal representation into a router configuration screen as well as in software configuration screens on TCP/IP program stacks operating on each subnet. Concerning the latter, you must then configure each station to indicate its subnet and host identifier so that each station obtains a full four-digit dotted decimal address.

Although the above example used octet boundaries for creating the subnet mask, this is not an addressing requirement. For example, you could assign the following mask to a network:

11111111 11111111 00001110 00001100

The only submask restriction is to assign 1's to at least all the network identifier positions, resulting in the ability to extend masking into the host identifier field if you desire to arrange the specific assignment of addresses to computers. However, doing so can make it more difficult to verify the correct assignment of addresses in routers and workstations. For this reason it is highly recommended that you should implement subnet masking on integral octet boundaries.

Domain Name Service

Addressing on a TCP/IP network occurs through the use of four decimal numbers ranging from 0 to 255, which are separated from one another by a dot. This dotted decimal notation represents a 32-bit address which consists of an assigned network number and a host number as previously described during our examination of IP addressing. Since numeric addresses are difficult to work with, TCP/IP also supports a naming convention based upon English words or mnemonics that are easier both to work with and to remember. The translation of English words or mnemonics to 32-bit IP addresses is performed by a Domain Name Server. Each network normally has at least one Domain Name Server, and communications established between such servers on TCP/IP networks connected to the Internet are referred to as a Domain Name Service (DNS).

The Domain Name Service is the naming protocol used in the TCP/IP protocol suite which enables IP routing to occur indirectly through the use of names instead of IP addresses. To accomplish this, DNS provides a domain name to IP address translation service.

A domain is a subdivision of a wide area network. When applied to the Internet (the collection of networks interconnected to one another), there are six top-level and seven pending domain names which were specified by the Internet Network

Table 2.5 Internet top-level domain names

Domain name	Assignment
Existing	
.COM	Commercial organization
.EDU	Educational organization
.GOV	Government agency
.MIL	Department of Defense
.NET	Networking organization
.ORG	Not-for-profit organization
Pending	
.firm	Business/commercial firms
.store	Goods for purchase
.web	World Wide Web related activities
.arts	Culture/entertainment organizations
.rec	Recreation/entertainment organizations
.info	Information/services
.nom	Individual/personal nomenclature

Information Center (InterNIC) at the time this book was prepared. These top level domains are listed in Table 2.5.

Under each top level domain the InterNIC will register subdomains which are assigned an IP network address. An organization receiving an IP network address can further subdivide their domain into two or more subdomains. In addition, instead of using dotted decimal notation to describe the location of each host, they can assign names to hosts as long as they follow certain rules and install a name server which provides IP address translation between named hosts and their IP addresses.

To illustrate the operation of a name server consider the network domain illustrated in Figure 2.32. In this example we will assume that a well-known government agency has a local area network with several computers that will be connected to the Internet. Each host address will contain the specific name of the host plus the names of all of the subdomains and domains to which it belongs. Thus, the computer 'WARRANTS' would have the official address:

warrants.telnet.fbi.gov

Similarly, the computer 'COPS' would have the address

cops.ftp.fbi.gov

In examining the domain naming structure illustrated in Figure 2.32 note that computers were placed in subdomains using common Internet application names, such as telnet and ftp. This is a common technique organizations use to make it easier for network users both within and outside the organization to remember mnemonics that represent specific hosts on their network.

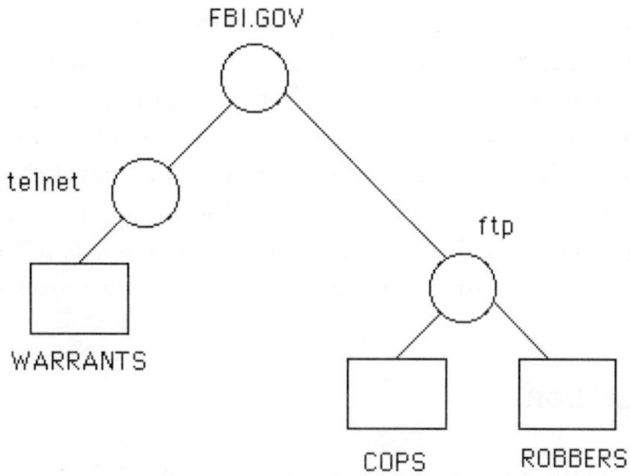

Figure 2.32 A domain-naming hierarchy

Since modern operating systems can support multiple applications, another common method of structuring domain names is based upon the use of the following format:

application.organization.top – level domain

Here applications can range from ftp and telnet servers to gopher and World Wide Web (WWW) servers. The organization can represent any registered entity, from Apple to IBM to Microsoft, while the top-level domain would be one of the domain names previously listed in the left-hand column of Table 2.5.

Although domain names provide a mechanism for identifying objects connected to wide area networks, hosts in a domain require network addresses to transfer information. Thus another host functioning as a name server is required to provide a name to address translation service.

Name server

The name server plays an important role in TCP/IP networks. In addition to providing a name-to-IP address translation service it must recognize that an address is outside its administrative zone of authority. For example, assume a host located on the domain illustrated in Figure 2.32 will use the address fred.microwear.com to transmit a message. The name server must recognize that that address does not reside in its domain and must forward the address to another name server for translation into an appropriate IP address. Since most domains are connected to the Internet via an Internet service provider, the name server on the domain illustrated in Figure 2.32 would have a pointer to the name server of the Internet Service Provider and forward the query to that name server. The Internet Service Provider's name server will either have an entry in its table in cache memory or forward the query to another higher level name server. Eventually a name server will be reached that has administrative authority over the domain containing the host name to resorve and will return an IP address through a reversed hierarchy to provide the originating name server with a response to its query. Most name servers cache the results of previous name queries which can considerably reduce off-domain or Internet DNS queries. In the event that a response is not received, possibly due to an incorrect name of the entry of a name no longer used, the local name server will generate a 'failure to resolve' message after a period of time that will be displayed on the requesting host's display.

TCP/IP configuration

The configuration of a station on a TCP/IP network normally requires the specification of four IP addresses as well as the station's host and domain names. To illustrate the configuration of a TCP/IP station, Figures 2.33 through 2.35 show the screen settings on a Microsoft Windows NT workstation used to configure the station as a participant on a TCP/IP network.

Figure 2.33 illustrates the Windows NT 4.0 Network Protocols tab with TCP/IP Protocol selected. Note that in the Network Protocols box the entry NWLink IPX/SPX Compatible Transport is shown. Windows NT has the ability to operate multiple protocol stacks to include NWLink which is Microsoft's implementation of the Novell IPX and SPX protocols. This capability enables you to use a Windows NT computer to access a Novell network.

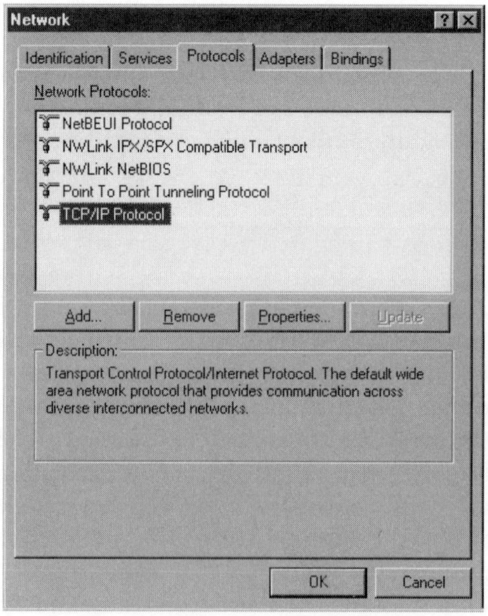

Figure 2.33 Using the Windows NT 4.0 Network protocols tab to select the use of the TCP/IP protocol

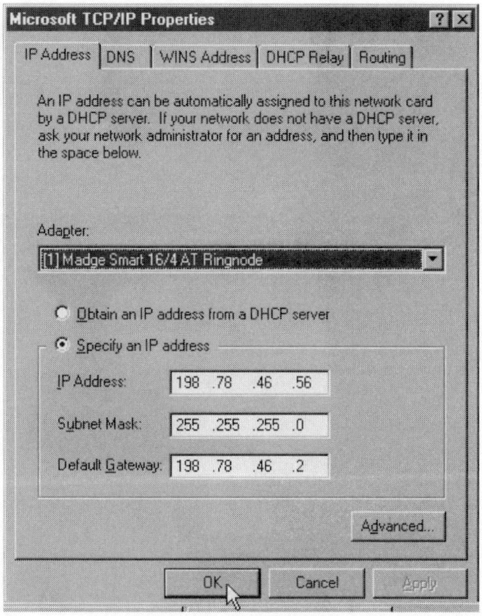

Figure 2.34 The Windows NT TCP/IP Properties display with entries for the IP address of the network interface, subnet mask, and default gateway

Clicking on the button labeled OK in Figure 2.33 results in the display of another box, labeled Microsoft TCP/IP Properties. Figure 2.34 illustrates the TCP/IP Properties dialog box with three address entries shown. These indicate the IP address of the interface assigned to the selected adapter card, the subnet mask and the IP address of the default gateway. Note that a computer can have multiple adapter cards, thus IP addresses are actually assigned to network interfaces. Also note that the term gateway dates from the time when such devices routed packets to other networks if their address was not on the local network. Thus, a more modern term for gateway is router .

After configuring the entries shown in Figure 2.34, you will require a few more entries. These include the address of the name server used to translate near-English mnemonics into IP addresses as well as the name of your computer and domain. To configure the address of the name server you would first click on the tab labeled DNS in Figure 2.34. This action will result in the display of the DNS dialog box which is shown in Figure 2.35.

The Windows NT DNS dialog box enables you to specify your host computer name and your domain name. These entries are optional; however, if you do not include them and your local DNS uses this configuration information, other TCP/IP users on either your network or a distant network will not be able to access your computer by entering your computer's near-English address name which in Figure 2.35 would be ftp.xyz.gov. Instead, users would have to know your numeric IP address.

The DNS entry area in Figure 2.35 allows you to specify up to four name servers in the order they should be searched. Many organizations operate two name servers, so the ability to enter four should suffice for most organizations.

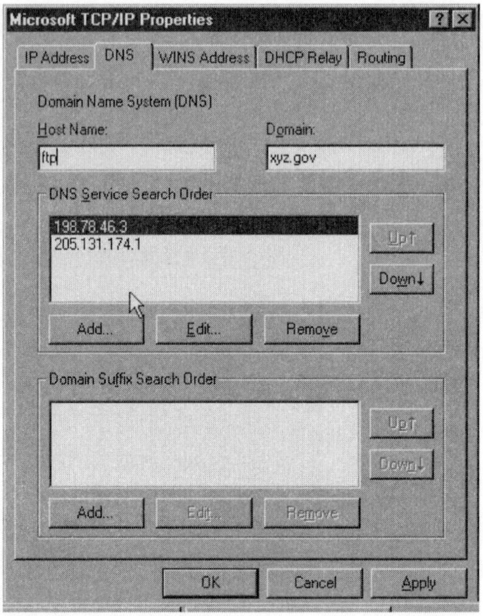

Figure 2.35 Using the Windows NT DNS Configuration dialox box to specify the station's name and domain name as well as two name server addresses

IPv6

In concluding this section on the Internet we will turn our attention to IPv6, the so-called next generation Internet protocol. Although the actual implementation of IPv6 is still a few years away, compliant equipment began to reach the marketplace during 1997 and Internet Service Providers can be expected to support this technology in the near future. Due to this, it is important to obtain an appreciation for the major characteristics of IPv6 and the migration issues involved in its use.

Evolution

The tremendous growth in the number of Internet users resulted in the recognition that the 32-bit address space of IPv4 had a limited life and required replacement. Although this recognition occurred during the late 1980s, it wasn't until the early 1990s that action began to take place. At a meeting of the Internet Society in Kobe, Japan, in June 1992 there were three proposals for a new IP. By December 1992 the number of IP replacement proposals reached seven and the Internet Engineering Steering Group (IESG) organized a directorate appropriately named IPng to review the new IP proposals and publish its recommendation. That recommendation was published in July 1994 at the Toronto IETF meeting and is documented in RFC 1752, *The Recommendation for the IP Next Generation Protocol*, published in January 1995. This new IP is version 6 of the Internet Protocol as the number 5 could not be used due to its prior allocation to an experimental protocol developed real-time data in parallel with IP. Thus, IPng is now known as IPv6 instead of IPv5.

Overview

IPv6 directly attacks the pending IP addressing problem by the use of a 128-bit address which replaces the 32-bit addresses of IPv4. This results in an increase of address space by a factor of 2^{96}. In addition to expanding the number of distinct IP addresses, the members of the IPng directorate simplified the IP header while providing a mechanism to improve the support of a variety of options to include those that may only be recognized as necessary to be developed many years from now, added the ability to label packets belonging to a particular type of traffic for which the originator requests special handling, and added a header extension capability which facilitates the support of encryption and authentication as well as a routing header whose use can be expected to enhance network performance.

Concerning network performance, one of the expected benefits of IPv6 over IPv4 is in the area of router performance. In IPv4 source routes are encoded in an optional header field whose contents must be checked by all routers, even if they do not represent a specific relay point in the source route. In comparison, under IPv6 routers are only required to examine the contents of the routing header if they recognize one of their own addresses in the destination field of the IPv6 main header. This method of operation required to support IPv6 should cut down on router CPU cycles, enabling a higher packet per second (PPS) processing rate to be obtained under normal operating conditions.

The best way to become familiar with IPv6 is by examining its header format. An appreciation of the simplicity of the IPv6 header and its ability to support a variety of options can be obtained by comparing that header to its predecessor, IPv4. Figure 2.36 illustrates the basic IPv6 header which consists of eight fields. This should be compared with the IPv4 header shown in Figure 2.28 from which it can be seen that six IPv4 fields have been eliminated. Perhaps less obviously, three fields have been renamed and two new fields incorporated into the IPv6 header.

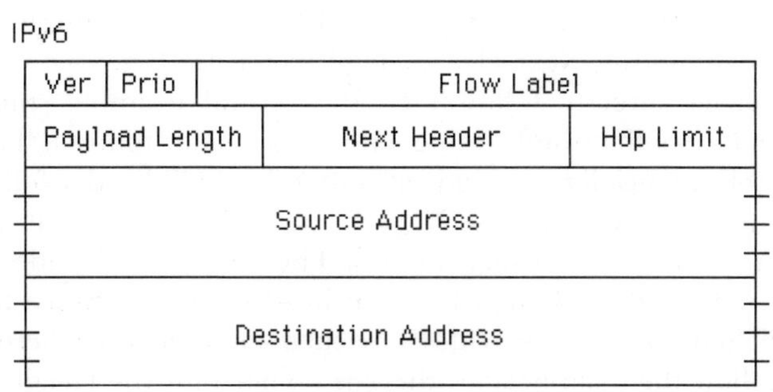

Figure 2.36 The IPv6 header. The only field that retains both its meaning and its position in IPv6 and IPv4 is the first, the version number (Ver)

The only field in the IPv6 header to retain both its meaning and its position is the version number (Ver). This 4-bit field is set to a value of 4 (0100 in binary) to identify an IPv4 packet and 6 (0110) to identify an IPv6 packet.

Although the IPv6 header does not contain any optional elements, it allows separate extension headers to be placed between the IP header and the transport layer header. For example, authentication and security encapsulation are performed using extension headers and identified by the 8-bit Next Header field which identifies the type of header immediately following the IPv6 header. Since each extension header other than the TCP header carries the header type of the following header, it becomes possible to create a 'daisy chain' of headers.

Currently, there are six extension headers defined in the IPv6 specification. These headers are listed in Table 2.6 in their recommended processing order when used in a daisy chain.

Table 2.6 IPv6 extension headers

Extension header	Description
Hop-by-hop options	Passes information, such as for management or debugging, to routers
Destination options	Carries information concerning one or more options to be performed by a receiver and action to be taken if the receiver does not recognize the option
Routing	Indicates a list of intermediate addresses through which a packet is relayed
Fragment	Contains information that enables a receiver to reassemble fragments
Authentication	Permits authentication of source addresses and enables a receiver to note that the contents of a packet were not altered during transmission
Encrypted security payload	Provides data confidentiality

Figure 2.37 illustrates three examples of the daisy chaining of IPv6 headers. The first example shows the use of an IPv6 header without an extension header. In this example, the Next Header field in the IPv6 header would indicate that the TCP header directly follows. In the second example, the IPv6 Next Header field indicates that the Routing extension header follows. The Next Header field in the Routing header then indicates that the TCP header follows the Routing header. In the third example, the Next header field in the Routing header indicates that the Authentication header follows. Then, the Next Header field in the Authentication header indicates that the TCP header follows. Since the use of the Next Header field under IPv6 provides a mechanism to denote whether or not extension headers follow the main header, there is no need for the Header Length field (IHL) which was used in IPv4. Similarly, the ability to use extension headers eliminated the necessity to retain the IPv4 variable length options field.

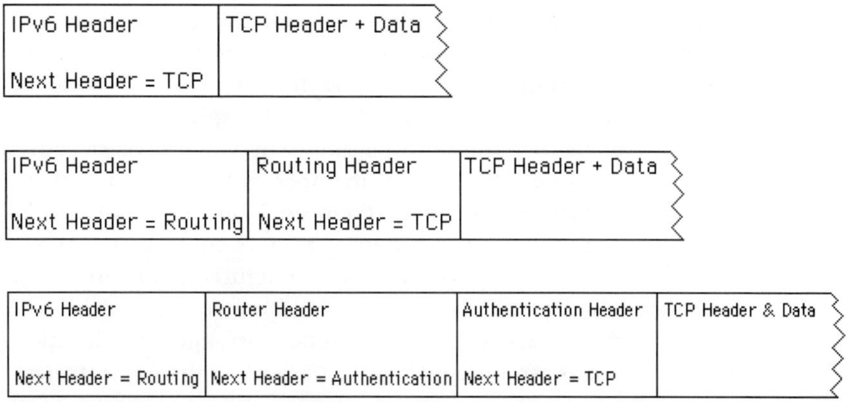

Figure 2.37 Creating a Daisy chain of headers. Through the chaining of headers specific information is easily added to packets to satisfy particular transmission requirements

Returning to a comparison of IPv6 and IPv4 headers, the Total Length field of IPv4 was replaced by the Payload Length field of IPv6. The two are similar but not equal, since the payload length represents the length of the data carried after the header while the Total Length field includes the length of the header.

Although the IPv4 Time to Live field was expressed as a number of seconds, in actuality its value is decreased by 1 as a packet flows through a router due to the difficulty associated with determining waiting times as a packet flows through a network. Recognizing the reality of using router hops as a method for decrementation, the Time to Live field was replaced by a Hop Limit field in IPv6.

Two new fields in the IPv6 header that were not included in IPv4 are Priority (Prio) and Flow Label. The 4-bit Priority field permits the originator to identify the priority of its packets. The 16 possible values of this field are divided into two ranges. Values 0 through 7 are used to denote the priority of traffic for which the originator is providing congestion control, while values 8 through 15 denote the priority of traffic that cannot back off in response to congestion. The latter represents real-time packets containing video, audio and similar data that is transmitted at a constant rate.

The 24-bit Flow Label field enables the identification of packets that require special handling and are routed from the same source to the same destination. This special handling can be conveyed by a control protocol, or by data within the flow's packets, such as the hop-by-hop header extension option. Through the use of the Priority and Flow Label headers, hosts can identify packets that require special handling by routers as well as the general method by which the packets should be processed with respect to other packets.

Addressing

There are three types of 128-bit IPv6 addresses—unicast, multicast, and anycast. A unicast address identifies a single interface. A multicast address identifies a group of interfaces. Thus, a packet transmitted to a unicast address is sent to a specific location while a packet sent to a multicast address are delivered to each member of the multicast group. The third type of IPv6 address, anycast, identifies a set of interfaces similar to multicast address; however, instead of being delivered to all members of the group, packets sent to an anycast address are delivered to only one interface—the nearest member of the anycast group.

Through the use of an anycast address a host does not have to know the specific address of a router since a group of routers could be assigned membership in an anycast group. Then, the host can be configured to use an anycast group address as its gateway address. This address would not require modification if at a later date the network was restructured to include rearranging the location of routers. This is because packets would continue to be sent to the nearest member of the anycast group. Based upon the use of anycast addresses, some of the reconfiguration problems associated with network restructuring now commonly experienced under IPv4 may be minimized under IPv6.

The format prefix

IPv6 allocates address space similar to the manner by which IPv4 addresses are grouped into classes based upon the composition of the first few bits of the address. This method of address space allocation is accomplished by a variable length field of unique bits denoted by the term Format Prefix. Table 2.7 lists the initial allocation of IPv6 address space to include the allocation assignment, binary prefix, and the fraction of the 128-bit address space assigned to the noted allocation.

Address notations and examples

Under the IPv6 design 128-bit addresses are written as eight 16-bit integers separated by colons. Each integer is represented by four hex digits, resulting in a reasonable expectation that you will have to enter 32 hex digits to specify an IP address. To facilitate the entry of 128-bit addresses, you can skip leading zeros in each hexadecimal integer sequence as well as use a double colon (: :) inside an address as a mechanism to replace a set of consecutive null 16-bit numbers. Both features can be expected to be used frequently during the initial stage of IPv6 deployment since only a portion of all 128 bits of the address need to used. This will more than likely result in addresses containing many zeros, enabling the ability to skip leading zeros in each hexadecimal component and use double colons to replace consecutive null numbers to facilitate the IP configuration process. For example, during what will probably be a lengthy transition period as the Internet moves from IPv4 to IPv6, IPv4-compatible IPv6 addresses will consist of a 32-bit IPv4 address in the low-ordered 32 bits of the IPv6 address space. Thus, such addresses transported in an IPv6 format will be prefixed with 96 zeros.

To illustrate the compaction of IPv6 addresses, consider the following 128-bit address created for illustrative purpose only:

504A : 0000 : 0000 : 0000 : 00FC : ABCD : 3A1F : 4D3A

Table 2.7 IPv6 initial address space allocation

Allocation assignment	Binary prefix	Fraction of address space
Reserved	0000 0000	1/256
Unassigned	0000 0001	1/256
Reserved for NSAP allocation	0000 001	1/128
Reserved for IPX allocation	0000 010	1/128
Unassigned	0000 011	1/128
Unassigned	0000 1	1/32
Unassigned	0001	1/16
Unassigned	001	1/8
Provider-based unicast address	010	
Unassigned	011	1/8
Reserved for geographically based unicast address	100	1/8
Unassigned	101	1/8
Unassigned	110	1/8
Unassigned	1110	1/16
Unassigned	1111 0	1/32
Unassigned	1111 10	1/64
Unassigned	1111 110	1/128
Unassigned	1111 1110 0	1/512
Link-local use address	1111 1110 10	1/1024
Site-local use address	1111 1110 11	1/1024
Multicast address	1111 1111	1/256

Since IPv6 enables leading zeros in each hexadecimal component to be skipped, you could enter the preceding address in a reduced form as follows:

504A : 0 : 0 : 0 : FC : ABCD : 3A1F : 4D3A

You can further simplify the above reduced form through the use of the IPv6 double-colon convention to eliminate one set of consecutive null 16-bit numbers. Thus, the reduced form shown above can be further reduced as follows:

504A :: FC : ABCD : 3A1F : 4D3A

Note that the further reduced IPv6 address containing the double-colon convention now consists of five specified 16-bit integers. Since each IPv6 address consists of eight 16-bit numbers, the double-colon represents eight munus five, or three, integers whose values are zero. This explains why it is a relatively simple process to expand a reduce address as well as why only one double-colon can be used within an IPv6 address Now that we have an appreciation for the types of IPv6 addresses and how they can be noted, let's turn our attention to a few IPv6 addresses. The first address we will look at is the IPv6 provider-based unicast address. This represents the first group of IPv6 addresses that will be allocated.

Figure 2.38 illustrates the format of a provider-based address. As you would expect from an examination of Table 2.7, the bit sequence 010 must be used as the address prefix for a provider-based address.

Prefix	Registry ID	Provider ID	Reserved	Subscriber ID	Reserved	Intrasubscriber
010	5 bits	16 bits	8 bits	24 bits	8 bits	64 bits

Figure 2.38 IPv6 provider-based unicast address format. The IPv6 provider-based unicast address is expected to be the first group of IPv6 addresses to be allocated

The Registry ID field identifies the Internet registry responsible for assigning the address, such as the InterNIC in North America, RIPE in Europe, and APNIC in Asia. The Provider ID field identifies the Internet Service Provider. This field is 16 bits in length and is followed by a reserved field of eight bits that could be used as an extension to the service provider. The 24-bit Subscriber ID identifies a unique user and the assignment of this portion of the provider-based address is the responsibility of the Internet Service Provider. Similar to the Provider ID field, the Subscriber ID field is followed by a reserved field eight bits in length that will initially be set to zero. This reserved field may be used as an extension to the Subscriber ID field. The remaining 64 bits in the provider-based address will be used in a manner similar to IPv4 Class A through C addresses in that they will identify a network and host. In actuality, since the ISP is identified by the Provider ID field, the Intrasubscriber field can be used by a network administrator who is a customer of the ISP to subdivide their assigned block of addresses into subnet and station-ID fields. The subnet portion would be used to identify different networks operated by the organization, while the Station ID field would identify stations on each network.

One of the more interesting aspects of IPv6 is the manner by which private networks will eventually be able to be connected to the Internet without requiring massive configuration changes as under IPv4. For example, under RFC 1918 the IETF reserved three blocks of addresses for networks that are not connected to the Internet. Although the intention of RFC 1918 is quite admirable, when an organization decides it's time to connect to the Internet they must obtain an appropriate group of addresses and then reconfigure IP addresses for each network device. Under IPv6 this address reassignment problem is handled far more gracefully due to the ability of allocate link-local use and site-local use addresses. Here the term link references a communications facility such as a frame relay or ATM network, a point-to-point leased line or a connection to an Ethernet, Token-Ring, or FDDI network. Thus, a link address represents an isolated device or transmission facility that has no router connection and is not currently connected to the Internet, while a link-local address provides a mechanism for connecting the network or device to the Internet without requiring a new address and an address reconfiguration. This capability is illustrated in the top portion of Figure 2.39 which illustrates the general format of a link-local address. Here the unique address field for connecting the link could represent a LAN MAC address. Since the prefix is 10 bits in length, the remainder of the address is 118 bits in length. Then, if the link-local address is used to connect an Ethernet network, the MAC address would be 48 bits in length. This would result in a link-local address having the 80-bit prefix FE80 :: followed by the 48-bit Ethernet MAC address. Thus, the use of a link-local address enables Ethernet, Token-Ring, and FDDI networks to be connected to the Internet without requiring the reconfiguration of network addresses.

Link-Local Address

1111111010	00....00	Unique address for link

Site-Local Address

1111111010	00....00	Unique address for link

Figure 2.39 Link-local and site-local addresses facilitate connecting private networks

The lower portion of Figure 2.39 illustrates the format of a site-local address under IPv6. This address would be assigned to a site that has routers but is not currently connected to the Internet via an ISP. When it's time to connect the site to the Internet, the router would be configured with a new prefix that would in effect generate a provider-based address through the inclusion of Registry ID, Provider ID, and Subscriber ID fields. Since no portion of the site-assigned part of the address would require changing, the connection of the site to the Internet is more gracefully accomplished under IPv6 than under IPv4.

Now that we have an appreciation for a few of the new types of addresses supported under IPv6, let's turn our attention to migration issues associated with moving to IPv6.

Migration issues

Recognizing reality, there will be no magic data for the 'mother of all cutovers' from IPv4 to IPv6. Instead, you can expect a gradual and incremental upgrade process to occur an you obtain new IPv6-compliant equipment for use. However, there is no cause for alarm for organizations that are happy with their current block of IP addresses and prefer the status quo. Such users will probably be able to continue using IPv4 for many years or until their existing equipment is gradually replaced by IPv6-compliant products and they elect to migrate. Thus, let's turn our attention to the two basic methods or migrating to IPv6 and how the migration process provides backward compatibility with IPv4.

Migration methods

There are two methods you can consider for migrating to IPv6—dual stacks and tunneling. Under the dual stack method each IP node becomes capable of supporting both IPv4 and IPv6 and is referred to as an IPv6/IPv4 node. Since the IPv6/IPv4 node can transmit and receive both IPv4 and IPv6 packets, it becomes capable of operating with both IPv4 and IPv6 nodes. Figure 2.40 illustrates a node operating a dual IPv6 and IPv4 stack. Here the IPv6/IPv4 node would be configured with a 32-bit address per interface to support IPv4 functions and 128-bit address per interface to support IPv6 functions.

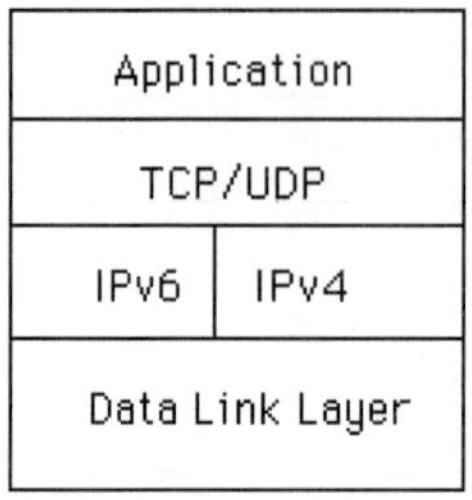

Figure 2.40 Through the use of a dual stack an IPv6/IPv4 node can communicate with both IPv4 only and IPv6 only nodes

The use of dual stack on a host provides a mechanism to gradually migrate a network to IPv6 on a host-by-host basis. In comparison, the use of a dual stack on a router could allow an organization to directly connect to an ISP that migrates to IPv6 as well as supports the graceful migration of hosts on the network by establishing dual stacks on those hosts. During the migration process, you will more than likely have to change your router's configuration information, update access control lists, and upgrade your Domain Naming System (DNS) to support larger IPv6 addresses. Although most IPv4 routing protocols can be used to route IPv6, the software upgrade may require additional RAM since routing tables are larger and dual stacks also require more memory. If you'are considering implementing dual stacks on hosts, you should check how memory is used and its effect on existing applications. Otherwise, it's quite possible that you may be unable to run your favorite application due to the manner by which dual stacks are implemented.

The second method for migrating to IPv6 is through tunneling. Tunneling is required because IPv6 is not backward compatible with IPv4 and IPv4-compatible devices cannot directly operate upon IPv6 packets. Thus, the movement of IPv6 packets via an IPv4 infrastructure requires tunneling which enables IPv6 packets to be transported via an IPv4 network, in effect encapsulating the IPv6 packet in an IPv4 packet. Since the Internet, as well as virtually all private IP-based networks, are currently based on an IPv4 infrastructure, tunneling enables IPv6 hosts and nodes to communicate with one another across what one day we will probably refer to as legacy IP networks.

Through the use of dual stacks you can obtain the ability to gradually migrate nodes on your network to IPv6. In comparison, tunneling enables those nodes to communicate with similar nodes across IPv4 networks. There are four tunnel configurations that can be established between routers and hosts—router to router, host to router, host to host, and router to host.

A router-to-router tunnel enables IPv6/IPv4 routers separated by an IPv4 network to tunnel IPv6 packets between themselves. If you upgrade a node to IPv6 on an IPv4 network, you would use a host-to-router tunnel to reach the router. Here the IPv6/IPv4 host would tunnel packets to the IPv6/IPv4 router. The host-to-host hunnel occurs when two IPv6/IPv4 hosts operating on an IPv4 network need to communicate with one another. The fourth tunnel configuration occurs when an IPv6/IPv4 router needs to communicate with an IPv6/IPv4 host via an IPv4 network. Then, the router encapsulates IPv6 packets in IPv4 packets and sends them to the IPv6/IPv4 host.

Although tunneling enables IPv6 and IPv4 to coexist across an IPv4 infrastructure, the additional header adds to network traffic and could be the proverbial straw that breaks the camel's back of heavily utilized network connection. Thus, some traffic monitoring and quick computation may be in order to determine if a WAN connection needs an upgrade to support the tunneling effort.

2.6 SNA AND APPN

To satisfy the requirements of customers for remote computing capability, mainframe computer manufacturers developed a variety of network architectures. Such architectures define the interrelationship of a particular vendor's hardware and software products necessary to permit communications to flow through a network to the manufacturer's mainframe computer.

IBM's System Network Architecture (SNA) is a very complex and sophisticated network architecture which defines the rules, procedures and structure of communications from the input–output statements of an application program to the screen display on a user's personal computer or terminal. SNA consists of protocols, formats and operational sequences which govern the flow of information within a data communications network linking IBM mainframe computers, minicomputers, terminal controllers, communications controllers, personal computers and terminals.

Since approximately 70% of the mainframe computer market belongs to IBM and plug-compatible systems manufactured by Amdahl and other vendors, SNA can be expected to remain as a connectivity platform for the foresseable future. This means that a large majority of the connections of local area networks to mainframe computers will require the use of gateways that support SNA operations.

As we will shortly note when examining SNA, it is a mainframe-centric, hierarchical structured networking architecture. While appropriate for most computer communications requirements of the 1980s, the growth in distributed processing and peer-to-peer communications represented a significant problem for network managers and LAN administrators that require access to mainframes in an SNA environment as well as the ability to support peer-to-peer communications. Recognizing this problem, IBM significantly revised SNA in the form of developing a new network architecture know as Advanced Peer-to-Peer Networking (APPN). After becoming familiar with SNA, we will then turn our attention to APPN in this section.

SNA concepts

An SNA network consists of one or more domains, where a domain refers to all of the logical and physical components that are connected to and controlled by one common point in the network. This common point of control is called the system services control point, which is commonly known by its abbreviation as the SSCP. There are three types of network addressable unit in an SNA network—SSCPs, physical units and logical units.

SSCP

The SSCP resides in the communications access method operating in an IBM mainframe computer, such as Virtual Telecommunications Access Method (VTAM), operating in a System/360, System/370, System/390, 4300 series or 308X computer, or in the system control program of an IBM minicomputer, such as System/3x or AS/400. The SSCP contains the network's address tables, routing tables and translation tables which it uses to establish connections between nodes in the network as well as to control the flow of information in an SNA network. Figure 2.41 illustrates single and multiple domain SNA networks.

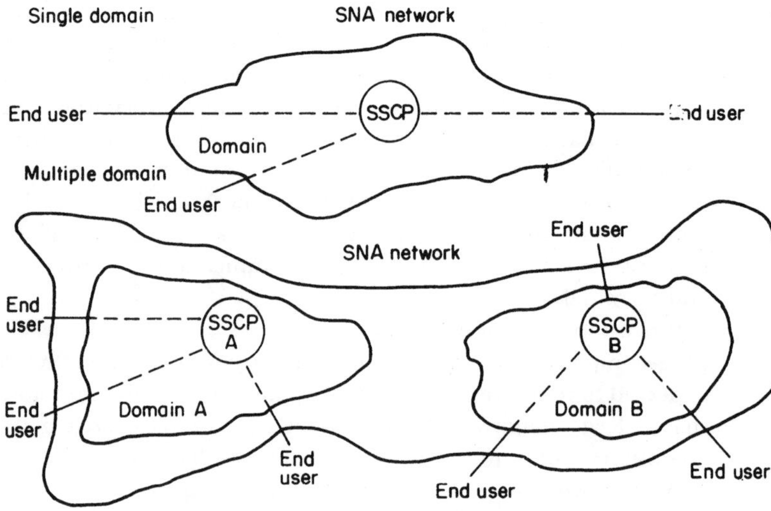

Figure 2.41 Single and multiple domain SNA networks

Network nodes

Each network domain will include one or more nodes, with an SNA network node consisting of a grouping of networking components which provides it with a unique characteristic. Examples of SNA nodes include cluster controllers, communications controllers and terminal devices, with the address of each device in the network providing its unique characteristic in comparison to a similar device contained in the network.

The physical unit

Each node in an SNA network contains a physical unit (PU) which controls the other resources contained in the node. The PU is not a physical device as its name appears to suggest, but rather a set of SNA components which provide service used to control terminals, controllers, processors and data links in the network. Those services include initializing nodes and managing the data flow between devices to include formatting and synchronizing data. In programmable devices, such as mainframe computers and communications controllers, the PU is normally implemented in software. In less intelligent devices, such as cluster controllers and terminals, the PU is typically implemented in read- only memory. In an SNA network each PU operates under the control of an SSCP. The PU can be considered to function as an entry point between the network and one or more logical units.

The logical unit

The third type of network-addressable unit in an SNA network is the logical unit, known by its abbreviation as the LU which represents a software module located in a devices memory. The LU is the interface or point of access between the end-user and an SNA network. Through the LU an end-user gains access to network resources and transmits and receives data over the network. Each PU can have one or more LUs, with each LU having a distinct address.

Multiple session capability

As an example of the communications capability of SNA, consider an end-user with an IBM PC and an SDLC communications adapter who establishes a connection to an IBM mainframe computer. The IBM PC is PU, with its display and printer considered to be LUs. After communications is established the PC user could direct a file to his or her printer by establishing an LU-to-LU session between the mainframe and printer while using the PC as an interactive terminal running an application program as a second LU-to-LU session. Thus, the transfer of data between PUs can represent a series of multiplexed LU-to-LU sessions, enabling multiple activities to occur concurrently.

SNA network structure

The structure of an SNA network can be considered to represent a hierarchy in which each device controls a specific part of the network and operates under the control of a device at the next higher level. The highest level in an SNA network is represented by a host or mainframe computer which executes a software module known as a communications access method. At the next lower level are one or more communications controllers, IBM's term for a front-end processor. Each communications controller executes a Network Control Program (NCP) which defines the operation of devices connected to the controllers, their PUs and LUs, operating rate, data code, and other communications-related functions such as the maximum packet size that can be transmitted. Connected to communications controllers are cluster controllers, IBM's term for a control unit. Thus, the third level in a SNA network can be considered to be represented by cluster controllers.

The cluster controllers support the attachment of terminals and printers which represent the lowest hierarchy of an SNA network. Figure 2.42 illustrates the SNA hierarchy and the Network Addressable Units (NAUs) associated with each hardware component used to construct an SNA network. Note that NAUs include lines connecting mainframes to communications controllers to cluster controllers, and cluster controllers to terminals and printers. In addition, NAUs also define application programs that reside in the mainframe. Thus, NAUs provide the mechanism for terminals to access specific programs via routing through hardware and transmission facilities that are explicitly identified.

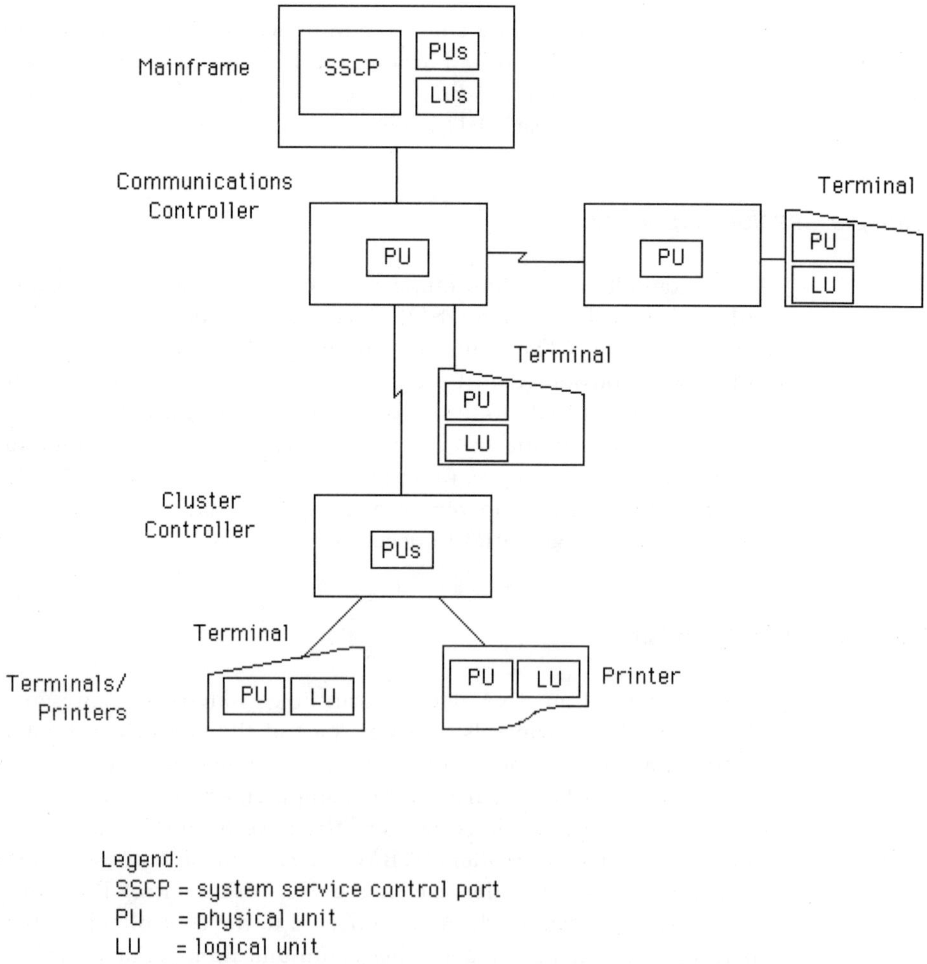

Figure 2.42 SNA hierarchical network structure. The structure of an SNA network is built upon a hierarchy of equipment, with mainframes connected to communications controllers and communications controllers connected to cluster controllers

Types of physical units

Table 2.8 lists five types of physical units in an SNA network and their corresponding node type. In addition, this table contains representative examples of hardware devices that can operate as a specific type of PU. As indicated in Table 2.8 the different types of PUs form a hierarchy of hardware classifications. At the lowest level, PU type 1 is a single terminal. PU type 2 is a cluster controller which is used to connect many SNA devices onto a common communications circuit. PU type 4 is a communications controller which is also known as a front-end processor. This device provides communications support for up to several hundred line terminations, where individual lines in turn can be connected to cluster controllers. At the top of the hardware hierarchy, PU type 5 is a mainframe computer.

Table 2.8 SNA PU summary

PU type	Node	Representative hardware
PU type 5	Mainframe	S/390, 43XX, 308X
PU type 4	Communications controller	3705, 3725, 3720, 3745
PU type 3	Not currently defined	N/A
PU type 2	Cluster controller	3274, 3276, 3174
PU type 1	Terminal	3180, PC with SNA adapter

The communications controller is also commonly referred to as a front-end processor. This device relieves the mainframe of most communications processing functions by performing such activities as sampling attached communications lines for data, buffering the data and passing it to the mainframe as well as performing error detection and correction procedures. The cluster controller functions similar to a multiplexer or data concentrator by enabling a mixture of up to 64 terminals and low-speed printers to share a common communications line routed to a communications controller or directly to the mainframe computer.

Multiple domains

Figure 2.43 illustrates a two-domain SNA network. By establishing a physical connection between the communications controller in each domain and coding appropriate software for operation on each controller, cross-domain data flow becomes possible. When cross-domain data flow is established terminal devices connected to one mainframe gain the capability to access applications operating on the other mainframe computer.

SNA was originally implemented as a networking architecture in which users establish sessions with application programs that operate on a mainframe computer within the network. Once a session is established a network control program (NCP) operating on an IBM communications controller, which in turn is connected to the IBM mainframe, would control the information flow between the user and the

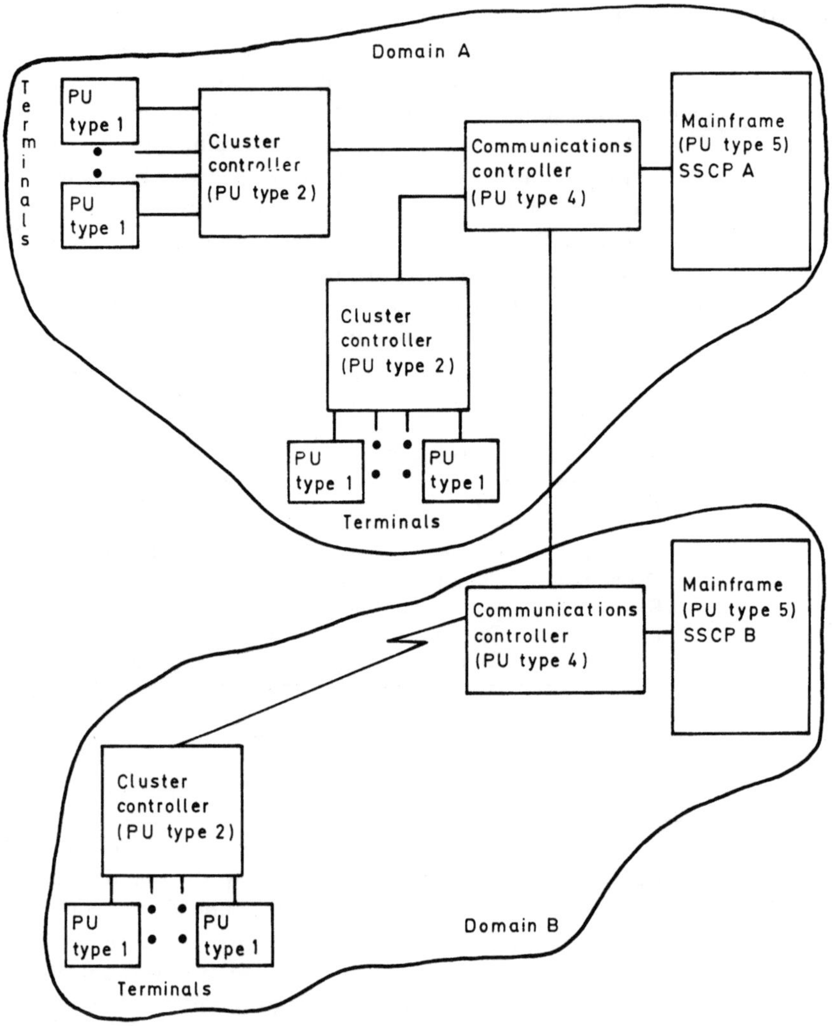

Figure 2.43 Two-domain SNA network

applications program. With the growth in personal computing many users no longer required access to a mainframe to obtain connectivity to another personal computer connected to the network. Thus, IBM modified SNA to permit peer-to-peer communications capability in which two devices on the network with appropriate hardware and software could communicate with one another without requiring access through a mainframe computer. In doing so, IBM introduced a PU2.1 node in 1987 in recognition of the growing requirement of its customers for a peer-to-peer networking capability. Unfortunately, it wasn't until the early 1990s that APPN became available and provided this networking capability, enabling communications between PU2.1 nodes without mainframe intervention through the use of LU6.2 sessions described later in this section.

SNA layers

SNA was developed as a connection-oriented communications service in 1974, predating the OSI Reference Model. A second version of SNA developed by IBM to include program-to-program communications capability differs from its original incarnation at the higher layers and more closery resembles the OSI Reference Model. Figure 2.44 provides a comparison between the current version of SNA and the ISO Reference Model which we will use for a discussion of SNA layers.

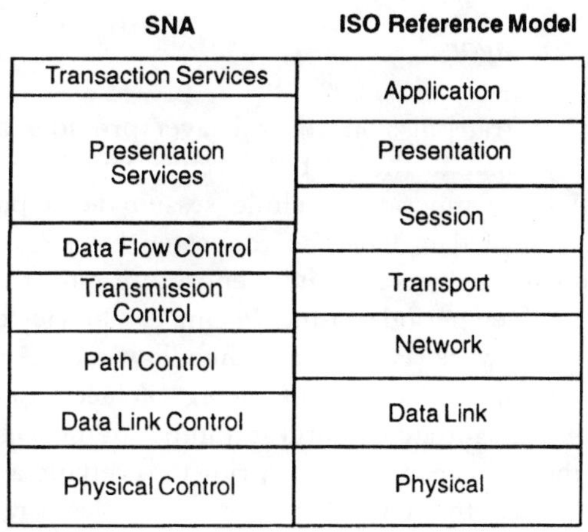

Figure 2.44 A comparison of SNA with the ISO Reference Model

Physical and data link layers

Similar to the OSI physical layer, SNA's physical control layer is concerned with the electrical, mechanical and procedural characteristics of the physical media and interfaces to the physical media. SNA's data link control layer is also quite similar to OSI's data link layer. Protocols defined by SNA include SDLC, System/370 channel, Token-Ring and X.25; however, only SDLC is used on a communications link in which master or primary stations communicate with secondary or slave stations. Some implementations of SNA using special software modules can also support Bisynchronous (BSC) communications, an IBM pre-SNA protocol still widely used as well as asynchronous communications.

Path control layer

Two of the major functions of the path control layer are routing and flow control. Concerning routing, since there can be many data links connected to a nose, path control is responsible for insuring that data is correctly passed through intermediate nodes as it flows from source to destination. At the beginning of an SNA session both sending and receiving nodes as well as all nodes between those nodes cooperate to select the most efficient route for the session. Since this route is only established for the duration of the session it is known as a virtual route. To increase the efficiency of transmission in an SNA network the path control layer at each node through which the virtual route is established has the ability to divide long messages into shorter segments for transmission by the data link layer. Similarly, path control may block short messages into larger data blocks for transmission by the data link layer. This enables the efficiency of SNA's transmission facility to be independent of the length of messages flowing on the network.

Transmission control layer

The SNA transmission control layer provides a reliable end-to-end connection service, similar to the OSI Reference Model transport layer. Other transmission control layer functions include session level pacing as well as encryption and decryption of data when so requested by a session. Here, pacing insures that a transmitting device does not send more data than a receiving device can accept during a given period of time. Pacing can be viewed as similar to the flow control of data in a network. However, unlike flow control which is essentially uncontrolled, NAUs negotiate and control pacing. To accomplish this the two NAUs at session end points negotiate the largest number of messages, known as a pacing group, that a sending NAU can transmit prior to receiving a pacing response from a receiving NAU. Here the pacing response enables the transmitting NAU to resume transmission. Session level pacing occurs in two stages along a session's route in an SNA network. One stage of pacing is between the mainframe NAU and the communications controller, while the second stage occurs between the communications controller and an attached terminal NAU .

Data flow control services

The data flow control services layer handles the order of communications wihin a session for error control and flow control. Here the order of communications is set by the layer controlling the transmission mode. Transmission modes available include full-duplex which permits each device to transmit any time, half-duplex flip-flop where devices can only transmit alternately and half-duplex contention, where one device is considered a master device and the slave cannot transmit until the master completes its transmission.

Presentation services layer

The SNA presentation services layer is responsible for the translation of data from one format to another. The layer also performs the connection and disconnection of sessions as well as updating the network configuration and performing network management functions. At this layer, the network addressable unit (NAU) services manager is responsible for formatting of data from an application to match the display or printer that is communicating with the application. Other functions performed at this layer include the compression and decompression of data to increase the efficiency of transmission on an SNA network.

Transaction service layer

The highest layer in SNA is the transaction services layer. This layer is responsible for application programs that implement distributed processing and management services, such as distributed data bases and document interchange as well as the control of LU-to-LU session limits.

SNA developments

The most significant development to SNA prior to the formal introduction of APPN can be considered to be the addition of new LU and PU subtypes to support what is known as Advanced Peer-to-Peer Communications (APPC) which represents the communications protocol of an APPN network. Previously, LU types used to define an LU-to-LU session were restricted to application-to-device and program-to-program sessions. LU1 through LU4 and LU7 are application-to-device sessions as indicated in Table 2.9, where LU4 and LU6 are program-to-program sessions.

Table 2.9 SNA LU session types

LU type	Session type
LU1	Host application and a remote batch terminal
LU2	Host application and a 3270 display terminal
LU3	Host application and a 3270 printer
LU4	Host application and SNA word processor or between two terminals via mainframe
LU6	Between applications programs typically residing on different mainframe computers
LU6.2	Peer-to-peer
LU7	Host application and a 5250 terminal

The addition to LU6.2 which operates in conjunction with PU 2.1 to support LU6.2 connections permits devices supporting this new LU to transfer data to any other device also supporting this LU without first sending the data through a mainframe computer. As new software products are introduced to support LU6.2 a more dynamic flow of data through SNA networks will occur, with many data links to mainframes that were previously heavily utilized or saturated gaining capacity as sessions between devices permit data flow to bypass the mainframe.

SNA sessions

All communications in SNA occur within sessions between NAUs. Here a session can be defined and a logical connection established between two NAUs over a specific route for a specific period of time, with the connection and disconnection of a session controlled by the SSCP. SNA defines four types of sessions; SSCP-to-PU, SSCP-to-LU, SSCP-to-SSCP and LU-to-LU. The first two types of sessions are used to request or exchange diagnostic and status information. The third type of session enables SSCPs in the same or different domains to exchange information. The LU-to-LU session can be considered as the core type of SNA session since all end-user communications take place over LU-to-LU sessions.

LU-to-LU sessions

In an LU-to-LU session one logical unit known as the Primary LU (PLU) becomes responsible for error recovery. The other LU, which normally has less processing power, becomes the secondary LU (SLU).

An LU-to-LU session is initiated by the transmission of a message from the PLU to the SLU. That message is known as a bind and contains information stored in the communications access method (known as VTAM) tables on the mainframe which identifies the type of hardware devices with respect to screen size, printer type, etc., configured in the VTAM table. This information enables a session to occur with supported hardware. Otherwise, the SLU will reject the bind and the session will not start.

Addressing

We previously discussed the concept of a domain which consists of an SSCP and the network resources it controls. Within a domain there exists a set of smaller network units that are known as subareas. In SNA terminology each host is a subarea as well as each communications controller and its peripheral nodes. The identification of an NAU in an SNA network consists of a subarea address and an element address within a subarea, having the format subarea:element. Here the subarea can be considered as being similar to an area code, as it identifies a portion of the network. Figure 2.45 illustrates the relationship between a domain and three subareas residing in that particular domain.

In SNA a subarea address is eight bits in length, while the element address within a subarea is also eight bits in length. This limits the number of subareas within a domain to 255 and also restricts the number of PUs and LUs within a subarea to 255. Each subarea address is shared by an SSCP and all of its LUs and PUs and represents a unique address within a domain. In comparison, element addresses are only unique within a subarea and can be duplicated. A third component of SNA addressing is a character-coded network name that is assigned to each component. Each name must be unique within a domain, and SSCPs maintain tables which map names to addresses. One of the key problems resulting from this addressing scheme

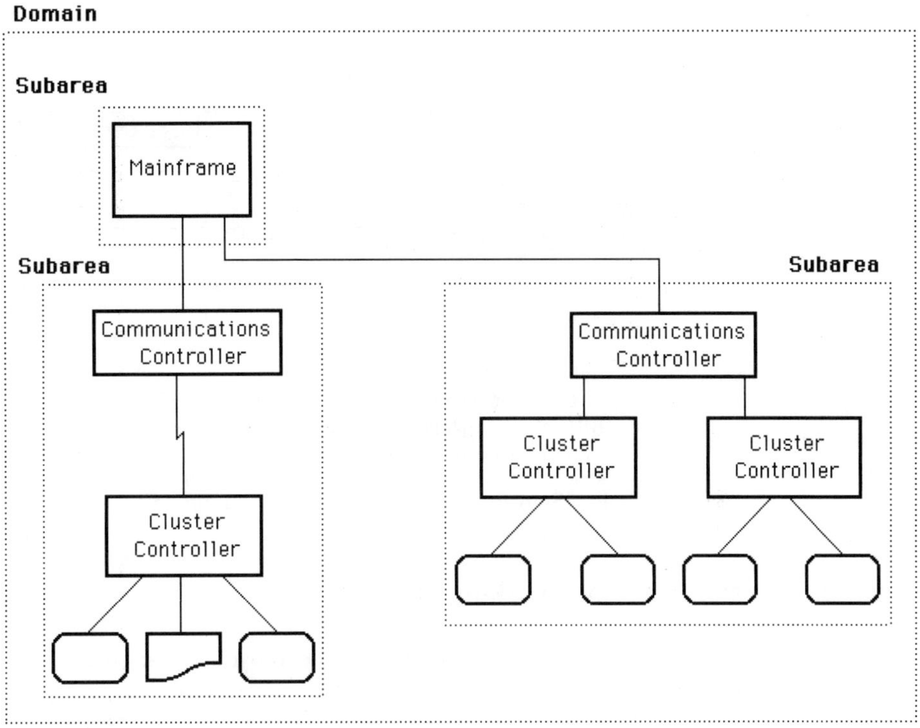

Figure 2.45 Relationship between a domain and its subareas. A subarea is a host or a communications controller and its peripheral nodes

is that unlike IP network addresses that are unique, SNA uses network names that can be duplicated by different organizations and makes SNA a 'non-routable' protocol, unlike TCP/IP which is a routable protcol.

The routing of packets between SNA subarea nodes occurs through the use of a sequence of message units created at different layers in the protocol stack. An application on an SNA node generates a Request Header (RH) which prefixes user data to form a Basic Transmission Unit (BTU) as shown at the top of Figure 2.46. At the path control layer, a Transmission Header (TH) is added as a prefix to the BTU to form a Basic Information Unit (BIU). The TH contains source and destination addresses that represent the sender and receiver of the packet, with subarea nodes examining the destination address within the TH of the BIU to make forwarding decisions.

Once a packet arrives at its destination subarea, routing must occur between the subarea and peripheral nodes in the destination subarea. That routing is based upon the Data Link Header (DLH) added to the BIU to form a Basic Link Unit (BLU), shown at the bottom of Figure 2.46. Now that we have a basic understanding of SNA, let's turn our attention to APPN.

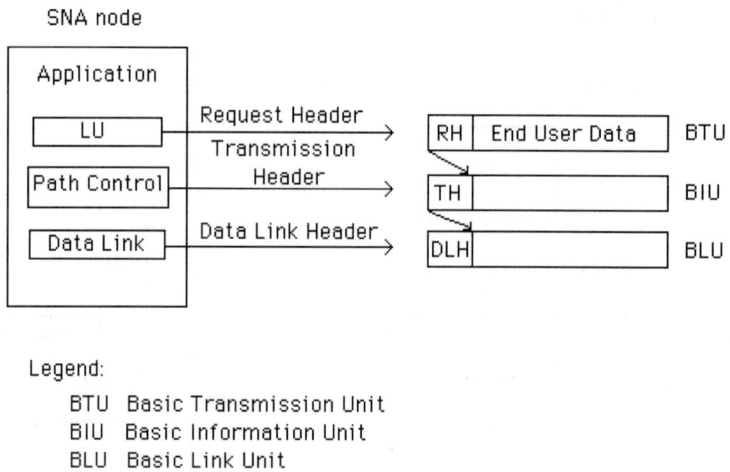

Figure 2.46 SNA routing based upon the headers in different message units

Advanced Peer-to-Peer Networking (APPN)

Although SNA represents one of the most successful networking strategies developed by a vendor, its centralized structure based upon mainframe-centered computing became dated in an evoloving era of client-server distributive computing. Recognizing the requirements of organizations to obtain peer-to-peer transmission capability instead of routing data through mainframes, IBM developed its Advanced Peer-to-Peer Networking (APPN) architecture during 1992 as a mechanism for computers ranging from PCs to mainframes to communicate as peers across local and wide area networks. The actual ability of programs on different computers to communicate with one another is obtained from special software known as Advanced Program-to-Program Communication (APPC) which represents a more marketable name for LU6.2 software. Since APPC enables the operation of APPN, we will first focus our attention upon APPC prior to examining the architecture associated with APPN.

APPC concepts

APPC represents a software interface between programs requiring communications with other programs and the network to which computers running those programs are connected. APPC represetns an open communications protocol that is available on a range of platforms to include PCs, mainframes, Macintosh, and UNIX-based systems as well as IBM 3174 control units and its 6611 Nways router series.

In its most basic structure, APPC can be considered to represent a stack existing above a network adapter but below the application using the adaptor. Figure 2.47 illustrates the general relationship of APPC to the software stack on two computers communicating with one another on a peer-to-peer basis. In this example, program A on computer 1 is shown communicating with program B on computer 2.

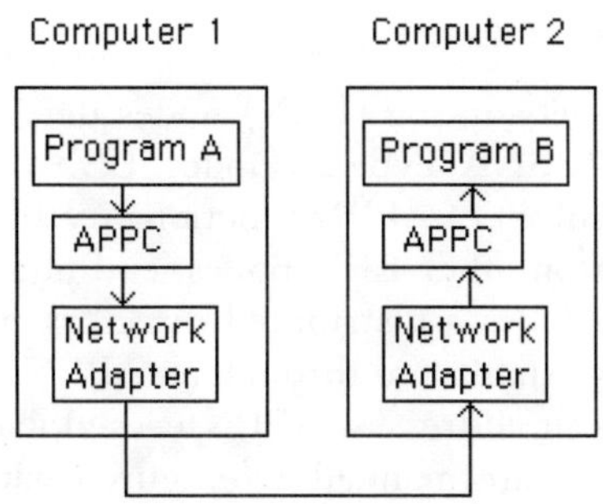

Figure 2.47 APPC software provides an interface between application programs and the network used

Under APPC terminology communication between two programs is referred to as a conversation which occurs according to a set of rules defined by the APPC protocol. Those rules specify how a conversation is established, how data is transmitted, and how the conversation is broken or deallocated.

Similar to most modern programming concepts, APPC supports a series of verbs that provide an application programming interface (API) between transaction programs and APPC software residing on a host. For example, the APPC verb ALLOCATE is used to initiate a conversation with another transaction program, the verb SEND_DATA enables the program to initiate the data transfer to its partner program, and the verb DEALLOCATE would be used to inform APPC to terminate the conversation previously established. Now that we have a general appreciation for the software that enables program-to-program communications, let's turn our attention to the architecture of APPN.

APPN architecture

APPN is a platform-independent network architecture which consists of three types of computers: low-entry networking (LEN) nodes, end nodes (ENs), and network nodes (NNs). Similar to SNA, APPN applications use network resources via logical unit (LU) software. LUs reside on each type of APPN node and application-to-application sessions occur between LUs, which in the case of APPN are LU6.2 LUs. APPN nodes can support a nearly unlimited number of LUs in comparison to the 255 SNA supports. In addition, APPN LUs can support multiple users, significantly inceasing its flexibility over SNA.

When an application program on one host requires communications with an application on another host, the first host tells its local LU to find a partner LU. The location of the partner LU is accomplished by a process in which several types of searches occur through an APPN network. Since those searches depend upon knowledge of the characteristics of the three types of APPN nodes, let's first turn our attention to the features and functions of those nodes.

LEN nodes

Low-entry networking (LEN) nodes data from the early 1980s when they were introduced as SNA Type 2.1 nodes. LEN nodes can be considered to represent the most basic subset of APPN functionality and have the ability to communicate with applications on other LEN nodes, end nodes, or network nodes.

APPN includes a distributed directory mechanism which enables routes to be dynamically established through an APPN network. However, unlike IP networks that use 32-bit addresses, APPN uses alphanumeric names.

LEN nodes are manually configured with a limited set of LUs. Thus, to use APPN directory services a LEN node requires the assistance of an adjacent APPN node, where adjacency is obtained through a LAN connection or a direct point-to-point link.

End nodes

End nodes (ENs) can be viewed as a more sophisticated type of LEN. In addition to supporting all of the functions of LEN nodes, end nodes know how to use APPN services, such as its directory services. To learn how to use such services, an end node identifies itself to the network when it is initially brought up. This identification process is accomplished by the end node registering its LUs with a network node server. Here the NN represents the third type of APPN node and is discussed in the next section. In comparison, LEN nodes do not perform this activity.

Network nodes

Network nodes (NNs) are the third component of an APPN network. NNs provide all of the functions associated with end nodes as well as routing and partner LU location services. Concerning routing, NNs work together to route information between such nodes, in effect providing a backbone transmission capability.

The partner LU location service depends upon network node searches when the partner is not registered by the NN serving the requestor. In such situations, the NN server will broadcast a search request to adjacent network nodes, requesting the location of the partner LU. This broadcasting will continue until the partner LU is located and a path or route is returned. Since broadcast searches are bandwidth intensive, NNs place directory entries they locate into cache memory which serves to limit broadcasts being propagated through an APPN network.

Operation

To illustrate the operation of an APPN network, let's examine a small network in which two end nodes are connected by a network node. Figure 2.48 illustrates an example of this network structure.

When the links between EN1 and NN1 and EN2 and NN1 are activated, the computers on each link automatically inform each other of their capabilities to include whether they are an end node or a network node, and ENs will register their capabilities with NNs. Thus, the NN will know the location and capability of

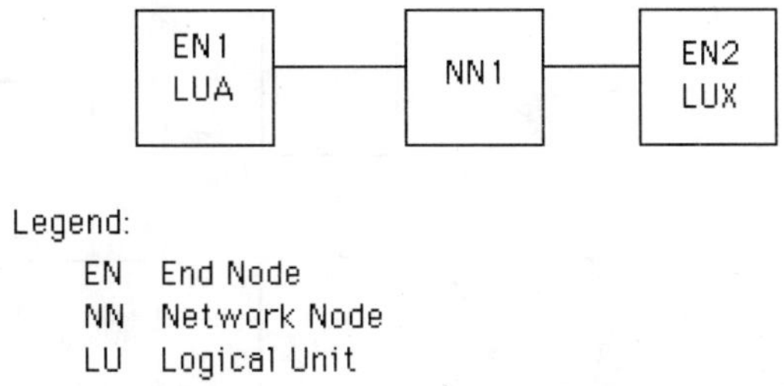

Figure 2.48 An APPN network consisting of two end nodes and a network node

both EN1 and EN2. When an application of EN1 needs to locate an LU in the network, such as LUX, it sends a request to its network node server, in this case NN1. Since NN1 is the server for EN1, both nodes establish a pair of control-point sessions to exchange APPN control information and EN1 registers its APPC LUs with NN1. Similarly, EN2 and NN1 establish a pair of control-point sessions when the link between those two nodes is brought up. Thus, NN1 knows how to get to EN1 and EN2 and which LUs are located at each node.

When EN1 asks NN1 to find LUX and determine a path through the network, NN1 checks its cache memory and notes that the only path available is EN1 to NN1 to EN2. NN1 passes this path information back to EN1 which enables the application operating on EN1 to establish an APPN session to LUX and initiate the exchange of information.

Now that we have an appreciation for basic APPN routing, let's examine a more complex example in which originating and destination LUs reside on end nodes separated from one another by multiple NNs. Figure 2.49 illustrates this more complex APPN network, consisting of four end nodes and three network nodes grouped together in a topology which allows multiple path routing between certain nodes.

Let's assume an APPN application on EN1 wants to initiate a conversation with an application on EN4. EN1 first requests NN1 to locate EN4 and determine which path through the network should be used. Since NN1 is not EN4's network node server, it will initally have no knowledge of EN4's location. Thus, NN1 will transmit a request to each adjacent network node in its quest to locate EN4. Since NN2 is the only network node adjacent to NN1, it passes the request to its adjacent nodes. Based upon the configuration shown in Figure 2.49, there is only one adjacent network node, NN3. Although EN4 is connected to both NN2 and NN3, an end node can have only one network node server. Thus, if we assume NN3 is the server for EN4, then NN2 has no knowledge of EN4 and does not respond on its behalf.

Next, upon locating EN4, NN3 queries EN4 to determine its existing communications links. Upon receipt of information from EN4 that has links to both NN2 and NN3, NN2 passes the information about EN4 to NN2 which passes the information back to NN1. NN1 uses the information received to determine which route to EN4 is best, selects an appropriate route, and passes the selected route back to EN1, allowing that end node to establish an APPC session to EN4.

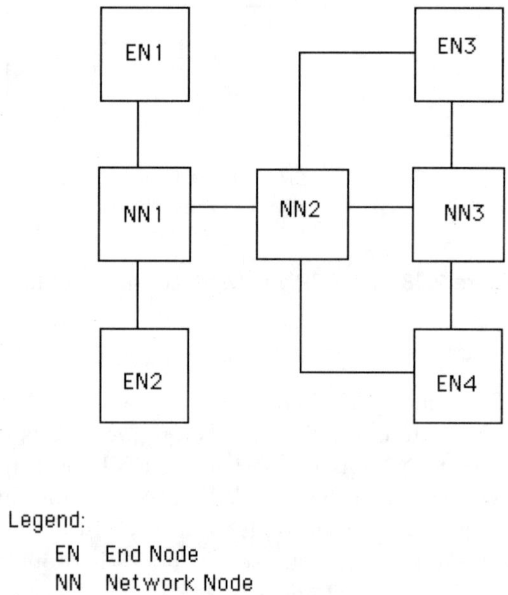

Figure 2.49 A more complex APPN network with multiple paths to some nodes

Route selection

Under APPN, routing is based upon network nodes maintaining a 'route addition resistance' value set up by network administrators and 'class of service' of data to be routed. Class-of-service routing enables different types of data to be routed via paths optimized for batch, interactive, batch-secure and interactive-secure. APPN uses eight values that are defined for each network link that are used in conjunction with the class of service to select an appropriate route. Value defined for each link include propagation delay, cost per byte, cost for connect time, effective capacity, and security. Using values defined for the links, a batch session might be routed on a path with high capacity and low cost, while an interactive session would probably be placed on a terrestrial link instead of a satellite link to minimize propagation delay.

APPN can be considered to represent a considerable enhancement to SNA as it provides efficient routing services that bypass the requirement of SNA date to flow in a hierarchical manner. However, APPN is similar to SNA in the fact that such networks do not have true network addresses, making pure routing between different networks difficult. In addition, the structure of APPN is similar to SNA with respect to its basic network layer operations which are illustrated in Figure 2.50. In examining Figure 2.50 note that APPN's network layer can be considered to represent APPC which converts LU service requests into frames for transport at the data link layer. Although SNA was originally limited to LLC Type 2 (LLC2) and SDLC transmission via a variety of physical layer interfaces, a number of conversion devices have been developed which extend both SNA and APPN transmission to frame relay. In addition, other products enable SNA and APPN data to be transported under a different network layer, a technique referred to as

<table>
<tr><td rowspan="2">Application</td><td colspan="4">Application Program</td></tr>
<tr><td colspan="4">LU Services</td></tr>
<tr><td>Network Layer</td><td colspan="4">APPC Path Control</td></tr>
<tr><td>Data Link Layer</td><td>LLC2
SNAP
Ethernet 2</td><td colspan="2">LLC2
SNAP</td><td>Frame Relay, SDLC</td></tr>
<tr><td>Physical Layer</td><td>Ethernet/
802.5</td><td>Token-
Ring</td><td>FDDI</td><td>V.24, V.35, RS232,
T1, E1, HSSI</td></tr>
</table>

Figure 2.50 The general structure of APPN

encapsulation or tunneling. In Chapter 5 when we examine LAN internetworking devices we will turn our attention to several gateway solutions, such as techniques to encapsulate SNA and APPN such that they can be routed over an IP network, a technique referred to as Data Link Switching (DLSw).

2.7 ATM

In concluding this chapter covering the fundamentals of wide area networks, we will focus our attention upon an emerging technology which represents a mechanism to transport voice, data, video, and images between private and public networks. This emerging technology is known as Asynchronous Transfer Mode, or ATM, and offers the ability to remove the barriers between local and wide area networks, providing a seamless interconnection for LAN internetworking between geographically separated areas. Although ATM represents a ubiquitous networking technology that can be used for both local and wide area networking, in this section we will primarily focus our attention upon its technology and use in a wide area networking environment. When we cover the fundamentals of local area networks in Chapter 3, we will then turn our attention to the operation of ATM in a LAN environment.

Overview

ATM represents a joining of packet switching and multiplexing technology to provide a transport facility for the movement of voice, data, video, and images over a common network infrastructure. Packet switching enables the use of network resources only when it becomes necessary to transmit information, while multiplexing enables communications from multiple sources to share the use of the ATM infrastructure by time. However, unlike conventional packet switching that divides messages into variable-length segments, ATM breaks all messages into fixed-length 53-byte cells. Through the use of fixed length cells, switching and multiplexing functions become more efficient. In addition, as we will shortly note, the use of relatively short cells makes it possible to transport a mixture of voice, video, data and image messages on a common infrastructure without the delay associated with transporting one type of message adversely affecting the receipt of cells containing another type of message.

Cell size

The ATM packet is relatively short, containing a 48-byte information field and a 5-byte header. This fixed-length packet, which is illustrated in Figure 2.51, is called a cell and the umbrella technology for the movement of cells is referred to as cell relay. ATM represents a specific type of cell relay service which is defined under broadband ISDN (BISDN) standards.

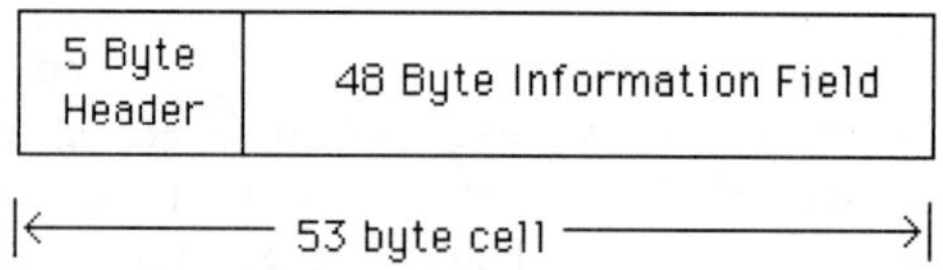

Figure 2.51 The ATM cell. The ATM cell is of fixed length, consisting of a 48-byte information field and a five-byte header

Benefits

The selection of a relatively short 53-byte ATM cell was based upon the necessity to minimize the effect of data transportation upon voice. For example, under frame relay or X.25 packet switching, both of which support variable-length information fields, a user could generate a burst of data which expands the frame length to a maximum value many times its minimum length. In doing so one user can effectively preclude other users from gaining access to the network until the first frame is completed. While this is a good design principle when all frames carry data, when digitized voice is transported the results can become awkward at the receiver. To illustrate this, assume your conversation is digitized and transported on a conventional packet switching network. As you say 'HELLO', 'HE' might be digitized and transported by a frame. Just prior to the letters 'LLO' leaving your mouth, assume a computer transmitted a burst of traffic that was transported by a long frame. Then, the digitization of 'LLO' would be stored temporarily until a succeeding frame could transport the data. At the receiver time gaps would appear, making the conversation sound awkward whenever bursts of data occurred. ATM is designed to eliminate this problem as cells are relatively short, resulting in cells transporting voice being able to arrive on a regular basis.

Another important advantage associated with the use of fixed-length cells concerns the design of switching equipment. This design enables processing to occur through the use of firmware instead of requiring more expensive software if variable-length cells were supported.

In addition to its relatively short cell length facilitating the integration of voice and data, ATM provides three additional benefits in the areas of scalabilitiy, transparency, and traffic classification.

Scalability

ATM cells can be transported on LANs and WANs at a variety of operating rates. This enables different hardware, such as LAN and WAN switches, to support a common cell format, a feature lacking with other communications technologies. Today, an ATM cell generated on a 25 Mpbs LAN can be transported from the LAN via a T1 line at 1.544 Mbps to a central office where it might be switched onto a 2.4 Gbps SONET network for transmission on the communications carrier infrastructure, the message being maintained in the same series of 53-byte cells, with only the operating rate scaled for a particular transport mechanism.

Transparency

The ATM cell is application transparent, enabling it to transport voice, data, images, and video. For this reason ATM enables networks to be constructed to support and type of application or application mix instead of requiring organizations to establish separate networks for different applications.

Traffic classification

When ATM was initially planned, it was recongnized that the ability to intermix voice, data, video, and images on a common network infrastructure would require the grouping of traffic into classes based upon the type of service they require. Once this was accomplished, a mechanism could be developed to handle traffic based upon its class of service. That mechanism, which we will cover later in this section, is known as the ATM Adaptation Layer (AAL) which is responsible for mapping information based upon its class of service into ATM cells. This mapping function includes the segmentation of data into cells for transport over the ATM infrastructure as well as the reassembly of cells received via the ATM infrastructure into their original data structure.

Four classes of traffic are defined by ATM. Each traffic class is defined through the use of three key attributes—the timing relationship between the source and destination, the variability of the bit rate, and its connection mode.

The timing relationship between the source and destination determines whether or not the receiver must be able to receive the originated data stream at the same rate at which it was originated. For example, a voice conversation which is digitized using PCM into a 64 kbps data stream must be 'read' by the receiving device at that rate to be correctly interpreted. In comparison, the transfer of a data file at 1.544 Mbps could be correctly received by a receiver operating at 64 kbps, although the reception of the file would take longer to occur than if the receiver operated at the same rate as the transmitter.

The second attribute, bit rate, defines whether the application requires a constant bit rate or can be supported via a variable bit rate. PCM-encoded voice would require a constant bit rate, while a variable bit rate would be suitable for the transfer of a file.

The third attribute, connection mode, determines whether the higher layer protocol above the ATM stack requires a connection to the distant location to be acknowledged prior to communications occurring. The connection mode can be connection-oriented or connectionless. Connection-oriented means a connection must be established prior to actual data transfer occurring, while connectionless refers to transmission occurring on a 'best effort' basis, with an acknowledgment flowing back only after transmission was initiated. Examples of connection-oriented applications include the public switched telephone network and such data transfer protocols as IBM's SNA and the X.25 packet transfer protocol. Examples of connectionless applications include the Ethernet LAN data link porotocol and TCP/IP that can operate on both LANs and WANs.

Table 2.10 indicates the relationship between four ATM classes of service defined by the ITU-T, their three key attributes, and the ATM Adaptation Layer (AAL) associated with the class of service.

Table 2.10 ATM classes of service

Class	Timing relationship	Bit rate	Connection mode	ATM Adaptation Layer
A	Required	Constant	Connection-oriented	AAL 1
B	Required	Variable	Connection-oriented	AAL 2
C	Not required	Variable	Connection-oriented	AAL3/4 and AAL5
D	Not required	Variable	Connectionless	AAL3/4 and AAL5

Classes A and B require a timing relationship between source and destination. Thus, they are both suitable for voice and video applications that cannot tolerate variable delays. In comparison, Classes C and D do not require a timing relationship. As such, they are oriented to supporting data transmission applications, with Class D supporting via simulation a connectionless communications mode of transmission which is commonly used on LANs.

Originally, five types of AALs were to be defined; however, AAL 3 and AAL 4 had a sufficient degree of commonality that enabled them to be merged. Later in this section when we examine the ATM protocol stack we will also examine each AAL. By associating such metrics as cell transmit delay, cell loss ratio, and cell delay variation to a traffic class, it becomes possible to provide a guaranteed quality of service (QoS) on a demand basis. This enables a traffic management mechanism to adjust network performance during periods of unexpected congestion to favor one or more classes of service based upon the metrics associated with each class. QoS provides the mechanism whereby constant bit-rate applications, such as voice, obtain priority over variable bit-rate applications. Although ATM enables a mixture of time-sensitive and time-insensitive applications to occur over a common network infrastructure, it is the QoS associated with each application which controls its ability to gain access to network resources.

The ATM protocol stack

Similar to other networking architectures, ATM is a layered protocol. The ATM protocol stack is illustrated in Figure 2.52 and consists of three layers—the ATM Adaptation Layer (AAL), the ATM Layer, and the Physical Layer. Both the AAL and Physical Layers are subdivided into two sublayers. Although the ATM protocol stack consists of three layers, as we will shortly note, those layers are essentially equivalent to the first two layers of the ISO Reference Model.

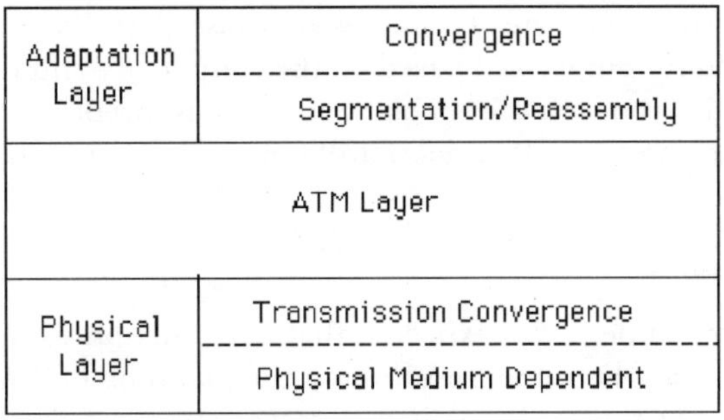

Figure 2.52 The ATM protocol stack

ATM Adaptation Layer (AAL)

As illustrated in Figure 2.52, the ATM Adaptation Layer consists of two sublayers—a convergence sublayer and a segmentation and reassembly sublayer. The function of the AAL is to adapt higher level data into formats compatible with ATM Layer requirements. To accomplish this task the ATM Adaptation Layer subdivides user information into segments suitable for encapsulation into the 48-byte information fields of cells. The actual adaptation process depends upon the type of traffic to be transmitted, although all traffic winds up in similar cells. There are five different AALs defined, referred to as AAL classes, with the relationship between traffic classes and AAL classes previously listed in Table 2.10. Although the ATM Adaptation Layer appears to be a network process, it is actually performed by network terminating equipment on the side of the User Network Interface (UNI). Through the inclusion of the AAL, the network does not have to be concerned with different types of traffic and is structured to facilitate the routing of cells between source and destination based upon information contained in each cell header.

Each AAL class represents a standard which defines how the data flow associated with a specific traffic class is converted into cells. Although not indicated in Table 2.10, the fifth AAL actually represents the lack of an adaptation processing standard, resulting in the mapping of a non-classified data stream into a sequence of 48-byte cell payloads. The lack of adaptation processing rules is referred to as AAL 0, the null adaptation layer, and requires source and destination equipment to assume responsibility for information formatting cell sequencing, cell loss and similar functions. An overview of the other AALs follows.

AAL 1

As indicated in Table 2.10, AAL 1 is associated with Class A traffic. Included in the AAL 1 process is the use of a timer to ensure that output and input are properly synchronized; however, AAL 1 does not differentiate between actual data and idle bits at the input interface.

AAL 2

AAL 2 is associated with Class B traffic that requires a timing relationship but uses a variable bit rate. Examples of Class B traffic are some types of video that are variably compressed based on the complexity and rate of change of moving images. At the time of this book revision was prepared AAL 2 was not standardized, requiring Class B services to be provided through either AAL 1 or AAL 3/4.

AAL 3/4

At one time it was expected that a distinction would be required between Class C and Class D traffic, resulting in the planning of AAL 3 for Class C and AAL 4 for Class D traffic. Later it was recognized that the two adaptation layers could be combined, resulting in AAL 3/4.

AAL 3/4 provides a data transport service for both connection-oriented and connectionless data. Information can be transferred in either a message or in a data streaming mode, with several optional delivery methods supported by this adaptation layer. When messages are presented to this layer, AAL 3/4 will perform sequencing to detect lost information. AAL 3/4 can also support the interleaving of multiple messages from the same source, which prevents short high-priority messages from being delayed behind longer messages that have a lower priority.

AAL 5

Originally referred to as the Simple and Efficient Adaptation Layer (SEAL), AAL 5 provides a similar service to AAL 3/4 but uses less overhead. AAL 5 can support both Class C and Class D traffic when the originator and destination can provide the ability to detect lost information and the interleaving of multiple messages is not required.

When receiving information, the ATM Adaptation Layer performs a reverse process. That is, it takes cells received from the network and reassembles them into a format the higher layers in the protocol stack understand. This process is known as reassembly. Thus, the segmentation and reassembly processes result in the name of the sublayer that performs those processes.

The ATM Layer

The ATM Layer provides the interface between the AAL and the Physical Layer. The ATM Layer is responsible for relaying cells both from the AAL to the Physical

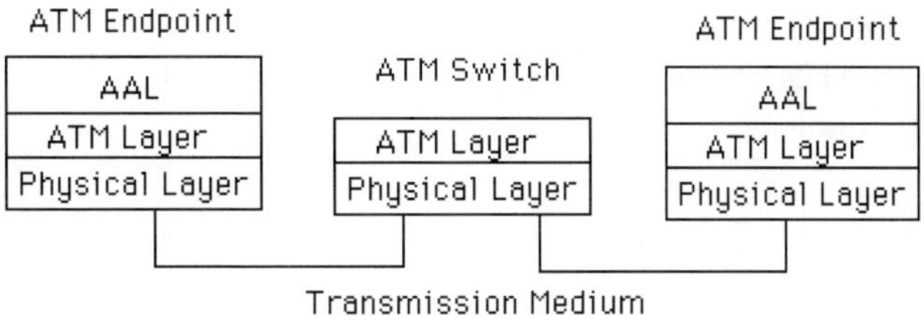

Figure 2.53 The ATM protocol stack within a network. The ATM Adaptation Layer is only required at endpoints within an ATM network

Layer and to the AAL from the Physical Layer. The actual method by which the ATM Layer performs this function depends upon its location within an ATM network. Since an ATM network consists of endpoints and switches, the ATM Layer can reside at either location. Similarly, a Physical Layer is required at both ATM endpoints and ATM switches.

Since a switch examines information within an ATM cell to make switching decisions, it does not perform any adaptation functions. Thus, the ATM switch operates at layers 1 and 2, while ATM endpoints operate at layers 1 through 3 of the ATM protocol stack as shown in Figure 2.53.

When the ATM Layer resides at an endpoint, it will generate idle or 'empty' cells whenever there is no data to send, a function not performed by a switch. Instead, in the switch the ATM Layer is concerned with facilitating switching functions, examining cell header information which enables the switch to determine where each cell should be forwarded to. For both endpoints and switches, the ATM Layer performs a variety of traffic management functions to include buffering incoming and outgoing cells as well as monitoring the transmission rate and conformance of transmission to service parameters that define a quality of service. At endpoints the ATM Layer also indicates to the AAL whether or not there was congestion during transmission, permitting higher layers to initiate congestion control.

The Physical Layer

Although Figures 2.52 and 2.53 illustrate an ATM Physical Layer, a specific physical layer is not defined within the protocol stack. Instead, ATM uses the interfaces to existing physical layers defined in other protocols, which enables organizations to construct ATM networks on different types of physical interfaces which in turn connect to different types of media. Thus, the omission of a formal physical layer specification results in a significant degree of flexibility which enhances the capability of ATM to operate on LANs and WANs.

ATM operation

As previously discussed, ATM represents a cell-switching technology that can operate at speeds ranging from T1's 1.544 Mbps to the gigabit per second rate of

SONET. In doing so the lack of a specific physical layer definition means that ATM can be used on many types of physical layers, which makes it a very versatile technology.

Components

A basic ATM network consists of endpoints and one or more switches, with switches used to examine the header in each cell and route cells received on one switch port onto another port towards its destination. The endpoint can be any device that supports the three layers of the ATM protocol stack; however, the most common type of endpoint is represented by network interface cards installed in LAN workstations.

ATM network interface cards

An ATM network interface card (NIC) is used to connect a LAN based workstation to an ATM LAN switch. The NIC converts data generated by the workstation into cells that are transmitted to the ATM LAN switch and converts cells received from the switch into a data format recognizable by the workstation.

ATM switches

An ATM switch is a multiport device which forms the basic inrastructure for an ATM network. By interconnecting ATM switches an ATM network can be constructed to span a building, city, country, or the globe.

The basic operation of an ATM switch is to route cells from an input port onto an appropriate output port. To accomplish this the switch examines fields within each cell header and uses that information in conjunction with table information maintained in the switch to route cells.

Network interfaces

ATM supports two types of basic interfaces: User-to-Network Interface (UNI) and Network-to-Node Interface (NNI).

User-to-Network Interface

The UNI represents the interface between an ATM switch and an ATM endpoint. Since the connection of a private network to a public network is also known as a UNI, the terms Public and Private UNI were used to differentiate between the two types of User-to-Network Interfaces. That is, a Private UNI references the connection between an endpoint and switch on an internal, private ATM network, such as an organizaiton's ATM based LAN. In comparison, a Public UNI would reference the interface between either a customer's endpoint or switch and a public ATM network.

Network-to-Node-Interface

The connection between an endpoint and a switch is similar than the connection between two switches. This results from the fact that switches communicate information concerning the utilization of their facilities as well as pass setup information required to support endpoint network requests.

The interface between switches is known as a Network-to-Node or Network-to-Network Interface (NNI). Similar to the UNI, there are two types of NNIs. A Private NNI describes the switch-to-switch interface on an internal network such as an organization's LAN. In comparison, a Public NNI describes the interface between public ATM switches, such as those used by communications carriers. Figure 2.54 illustrates the four previously described ATM network interfaces.

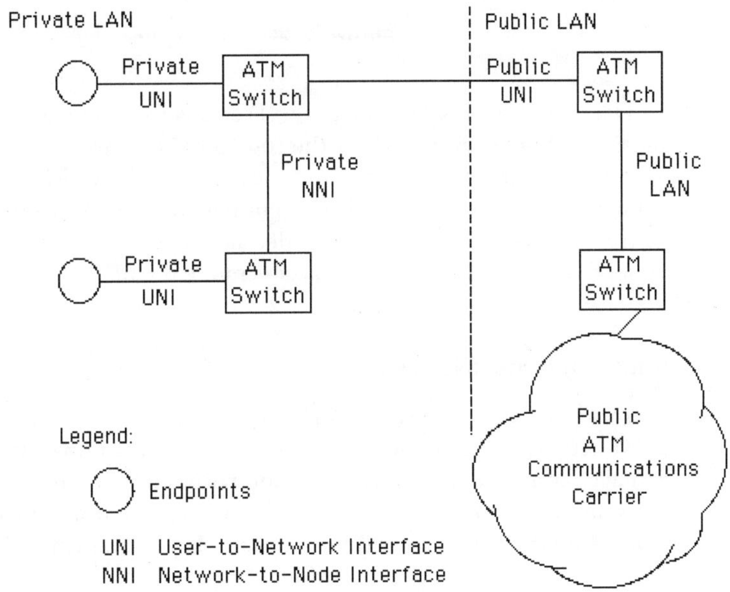

Figure 2.54 ATM network interfaces

The ATM cell header

The structure of the ATM cell is identical in both public and private ATM networks, with Figure 2.55 illustrating the fields within the five-byte cell header. As we will soon note, although the cell header fields are identical throughout an ATM network, the use of certain fields depends upon the interface or the presence or absence of data being transmitted by an endpoint.

Generic Flow Control field

The Generic Flow Control (GFC) field consists of the first four bits of the first byte of the ATM cell header. This field is used to control the flow of traffic across

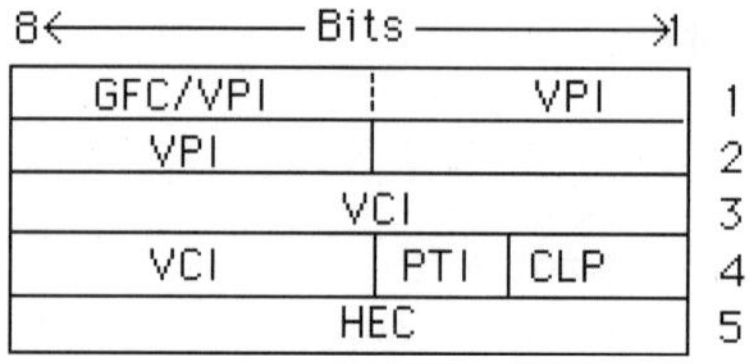

GFC Generic Flow Control
VPI Virtual Path Identifier
VCI Virtual Channel Identifier
PTI Payload Type Identifier
CLP Cell Loss Priority
HEC Header Error Check

Figure 2.55 The ATM cell header

the User-to-Network Interface (UNI) and is used only at the UNI. When cells are transmitted between switches, the four bits become an extension of the VPI field, permitting a larger VPI value to be carried in the cell header.

Two modes of GFC based flow-control are specified—uncontrolled access and controlled access. When uncontrolled access is specified, all bits in the GFC field are set to zero. In the controlled access mode the field is set when congestion occurs.

Virtual Path Identifier fields

The Virtual Path Identifier (VPI) identifies a path between two locations in an ATM network that provides transportation for a group of virtual channels, where a virtual channel represents a connection between two communicating ATM devices. Thus, a virtual path can be considered to represent a bundle of channels between two endpoints. When an endpoint has no data to transmit, the VPI field is set to all zeros to indicate an idle condition. As previously explained, when transmission occurs between switches, the GFC field is used to support an extended VPI value.

Virtual Channel Identifier field

The Virtual Channel Identifier (VCI) can be considered to represent the second part of the two-level routing hierarchy used by ATM, where a group of virtual channels (VCs) are used to form a virtual path (VP). The virtual channel identifier identifies a virtual channel, which, in turn, represents the flow of a single network connection data flow between two ATM endpoints. ATM standards defines the virtual channel as a unidirectional connection. Thus, two virtual channels must be established between ATM endpoints to support a bidirectional data flow.

Figure 2.56 illustrates the relationship between virtual paths and virtual channels. Here the virtual channel represents a connection between two communicating ATM entities, such as an endpoint to a central office switch, or between two

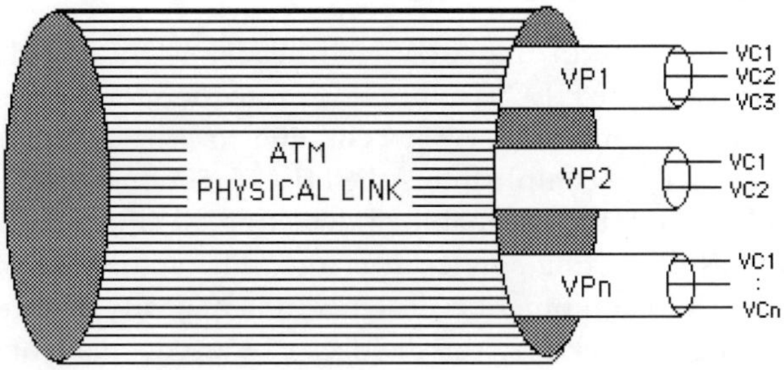

Figure 2.56 Relationship between virtual paths and virtual channels

switches. The virtual channel can represent a single ATM link or a concatenation of two or more links, with communications on the channel occurring in cell sequence order at a predefined quality of service. In comparison, each virtual path represents a group of VCs transported between two points that can flow over one or more ATM links. Although VCs are associated with a VP, they are neither unbundled nor processed. Thus, the purpose of a virtual path is to provide a mechanism for bundling traffic routed towards the same destination. This technique enables ATM switches to examine the VPI field within the cell header to make a decision concerning the relaying of the cell instead of having to examine the entire three-byte address formed by the VPI and the VCI. When an endpoint is in an idle condition, the VPI field is set to all zeros. Although the VCI field will also be set to all zeros to indicate the idle condition, other non-zero VCI values are reserved for use with a VPI zero value to indicate certain predefined conditions. For example, the VPI field value of zero with a VCI field value of 5 is used to transmit a signaling/connection request.

Payload Type Identifier field

The Payload Type Identifier (PTI) field consists of three bits in the fourth byte of the cell header. This field is used to identify the type of information carried by the cell. Values 0–3 are reserved to identify various types of user data, 4 and 5 denote management information, while 6 and 7 are reserved for future use.

Cell Loss Priority field

The last bit in the fourth byte of the cell header represents the Cell Loss Priority (CLP) field. This bit is set by the AAL Layer and used by the ATM Layer throughout an ATM network as an indicator of the importance of the cell. If the CLP is set to 1, it indicates the cell can be discarded by a switch experiencing congestion. If the cell should not be discarded because of the necessity to support a predifined quality of service (QoS), the AAL Layer will set the CLP bit to 0. The CLP bit can also be set by the ATM Layer if a connection exceeds the quality of service (QoS) level agreed to during the initial communications handshaking process when setup information is exchanged.

Header Error Check field

The last byte in the ATM cell header is the Header Error Check (HEC). The purpose of the HEC is to provide protection for the first four bytes of the cell header against the misdelivery of cells due to errors affecting the addresses within the header. To accomplish this the HEC functions as an error detecting and correcting code. The HEC is capable of detecting all single and certain multiple bit errors as well as correcting single-bit errors. The actual use of this field will depend upon how ATM equipment is designed. If a majority of anticipated errors are expected to be single-bit errors, this field can be used for error correction; however, its use introduces a risk of generating unwanted erroneous traffic if a mistake is made in the correction process when a number of bits are in error.

ATM connections and cell switching

Now that we have a basic understanding of the ATM cell header to include the virtual path and virtual channel identifiers, we can turn our attention to the methods used to establish connections between endpoints as well as to how connection identifiers are used for cell switching to route cells to their destination.

Connections

In comparison to most LANs that are connectionless, ATM is a connection-oriented communications technology. This means that a connection has to be established between two ATM endpoints prior to actual data being transmitted between the endpoints. The ATM connection can be established as a Permanent Virtual Circuit (PVC) or as a Switched Virtual Circuit (SVC).

A PVC can be considered as being similar to a leased line, with routing established for long-term use. Once a PVC is established, no further network intervention is required any time a user wishes to transfer data between endpoints connected via a PVC. In comparison, a SVC can be considered as being similar to a telephone call made on the switched telephone network. That is, the SVC requires network intervention to establish the path linking endpoints each time a SVC occurs.

Both PVCs and SVCs include the 'V' (Virtual) as they represent virtual rather than permanent or dedicated connections. This means that through statistical multiplexing, an endpoint can receive calls from one or more distant endpoints.

Cell switching

The VPI and VCI fields within a cell header can be used individually or collectively by a switch to determine the output port for relaying or transferring a cell. To determine the output port, the ATM switch first reads the incoming VPI, VCI, or both fields, with the field read dependent upon the location of the switch in the network. Next, the switch will use the connection identifier information to perform a table lookup operation. That operation uses the current connection identifier as a match criteron to determine the output port the cell will be routed onto as well as a new connection identifier to be placed into the cell header. The new connection identifier is then used for routing between the next pair of switches or from a switch to an endpoint.

Types of switches

There are two types of ATM switches, the differences being related to the type of header fields read for establishing cross-connections through the switch. A switch limited to reading and substituting VPI values is commonly referred to as a VP switch. This switch operates relatively fast. A switch that reads and substitutes both VPI and VCI values is commonly referred to as a Virtual Channel (VC) switch. A VC switch generally has a lower cell operating rate than a VP switch as it must examine additional information in each cell header. You can consider a VP switch as being similar to a central office switch, while a VC switch would be similar to end-office switches.

Using connection identifiers

To illustrate the use of connection identifiers in cell switching, consider Figure 2.57, which illustrates a three-switch ATM network with four endpoints. When switch 1 receives a cell on port 2 with VPI = 0, VCI = 10, it uses the VPI and VCI values to perform a table lookup, assigning VPI = 1, VCI = 12 for the cell header and switching the cell onto port 1. Similarly, when switch 1 receives a cell on port 3 with VPI = 0, VCI = 18 its table lookup operation results in the assignment of VPI = 1, VCI = 15 to the cell's header and the forwarding of the cell onto port 1. If we assume switch 2 is a VP switch, it reads and modifies only the VPI, thus, the VCIs are shown exiting the switch with the same values they had upon entering the switch. At switch 3, the VPC is broken down, with virtual channels assigned to route cells to endpoints C and D that were carried in a common virtual path from switch 1 to switch 3.

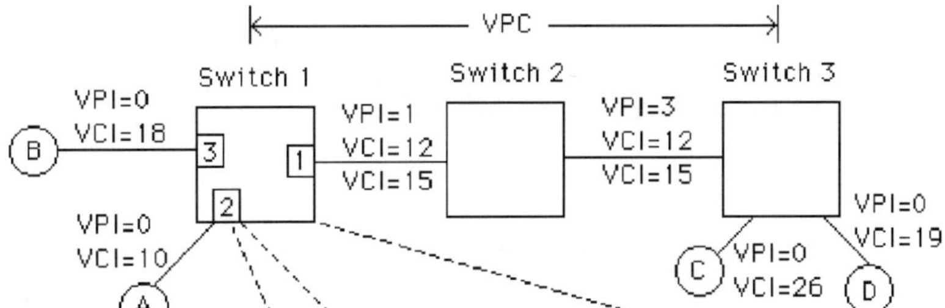

Current VPI/VCI	Port #	New VPI/VCI
0,10	1	1,12
0,18	1	1,15

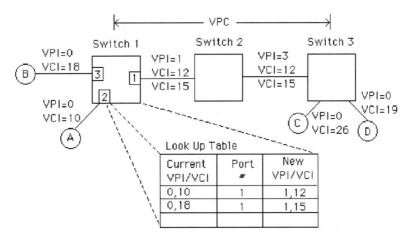

Figure 2.57 Cell switching example

The assignment of VPI and VCI values is an arbitrary process which considers those already in use, with the lookup tables being created when a connection is established through the network. That connection results from an ATM endpoint requesting a connection setput via the User-to-Network Interface through the use of a signaling protocol which contains an address within the cell. That address can be in one of three formats. One known as E.164 is the same as that used in public telephone networks, while the other two address formats include domain identifiers that allow address fields to be assigned by different organizations. The actual signaling method is based upon the signaling protocol used in ISDN and enables a quality of service to be negotiated and agreed to during the connection setup process. The quality of service is based upon metrics assigned to different traffic classes, permitting an endpoint to establish several virtual connections where each connection transports different types of data with different performance characteristics assigned to each connection.

ATM is a complex evolving technology which provides the potential to integrate voice, data, video, and images across LANs and WANs. Although it is still in a state of infancy, several communications carriers have gone on record to commit the expenditure of a considerable amount of money to implement the technology into their infrastructure for WAN operations. A similar commitment is not possible with respect to LANs as each organization is the ultimate controller of their destiny. Like all technologies, ATM will compete based upon price and performance. Since such high speed technologies as Fast Ethernet offer 100 Mbps operations at a fraction of the current cost of ATM equipment, the adaptation of ATM on LANs may become more of a gradual evolutionary process than a technological revolution.

REVIEW QUESTIONS

2.1.1 Define the term 'wide area network'.

2.1.2 Describe four types of wide area network transmission facilities.

2.2.1 What is the most commonly used circuit switched network?

2.2.2 Define the term 'switched virtual call'.

2.2.3 What is the key difference between LAN and WAN circuit switching?

2.2.4 Describe how a communications carrier would employ FDM between two carrier offices.

2.2.5 How many voice conversations can an ITU FDM standard mastergroup transport?

2.2.6 What is the actual data transmission capability of a T1 circuit that operates at 1.544 Mbps?

2.2.7 Describe the functions performed by the three key components of a digital channel bank.

2.2.8 What is the data rate generated by the Pulse Code Modulation process used to construct a digitized voice signal?

2.2.9 Describe the functions of the three primary components of a channel bank.

2.2.10 Discuss the difference between transmitting voice and data with respect to the delay each can tolerate when flowing through a network.

2.3.1 Describe three types of private networks that can be constructed through the use of leased lines.

2.3.2 List two types of analog and four types of digital leased lines you can consider when constructing a private network.

2.3.3 What is the key difference between multiplexing and routing?

2.3.4 What are the functions of a router's LAN and serial ports?

2.4.1 What was the rationale for the development of packet swithcing networks?

2.4.2 What is the primary function of a packet switch?

2.4.3 What is the difference between ITU Recommendations X.25, X.28, X.29, and X.3?

2.4.4 Describe two methods by which a non-packet mode device can be connected to an X.25 network.

2.4.5 What is the difference between a datagram packet network and a virtual circuit based network with respect to packet flow?

2.4.6 What is the primary function of PAD?

2.4.7 What are some of the functions performed by layer 3 operations on an X.25 packet network?

2.4.8 How are logical channel numbers used in an X.25 packet network?

2.4.9 Describe the potential use of receiver ready (RR) and receiver not ready (RNR) packets.

2.4.10 Compare X.25 to frame relay by discussing the role of flow control and error checking for each network.

2.4.11 What is the difference between a PAD and a FRAD?

2.4.12 What is a Committed Information Rate (CIR)?

2.4.13 Under what conditions will a frame relay network discard frames?

2.4.14 What is the purpose of a frame relay data link connection identifier (DLCI)? What is the equivalent DLCI on an X.25 network?

2.4.15 Assume congestion occurred on a frame relay network. What two bits in the frame relay header would be set when frames flow through a switch experiencing congestion?

2.4.16 How does a frame relay network handle errored frames?

2.4.17 What is the relationship between the CIR, the measurement interval (T_c) and the committed burst size (B_c) on a frame relay network?

2.4.18 What value does the sum of the committed burst size (B_c) and excess burst size (B_e) represent?

2.4.19 When voice is transported over a frame relay network, why is it important to place frames transporting voice in a FRADs queue ahead of frames transporting data?

2.5.1 What is the only protocol suite supported on the Internet?

2.5.2 What are two key differences between TCP and UDP?

2.5.3 Is TCP/IP a WAN, a LAN, or a combined WAN and LAN protocol?

2.5.4 What is the primary purpose of data fragmentation?

2.5.5 What layers in the TCP/IP protocol suite are responsible for the division of a message into datagrams and message fragmentation?

2.5.6 What is the purpose of the Address Resolution Protocol (ARP)?

2.5.7 How would you identify TCP segments containing a file transfer and a Telnet session flowing on the same transmission path?

2.5.8 What field in the TCP header provides a flow control mechanism?

2.5.9 Explain how the altering of a TCP windows field value can be used to control the flow of information.

2.5.10 Explain the relationship between port addresses contained in the TCP or UDP header and IP address with respect to the transmission and delivery of different application data between two computers.

2.5.11 How is a datagram prevented from wandering the Internet forever?

2.5.12 How does an IP datagram indicate the higher layer protocol used to create a message carried in the datagram?

2.5.13 Discuss the relationship between the number of network identifiers that can be defined on Class A, Class B, and Class C addresses.

2.5.14 Explain why a maximum of 254 devices can be assigned addresses on a Class C network even though the host portion of a Class C address is 8 bits in length.

2.5.15 What device is responsible for providing a translation between a device's near-English domain name and its IP address?

2.5.16 Create an example of a domain name based address for an ftp server operated by a commercial organization called Microware.

2.5.17 What are the three types of IPvc6 addresses?

2.5.18 Reduce the IPv6 address

301C : 0000 : 0000 : 0000 : 000A : FFBA : 000E : 1234

through the use of leading zero suppression and a double colon.

2.5.19 Describe two methods that can be used to migrate to IPv6.

2.6.1 Discuss the role of the SSCP, PU and LU in an SNA network.

2.6.2 Describe how an IBM PC or compatible computer connected to an SNA network can provide a simultaneous interactive display and printing capability.

2.6.3 What is a domain?

2.6.4 What is the function of an IBM communications controller?

2.6.5 What does peer-to-peer communications mean?

2.6.6 What is the function of pacing?

2.6.7 Describe the function of APPN end nodes, network nodes, and low-entry networking nodes.

2.6.8 Discuss the importance of APPN's 'class of service' route selection.

2.7.1 Discuss two benefits obtained from the use of the relatively short fixed sized cell used in ATM.

2.7.2 When referring to ATM, what does the term 'scalability' reference?

2.7.3 What feature enables ATM to adjust network performance during unexpected congestion?

2.7.4 What is the function of the ATM Adaptation Layer?

2.7.5 Why is the ATM Adaptation Layer not required at an ATM switch?

2.7.6 What is the function of the ATM Layer?

2.7.7 Where in the ATM network are idle or empty cells generated?

2.7.8 What is the advantage associated with the absence of a specific Physical Layer being defined in the ATM protocol stack?

2.7.9 What are some of the functions of an ATM network interface card?

2.7.10 What are the two key functions performed by a LAN switch that has an ATM NIC?

2.7.11 Discuss the differences between a LAN switch and an ATM switch.

2.7.12 Discuss the difference between the User-to-Network Interface and the Network-to-Node Interface.

2.7.13 What is the purpose of a virtual path?

2.7.14 What is the purpose of the Cell Loss Priority field in the ATM cell header?

2.7.15 What is the purpose of the Header Error Check field in the ATM cell header?

2.7.16 What are the two types of connections supported by ATM?

2.7.17 What is the difference between a VP switch and a VC switch with respect to their operating rate?

3

LOCAL AREA NETWORKS

To obtain an appreciation for the role of different networking devices we must have a firm understanding of the operational characteristics of both wide area and local area networks. Since the first two chapters in this book were oriented towards wide area network concepts and different types of WANs, we will now turn our attention to local area networks. In doing so we will first briefly examine the origins and major benefits derived from the utilization of local area networks and their relationship to typical network applications. In doing so we will compare and contrast LANs and WANs to obtain a better understanding of the similarities and differences between the two types of networks. Next, we will look at the major areas of local area network technology and the effect these areas have upon the efficiency and operational capability of such networks. Here, our examination will focus upon network topology, transmission media and the major access methods employed in LANs. Using the previous material as a base, we will then focus our attention upon the operation of several types of local area networks.

3.1 OVERVIEW

This chapter, in conjunction with the first two chapters in this book and along with Chapter 4, provides the foundation for Chapter 5 which is focused upon the detailed operation and utilization of internetworking devices. Those devices connect LANs together, both directly when they are located in close proximity to one another, and indirectly via a wide area network when they are located in different geographical areas. Thus, obtaining an understanding of WANs and LANs provides the foundation for discussing LAN internetworking devices presented in Chapter 5. In that chapter we will examine the use of bridges, routes, switches, and gateways. Since an explanation of some of the functions and methods of utilization of those internetworking devices requires the knowledge of digital networking equipment and techniques presented in Chapter 4, the chapter covering internetworking devices follows that chapter.

Although several types of Ethernet and Token-Ring LANs are covered in this chapter, ATM may be conspicuous by its absence. ATM represents both a LAN and a WAN technology, and an overview of ATM with respect to its use on a wide area network was provided in Chapter 2. Since the use of ATM directly to the desktop is similar for both LANs and WANs, its coverage in this chapter would duplicate information previously presented in Chapter 2. However, since the use of ATM as a LAN backbone requires a process referred to as LAN Emulation (LANE), we will examine that process in Chapter 5 when we cover LAN internetworking devices.

Origin

The origin of local area networks can be traced, in part, to IBM terminal equipment introduced in 1974. At that time, IBM introduced a series of terminal devices designed for use in transaction-processing applications for banking and retailing. What was unique about those terminals was their method of connection; a common cable that formed a loop provided a communications path within a localized geographical area. Unfortunately, limitations in the data transfer rate, incompatibility between individual IBM loop systems, and other problems precluded the widespread adoption of this method of networking. The economics of media-sharing and the ability to provide common access to a centralized resource were, however, key advantages, and they resulted in IBM and other vendors investigating the use of different techniques to provide a localized communications capability between different devices. In 1977, Datapoint Corporation began selling its Attached Resource Computer Network (Arcnet), considered by most people to be the first commercial local area networking product. Since then, hundreds of companies have developed local area networking products, and the installed base of terminal devices connected to such networks has increased exponentially. They now number in the tens of millions.

Comparison to WANs

Local area networks can be distinguished from wide area networks by geographic area of coverage, data transmission and error rates, ownership, government regulation, data routing and, in many instances, the type of information transmitted over the network.

Geographic area

The name of each network provides a general indication of the scope of the geographic area in which it can support the interconnection of devices. As its name implies, a LAN is a communications network that covers a relatively small local area. This area can range in scope from a department located on a portion of a floor in an office building, to the corporate staff located on several floors in the building, to several buildings on the campus of a university.

Regardless of the LAN's area of coverage, its geographic boundary will be restricted by the physical transmission limitations of the local area network. These limitations include the cable distance between devices connected to the LAN and the total length of the LAN cable. In comparison, a wide area network can provide communications support to an area ranging in size from a town or city to a state, country, or even a good portion of the entire world. Here, the major factor governing transmission is the availability of communications facilities at different geographic areas that can be interconnected to route data from one location to another.

Data transmission and error rates

Two additional areas that differentiate LANs from WANs and explain the physical limitation of the LAN geographic area of coverage are the data transmission rate and the error rate for each type of network. LANs normally operate at a megabit-per-second rate, typically ranging from 4 Mbps to 16 Mbps, with several LANs operating at 100 Mbps and the recently introduced Gigabit Ethernet network operating at 1 Gbps. In comparison, the communications facilities used to construct a major portion of most WANs provide a data transmission rate at or under the T1 and E1 data rates of 1.544 Mbps and 2.048 Mbps.

Since LAN cabling is primarily within a building or over a small geographical area, it is relatively safe from natural phenomena, such as thunderstorms and lightning. This safety enables transmission at a relatively high data rate, resulting in a relatively low error rate. In comparison, since wide area networks are based on the use of communications facilities that are much farther apart and always exposed to the elements, they have a much higher probability of being disturbed by changes in the weather, electronic emissions generated by equipment, or such unforeseen problems as construction workers accidentally causing damage to a communications cable. Because of these factors, the error rate on WANs is considerably higher than the rate experienced on LANs. On most WANs you can expect to experience an error rate between 1 in a million (10^6) and 1 in 10 million (10^7) bits. In comparison, the error rate on a typical LAN may exceed that range by one or more orders of magnitude, resulting in an error rate from 1 in 10 million (10^7) to 1 in 100 million (10^8) bits.

Ownership

The construction of a wide area network requires the leasing of transmission facilities from one or more communications carriers. Although your organization can elect to purchase or lease communications equipment, the transmission facilities used to connect diverse geographical locations are owned by the communications carrier. In comparison, an organization that installs a local area network normally owns all of the components used to form the network, including the cabling used to form the transmission path between devices.

Regulation

Since wide area networks require transmission facilities that may cross local, state, and national boundaries, they may be subject to a number of governmental regulations at the local, state, and national levels. In comparison, regulations affecting local area networks are primarily in the areas of building codes. Such codes regulate the type of wiring that can be installed in a building and whether the wiring must run in a conduit.

Data routing and topology

In a local area network, data is routed along a path that defines the network. That path is normally a bus, ring, tree, or star structure, and data always flows on that structure. The topology of a wide area network can be much more complex. In fact, many wide area networks resemble a mesh structure, including equipment used to reroute data in the event of communications circuit failure or excessive traffic between two locations. Thus, the data flow on a wide area network can change, while the data flow on a local area network primarily follows a single basic route.

Type of information carried

The last major difference between local and wide area networks is in the type of information carried by each network. Many wide area networks support the simultaneous transmission of voice, data, and video information. In comparison, most local area networks are currently limited to carrying data. In addition, although all wide area networks can be expanded to transport voice, data, and video, many local area networks are restricted by design to the transportation of data. Table 3.1 summarizes the similarities and differences between local and wide area networks.

Table 3.1 Comparing LANs and WANs

Characteristic	Local area network	Wide area network
Geographic area of coverage	Localized to a building, group of buildings, or campus	Can span an area ranging in size from a city to the globe
Data transmission rate	Typically 4 Mbps to 16 Mbps, with relatively new copper fiber optic-based networks operating at 100 Mbps and 1 Gbps	Normally operate at or below T1 and E1 transmission rates of 1.544 Mbps and 2.048 Mbps
Error rate	1 in 10^7 to 1 in 10^8	1 in 10^6 to 1 in 10^7
Ownership	Usually with the implementor	Communications carrier retains ownership of line facilities
Data routing	Normally follows fixed route	Switching capability of network allows dynamic alteration of data flow
Topology	Usually limited to bus, ring, tree, and star	Virtually unlimited design capability
Type of information carried	Primarily data	Voice, data, and video commonly integrated

Utilization benefits

In its simplest form, a local area network is a cable that provides an electronic highway for the transportation of information to and from different devices connected to the network. Because a LAN provides the capability to route data between devices connected to a common network within a relatively limited distance, numerous benefits can accrue to users of the network. These can include the ability to share the use of peripheral devices, thus obtaining common access to data files and programs, the ability to communicate with other people on the LAN by electronic mail, and the ability to access the larger processing capability of mainframes or minicomputers through the common gateways that link a local area network to larger computer systems.

Peripheral sharing

Peripheral sharing allows network users to access relatively expensive color laser printers, CD-ROM systems, and other devices that may be needed only a small portion of the time a workstation is in operation. Thus, users of a LAN can obtain access to resources that would probably be too expensive to justify for each individual workstation user.

Common software access

The ability to access data files and programs from multiple workstations can substantially reduce the cost of software. In addition, shared access to database information allows network users to obtain access to updated files on a real-time basis.

Electronic mail

One popular type of application program used on LANs enables users to transfer messages electronically. Commonly referred to as *electronic mail* or *e-mail*, this type of application program can be used to supplement and, in many cases, eliminate the need for paper memoranda.

Gateway access to mainframes

For organizations with mainframe or minicomputers, a local area network gateway can provide a common method of access to those computers. Without the use of a LAN gateway, each personal computer requiring access to a mainframe or minicomputer would require a separate method of access. This might increase both the complexity and the cost of providing access.

3.2 TECHNOLOGICAL CHARACTERISTICS

Although a local area network is a limited distance transmission system, the variety of options available for constructing such networks is anything but limited. Many of the options available for the construction of local area networks are based on the technological characteristics that govern their operation. These characteristics include different topologies, signaling methods, transmission media, access methods used to transmit data on the network, and the hardware and software required to make the network operate.

Topology

The *topology* of a local area network is the structure or geometric layout of the cable used to connect stations on the network. Unlike conventional data communications networks, which can be configured in a variety of ways with the addition of hardware and software, most local area networks are designed to operate based upon the interconnection of stations that follow a specific topology. The most common topologies used in LANs include the loop, bus, ring, star, and tree, as illustrated in Figure 3.1.

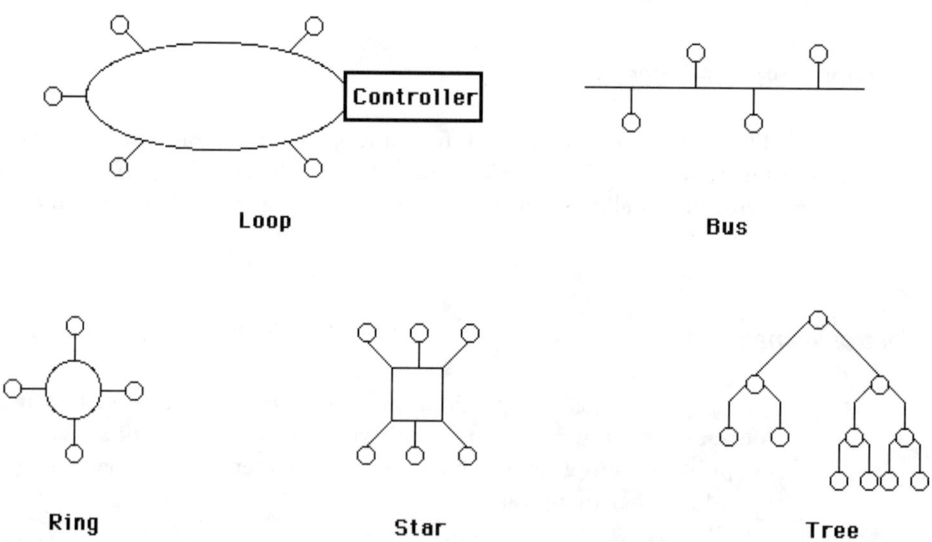

Figure 3.1 LAN topology. The five most common geometric layouts of local area network cabling form a loop, bus, ring, star, or tree structure

Loop

As previously mentioned in this chapter, IBM introduced a series of transaction-processing terminals in 1974 that communicated through the use of a common controller on a cable formed into a loop. This type of topology is illustrated at the top of Figure 3.1.

Since the controller employed a poll-and-select access method, terminal devices connected to the loop required a minimum of intelligence. Although this reduced the cost of terminals connected to the loop, the controller lacked the intelligence to distribute the data flow evenly among terminals. A lengthy exchange between two terminal devices or between the controller and a terminal would thus tend to bog down this type of network structure. A second problem associated with this network structure was the centralized placement of network control in the controller. If the controller failed, the entire network would become inoperative. Due to these problems, the use of loop systems is restricted to several niche areas, and they are essentially considered a derivative of a local area network.

Bus

In a bus topology structure, a cable is usually laid out as one long branch, onto which other branches are used to connect each station on the network to the main data highway. Although this type of structure permits any station on the network to talk to any other station, rules are required for recovering from such situations as when two stations attempt to communicate at the same time. Later in this chapter, we will examine the relationships among the network topology, the method employed to access the network, and the transmission medium employed in building the network.

Ring

In a ring topology, a single cable that forms the main data highway is shaped into a ring. As with the bus topology, branches are used to connect stations to one another via the ring. A ring topology can thus be considered to be a looped bus. Typically, the access method employed in a ring topology requires data to circulate around the ring, with a special set of rules governing when each station connected to the network can transmit data.

Star

The fourth major local area network topology is the star structure. In a star network, each station on the network is connected to a network controller. Access from any one station on the network to any other station is accomplished through the network controller. Here, the network controller can be viewed as functioning similarly to a telephone switchboard, since access from one station to another station on the network can occur only through the central device.

Tree

A tree network structure represents a complex bus. In this topology, the common point of communications at the top of the structure is known as the *headend*. From the headend, feeder cables radiate outward to nodes, which in turn provide workstations with access to the network. There may also be a feeder cable route to additional nodes, from which workstations gain access to the network.

Mixed topologies

Some network are a mixture of topologies. For example, as previously discussed, a tree structure can be viewed as a series of interconnected buses. Another example of the mixture of topologies is a type of Ethernet known as 10BASE-T. That network can actually be considered a star–bus topology, in which workstations are first connected to a common device known as a hub, which in turn can be connected to other hubs to expand the network.

Comparison of topologies

Although there are close relationships among the topology of the network, its transmission media, and the method used to access the network, we can examine topology as a separate entity and make several generalized observations. First, in a star network, the failure of the network controller will render the entire network inoperative. This is because all data flow on the network must pass through the network controller. On the positive side, the star topology normally consists of telephone wires routed to a switchboard. A local area network that can use in-place twisted-pair telephone wires in this way is simple to implement and usually very economical.

In a ring network, the failure of any node connected to the ring normally inhibits data flow around the ring. Due to the fact that data travels in a circular path on a ring network, any cable break has the same effect as the failure of the network controller in a star-structured network. Since each network station is connected to the next network station, it is usually easy to install the cable for a ring network. In comparison, a star network may require cabling each section to the network controller if existing telephone wires are not available, and this can result in the installation of very long cable runs.

In a bus-structured network, data is normally transmitted from a single station to all other stations located on the network, with a destination address appended to each transmitted data block. As part of the access protocol, only the station with the destination address in the transmitted data block will respond to the data block. This transmission concept means that a break in the bus may be limited to affecting only network stations on one side of the break that wish to communicate with stations on the other side of the break. Thus, unless a network station functioning as the primary network storage device becomes inoperative, a failure in a bus-structured network is usually less serious than a failure in a ring network. However, some local area networks, such as Token-Ring and FDDI, were designed to overcome the effect of certain types of cable failures. Token-Ring networks include a backup path which, when manually placed into operation, may be able to overcome the effect of a cable failure between hubs (referred to as *multistation access units or MAUs*). In an FDDI network, a second ring can be activated automatically as part of a self-healing process to overcome the effect of a cable break.

A tree-structured network is similar to a star-structured network in that all signals flow through a common point. In the tree-structured network the common signal point is the headend. Failure of the headend renders the network inoperative. This network structure requires the transmission of information over relatively long distances. For example, communications between two stations located at opposite ends of the network would require a signal to propagate twice the length of the longest network segment. Due to the propagation delay associated with the transmission of any signal, the use of a tree structure may result in a response time delay for transmissions between the nodes that are most distant from the headend.

Signaling methods

The signaling method used by a local area network refers to both the way data is encoded for transmission and the frequency spectrum of the media. To a large degree, the signaling method is related to the use of the frequency spectrum of the media.

Broadband versus baseband

Two signaling methods used by LANs are broadband and baseband. In *broadband signaling*, the bandwidth of the transmission medium is subdivided by frequency to form two or more subchannels, with each subchannel permitting data transfer to occur independently of data transfer on another subchannel. In *baseband signaling* only one signal is transmitted on the medium at any point in time.

Broadband is more complex than baseband, because it requires information to be transmitted via the modulation of a carrier signal, thus requiring the use of special types of modems.

Figure 3.2 illustrates the difference between baseband and broadband signaling with respect to channel capacity. It should be noted that although a twisted-pair wire system can be used to transmit both voice and data, data transmission is baseband since only one channel is normally used for data. In comparison, a broadband system on coaxial cable can be designed to carry voice and several subchannels of data, as well as fax and video transmission.

Broadband signaling

A broadband local area network uses analog technology, in which high frequency (HF) modems operating at or above 4 kHz place carrier signals onto the transmission medium. The carrier signals are then modified—a process known as *modulation*, which impresses information onto the carrier. Other modems connected to a broadband LAN reconvert the analog signal block into its original digital format—a process known as *demodulation*.

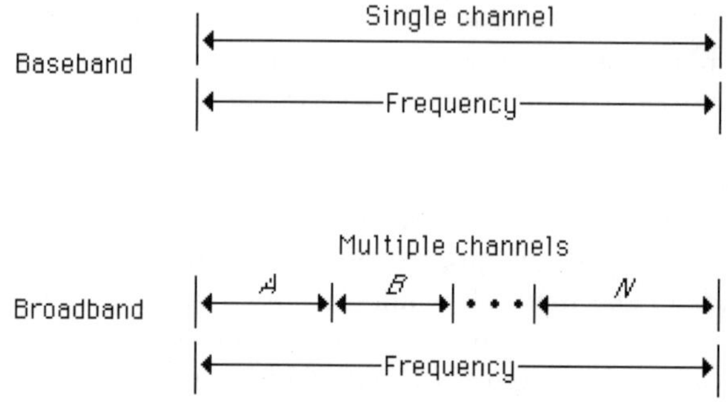

Figure 3.2 Baseband versus broadband signaling. In baseband signaling the entire frequency bandwidth is used for one channel. In comparison, in broadband signaling the channel is subdivided by frequency into many subchannels

The most common modulation method used on broadband LANs is *frequency shift keying (FSK)*, in which two different frequencies are used, one to represent a binary 1 and another to represent a binary 0. Another popular modulation method uses a combination of amplitude and phase shift changes to represent pairs of bits. Referred to as *amplitude modulation phase shift keying (AM PSK)*, this method of analog signaling is also known as *duobinary signaling* because each analog signal represents a pair of digital bits.

Because it is not economically feasible to design amplifiers that boost signal strength to operate in both directions, broadband LANs are unidirectional. To provide a bidirectional information transfer capability, a broadband LAN uses one channel for inbound traffic and another channel for outbound traffic. These channels can be defined by differing frequencies or obtained by the use of a dual cable.

Baseband signaling

In comparison to broadband local area networks, which use analog signaling, baseband LANs use digital signaling to convey information.

To understand the digital signalnig methods used by most baseband LANs, let us first review the method of digital signaling used by computers and terminal devices. In that signaling method a positive voltage is used to represent a binary 1, while the absence of voltage (0 volts) is used to represent a binary 0. If two successive 1 bits occur, two successive bit positions then have a similar positive voltage level or a similar zero voltage lever. Since the signal goes from 0 to some positive voltage and does not return to 0 between successive binary 1's, it is referred to as a *unipolar non-return to zero (NRZ)* signal. This signaling technique is illustrated in Figure 3.3a.

Although unipolar non-return to zero signaling is easy to implement, its use for transmission has several disadvantages. One of the major disadvantages associated with this signaling method involves determining where one bit ends and another begins. Overcoming this problem requires synchronization between a transmitter and receiver by the use of clocking circuitry, which can be relatively expensive.

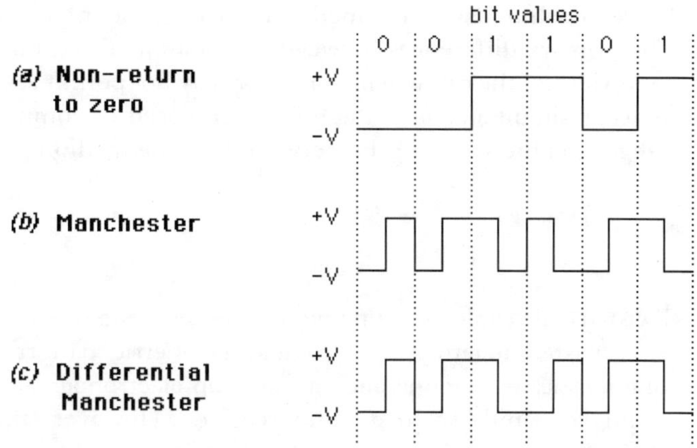

Figure 3.3 Baseband signaling techniques. In Manchester coding a timing transition occurs in the middle of each bit and the line code maintains an equal amount of positive and negative voltage. Under differential Manchester coding, the direction of the signal's voltage transition changes whenever a binary 1 is transmitted

To overcome the need for clocking, baseband LANs use *Manchester or Differential Manchester coding*. In Manchester coding a timing transition always occurs in the middle of each bit, while an equal amount of positive and negative voltage is used to represent each bit. This coding technique provides a good timing signal for clock recovery from received data, due to its timing transitions. In addition, since the Manchester code always maintains an equal amount of positive and negative voltage, it prevents direct current (DC) voltage buildup, enabling repeaters to be spaced further apart from one another.

Figure 3.3(b) illustrates an example of Manchester coding. Note that a low to high voltage transition represents a binary 1, while a high to low voltage transition represents a binary 0. Differential Manchester coding is illustrated in Figure 3.3(c). The difference between Manchester and Differential Manchester coding occurs in the method by which binary 1's are encoded. In Differential Manchester coding, the direction of the signal's voltage transition changes whenever a binary 1 is transmitted, but remains the same for a binary 0. The IEEE 802.3 standard specifies the use of Manchester coding for baseband Ethernet operating at data rates up to 10 Mbps. The IEEE 802.5 standard specifies the use of Differential Manchester coding for Token-Ring networks at the physical layer to transmit and detect four distinct symbols—a binary 0, a binary 1, and two non-data symbols.

Transmission medium

The transmission medium used in a local area network can range in scope from twisted-pair wire, such as is used in conventional telephone lines, to coaxial cable, fiber optic cable, and electromagnetic waves such as those used by FM radio and infrared. Here, the latter two methods are used for the construction of wireless

LANs. Each transmission medium has a number of advantages and disadvantages. The primary differences between media are their cost and ease of installation; the bandwidth of the cable, which may or may not permit several transmission sessions to occur simultaneously; the maximum speed of communications permitted; and the geographic scope of the network that the medium supports.

Twisted-pair

Twisted-pair cable, as its name implies, consists of one or more pairs of insulated wire twisted together in a regular geometric pattern. Since a length of wire functions as an antenna and can pick up electromagnetic emissions, the wire pair would act similarly to a radio receiver. However, the twists are designed to produce a counterbalance to the receipt of such emissions, since the electromagnetic fields of the pair of wires have opposite polarities and intensities and cancel each other out, reducing their potential to cause transmission errors.

UTP and STP

Twisted-pair cables usually consist of 4, 8 or 12 wires producing 2, 4 or 6 pairs. A common 25-pair cable used to consolidate smaller pair bundles is primarily used for telephone wiring and is normally unsuitable for LAN applications. When twisted-pairs are shielded with foil or another substance to reduce the effect of electromagnetic emissions the wiring is referred to as shielded twisted-pair (STP). Thus, non-shielded twisted-pair is referred to as unshielded twisted-pair (UTP).

Conductors

The conductors in twisted-pair wiring are referenced with respect to their thickness using American Wire Gauge (AWG) numbering. The most common AWG conductors used in twisted-pair are 19, 22, 24 and 26; however, the AWG number is inversely proportional to the thickness of the conductor. As you might expect, a thin conductor has more resistance to data flow than a thicker wire. This is illustrated by the entries in Table 3.2 which denote the resistance for four common AWG cable pairs. As expected, a lower wire gauge has a thicker conductor which results in a lower resistance.

Table 3.2 American Wire Gauge conductor resistances

AWG	Ohms/1000 feet
19	16.1
22	32.4
24	51.9
26	83.5

Cabling standards: general

There are two general types of cabling standards by which many types of twisted-pair wiring can be categorized—de facto and de jure (from the Latin phrases *de facto* and *dejure*).

De facto standards represent a commonly used set of cabling rules and requirements which were developed by one or more vendors and do not carry the backing of a standards-making organization. Examples of defacto cabling 'standards' include the original Ethernet cabling, which specified the type of coaxial cable to be used in developing a LAN, and IBM's cabling system. The latter specifies a variety of twisted-pair and optical cables for use with the development of Token-Ring networks.

A de jure standard represents a standard developed by a standards-making organization. In the area of twisted-pair cabling the Electronics Industry Association and the Telecommunications Industry Association (EIA/TIA) working together developed several standards which have also been adopted by the American National Standards Institute (ANSI). Probably the most important standard is EIA/TIA-568, Commercial Building Telecommunications Wiring Standard. This standard contains detailed specifications on the electrical and physical characteristics of twisted-pair coaxial and optical fiber cable as well as guidelines concerning the cabling of new and existing buildings. In discussion cabling standards I will review the characteristics of IBM's cabling system and the TIA/EIA-568 standard, focusing my discussion towards the twisted-pair cabling used by each standard.

IBM cabling system

The IBM cabling system was introduced in 1984 as a mechanism to support the networking requirements of office environments. By defining standards for cables, connectors, faceplates, distribution panels, and other facilities IBM's cabling system was designed to support the interconnection of personal computers, conventional terminals, mainframe computers, and office systems. In addition, this system permits devices to be moved from one location to another or added to a network through a simple connection to the cabling system's wall plates or surface mounts.

The IBM cabling system specifies seven different cabling categories. Depending upon the type of cable selected you can install the selected wiring indoors, outdoors, under a carpet, or in ducts and other air spaces.

The IBM cabling system uses wire which conforms to the American Wire Gauge or AWG. As previously discussed, as the wire diameter gets larger the AWG number decreases, in effect resulting in an inverse relationship between wire diameter and AWG. The IBM cabling system uses wire between 22 AWG (0.644 mm) and 26 AWG (0.405 mm). Since a larger diameter wire has less resistance to current flow than a smaller diameter wire, a smaller AWG permits cabling distances to be extended in comparison to a higher AWG cable.

Type 1

The IBM cabling system Type 1 cable contains two twisted-pairs of 22 AWG conductors. Each pair is shielded with a foil wrapping and both pairs are surrounded by an outer braided shield or with a corrugated metallic shield. One pair of wires uses shield colors of red and green, while the second pair of wires uses shield colors of orange and black. The braided shield is used for indoor wiring, while the corrugated metallic shield is used for outdoor wiring. Type 1 cable is available in two different designs—plenum and non-plenum. Plenum cable is installed without the use of a conduit while non-plenum cable requires a conduit. Type 1 cable is typically used to connect a distribution panel or multistation access unit and the faceplate or surface mount at a workstation.

Type 2

Type 2 cable is actually a Type 1 indoor cable with the addition of four pairs of 22 AWG conductors for telephone usage. For this reason Type 1 cable is also referred to as data-grade twisted-pair cable, while Type 2 cable is known as two data-grade and four-grade twisted-pair. Due to its voice capability, Type 2 cable can support PBX interconnections. Like Type 1 cable, Type 2 cable supports plenum and non-plenum designs. Type 2 cable is not available in an outdoor version.

Type 3

Type 3 cable is conventional twisted-pair telephone wire, with a minimum of two twists per foot. Both 22 AWG and 24 AWG conductors are supported by this cable type. One common use of Type 3 cable is to connect PCs to hubs in a Token-Ring network.

Type 5

Type 5 cable is fiber optic cable. Two 100/140 μm optical fibers are contained in a Type 5 cable. This cable is suitable for indoor, non-plenum installation or outdoor aerial installation. Due to the extended transmission distance obtainable with fiber optic cable, Type 5 cable is used in conjunction with the IBM 8219 Token-Ring Network Optical Fiber Repeater to interconnect two hubs up to 6600 feet (2 km) from one another.

Type 6

Type 6 cable contains two twisted-pairs of 26 AWG conductors for data communications. It is available for non-plenum applications only and its smaller diameter than Type 1 cable makes it slightly more flexible. The primary use of Type 6 cable is for short runs as a flexible path cord. This type of cable is often used to connect an adapter card in a personal computer to a faceplate which, in turn, is connected to a Type 1 or Type 2 cable which forms the backbone of a network.

Type 8

Type 8 cable is designed for installation under a carpet. This cable contains two individually shielded, parallel pairs of 26 AWG conductors with a plastic ramp designed to make under-carpet installation as unobtrusive as possible. Although Type 8 cable can be used in a manner similar to Type 1, it provides only half of the maximum transmission distance obtainable through the use of Type 1 cable.

Type 9

Type 9 cable is essentially a low-cost version of Type 1 cable. Like Type 1, Type 9 cable consists of two twisted-pairs of data cable; however, 26 AWG conductors are used in place of the 22 AWG wire used in Type 1 cable. As a result of the use of a smaller diameter cable, transmission distances on Type 9 cable are approximately two-thirds those obtainable through the use of Type 1 cable. The color coding on the shield of Type 9 cable is the same as that used for Type 1 cable.

Summary of cable types

All seven types of cables defined by the IBM cabling system can be used to construct Token-Ring networks. However, the use of each type of cable has a different effect upon the ability to connect devices to the network, the number of devices that can be connected to a common network, the number of wiring closets in which hubs can be installed to form a ring, and the ability of the cable to carry separate voice conversations. The latter capability enables a common cable to be routed to a user's desk where a portion of the cable is connected to their telephone while another portion of the cable is connected to their computer's Token-Ring adapter card.

Table 3.3 summarizes the performance characteristics of the cables defined by the IBM cabling system. The drive distance entry indicates the relative relationship between different types of cables with respect to the maximum cabling distance between a workstation and a hub as well as between hubs. Type 1 cable provides a maximum drive distance of 100 m between a workstation and hub and 300 m between hubs for a network operating at 4 Mbps. Other drive distance entries in Table 3.3 are relative to the drive distance obtainable when Type 1 cable is used.

Table 3.3 IBM cabling system cable performance characteristics

	Cable type						
Performance characteristics	1	2	3	5	6	8	9
Drive distance (relative to type 1)	1.0	1.0	0.45	3.0	0.75	0.5	0.66
Data rate (Mbps)	16	16	4*	250	16	16	16
Maximum devices per ring	260	260	72	260	96	260	260
Maximum closets per ring	12	12	2	12	12	12	12
Voice support	no	yes	yes	no	no	no	no

*Note: Although 16-Mbps operations are not directly supported by Type 3 cable, its use is quite common when drive distances are very short.

Connectors

The IBM cabling system includes connectors for terminating both data and voice conductors. The data connector has a unique design based upon the development of a latching mechanism which permits it to mate with another, identical connector.

Figure 3.4 illustrates the IBM cabling system data connector. Its design makes it self-shorting when disconnected from another connector. This provides a Token-Ring network with electrical continuity when a station is disconnected. Unfortunately, the data connector is expensive in comparison with RJ telephone connectors, with a typical retial price between $3 and $5, whereas an RJ telephone connector can be purchased for 10 cents or so.

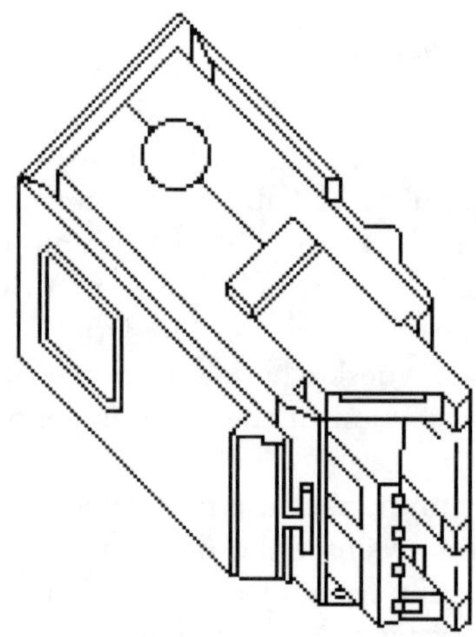

Figure 3.4 IBM cabling system data connector

Due to the cost of data connectors and cable, acceptance of the IBM cabling system by end-users has been slow. Instead of being designed for IBM data connectors, many hub vendors as well as network adapter manufacturers design their products to use less expensive and far more available RJ connectors. Other vendors provide both an IBM data connector and an RJ telephone connector on their hubs, permitting users to select the type of connector they wish to use.

EIA/TIA-568 cabling standard

The Electronics Industry Association/Telecommunications Industries Association 'Commercial Building Telecommunications Standard', commonly referred to as EIA/TIA-568, was ratified in 1992. This standard specifies a variety of building cabling parameters, ranging from backbone cabling used to connect a building's telecommunications closets to an equipment room, to horizontal cabling used to cable individual users to the equipment closet. The standard defines the performance characteristics of both backbone and horizontal cables as well as different types of connectors used with different types of cable.

Backbone cabling

Four types of media are recognized by the EIA/TIA-568 standard for backbone cabling. Table 3.4 lists the media options supported by the EIA/TIA-568 standard for backbone cabling.

Table 3.4 EIA/TIA-568 backbone cabling media options

Media type	Maximum cable distance
100 ohm UTP	800 meters (2624 feet)
150 ohm STP	700 meters (2296 feet)
50 ohm thick coaxial cable	500 meters (1640 feet)
62.5/125 μm multimode optical fiber	2000 meters (6560 feet)

Horizontal cabling

Horizontal cabling under the EIA/TIA-568 standard consists of cable which connects equipment in a telecommunications closet to a user's work area. The media options supported for horizontal cabling are the same as those specified for backbone cabling with the exception of coaxial cable for which 50 ohm thin cable is specified; however, cabling distances are restricted to 90 meters in length from equipment in the telecommunications closet to a telecommunications outlet. This permits a patch cord or drop cable up to 10 meters in length to be used to connect a user workstation to a telecommunications outlet, resulting in the total length of horizontal cabling not exceeding the 100 meter restriction associated with many LAN technologies that use UTP cabling.

UTP categories

One of the more interesting aspects of the EIA/TIA-568 standard is its recognition that different signaling rates require different cable characteristics. This resulted in the EIA/TIA-568 standard classifying UTP cable into five categories. Those categories and their suitability for different type of voice and data applications are indicated in Table 3.5.

Table 3.5 EIA/TIA-568 UTP cable categories

Category 1	Voice or low speed data up to 56 kbps; not useful for LANs
Category 2	Data rates up to 1 Mbps
Category 3	Supports transmission up to 16 MHz
Category 4	Supports transmission up to 20 MHz
Category 5	Supports transmission up to 100 MHz

In examining the entries in Table 3.5 note that categories 3 through 5 support transmission with respect to indicated signaling rates. This means that the ability of those categories of UTP to support different types of LAN transmission will depend upon thee signaling method used by different LANs. For example, consider a LAN encoding technique which results in six bits encoded into four signaling elements that have a 100 MHz signaling rate. Through the use of category 5 cable a data transmission rate of 150 Mbps $((6/4) \times 100)$ could be supported.

Cateogory 3 cable is typically used for Ethernet and 4 Mbps Token-Ring LANs. Category 4 is normally used for 16 Mbps Token-Ring LANs, while category 5 cable supports 100 Mbps Ethernet LANs, such as 100VGAny-LAN and 100BASE-T as well as ATM to the desktop at both 25 Mbps and 155 Mbps operating rates.

UTP specifications: general

The requirement to qualify a segment of installed cable and attached connectors resulted in the EIA/TIA-568 standard defining a series of link performance parameters. These parameters cover attenuation and Near End CrossTalk (NEXT) and are specified for UTP cable categories in Annex E to the standard. Thus, let's turn our attention to the manner by which attenuation and NEXT are measured prior to examining the specification limits for those two parameters on different types of cable.

UTP specifications: attenuation

Attenuation represents the loss of signal power as a signal propagates from a transmitter at one end of the cable towards a receiving device located at the distant end of the cable. Attenuation is measured in decibels (dB) as indicated below:

$$\text{Attenuation} = 20 \log_{10}\left(\frac{\text{Transmit voltage}}{\text{Receive voltage}}\right)$$

For those of us a little rusty with logarithms let's examine a few examples of attenuation computations. First, let's assume the transmit voltage was 100, while the receive voltage was 1. Then,

$$\text{Attenuation} = 20 \log_{10}\left(\frac{100}{1}\right) = 20 \log_{10} 100$$

The value of $\log_{10}100$ can be obtained by determining 10 to the appropriate power to equal 100. Since the answer is 2 ($10^2 = 100$), $\log_{10}100$ has the value of 2 and 20 $\log_{10}100$ then has a value of 40.

Now assume the transmit voltage was 10 while the receive voltage was 1. Then,

$$\text{Attenuation} = 20 \log_{10}\left(\frac{10}{1}\right) = 20 \log_{10} 10$$

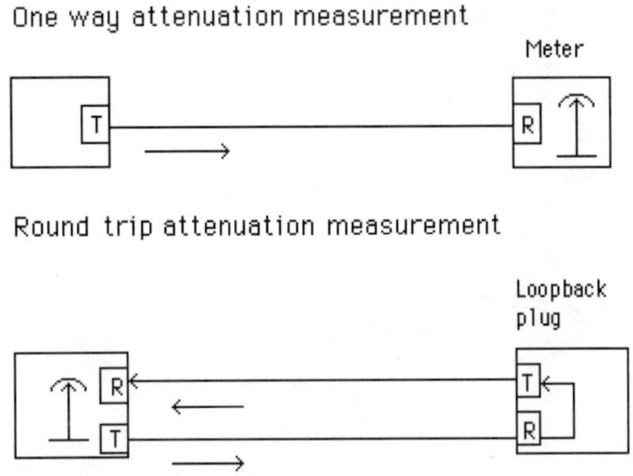

Figure 3.5 Measuring attenuation. A one-way attenuation measurement requires the use of a meter or measuring device at the distant end. In comparison, a round-trip attenuation measurement can be accomplished through the use of a loopback plug at the distant end which ties transmit (T) and receive (R) wire pairs together

Since the value of $\log_{10}10$ is 1 ($10^1 = 10$), $20\log_{10}10$ has a value of 20. Note that a comparison of the two examples indicates that a lower level of signal power loss results in a lower level of attenuation. Thus, the lower the attenuation the lower the signal loss.

There are two methods by which attenuation can be measured—one-way and round-trip. Figure 3.5 compares each method of attenuation measurement.

UTP specifications: Near End CrossTalk (NEXT)

Crosstalk represents the electromagnetic interference caused by a signal on one wire pair being emitted onto another wire pair, resulting in noise. Figure 3.6 illustrates the generation of crosstalk due to the flow of current on one wire pair resulting in the creation of a magnetic field. The magnetic field induces a signal on the adjacent wire pair which represents noise. Since transmit and receive pairs are twisted and the transmit signal is strongest at its source, the maximum level of interference occurs at the cable connector and decreases as the signal traverses the cable. Thus, crosstalk is measured at the near end, hence the term NEXT.

NEXT results in an induced or coupled signal flowing from the transmit pair to the receive pair even though both pairs are not interconnected. Mathematically, NEXT is defined in decibels (dB) as follows:

$$\text{NEXT} = 20\log_{10}\left(\frac{\text{Transmit voltage}}{\text{Coupled voltage}}\right)$$

Here the transmit voltage represents the power placed on the transmit pair, while the coupled signal is measured on the receive pair at the location where the transmit voltage was generated. Note that a larger dB NEXT measurement is better as it indicates a lower level of crosstalk. This is the opposite of attenuation,

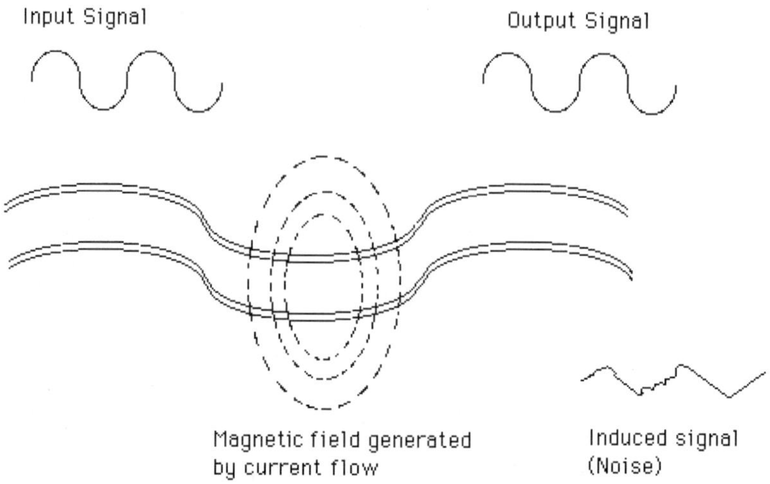

Figure 3.6 Crosstalk

since a lower attenuation reading indicates less signal loss and is better than a higher reading for that parameter.

Table 3.6 indicates the EIA/TIA-568 specification limits for categories 3, 4 and 5 UTP cable and for IBM cabling system types 1, 2, 6 and 9 cable. These categories represent the primary types of UTP and STP cable used for local area network data transmission.

In examining the entries in Table 3.6 note that the attenuation and NEXT of a cable must be measured over a range of frequencies. That range is based upon the cable category. For example, since category 3 cable is designed to support signaling rates up to 16 MHz, attenuation and NEXT should be measured up to and including the highest signaling rate supported by that type of cable.

Coaxial cable

Coaxial cable consists of a central conductor copper wire which is then covered by an insulator known as a dielectric. An overlapping woven copper mesh surrounds the dielectric and the mesh, in turn, is covered by a protective jacket which can consist of polyethylene or aluminum. Figure 3.7 illustrates the composition of a typical coaxial cable; however, it should be noted that over 100 types of coaxial cable are currently marketed. The key differences between such cables involve the number of conductors contained in the cable, the dielectric employed and the type of protective jacket and material used to provide strength to the cable which allows it to be pulled through conduits without breaking.

Two basic types of coaxial cable are used in local area networks, with the type of cable based upon the transmission technique employed: baseband or broadband signaling. Both cable types are much more expensive than twisted-pair wire; however, the greater frequency bandwidth of coaxial cable permits higher data rates for longer distances than are obtainable over twisted-pair wire.

Table 3.6 Attenuation and Next limits in decibels (dB)

(a) EIA/TIA-568

Frequency (MHz)	Category 3		Category 4		Category 5	
	Attenuation	NEXT	Attenuation	NEXT	Attenuation	NEXT
1.0	4.2	39.1	2.6	53.3	2.5	60.3
4.0	7.3	29.3	4.8	43.3	4.5	50.6
8.0	10.2	24.3	6.7	38.2	6.3	45.6
10.0	11.5	22.7	7.5	36.6	7.0	44.0
16.0	14.9	19.3	9.9	33.1	9.2	40.6
20.0	—	—	11.0	31.4	10.3	39.0
25.0	—	—	—	—	11.4	37.4
31.2	—	—	—	—	12.8	35.7
62.5	—	—	—	—	18.5	30.6
100.0	—	—	—	—	24.0	27.1

(b) IBM cabling system

Frequency (MHz)	Attenuation (dB/km)				NEXT (dB/km)			
Cable type (P=plenum)	1P/2P	6P/9P	1/2	6/9	1P/2P	6P/9P	1/2	6/9
4	22	33	22	33	−58	−52	—	—
8	31.1	46.7	N/S	N/S	−54.9	−48.9	—	—
10	34.8	52.2	N/S	N/S	−53.5	−47.5	−40*	−34*
16	44.0	66.0	45	66	−50.4	−44.4	N/S	N/S
20	49.2	73.8	N/S	N/S	−49.0	−43.0	N/S	N/S
25	61.7	93.3	N/S	N/S	−47.5	−41.5	N/S	N/S
31.25	68.9	104.3	N/S	N/S	−46.1	−40.1	N/S	N/S
62.5	97.5	147.5	N/S	N/S	−41.5	−35.5	N/S	N/S
100.0	123.3	186.6	128	N/S	−38.5	−32.5	N/S	N/S
300.0	209.2	323.2	N/S	N/S	−31.3	−25.3	N/S	N/S

* 12–20 MHz

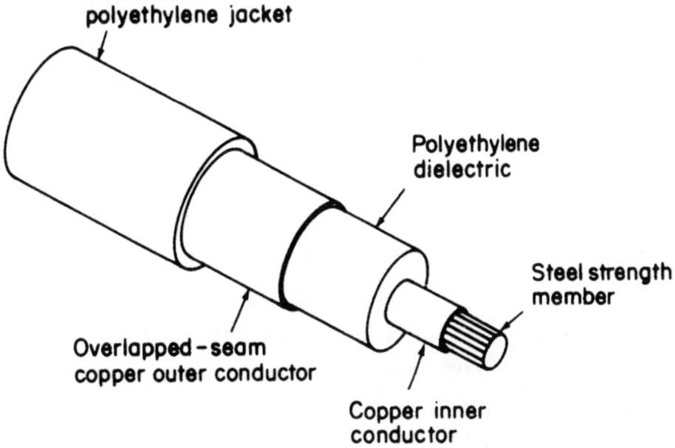

Figure 3.7 Coaxial cable

Normally, 50 Ω coaxial cable is used in baseband networks, while 75 Ω cable is used in broadband networks. The latter coaxial is identical to that used in cable television (CATV) applications, including the coaxial cable used in one's home. Data rates on baseband networks using coaxial cable range upward to between 50 and 100 Mbps. With broadband transmissions, data rates up to and including 400 Mbps are obtainable.

Hardware interface

A coaxial cable with a polythylene jacket is normally used for baseband signaling. Data is transmitted from stations on the network to the baseband cable in a digital format and the connection from each station to the cable is accomplished by the use of a simple coaxial T-connector. Figure 3.8 illustrates the hardware interface designed to connect a personal computer to a coaxial cable of a typical baseband local area network. Here the network adapter card is a hardware device that contains the logic to control network access and is inserted into one of the expansion slots in the system unit of a PC. At the rear of the computer's system unit a short section of coaxial cable is used to connect the network adapter card to baseband cable via a T-connector.

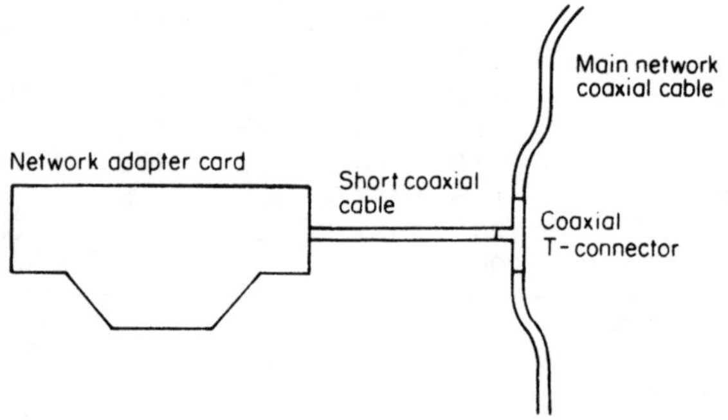

Figure 3.8 Hardware interface to coaxial cable. The network adapter card is installed in the system unit of the PC and connected to the main coaxial cable of the network via a short coaxial cable interfaced to a T-connector

Since data on a baseband network travels in a digital form, those signals can be easily regenerated by the use of a device known as a repeater or data regenerator. This is a low-cost device that is constructed to look for a pulse rise; upon detecting the occurrence of the rise, it will disregard the entire pulse and regenerate an entirely pulse. You can thus install low-cost repeaters into a baseband coaxial network to extend the distance transmission can occur on the cable. Typically, a coaxial cable baseband system can cover a network of several miles and may contain hundreds to thousands of stations.

Broadband coaxial cable

To obtain independent subchannels derived by frequency on coaxial cable broadband transmission requires a method to translate the digital signals from PCs and other workstations into appropriate frequencies. This translation process is accomplished by the use of radio-frequency (RF) modems which modulate the digital data into analog signals and convert or demodulate received analog signals into digital signals. Since signals are transmitted at one frequency and received at a different frequency, a 'head end' or frequency translator is also required for broadband transmission on coaxial cable. This device is also known as a remodulator as it simply converts the signals from one subchannel to another subchannel.

The requirement for modems and frequency translators normally makes broadband transmission more expensive than baseband. Although the ability of broadband to support multiple channels provides it with an aggregate data transmission capacity that exceeds baseband, in general, baseband transmission permits a higher per-channel data flow. While this is an important consideration for mainframe-to-mainframe communications when massive amounts of data must be moved, for most personal computer file transfer operations the speed of either baseband or boradband transmission should be sufficient. This fact may be better understood by comparing the typical transmission rates obtainable on baseband and broadband networks to drive a high-speed dot matrix printer and the differences between the time required to trnasmit data on the network and the time required to print the data.

Typical transmission speeds on commonly employed baseband and broadband networks range from 2 to 16 Mbps. In comparison, a high-speed dot matrix printer operating at 120 cps would require approximately 200 s to print 1 second's worth of data transmitted at 2 Mbps and 1600 s to print 1 second's worth of data transmitted at 16 Mbps.

Fiber optic cable

Fiber optic cable is a transmission medium for light energy and as such provides a very high bandwidth, permitting data rates ranging up to billions of bits per second. The fiber optic cable consists of a thin core of glass or plastic which is surrounded by a protective shield. Several shielded fibers in turn are bundled in a jacket with a central member of aluminum or steel employed for tensile strength.

Digital data represented by electrical energy must be converted into light energy for transmission on a fiber optic cable. This is normally accomplished by a low-power laser or through the use of a light emitting diode and appropriate circuitry. At the receiver, light energy must be reconverted into electrical energy. Normally, a device known as a photodetector, as well as appropriate circuitry to regenerate the digital pulses and an amplifier, are used to convert the received light energy into its original digital format.

In addition to the high bandwidth of fiber optic cables, they offer users several additional advantages in comparison to conventional tarnsmission mediums. Since data travels in the form of light, it is immune to electrical interference and building codes that may require expensive conduits to be installed for conventional cables are usually unnecessary. Similarly, fiber optic cable can be installed through areas where the flow of electricity could be dangerous since only light flows through such cables.

Since most fibers only provide a single, unidirectional transmission path a minimum of two cables is normally required to connect all transmitters to all receivers on a network built using fiber optic cable. Due to the higher cost of fiber optic cable than coaxial or twisted-pair, the dual cable requirement of fiber cables can make them relatively expensive in comparison to other types of cable. In addition, it is very difficult to splice such cable, which usually means skilled installers are required to implement a fiber optic-based network. Similarly, once this type of network is installed, it is difficult to modify the network.

Currently, the cost of the cable, difficulty of installation and modification make the utilization of fiber optic-based local area networks impractical for many commercial applications. Today, the primary use of fiber optic cable is to extend the distance between workstations on a network or to connect two distant networks to one another. The device used to connect a length of fiber optic cable into the LAN or between LANs is a fiber optic repeater. The repeater converts the electrical energy of signals flowing on the LAN into light energy for transmission on the fiber optic cable. At the end of the fiber optic cable, a second repeater converts light energy back into electrical energy. With the cost of the fiber optic cable declining and improvements expected to simplify the installation and modification of networks using this type of cable, the next few years may witness a profound movement toward the utilization of this transmission medium throughout local area networks.

Access method

If the topology of a local area network can be compared to a data highway, then the access method might be viewed as the set of rules that enable data from one workstation to successfully reach its destination via the data highway. Without such rules, it is quite possible for two messages sent by two differnet workstations to collide, with the result that neither message reaches its destination. Two common access methods primarily employed in local area networks are Carrier-Sense Multiple Access with Collision Detection (CSMA/CD) and token passing. Each of these access methods is uniquely structured to address the previously mentioned collision and data destination problems.

Prior to discussing how access methods work, let us first examine the two basic types of devices that can be attached to a local area network to gain an appreciation for the work that the access method must accomplish.

Listeners and talkers

We can categorize each device by its operating mode as being a *listener* or a *talker*. Some devices, like printers, only receive data, and thus operate only as listeners. Other devices, such as personal computers, can either transmit or receive data and are capable of operating in both modes. In a baseband signaling environment where only one channel exists, or on an individual channel on a broadband system, if several talkers wish to communicate at the same time, a collision will occur. Therefore, a scheme must be employed to define when each device can talk and, in the event of a collision, what must be done to keep it from happening again.

For data to reach its destination correctly, each listener must have a unique address, and its network equipment must be designed to respond to a message on the net only when it recognizes its address. The primary goals in the design of an access method are to minimize the potential for data collision, to provide a mechanism for corrective action when data collides, and to ensure that an addressing scheme is employed to enable messages to reach their destination.

Carrier-Sense Multiple Access with Collision Detection (CSMA/CD)

CSMA/CD can be categorized as a *listen then send* access method. CSMA/CD was one of the earliest access techniques to be developed and is the technique used in Ethernet.

Under the CSMA/CD concept, when a station has data to send, it first listens to determine if any other station on the network is talking. The fact that the channel is idle is determined in one of two ways, based on whether the network is broadband or baseband.

In a broadband network, the fact that a channel is idle is determined by *carrier-sensing*, or noting the absence of a carrier tone on the cable.

In a baseband Ethernet network, one channel is used for data transmission and there is no carrier to monitor. Instead, baseband Ethernet encodes data using a Manchester code in which a timing transition always occurs in the middle of each bit as previously illustrated in Figure 3.3. Although baseband Ethernet does not transmit data via a carrier, the continuous transitions of the Manchester code can be considered as equivalent to a carrier signal. Carrier-sensing on a baseband network is thus performed by monitoring the line for activity.

In a CSMA/CD network, if the channel is busy, the station will wait until it becomes idle before transmitting data. Since it is possible for two stations to listen at the same time and discover an idle channel, it is also possible that the two stations could then transmit at the same time. When this situation arises, a collision will occur. Upon sensing that a collision has occurred, a delay scheme will be employed to prevent a repetition of the collision. Typically, each station will use either a randomly generated or a predefined time-out period before attempting to retransmit the message that collided. Since this access method requires hardware capable of detecting the occurrence of a collision, additional circuitry required to perform collision detection adds to the cost of such hardware.

Figure 3.9 illustrates a CSMA/CD bus-based local area network. Each workstation is attached to the transmission medium, such as coaxial cable, by a device known as a *bus interface unit (BIU)*. To obtain an overview of the operation of a CSMA/CD network, assume that station A is currently using the channel and stations C and D wish to transmit. The BIUs connecting stations C and D to the network would listen to the channel and note it is busy. Once station A completes

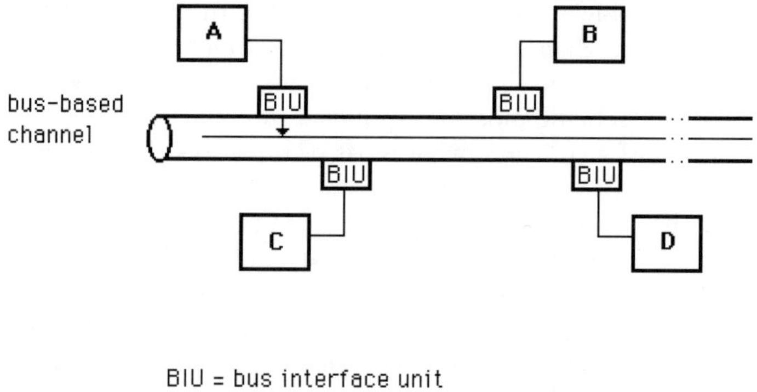

Figure 3.9 CSMA/CD network operation. In a CSMA/CD network, as the distance between workstations increases, the resulting increase in propagation delay time increases the probability of collisions

its transmission, stations C and D attempt to gain access to the channel. Since station A's signal takes longer to propagate down the cable to station D than to station C, C's BIU notices that the channel is free slightly before station D's BIU. However, as station C gets ready to transmit, station D now assumes the channel is free. Within an infinitesimal period of time, C starts transmission, followed by D, resulting in a collision. Here, the collision is a function of the propagation delay of the signal and the distance between two competing stations. CSMA/CD networks therefore work better as the main cable length decreases.

The CSMA/CD access technique is best suited for networks with intermittent transmission, since an increase in traffic volume causes a corresponding increase in the probability of the cable being occupied when a station wishes to talk. In addition, as traffic volume builds under CSMA/CD, throughput may decline, since there will be longer waits to gain access to the network, as well as additional time-outs required to resolve collisions that occur.

Token passing

In a *token passing* access method, each time the network is turned on, a token is generated. The token, consisting of a unique bit pattern, travels the length of the network, either around a ring or along the length of a bus. When a station on the network has data to transmit, it must first seize a free token. On a Token-Ring network, the token is then transformed to indicate that it is in use. Information is added to produce a frame, which represents data being transmitted from one station to another. During the time the token is in use, other stations on the network remain idle, eliminating the possibility of collisions. Once the transmission is completed, the token is converted back into its original form by the station that transmitted the frame, and becomes available for use by the next station on the network.

Figure 3.10 illustrates the general operation of a token-passing Token-Ring network using a ring topology. Since a station on the network can only transmit

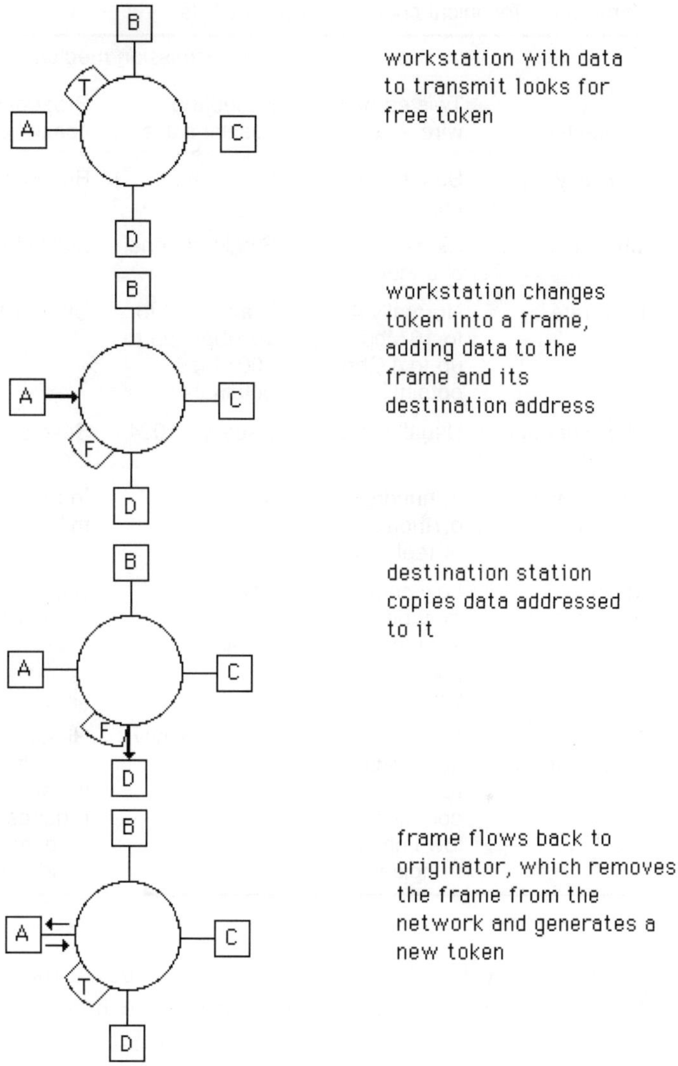

Figure 3.10 Token-Ring operation

when it has a free token, token passing eliminates the requirement for collision detection hardware. Due to the dependence of the network upon the token, the loss of a station can bring the entire network down. To avoid this, the design characteristics of Token-Ring networks include circuitry that automatically removes a failed or failing station from the network as well as other self-healing features. This additional capability is costly; Token-Ring adapter cards in 1998 were typically priced at two to three times the cost of an Ethernet adapter card.

Due to the variety of transmission media, network structures, and access methods, there is no one best network for all users. Table 3.7 provides a generalized

Table 3.7 Technical characteristics of LANs

	Transmission medium			
Characteristic	Twisted-pair wire	Baseband coaxial cable	Broadband coaxial cable	Fiber optic cable
Topology	Bus, star or ring	Bus or ring	Bus or ring	Bus, ring or star
Channels	Single channel	Single channel	Multi-channel	Single, multi-channel
Data rate	Normally 4 to 16 Mbps up to 1 Gbps obtainable	Normally 2 to 10 Mbps, up to 100 Mbps obtainable	Up to 400 Mbps	Up to Gbps
Maximum nodes on net	Usually <255	Usually <1024	Several thousand	Several thousand
Geographical coverage	In hundreds or/thousands of feet	In miles	In tens of miles	In tens of miles
Major advantages	Low cost, may be able to use existing wiring	Low cost, simple to install	Supports voice, data, video applications simultaneously	Supports voice, data video applications simultaneously
Major disadvantages	Limited bandwidth, requires conduits, low immunity to noise	Low immunity to noise	High cost, difficult to install, requires RF modems and headend	Cable cost, difficult to splice

comparison of the advantages and disadvantages of the technical characteristics of local area networks, using the transmission medium as a frame of reference.

Now that we have a general appreciation for the characteristics of CSMA/CD and token-passing networks, we will turn our attention to obtaining specific information about each type of network. However, prior to doing so we will briefly examine the work of the IEEE and its role in the developement of LAN standards. As we will shortly note, the work of the IEEE in subdividing the data link layer facilitated the development of a standard method of control over LANs with different access methods.

3.3 IEEE 802 STANDARDS

The Institute of Electrical and Electronic Engineers (IEEE) Committee 802 was formed at the beginning of the 1980s to develop standards for emerging technologies. The IEEE fostered the development of local area networking equipment from different vendors that can work together. In addition, IEEE LAN standards provided a common design goal for vendors to access a relatively larger market than if proprietary equipment were developed. This, in turn, enabled economies of scale to lower the cost of products developed for larger markets.

802 committees

Table 3.8 lists the organization of IEEE 802 committees involved in local area networks. In examining the lists of committees in Table 3.8 it is apparent that the IEEE early on noted that a number of different systems would be required to satisfy the requirements of a diverse end-user population. Accordingly, the IEEE adopted the CSMA/CD, Token Bus, and Token Ring as standards 802.3, 802.4, and 802.5, respectively.

Table 3.8 IEEE Series 802 committees

802. 1	High Level Interface
802. 2	Logical Link Control
802. 3	CSMA/CD
802. 3z	Gigabit Ethernet
802. 4	Token-Passing Bus
802. 5	Token-Passing Ring
802. 6	Metropolitan Area Networks
802. 7	Broadband Technical Advisory Group
802. 8	Fiber Optic Technical Advisory Group
802. 9	Integrated Voice and Data Networks
802. 10	Network Security
802. 11	Wireless LANs
802. 12	100VG-AnyLAN

The IEEE Committee 802 published draft standards for CSMA/CD and Token Bus local area networks in 1982. Standard 802.3, which describes a baseband CSMA/CD network similar to Ethernet, was published in 1983. Since then, several addenda to the 802.3 standard have been adopted to govern the operation of CSMA/CD on different types of media. These addenda include 10BASE-2 which defines a 10 Mbps baseband network operating on thin coaxial cable; 1BASE-5, which defines a 1 Mbps baseband network operating on twisted-pair; 10BASE-T, which defines a 10 Mbps baseband network operating on twisted-pair; and 10BROAD-36, which defines a broadband 10 Mbps network that operates on thick coaxial cable.

The next standard published by the IEEE was 802.4, which describes a token-passing bus-oriented network for both baseband and broadband transmission. This standard is similar to the Manufacturing Automation Protocol (MAP) standard developed by General Motors.

The third LAN standard published by the IEEE was based upon IBM'S specifications for its Token-Ring network. Known as the 802.5 standard, it defines the operation of Token-Ring networks on shielded twisted-pair cable at data rates of 1 and 4 Mbps. That standard was modified to acknowledge three IBM enhancements to Token-Ring network operations. These enhancements include the 16 Mbps operating rate, the ability to release a token early on a 16 Mbps network, and a bridge routing protocol known as source reouting.

In late 1992 Graud Junction proposed to the IEEE a method for operating Ethernet at 100 Mbps. At approximately the same time, AT&T and Hewlett-Packard proposed a different method to the IEEE which was originally named 100BaseVG, with VG referencing voice-grade twisted-pair cable. IBM joined AT&T and HP, adding support for Token-Ring to 100BaseVG, resulting in the proposed standard having its name changed to 100VG-AnyLAN in recognition of its ability to support either Ethernet or Token-Ring. Because of the merits associated with each proposed standard, the IEEE approved both in 1995. 100BASE-T, also commonly referred to as Fast Ethernet, was approved as an update to 802.3. The specification for 100VG-AnyLAN was approved as 802.12.

At the time this book was prepared a standard covering the CSMA/CD protocol operating at 1 Gbps was being developed. Referred to as Gigabit Ethernet, this proposed standard, which is being worked on by the IEEE 802.3z committee, should be promulgated by mid-1998.

Data link subdivision

One of the more interesting facets of IEEE 802 standards is the subdivision of the ISO Open Systems Interconnection model's data link layer into a minimum of two sublayers: Logical Link Control (LLC) and Medium Access Control (MAC). Figure 3.11 illustrates the relationship between IEEE 802 local area network standards and the first three layers of the OSI Reference Model.

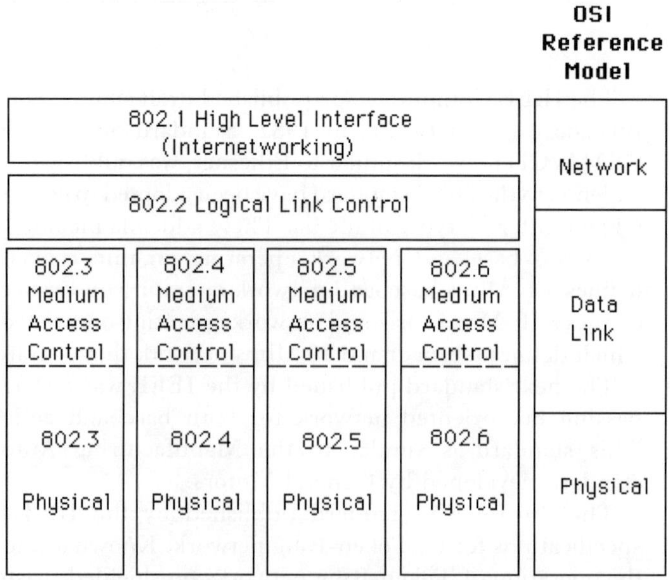

Figure 3.11 Relationship between IEEE standards and the OSI Reference Model

The separation of the data link layer into two entities provides a mechanism for regulating access to the medium that is independent of the method for establishing, maintaining and terminating the logical link between workstations. The method of regulating access to the medium is defined by the Medium Access Control portion of each local area network standard. This enables the Logical Link Control standard to be applicable to each type of network.

Medium Access Control

The Medium Access Control (MAC) sublayer is responsible for controlling access to the network. To accomplish this, it must ensure that two or more stations do not attempt to transmit data onto the network simultaneously. For Ethernet networks, this is accomplished through the use of the CSMA/CD access protocol.

In addition to network access control, the MAC sublayer is responsible for the orderly movement of data onto and off of the network. To accomplish this, the MAC sublayer is responsible for MAC addressing, frame type recognition, frame control, frame copying, and similar frame-related functions.

The MAC address represents the physical address of each station connected to the network. That address can belong to a single station, can represent a predefined group of stations (group address), or can represent all stations on the network (broadcast address). Through MAC addresses, the physical source and destination of frames are identified.

Frame type recognition enables the type and format of a frame to be recognized. To ensure that frames can be processed accurately, frame control prefixes each frame with a preamble, which consists of a predefined sequence of bits. In addition, a frame check sequence (FCS) is computed by applying an algorithm to the contents of the frame; the results of the operation are placed into the frame. This enables a receiving station to perform a similar operation. Then, if the locally computed FCS matches the FCS carried in the frame, the frame is considered to have arrived without error.

Logical Link Control

Logical Link Control (LLC) frames are used to provide a link between network layer protocols and Medium Access Control. This linkage is accomplished through the use of Service Access Points (SAPs), which operate in much the same way as a mailbox. That is, both network layer protocols and Logical Link Control have access to SAPs and can leave messages for each other in them.

Like a mailbox in a post office, each SAP has a distinct address. For the Logical Link Control, a SAP represents the location of a network layer process, such as the location of an application within a workstation as viewed from the network. From the network layer perspective, a SAP represents the place to leave messages concerning the network services requested by an application.

LLC frames contain two special address fields, known as the Destination Services Access Point (DSAP) and the Source Services Access Point (SSAP). The former specifies the receiving network layer process, while the latter specifies the sending network layer process. Both DSAP and SSAP addresses are assigned by the IEEE. When we turn our attention to specific types of LANs later in this chapter, we will also examine how DSAP and SSAP information is carried within a LAN frame.

Physical layer subdivision

Similar to the manner by which the IEEE subdivided the data link layer, certain IEEE LANs have their physical layer subdivided. This subdivision resulted from the recognition that different coding schemes were better suited to certain types of media. Thus, the subdivision of the physical layer enables different coding schemes to be used with different media for a common type of LAN. When applicable in later sections of this chapter, we will examine the subdivision of the physical layer.

3.4 ETHERNET NETWORKS

From the title of this section, it is apparent that there is more than one type of Ethernet network. From a network access perspective, there is actually only one Ethernet network. However, the CSMA/CD access protocol used by Ethernet, as well as its general frame format and most of its operating characteristics, were used by the Institute of Electrical and Electronic Engineers (IEEE) to develop a series of Ethernet-type networks under the IEEE 802.3 umbrella. Thus, this section will focus on different types of Ethernet networks by closely examining the components and operating characteristics of Ethernet and then comparing its major features to the different networks defined by the IEEE 802.3 standard. Once this is accomplished, we will focus our attention on the wiring, topology, and hardware components associated with each type of IEEE 802.3 Ethernet network, as well as the Ethernet frame used to transport data.

Original network components

The 10 Mbps Ethernet network standard originaly developed by Xerox, Digital Equipment Corporation, and Intel was based on the use of five hardware components. Those components include a coaxial cable, a cable tap, a transceiver, a transceiver cable, and an interface board (also known as an Ethernet controller). Figure 3.12 illustrates the relationship between Ethernet components.

Coaxial cable

One of the problems faced by the designers of Ethernet was the selection of an appropriate medium. Although twisted-pair wire is relatively inexpensive and easy to use, the short distances between twists serve as an antenna for receiving

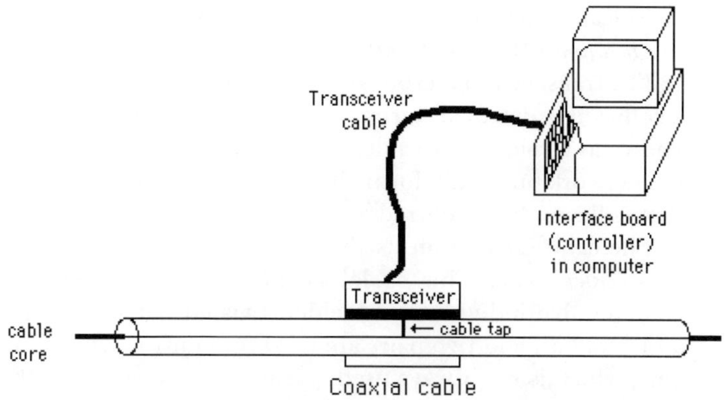

Figure 3.12 Ethernet hardware components. When thick coaxial cable is used for the bus, an Ethernet cable connection is made with a transceiver cable and a transceiver tapped into the cable

electromagnetic and radio frequency interference in the form of noise. Thus, the use of twisted-pair cable restricts the network to relatively short distances. Coaxial cable, however, has a dielectric shielding the conductor. As long as the ends of the cable are terminated, coaxial cable can transmit over greater distances than twisted-pair cable. Thus, the initial selection for the Ethernet transmission medium was coaxial cable.

There are two types of coaxial cable that can be used to form the main Ethernet bus. The first type of coaxial cable specified for Ethernet was a relatively thick 50-ohm cable, which is normally colored yellow and is commonly referred to as 'thick' Ethernet. This cable has a marking every 2.5 meters to indicate where a tap should occur, if one is required to connect a station to the main cable at particular location. These markings represent the minimum distance one tap must be separated from another on an Ethernet network.

A second type of coaxial cable used with Ethernet is smaller and more flexible; however, it is capable of providing a transmission distance only one-third of that obtainable on thick cable. This lighter and more flexible cable is referred to as 'thin' Ethernet and also has an impedance of 50 ohms.

Two of the major advantages of thin Ethernet over thick cable are its cost and its use of BNC connectors. Thin Ethernet is significantly less expensive than thick Ethernet. Thick Ethernet requires connections via taps, whereas the use of thin Ethernet permits connections to the bus via industry-standard BNC connectors that form T-junctions.

Transceiver and transceiver cable

Transceiver is a shortened form of *transmitter–receiver*. This device contains electronics to both transmit onto and receive signals carried by the coaxial cable. The transceiver contains a tap that, when pushed against the coaxial cable, penetrates the cable and makes contact with the core of the cable. In books and technical literature the transceiver, its tap, and its housing are often referred to as the *medium attachment unit (MAU)*.

The transceiver is responsible for carrier detection and collision detection. When a collision is detected during a transmission, the transceiver places a special signal, known as a *jam*, on the cable. This signal is of sufficient duration to propagate down the network bus and inform all of the other transceivers attachment to the bus node that collision has occurred.

The cable that connects the interface board to the transceiver is known as the *transceiver cable*. This cable can be up to 50 meters (165 feet) in lengths, and contains five individually shielded twisted-pairs. Two pairs are used for data in and data out, and two pairs are used for control signals in and out. The remaining pair, which is not always used, permits the power from the computer in which the interface board is inserted to power the transceiver.

Since collision detection is a critical part of the CSMA/CD access protocol, the original version of Ethernet was modified to inform the interface board that the transceiver collision circuitry is operational. This modification resulted in each transceiver's sending a signal to the attached interface board after every transmission, informing the board that the transceiver's collision circuitry is operational. This signal is sent by the transceiver over the collision pair of the transceiver cable, and must start within 0.6 microseconds after each frame is transmitted. The duration of the signal can vary between 0.5 and 1.5 microseconds. Known as the *Signal Quality Error* and also referred to as the *SQE or heartbeat*, this signal is supported by Ethernet Version 2.0, published as an standard in 1982, and by the IEEE 802.3 standard.

Interface board

The *interface board*, or *network interface card (NIC)*, is inserted into an expansion slot within a computer, and is responsible for transmitting frames to and receiving frames from the transceiver. This board contains several special chips, including a controller chip that assembles data into an Ethernet frame and computes the cyclic redundancy check used for error detection. Thus, this board is also referred to as an *Ethernet controller*.

Repeaters

A repeater is a device that receives, amplifies, and retransmits signals. Since a repeater operates at the physical layer, it is transparent to data and simply regenerates signals. Figure 3.13 illustrates the use of a repeater to connect two Ethernet cable segments. As indicated, a transceiver is tapped to each cable segment to be connected, and the repeater is cabled to the transceiver. When used to connect cable segments, a repeater counts as one station on each connected segment. Thus, a segment capable of supporting up to 100 stations can support only 99 additional stations when a repeater is used to connect cable segments.

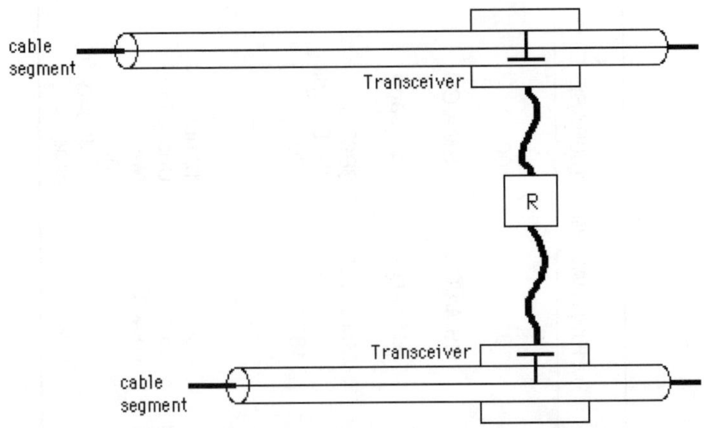

Figure 3.13 Using a repeater. Cable segments can be joined together by a repeater (R) to expand the network. The repeater counts as a station on each cable segment

IEEE 802.3 networks

The IEEE 802.3 standard is based on Ethernet. However, it has several significant differences, particularly its support of multiple physical layer options, which include 50 and 75 ohm coaxial cable, unshielded twisted-pair wire, an operating rate of 100 Mbps for two recently standardized versions of Ethernet, and a pending standard referred to as Gigabit Ethernet that will support data transmission at 1 Gbps. Other differences between various types of IEEE 802.3 networks and Ethernet include the data rates supported by some 802.3 networks, their method of signaling, the maximum cable segment lengths permitted prior to the use of repeaters, and their network topologies.

Network names

The standards thad define IEEE 802.3 networks have been given names that generally follow the form '*s type-l*'. Here , *s* refers to the speed of the network in Mbps, *type* is BASE for baseband and BROAD for broadband, and *l* refers to the maximum segment length in 100-meter multiples. Thus, 10BASE-5 refers to an IEEE 802.3 baseband network that operates at 10 Mbps and has a maximum segment length of 500 meters. One exception to this general form is 10BASE-T, which is the name for an IEEE 802.3 network that operates at 10 Mbps using unshielded twisted-pair (UTP) wire.

Table 3.9 compares the operating charateristics of six currently defined IEEE 802.3 networks to Ethernet.

10BASE-5

An examination of the operating characteristics of Ethernet and 10BASE-5 indicates that these networks are the same.

Table 3.9 Ethernet and IEEE 802.3 network characteristics

Operational characteristics	Ethernet	10BASE-5	10BASE-2	1BASE-5	10BASE-T	10BROAD-36	100BASE-T
Operating rate (Mbps)	10	10	10	1	10	10	100
Access protocol	CSMA/CD	CSMA/CD	CSMA/CD	CSMA/CD	CSMA/CD	CSMA/CD	CSMA/CD
Type of signaling	Baseband	Baseband	Baseband	Baseband	Baseband	Broadband	Baseband
Data encoding	Manchester	Manchester	Manchester	Manchester	Manchester	Manchester	8B6T or 4B5B coding
Maximum segment length (meters)	500	500	185	250	100	1800	100
Stations/segment	100	100	30	12/hub*	12/hub*	100	12/hub*
Medium	50 ohm coaxial (thick)	50 ohm coaxial (thick)	50 ohm coaxial (thin)	Unshielded twisted pair	Unshielded twisted pair	75 ohm coaxial	Unshielded twisted pair (Category 5)
Topology	Bus	Bus	Bus	Star	Star	Bus	Star

* number of hub ports typically varies from 4 to 8, 12, 16 or 24 depending upon manufacturer

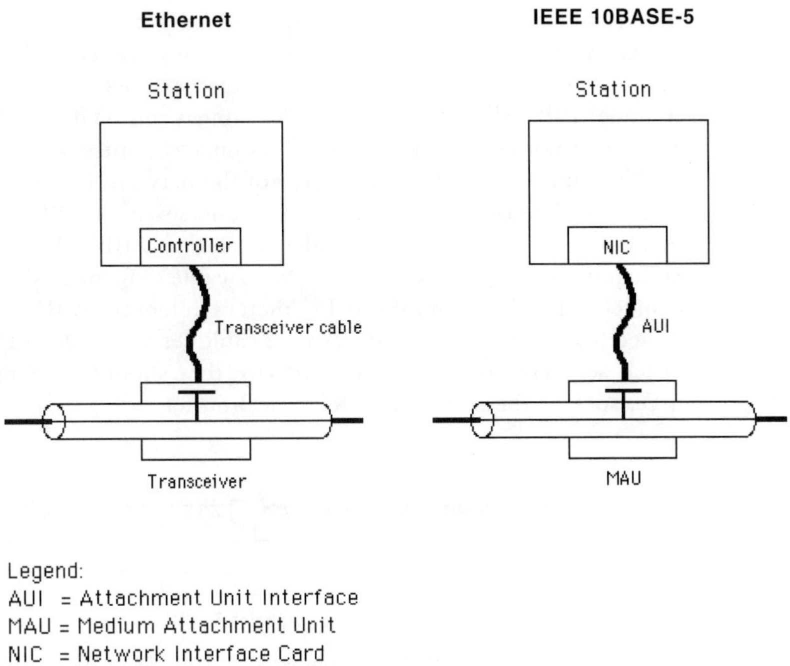

Figure 3.14 Ethernet and 10BASE-5 media interface differences

Figure 3.14 illustrates the major terminology changes between Ethernet and the IEEE 802.3 10BASE-5 network. These changes are in the media interface: the transceiver cable is referred to as the *Attachment Unit Interface (AUI)*, and the transceiver, including its tap and housing, is referred to as the *Medium Attachment Unit (MAU)*. The Ethernet controller, also known as an interface board, is now known as the *Network Interface Card (NIC)*.

Both Ethernet and the IEEE 802.3 10BASE-5 standards support a data rate of 10 Mbps and maxium cable segment length of 500 meters. 10BASE-5, like Ethernet, requires a minimum spacing of 2.5 meters between MAUs and supports a maximum of five segments in any end-to-end path through the traversal of up to four repeaters in any path. Within any path, no more than three cable segments can be *populated*—have stations attached to the cable—and the maximum number of attachments per segment is limited to 100.

10BASE-2

10BASE-2 is a smaller and less expensive version of 10BASE-5. This standard uses a thinner RG-58 coaxial cable, thus earning the names of 'cheapnet' and 'thinnet', as well as 'thin Ethernet'. Although 10BASE-2 cable is both less expensive and easier to use than 10BASE-5 cable, it cannot carry signals as far as 10BASE-5 cable.

Under the 10BASE-2 standard, the maximum cable segment length is reduced to 185 meters (607 feet), with a maximum of 30 stations per segment. Another difference between 10BASE-5 and 10BASE-2 concerns the integration of transceiver electronics into the network interface card under the 10BASE-2 standard. This permits the NIC to be directly cabled to the main trunk cable. In fact, under 10BASE-2 the thin Ethernet cable is routed directly to each workstation location and routed through a BNC T-connector, one end of which is pressed into the BNC connector built into the rear of the network interface card.

Figure 3.15 illustrates the cabling of a one-segment 10BASE-2 network, which can support a maximum of 30 nodes or stations. BNC barrel connectors can be used to join two lengths of thin 10BASE-2 cable to form a cable segment, as long as the joined cable does not exceed 185 meters in length. A BNC terminator must be attached to each end of each 10BASE-2 cable segment. One of the two terminators on each segment contains a ground wire that should be connected to a ground source, such as the screw on an electrical outlet.

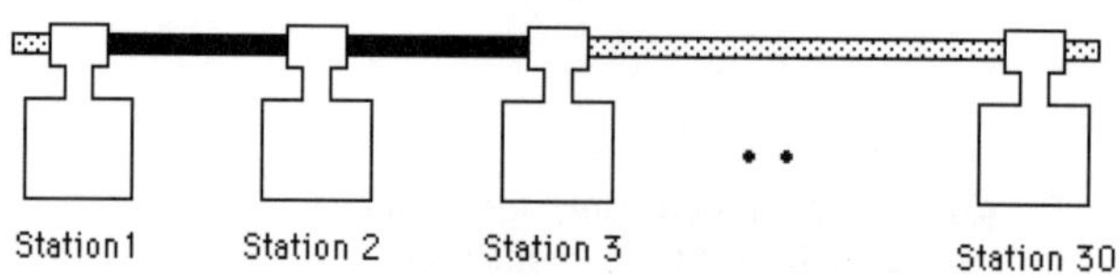

Figure 3.15 Cabling a 10BASE-2 network. A 10BASE-2 cable segment cannot exceed 185 meters and is limited to supporting up to 30 nodes or stations

10BROAD-36

10BROAD-36 is the only broadband network based on the CSMA/CD access protocol standardized by the IEEE. Unlike a baseband network, in which Manchester-coded signals are placed directly onto the cable, the 10BROAD-36 standard requires the use of radio frequency (RF) modems. Those modems modulate non-return to zero (NRZ) coded signals for transmission on one channel at a specified frequency, and demodulate received signals by listening for tones on another channel at a different frequency.

A 10BROAD-36 network is constructed with a 75-ohm coaxial cable, similar to the cable used in modern cable television (CATV) systems. Under the IEEE 802.3 broadband standard, either single or dual cables can be used to construct a network. If a single cable is used, the end of the cable (referred to as the *headend*) must be terminated with a frequency translator. That translator converts the signals received on one channel to the frequency assigned to the other channel, retransmitting the signal at the new frequency. Since the frequency band for a transmitted signal is below the frequency band 10BROAD-36 receivers scan, we say the frequency translator *upconverts* a transmitted signal and retransmits it for reception by other stations on the network. If two cables are used, the headend simply functions as a relay point, transferring the signal received on one cable onto the second cable.

A broadband transmission system has several advantages over a baseband system. Two of the primary advantages of broadband are its ability to support multiple transmissions occurring on independent frequency bands simultaneously,

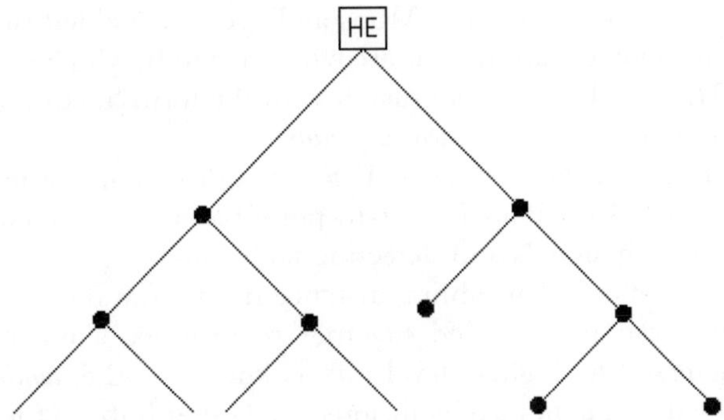

Figure 3.16 Broadband tree topology support. The headend (HE) receives transmissions at one frequency and regenerates the received signals back onto the network at another predefined frequency

and its ability to support a tree structure topology carrying multiple simultaneous transmissions. Using independent frequency bands, you can establish several independent networks. In fact, each network can be used to carry voice, data, and video over a common cable. As for topology, broadband permits the use of a tree structure network, such as the structure shown in Figure 3.16. In this example, the top of the tree would be the headend; it would contain a frequency translator, which would regenerate signals received at one frequency back onto the cable at another predefined frequency.

The higher noise immunity of 75-ohm coaxial cable permits a 10BROAD-36 network to span 3600 meters, making this medium ideal for linking buildings on a campus. In addition, the ability of 10BROAD-36 to share channel space on a 75-ohm coaxial cable permits organizations that have an existing CATV system, such as one used for video security, to use that cable for part or all of their broadband CSMA/CD network. Although these advantages can be significant, the cost associated with RF modems limited the use of broadband primarily to campus environments that require an expanded network span. In addition, the rapid development and acceptance of 10BASE-T and 100BASE-T and the decline in the cost of fiber optic cable to extend the span of twisted-pair networks have severely limited what was once widely anticipated to be a promising future for 10BROAD-36 networks.

1BASE-5

The 1BASE-5 standard was based on AT&T's low-cost CSMA/CD network known as StarLan. Thus, 1BASE-5 is commonly referred to as StarLan, although AT&T uses that term to refer to CSMA/CD networks operating at both 1 and 10 Mbps using unshielded twisted-pair cable. The latter is considered the predecessor to 10BASE-T.

The 1BASE-5 standard differs significantly from Ethernet and 10BASE-5 standards in its use of media and topology, and in its operating rate. The 1BASE-5 standard operates at 1 Mbps and uses unshielded twisted-pair (UTP) wiring in a star topology; all stations are wired to a hub, which is known as a *multiple access unit (MAU)*. To avoid confusion with the term 'medium access unit', we will refer to this wiring concentrator as a *hub*.

Each station in a 1BASE-5 network contains a network interface card (NIC), cabled via UTP on a point-to-point basis to a hub port. The hub is responsible for repeating signals and detecting collisions.

The maximum cabling distance from a station to a hub is 250 meters; up to five hubs can be cascaded together to produce a maximum network span of 2500 meters. The highest level hub is known as the *header hub* and is responsible for broadcasting news of collisions to all other hubs in the network. These hubs, which are known as *intermediate hubs*, are responsible for reporting all collisions to the header hub.

AT&T's 1 Mbps StarLan network, along with other 1BASE-5 systems, initially received a degree of acceptance for use in small organizations. However, the introduction of 10BASE-T, which provided an operating rate 10 times that obtainable under 1BASE-5, severely limited the further acceptance of 1BASE-5 networks.

10BASE-T

In the late 1980s, a committee of the IEEE recognized the requirement of organizations to transmit Ethernet at a 10 Mbps operating rate over low-cost and readily available unshielded twisted-pair cable. Although several vendors had already introduced equipment that permitted Ethernet signaling via UTP cabling, such equipment was based on proprietary designs and was not interperable. Thus, a new task of the IEEE was to develop a standards for 802.3 networks operating at 10 Mbps using UTP cable. The resulting standard was approved by the IEEE as 802.3i in September 1990, and is more commonly known as 10BASE-T.

The 10BASE-T standard supports an operating rate of 10 Mbps at a distance of up to 100 meters (328 feet) over UTP Category 3 cable without the use of a repeater. The UTP cable requires two pairs of twisted wire. One pair is used for transmitting, while the other pair is used for receiving. Each pair of wires is twisted together, and each twist is 180 degrees. Any electromagnetic interference (EMI) or radio frequency interference (RFI) is therefore received 180 degrees out of phase; this theoretically cancels out EMI and RFI noise while leaving the network signal. In reality, the wire between twists acts as an antenna and receives a degree of noise.

Network components

A 10BASE-T network can be constructed with network interface cards, UTP cable, and one or more hubs. Each NIC is installed in the expansion slot of a computer and wired on a point-to-point basis to a hub port. When all of the ports on a hub are used, one hub can be connected to another to expand the network, resulting in a physical star, logical bus network structure.

Network interface cards

Most 10BASE-T network interface cards contain multiple connectors, which enable the card to be used with different types of 802.3 networks. For example, most modern NICs include an RJ-45 jack as well as BNC and DB-15 connectors. The RJ-45 jack supports the direct attachment of the NIC to a 10BASE-T network, while the BNC connector permits the NIC to be mated to a 10BASE-2 T-connector. The DB-15 connector enables the NIC to be cabled to a transceiver, and is more commonly referred to as the NIC's *attachment unit interface (AUI)* port.

Figure 3.17 illustrates an example of an Ethernet multiple media network interface card. Not only does this type of card simplify the manufacturer's inventory, but in addition it provides end-users with the ability to easily migrate from one wiring infrastructure to another without having to replace network interface cards.

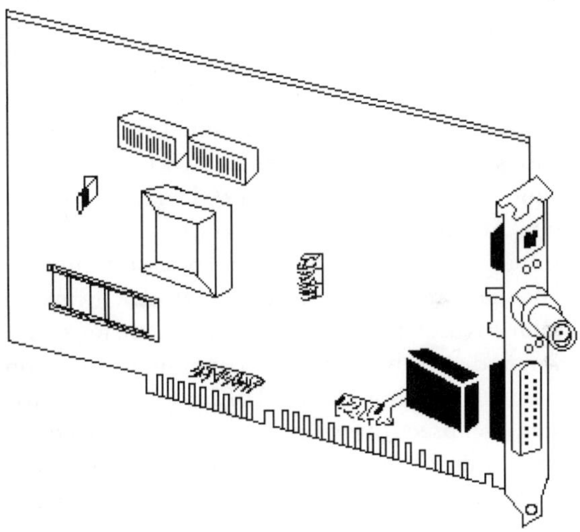

Figure 3.17 Multiple media network interface card. Some multiple media network interface cards, such as the one illustrated, support the direct attachment to UTP and thin coaxial cable, while the DB-15 connector permits the card to be cabled to a transceiver connected to thick coaxial cable

The use of the RJ-45 connector can be considered to be forward looking as it allowed different features to be developed without changing the connector. 10BASE-T uses four of the eight RJ-45 pins, with Table 3.10 providing a comparison of 10BASE-T pin numbers to RJ-45 pin numbers as well as indicating the signal names used with 10BASE-T UTP cable. Other versions of Ethernet as well as other types of LANs use additional pins in the RJ-45 connector. For example, a full-duplex version of Ethernet requires the use of eight pins. Thus, the original selection of the RJ-45 connector proved to be a wise choice.

Table 3.10 10BASE-T wiring

10BASE-T pin #	RJ-45 pin #	10BASE-T signal name
1	1	Transmit Data +
2	2	Transmit Data –
3	3	Receive Data +
—	4	Not used
—	5	Not used
6	6	Receive Data –
—	7	Not used
—	8	Not used

Hub

The wiring hub in a 10BASE-T network functions as a multiport repeater: it receives, retimes, and regenerates signals received from any attached station. The hub also functions as a filter: it discards severely distorted frames.

A 10BASE-T hub tests the integrity of the link from each hub port to a connected station by transmitting a special signal to the station. If the device doesn't respond, the hub will automatically shut down the port, and may illuminate a status light-emitting diode (LED) to indicate the status of each port.

Hubs monitor, record, and count consecutive collisions that occur on each individual station link. Since an excessive number of consecutive collisions will prevent data transfer on all of the attached links, hubs are required to cut off or partition any link on which too many collisions occurred. This partitioning enables the remainder of the network to operate in situations where a faulty NIC transmits continuously. Although the IEEE 802.3 standard does not specify a maximum number of consecutive collisions, the standard does specify that partitioning can be initiated after 30 or more consecutive collisions occur. Thus, some hub vendors initiate partitioning when 31 consecutive collisions occur, while other manufacturers use a higher value.

Although a wiring hub is commonly referred to as a *concentrator*, this term is not technically correct. A 10BASE-T wiring hub is a self-contained unit that typically includes 8, 10, or 12 RJ-45 ports for direct connection to stations, and a BNC and/or DB-15 AUI port for expanding the hub to other network equipment. The BNC and AUI ports enable the 10BASE-T hub to be connected to 10BASE-2 and 10BASE-5 networks, respectively. For the latter, the AUI port is cabled to a 10BASE-5 MAU (transceiver), which is tapped into thick 10BASE-5 coaxial cable. One 10BASE-T hub can be connected to another with a UTP link between RJ-45 ports on each hub.

Figure 3.18 illustrates the connectors on a typical 10BASE-T hub. On some hubs, one RJ-45 jack is labeled uplink/downlink for use in cascading hubs, while other vendors permit any RJ-45 port to be used for connecting hubs.

Unlike a hub, a concentrator consists of a main housing into which modular cards are inserted. Although some modular cards may appear to represent hubs, and do indeed function as 10BASE-T hubs, the addition of other modules permits the

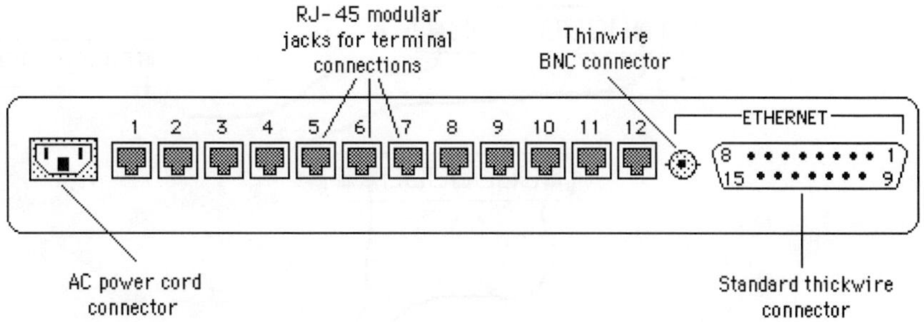

Figure 3.18 10BASE-T hub connectors. In addition to 8, 10, or 12 RJ-45 modular jacks for terminal connections, most 10BASE-T hubs contain a BNC and DB-15 port to permit attachment to thin and thick backbone networks

network to be easily expanded from one location and allows additional features to be supported. For example, the insertion of a fiber optic inter-repeater module permits concentrators to be interconnected over relatively long distances of approximately 3 km.

Expanding a 10BASE-T network

A 10BASE-T network can be expanded with additional hubs once the number of stations serviced uses up the hub's available terminal ports. In expanding a 10BASE-T network, the wiring that joins each hub together is considered to represent a cable segment, while each hub is considered as a repeater. Under the 802.3 specification, no two stations can be separated by more than four hubs connected together by five cable segments.

Figure 3.19 illustrates the expansion of a 10BASE-T network through the use of five hubs. Note that the connection between station A and station B traverses five segments and four hubs, and so does not violate IEEE 802.3 connection rules.

100BASE-T

The standardization of 100BASE-T, commonly known as Fast Ethernet, required an extension of previously developed IEEE 802.3 standards. In the definition process of standardization development, both the Ethernet Media Access Control (MAC) and physical layer required adjustments to permit 100 Mbps operational support. For the MAC layer, scaling its speed to 100 Mbps from the 10BASE-T 10 Mbps operational rate required a minimal adjustment, since in theory the 10BASE-T MAC layer was developed independent of the data rate. For the physical layer, more than a minor adjustment was required since Fast Ethernet was designed to support three types of media, resulting in three new names which fall under the 100BASE-T umbrella. 100BASE-T4 uses three wire pairs for data transmission and a fourth for collision detection resulting in T4 being appended to the 100BASE mnemonic. 100BASE-TX uses two pairs of category 5 UTP with one pair employed for transmission while the second is used for both collision detection and reception of data. The third 100BASE-T standard is 100BASE-FX which represents the use of fiber optic media.

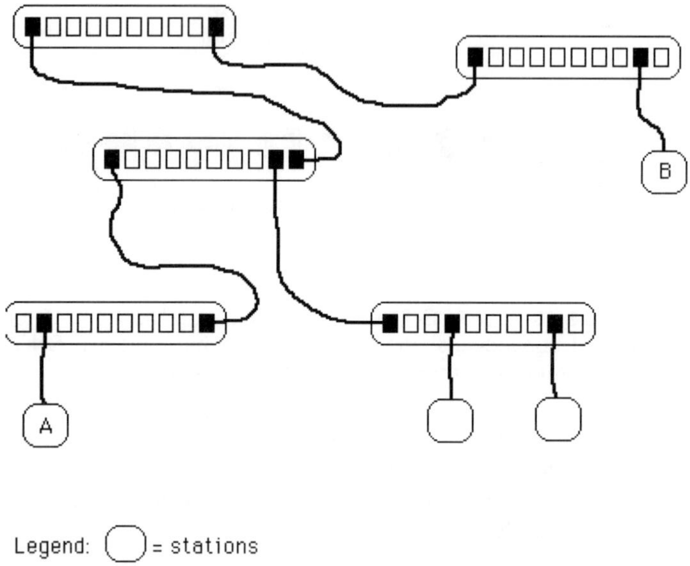

Figure 3.19 Expanding a 10BASE-T network. No two stations can be separated by more than four hubs in a 10BASE-T network

Using work developed in the standardization process of FDDI in defining 125 Mbps full-duplex signaling to accommodate optical fiber, UTP and STP through Physical Media Dependent (PMD) sublayers, Fast Ethernet borrowed this strategy. Since a mechanism was required to map the PMD's continuous signaling system to the start–stop 'half-duplex' system used at the Ethernet MAC layer, the physical layer was subdivided. This subdivision is illustrated in Figure 3.20. The PMD sublayer supports the appropriate media to be used, while the convergence sublayer (CS), which was later renamed the physical coding sublayer, performs the mapping between the PMD and the Ethernet MAC layer.

Although Fast Ethernet represents a tenfold increase in the LAN operating rate from 10BASE-T, to ensure proper collision detection the 100BASE-T network span was reduced to 250 meters, with a maximum of 100 meters permitted between a network node and a hub. The smaller network diameter reduces potential propagation delay. When coupled with a tenfold operating rate increase and no change in network frame size, the ratio of frame duration to network propagation delay for a 100BASE-T network is the same as for a 10BASE-T network.

Physical layer

The physical layer subdivision previously illustrated in Figure 3.20, as indicated in the title of the figure, presents an overview of the true layer subdivision. In actuality, a number of changes were required at the physical layer to obtain a 10 Mbps operating rate. Those changes include the use of three wire pairs for data (the fourth is used for collision detection), 8B6T ternary coding (for 100BASE-T4) instead of Manchester coding, and an increase in the clock signaling speed from 20 MHz to 25 MHz.

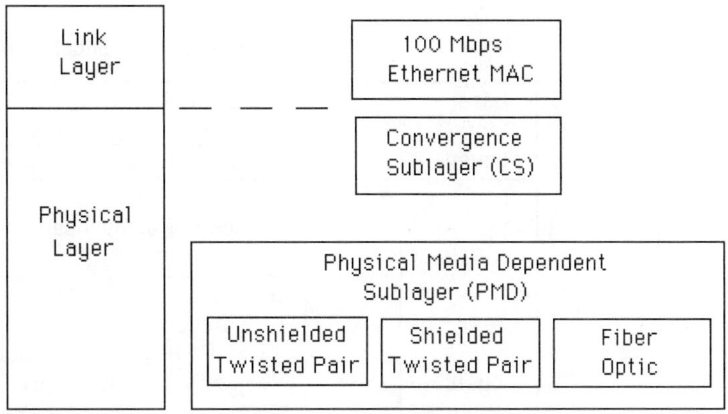

Figure 3.20 Fast Ethernet (100BASE-T) physical layer subdivision overview

When the specifications for Fast Ethernet were being developed it was recognized that the physical signaling layer would incorporate medium-dependent functions if support was extended to two-pair cable (100BASE-TX) operations. To separate medium-dependent interfaces to accommodate multiple physical layers, a common interface referred to as the Medium Independent Interface (MII) was inserted between the MAC layer and the physical encoding sublayer. The MII represents a common point of interoperability between the medium and the MAC layer. The MII can support two specific data rates, 10 Mbps and 100 Mbps, permitting older 10BASE-T nodes to be supported at Fast Ethernet hubs. To reconcile the MII signal with the MAC signal, a reconciliation sublayer was added under the MAC layer, resulting in the subdivision of the link layer into three parts—a Logical Link Control sublayer, a Media Access Control sublayer and a reconciliation sublayer. The top portion of Figure 3.21 illustrates this subdivision.

That portion of Fast Ethernet below the MII, which is the new physical layer, is now subdivided into three sublayers. The lower portion of Figure 3.21 illustrates the physical sublayers for 100BASE-T4 and 100BASE-TX.

The physical coding sublayer performs the data encoding, transmit, receive and carrier sense functions. Since the data coding method differs between 100BASE-T4 and 100BASE-TX, this difference requires distinct physical coding sublayers for each version of Fast Ethernet.

The physical medium attachment (PMA) sublayer maps messages from the physical coding sublayer (PCS) onto the twisted-pair transmission media and vice versa.

The medium dependent interface (MDI) sublayer specifies the use of a standard RJ-45 connector. Although the same connector is used for 100BASE-TX, the use of two pairs of cable instead of four results in different pin assignments.

100BASE-T4

Figure 3.22 illustrates the RJ-45 pin assignments of wire pairs used by 100BASE-T4. Note that wire pairs D1 and D2 are unidirectional. As indicated in Figure 3.22,

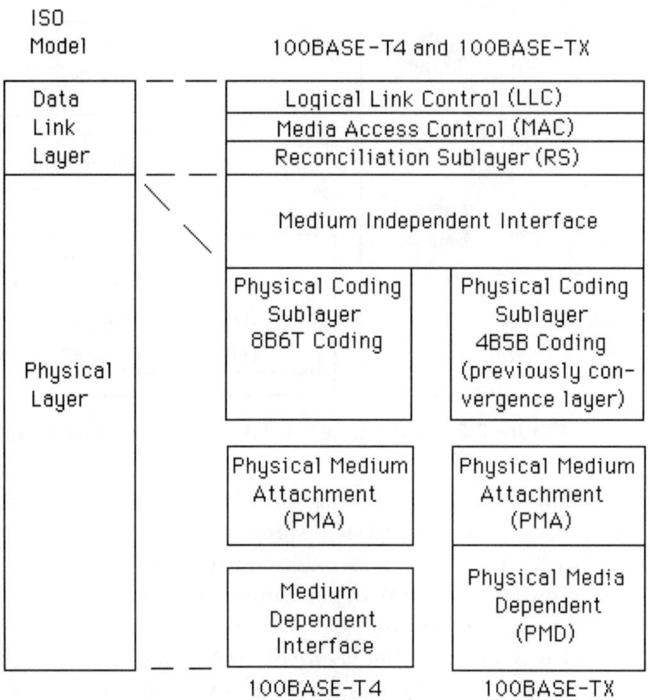

Figure 3.21 100BASE-T4 and 100BASE-TX physical and link layers

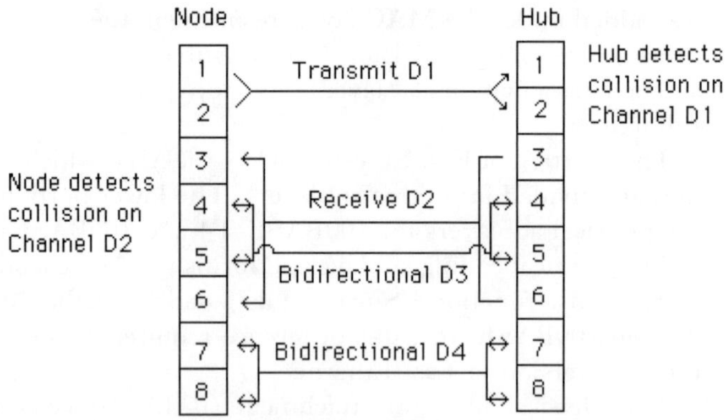

Figure 3.22 100BASE-T4 pin assignments

three wire pairs are available for data transmission and reception in each direction, while the fourth pair is used for collision detection.

The 100BASE-T4 physical coding sublayer implements 8B6T block coding. Under this coding technique, each block of eight input bits is transformed into a unique code group of six ternary symbols. Figure 3.23 provides an overview of the 8B6T coding process used by 100BASE-T4.

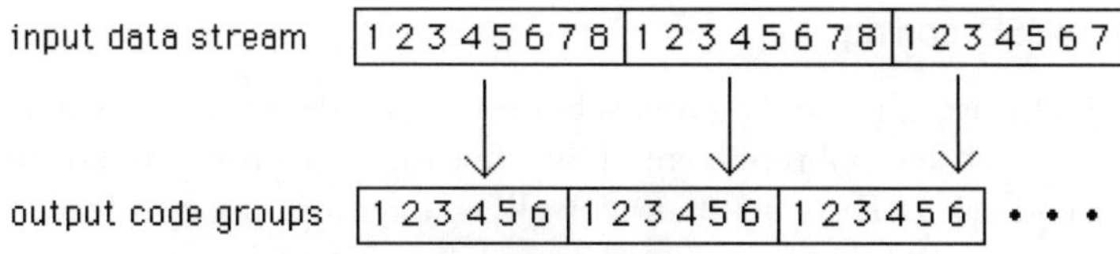

Figure 3.23 8B6T coding process

The output code groups resulting from 8B6T coding flow out to three parallel channels that are placed on three twisted pairs. Thus, the effective data rate of each pair is 100/3 Mbps or 33.33 Mbps. Since six bits are represented by eight bit positions, the signaling rate or baud rate on each cable pair becomes 33 Mbps × 6/8 or 25 MHz, which is the clock rate used at the MII sublayer.

100BASE-TX

100BASE-TX represents 100BASE-T which supports the use of two pairs of category 5 UTP cabling with RJ-45 connectors. A 100BASE-TX network requires a hub, and the maximum cable run is 100 meters from hub port to node, with a maximum network diameter of 250 meters.

Figure 3.24 illustrates the cabling of two pairs of UTP wires between a hub and node to support 100BASE-TX transmission. One pair of wires is used for transmission, while the second pair is used for collision detection and reception of data. The use of a 125 MHz frequency requires the use of a 'data grade' cable. Thus, 100BASE-TX is based upon the use of category 5 UTP.

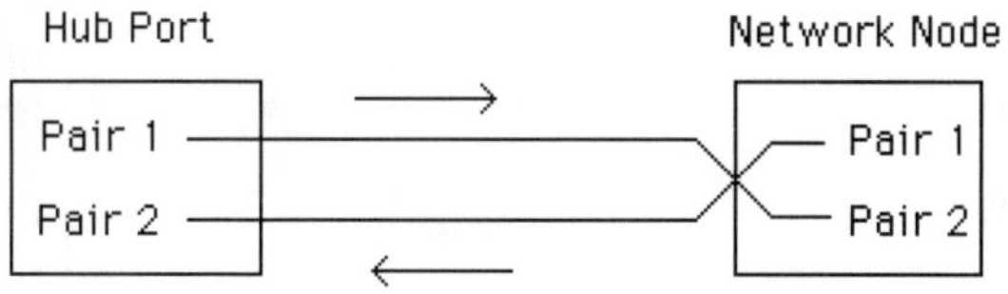

Figure 3.24 100BASE-TX cabling

Although the 100BASE-TX physical layer structure resembles the 100BASE-T4 layer, there are significant differences between the two to accommodate the differences in media used. At the physical coding sublayer, the 100 Mbps start–stop bit stream from the MII first converted to a full-duplex 125 Mbps bit stream. This conversion is accomplished by the use of the FDDI PMD as the 100BASE-TX PMD. Next, the data stream is encoded using a 4B5B coding scheme. The 100BASE-TX PMD decodes symbols from the 125 Mbps continuous bit stream and converts the stream to 100 Mbps start–stop data bits when the data flow is reversed.

4B5B coding

The use of a 4B5B coding scheme enables data and control information to be carried in each symbol represented by a five bit code group. In addition, an inter-stream fill code (IDLE) is defined, as well as a symbol used to force signaling errors. Since four data bits are mapped into a five bit code, only 16 symbols are required to

Table 3.11 4B5B 100BASE-TX code groups

Code group 4320	Name	TXD/RD 3210	Interpretation
Data			
11110	0	0000	Data 0
01001	1	0001	Data 1
10100	2	0010	Data 2
10101	3	0011	Data 3
01010	4	0100	Data 4
01011	5	0101	Data 5
01110	6	0110	Data 6
01111	7	0111	Data 7
10010	8	1000	Data 8
10011	9	1001	Data 9
10110	A	1010	Data A
10111	B	1011	Data B
11010	C	1100	Data C
11011	D	1101	Data D
11100	E	1110	Data E
11101	F	1111	Data F
Idle			
11111	I		IDLE: used as inter-stream fill code
Control			
11000	J		Start-of-stream Delimiter, Part 1 of 2; always used in pairs with K
10001	K		Start-of-stream Delimiter, Part 2 of 2; always used in pairs with J
01101	T		End-of-stream Delimiter, Part 1 of 2; always used in pairs with R
00111	R		End-of-stream Delimiter, Part 2 of 2; always used in pairs with T
Invalid			
00100	V		Transmit error; used to force signaling errors
00000	V		Invalid code
00001	V		Invalid code
00010	V		Invalid code
00011	V		Invalid code
00101	V		Invalid code
00110	V		Invalid code
01000	V		Invalid code
01100	V		Invalid code
10000	V		Invalid code
11001	V		Invalid code

represent data. The remaining symbols not used for control or to denote an IDLE condition are not used by 100BASE-TX and are considered as invalid.

Table 3.11 lists the 4B5B 100BASE-TX code groups. Note that all 1's indicate an idle condition.

100BASE-FX

100BASE-FX represents the third 100BASE-T wiring scheme, defining Fast Ethernet transmission over fiber optic media. 100BASE-FX requires the use of two-strand 62.5/125 μm multimode fiber media and supports the 4B5B coding scheme, identical to the one used by 10BASE-TX. Although 100BASE-FX standards were in the process of being developed when this book was written, the use of fiber optic can be expected to result in longer cable runs than are permissible with UTP. This should enable 100BASE-FX to be used for connections between bridges, routers and switches.

Network utilization

Since 100BASE-T4 and 100BASE-TX preserve the 10BASE-T MAC layer, both standards are capable of interoperability with existing 10BASE-T networks as well as with other low speed Ethernet technology. Through the use of NWay autosensing logic, Fast Ethernet adapters, hub and switch ports can determine if attached equipment can transmit at 10 or 100 Mbps and adjust to the operating rate of the distant device.

NWay is a cable and transmission autosensing scheme proposed by National Semiconductor to the IEEE 802.3 standards group in May 1994. The NWay autosensing scheme permits Ethernet circuits to detect both the cable type and speed of incoming Ethernet data as well as enabling Ethernet repeaters to configure themselves for correct network operations. Since NWay can detect 10 Mbps versus 100 Mbps operations, as well as half- and full-duplex transmission, it permits Ethernet circuits to be developed to automatically adjust to the operating rate and cabling scheme used. This in turn can be expected to simplify the efforts of network managers and administrators, since products incorporating NWay are self-configurable and will not require the setting of DIP switches or software parameters.

There are several network configurations you can consider for Fast Ethernet operations. First, you can construct a 100BASE-TX network similar to a 10BASE-T network, using an appropriate hub and workstations that support the specific network type. This type of network, commonly referred to as shared-media hub-based network, should be considered when most, if not all, network users access servers to perform graphic-intensive or similar bandwidth-intensive operations. Figure 3.25 illustrates a Fast Ethernet shared-media hub network configuration.

A more common use for Fast Ethernet during 1998 was its incorporation into one or more ports in an Ethernet switch. An Ethernet switch can be viewed as a sophisticated hub which is designed to transmit packets arriving on one input port

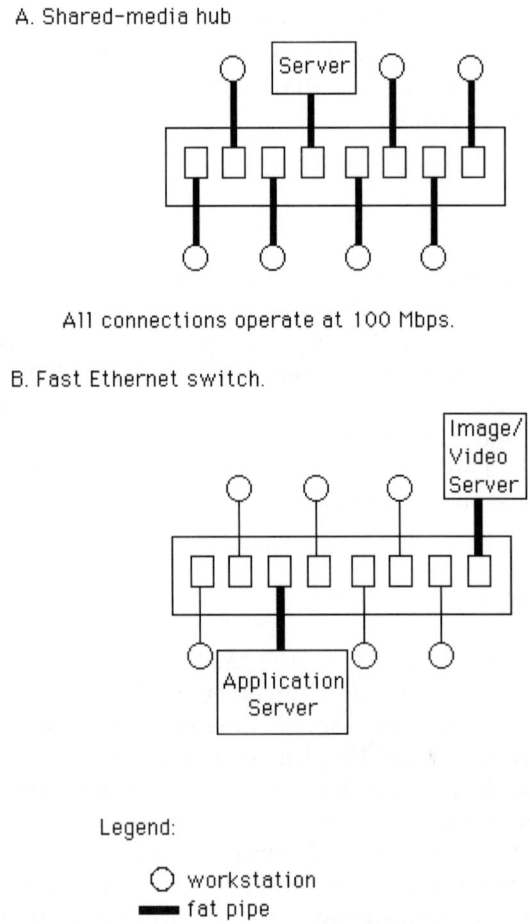

Figure 3.25 Fast Ethernet network applications

to a predefined output port. This is accomplished by the switch reading the destination address in each frame and comparing that address to a table of learned address–port relationships. Most Ethernet switches contain one or two 100 Mbps operating ports, while the other ports are conventional 10BASE-T ports. Typically you would connect servers that have a heavy workload to 100 Mbps Fast Ethernet ports, such as an image/video server and a database server. Figure 3.25B illustrates the use of a Fast Ethernet switch. In examining Figures 3.25A and B, note that the heavily shaded connections to all workstations in Figure 3.25A and to the servers in Figure 3.25B represent 100 Mbps Fast Ethernet ports requiring Fast Ethernet adapter cards to be installed in each server or workstation connected to a 100 Mbps port. A common term used to reference the 100 Mbps connection is 'fat pipe'.

Since a Fast Ethernet port provides downward compatibility with 10BASE-T, you can interconnect conventional 10BASE-T hubs to a Fast Ethernet shared-media hub or Fast Ethernet switch.

Gigabit Ethernet

In July 1996 the IEEE created the 802.3z task force and assigned it the task of completing a standard for Ethernet operating at 1 Gbps by 1998. At the time this book was prepared, the IEEE was working on half-duplex shared bandwidth and full-duplex point-to-point versions of this high speed version of Ethernet. Due to its 1 Gbps operating rate, the term Gigabit Ethernet was commonly used to reference both versions of the developing standard as well as different media versions of 1 Gbps Ethernet.

Currently, Gibabit Ethernet is being developed to operate over three types of media: multi-mode and single-mode fiber, and copper. The 802.3z task force was in the process of developing a series of four specifications which use different types of transmission to achieve different transmission distances on different types of media. Table 3.12 provides a summary of the Gigabit Ethernet specifications being developed by the IEEE.

Table 3.12 IEEE Gigabit Ethernet specifications

Specification	Transmission facility	Utilization
1000BASE-LX	Long-wavelength laser transceivers	Support transmission up to 550 m on multimode fiber or 3000 m on single-mode fiber
1000BASE-SX	Short-wavelength laser transceiver operating on multimode fiber	Support transmission up to 3000 m on 62.5 μm multimode fiber or up to 550 m on 50 μm multimode fiber
1000BASE-CX	Short-haul copper jumpers	Transmit up to 25 m among devices located in a room or rack
1000BASE-T	UTP category 5	Transmit up to 100 m on four-pair category 5 UTP

During 1997 and early 1998 several vendors introduced Gigabit Ethernet products without waiting for standards to be finalized. Such products were primarily designed to serve as a backbone hub to interconnect lower speed Ethernet networks, or as a switch designed to alleviate bottlenecks on networks transmitting a high proportion of graphic-intensive applications, such as weather maps. Since few networks actually require a 1 Gbps transmission capability, most implementations of Gigabit Ethernet for the next few years will more than likely serve as a backbone for highly utilized 10BASE-T and 100BAST-T networks.

From a technical perspective, Gigabit Ethernet represents an extension to the 10 and 100 Mbps IEEE 802.3 standards. The half-duplex version of Gigabit Ethernet is designed for shared bandwidth operations and uses the CSMA/CD access protocol; however, the minimum frame length, which is 64 bytes for 100BASE-T and 100BASE-T networks, required an expansion to 512 bytes by a technique referred to as carrier extension. Carrier extension is required to support a network diameter of 200 meters. Otherwise, the use of a 64-byte minimum length frame which provides a collision detection window of 51.2 μs at 10 Mbps and 5.12 μs at 100 Mbps would result in a reduction of the window to 0.512 μs at 1 Gbps. This would result in a round-trip delay that would restrict the network diameter of a Gigabit network to a maximum span under a few dozen meters, insufficient for supporting horizontal wiring to desktops distributed over a building floor. Through padding of frames under 512 bytes in length to 512 bytes by the technique referred to as carrier extension, individual desktops can be up to 100 m from a hub, and the network diameter is extended to 200 m.

Although carrier extension enables greater transmission distances, it also adversely effects network utilization since pads are added to short frames to produce a minimum 512-byte frame. For example, a 64-byte frame would be padded with 448 bytes of carrier extension symbols that would reduce the information carrying capacity of the frame to 64/512, or 12.5%. Recognizing this problem, a second technology was added to Gigabit Ethernet. This technology is referred to as packet bursting and enables stations to transmit several frames instead of a single frame if the first frame is successfully transmitted. Since only the first frame will require a carrier extension if its length is less than 512 bytes, packet bursting in effect averages the extension over a number of frames.

The full-duplex version of Gigabit Ethernet uses a four-wire pair on point-to-point links, avoiding collisions. For this reason there is no need for the use of carrier extension, nor for packet bursting, which are required on a half-duplex connection.

Frame composition

Figure 3.26 illustrates the general frame composition of Ethernet and IEEE 802.3 frames. You will note that they differ slightly. An Ethernet frame contains an eight-byte preamble, while the IEEE 802.3 frame contains a seven-byte preamble followed by a one-byte start-of-frame delimiter field. A second difference between the composition of Ethernet and IEEE 802.3 frames concerns the two-byte Ethernet type field. That field is used by Ethernet to specify the protocol carried in the frame, enabling several protocols to be carried independently of one another. Under the IEEE 802.3 frame format, the type field was replaced by a two-byte length field, which specifies the number of bytes that follow that field as data.

Now that we have an overview of the structure of Ethernet and 802.3 frames, let's probe deeper and examine the composition of each frame field. We will take advantage of the similarity between Ethernet and IEEE 802.3 frames to examine the fields of each frame on a composite basis, noting the differences between the two when appropriate.

Ethernet

Preamble	Destination Address	Source Address	Type	Data	Frame Check Sequence
8 bytes	6 bytes	6 bytes	2 bytes	46-1500 bytes	4 bytes

IEEE 802.3

Preamble	Start of Frame Delimiter	Destination Address	Source Address	Length	Data	Frame Check Sequence
7 bytes	1 byte	2/6 bytes	2/6 bytes	2 bytes	46-1500 bytes	4 bytes

Figure 3.26 Ethernet and IEEE 802.3 frame formats

Preamble field

The preamble field consists of eight (Ethernet) or seven (IEEE 802.3) bytes of alternating 1 and 0 bits. The purpose of this field is to announce the frame and to enable all receivers on the network to synchronize themselves to the incoming frame. In addition, this field by itself (under Ethernet) or in conjunction with the start-of-frame delimiter field (under the IEEE 802.3 standard) ensures there is a minimum spacing period of 9.6 us at 10 Mbps between frames for error detection and recovery operations.

Start of frame delimiter field

This field is applicable only to the IEEE 802.3 standard, and can be viewed as a continuation of the preamble. In fact, the composition of this field continues in the same manner as the format of the preamble, with alternating 1 and 0 bits used for the first six bit positions of this one-byte field. The last two bit positions of this field are 11—this breaks the synchronization pattern and alerts the receiver that frame data follows.

Destination address field

The destination address identifies the recipient of the frame. Although this may appear to be a simple field, in reality its length can vary between IEEE 802.3 and Ethernet frames. In addition, each field can consist of two or more subfields, whose settings govern such network operations as the type of addressing used on the LAN, and whether or not the frame is addressed to a specific station or more than one station. To obtain an appreciation for the use of this field, let's examine how

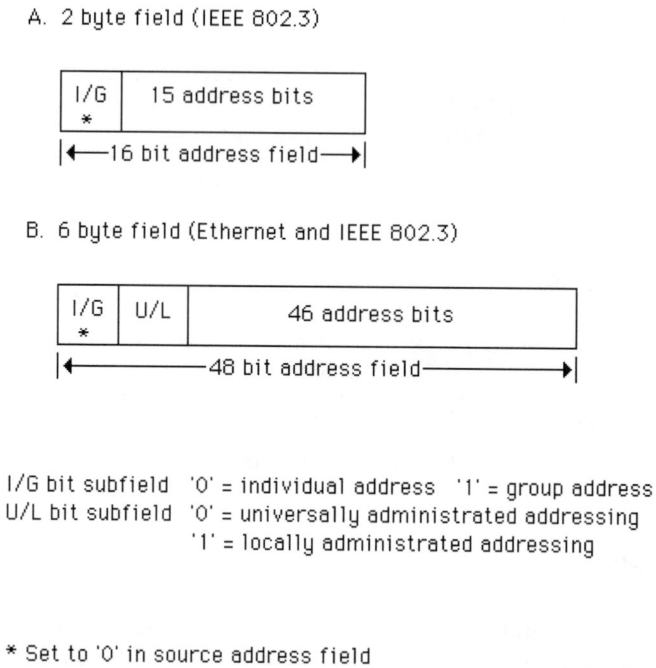

Figure 3.27 Source and destination address field formats

this field is used under the IEEE 802.3 standard as one of the two field formats applicable to Ethernet.

Figure 3.27 illustrates the composition of the source and destination address fields. As indicated, the two-byte source and destination address fields are applicable only to IEEE 802.3 networks, while the six-byte source and destination address fields are applicable to both Ethernet and IEEE 802.3 networks. A user can select either a two- or a six-byte destination address field; however, with IEEE 802.3 equipment, all stations on the LAN must use the same addressing structure. Today, almost all 802.3 networks use six-byte addressing.

I/G subfield

The one-bit I/G subfield is set to 0 to indicate that the frame is destined to an individual station, or 1 to indicate that the frame is addressed to more than one station—a *group address*. One special example of a group address is the assignment of all 1's to the address field. Hex 'FFFFFFFFFFFF' is recognized as a broadcast address, and each station on the network will receive and accept frames with that destination address.

When a destination address specifies a single station, the address is referred to as a *unicast address*. A group address that defines multiple stations is known as a *multicast address*, while a group address that specifies all stations on the network is, as previously mentioned, referred to as a *broadcast address*.

U/L subfield

The U/L subfield is applicable only to the six-byte destination address field. The setting of this field's bit position indicates whether the destination address is an address that was assigned by the IEEE (universally administered) or assigned by the organization via software (locally administered).

Universal versus locally administered addressing

Each Ethernet Network Interface Card (NIC) contains a unique address burned into its read-only memory (ROM) at the time of manufacture. To ensure that this universally administered address is not duplicated, the IEEE assigns blocks of addresses to each manufacturer. These addresses normally include a three-byte prefix, which identifies the manufacturer and is assigned by the IEEE, and a three-byte suffix, which is assigned by the adapter manufacturer to its NIC. For example, the prefix hex 02608C identifies an NIC manufactured by 3Com, while a prefix of hex 08002B identifies an NIC manufactured by Digital Equipment Corporation.

Although universally administered addressing eliminates the potential for duplicate network addresses, it does not provide the flexibility obtainable from locally administered addressing. For example, under locally administered addressing, you can configure mainframe software to work with a predefined group of addresses via a gateway PC. Then, as you add new stations to your LAN, you simply use your installation program to assign a locally administered address to the NIC instead of using its universally administered address. As long as your mainframe computer has a pool of locally administered addresses that includes your recent assignment, you do not have to modify your mainframe communications software configuration. Since the modification of mainframe communications software typically requires recompiling and reloading, the attached network must become inoperative for a short period of time. Because a large mainframe may service hundreds to thousands of users, such changes are normally performed late in the evening or at a weekend. Thus, the changes required for locally administered addressing are more responsive to users than those required for universally administered addressing.

Source address field

The source address field identifies the station that transmitted the frame. Like the destination address field, the source address can be either two or six bytes in length.

The two-byte source address is supported only under the IEEE 802.3 standard and requires the use of a two-byte destination address; all stations on the network must use two-byte addressing fields. The six-byte source address field is supported by both Ethernet and the IEEE 802.3 standard. When a six-byte address is used, the first three bytes represents the address assigned by the IEEE to the manufacturer for incorporation into each NIC's ROM. The vendor then normally assigns the last three bytes for each of its NICs.

Type field

The two-byte type field is applicable only to the Ethernet frame. This field identifies the higher-level protocol contained in the data field. Thus, this field tells the receiving device how to interpret the data field.

Under Ethernet, multiple protocols can exist on the LAN at the same time. Xerox served as the custodian of Ethernet address ranges licensed to NIC manufacturers and defined the protocols supported by the assignment of type field values. Under the IEEE 802.3 standard, the type field was replaced by a length field, which precludes compatibility between pure Ethernet and 802.3 frames.

Length field

The two-byte length field, applicable to the IEEE 802.3 standard, defines the number of bytes contained in the data field. Under both Ethernet and IEEE 802.3 standards, the minimum size frame must be 64 bytes in length from preamble through FCS fields. This minimum size frame ensures that there is sufficient transmission time to enable Ethernet NICs to detect collisions accurately based on the maximum Ethernet cable length specified for a network and the time required for a frame to propagate the length of the cable. Based on the minimum frame length of 64 bytes and the possibility of using two-byte addressing fields, this means that each data field must be a minimum of 46 bytes in length.

Data field

As previously discussed, the data field must be a minimum of 46 bytes in length to ensure that the frame is at least 64 bytes in length. This means that the transmission of one byte of information must be carried within a 46-byte data field; if the information to be placed in the field is less than 46 bytes long, the remainder of the field must be padded. Although some publications subdivide the data field to include a PAD subfield, the latter actually represents optional fill characters that are added to the information in the data field to ensure a length of 46 bytes. The maximum length of the data field in 1500 bytes.

Frame check sequence field

The frame check sequence (FCS) field, applicable to both Ethernet and the IEEE 802.3 standard, provides a mechanism for error detection. Each transmitter computes a cyclic redundancy check (CRC) that covers both address fields, the type/length field, and the data field. The transmitter then places the computed CRC in the four-byte FCS field.

The CRC treats the previously mentioned fields as one long binary number. The n bits to be covered by the CRC are considered to represent the coefficients of a polynomial $M(X)$ of degree $n-1$. Here, the first bit in the destination address field corresponds to the X^{n-1} term, while the last bit in the data field corresponds to the X^0 term. Next, $M(X)$ is multiplied by X^{32} and the result of that multiplication process is divided by the following polynomial:

$$G(X) = X^{32} + X^{26} + X^{23} + X^{22} + X^{16} + X^{12} + X^{11} + X^{10} + X^8 + X^7 + X^5 + X^4 + X^2 + X + 1$$

Note that the term X^n represents the setting of a bit to a 1 in position n. Thus, part of the generating polynomial $X^5 + X^4 + X^2 + X^1$ represents the binary value 11011.

This division produces a quotient and a remainder. The quotient is discarded, and the remainder becomes the CRC value placed in the four-byte FCS field. This 32-bit CRC reduces the probability of an undetected error to 1 bit in every 4.3 billion, or approximately 1 bit in 2^{32-1} bits.

Once a frame reaches its destination, the receiver uses the same polynomial to perform the same operation upon the received data. If the CRC computed by the receiver matches the CRC in the FCS field, the frame is accepted. Otherwise, the receiver discards the received frame, as it is considered to have one or more bits in error. The receiver will also consider a received frame to be invalid and discard it under two additional conditions. Those conditions occur when the frame does not contain an integral number of bytes, or when the length of the data field does not match the value contained in the length field. The latter condition, obviously, is only applicable to the 802.3 standard, since an Ethernet frame uses a type field instead of a length field.

Media Access Control (MAC) overview

As previously discussed, under the IEEE 802 series of standards the data link layer of the OSI Reference Model was subdivided into two sublayers—Logical Link Control (LLC) and Media a Access Control (MAC). The frame formats examined in Figure 3.27 represent the manner in which LLC information is transported in an Ethernet frame. Directly under the LLC sublayer is the MAC sublayer. The MAC sublayer is responsible for checking the channel and transmitting data if the channel is idle, checking for the occurrence of a collision, and taking a series of predefined steps if a collision is detected. Thus, this layer provides the required logic to control the network.

Logical Link Control (LLC) overview

The Logical Link Control (LLC) sublayer was defined under the IEEE 802.2 standard to make the method of link control independent of a specific access method. Thus, the 802.2 method of link control spans Ethernet (IEEE 802.3), Token Bus (IEEE 802.4), and Token-Ring (IEEE 802.5) local area networks. Functions performed by the LLC include generating and interpreting commands to control the flow of data, including recovery operations for when a transmission error is detected.

Link control information is carried within the data field of an IEEE 802.3 frame as an LLC Protocol Data Unit. Figure 3.28 illustrates the relationship between the IEEE 802.3 frame and the LLC Protocol Data Unit.

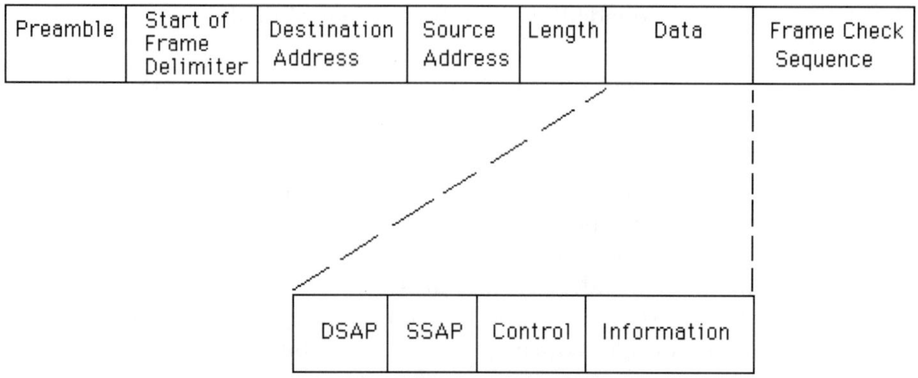

Figure 3.28 Formation of LLC protocol data unit. Control information is carried within a MAC frame

Service Access Points (SAPs) function much like a mailbox. Since the LLC layer is bounded below by the MAC sublayer and bounded above by the network layer, SAPs provide a mechanism for exchanging information between the LLC layer and the MAC and network layers. For example, from the network layer perspective, a SAP represents the place to leave messages about the services requested by an application. The Destination Services Access Point (DSAP) is one byte in length and is used to specify the receiving network layer process. The Source Service Access Point (SSAP) is also one byte in length. The SSAP specifies the sending network layer process. Both DSAP and SSAP addresses are assigned by the IEEE. For example, hex address 'FF' represents a DSAP broadcast address.

Type and classes of service

Under the 802.2 standard, there are three types of service available for sending and receiving LLC data. These types are discussed in the next three paragraphs. Figure 3.29 provides a visual summary of the operation of each LLC service type.

Type 1

Type 1 is an unacknowledged connectionless service. The term *connectionless* refers to the fact that transmission does not occur between two devices as if a logical connection were established. Instead, transmission flows on the channel to all stations; however, only the destination address acts upon the data. As the name of this service implies, there is no provision for the acknowledgement of frames. Neither are there provisions for flow control or for error recovery. Therefore, this is an unreliable service.

Despite those shortcomings, Type 1 is the most commonly used service, since most protocol suites use a reliable transport mechanism at the transport layer, thus eliminating the need for reliability at the link layer. In addition, by eliminating the

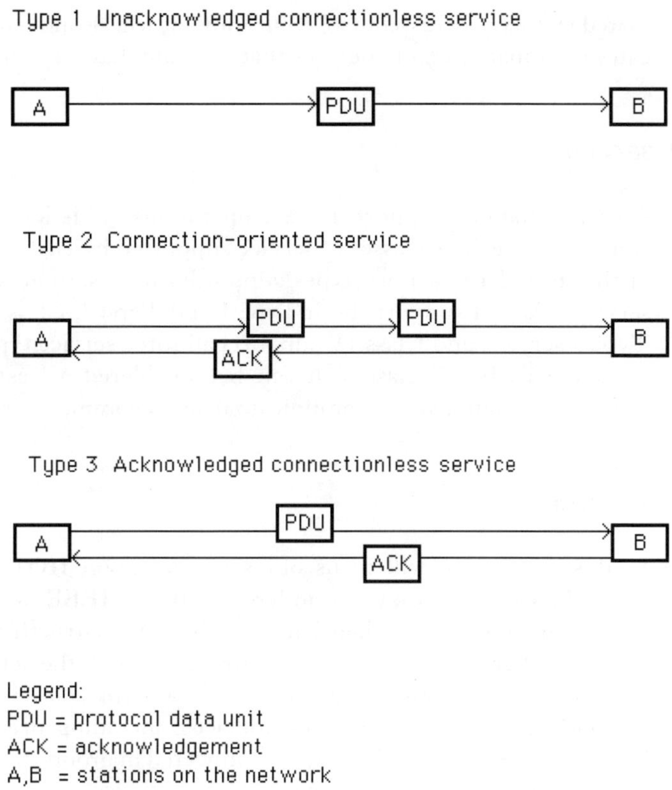

Figure 3.29 Logical Link Control service types

time needed to establish a virtual link and the overhead of acknowledgements, a Type 1 service can provide a greater throughput than other LLC service types.

Type 2

The type 2 connection-oriented service requires that a logical link be established between the sender and the receiver prior to information transfer. Once the logical connection is established, data will flow between the sender and receiver until either party terminates the connection. During data transfer, a Type 2 LLC service provides all of the functions lacking in a Type 1 service, with a sliding window used for flow control.

Type 3

The type 3 acknowledged connectionless service contains provision for the setup and disconnection of transmission; it acknowledges individual frames using the stop-and-wait flow control method. Type 3 service is primarily used in an automated factory process-control environment, where one central computer communicates with many remote devices that typically have a limited storage capacity.

Classes of service

All LLC stations support Type 1 operations. This level of support is known as Class I service. The classes of service supported by LLC indicate the combinations of the three LLC service types supported by a station. Class I supports Type 1 service, Class II supports both Type 1 and Type 2, Class III supports Type 1 and Type 3 service, and Class IV supports all three service types. Since service Type 1 is supported by all classes, it can be considered a least common denominator, enabling all stations to communicate using a common form of service.

3.5 TOKEN-RING

In this section we will focus our attention upon IBM's Token-Ring local area network. This network was standardized by the IEEE as the IEEE 802.5 standard.

Although the term 'Token-Ring' implies a ring structure, in actuality this type of LAN is either a star or a star–ring structure, with the actual topology based upon the number of stations to be connected. The term 'star' is derived from the fact that a grouping of stations and other devices, including printers, plotters, repeaters, bridges, routers, and gateways, are connected in groups to a common device called a Multistation Access Unit (MAU).

Figure 3.30 illustrates a single ring formed through the use of one MAU in which up to eight devices are interconnected. Thus, for a very small Token-Ring

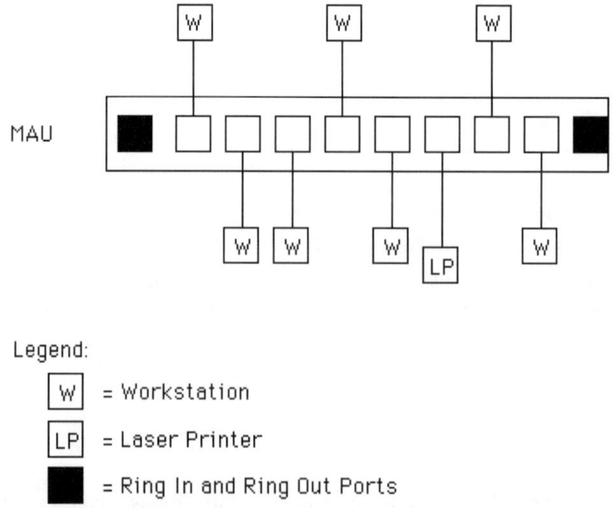

Figure 3.30 Single-ring LAN. A single-ring LAN can support up to eight devices through their attachment to a common MAU

LAN consisting of a mixture of eight or fewer devices, the structure actually resembles a star.

When IBM introduced its 4 Mbps Token-Ring network, its first MAU, known as the 8228, was a 10-port device, of which two ports were used for Ring-In (RI) and Ring-Out (RO) connectors which enable multiple MAUs to be connected to one another to expand the network. The remaining eight ports on the 8228 are designed for connecting devices to the ring. Since IBM's eight-port 8228 reached the market, other vendors have introduced similar products with different device support capacities. You can now commonly obtain MAUs that support 4, 8, and 16 devices.

In examining Figure 3.30, note that an eight-port MAU is illustrated, which enables up to eight devices to be interconnected to a Token-Ring LAN. The MAU can be considered the main ring path, as data will flow from one port to another via each device connected to the port. If you have more than eight devices, you can add additional MAUs, interconnecting the MAUs via the Ring-In and Ring-Out ports located at each side of each MAU. When this interconnection occurs, by linking two or more MAUs together you form a star–ring topology, as illustrated in Figure 3.31. In this illustration, the stations and other devices form a star structure, while the interconnection of MAUs forms the ring; hence you obtain a star–ring topology.

Redundant versus non-redundant main ring paths

When two or more MAUs are interconnected, the serial path formed by those interconnections is known as the main ring path. Connections between MAUs can be accomplished through the use of one or two pairs of wiring. One pair will be used as the primary data path, while the other pair functions as a backup data path.

Figure 3.32 (a) illustrates the formation of a ring consisting of two MAUs in which both primary and backup paths are established to provide a redundant main ring path. If one of the cables linking the MAUs becomes disconnected, cut, or crimped, the network can continue to operate since the remaining wiring pair provides a non-redundant main ring path capability as shown in Figure 3.32 (b).

The backup capability provided by redundant main ring paths is established through the use of loopback plugs or a built-in MAU self-shorting feature, both of which are discussed in additional detail later in this chapter. Since it is both difficult and tedious to draw wire pairs, in the remainder of this section we will use a single line to indicate wiring pairs between MAUs and MAUs and between stations connected to MAUs.

Cabling and device restrictions

The type of cable or wiring used to connect devices to MAUs and to interconnect MAUs is a major constraint that governs the size of a Token-Ring network. In Section 3.2 we examined the performance characteristics of IBM cabling system cable. Now we will examine some of the constraints associated with the use of different types of IBM cabling system cable.

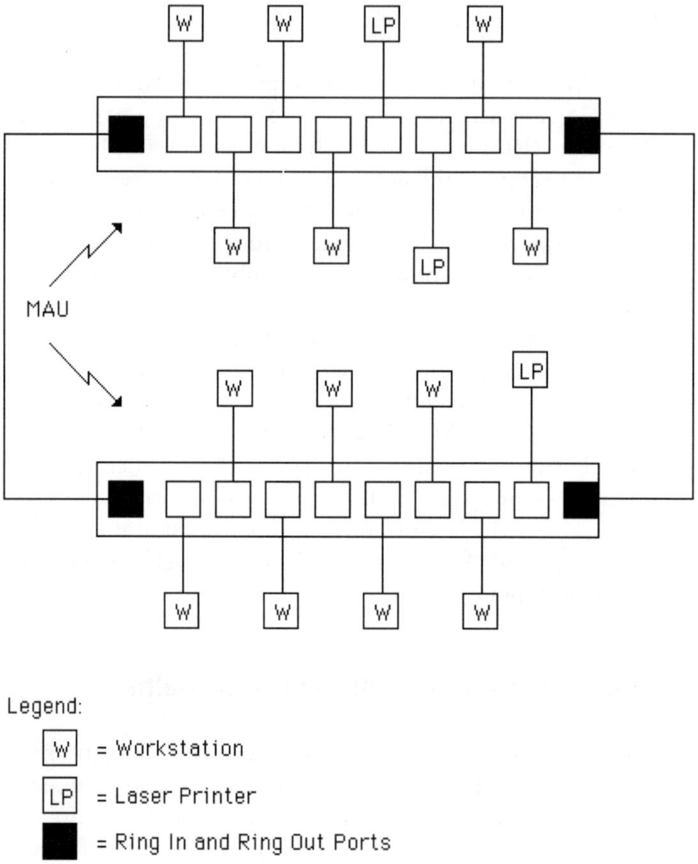

Figure 3.31 Developing the star–ring

There are several cabling and device constraints you must consider when designing a Token-Ring network to ensure the network will work correctly. First, you must consider the maximum cabling distance between each device and an MAU that will service the device. The cable between the MAU and the device is referred to as a lobe, with the maximum lobe distance being 100 m (330 feet) at both 4 and 16 Mbps. This means that you must consider the lobe distance in conjunction with the cabling distance restrictions between MAUs if you have more than eight devices to be connected to a Token-Ring LAN. In addition, for larger networks, you must also consider restrictions on the number of MAUs in the network and their placement in wiring closets, as well as a parameter known as the adjusted ring length, since they collectively govern the maximum number of devices that can be supported.

Intra-MAU cabling distances

Table 3.13 lists the maximum intra-MAU cabling distances permitted on a Token-Ring network for the two most commonly used types of IBM cables. Those

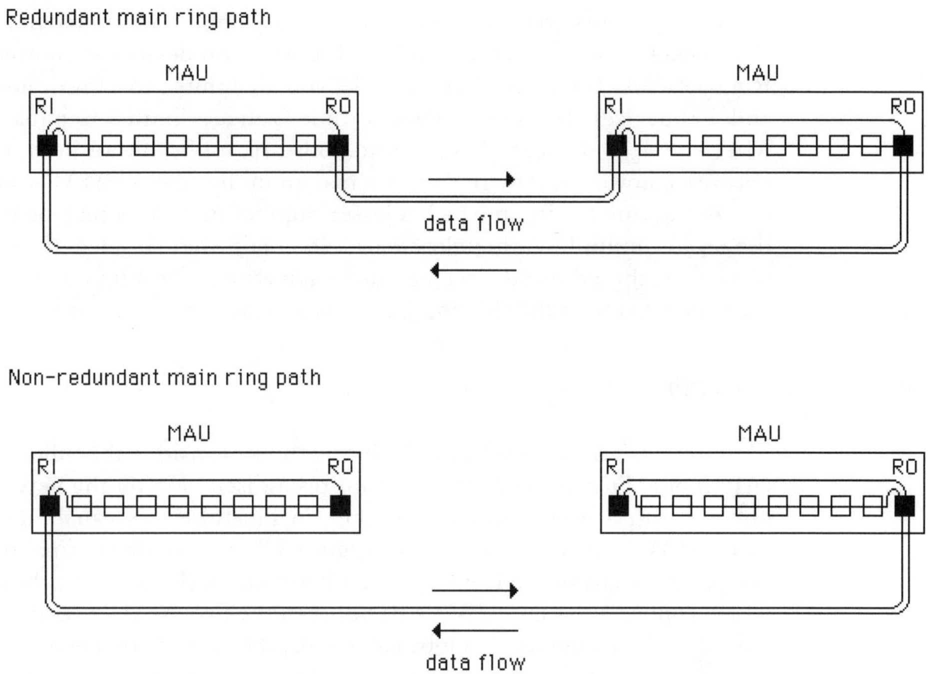

Figure 3.32 (a) Redundant and (b) non-redundant main ring paths

Table 3.13 Intra-MAU cabling distance (feet)

	Type of cable	
Operating rate	Type 1	Type 3
4 Mbps	330	1000
16 Mbps	330	250

distances can be extended through the use of repeaters; however, their use adds to both the complexity and the cost of the network.

As indicated in Table 3.13, the cabling distance between MAUs depends upon both the operating rate of the LAN—4 Mbps or 16 Mbps—and the type of cable used. Type 1 is a double-shielded pair cable, while Type 3 is non-shielded twisted-pair telephone wire.

In examining the entries in Table 3.13, it may appear odd that the maximum intra-MAU cable distance is the same for both 4 and 16 Mbps networks when Type 1 cable is used. This situation occurred because, at the time IBM set a 100 meter recommended limit for a 4 Mbps network, the company took into consideration the need to reuse the same cabling when customers upgraded to a 16 Mbps operating rate. When using Type 3 cable, distances are shorter since signal line noise increases in proportion to the square root of frequency. Thus, upgrading the operating rate from 4 to 16 Mbps with Type 3 cable decreases the maximum permissible distance.

As you plan to extend your network to interconnect additional devices, you must also consider the maximum number of MAUs and devices supported by a Token-Ring network. If you use Type 1 cable, you are limited to a maximum of 33 MAUs and 260 devices. If you use Type 3 cable, you are limited to a maximum of nine MAUs and 72 devices. These limitations are applicable to both 4 and 16 Mbps networks; however, they represent a maximum number of MAUs and devices and do not indicate reality in which a lesser number of MAUs may be required due to the use of multiple wiring closets or a long adjusted ring length. Due to the role played by the adjusted ring length in governing the number of MAUs, let us examine what an adjusted ring length is and how it functions as a constraint.

Adjusted ring length

To fully understand the reason why we must consider the adjusted ring length (ARL) of a Token-Ring network requires a discussion on the network's ability to operate with a faulty cabling section. To illustrate this capability, consider the three-MAU network illustrated in Figure 3.33 (a). Under normal network operation, data is transmitted from RO to RI between MAUs and over the main ring path which connects the last MAU's RO port to the first MAU's RI port.

If an attached device or a lobe cable fails, the lack of voltage on the MAU's port causes the port to be bypassed, permitting information from other stations to flow on the main ring path. If an MAU or a cable interconnecting two MAUs fails, the previously described built-in backup capability of MAUs permits the network to continue operating. This backup capability permits the Token-Ring to be re-established in one of two ways, dependent upon the capability of the MAUs used in the network. Some MAUs have a 'self-shorting' capability, which means that RI and RO connectors are joined together without requiring the use of a cable or plug to complete a ring. Other MAUs require the use of a cable or plug between the RI and RO connectors. By using the 'self-shorting' capability, or using a cable or plug in the input and output connectors of the MAUs located at both ends of a failed cable, the ring can be reconfigured for operation. Figure 3.33(b) illustrates the reconfigured ring. Note that the total cable length of the main ring path and available cable segments represents the adjusted ring length.

In actuality, the adjusted ring length is the total ring length (main ring path plus all cable segments) less the shortest cable segment between MAUs. The reason we subtract the shortest cable segment is that doing so provides the longest total cable distance for a reconfigured ring. It is that distance that a signal must be capable of flowing around a ring without excessive distortion adversely affecting the signal. For example, suppose the main ring path is 300 feet and cable segments A and B are 200 and 150 feet, respectively. Then, the total drive distance is 300 + 200 + 150, or 650 feet, and the adjusted ring length is 650 − 150, or 500 feet.

Other ring size considerations

The adjusted ring length as well as the length of the main ring path are two constraints that govern the size of a Token-Ring network. Other constraints include

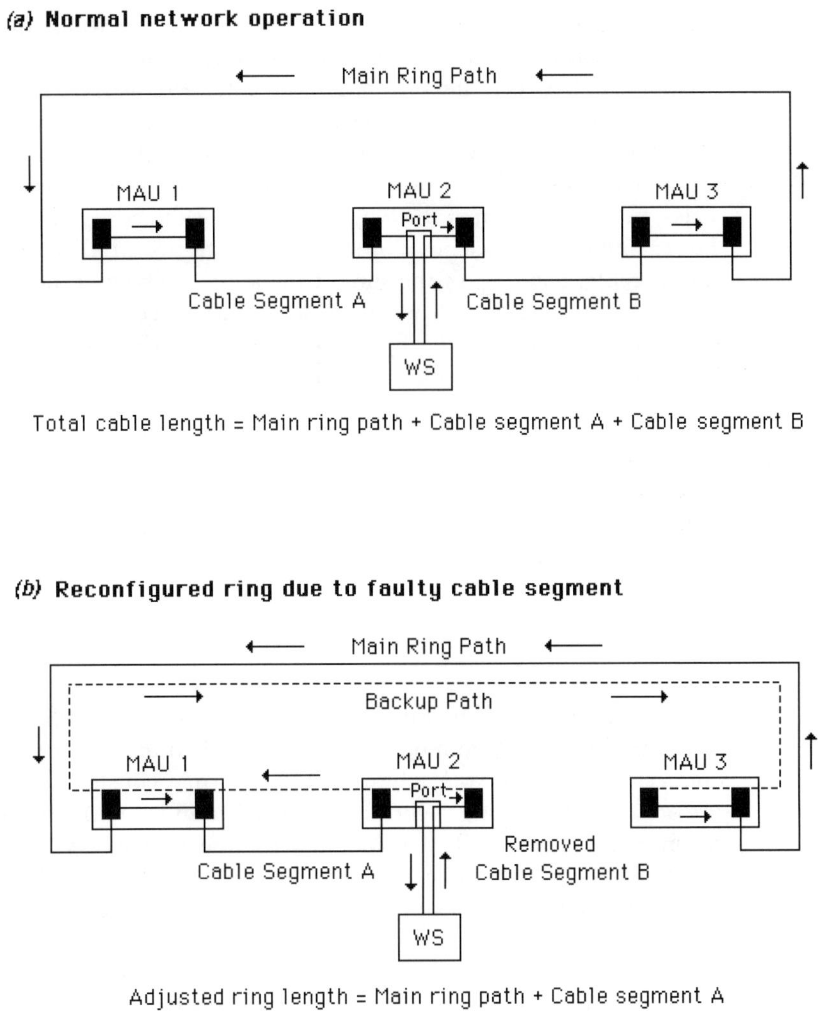

Figure 3.33 Computing the adjusted ring length

the operating rate of the ring, the length of the longest lobe, the type and number of MAUs in the network, and the number of distinct locations where the MAUs are installed—the latter referred to by IBM as wiring closets. The type of MAU is equivalent to the type of cabling used. A Type 1 MAU is cabled using Type 1 (double-shielded pair) cable, while a Type 3 MAU uses Type 3 (non-shielded twisted-pair telephone wire) cable.

When Type 1 MAUs are used for 4 or 16 Mbps operation and all lobe cables terminate in a single wiring closet, you can interconnect up to 33 IBM 8228 MAUs, which can serve 260 devices, with each device cabled using up to a 100 meter lobe cable. Here, the number of 8228s is given by the following formula:

$$8228\text{s} = \frac{\text{Int (device} + 0.5)}{8} \leq 33$$

For 2 to 12 wiring closets, IBM provides a series of charts in the firm's *Token-Ring Network, Introduction and Planning Guide,* which supplies details about the relationship between the lobe length, adjusted ring length, number of MAUs, and wiring closets when Type 1 cable is used. Those charts are two-dimensional arrays with the number of wiring closets listed across the horizontal heading and the number of MAUs listed down the vertical heading. Each entry in the array provides, in feet, the sum of the longest lobe cable and adjusted ring length permitted. For example, a two-wiring-closet network with two MAUs can have a total of 1192 feet of cable, made up of the sum of the longest lobe and the adjusted ring length. For a network with two wiring closets and three MAUs, the maximum cable distance decreases to 1163 feet, while a three-wiring-closet three-MAU network has a maximum cable distance of 1148 feet.

Transmission formats

Three types of transmission formats are supported on a Token-Ring network—token, abort, and frame.

Token

The token format as illustrated in Figure 3.34(a) is the mechanism by which access to the ring is passed from one computer attached to the network to another device connected to the network. Here the token format consists of three bytes, of which the starting and ending delimiters are used to indicate the beginning and end of a token frame. The middle byte of a token frame is an access control byte. Three bits are used as a priority indicator, three bits are used as a reservation indicator, while one bit is used for the token bit and another bit position functions as the monitor bit.

When the token bit is set to a binary 0, it indicates that the transmission is a token. When it is set to a binary 1, it indicates that data in the form of a frame is being transmitted.

Abort

The second Token-Ring frame format signifies an abort token. In actuality, there is no token, since this format is indicated by a starting delimiter followed by an ending delimiter. The transmission of an abort token is used to abort a previous transmission. The format of an abort token is illustrated in Figure 3.34(b).

Frame

The third type of Token-Ring frame format occurs when a station seizes a free token. At that time the token format is converted into a frame which includes the addition of frame control, addressing data, an error detection field, and a frame

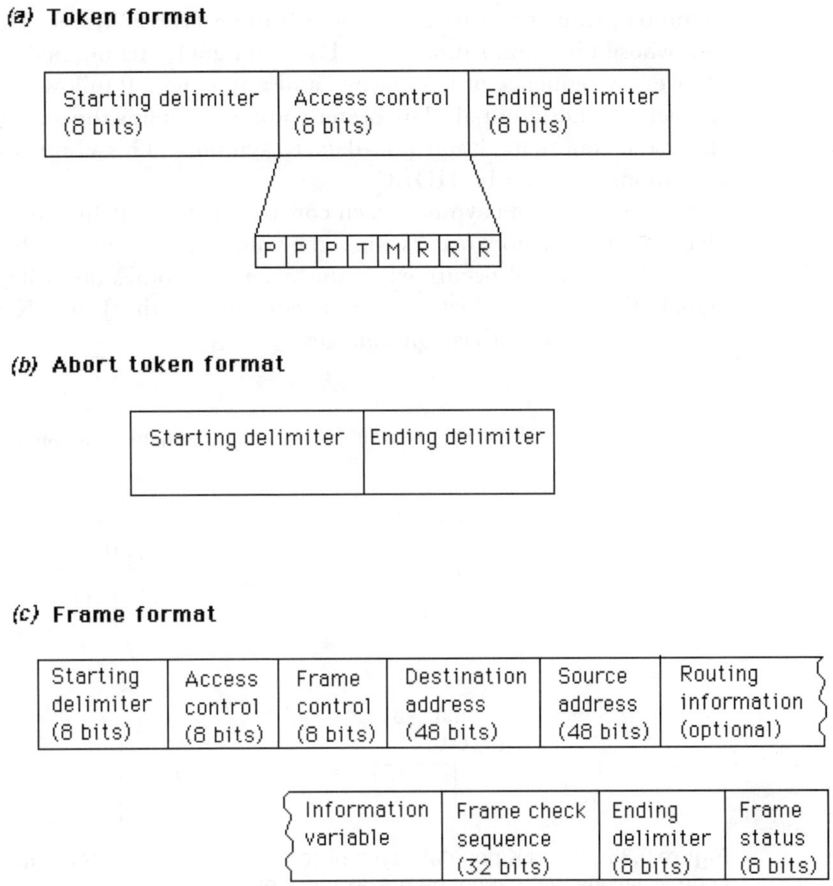

Figure 3.34 (a) Token format (P: priority bits, T: token bit, M: monitor bit, R: reservation bits); (b) abort token format; (c) frame format

status field. The format of a Token-Ring frame is illustrated in Figure 3.34(c). By examining each of the fields in the frame, we will also examine the token and token abort frames, due to the commonality of fields between each frame.

Starting/ending delimiters

The starting and ending delimiters mark the beginning and end of a token or frame. Each delimiter consists of a unique code pattern which identifies it to the network.

Non-data symbols

Under Manchester and Differential Manchester coding there are two possible code violations that can occur. Each code violation produces what is known as a non-data symbol and is used in the Token-Ring frame to denote starting and ending delimiters similar to the use of the flag in an HDLC frame. However, unlike the flag whose bit composition 01111110 is uniquely maintained by inserting a 0 bit after every sequence of five set bits and removing a 0 following every sequence of five set bits, Differential Manchester coding maintains the uniqueness of frames by the use of non-data J and non-data K symbols. This eliminates the bit-stuffing operations required by HDLC.

The two non-data symbols each consists of two half-bit times without a voltage change. The J symbol occurs when the voltage is the same as that of the last signal, while the k symbol occurs when the voltage becomes opposite to that of the last signal. Figure 3.35 illustrates the occurrence of the J and K non-data symbols based upon different last-bit voltages.

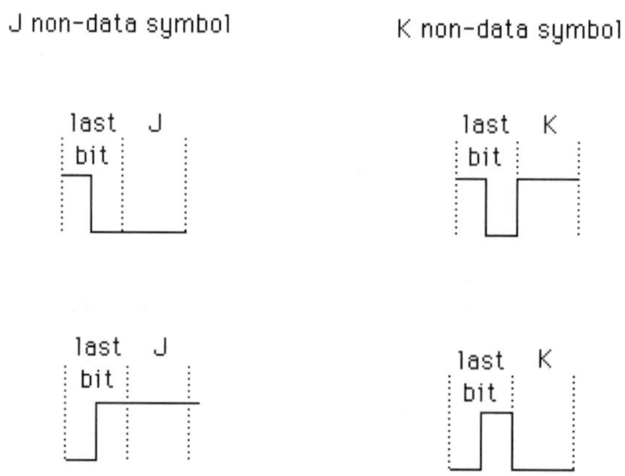

Figure 3.35 J and K non-data symbol composition. J and K non-data symbols are distinct code violations that cannot be mistaken for data

The start delimiter field marks the beginning of a frame. The composition of this field is the bits and non-data symbols JK0JK000. The end delimiter field marks the end of a frame as well as denoting whether or not the frame is the last frame of a multiple frame sequence using a single token or if there are additional frames following this frame. The format of the end delimiter field is JK1JK1IE, where I is the intermediate frame bit. If I is set to 0, this indicates it is the last frame transmitted by a station. If I is set to 1, this indicates that additional frames follow this frame. E is an Error-Detected bit. The E bit is initially set to 0 by the station transmitting a frame, token, or abort sequence. As the frame circulates the ring, each station checks the transmission for errors. Upon detection of a Frame Check Sequence (FCS) error, inappropriate non-data symbol, illegal framing, or another type of error, the first station detecting the error will set the E bit to a value of 1. Since stations keep track of the number of times they set the E bit to a value of 1, it becomes possible to use this information as a guide to locating possible cable errors. For example, if one workstation accounted for a very large percentage of E bit settings in a 72-station network, there is a high degree of probability that there is a problem with the lobe cable to that workstation. The problem could be a crimped cable or a loose connector and represents a logical place to commence an investigation in an attempt to reduce E bit errors.

Access control

The second field in both token and frame formats is the access control byte. As illustrated in Figure 3.34(a), this byte consists of four subfields and serves as the controlling mechanism for gaining access to the network. When a free token circulates the network, the access control field represents one-third of the length of the frame since it is prefixed by the start delimiter and suffixed by the end delimiter.

The lowest priority that can be specified by the priority bits in the access control byte is 0 (000), while the highest is seven (111), providing eight levels of priority. Table 3.14 lists the normal use of the priority bits in the access control field. Workstations have a defauld priority of three, while bridges have a default priority of four.

Table 3.14 Priority bit settings

Priority bits	Priority
000	Normal user priority, MAC frames that do not require a token and response-type MAC frames
001	Normal user priority
010	Normal user priority
011	Normal user priority and MAC frames that require tokens
100	Bridge
101	Reserved
110	Reserved
111	Specialized station management

To reserve a token, a workstation inserts its priority level in the priority reservation subfield. Unless another station with a higher priority bumps the requesting station, the reservation will be honored and the requesting station will obtain the token. If the token bit is set to 1, this serves as an indication that a frame follows instead of the ending delimiter.

A station that needs to transmit a frame at a given priority can use any available token that has a priority level equal to or less than the priority level of the frame to be transmitted. When a token of equal or lower priority is not available, the ring station can reserve a token of the required priority through the use of the reservation bits. In doing so, the station must follow two rules. First, if a passing token has a higher priority reservation than the reservation level desired by the workstation, the station will not alter the reservation field contents. Secondly, if the reservation bits have not been set or indicate a lower priority than that desired by the station, the station can now set the reservation bits to the required priority level.

Once a frame is removed by its originating station, the reservation bits in the header will be checked. If those bits have a non-zero value, the station must release a non-zero priority token, with the actual priority assigned based upon the priority used by the station for the recently transmitted frame, the reservation bit settings received upon the return of the frame, and any stored priority.

On occasion, the Token-Ring protocol will result in the transmission of a new token by a station prior to that station having the ability to verify the settings of the access control field in a returned frame. When this situation arises, the token will be issued according to the priority and reservation bit settings in the access control field of the transmitted frame.

The monitor bit

The monitor bit is used to prevent a token with a priority exceeding zero or a frame from continuously circulating on the Token-Ring. This bit is transmitted as a 0 in all tokens and frames, except for a device on the network which functions as an active monitor and thus obtains the capability to inspect and modify that bit.

When a token or frame is examined by the active monitor, it will set the monitor bit to a 1 if it was previously found to be set to 0. If a token or frame is found to have the monitor bit already set to 1, this indicates that the token or frame has already made at least one revolution around the ring and an error condition has occurred, usually caused by the failure of a station to remove its transmission from the ring, or the failure of a high-priority station to seize a token. When the active monitor finds a monitor bit set to 1, it assumes an error condition has occurred. The active monitor then purges the token or frame and releases a new token onto the ring. Now that we have an understanding of the role of the monitor bit in the access control field and the operation of the active monitor on that bit, let's focus our attention upon the active monitor.

Active monitor

The active monitor is the device that has the highest address on the network. All other stations on the network are considered as standby monitors and watch the active monitor.

As previously explained, the function of the active monitor is to determine if a token or frame is continuously circulating the ring in error. To accomplish this, the active monitor sets the monitor count bit as a token or frame goes by. If a destination workstation fails or has its power turned off, the frame will circulate back to the active monitor, where it is then removed from the network. In the event of the active monitor failing or being turned off, the standby monitors watch the active monitor by looking for an active monitor frame. If one does not appear within seven seconds, the standby monitor that has the highest network address then takes over as the active monitor.

Frame control

The frame control field informs a receiving device on the network of the type of frame that was transmitted and how it should be interpreted. Frames can either be

Logical Link Control (LLC) or reference physical link functions according to the IEEE 802.5 Media Access Control (MAC) standard. A MAC frame carries network control information and responses, while an LLC frame carries data.

The eight-bit frame control field has the format FFZZZZZZ, where FF are frame definition bits. The top part of Table 3.15 indicates the possible settings of the frame bits and the assignment of those settings. The ZZZZZZ bits convey MAC buffering information when the FF bits are set to 00. When the FF bits are set to 01 to indicate an LLC frame, the ZZZZZZ bits are split into two fields, designated rrrYYY. Currently, the rrr bits are reserved for future use and are set to 000. The YYY bits indicate the priority of the LLC data. The lower portion of Table 3.15 indicates the value of the Z bits when used in MAC frames to notify a Token-Ring adapter that the frame is to be expressed buffered.

Table 3.15 Frame control field subfields

F bit settings	Assignment
00	MAC frame
01	LLC frame
10	Undefined (reserved for future use)
11	Undefined (reserved for future use)

Z bit settings	Assignment*
000	Normal buffering
001	Remove ring station
010	Beacon
011	Claim token
100	Ring purge
101	Active monitor present
110	Standby monitor present

* When F bits are set to 00, Z bits are used to notify an adapter that the frame is to be express buffered.

Destination address

Although the IEEE 802.5 standard is similar to the 802.3 standard in its support of 16-bit and 48-bit address fields, almost all implementations of Token-Ring use 48-bit addresses. The destination address field is made up of five subfields as illustrated in Figure 3.36. The first bit in the destination address identifies the destination as an individual station (bit set to 0) or as a group (bit set to 1) of one or more stations. The latter provides the capability for a message to be broadcast to a group of stations.

Universally and locally administered addresses

Similar to the IEEE 802.3 standard, universally administered addresses are assigned in blocks of numbers by the IEEE to each manufacturer of Token-Ring

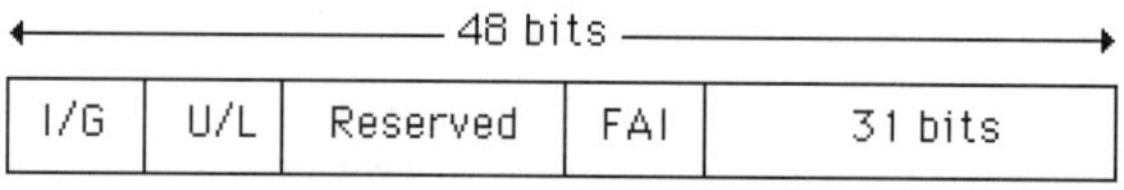

Figure 3.36 Destination address subfields (I/G: individual or group bit address identifier; U/L: universally or locally administered bit identifier; FAI: functional address indicator). The reserved field contains the manufacturer's identification in 22 bits represented by 6 hex digits

equipment, with the manufacturer encoding a unique address into each adapter card. Locally administered addressing permits users to temporarily override universally administered addressing and can be used to obtain addressing flexibility.

Functional address indicator

The functional address indicator subfield in the destination address identifies the function associated with the destination address, such as a bridge, active monitor, or configuration report server.

The functional address indicator indicates a functional address when set to 0 and the I/G bit position is set to 1, the latter indicating a group address. This condition can only occur when the U/L bit position is also set to 1 and results in the ability to generate locally administered group addresses that are called functional addresses.

Source address

The source address field always represents an individual address which specifies the adapter card responsible for the transmission. The source address field consists of three major subfields as illustrated in Figure 3.37. When locally administered addressing occurs, only 24 bits in the address field are used since the 22 manufacturer identification bit positions are not used.

The routing information bit identifier identifies the fact that routing information is contained in an optional routing information field. This bit is set when a frame is routed across a bridge using IBM's source routing technique which is described in detail in Chapter 5.

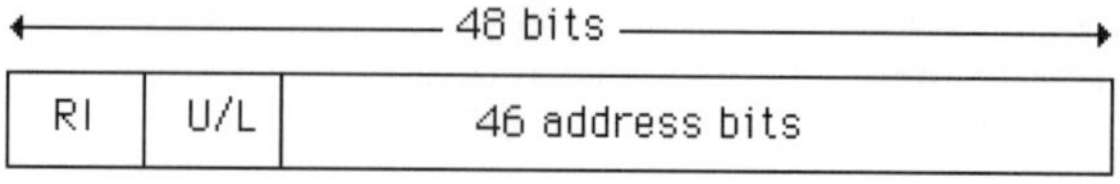

Figure 3.37 Source address field (RI: routing information bit identifier; U/L: universally or locally administered bit identifier). The 46 address bits consist of 22 manufacturer identification bits and 24 universally administered bits when the U/L bit is set to 0. If set to 1, a 31-bit locally administered address is used with the manufacturer's identification bits set to 0

Routing information

The routing information field is optional and is included in a frame when the RI bit of the source address field is set. Figure 3.38 illustrates the format of the optional routing information field. If this field is omitted, the frame cannot leave the ring it was originated on under IBM's source routing bridging method. Under transparent bridging, the frame can be transmitted onto another ring. Both source routing bridging and transparent bridging are covered in detail in Chapter 5. The routing information field is of variable length and contains a control subfield and one or more two-byte route designator fields when included in a frame, as the latter are required to control the flow of frames across one or more bridges.

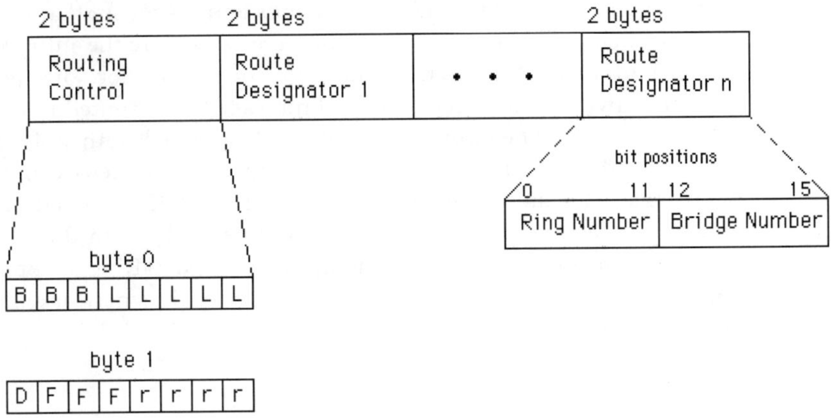

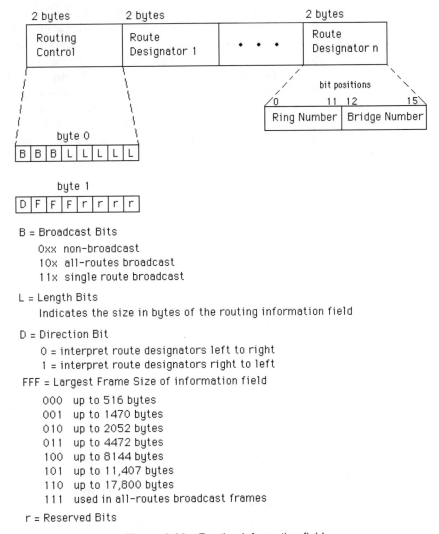

B = Broadcast Bits
- 0xx non-broadcast
- 10x all-routes broadcast
- 11x single route broadcast

L = Length Bits
- Indicates the size in bytes of the routing information field

D = Direction Bit
- 0 = interpret route designators left to right
- 1 = interpret route designators right to left

FFF = Largest Frame Size of information field
- 000 up to 516 bytes
- 001 up to 1470 bytes
- 010 up to 2052 bytes
- 011 up to 4472 bytes
- 100 up to 8144 bytes
- 101 up to 11,407 bytes
- 110 up to 17,800 bytes
- 111 used in all-routes broadcast frames

r = Reserved Bits

Figure 3.38 Routing information field

The maximum length of the routing information field (RIF) supported by IBM is 18 bytes. Since each RIF field must contain a two-byte routing control field, this leaves a maximum of 16 bytes available for use by up to eight route designators. As illustrated in Figure 3.38, each two-byte route designator consists of a 12-bit ring number and a four-bit bridge number. Thus, a maximum total of 16 bridges can be used to join any two rings in an Enterprise Token-Ring network.

Information field

The information field is used to contain Token-Ring commands and responses as well as carry user data. The type of data carried by the information field depends upon the F bit settings in the frame type field. If the F bits are set to 00, the information field carries MAC commands and responses that are used for network management operations. If the F bits are set to 01, the information field carries LLC or user data. Such data can be in the form of portions of a file being transferred on the network or an electronic mail message being routed to another workstation on the network. The information field is of variable length and can be considered to represent the higher level protocol enveloped in a Token-Ring frame.

In the IBM implementation of the IEEE 802.5 Token-Ring standard, the maximum length of the information field depends upon the Token-Ring adapter used and the operating rate of the network. Token-Ring adapters with 64 Kbytes of memory can handle up to 4.5 Kbytes on a 4 Mbps network and up to 18 Kbytes on a 16 Mbps network.

Frame check sequence

The frame check sequence field contains four bytes which provide the mechanism for checking the accuracy of frames flowing on the network. The cyclic redundancy check data included in the frame check sequence field covers the frame control, destination address, source address, routing information, and information fields. If an adapter computes a cyclic redundancy check that does not match the data contained in the frame check sequence field of a frame, the destination adapter discards the frame information and sets an error bit (E bit) indicator. This error bit indicator actually represents a ninth bit position of the ending delimiter and serves to inform the transmitting station that the data was received in error.

Frame status

The frame status field serves as a mechanism to indicate the results of a frame's circulation around a ring to the station that initiated the frame. Figure 3.39 indicates the format of the frame status field. The frame status field contains three subfields that are duplicated for accuracy purposes, since they reside outside CRC checking. One field (A) is used to denote whether an address was recognized, while a second field (C) indicates whether the frame was copied at its destination. Each of

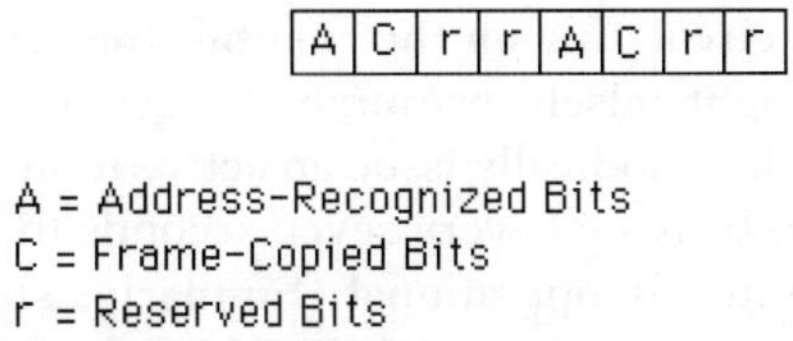

Figure 3.39 Frame status field. The frame status field denotes whether the destination address was recognized and whether the frame was copied. Since this field is outside CRC checking its subfields are duplicated for accuracy

these fields is one bit in length. The third field, which is two bit positions in length (rr), is currently reserved for future use.

Medium Access Control

As previously discussed, a MAC frame is used to transport network commands and responses. As such, the MAC layer controls the routing of information between the LLC and the physical network. Examples of MAC protocol functions include the recognition of adapter addresses, physical medium access management, and message verification and status generation.

A MAC frame is indicated by the setting of the first two bits in the frame control field to 00. When this situation occurs, the contents of the information field which carries MAC data is known as a vector. Table 3.16 lists currently defined vector identifier codes for six MAC control frames defined under the IEEE 802.5 standard.

Table 3.16 Vector identifier codes

Code value	MAC frame meaning
010	Beacon (BCN)
011	Claim token (CL_TK)
100	Purge MAC frame (PRG)
101	Active monitor present (AMP)
110	Standby monitor present (SMP)
111	Duplicate address test (DAT)

MAC control

As discussed earlier in this section, each ring has a station known as the active monitor which is responsible for monitoring tokens and taking action to prevent the endless circulation of a token on a ring. Other stations function as standby monitors and one such station will assume the functions of the active monitor if that device should fail or is removed from the ring. For the standby monitor with the highest network address to take over the functions of the active monitor, the standby monitor needs to know there is a problem with the active monitor. If no frames are circulating on the ring but the active monitor is operating, the standby monitor might falsely presume the active monitor has failed. Thus, the active monitor will periodically issue an active monitor present (AMP) MAC frame. This frame must be issued every seven seconds to inform the standby monitors that the active monitor is operational. Similarly, standby monitors periodically issue a standby monitor present (SMP) MAC frame to denote they are operational.

If an active monitor fails to send an AMP frame within the required time interval, the standby monitor with the highest network address will continuously transmit claim token (CL_TK) MAC frames in an attempt to become the active monitor. The standby monitor will continue to transmit CL_TK MAC frames until one of three conditions occurs:

- A MAC CL_TK frame is received and the sender's address exceeds the standby monitor's station address.
- A MAC beacon (BCN) frame is received.
- A MAC purge (PRG) frame is received.

If one of these conditions occurs, the standby monitor will cease its transmission of CL_TK frames and resume its standby function.

Purge frame

If a CL_TK frame issued by a standby monitor is received back without modification and neither a beacon nor a purge frame is received in response to the CL_TK frame, the standby monitor becomes the active monitor and transmits a purge MAC frame. The purge frame is also transmitted by the active monitor each time a ring is initialized or if a token is lost. Once a purge frame is transmitted, the transmitting device will place a token back on the ring.

Beacon frame

In the event of a major ring failure, such as a cable break or the continuous transmission by one station (known as jabbering), a beacon frame will be transmitted. The transmission of BCN frames can be used to isolate ring faults. For an example of the use of a beacon frame, consider Figure 3.40 in which a cable fault results in a ring break. When a station detects a serious problem with the ring, such as the failure to receive a frame or token, it transmits a beacon frame. That frame defines a failure domain which consists of the station reporting the failure via the transmission of a beacon and its nearest active upstream neighbor (NAUN), as well as everything between the two.

If a beacon frame makes its way back to the issuing station, that station will remove itself from the ring and perform a series of diagnostic tests to determine if it should attempt to reinsert itself into the ring. This procedure ensures that a ring error caused by a beaconing station can be compensated for by having that station remove itself from the ring. Since beacon frames indicate a general area where a failure occurred, they also initiate a process known as auto-reconfiguration. The

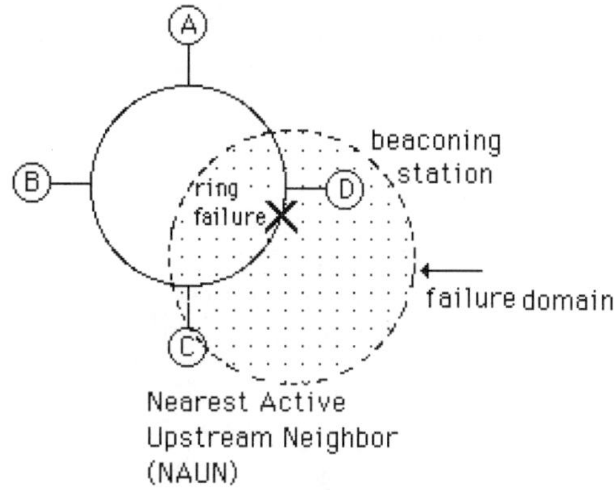

Figure 3.40 Beaconing. A beaconing frame indicates a failure occurring between the beaconing station and its nearest active upstream neighbor—an area referred to as a failure domain

first step in the auto-reconfiguration process is the diagnostic testing of the beaconing station's adapter. Other steps in the auto-reconfiguration process include diagnostic tests performed by other nodes located in the failure domain in an attempt to reconfigure a ring around a failed area.

Duplicate address test frame

The last type of MAC command frame is the duplicate address test (DAT) frame. This frame is transmitted during a station initialization process when a station joins a ring. The station joining the ring transmits a MAC DAT frame with its own address in the frame's destination address field. If the frame returns to the originating station with its address-recognized (A) bit in the frame control field set to 1, this means that another station on the ring is assigned that address. The station attempting to join the ring will send a message to the ring network manager concerning this situation and will not join the network.

Logical Link Control

In concluding this section, we will examine the flow of information within a Token-Ring network at the Logical Link Control (LLC) sublayer. Similar to its use on Ethernet, the LLC sublayer is responsible for performing routing, error control, and flow control. In addition, this sublayer is responsible for providing a consistent view of a LAN to upper OSI layers, regardless of the type of media and protocols used on the network.

Figure 3.41 illustrates the format of an LLC frame which is carried within the information field of the Token-Ring frame. As previously discussed in this section, the setting of the first two bits in the frame control field of a Token-Ring frame to 01 indicates that the information field should be interpreted as an LLC frame. The portion of the Token-Ring frame which carries LLC information is known as a

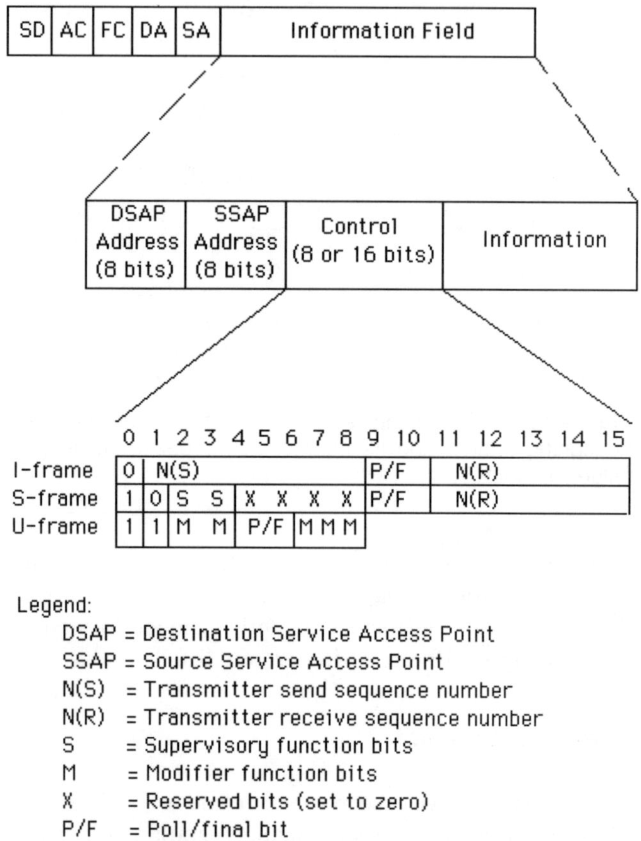

Figure 3.41 Logical Link Control frame format

protocol data unit and consists of either three or four fields, depending upon the inclusion or omission of an optional information field. The control field is similar to the control field used in the HDLC protocol and defines three types of frames—information (I-frame) are used for sequenced messages, supervisory (S-frames) are used for status and flow control, while unnumbered (U-frames) are used for unsequenced, unacknowledged messages. Both Service Access Points (SAPs) and Destination Service Access Points (DSAPs) function in the same manner as previously covered in our examination of Ethernet. In addition, Token-Ring, like Ethernet, supports the same types of connectionless and connection-oriented services at the data link layer.

REVIEW QUESTIONS

3.1.1 Discuss the differences between LANs and WANs with respect to their geographic area of coverage, data transmission and error rates, ownership, government regulation, and data routing.

3.1.2 Discuss four reasons that can justify the use of a LAN.

3.2.1 Describe the advantages and disadvantages associated with five LAN topologies.

3.2.2 Discuss the effect of a cable break on a star, bus, and ring-based LAN.

3.2.3 What is the difference between broadband and baseband signaling?

3.2.4 Discuss the rationale for using Manchester or Differential Manchester coding for LAN signaling.

3.2.5 What is the purpose of placing twists in a pair of wires?

3.2.6 What is the relationship between the American Wire Gauge (AWG) of a wire and its resistance?

3.2.7 What is the relationship between cabling distance and AWG values?

3.2.8 What is the total length of horizontal cabling permitted under the EIA/TIA-568 standard?

3.2.9 If a signal with transmitted voltage of 1000 mW is received at a voltage of 10 mW, what is the attenuation on the circuit path?

3.2.10 What is the general relationship between signal power loss and attenuation?

3.2.11 Which has a greater effect upon communications—near or far end crosstalk? Why?

3.2.12 What is the key function performed by a headend on a broadband network?

3.2.13 Compare and contrast CSMA/CD and token-passing access methods.

3.2.14 How can a collision occur on a CSMA/CD network?

3.2.15 How is the state of a communications channel (active or idle) determined on broadband and baseband networks?

3.3.1 What is an advantage associated with the subdivision of the data link layer by the IEEE?

3.3.2 Why did the IEEE subdivide the physical layer for certain types of LANs?

3.4.1 What are the two types of coaxial cable used to construct Ethernet networks? What are the advantages and disadvantages associated with the use of each type of cable?

3.4.2 What is the purpose of a jam signal?

3.4.3 What are the key functions performed by an Ethernet transceiver?

3.4.4 What is a 10BASE-5 network?

3.4.5 What is a 10BASE-2 network?

3.4.6 What are the key differences between the original Ethernet network and the IEEE 10BASE-5 network?

3.4.7 Which version of Ethernet requires the use of RF modems?

3.4.8 Discuss the three key components required to construct the physical infrastructure for a 10BASE-T network.

3.4.9 Describe the functions performed by a 10BASE-T hub.

3.4.10 What is the constraint on stations on a 10BASE-T network when the network is expanded via the interconnection of hubs?

3.4.11 What is the difference between 100BASE-T4 and 100BASE-TX with respect to cable use and signaling method employed?

3.4.12 Why does the use of 100BASE-TX require the use of category 5 UTP?

3.4.13 Why is NWay important?

3.4.14 Describe two 100BASE-T applications.

3.4.15 Why do Ethernet 10 Mbps operations require a minimum period of 9.6 ms between frames?

3.4.16 What does a destination address of all F's in a frame denote?

3.4.17 What is the difference between universally administered addressing and locally administered addressing?

3.4.18 Why must an Ethernet frame have a minimum length of 64 bytes?

3.4.19 What is meant by the term connectionless transmission?

3.5.1 Draw a two-MAU Token-Ring network illustrating ring-in and ring-out connections.

3.5.2 How does a Token-Ring MAU provide a redundant ring path?

3.5.3 Discuss two constraints associated with the design of a Token-Ring network.

3.5.4 What are the two methods by which a break in a Token-Ring network can be compensated for?

3.5.5 Suppose a main ring path is 380 feet while two segments are 150 and 160 feet in length. What is the adjusted ring length of this network?

3.5.6 What is the purpose of an abort token?

3.5.7 What are J and K symbols?

3.5.8 What is the purpose of the monitor bit? Who is responsible for monitoring the monitor bit?

3.5.9 What is the function of the Token-Ring functional address indicator?

3.5.10 How is a MAC frame distinguished from an LLC frame on a Token-Ring network?

3.5.11 What is the purpose of the E bit in a Token-Ring frame?

3.5.12 What is the purpose of a beacon frame?

3.5.13 What is a failure domain?

3.5.14 What is the purpose of a duplicate address test frame?

4

WIDE AREA NETWORK TRANSMISSION EQUIPMENT

One method of categorizing data communications components is by the function or group of functions they are designed to perform. In this chapter, the operation and utilization of networking devices designed primarily to provide data transmission via a wide area network communications medium will be covered. Specific devices which will be explored in this chapter include a variety of analog and digital transmission products that can be used to transmit data over analog or digital wide area network transmission facilities. Analog transmission devices covered in this chapter range in scope from the nearly obsolete acoustic coupler to a half-dozen types of modems, including state-of-the-art cable and digital subscriber line modems. Digital transmission devices covered in this chapter include different types of data and channel service units and a device that permits the extension of parallel transmission from a computer to peripheral units located at a distance from a computational facility.

4.1 ACOUSTIC COUPLERS

An acoustic coupler is in essence a modem which permits data transmission through the utilization of the handset of an ordinary telephone. Similar in functioning to a modem, an acoustic coupler is a device which will accept a serial asynchronous data stream from terminal devices, modulates that data stream into the audio spectrum, and then transmit the audio tones over a switched or dial-up telephone connection.

Acoustic couplers are equipped with built-in cradles or fittings into which a conventional telephone headset is placed. Through the process of acoustic coupling, the modulated tones produced by the acoustic coupler are directly picked up by the attached telephone headset. Likewise, the audible tones transmitted over a telephone line are picked up by the telephone earpiece and demodulated by the acoustic coupler into a serial data stream which is acceptable to the attached terminal.

Acoustic couplers normally use two distinct frequencies to transmit information, while two other frequencies are employed for data reception. A frequency from each pair is used to create a mark tone which represents an encoded binary one from the digital data stream, while another pair of frequencies generates a space tone which represents a binary zero. This utilization of two pairs of frequencies permits full-duplex transmission to occur over the two-wire switched telephone network.

Since acoustic couplers enable any conventional telephone to be used for data transmission purposes, the coupler does not have to be physically wired to the line. This permits considerable flexibility in selecting a transmission location, which can include pay telephones in airports and hard-wired telephones in hotel rooms. Acoustic couplers are manufactured both as separate units and as built-in units to data terminals, as shown in Figure 4.1. Due to the significant replacement of portable terminals by laptop computers with built-in modems with modular jacks, the market for both portable terminals and acoustic couplers has greatly diminished. However, if you read some mobile computing trade magazines you will still note advertisements for battery-powered acoustic couplers as they permit the transmission mobility many travelers require.

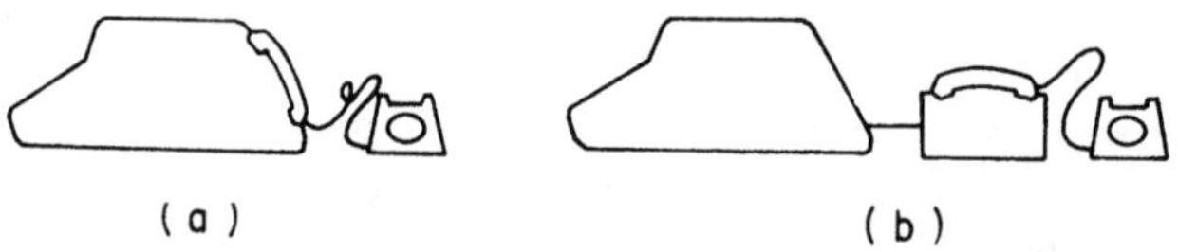

Figure 4.1 Varying coupler connections: (a) terminal with built-in coupler; (b) terminal connected to coupler

US and European compatibility

Since acoustic couplers are normally employed to permit portable data processing devices to communicate with data-processing facilities, and since a large portion of low-speed modems at such facilities in the United States were originally furnished by AT&T and its operating companies prior to its break-up into independent organizations, most manufacturers of acoustic couplers designed them to be compatible with low-speed 'Bell System' modems. Here the term 'Bell System' refers to the operating characteristics of modems that were manufactured by Western Electric for use by AT&T operating companies prior to those operating companies becoming independent organizations.

In Europe, most acoustic couplers are designed to be compatible with ITU recommendations that govern the operation of low-speed modems. To understand the differences between low speed Bell System and ITU modems, we will examine acoustic couplers that operate at data rates between 0 and 450 bps. In the United States, such couplers are compatible with Bell System 103 and 113 type modems while in Europe such couplers are compatible with the ITU V.21 recommendation. Table 4.1 lists the operating frequencies of acoustic couplers designed to operate with Bell System 103/113 type modems and modems that follow the ITU V.21 recommendation.

Table 4.1 Acoustic coupler modem compatibility (operating frequencies in Hz)

	Bell System 103/113 type		ITU V.21	
	Originate	Answer	Originate	Answer
Transmit				
Mark	1270	2225	980	1650
Space	1070	2025	1180	1850
Receive				
Mark	2225	1270	1650	980
Space	2025	1070	1850	1180

Couplers like low-speed modems must operate in one of two modes—originate or answer. This operational mode should not be confused with a transmission mode of simplex, half-duplex, or full-duplex. What the operational mode refers to is the frequency assignments for transmitting marks and spaces. Thus, from Table 4.1, an acoustic coupler compatible with a Bell System 103/113 type modem would transmit a tone at 1270 Hz to represent a mark and a tone at 1070 Hz to represent a space when it is in the originate mode of operation. To communicate effectively, the device (modem or coupler) at the other end of the line must be in the answer mode of operation. If so, then it would receive a mark at 1270 Hz and a space at 1070 Hz, ensuring that the tones transmitted by the originate mode device would be heard by the receive mode device. This explains why two terminal operators, each with an originate mode coupler, could not communicate with one another. This communications incompatibility results from the fact that one coupler would transmit a mark at 1270 Hz, while the other coupler would be set to receive the mark at 2225 Hz. Thus, the second coupler would never hear the tone originated by the first coupler.

By convention, originate mode couplers and modems are connected to terminals while answer mode devices are connected to computer ports, since terminals original calls and computers typically answer such calls. Some couplers can be obtained with an originate/answer mode switch. By changing the position of the switch, you change the coupler's operating frequency assignments.

In Figure 4.2 the frequency assignments of couplers designed to be compatible with Bell System 103/113 type modems is graphically illustrated. Note that 1170 Hz and 2125 Hz are the channel center frequencies and two independent data channels are derived by frequency, permitting full-duplex transmission to occur over the two-wire public switched telephone network.

Returning to Table 4.1, note that the operating frequencies of Bell System 103/113 type modems are completely different from modems designed to operate according to the V.21 recommendation. This frequency incompatibility explains why, for example, an American traveling in Europe will often be unable to use his or her portable personal computer to communicate with either a public packet network or his or her company's mainframe computer located in Europe.

Originally acoustic couplers were developed to transmit and receive data at 300 bps. Today, a few vendors market devices that operate at 1200 bps. Such couplers

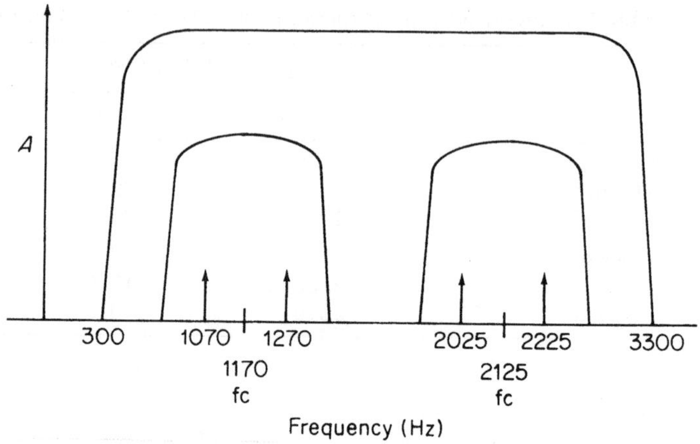

Figure 4.2 Bell System 103/113 frequency spectrum

are compatible with either Bell System 202 or 212A or ITU V.22 modems and their method of modulation will be described in Section 4.2 where modems are covered.

Operation

The operation of an acoustic coupler is a relatively simple process. An operator wishing to transmit data dials a telephone access number and upon establishing the proper connection by hearing a high-pitched tone, places the telephone headset into the coupler. Although usage varies by numerous applications, the prevalent utilization of acoustic couplers is in obtaining access to a packet network.

In a packet network, a group of dial-in computer telephone access numbers are interfaced to rotary which enables users to dial the low telephone number of the group and automatically 'step' or bypass currently busy numbers. Each telephone line is then connected to a modem on a permanent basis, and the modem in turn is connected to a computer port or channel. An automatic answering device in each modem automatically answers the incoming call and in effect establishes a connection from the user who dialed the number to the computer port, as shown in Figure 4.3. Once access is established to a packet network, the operator enters a routing code to establish a virtual path through the network to a corporate computer connected to the packet network.

Problems in Usage

A disadvantage associated with the use of acoustic couplers is a reduction of transmission rates when compared to rates which can be obtained by using modems. Owing to the properties of carbon microphones in telephone headsets, the frequency band that can be passed is not as wide as the band modems can pass. Although typical data rates of acoustic couplers vary between 110 and 300 bps,

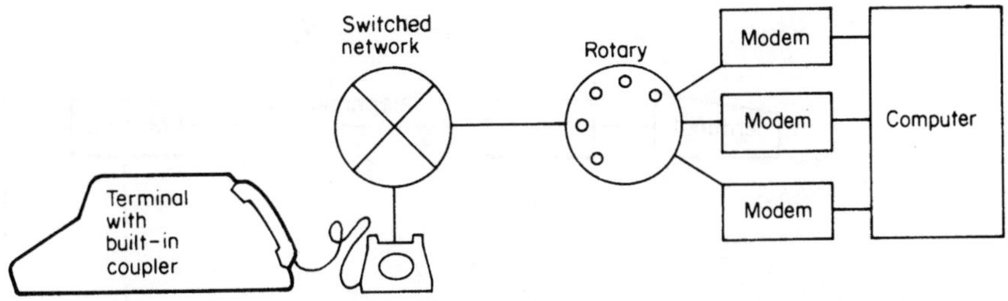

Figure 4.3 Network access in a time-sharing environment. After dialing the computer access number the terminal user places the telephone headset into the cradle of the acoustic coupler

some units manufactured permit transmission at 450, 600 and 1200 bps. For usage with slow-speed terminals, the acoustic coupler can be viewed as a low-cost alternative to a modem while increasing user transmission location flexibility.

One item which may warrant user attention is the placement of a piece of cotton inside the earpiece behind the receiver of the telephone. Although the placement of cotton at this location is normally done by most telephone companies, this should be checked, since the cotton keeps speaker and receiver noise from interfering with each other and acts to prevent transmitted data from interfering with received data.

An easily resolved problem is the placement of the telephone headset into the coupler. On many occasions users have hastily placed the handset only partially into the coupler, and this will act to reduce the level of signal strength necessary for error-free transmission.

4.2 MODEMS

Today, despite the introduction of a number of all-digital transmission facilities by most communications carriers, the analog telephone system remains the primary facility utilized for data communications. Since terminals and computers produce digital pulses, whereas telephone circuits are designed to transmit analog signals which fall within the audio spectrum used in human speech, a device to convert the digital data pulses of terminals and computers to analog tones that are carried on telephone circuits becomes necessary to transmit data over such circuits. Such a device is called a modem, which derives its meaning from a contraction of the two main functions of such a unit—modulation and demodulation. Although modem is the term most frequently used for such a device that performs modulation and demodulation, 'data set' is another common term whose use is synonymous in meaning.

In its most basic form a modem consists of a power supply, a transmitter, and a receiver. The power supply provides the voltage necessary to operate the modem's circuitry. In the transmitter a modulator and amplifier, as well as filtering, wave-shaping, and signal control circuitry convert digital direct current pulses; these

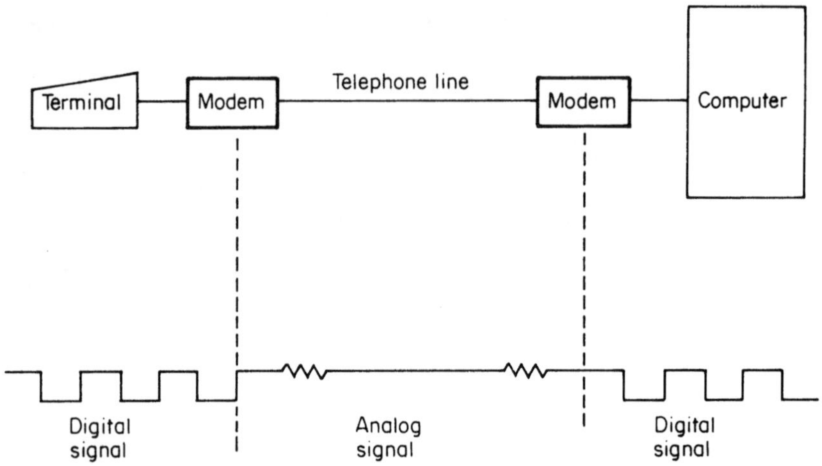

Figure 4.4 Signal conversion performed by modems. A modem converts a digital signal to an analog tone (modulation) and reconverts the analog tone (demodulation) into its original digital signal

pulses, originated by a computer or terminal, are converted into an analog, wave-shaped signal which can be transmitted over a telephone line. The receiver contains a demodulator and associated circuitry which reverse the process by converting the analog telephone signal back into a series of digital pulses that is acceptable to the computer or terminal device. This signal conversion is illustrated in Figure 4.4.

Basic components

In Figure 4.5 the basic components of a modem are indicated in a block diagram format, with those components associated with the transmitter at the top portion of the illustration, while those components associated with the receiver are located in the lower portion of the figure. Prior to examining each of the components, readers should note that Figure 4.5 represents a 'general' modem and those components indicated by dashed lines are only applicable to synchronous devices. In addition, other components, including a microprocessor, ROM and RAM, which can be included to provide a modem with intelligence have been purposely omitted from Figure 4.5 to enable us to focus our attention upon the modulation and demodulation of data by the transmitter and receiver in each device.

Modem transmitter section

The key components of a modem's transmitter section include a data encoder, scrambler, modulator, amplifier, filter, timing source and transmit control circuits. Of these components, the scrambler and transmit clock provided by the timing source are used only in synchronous modems.

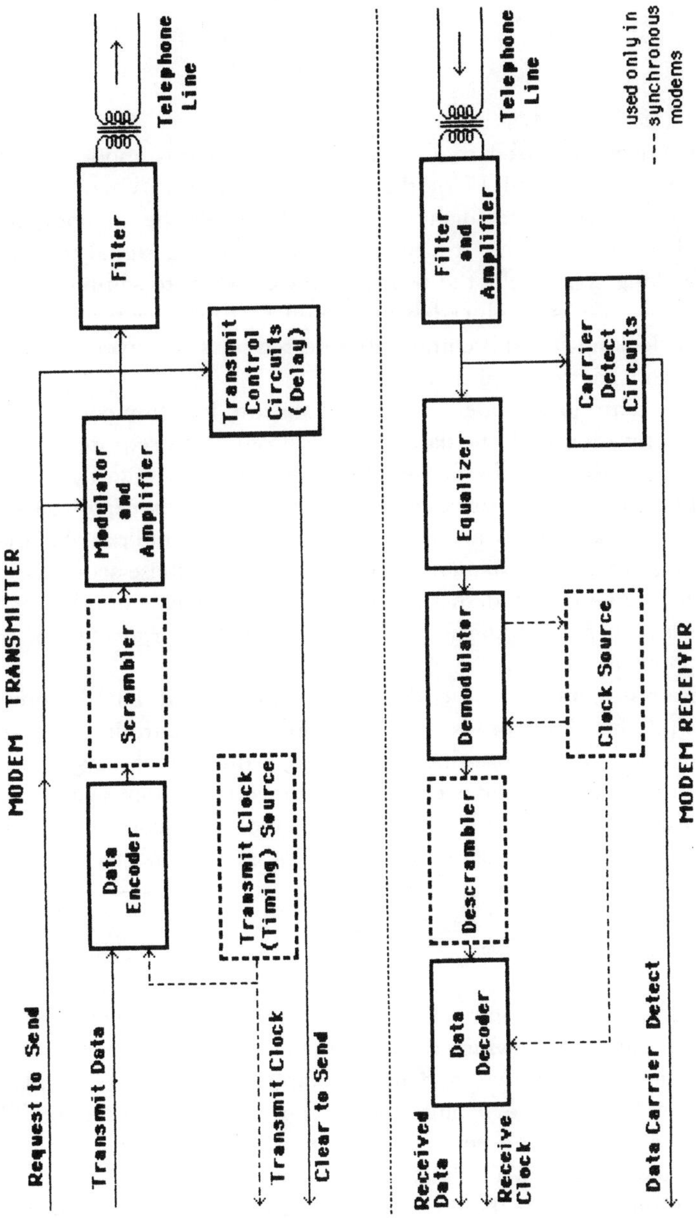

Figure 4.5 Basic components of a modem

The data encoder is an option built into many modems. The encoder is used in conjunction with some modulation schemes, enabling each signal change to represent more than one bit of information. We will discuss the use of data encoders and decoders later in this chapter when we examine the difference between a bit and a baud.

Scramblers

As previously discussed in Chapter 1, synchronous modems provide clocking signals on pins 15 and 17 of the RS-232 interface. When a modem receives a modulated synchronous data stream and passes the demodulated data to an attached terminal device, it also provides a clocking signal to the data terminal. This clocking signal tells the terminal device when to sample pin 3, the received data circuit, and is produced by the modem from the received data. Thus, the received clocking signal is commonly referred to as a derived clocking signal as it is derived from the received data.

For a synchronous modem's received clock to function correctly it must remain in synchronization with the data being received. This requires a sufficient number of changes in the composition of the data, e.g., 0 to 1 and 1 to 0, to permit the receiving modem's circuitry to derive timing from the received data. Since the data stream can consist of any arbitrary bit pattern, it is quite possible that the data will randomly contain long sequences of 0s or 1s. When these sequences occur the data will not provide the modem's receiver with a sufficient number of transitions for clock recovery, a condition which resulted in the incorporation of scramblers into synchronous modems.

A scrambler modifies the data to be modulated based upon a predefined algorithm. This algorithm is normally implemented through the use of a feedback shift register, which examines a defined sequence of bits and modifies their composition to ensure that every possible bit combination is equally likely to occur. At the receiving modem a descrambler employs the inverse of the predefined algorithm to restore the data into its original serial data stream.

Modulator, amplifier and filter

The modulator acts upon a serial data stream by using the composition of the data to alter the carrier tone which the modem places on the communications line. When a connection between two modems is established one modem will 'raise' a carrier tone that is heard by the distant modem. The reader is referred to 'The modulation process' in this section for information concerning the operation of different methods used to modulate data.

The amplifier boosts the level of the modulated signal for transmission onto the telephone line while the filter limits the frequencies of the tones placed on the line to comply with federal regulations. At the receiver, modulated tones received from the telephone line are filtered to remove extraneous tones caused by noise and then amplified to boost the received signal level.

Equalizer

The equalizer illustrated in the modem receiver section in Figure 4.5 is designed to measure the characteristics of a received analog signal and to adjust itself to that signal. In doing so the equalizer minimizes the effect of attenuation and delay upon the various components of a transmitted signal. To accomplish this task the modem's transmitter will prefix each transmission with a short 'training' signal whenever the direction of transmission changes. This training signal represents a predefined modulation of the carrier whose ideal reception characteristics are known by the equalizer in the receiver of the distant modem. Thus, the equalizer will be adjusted by the receiving modem until the best possible signal is received.

To obtain an appreciation for the operation and utilization of equalizers, let us first focus our attention upon several basic data channel parameters and the method by which communications carriers create a telephone channel. This will provide us with the foundation for discussing several analog signal impairments associated with transmission on the switched telephone network and the method by which equalizers can be used to compensate for those impairments.

Bandwidth

Bandwidth is a measurement of the width of a range of frequencies such that

$$B = f_2 - f_1$$

where B is the bandwidth, f_2 the highest frequency and f_1 the lowest frequency in a range of frequencies. Figure 4.6 illustrates the bandwidth of a telephone channel in comparison to the audio spectrum heard by the human ear. Here the unit hertz (Hz) is used to represent a cycle per second.

The 3000 Hz bandwidth which forms a telephone channel is commonly referred to as the passband of the channel. The term passband references a contiguous portion of an area in the frequency spectrum which permits a predefined range of frequencies to pass. Thus, the passband of a telephone channel permits frequencies between 300 and 3300 Hz to pass.

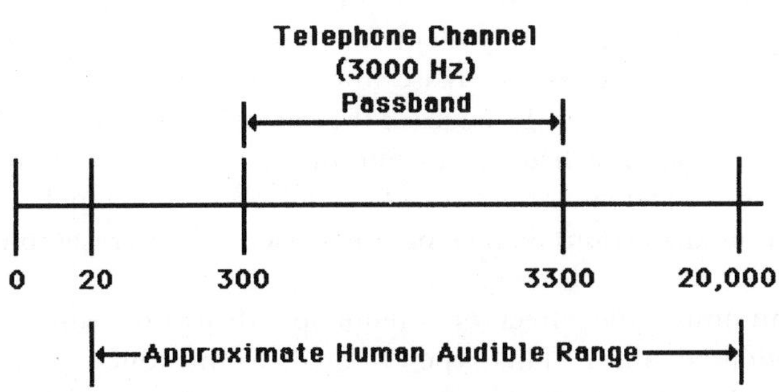

Figure 4.6 Bandwidth of a telephone channel

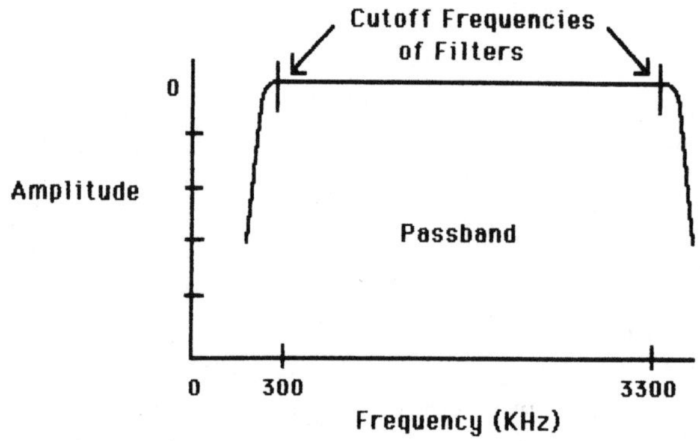

Figure 4.7 Telephone channel passband creation

The rationale for the telephone channel passband is economics. Frequencies under 300 Hz and above 3300 Hz are essentially not required to understand a telephone conversation even though it precludes a soprano from being fully appreciated at the other end of the telephone connection. By transmitting only 3000 Hz instead of the 20 000 Hz that the human ear can hear the bandwidth required for each call is reduced by a factor of approximately six. This bandwidth reduction enabled telephone companies to more efficiently employ frequency division multiplexing, a technique which allows many voice calls to be simultaneously carried on a common circuit routed between telephone company offices.

To construct a telephone channel passband the telephone company uses low and high-pass filters which are designed to permit either all signals up to a predefined frequency or all signals under a predefined frequency to pass through the channel. As a result of the use of filters the amplitude-frequency response becomes rounded at the cut-off frequencies at which the filters operate and then begin to approach large negative values as the filters' attenuation becomes more pronounced. Figure 4.7 illustrates how the use of filters results in the creation of a passband on a telephone channel.

Ideally, all frequencies across the passband of a telephone channel should undergo the same amount of attenuation as illustrated by the straight line between the cut-off frequencies shown in Figure 4.7. Unfortunately, high frequencies lose their strength more rapidly than low frequencies, which results in attenuation increasing as frequencies increase towards the end of the passband. In addition, attenuation increases as the edges of the operating frequencies of bandpass filters on a channel are approached. As a result of the two previously mentioned factors the amplitude-frequency response of a telephone channel which indicates the attenuation distortion that signals experience will resemble that illustrated in Figure 4.8.

To minimize the effect of attenuation distortion some modems include an attenuation equalizer. This type of equalizer introduces a variable gain at frequencies within the passband which compensate for the differences in attenuation between high and low frequencies as well as the increased attenuation at the edges

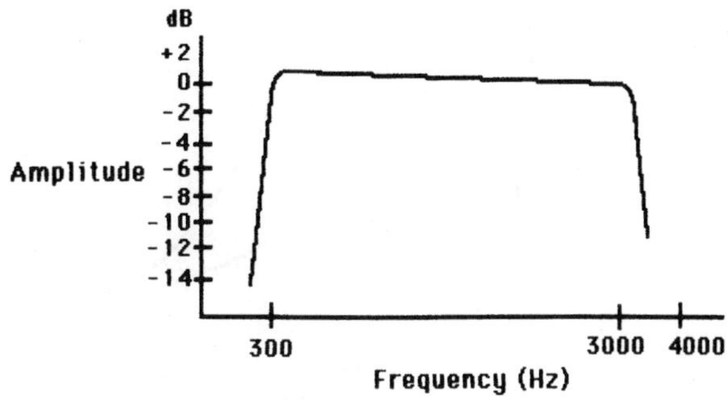

Figure 4.8 Typical amplitude–frequency response across a voice channel

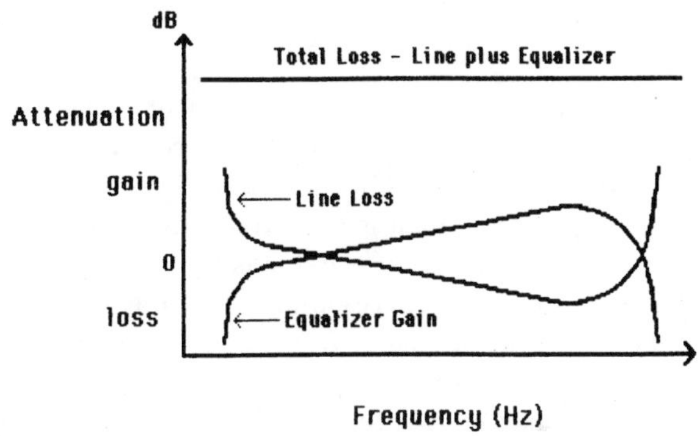

Figure 4.9 Using an equalizer to correct attenuation distortion

of the passband. Figure 4.9 illustrates the operation of an attenuation equalizer whose use results in a near uniform signal level across the passband.

Delay distortion

A second type of distortion which affects the recovery of information from a received signal is delay distortion. In a distortionless channel, all frequencies pass through the channel at the same speed. This results in the frequency and phase of the signal having a constant linear relationship with respect to time and ensures that the transmission of one signal will not interfere with the reception of a previously transmitted signal. Unfortunately, all channels except perhaps those in the laboratory have a degree of distortion. When distortion occurs, the relationship between the phase and frequency of a signal becomes non-linear. As the level of distortion increases, the relationship between the phase and frequency of a signal degenerates further. This degeneration, which is called phase delay, is measured at a particular point on the frequency spectrum by dividing the phase of the signal by

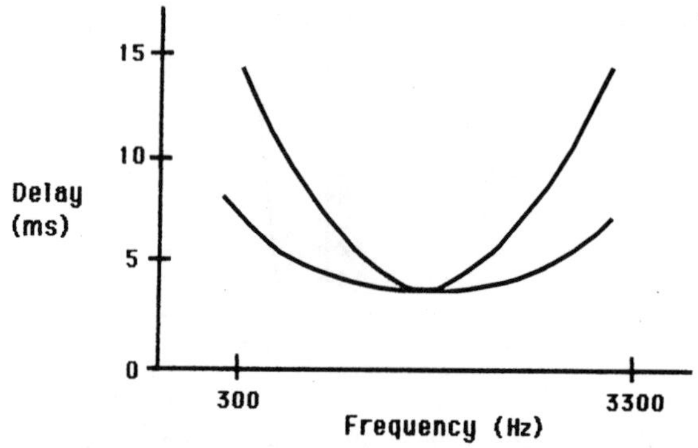

Figure 4.10 Typical envelope delay curves

its frequency. The direct measurement of phase delay is not practical due to the requirement to have an absolute phase reference and keep track of phase changes over multiples of 360°. This resulted in the use of the slope of the phase plotted against frequency, which is known as envelope delay.

Mathematically, envelope delay is the first derivative of phase delay. The shape of the envelope delay curve obtained by measuring delays at different frequencies reflects the degree of change in the slope of the phase versus frequency curve. This delay change varies based upon the transmission distance. Figure 4.10 illustrates two typical delay curves for signals transmitted on a telephone channel, with the steeper curve representing the envelope delay on a longer distance circuit than the flatter curve.

To illustrate the potential effect of envelope delay upon communications, assume a modem transmits one of two tones (f_1) to represent a binary zero and the second tone (f_2) to represent a binary one. This is one of the earliest methods used for modulation, a technique referred to as frequency shift keying or FSK. If the envelope delay curve is not symmetrical, consider what can happen if tone f_1 is transmitted, followed by tone f_2. Due to different delays associated with different frequencies there now exists the possibility that the delay in f_1 being received could result in that tone reaching the receiving modem at the same time that tone f_2, which represents a different binary value, is received. This could cause one received signal to be superimposed upon the second signal by time, resulting in one tone distorting the other tone.

Although all communications circuits will exhibit a degree of delay, it is important to flatten the delay time across the passband to minimize the potential for one tone of a signal to be superimposed on another tone by time. Some modems are designed with delay equalizers which introduce a delay approximately inverse to that exhibited by the telephone channel. Through the use of a delay equalizer the delay time associated with frequencies within the passband can be made relatively flat as illustrated in Figure 4.11. Doing so reduces the potential of one tone interfering with another, a condition formally referred to as intersymbol interference.

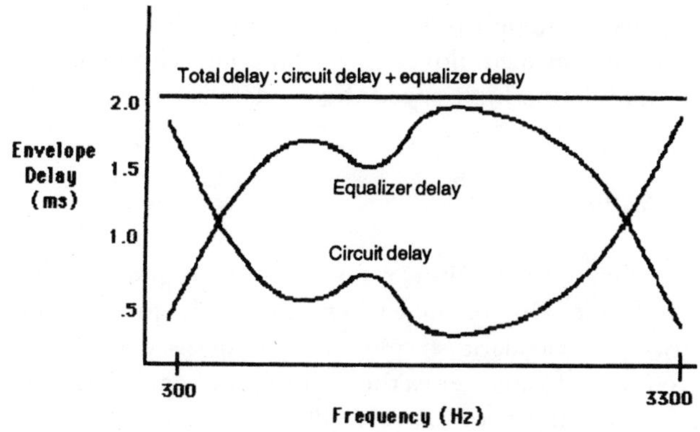

Figure 4.11 Using a delay equalizer

The modulation process

The modulation process alters the characteristics of a carrier signal. By itself, a carrier is a repeating signal that conveys no information. However, when the carrier is changed by the modulation process information is impressed upon the signal. For analog signals, the carrier is a sine wave, represented by

$$a = A \sin(2\pi f t + \phi)$$

where a is the instantaneous value of voltage at time t, A the maximum amplitude, f the frequency and ϕ the phase.

The carrier's characteristics that can be altered are thus the carrier's amplitude for amplitude modulation (AM), the carrier's frequency for frequency modulation (FM), and the carrier's phase angle for phase modulation (ϕM).

Amplitude modulation

The simplest method of employing amplitude modulation is to vary the magnitude of the signal from a zero level to represent a binary zero to a fixed peak-to-peak voltage to represent a binary one. Figure 4.12 illustrates the use of amplitude modulation to encode a digital data stream into an appropriate series of analog

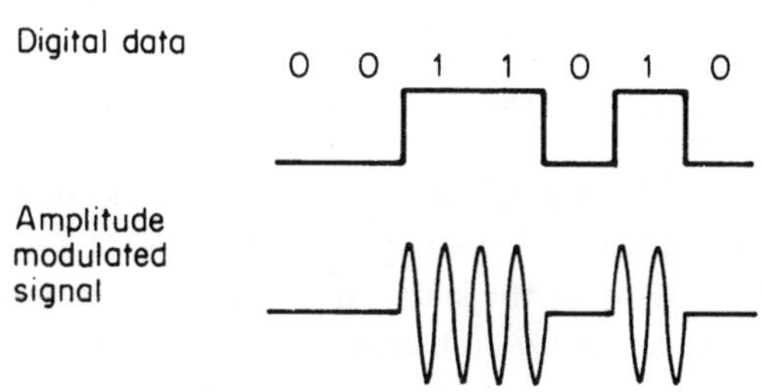

Figure 4.12 Amplitude modulation

signals. Although pure amplitude modulation is normally used for very low data rates, it is also employed in conjunction with phase modulation to obtain a method of modulating high-speed digital data sources.

Frequency modulation

Frequency modulation refers to how frequently a signal repeats itself at a given amplitude. One of the earliest uses of frequency modulation was in the design of low-speed acoustic couplers and modems where the transmitter shifted from one frequency to another as the input digital data changed from a binary one to a binary zero or from a zero to a one. This shifting in frequency is known as frequency shift keying (FSK) and is primarily used by modems operating at data rates up to 300 bps in a full-duplex mode of operation and up to 1200 bps in a half-duplex mode of operation. Figure 4.13 illustrates frequency shift keying frequency modulation.

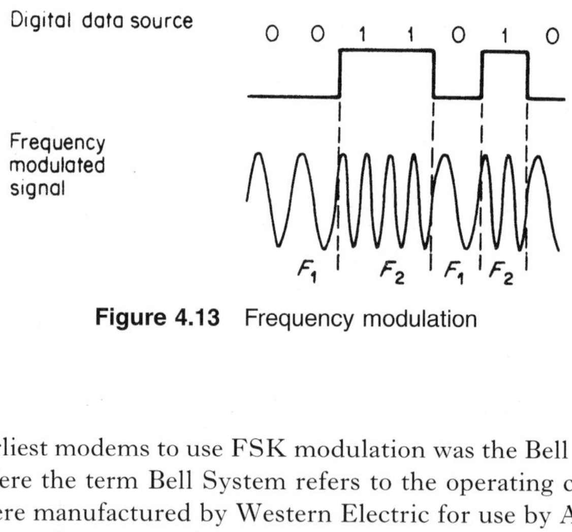

Figure 4.13 Frequency modulation

One of the earliest modems to use FSK modulation was the Bell System 103/113 type modem. Here the term Bell System refers to the operating characteristics of modems that were manufactured by Western Electric for use by AT&T operating companies prior to those operating companies becoming independent organizations. A large portion of low-speed modems in the United States were furnished by AT&T and its operating companies prior to the court-ordered divestiture of those companies. Due to this large installed base of older devices, many modern modems marketed today are designed to be compatible with earlier modems whose operating characteristics can be considered to be *de facto* standards.

The Bell System 103/113 type modem was designed to operate in one of two modes—originate or answer. This operational mode refers to the frequency assignment used by the modem for transmitting marks and spaces and should not be confused with the half- or full-duplex transmission modes that reference the modem's ability to transmit and receive data alternately or simultaneously. By splitting a two-wire circuit into two separate transmission paths by frequency, full-duplex transmission becomes possible on a half-duplex line facility.

Table 4.1 previously listed the operating frequencies of the Bell System 103/113 type modem. The originate-mode modem is normally connected to a terminal device that originates calls, whereas the answer-mode modem is normally connected to computer ports that answer calls occurring over the switched telephone network (PSTN).

In Figure 4.2 the frequency assignments of modems designed to be compatible with Bell System 103/113 type modems is graphically illustrated. Note that 1170 Hz and 2125 Hz are the channel center frequencies and two independent data channels are derived by frequency, permitting full-duplex transmission to occur over the two-wire public switched telephone network.

Phase modulation

Phase modulation is the process of varying the carrier signal with respect to the origination of its cycle as illustrated in Figure 4.14. Several forms of phase modulation are used in modems to include single and multiple-bit phase-shift keying (PSK) and the combination of amplitude and multiple-bit phase-shift keying.

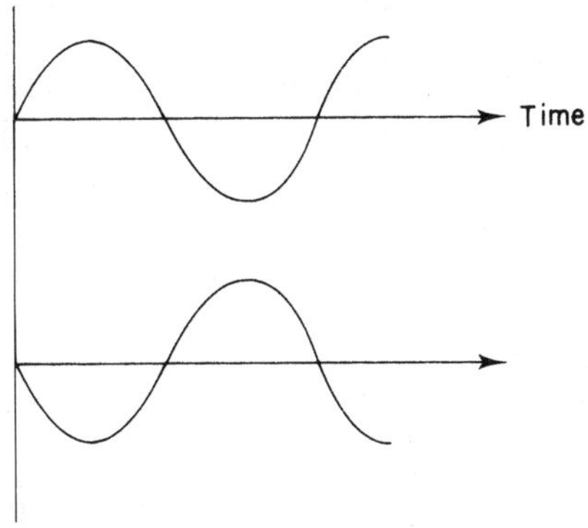

Figure 4.14 Phase modulation. Phase is the position of the wave form of a signal with respect to the origination of the carrier cycle. In this illustration, the bottom wave is 180 degrees out of phase with a normal sine wave illustrated at the top

In single-bit, phase-shift keying, the transmitter simply shifts the phase of the signal to represent each bit entering the modem. Thus, a binary one might be represented by a 90° phase change while a zero bit could be represented by a 270° phase change. Due to the variance of phase between two-phase values to represent binary ones and zeros this technique is known as two-phase modulation.

Prior to discussing multiple-bit, phase-shift keying let us examine the difference between the data rate and signaling speed. This will enable us to understand the rational for the utilization of multiple-bit, phase-shift keying, where two or more bits are grouped together and represented by one phase shift in a signal.

Bps versus baud

Bits per second is the number of binary digits transferred per second and represents the data transmission rate of a device. Baud is the signaling rate of a device such as a modem. If the signal of the modem changes with respect to each bit entering the device, then 1 bps = 1 baud. Suppose a modem is constructed such that one signal change is used to represent two bits. Then the baud rate would be one-half the bps rate.

When one baud is used to represent two bits the encoding technique is known as dibit encoding. Similarly, the process of using one baud to represent three bits is known as tribit encoding and the bit rate is then one-third of the baud rate. Both dibit and tribit encoding are known as multilevel coding techniques and are commonly implemented using phase modulation.

Voice circuit parameters

Bandwidth is a measurement of the width of a range of frequencies. As previously explained, a voice-grade telephone channel has a passband, which defines its slot in the frequency spectrum, which ranges from 300 to 3300 Hz. The bandwidth of a voice-grade telephone channel is thus 3300 – 300 or 3000 Hz.

As data enters a modem it is converted into a series of analog signals, with the signal change rate of the modem known as its baud rate. In 1928, Nyquist developed the relationship between the bandwidth and the baud rate on a circuit as

$$B = 2W$$

where B is the baud rate and W the bandwidth in Hz.

For a voice-grade circuit with a bandwidth of 3000 Hz, this relationship means that data transmission can only be supported at baud rates lower than 6000 symbols or signaling elements per second, prior to one signal interfering with another and causing intersymbol interference.

Since any oscillating modulation technique immediately halves the signaling rate, this means that most modems are limited to operating at one-half of the Nyquist limit. Thus, in a single-bit, phase-shift keying modulation technique, where each bit entering the modem results in a phase shift, the maximum data rate obtainable would be limited to approximately 3000 bps. In such a situation the bit rate would equal the baud rate, since there would be one signal change for each bit.

To overcome the Nyquist limit required engineers to design modems that first group a sequence of bits together, examined the composition of the bits, and then implemented a phase shift based upon the value of the grouped bits. This technique is known as multiple bit, phase-shift keying or multilevel, phase-shift keying. Two-bit codes called dibits and three-bit codes known as tribits are formed and transmitted by a single phase shift from a group of four or eight possible phase states.

Most modems operating at 600 to 4800 bps employ multilevel, phase-shift keying modulation. Some of the more commonly used phase patterns employed by modems using dibit and tribit encoding are listed in Table 4.2.

Table 4.2 Common phase-angle values used in multilevel, phase-level keying

Bits transmitted	Possible phase-angle values (degrees)		
dibit			
00	0	45	90
01	90	135	0
10	180	225	270
11	270	315	180
tribit			
000	0	22.5	45
001	45	67.5	0
010	90	112.5	90
011	135	157.5	135
100	180	202.5	180
101	225	247.5	225
110	270	292.5	270
111	315	337.5	315

Combined modulation techniques

Since the most practical method to overcome the Nyquist limit is obtained by placing additional bits into each signal change, modem designers have combined modulation techniques to obtain very high-speed data transmission over voice-grade circuits. One combined modulation technique commonly used involves both amplitude and phase modulation. This technique is known as quadrature amplitude modulation (QAM) and results in four bits being placed into each signal change, with the signal operating at 2400 (baud), causing the data rate to become 9600 bps.

The first implementation of QAM involved a combination of phase and amplitude modulation, in which 12 values of phase and three values of amplitude are employed to produce 16 possible signal states as illustrated in Figure 4.15. One of the earliest modems to use QAM in the United States was the Bell System 209, which modulated a 1650 Hz carrier at a 2400 baud rate to effect data transmission at 9600 bps. Today, most 9600 bps modems manufactured adhere to the ITU V.29 standard. The V.29 modem uses a carrier of 1700 Hz which is varied in both phase and amplitude, resulting in 16 combinations of eight phase angles and four amplitudes. Under the V.29 standard, fallback data rates of 7200 and 4800 bps are specified.

In addition to combining two modulation techniques, QAM also differs from the previously discussed modulation methods by its use of two carrier signals. Figure 4.16 illustrates a simplified block diagram of a modem's transmitter employing QAM. The encoder operates upon four bits from the serial data stream and causes both an in-phase (IP) cosine carrier and a sine wave that serves as the quadrature component (QC) of the signal to be modulated. The IP and QC signals are then summed and result in the transmitted signal being changed in both amplitude and

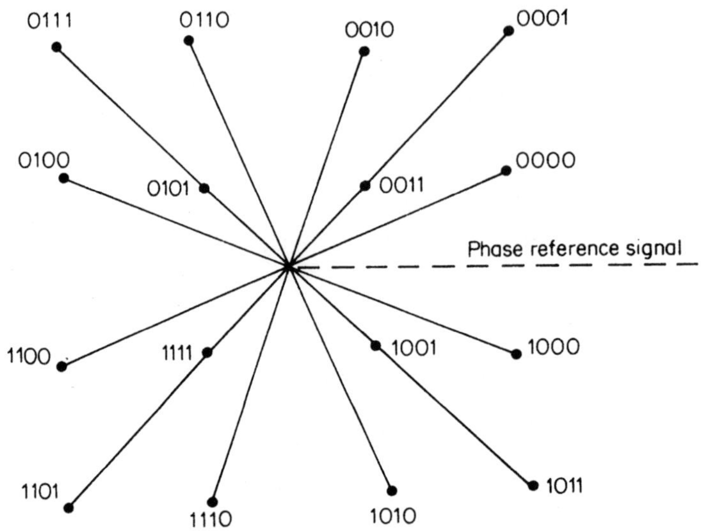

Figure 4.15 Quadrature amplitude modulation produces 16 signal states from a combination of 12 angles and three amplitude levels

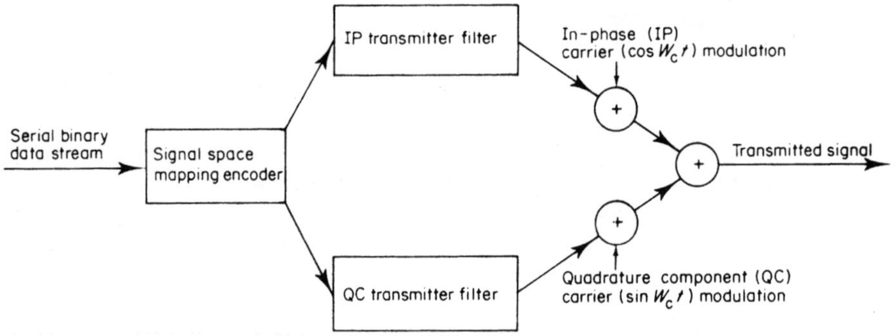

Figure 4.16 QAM modem transmitter

phase, with each point placed at the *x*-*y* coordinates representing the modulation levels of the cosine carrier and the sine carrier.

If you plot the signal points previously illustrated in Figure 4.15 which represent all of the data samples possible in that particular method of QAM, the series of points can be considered to be the signal structure of the modulation technique. Another popular term used to describe these points is the constellation pattern. By an examination of the constellation pattern of a modem, it becomes possible to predetermine its susceptibility to certain transmission impairments. As an example, phase jitter which causes signal points to rotate about the origin can result in one signal being misinterpreted for another, which would cause four bits to be received in error. Since there are 12 angles in the QAM method illustrated in Figure 4.15, the minimum rotation angle is 30°, which provides a reasonable immunity to phase jitter.

Other modulation techniques

By the late 1980s several vendors were offering modems that operated at data rates up to 19 200 bps over leased voice-grade circuits. Originally, modems that operated at 14 400 bps employed a quadrature amplitude modulation technique, collecting data bits into a 6-bit symbol 2400 times per second, resulting in the transmission of a signal point selected from a 64-point signal constellation. The signal pattern of one vendor's 14 400 bps modem is illustrated in Figure 4.17. Note that this particular signal pattern appears to form a hexagon and according to the vendor was used since it provides a better performance level with respect to signal-to-noise (S/N) ratio and phase jitter than conventional rectangular grid signal structures. However, in spite of hexagonal packed signal structures, it should be obvious that the distance between signal points for a 14 400 bps modem are closer than the resulting points for a 9600 bps modem. This means that a 14 400 bps conventional QAM modem is more susceptible to transmission impairments and the overall data throughput under certain situations can be less than that obtainable with 9600 bps modems. Figure 4.18 illustrates the typical throughput variance of 9600 and 14 400

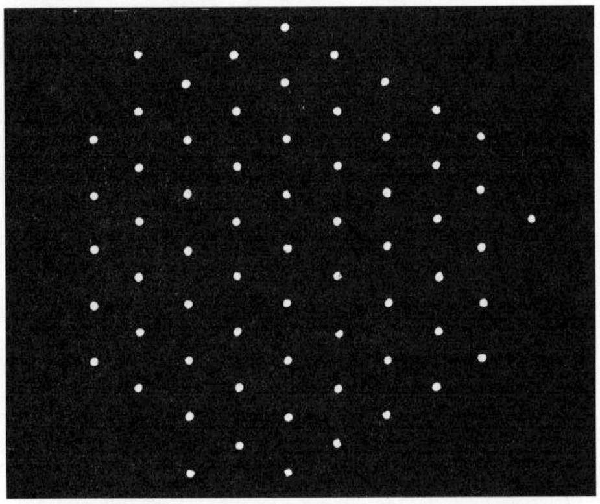

Figure 4.17 14 400 bps hexagonal signal constellation pattern

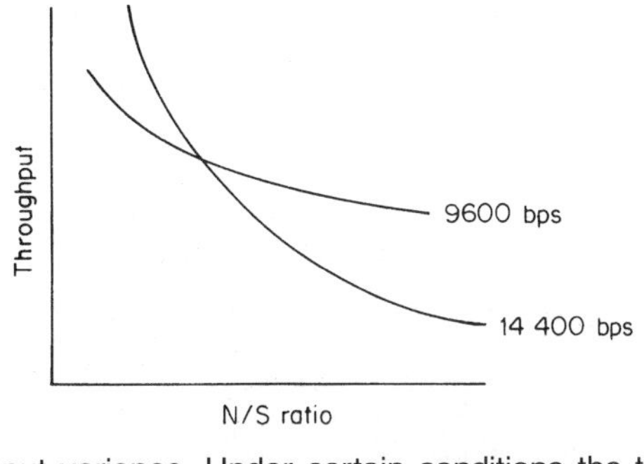

Figure 4.18 Throughput variance. Under certain conditions the throughput obtained by using 9600 bps modems can exceed the throughput obtained when using 14 400 bps devices

bps modems with respect to the ratio of noise to the strength of the signal (N/S) on the circuit. From this illustration, it should be apparent that 14 400 bps modems using conventional quadrature amplitude modulation should only be used on high-quality circuits.

Modems that transmit data at 16 000 bps are very similar to 14 400 bps devices, with the major difference being in the baud rate. Thus, most 16 000 bps modems encode data into 6-bit symbols and transmit the signals 2667 times per second. This method also employs a total of 64 signal points; however, the baud rate is increased from 2400 to 2667 to obtain the higher data transfer rate.

Trellis coded modulation

Due to the susceptibility of conventional QAM modems to transmission impairments, a new generation of modems based upon Trellis coded modulation (TCM) was developed. Such modems tolerate more than twice as much noise power as conventional QAM modems, permitting bidirectional data rates up to 33.6 kbps to be obtained over the switched telephone network.

To understand how TCM provides a higher tolerance to noise and other line impairments, to include phase jitter and distortion, let us consider what happens when a line impairment occurs when conventional QAM modems are used. Here the impairment causes the received signal point to be displaced from its appropriate location in the signal constellation. The receiver then selects the signal point in the constellation that is closest to what it received. Obviously, when line impairments are large enough to cause the received point to be closer to a signal point that is different from the one transmitted, an error occurs. To minimize the possibility of such errors, TCM employs an encoder that adds a redundant code bit to each symbol interval.

In actuality, at 14 400 bps the transmitter converts the serial data stream into 6-bit symbols and encodes two of the six bits employing a binary convolutional encoding scheme as illustrated in Figure 4.19. The encoder adds a code bit to the two input bits, forming three encoded bits in each symbol interval. As a result of this encoding operation, three encoded bits and four remaining data bits are then

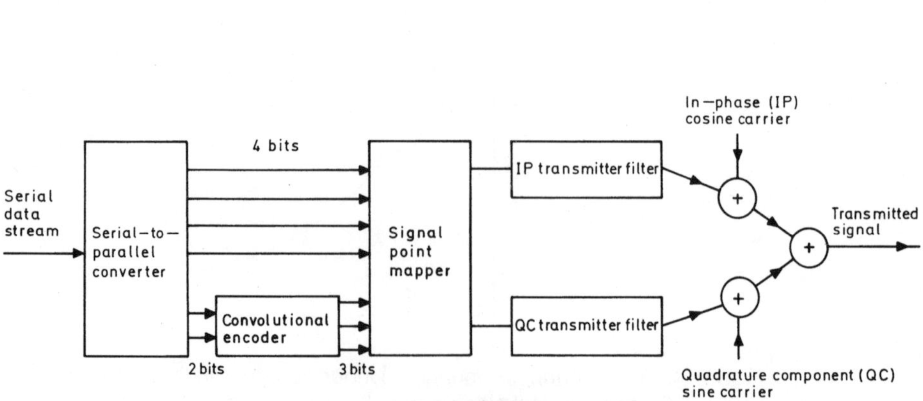

Figure 4.19 Trellis coded modulation

mapped into a signal point which is selected from a 128-point (2^7) signal constellation.

The key to the ability of TCM to minimize errors at high data rates is the employment of forward error correcting (FEC) in the form of convolutional coding. With convolutional coding, each bit in the data stream is compared with one or more bits transmitted prior to that bit. The value of each bit, which can be changed by the convolutional encoder, is therefore dependent upon the value of other bits. In addition, a redundant bit is added for every group of bits compared in this manner. The following examination of the formation of a simple convolutional code clarifies how the convolutional encoder operates.

Convolutional encoder operation

Assuming that a simple convolutional code is formed by the modulo 2 sum of the two most recent data bits, then two output bits will be produced for each data bit—a data bit and a parity bit. If we also assume that the first output bit from the encoder is the current data bit then the second output bit is the modulo 2 sum of the current bit and its immediate predecessor. Figure 4.20 illustrates the generation of this simple convolutional code.

Because each parity bit is the modulo 2 sum of the two most recent data bits, the relationship between the parity bits and the data bits becomes:

$$P_i = b_i + b_{i-1} \qquad i = 1, 2, 3 \ldots$$

If the composition of the first four data bits entering the encoder was 1101 ($b_4b_3b_2b_1$), the four parity bits are developed as follows:

$$\mathrm{P}_1 = \mathrm{b}_1 + \mathrm{b}_0 = 1 + 0 = 1$$
$$\mathrm{P}_2 = \mathrm{b}_2 + \mathrm{b}_1 = 0 + 1 = 1$$
$$\mathrm{P}_3 = \mathrm{b}_3 + \mathrm{b}_2 = 1 + 0 = 1$$
$$\mathrm{P}_4 = \mathrm{b}_4 + \mathrm{b}_3 = 1 + 1 = 0$$

Thus, the four-bit sequence 1101 is encoded as 01111011.

The preceding example also illustrates how dependencies can be constructed. In actuality, there are several trade-offs in developing a forward error correction scheme based upon convolutional coding. When a bit is only compared with a previously transmitted bit the number of redundant bits required for decoding at the receiver is very high. The complexity of the decoding process is, however, minimized. When the bit to be transmitted is compared with a large number of previously transmitted bits the number of redundant bits required is minimized. The processing required at both ends, however, increases in complexity.

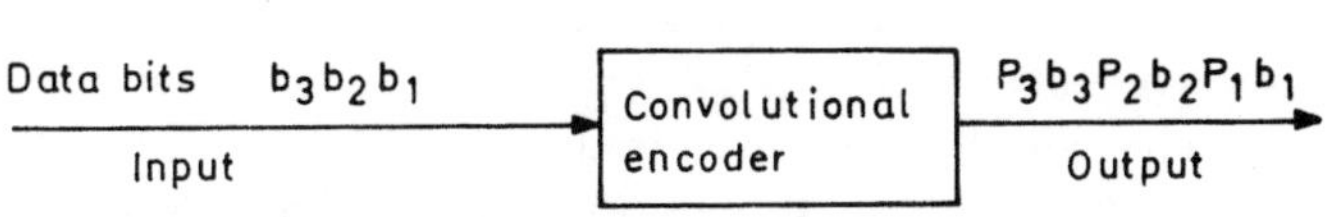

Figure 4.20 Simple convolutional code generation

In a 14 400 bps TCM modem, the signal point mapper uses the three encoded bits to select one of eight (2^3) subsets consisting of 16 points developed from the four data bits. This encoding process ensures that only certain points are valid. At the receiving modem, the decoder compares the observed sequence of signal points and selects the valid point closest to the observed sequence. The encoder makes this selection process possible by generating redundant information that establishes dependencies between successive points in the signal constellation. At the receiving modem, the decoder uses an alogorithm that compares previously received data with currently received data. The convolutional decoding algorithm then enables the modem to select the optimum signal point. Because of this technique, a TCM modem is twice as immune to noise as a conventional QAM modem. In addition, the probability of an error occurring when a TCM modem is used is substantially lower than when an uncoded QAM modem is used.

TCM modem developments

The first TCM modems marketed in 1984 operated at 14.4 kbps. Since then, a number of modems that use more complex TCM techniques have been introduced. These modems operate synchronously at 19.2 kbps, 24.4 kbps, 28.8 kbps and 33.6 kbps. By 1998, the application of TCM to modems enabled network users to more than double the transmission speed on analog circuits from that achievable only a few years before.

Echo cancellation

Prior to 1984, full-duplex transmission was achieved on the switched telephone network by splitting the available bandwidth on the line into two separate channels, each of which was used to transmit and receive data simultaneously. While this method of frequency division of the channel enabled full-duplex transmission at data rates up to 2400 bps, higher data rates using this modulation technique required more bandwidth than was available on the line. This constraint resulted in modem engineers developing a new modulation technique to achieve 9600 bps full-duplex transmission on the switched telephone network that was based upon the use of echo cancellation technology. Two other techniques used by modem designers that simulate full-duplex transmission are asymmetric transmission and 'ping-pong' transmission. Both of these transmission techniques are discussed later in this section.

Using echo cancellation, both the sending and receiving modem use the same frequency, which would normally result in the occurrence of interference between transmitted and received signals. By the use of echo cancellation technology it becomes possible for the modem's receiver to cancel out the effect of its own transmitted signal, enabling the modem to obtain the ability to distinguish the received signal.

Echo cancellation is used in the ITU V.32, V.32 bis, V.33 and V.34 modems that are described later in this section. The V.32 modem uses QAM to encode 4 bits into one signal change or baud, operating at 2400 baud to provide 9600 bps full-duplex transmission through the use of echo cancellation. In addition, trellis coding of data is an optional mode of operation for V.32 modems which, when in effect, results in the probability of a bit error occurring less than that of most modems operating at significantly lower data rates than a V.32 modem.

Types of modems and features

Modems can be categorized based upon a large number of features, their physical construction, the type of data they can modulate, and the type of telephone facility they can operate upon. To become familiar with the basic types of modems, we will examine many of the more popular features that can be used to define different types of this communications device.

Mode of transmission

If the transmitter or the receiver of the modem is such that the modem can send or receive data in one direction only, the modem will function as a simplex modem. If the operations of the transmitter and receiver are combined so that the modem may transmit and receive data alternately, the modem will function as a half-duplex modem. In the half-duplex mode of operation, the transmitter must be turned off at one location, and the transmitter of the modem at the other end of the line must be turned on before each change in transmission direction. The time interval required for this operation is referred to as turnaround time. If the transmitter and receiver operate simultaneously, the modem will function as a full-duplex modem. This simultaneous transmission in both directions can be accomplished by the use of echo cancellation, splitting the telephone line's bandwidth into two channels on a two-wire circuit, or by the utilization of two two-wire circuits, such as are obtained on a four-wire leased line.

Transmission technique

Modems are designed for asynchronous or synchronous data transmission. Asynchronous transmission is also referred to as start-stop transmission and is usually employed by unbuffered terminal devices where the time between character transmission occurs randomly.

In asynchronous transmission, the character being transmitted is initialized by the character's start bit as a mark-to-space transition on the line and terminated by the character's stop bit which is converted to a 'space/marking' signal on the line. The digital pulses between the start and stop bits are the encoded bits which determine the type of character which was transmitted. Between the stop bit of one character and the start bit of the next character, the asynchronous modem places the line in the 'marking' condition. Upon receipt of the start bit of the next character the line is switched to a mark-to-space transition, and the modem at the other end of the line starts to sample the data.

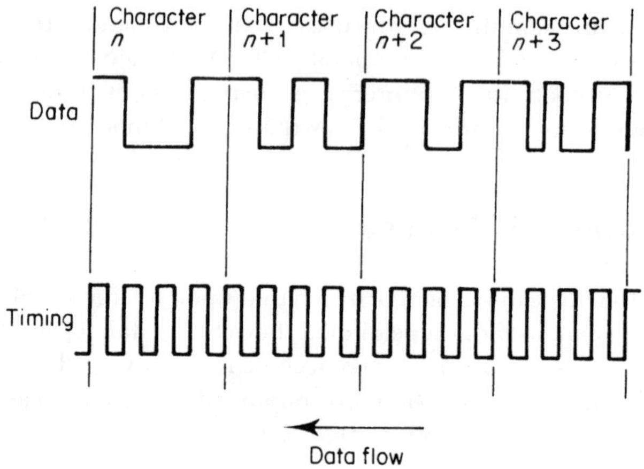

Figure 4.21 Synchronous timing signals. The timing signal is used to place the bits that form each character into a unique time period

The marking and spacing conditions are audio tones produced by the modulator of the modem to denote the binary data levels. These tones are produced at predefined frequencies, and their transition between the two states as each bit of the character is transmitted defines the character.

Synchronous transmission permits more efficient line utilization since the bits of one character are immediately followed by the bits of the next character, with no start and stop bits required to delimit individual characters. In synchronous transmission, groups of characters are formed into data blocks, with the length of the block varying from a few characters to a thousand or more. In synchronous transmission, the individual bits of each of the characters within each block are identified based upon a transmitted timing signal which is usually provided by the modem and which places each bit into a unique time period. This timing or clock signal is transmitted simultaneously with the serial bit stream as shown in Figure 4.21.

Line use classification

Modems can be classified into many categories to include the mode of transmission and transmission technique as well as by the application features they contain and the type of lines they are built to service. Generaly, modems can be classified into four line-servicing groups: subvoice or narrowband lines, voice-grade lines, wideband lines, and dedicated lines. Subvoice-band modems require only a portion of the voice-grade channel's available bandwidth and are commonly used with equipment operating at speeds up to 300 bps. On narrow-band facilities, modems can operate in the full-duplex mode by using one-half of the available bandwidth for transmission in each direction and use an asynchronous transmission technique.

Modems designed to operate on voice-grade facilities may be asynchronous or synchronous, half-duplex or full-duplex. Asynchronous transmission is normally employed at speeds up to and including 33.6 kbps. Although a leased, four-wire line will permit full-duplex transmission at high speeds without requiring the use of echo cancellation, transmission via the switched telephone network at high data rates requires the use of this technology to obtain a full-duplex operational capability.

Wideband modems, which are also referred to as group-band modems since a wideband circuit is a grouping of lower-speed lines, permit users to transmit synchronous data at speeds in excess of 48 800 bps. Although wideband modems are primarily used for computer-to-computer transmission applications, they are also used to service multiplexers which combine the transmission of many low- or medium-speed terminals to produce a composite higher transmission speed. Dedicated or limited-distance modems, which are also known by such names as shorthaul modems and modem bypass units, operate on dedicated solid conductor twisted-pair or coaxial cables, permitting data transmission at distances ranging up to 15 to 20 miles, depending upon the modem's operating speed and the resistance of the conductor.

Intelligence

Until the 1970s, modems could be categorized as dumb or non-intelligent devices. This categorization referenced their inability to perform different functions based upon a request initiated from a computer or terminal operator whose equipment was connected to the modem.

The incorporation of microprocessor technology and addition of random access memory (RAM), read only memory (ROM), and erasable programmable read only memory (EPROM) to modem circuit boards allowed manufacturers to add intelligence to their products. Through built-in routines in ROM and the ability to recognize and act upon commands and events, modems can perform such functions as automatically performing dialing operations, negotiating the method of modulation to be used to communicate with a distant modem, performing error detection and correction operations to insure data integrity, responding to status requests to report the 'state of its health', and many other functions.

Method of fabrication

Modems are manufactured in three different configurations—standalone, fabricated on adapter cards, and fabricated on rack-mount cards. The standalone modem is a self-contained device that includes one or more circuit boards contained in a common housing.

Modems fabricated on adapter cards are designed for insertion into the system unit of specific personal computers, such as an IBM PC or Apple Macintosh. Unlike an external modem that requires a separate power supply, the internal modem obtains its power from the personal computer. This eliminates the necessity of having an additional outlet for powering the modem.

Two additional differences between internal and external modems concern their cabling and desk space utilization. External modems must be cabled to a personal computer, whereas, an internal modem can use the bus in the system unit of the PC for data transfer between the modem and the computer. Concerning desk space or what many persons call footprint, since an internal modem fits into the system unit of a personal computer it requires no additional footprint or desk space.

Both internal and external modems are designed to operate as singular devices. In comparison, rack-mount modem cards are designed for insertion into a rack which is designed to provide both power and control to a common group of modems. Rack-mount modem cards are normally installed at data processing centers or at locations where communications equipment is located to support the transmission requirements of many users to a remote location, such as between dial-in lines and the ports on a multiplexer.

Reverse and secondary channels

To eliminate turnaround time when transmission is over the two-wire switched network or to relieve the primary channel of the burden of carrying acknowledgement signals on four-wire dedicated lines, modem manufacturers developed a reverse channel which is used to provide a path for the acknowledgement of transmitted data at a slower speed than the primary channel. This reverse channel can be used to provide a simultaneous transmission path for the acknowledgement of data blocks transmitted over the higher speed primary channel at up to 150 bps.

A secondary channel is similar to a reverse channel. It can, however, be used in a variety of applications which include providing a path for a high-speed terminal and a low-speed terminal. When a secondary channel is used as a reverse channel, it is held at one state until an error is detected in the high-speed data transmission and is then shifted to the other state as a signal for retransmission. Another application where a secondary channel can be utilized is when a location contains a high-speed, synchronous terminal and a slow-speed, asynchronous terminal such as a Teletype. If both devices are required to communicate with a similar distant location, one way to alleviate dual line requirements as well as the cost of extra modems to service both devices is by using a pair of modems that have secondary channel capacity, as shown in Figure 4.22. Although a reverse channel is usable on both two-wire and

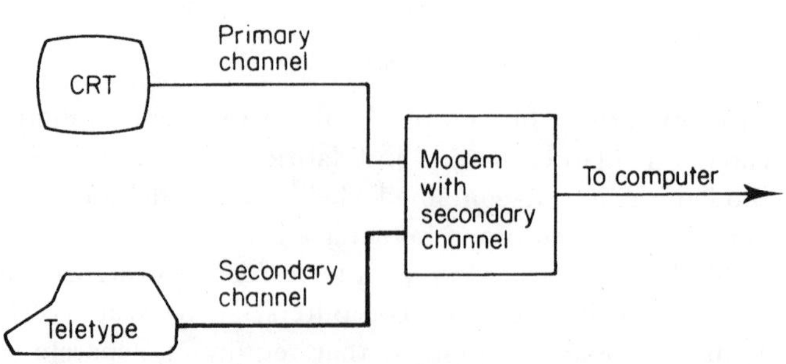

Figure 4.22 Secondary channel operation. Two terminals can communicate with a distant location by sharing a common line through the use of a modem with a secondary channel

four-wire telephone lines, the secondary channel technique is usable only on a four-wire circuit. A secondary channel modem derives two channels from the same line, a wide one to carry synchronous data and a narrow channel to carry asynchronous teletype-like data. Some modems with the secondary channel option can actually provide two slow-speed channels as well as one high-speed channel, with the two slow-speed channels being capable of transmitting asynchronous data up to a composite speed of 150 bps.

Equalization

Inconsistencies inherent in a transmission medium designed for voice rather than data transmission required modem manufacturers to build equalizers into their products to compensate for those inconsistencies produced by the telephone circuit, amplifiers, switches, and relays, as well as other equipment that data may be transmitted across in establishing a connection between two or more points. An equalizer is basically an inverse filter which is used to correct amplitude and delay distortion which, if uncorrected, could lead to intersymbol interference during transmission. A well designed equalizer matches line conditions by maintaining certain of the modem's electrical parameters at the widest range of marginal limits in order to take advantage of the data rate capability of the line while eliminating intersymbol interference. The design of the equalizer is critical, since if the modem operates too near or outside these marginal limits, the transmission error rate will increase. There are three basic methods for achieving equalization. The first method, the utilization of fixed equalizers, is typically accomplished by using marginally adjustable high-Q filter sections. Modems with transversal filters use a tapped delay line with manually adjustable variable tap gains, while automatic equalization is usually accomplished by a digital transversal filter with automatic tap gain adjustments. The faster the modem's speed, the greater the need for equalization and the more complex the equalizer. Most modems with rated speeds up to 4800 bps incorporate non-adjustable, fixed equalizers which have been designed to match the average line conditions that have been found to occur on the dial-up network. Most modems with fixed or non-adjustable equalizers are thus designed for a normal, randomly routed call between two locations over the dial-up network. If the modem is equipped with a signal-quality light which indicates an error rate that is unacceptable, or if the operator encounters difficulty with the connection, the problem can be alleviated by simply disconnecting the call and dialing a new call, which should reroute the connection through different points on the dial-up network.

Manual equalization

Manually adjusted equalization was originally employed on some 4800 bps modems used for transmission over leased lines, with the parameters being tuned or preset at installation time, and re-equalization usually not required unless the lines are reconfigured. Primarily designed to operate over unconditioned leased telephone lines, manually equalized modems allow the user to eliminate the monthly expense associated with line conditioning. Due to the incorporation of microprocessors into modems for signal processing they were soon employed to perform automatic equalization. This resulted in most modern modems incorporating automatic equalization.

Automatic equalization

Automatic equalization is used on most 4800 bps and above modems designed for operation over the switched telephone network and on all 7200 bps and above modems which are primarily designed to operate over dedicated lines but which can operate over the switched network in a fallback operational mode. With automatic equalization, a certain initialization time is required to adapt the modem to existing line conditions. This initialization time becomes important during and after line outages, since long initial equalization times can extend otherwise short dropouts unnecessarily. Recent modem developments have shortened the initial equalization time to between 15 and 25 ms, whereas only a few years ago a much longer time was commonly required. After the initial equalization, the modem continuously monitors and compensates for changing line conditions by an adaptive process. This process allows the equalizer to 'track' the frequently occurring line variations that occur during data transmission without interrupting the traffic flow. On one 9600 bps modem, this adaptive process occurs 2400 times a second, permitting the recognition of variations as they occur.

Synchronization

For synchronous communications, generally in speeds exceeding 1800 bps, the start–stop bits characteristic of asynchronous communications can be eliminated. Bit synchronization is necessary so that the receiving modem samples the link at the exact moment that a bit occurs. The receiver clock is supplied by the modem in phase coherence with the incoming data bit stream, or more simply stated, tuned to the exact speed of the transmitting clock. The transmitting clock can be supplied by either the modem (internal) or the terminal (external).

The transmission of synchronous data is generally under the control of a master clock which is the fastest clock in the system. Any slower data clock rates required are derived from the master clock by digital division logic, and those clocks are referred to as slave clocks. For instance, a master clock oscillating at a frequency of 96 kHz could be used to derive 9.6 kpbs (1/10), 4.8 kbps (1/20), and 2.4 kbps (1/40) clock speeds.

Multiport capability

Modems with a multiport capability offer a function similar to that provided by a multiplexer. In fact, multiport modems contain a limited function time division multiplexer (TDM) which provides the user with the capability of transmitting

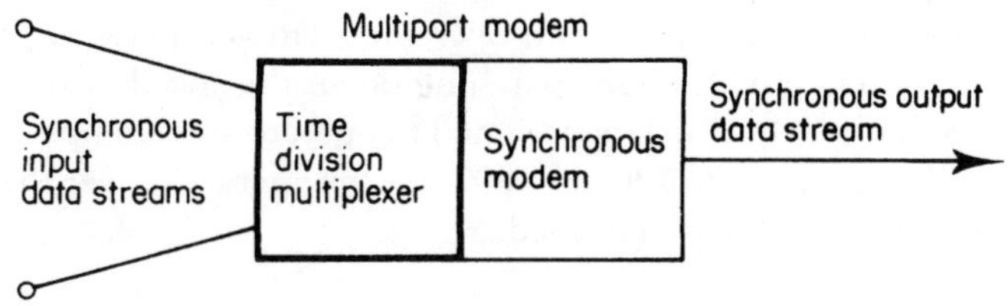

Figure 4.23 Multiport modem. Containing a limited function time division multiplexer, a multiport modem combines the input of a few synchronous input data streams for transmission

more than one synchronous data stream over a single transmission line, as shown in Figure 4.23.

In contrast with typical multiplexers, the limited function multiplexer used in a multiport modem combines only a few high-speed synchronous data streams, whereas multiplexers can normally concentrate a mixture of asynchronous and synchronous, high- and low-speed data streams. A further description, as well as application examples, will be found in Section 4.4.

Security capability

To provide an additional level of network protection for calls originated over the PSTN several vendors market 'security' modems. Some of these modems contain a buffer area into which a network administrator enters authorized passwords and associated telephone numbers. When a potential network user dials the telephone number assigned to the security modem that device prompts the person to enter his password. If a valid match between the entered and previously stored password occurs the security modem disconnects the line and then dials the telephone number associated with the password. A modem with a security capability thus provides a mechanism to verify the originator of calls over the PSTN by his telephone number. A second type of security modem verifies the location of the call originator. To do so this type of modem is first configured by associating callback telephone numbers to passwords. Once a caller establishes a connection with this type of security modem and enters his or her password, the distant modem breaks the connection and initiates a callback using the telephone number associated with the password. Based upon this type of operation, this feature is referred to as callback security. A further description of security modems is contained in Section 4.6.

Multiple speed selection capability

For data communication systems which require the full-time service of dedicated lines but need to access the switched network if the dedicated line should fail or degrade to the point where it cannot be used, dial backup capability for the modems used is necessary. Since transmission over dedicated lines usually occurs at a higher speed than one can obtain over the switched network, one method of facilitating dial backup is through switching down the speed of the modem. Thus, a multiple speed modem which is designed to operate at 9600 bps over dedicated lines may be switched down to 7200 or 4800 bps for operation over the dial-up network until the dedicated lines are restored.

Voice/data capability

Many modems can now be obtained with a voice/data option which provides the user with a voice communication capability over the same line which is used for data transmission. Depending upon the modem, this voice capability can be either alternate voice/data or simultaneous voice/data. The user may thus communicate with a distant location at the same time as data transmission is occurring, or the user may transmit data during certain times of the day and use the line for voice communications at other times. Voice/data capability can also be used to minimize normal telephone charges when data transmission sequences require voice coordination.

Modem operations and compatibility

Many modem manufacturers describe their product offerings in terms of compatibility or equivalency with modems manufactured by Western Electric for the Bell System, prior to its breakup into independent telephone companies, or with International Telecommunications Union (ITU) recommendations. The International Telecommunications Union, which is based in Geneva, is an agency of the United Nations which developed a series of modem standards for recommended use. These recommendations are primarily adapted by the Post, Telephone and Telegraph (PTT) organizations that operate the telephone networks of many countries outside the United States; however, owing to the popularity of certain ITU recommendations, they have also been followed in designing certain modems for operation on communications facilities within the USA. The following examination of the operation and compatibility of the major types of Bell System and ITU modems is based upon their operating rate.

300 bps

Modems operating at 300 bps use a frequency shift keying (FSK) modulation technique as previously described during the discussion of acoustic couplers in Section 4.1. In this technique the frequency of the carrier is alternated to one of two frequencies, one frequency representing a space or zero bit while the other frequency represents a mark or a one bit. Table 4.3 lists the frequency assignment for Bell System 103/113 and ITU V.21 modems which represent the two major types of modems that operate at 300 bps.

Table 4.3 Frequency assignments (Hz) for 300 bps modems

Major modem types	Originate	Answer
Bell System (103/113 type)	Mark 1270 Space 1070	2225 2025
ITU V.21	Mark 980 Space 1180	1650 1850

Bell System 103 and 113 series modems are designed so that one channel is assigned to the 1070–1270 Hz frequency band while the second channel is assigned to the 2025–2225 Hz frequency band. Modems that transmit in the 1070–1270 Hz band but receive in the 2025–2225 Hz band are designated as originate modems, while a modem which transmits in the 2025–2225 Hz band but receives in the 1070-1270 Hz band is designated as an answer modem. When using such modems, their correct pairing is important, since two originate modems cannot communicate with each other.

Bell System 113A modems are originate only devices that should normally be used when calls are to be placed in one direction. This type of modem is mainly used to enable teletype-compatible terminals to communicate with time-sharing systems where such terminals only originate calls. Bell System 113B modems are answer only and are primarily used at computer sites where users dial in to establish communications.

Modems in the 103 series, which includes the 103A, E, F, G and J modems, can transmit and receive in either the low or the high band. This ability to switch modes is denoted as 'originate and answer', in comparison to the Bell 113A which operates only in the originate mode and the Bell 113B which operates only in the answer mode.

As indicated in Table 4.3, modems operating in accordance with the ITU V.21 recommendation employ a different set of frequencies for the transmission and reception of marks and spaces. Thus, Bell System 103/113 type modems and V.21 devices can never communicate with one another.

The Bell System 103/113 type modem was used extensively in North America, and V.21 devices are primarily used in Europe. This incompatibility explains many problems that transportable and lap-top computer users experience when traveling internationally. An American traveling to Europe with a Bell 103/113 type modem built into the computer is thus precluded from communicating at 300 bps with modems in Europe. Similarly, Europeans traveling to North America with a V.21 type modem in their computers cannot communicate at 300 bps, because modems in North America normally are Bell System 103/113 compatible at that data rate.

The two pairs of frequencies used by the modems listed in Table 4.3 permit the bandwidth of a communications channel to be split into two subchannels by frequency. This technique was illustrated in Figure 4.2 for Bell System 103/113 modems. Since each subchannel can permit data to be transmitted in a direction opposite that transmitted on the other subchannel, this technique permits full-duplex transmission to occur on the switched telephone network which is a two-wire circuit that normally can only support half-duplex transmission.

Echo suppression

One of the problems associated with designing full-duplex modems for operation on the switched telephone network is the effect of echo suppressors upon modem operations. To understand how modem designers overcome the effect of echo suppressors, let us first examine why they are required. Then we can note how modem designers can disable the echo suppressors in the path established by a PSTN call as well as the effect of echo suppression upon both full- and half-duplex modems.

Although the switched telephone network is considered to be a two-wire network, in actuality, the routing of each call that goes through more than one telephone office will be via both two-wire and four-wire paths. Each local loop from a subscriber to the local telephone company serving office is a two-wire path; however, each connection between telephone offices is a four-wire path as illustrated in Figure 4.24.

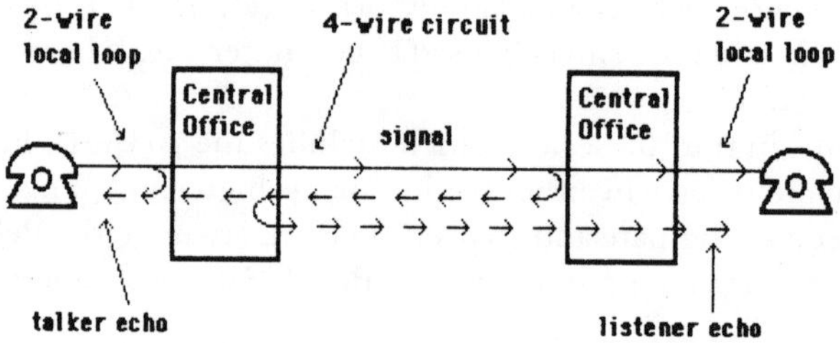

Figure 4.24 Talker and listener echoes

The actual conversion of signals from a two-wire circuit to a four-wire circuit results in an impedance mismatch. This impedance mismatch causes a portion of signal energy to be reflected back toward the originator and is referred to as 'talker', 'local', or 'near-end' echo. If this echo encounters another impedance mismatch as it flows back to the originator, another echo will be produced which now flows in the same direction as the desired signal. This doubly reflected echo, which would be heard by the listener in a telephone conversation, is called a 'listener' echo.

To minimize the effect of echoes, communications carriers added echo suppressors to their network. Echo suppressors are signal-activated devices which, unless disabled, insert a high degree of attenuation in the return echo path during the time a signal flows in the opposite path. While the use of echo suppressors provides a better quality voice circuit, their operation normally limits data transmission to one direction at a time. In addition, since they require a small amount of time to disable every time the direction of transmission is reversed, their suppression in a half-duplex environment will adversely affect data transmission throughput.

Disabling echo suppressors

One method to obtain full-duplex transmission on the PSTN is to use different frequencies for each transmission direction, similar to that described for FSK 300 bps modems. To accomplish simultaneous transmission in both directions requires the disabling of echo suppressors that lie in the switched network path established for the call. Echo suppressor disabling is accomplished by modems generating a signal in the 2010-2240 Hz band for at least 400 ms. Once echo suppressors have been disabled by an echo suppressor tone (usually 2100 Hz in North America), the signal energy of the modulated carrier in the 300-3300 Hz band in either direction of transmission is sufficient to keep the suppressors disabled as long as power is not interrupted for more than a period of 100 ms.

300 to 1800 bps

There are several Bell System and ITU V series modems that operate in the range between 300 to 1800 bps. Some of these modems such as the Bell System 212A and ITU V.22 devices can operate at either of two speeds; and other modems such as the Bell System 202 and the ITU V.23 only operate at one data rate. We will examine these modems in pairs, enabling their similarities and differences to be compared.

Bell System 212A and V.22 modems

In the late 1970s and early 1980s, the Bell System 212A and ITU V.22 modems represented the largest base of installed devices used for data transmission on the switched telephone network. Although both types of modems were rapidly replaced by more modern and higher operating rate devices, they can be expected to remain in use at many locations through the 1990s. Most higher-speed modems manufactured today are thus downward compatible with Bell System 212A and ITU V.22 devices.

The Bell System 212A modem permits either asynchronous or synchronous transmission over the public switched telephone network. The 212A contains a 103-type modem for asynchronous transmission at speeds up to 300 bps. At this data rate FSK modulation is employed, using the frequency assignments previously indicated in Table 4.3. At 1200 bps, differential phase shift keyed (DPSK) modulation is used which permits the modem to operate either asynchronously or synchronously. The phase shift encoding of the 212A type modem is illustrated in Table 4.4.

The actual phase shifts listed in Table 4.4 occur with respect to the phase angle of the previous phase shift. Thus, the name for this modulation technique is differential phase-shift keying or DPSK. In comparison, since no phase shift occurs at

Table 4.4 212A type modem phase shift encoding

Dibit	Phase shift (degrees)
00	90
01	0
10	180
11	270

300 bps the FSK modulation scheme used at that data rate is sometimes referred to as continuous phase FSK modulation.

Although a 212A type modem is designed to accept asynchronous data, transmission between two 212As at 1200 bps is actually synchronous. For this reason a 212A modem is not transparent to the binary data stream and must therefore be able to be set to support the correct character length.

All 212A type modems operating at 1200 bps support 9- and 10-bit characters, while some vendor modems also support 8- and 11-bit characters. Since the character bit length includes start and stop bits, the universal support of 10-bit characters permits the transmission of 7-bit ASCII plus parity and 8-bit extended ASCII and EBCDIC characters by all 212A modems.

When operating at 1200 bps the 212A modem uses 1200 Hz originate and 2400 Hz answer carrier frequencies. This enables the 3000 Hz bandwidth of the telephone channel to be subdivided into two channels by frequency, enabling full-duplex transmission to occur on the PSTN.

One advantage in the use of this modem is that it was the first device developed to permit the reception of transmission at two different transmission speeds. Before the operator initiates a call, he or she selects the operating speed at the originating set. The manner in which the operating speed is selected depends upon the type of 212A modem used. If the modem is what is now commonly referred to as a 'dumb' modem the operator selects the higher operating speed by pressing a 'HS' (high speed) button on the front panel of the modem. If the modem is an intelligent modem built to respond to software commands the operator can either use a communications program or send a series of commands through the serial port of a personal computer or terminal connected to the modem to set its operating speed. Due to the substantial use of intelligent modems with personal computers these modems will be reviewed as a separate entity later in this chapter. When the call is made, the answering 212A modem automatically switches to that operating speed. During data transmission, both modems remain in the same speed mode until the call is terminated, when the answering 212A can be set to the other speed by a new call. The dual-speed 212A permits both terminals connected to Bell System 100 series data sets operating at up to 300 bps or terminals connected to other 212A modems operating at 1200 bps to share the use of one modem at a computer site and thus can reduce central computer site equipment requirements.

The V.22 standard is for modems that operate at 1200 bps on the PSTN or leased circuits and has a fallback data rate of 600 bps. The modulation technique

Table 4.5 V.22 modulation phase shift as opposed to bit patterns

Dibit values 1200 bps)	Bit values (600 bps)	Phase change modes 1,2,3,4	Phase change mode 5
00	0	90	270
01	—	0	180
11	1	270	90
10	—	180	0

employed is 4-phase PSK at 1200 bps and 2-phase PSK at 600 bps, with five possible operational modes specified for the modem at 1200 bps. Table 4.5 lists the V.22 modulation phase shifts with respect to the bit patterns entering the modem's transmitter. Modes 1 and 2 are for synchronous and asynchronous data transmission at 1200 bps respectively, while mode 3 is for synchronous transmission at 600 bps. Mode 4 is for asynchronous transmission at 600 bps while mode 5 represents an alternate phase change set for 1200 bps asynchronous transmission.

In comparing V.22 modems to the Bell System 212A devices it should be apparent that they are totally incompatible at the lower data rate, since both the operating speed and modulation techniques differ. At 1200 bps the modulation techniques used by a V.22 modem in modes one through four are exactly the same as that used by a Bell System 212A device. Unfortunately, a Bell 212A modem that answers a call sends a tone of 2225 Hz on the line that the originating modem is supposed to recognize. This frequency is used because of the construction of the switched telephone network in the United States and other parts of North America. Under V.22, the answering modem first sends a tone of 2100 Hz since this frequency is more compatible with the design of European switched telephone networks. Then, the V.22 modem sends a 2400 Hz tone that would not be any better except that the V.22 modem also sends a burst of data whose primary frequency is about 2250 Hz, which is close enough to the Bell standard of 2225 Hz that many Bell 212A-type modems will respond. Thus, some Bell 212A modems can communicate with V.22 modems at 1200 bps while other 212-type modems may not be able to communicate with V.22 devices, with the ability to successfully communicate being based upon the tolerance of the 212 type modem to recognise the V.22 modem's data burst at 2250 Hz.

Bell System 202 series modems

Bell System 202 series modems were introduced during the early 1960s and provided users with high-speed transmission for their time. Today most 202 type modems have been replaced by the use of higher operating devices, however, some third-party vendors continue to manufacture 202 compatible products.

Bell System 202 series modems are designed for speeds up to 1200 or 1800 bps. One model in the 202 series, the 202C modem, can operate on either the switched network or on leased lines, in the half-duplex mode on the former and the full-duplex mode on the latter. The 202C modem can operate half-duplex or full-duplex on leased lines. This series of modems uses frequency shift keyed (FSK) modulation, and the frequency assignments are such that a mark is at 1200 Hz and a space at 2200 Hz. When either modem is used for transmission over a leased four-wire circuit in the full-duplex mode, modem control is identical to the 103 series modem in that both transmitters can be strapped on continuously which alleviates the necessity of line turnarounds.

Since the 202 series modems do not have separate bands, on switched network utilization half-duplex operation is required. This means that both transmitters (one in each modem) must be alternately turned on and off to provide two-way communication.

The time required to turn off the modem's transmitter and turn on its receiver is referred to as its turnaround delay time. While not significant on an individual basis, when transmitting and receiving small amounts of data the cumulative turnaround delay time will adversely affect performance. The use of 202 series modems on the switched network has thus been essentially replaced by 212A, V.22 bis, and other higher operating rate full-duplex modems whose use eliminates adverse performance due to turnaround delays.

The Bell 202 series modems have a 5 bps reverse channel for switched network use, which employs amplitude modulation for the transmission of information. The channel assignments used by a Bell System 202 type modem are illustrated in Figure 4.25, where the 387 Hz signal represents the optional 5 bps AM reverse channel. Owing to the slowness of this reverse channel, its use is limited to status and control function transmission. Status information such as 'ready to receive data' or 'device out of paper' can be transmitted on this channel. Also owing to the slow transmission rate, error detection of received messages and an associated NAK and request for retransmission is normally accomplished on the primary channel since even with the turnaround time, it can be completed at almost the same rate one obtains in using the reverse channel for that purpose. Non-Bell 202-equivalent modems produced by many manufacturers provide reverse channels of 75 to 150 bps which can be utilized to enhance overall system performance. Reverse keyboard-entered data as well as error detection information can be practically transmitted over such a channel.

While a data rate of up to 1800 bps can be obtained with the 202D modem, transmission at this speed requires that the leased line be conditioned for transmission by the telephone company. The 202S and 202T modems are additions to

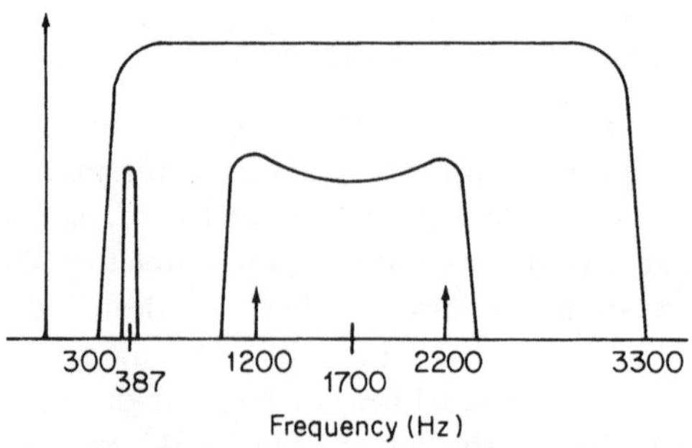

Figure 4.25 Bell System 202-type modem channel assignments

the 202 series and are designed for transmission at 1200 and 1800 bps over the switched network and leased lines, respectively. At speeds in excess of 1400 bps, the 202T requires line conditioning when interfaced to either two- or four-wire circuits, whereas for a two-wire circuit, conditioning is required at speeds in excess of 1200 bps when an optional reverse channel is used.

V.23 modems

The V.23 standard is for modems that transmit at 600 or 1200 bps over the PSTN. Both asynchronous and synchronous transmission is supported by using FSK modulation; and, an optional 75 bps backward or reverse channel can be used for error control. Figure 4.26 illustrates the channel assignments for a V.23 modem. In comparing Figure 4.26 with Figure 4.25, it is obvious that Bell System 202 and V.23 modems are incompatible with each other.

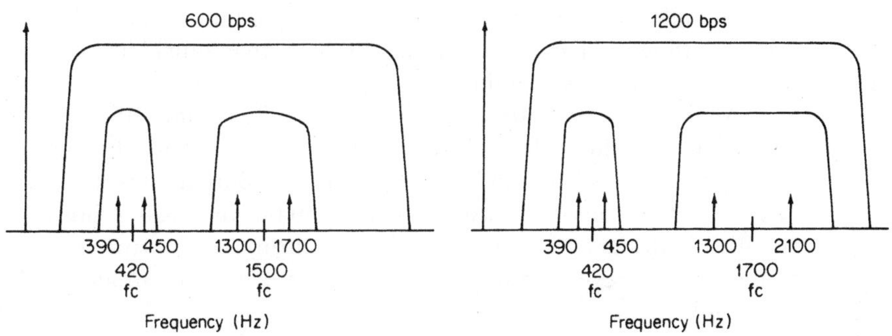

Figure 4.26 V.23 channel assignments

The V.23 reverse channel uses a frequency of 390 Hz to represent a binary 1 and a frequency of 450 Hz to represent a binary 0. In this modem's first mode of operation, which governs transmission up to 600 bps, a binary 1 is represented by a 1300 Hz tone while a binary 0 is represented by a 1700 Hz tone. In the modem's second mode of operation, which governs data transmission at 1200 bps, a binary 1 is represented by a frequency of 1300 Hz while a binary 0 is represented by a frequency of 2100 Hz.

Unlike the Bell System 202 that has been essentially replaced by more modern modems, the 75 bps reverse channel of the V.23 modem has prolonged its life as it makes the modem well suited to support many applications, including Videotext transmission requirements in Europe. When used in this manner, a V.23 modem is built into television sets and special terminal devices. The reverse channel which operates at 75 bps is used to transmit screen selection information to the computer controlling the Videotext system. Since the screen selection information may consist of a few alphanumeric characters, the 75 bps data rate is normally more than sufficient to support the transmission of data to the computer. The primary channel which operates at 1200 bps is then used as a reverse channel with respect to the requester providing the transmission path for data sent from the computer to the user's screen.

2400 bps

Examples of modems that operate at 2400 bps include the Bell System 201, the ITU V.26 series, and the V.22 bis modem. The Bell System 201 and ITU V.26 series modems are designed for synchronous bit serial transmission at a data rate of 2400 bps, while the V.22 bis standard governs 2400 bps asynchronous transmission.

Bell System 201B/C

The Bell System 201 series was originally introduced during the mid-1960s to support what was then considered to be high-speed synchronous data transmission. Although 201 modems are no longer manufactured by Western Electric, compatible modems are manufactured by that subsidiary of AT&T as well as by several third-party vendors.

Members of the 201 series include the 201B and 201C models. Both of these modems use dibit phase shift keying modulation, with the phase shifts based upon the dibit values listed in Table 4.6. The 201B modem is designed for half- or full-duplex synchronous transmission at 2400 bps over leased lines. In comparison, the 201C is designed for half-duplex, synchronous transmission over the PSTN. A more modern version of the 201C is AT&T's 2024A modem, which is compatible with the 201C.

Table 4.6 Bell System 201 B/C phase as opposed to bit pattern

Dibit values	Phase shift (degrees)
00	225
10	315
11	45
10	135

V.26 modem

The V.26 standard specifies the characteristics for a 2400 bps synchronous modem for use on a four-wire leased line. Modems operating according to the V.26 standard employ differential phase shift keying, using one of two recommended coding schemes. The phase change based upon the dibit values for each of the V.26 coding schemes is listed in Table 4.7. The use of each phase change pattern results in a modulation rate of 1200 baud providing a data signaling rate of 2400 bps. Under the V.26 standard a reverse channel can be used. This channel has the same specifications as the V.23 reverse channel.

Table 4.7 V.26 modulation phase as opposed to bit pattern

	Phase change	
Dibit values	Pattern A	Pattern B
00	0	45
01	90	135
11	180	225
10	270	315

Two similar recommendations to V.26 are V.26 bis and V.26 ter. The V.26 bis recommendation defines a dual speed 2400/1200 bps modem for use on the PSTN. At 2400 bps the modulation and coding method is the same as the V.26 recommendation for pattern B listed in Table 4.7. At the reduced data rate of 1200 bps a two-phase shift modulation scheme is employed, with a binary zero represented by a 90° phase shift while binary one is represented by a 270° phase shift. The V.26 bis recommendation also includes an optional reverse or backward channel that can be used for data transfer up to 75 bps. When employed, frequency shift keying is used to obtain this channel capacity, with a mark or one bit represented by a 390 Hz signal and a space or zero bit represented by a 450 Hz signal.

The V.26 ter recommendation uses the same phase shift scheme as the V.26 modem, but incorporates an echo-canceling technique that allows transmitted and received signals to occupy the same bandwidth. The V.26 ter modem is thus capable of operating in full duplex at 2400 bps on the PSTN. Echo canceling will be described later in this chapter when the V.32 modem is examined. Although the V.26 ter modem is popular in France, its use has been superseded by the V.22 bis modem in most countries.

V.22 bis

The ITU V.22 bis recommendation governs modems designed for 2400 bps full-duplex transmission on the PSTN and two-wire leased lines. When operating at 2400 bps, a V.22 bis modem can accept either asynchronous or synchronous data; however, transmission between modems occurs synchronously using quadrature amplitude modulation at 600 baud. Since each baud represents four bits, this results in a 2400 bps data rate.

Similar to 212A and V.22 modems, a V.22 bis modem uses carrier frequencies of 1200 and 2400 Hz for each channel obtained by frequency division. Unlike those modems that use DPSK modulation, the V.22 bis when operating at 2400 bps uses QAM. In this modulation technique the data to be transmitted is divided into groups of four consecutive bits known as quadbits. The first two bits are encoded as a phase change relative to the quadrant occupied by the preceding signal element. Table 4.8 indicates these phase quadrant changes. The last two bits of each quadbit define one of four signaling elements associated with each new quadrant. Figure 4.27 illustrates the signal constellation of a V.22 bis modem in which all possible

Table 4.8 V.22 bis dibit to phase quadrant change encoding

First two bits in quadbit (2400 bps) or dibit values (1200 bps)	Phase quadrant change (degrees)	
00	1→2 2→3 3→4 4→1	90
01	1→1 2→2 3→3 4→4	0
11	1→4 2→1 3→2 4→3	270
10	1→3 2→4 3→1 4→2	180

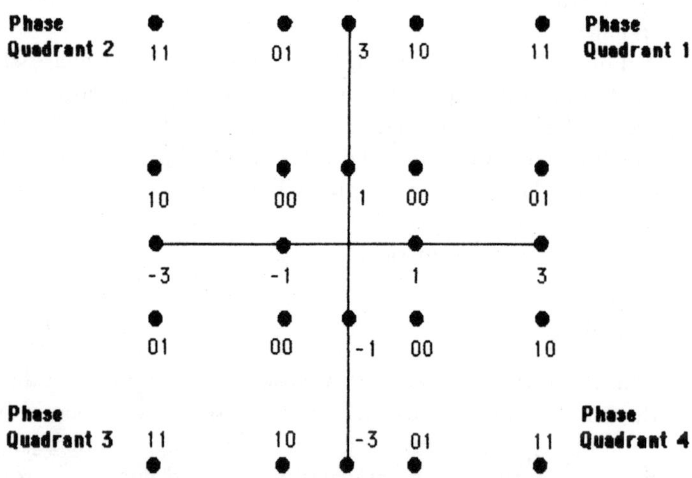

Figure 4.27 V.22 bis signal constellation

signal points are shown. Note that the dibit values in each phase quadrant represent the last two bits in a quadbit.

When operating at 1200 bps a V.22 bis modem follows the modulation scheme used by a V.22 device. That is, it uses DPSK modulation in which the value of each dibit is used to generate a phase change relative to the previous phase change as indicated in Table 4.5.

Since V.22 bis defines operations at 1200 bps to follow the V.22 format, when using a V.22 bis modem manufactured in Europe communications capability with

Bell System 212A type modems at that data rate may not always be possible due to the answer tone incompatibility usually encountered between modems following Bell System specifications and ITU recommendations. In addition, V.22 bis modems manufactured in the United States are usually not compatible with such modems manufactured in Europe at fallback data rates. This is because V.22 bis modems manufactured in Europe that are fully compliant with the recommendation only support four modes of operation to include asynchronous and synchronous transmission at 2400 and 1200 bps. This precludes the operation of those modems at 300 bps. At 1200 bps, the incompatibility between most European telephone networks (which are designed to accept only 2100 Hz answer tones) and the US telephone network (which usually accepts an answer tone between 2100 and 2225 Hz) may preclude communications between a US and a European manufactured V.22 bis modem at 1200 bps.

Since most US manufactured V.22 bis modems are fully compatible with Bell System 212A devices, this means that such modems can also operate at 300 bps using FSK modulation. At that data rate, a US manufactured V.22 bis modem would be incompatible with a European manufactured V.22 bis modem.

In spite of the previously mentioned problems, V.22 bis modems became a *de facto* standard for use with personal computers communicating over the PSTN during the early 1990s. This is due to several factors, to include the manufacture in the United States of V.22 bis modems that are Bell System 212A compatible, permitting PC users with such modems to be able to communicate with other PCs and mainframe computers connected to either 212A or 103/113 type modems. In addition, at 2400 bps, US V.22 bis modems can communicate with European V.22 bis modems, in effect, providing worldwide communications capability over the PSTN.

4800 bps

Modems that operate at 4800 bps were originally considered high-speed devices. In the late 1990s this operating rate is more representative of a low to medium speed.

The two modems examined in this section represent devices that were originally marketed during the early 1970s. Although the modulation method employed by both modems covered in this section can be considered as dated, the use of each modem can be expected to continue in a limited manner for the foreseeable future based upon reasons explained later in this section.

Bell System 208 and V.27 modems

The Bell System 208 series and ITU V.27 modems represent the most common types of modems designed for synchronous data transmission at 4800 bps. The Bell System 208 Series modems use a quadrature amplitude modulation technique. The 208A modem is designed for either half-duplex or full-duplex operation at 4800 bps over leased lines. The 208B modem is designed for half-duplex operation at 4800 bps on the switched network. The V.27 modem is designed for full- or half-duplex operations on leased lines. Later versions of the 208A were offered by

AT&T as the 2048A and 2048C models, which were designed for four-wire leased line operation. The 2048C has a start-up time less than one half of the 2048A, which makes it more suitable for operations on multidrop lines. This is because a requirement by any remote modem in a multidrop network for retraining can be satisfied quicker than by the use of a 2048A modem. Other third-party vendors introduced 208B type modems with rapid equalization times that provide enhanced throughput capability for switched network transmission. This primarily occurs during interactive transmission since the half-duplex operation of the 208B on the switched network results in numerous line turnarounds. By shortening the time required for the modems to re-equalize each time the direction of transmission changes the modem can begin to modulate data quicker after a line turnaround, in effect, boosting data throughput.

Both Bell 208 type modems and ITU V.27 modems pack data three bits at a time, encoding them for transmission as one of eight phase angles. Unfortunately, since each type of modem uses different phase angles to represent a tribit value, they cannot talk to each other. Table 4.9 lists the V.27 modulation phase shifts with respect to each of the eight possible tribit values.

Table 4.9 V.27 modulation phase drift versus bit pattern

Tribit values	Phase change (degrees)
001	0
000	45
010	90
011	135
111	180
110	225
100	270
101	315

During the 1970s to the mid-1980s, the primary use of 208 and V.27 modems was to support the transmission requirements of synchronous remote batch terminals (RBTs) that were used to communicate with mainframe computers. The use of RBTs allowed users located remote from a mainframe to enter jobs for processing as well as pull system output (Sysout) to the RBT's printer. In addition, many RBTs had tape and disk storage that allowed jobs to be batched for transmission as well as for Sysout to be queued to tape or disk for printing if the printer was being used for local processing.

From the early 1980s, personal computers began to be used as RBT replacements. Several manufacturers introduced synchronous communications adapter cards for installation in the system unit of PCs which allow them to be connected to synchronous modems, such as Bell System 208 and ITU V.27 devices. Other manufacturers offer internal 208 and V.27 type modems which permit PCs to communicate synchronously at 4800 bps over the PSTN or on a leased line by simply connecting a modular plug into the jack built into the rear of the adapter card.

In the late 1980s several third-party vendors introduced 208 type modems with enhanced capabilities. One model, which is manufactured by several vendors, is a 208A/B device that can be used on either the PSTN or on leased lines by simply changing a connector in the modem. A second 208 type modem manufactured by third-part vendors can be obtained with an optional asynchronous to synchronous converter. Through the use of this converter, which also functions as a speed converter, you can use the modem to transmit asynchronous data rates of 1200, 2400 and 4800 bps.

9600 bps

Three common modems that are representative of devices that operate at 9600 bps are the Bell System 209, and the ITU V.29 and V.32 modems.

Bell System 209 and V.29 modems

Modems equivalent to the Bell System 209 and ITU V.29 devices are designed to operate in a full-duplex, synchronous mode at 9600 bps over private lines. The Bell System 209A modem operates by employing a quadrature amplitude modulation technique as previously illustrated in Figure 4.15. Included in this modem is a built-in synchronous multiplexer which will combine up to four data rate combinations for transmission at 9600 bps. The multiplexer combinations are shown in Table 4.10. The use of a multiplexer incorporated into a modem is discussed more thoroughly in Section 4.4. A newer version of the 209A that was offered by AT&T is the 2096A. This modem is noteworthy because it has an EIA RS-449/423 interface with RS-232-C/D compatibility.

Table 4.10 Bell 209A multiplexer combinations

2400–2400–2400–2400 bps
4800–2400–2400 bps
4800–4800 bps
7200–2400 bps
9600 bps

With the exception of Bell System 209-type modems, a large majority of 9600 bps devices manufactured throughout the world adhere to the ITU V.29 standard. The V.29 standard governs data transmission at 9600 bps for full- or half-duplex operation on leased lines, with fallback data rates of 7200 and 4800 bps allowed. At 9600 bps the serial data stream is divided into groups of four consecutive bits. The first bit in the group is used to determine the amplitude to be transmitted while the remaining three bits are encoded as a phase change, with the phase changes identical to those of the V.27 recommendation listed in Table 4.9.

Table 4.11 lists the relative signal element amplitude of V.29 modems, based upon the value of the first bit in the quadbit and the absolute phase which is determined from bits two through four. Thus, a serial data stream composed of the bits 1 1 0 0 would have a phase change of 270° and its signal amplitude would be 5. The resulting signal constellation pattern of V.29 modems is illustrated in Figure 4.28.

Table 4.11 V.29 signal amplitude construction

Absolute phase (degrees)	First bit	Relative signal element amplitude
0, 90, 180, 270	0	3
	1	5
45, 135, 225, 315	0	$\sqrt{2}$
	1	$3\sqrt{2}$

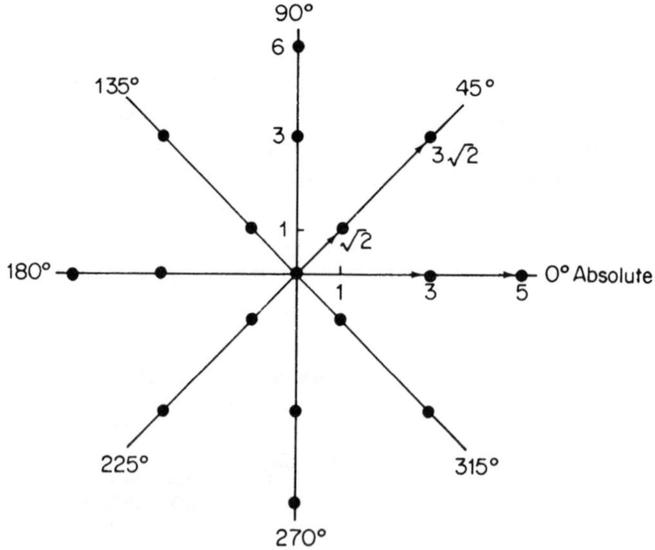

Figure 4.28 V.29 signal constellation pattern

V.29 variants

The basic chip set which provides the functionality of a V.29 modem, which is called a V.29 data pump, is used by many vendors as the basis for developing non-standard modems. Some of these non-standard modems are designed to simulate full-duplex transmission when used on the PSTN by the formation of asymmetrical channels or implementation of a 'ping-pong' scheme is described later in this section. When an asymmetrical channel scheme is used the primary channel employs QAM while the secondary channel uses FSK modulation. One channel will thus operate at 9600 bps while the other channel normally operates at 300 bps.

The primary use of asymmetrical and 'ping-pong' modulation schemes based upon the V.29 data pump is to provide an economical high-speed data transmission capability for personal computer users communicating over the PSTN. Since the vast majority of personal computers have a built-in asynchronous serial port, to support asynchronous data the modified V.29 modems include an asynchronous to synchronous converter. This allows a modem originally designed for synchronous transmission to be used to support asynchronous data transmission requirements.

The driving force behind the development of non-standard modems based upon the use of V.29 data pumps was economics. When these modems reached the market in the late 1980s they had a retail price ranging between $995 and $1295. In comparison, the only true full-duplex 9600 bps modem designed for use on the PSTN, the V.32 modem, had a retail price exceeding $3500. The retail price of V.32 modems significantly declined to the point where their cost became close to V.29 data pump based modems. Due to the significant reduction in the cost differential between V.29 data pump based modems and V.32 modems, the major rationale was removed for the selection of the former category of communications equipment. A secondary reason for the failure of V.29 data pump based modems to achieve a significant installed base resulted from their incompatibility with other vendor products when operating in their 'fast' or high-speed transmission mode. This was because the techniques used to implement asymmetrical transmission or the 'ping-pong' technique differed among vendors, in effect, forcing users to communicate at a lower speed as a V.22 bis device if they wanted to communicate with a modem manufactured by a different vendor.

Although V.29 based asymmetrical modems failed to obtain the market they were developed for, their transmission concept is incorporated into several types of modern digital subscriber line modems which are described in detail later in this chapter.

V.32 modem

V.32 is based upon a modified quadrature amplitude modulation technique and was designed to permit full-duplex 9600 bps transmission over the switched telephone network.

A V.32 modem establishes two high-speed channels in the opposite direction of one another as illustrated in Figure 4.29. Each of these channels shares approximately the same bandwidth, with an echo canceling technique employed to enable transmitted and received signals to occupy the same bandwidth. This is made possible by designing intelligence into the modem's receiver that permits it to cancel out the effects of its own transmitted signal enabling the modem to distinguish its sending signal from the signal being received.

Under the V.32 recommendation synchronous data signaling rates of 2400, 4800 and 9600 bps are supported for asynchronous data entering the modem at those rates. This support is accomplished through the use of an asynchronous to

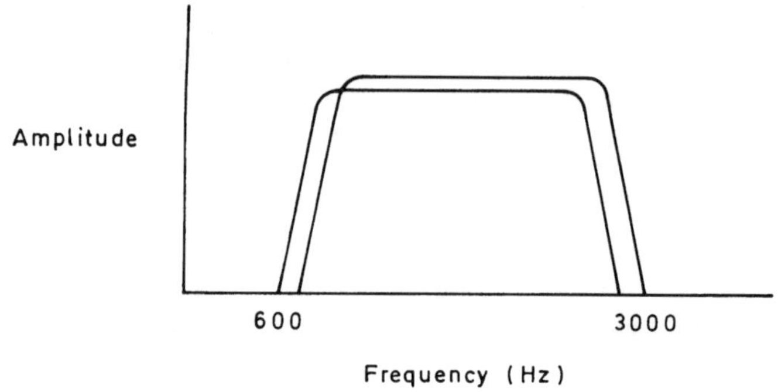

Figure 4.29 V.32 channel derivation

synchronous converter built into the modem. A V.32 modem uses a carrier frequency of 1800 Hz and has a modulation rate of 2400 bauds to support a data transfer rate of 9600 bps. At 9600 bps, the V.32 recommendation specifies two alternative modulation schemes—non-redundant coding and trellis coding.

The use of non-redundant coding results in a 16-point constellation pattern, while trellis coding results in a 32-point constellation pattern. Readers should note that while all V.32 modems must be capable of 'interworking' using the 16-point constellation pattern not all V.32 modems will include trellis coding. Users that require the better immunity to impairments thus afforded by trellis coding should ensure that the V.32 modem they are considering supports that coding alternative.

Non-redundant coding

Under the non-redundant coding technique, the data to be transmitted at 9600 bps is divided into groups of four consecutive data bits. The value of the first two bits is used in conjunction with the value of the two bits last output to generate the value for the next two output bits. These two output bits are then used with the value of bits three and four of the quadbit to select an appropriate signal point. Table 4.12 indicates the dependencies of the output dibit selection upon the value of the input dibit and the previous output dibits. Note that the value of the two input bits ($Q1_n$ and $Q2_n$) and the value of the previous dibit outputs ($Y1_{n-1}$ and $Y2_{n-1}$) are used to determine the phase quadrant change where the signal point will be located; however, this does not actually locate the point in the quadrant. To perform the latter operation the dibit outputs ($Y1_n$ and $Y2_n$) are used in conjunction with the values of the third and fourth bits in the quadbit input to select a position in the quadrant.

To select a quadrant position, the value of the dibit output ($Y1_n$ and $Y2_n$) is used in conjunction with the value of the third and fourth input (Q3 and Q4) bits. Table 4.13 indicates the selection of the non-redundant coding axis position based upon the values of the dibit output and the second dibit input.

To illustrate the use of Tables 4.12 and 4.13, assume a quadbit input has the value 0001 and the previous dibit output was 01. Since $Q1_n$ and $Q2_n$ have the value 00 while $Y1_{n-1}$ and $Y2_{n-1}$ have the value 01, from Table 4.12 the outputs ($Y1_n$ and

Table 4.12 Differential quadrant coding for 4800 bps and non-redundant coding at 9600 bps

Inputs		Previous outputs		Phase quadrant change (degrees)	Outputs		Signal state for 4800 bps
$Q1_n$	$Q2_n$	$Y1_{n-1}$	$Y2_{n-1}$		$Y1_n$	$Y2_n$	
0	0	0	0	+90	0	1	B
0	0	0	1	1	1	1	C
0	0	1	0		0	0	A
0	0	1	1		1	0	D
0	1	0	0	0	0	0	A
0	1	0	1		0	1	B
0	1	1	0		1	0	D
0	1	1	1		1	1	C
1	0	0	0	+180	1	1	C
1	0	0	1		1	0	D
1	0	1	0		0	1	B
1	0	1	1		0	0	A
1	1	0	0	+270	1	0	D
1	1	0	1		0	0	A
1	1	1	0		1	1	C
1	1	1	1		0	1	B

Table 4.13 V.32 non-redundant coding signal-state mappings for 9600 bps

Coded inputs				Non-redundant coding axis position	
Y_1	Y_2	Q3	Q4	X	Y
0	0	0	0	−1	−1
0	0	0	1	−3	−1
0	0	1	0	−1	−3
0	0	1	1	−3	−3
0	1	0	0	1	−1
0	1	0	1	1	−3
0	1	1	0	3	−1
0	1	1	1	3	−3
1	0	0	0	−1	1
1	0	0	1	−1	3
1	0	1	0	−3	1
1	0	1	1	−3	3
1	1	0	0	1	1
1	1	0	1	3	1
1	1	1	0	1	3
1	1	1	1	3	3

$Y2_n$) become 11 and the phase quadrant change will be 90°. From Table 4.13, the axis position will be 3,1 (Y1 = Y2 = 1, Q3 = 0, Q4 = 1), which will place the signal in the lower right position of the quadrant.

Figure 4.30 illustrates the 16-point constellation pattern generated by a V.32 modem using non-redundant coding. The four points that are circled represent valid signal points when the modem operates at 4800 bps. When operating at 4800 bps using non-redundant coding the modem operates on two bits at time, comparing the input bit values to the previous output dibit values to select one of four signal points.

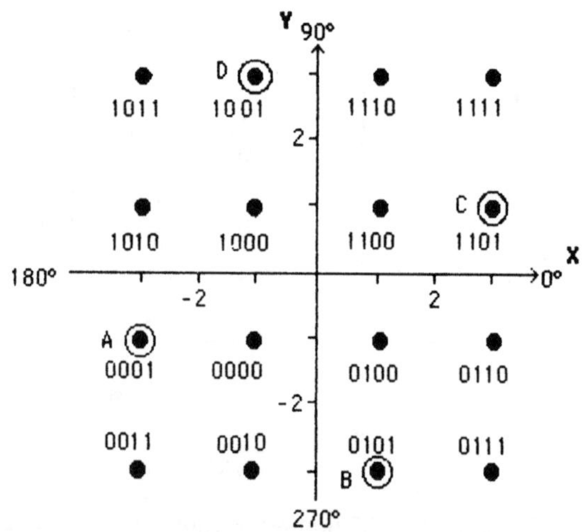

Figure 4.30 V.32 16-point signal constellation. The binary numbers denote $Y1_n$ $Y2_n$ $Q3_n$ $Q4_n$

Trellis coding

When a V.32 modem employs trellis coding, the input data stream is also divided into groups of four bits. Similar to the non-redundant coding method, the first two bits in each group ($Q1_n$ and $Q2_n$) are differentially encoded into $Y1_n$ and $Y2_n$ based upon the previous dibit value output. Table 4.14 indicates the differential encoding used by a V.32 modem operating at 9600 bps employing trellis coding.

Unlike non-redundant coding, under trellis coding the two differentially encoded bits ($Y1_n$ and $Y2_n$) are used as input to a convolutional encoder. The use of the convolutional encoder results in the generation of three bits based upon the two-bit input. Two of the three bits are the differentially encoded bits ($Y1_n$ and $Y2_n$) which are passed by the encoder. The third bit ($Y0_n$) is a redundant bit produced by the convolutional encoding process whose value is based upon the value of $Y1_n$ and $Y2_n$.

At 9600 bps the two passed through output bits ($Y1_n$ and $Y2_n$) and the redundant bit generated by the encoder ($Y0_n$) are used in conjunction with the value of the third and fourth bits in each quadbit to select one of 32 signal state mapping points.

Table 4.14 V.32 differential encoding for use with trellis-coded alternative at 9600 bps

Inputs		Previous outputs		Outputs	
$Q1_n$	$Q2_n$	$Y1_{n-1}$	$Y2_{n-1}$	$Y1_n$	$Y2_n$
0	0	0	0	0	0
0	0	0	1	0	1
0	0	1	0	1	0
0	0	1	1	1	1
0	1	0	0	0	1
0	1	0	1	0	0
0	1	1	0	1	1
0	1	1	1	1	0
1	0	0	0	1	0
1	0	0	1	1	1
1	0	1	0	0	1
1	0	1	1	0	0
1	1	0	0	1	1
1	1	0	1	1	0
1	1	1	0	0	0
1	1	1	1	0	1

Table 4.15 indicates the trellis coding signal points while Figure 4.31 illustrates the constellation pattern formed by a plot of all 32 signal points. When a V.32 modem operates at 4800 bps, as previously explained the device operates upon dibits instead of quadbits. When this occurs, the bits $Q1_n$ and $Q2_n$ are differentially encoded into $Y1_n$ and $Y2_n$ according to Table 4.12 which results in a constellation pattern of four signal points indicated in Figure 4.31 by the letters A, B, C, and D.

Since the V.32 recommendation was promulgated in 1984 numerous modem manufacturers have used the V.32 modulation scheme as a platform for adding enhancements to this modem. These enhancements include error detection and correction capability, data compression, callback security, and other features.

14 400 bps

Currently, the only standardized 14 400 bps modems are the ITU V.32 bis and V.33 recommendations.

V.32 bis modem

The V.32 bis recommendation was promulgated in 1981 and represents the ITU standard for modems operating on the switched telephone network at data rates up to 14 400 bps. Although this modem has been available for over a decade, until 1992 the average retail price of this modem exceeded $1000 and limited its widespread use. Since then, manufacturers have reduced the retail price of a V.32 bis modem to under $100, which has significantly increased its acquisition by personal computer

Table 4.15 V.32 trellis coding at 9600 bps

Coded inputs					Trellis coding	
(YO)	Y1	Y2	Q3	Q4	Re	Im
0	0	0	0	0	−4	1
	0	0	0	1	0	−3
	0	0	1	0	0	1
	0	0	1	1	4	1
	0	1	0	0	4	−1
	0	1	0	1	0	3
	0	1	1	0	0	−1
	0	1	1	1	−4	−1
	1	0	0	0	−2	3
	1	0	0	1	−2	−1
	1	0	1	0	2	3
	1	0	1	1	2	−1
	1	1	0	0	2	−3
	1	1	0	1	2	1
	1	1	1	0	−2	−3
	1	1	1	1	−2	1
1	0	0	0	0	−3	−2
	0	0	0	1	1	−2
	0	0	1	0	−3	2
	0	0	1	1	1	2
	0	1	0	0	3	2
	0	1	0	1	−1	2
	0	1	1	0	3	−2
	0	1	1	1	−1	−2
	1	0	0	0	1	4
	1	0	0	1	−3	0
	1	0	1	0	1	0
	1	0	1	1	1	−4
	1	1	0	0	−1	−4
	1	1	0	1	3	0
	1	1	1	0	−1	0
	1	1	1	1	−1	4

users in the home, business, government, and academia. Although more modern modems provide operating rates up to 33.6 kbps, for the next few years those modems can be expected to have a retail cost up to twice that of a V.32 bis modem. Thus, the popularity of V.32 bis modems can be expected to continue for the foreseeable future.

The V.32 bis recommendation is very similar to the V.32 recommendation in several key areas. Both recommendations specify the use of echo cancellation to obtain a full duplex transmission capability on the two-wire switched telephone network, and both specify the use of trellis coding, which significantly reduces the probability of transmission errors by improving the signal-to-noise ratio of

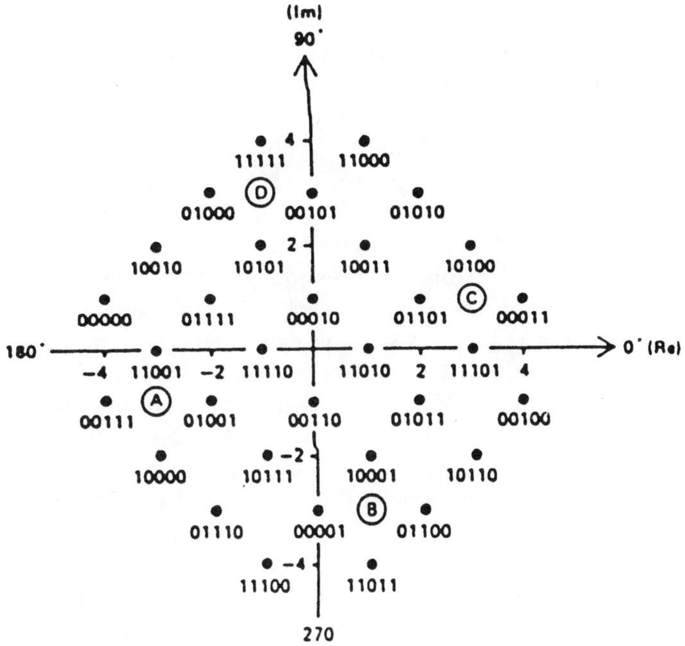

Figure 4.31 V.32 trellis coding signal constellation pattern

Table 4.16 Comparison of features of V.32 and V.32 bis

Feature	V.32	V.32 bis
Operating rates	9600 7200 (optional) 4800	14 400 12 000 9600 7200 4800
Encoded bits/symbol and signal constellation points:		
14 400 bps	—	6 (+TCM) 128
12 000 bps	—	5 (+TCM) 64
9600 bps	4 (+TCM)32	—
Fallback	yes	yes
Fall-forward	no	yes
Retrain	15 seconds	10 seconds

transmission. Major differences between the V.32 bis and V.32 recommendations are in the areas of operating rates supported, the number of encoded bits used per signal change and the resulting constellation pattern, the support of alternative operating rates during a transmission session, and the method and time required for retraining. Table 4.16 summarizes the major differences between the V.32 and V.32 bis recommendations.

As indicated in Table 4.16, the V.32 bis modem supports two operating rates above the maximum operating rate of the V.32 modem. Since those operating rates reflect the data transfer capability of each modem without the effect of data compression, the addition of that feature to each modem significantly increases the difference in throughput achievable through the use of each modem. For example, a V.32 modem that supports V.42 bis data compression, which provides an average compression ratio of 4:1, results in a throughput of 38.4 kbps when the modem operates at 9600 bps. In comparison, a V.32 bis modem operating at 14 400 bps employing V.42 bis compression provides a throughput of 57.6 kbps when the average compression ratio is 4:1. Thus, the operating rate difference of 4800 bps between a V.32 and a V.32 bis can expand to a throughput difference of almost 20 kbps when the effect of compression is considered.

Both V.32 and V.32 bis modems employ trellis coding at operating rates of 9600 bps and above. Each modem operates at 2400 baud and packs either four, five, or six data bits plus the Trellis Coded Modulation (TCM) bit into each signal change. The constellation pattern thus increases from 32 signal points for a V.32 modem operating at 9600 bps to 64 and 128 signal points for a V.32 bis modem operating at 12 000 and 14 400 bps, respectively.

At an operating rate of 14.4 kbps the V.32 bis modem divides data to be transmitted into groups of six consecutive bits. The first two bits in time ($Q1_n$ and $Q2_n$) in each group are differently coded into Y1 and Y2 in a similar manner to that previously described for the coding used by a V.32 modem. However, the outputs based on similar inputs and similar previous output are changed. Table 4.17 provides a summary of the differential encoding used by a V.32 bis modem operating at 14.4 kbps. You can note the differences between the coding of bits $Y1_n$

Table 4.17 V.32 bis differential dncoding for use with trellis coding

Inputs		Previous Outputs		Outputs	
$Q1_n$	$Q2_n$	$Y1_{n-1}$	$Y2_{n-2}$	$Y1_n$	$Y2_n$
0	0	0	0	0	0
0	0	0	1	0	1
0	0	1	0	1	0
0	0	1	1	1	1
0	1	0	0	0	1
0	1	0	1	0	0
0	1	1	0	1	1
0	1	1	1	1	0
1	0	0	0	1	0
1	0	0	1	1	1
1	0	1	0	0	1
1	0	1	1	0	0
1	1	0	0	1	1
1	1	0	1	1	0
1	1	1	0	0	0
1	1	1	1	0	1

and $Y2_n$ from the coding performed by a V.32 modem by comparing the entries in the two rightmost columns of Table 4.17 to the sixth and seventh columns in Table 4.12.

Once the two differentially encoded bits ($Y1_n$ and $Y2_n$) are computed, they are used as input to a convolutional encoder. That encoder generates a redundant bit ($Y0_n$) that is used for the six information-carrying bits ($Y1_n$, $Y2_n$, $Q3_n$, $Q4_n$, $Q5_n$, and $Q6_n$) for mapping into predefined coordinates of the transmitted signal. Figure 4.32 illustrates the signal constellation and mapping for the V.32 bis modem using trellis-coded modulation at 14.4 kbps. Note that the seven-bit binary numbers in the signal constellation reference the sequence of bits $Q6_n$, $Q5_n$, $Q4_n$, $Q3_n$, $Y2_n$, $Y1_n$, and $Y0_n$. Also note that the letters A, B, C, and D reference synchronization signal elements.

Figure 4.32 V.32 bis signal constellation and mapping for trellis-coded modulation at 14.4 kbps. A, B, C and D reference synchronizing signal elements

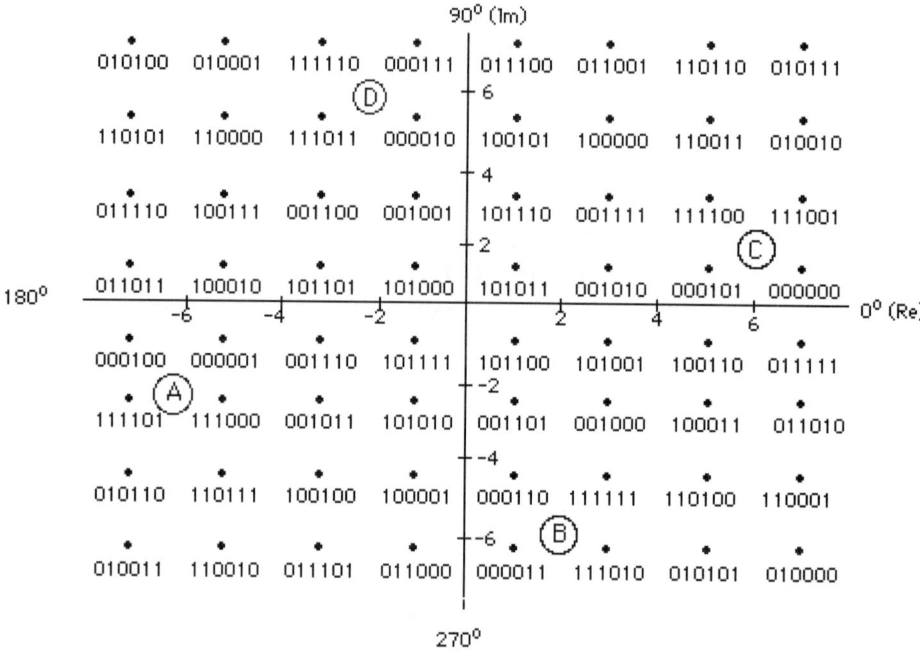

Figure 4.33 V.32 bis signal constellation and mapping for trellis-coded modulation at 12.0 kbps. A, B, C and D reference synchronizing signal elements

In addition to using trellis coding at 14.4 kbps, a V.32 bis modem also supports that coding method at 12.0 kbps, for which Figure 4.33 illustrates the constellation pattern and mapping. At that data rate the convolutional encoder produces a redundant bit that is used with five information-carrying bits ($Y1_n$, $Y2_n$, $Q3_n$, $Q4_n$ and $Q5_n$). Thus, the six binary numbers at each position in the signal constellation reference the bit sequence $Q5_n$, $Q4_n$, $Q3_n$, $Y2_n$, $Y1_n$, and $Y0_n$. The locations labeled A, B, C, and D again reference synchronizing signal elements.

The fallback feature in Table 4.16 refers to the ability of a modem to change its operating rate downward automatically when it encounters a predefined signal-to-noise ratio that would result in an unacceptable error rate if transmission at the current operating rate were maintained. Unfortunately, under the V.32 recommendation there was no provision for restoring the original operating rate if line quality improved after a fallback. Under V.32 bis an automatic fall-forward capability is included in the recommendation. This feature enables a V.32 bis modem to return to a higher operating rate if line quality improves.

Some modem vendors have implemented an enhanced fall-forward capability that can improve the transmission capability of the modem. For example, U.S. Robotics developed a feature known as Adaptive Speed Leveling for its V.32 bis modems. This feature enables the data rate of each direction of transmission to vary. Thus, if the incoming data rate is lowered due to noise encountered in one direction or another impairment, the outgoing data can still be transmitted at the highest operating rate, and vice versa. In comparison, most V.32 bis modem manufacturers implement an auto fall-forward capability symmetrically. The last major difference between V.32 and V.32 bis modems concerns retraining.

The V.32 modem has a 15-second retrain time, which minimizes the effect of a retrain upon modem throughput.

V.33 modem

The V.33 modem can be viewed as a less complex extension of V.32 technology. This is because, although the operating rate of the V.32 bis modem increased to 14.4 kbps from the 9600 bps operating rate of the V.32 modem, the V.33 achieves full-duplex transmission without the use of echo cancellation. This is possible since the V.33 modem is designed to operate on four-wire leased lines which permits the use of two two-wire signal paths.

V.32-compatible modems can operate at data rates of 14.4, 12.0, and 9.6 kbps. When operating at 14.4 kbps, the V.33 modem uses QAM modulation, assigning six data bits to each signal change and a seventh bit for trellis coding. This results in a constellation pattern of 128 signal points, which is the same as the constellation pattern of the V.32 bis modem illustrated in Figure 4.32. Although the V.33 standard is specific in its requirement for operation on four-wire leased lines, a few vendors developed proprietary half-duplex versions of the V.33 modem for use on the two-wire switched network. Those modems never achieved popularity, as most switched network users prefer using V.32 bis modems to obtain a full-duplex transmission capability at an operating rate of 14.4 kbps. In addition, the V.32 bis modem provides those modem users with the ability to communicate with a large potential audience in comparison to the small base of installed modems using a proprietary half-duplex V.33 modulation scheme.

19200 bps

When the V.32 bis modem was ratified in 1991, the ITU formed a study group that was subsequently referred to as vFast. That study group was tasked to look at higher-speed modem standards. The work of that study group was relatively long and resulted in the ratification of the V.34 standard in late 1994 and a modification to that standard in late 1995. While waiting for the standards to be promulgated, several vendors introduced interim technologies in an effort to build market penetration as well as to provide users with access to higher modem operating rates. Three of those interim technologies are referred to as vFast, V.FC, and V.32 terbo.

v.Fast represents the interim standard that preceded V.34. Although similar to V.34, there are some differences between v.Fast and V.34 modems.

The V.FC modem, with FC representing a mnemonic for Fast Class, represents a proprietary modulation scheme developed by Rockwell Semiconductor for communication at data rates up to 28.8 kbps. Early adopters of Rockwell's V.FC products were provided with hardware-based methods to upgrade to full V.34 compliance, and Rockwell now manufactures V.34-compliant chip sets.

AT&T took a different approach from Rockwell, developing an extension of V.32 bis technology in their V.32 terbo modem, which is designed to operate at data rates up to 19.2 kbps. The AT&T approach requires significantly less digital signal processing capability to implement than a V.34 modem, which requires the use of a digital signal processor capable of providing between 35 and 40 million instructions per second (mips). Similar to Rockwell, Paradyne Corporation, which until recently was part of AT&T now manufactures V.34 chip sets, although the V.32 terbo modulation capability is also included in many vendor high-speed modem products to provide transmission compatibility with the base of previously manufactured V.32 terbo modems. Since the V.32 terbo modem represents an extension of V.32 bis technology and provides a maximum operating rate of 19 200 bps, we will focus our attention on that modem in this section.

V.32 terbo

As previously discussed, V.32 terbo represents a relatively simple change to the 14.4 kbps V.32 bis standard. For this reason many people consider V.32 terbo as an enhancement of V.32 bis rather than as a competitor to more complex and higher operating rate V.34 modems.

Although the V.32 terbo modem was championed by AT&T, its specification was not formally adopted by any standards body. In spite of this, its technology was incorporated into products from several modem manufacturers, both as the highest operating rate of a modem and as one of several modulation methods supported by a V.34 modem to provide downward compatibility with V.32 terbo modems. With a retail price only $30 more than a 14.4 kbps modem, the V.32 terbo modem provides a respectable level of price performance. In addition, unlike V.34 modems which have difficulty in reaching their highest-rated speed over the PSTN and fall back to a lower operating rate, a V.32 terbo modem in many cases can be expected to provide a more consistent level of performance.

Operation

The key to the design of the V.32 terbo modem is its modification of the V.32 bis standard to reflect two new data signaling rates: 16.8 and 19.2 kbps. All other characteristics of the V.32 terbo modem are identical to the V.32 bis standard. Thus, a V.32 terbo modem uses two-dimensional trellis coding and echo cancellation.

Signal element coding

At 14 400, 16 800, and 19 200 bps, the scrambled data stream is divided into groups of six, seven, or eight consecutive data bits, respectively. The first two bits in time Q1 and Q2 are differentially encoded and trellis encoded following recommendation V.32 bis to generate three trellis-encoded bits, referred to as Y0, Y1, and Y2. These trellis bits and all remaining information bits, Q3 through Q8, are mapped into the coordinates of the signal elements to form the constellation pattern at a particular operating rate. Figures 4.34 and 4.35 illustrate the signal space diagram and mapping for V.32 terbo operations at 16.8 kbps and 19.2 kbps, respectively. At 14.4 kbps, the signal space diagram is the same as that of the V.32 bis recommendation for operation at 14.4 kbps.

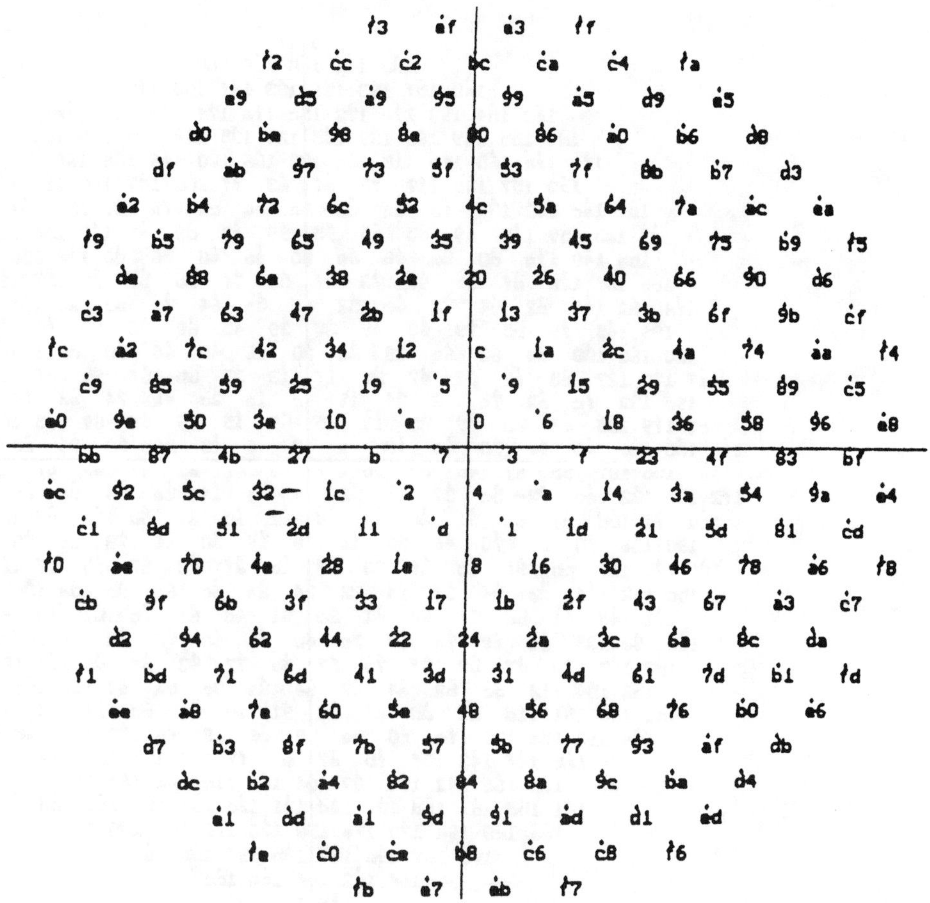

Figure 4.34 V. 32 terbo signal space diagram and mapping for 16 800 bps. Hexadecimal numbers refer to Q7Q6Q5Q4 and Q3Y2Y1Y0. (Diagram courtesy of AT&T Paradyne)

In examining the constellation patterns illustrated in Figures 4.34 and 4.35, note that each signal element is identified in hexadecimal format. Y0 is the least significant bit in the least significant hexadecimal digit. Q4 is the least significant bit in the second hexadecimal digit. At 19 200 bps, Q8 is the least significant bit in the third hexadecimal digit. Also note that the 16 800 bps signal space diagram is a subset of the 19 200 bps signal space diagram.

Non-linear encoding

At the new data rates of 16 800 and 19 200 bps, the V.32 terbo signal space diagrams are modified by non-linear encoding. Non-linear encoding is enabled only during data transmission at these rates and not during other sequence states. Non-linear encoding is disabled during startup and retraining.

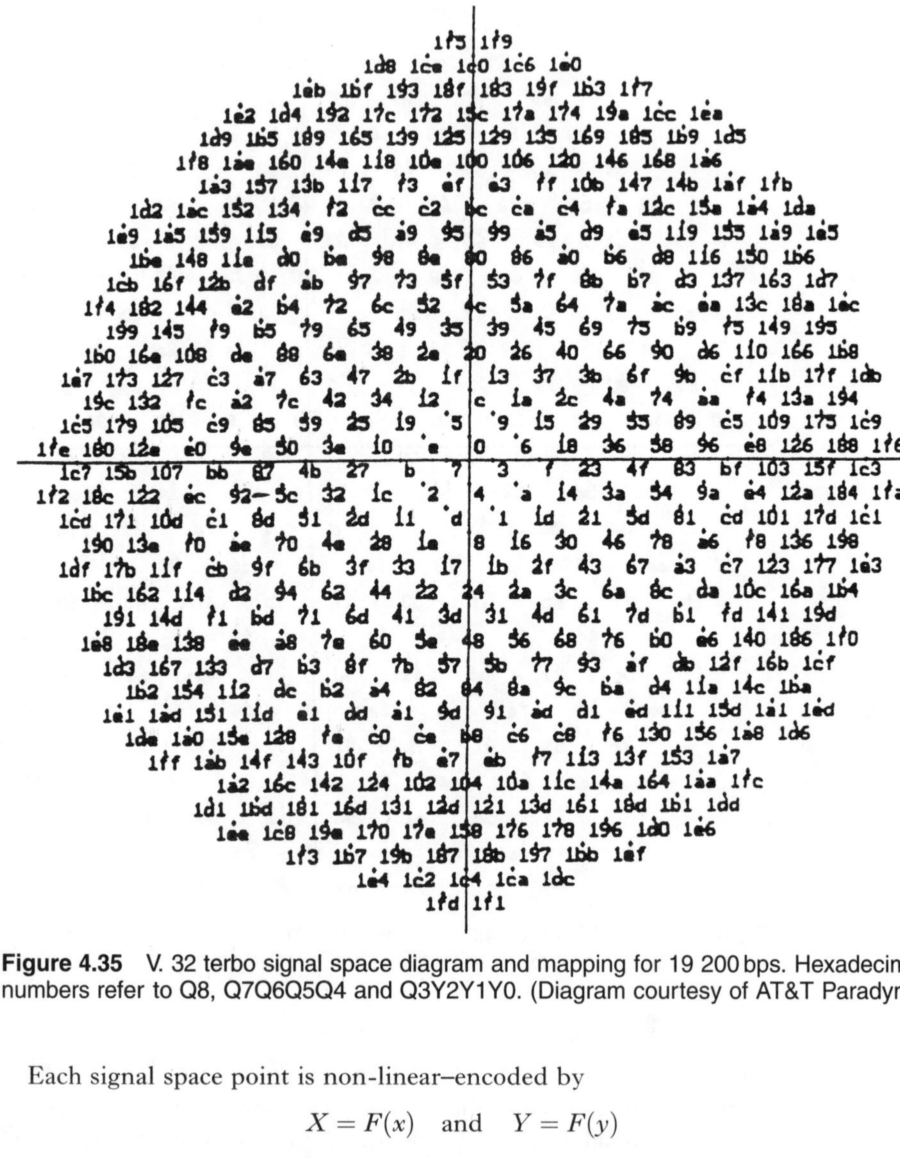

Figure 4.35 V. 32 terbo signal space diagram and mapping for 19 200 bps. Hexadecimal numbers refer to Q8, Q7Q6Q5Q4 and Q3Y2Y1Y0. (Diagram courtesy of AT&T Paradyne)

Each signal space point is non-linear–encoded by

$$X = F(x) \quad \text{and} \quad Y = F(y)$$

where the non-linear encoding projection function is

$$F = K(16\,384 + 2731P + 137P^2 + 3P^3)/16\,384$$

and $P = C(X^2 + Y^2)$. Note that $C = g/R^2$ is a constant, where the non-linear encoding gain is $G = 0.8$, and R^2 is the square of the radius of the signal space at the data signaling rate of 19 200 bps. The transmitter gain correction factor is

$$K = 1 - 1697g/16384$$

All data rates require scaling of the signal space diagrams to achieve equal transmission power. It is recommended that you perform non-linear encoding after this power adjustment. In this case, the radius of the signal space is approximately the same at 16 800 and 19 200 bps. The radius at 19 200 bps can be used to calculate C at both data rates so that identical equations are used.

Non-linear decoding

At the new data rates of 16 800 and 19 200 bps, the signal space diagrams are modified by non-linear decoding at the output of the equalizer in the receiver. Non-linear decoding is enabled only during data reception at these rates and not during other sequence states. Non-linear decoding is disabled during startup and retraining.

The reciprocal non-linear decoding operation is performed at the equalizer output:

$$X = R(x) \quad \text{and} \quad Y = R(y)$$

using the reciprocal function

$$R = 1/K + (2731P + 1229P^2 + 731P^3 + 264P^4)/16\,384$$

where $P = C(X^2 + Y^2)/K^2$ and the constants C and K are as defined in the non-linear encoding section. K is the transmitter gain correction factor. The recovered points x and y are decoded using conventional techniques.

Note that in V.32 terbo the non-linear decoder is only active during data transmission at data rates of 16 800 and 19 200 bps. The non-linear decoder is not active during training. Training is completed before the rate sequence is detected. For this reason, the transmitter gain correction factor K is incorporated in the calculation of the non-linear decoder. The equalizer would remove this constant gain factor if the non-linear decoder were active during training.

The V.32 terbo modem can provide several advantages over the higher operating rate of V.34 and V.34 bis standards that can enable it to serve a niche market for a considerable period of time. Those advantages are based on the problems many V.34 and V.34 bis users will experience when attempting to operate those modems at data rates beyond 19.2 kbps. In addition, from a practical perspective, many computers cannot support the DTE rate required to transmit data into a very-high-speed modem when the modem is in its compression mode of operation. For example, assuming a 4:1 compression ratio, you would want to set the speed of a computer port to 76.8 kbps (19.2×4) to work with a V.32 terbo modem. In comparison, you would want to set the port speed to 134.4 kbps (33.6×4) when using a V.34 bis modem. When working in Windows it is often difficult to achieve an average DTE speed greater than 57.6 kbps, which means that the use of a V.34 bis modem may not result in any noticeable improvement over a V.32 terbo modem unless the transmitted data is non-compressible. Thus, the key advantages of a V.32 terbo modem include its cost and practicality of use.

28 800 bps

There are three modems that have a maximum operating rate of 28 800 bps. Those modems include the two pre-standardization versions of the V.34 recommendation, referred to as F.FC and v.Fast, as well as the V.34 modem. To eliminate a potential degree of confusion, readers should note that the first generation of V.34 modems were limited to a maximum operating rate of 28 800 bps. During 1997 a revision to the V.34 recommendation, instead of an anticipated new recommendation, which many persons referred to as V.34 bis, raised the maximum transmission rate of this modem to 33.6 kbps. Thus, the V.34 modem will be discussed under both 28 800 bps and 33 600 bps headings.

V.FC modem

While waiting for the ITU-T study group to complete its work on finalizing the V.34 standard, Rockwell Semiconductor, a division of Rockwell International, developed a modem to provide customers with a 28.8 kbps transmission capability. This modem, which obtained the mnemonic v.FC for v.Fast Class, represents Rockwell's proprietary technology that was based on the basic features of the proposed ITU v.Fast recommendation during the March–April 1993 time frame. Although the resulting V.FC modem was manufactured as a chipset by Rockwell and incorporated into several modem manufacturer products, the resulting V.34 standard differs from the V.FC specifications enough to preclude interoperability at 28.8 kbps unless the V.34 modem includes V.FC compatibility. Otherwise, the V.34 and V.FC modems will negotiate a V.32 bis connection at 14.4 kbps.

A V.FC modem provides a maximum operating rate of 28.8 kbps as well as V.FC modulation fallback rates of 26.4 kbps and 24.0 kbps. In addition, the V.FC chipset includes full and transparent downward compatibility with V.32 bis (14.4 kbps), V.32 (9.6 kbps), and V.22 bis (2.4 kbps).

Unlike a V.34 modem that can fall back through a range of operating rates, a V.Fc modem will not fall back to V.32 bis. Other differences between a V.FC and a V.34 modem are in the areas of line probing, precoding, and trellis coding, which are discussed when the V.34 standard is described later in this section.

Although the V.FC modem filled an important niche as users waited for the V.34 standard, most 28.8 kbps modems now manufactured are based on the V.34 standard. To provide compatibility with the large base of V.FC modems, many vendors include a V.FC compatibility mode in their products, an important feature you may wish to consider.

v.Fast modem

v.Fast represents the interim standard that preceded V.34. In the haste to get modems to the marketplace as well as for semiconductor firms to develop chipsets for modem manufacturers, several versions of v.Fast were implemented. In fact, the previously described V.FC can be considered to represent v.Fast. Needless to say, different implementations of v.Fast caused some degree of compatibility problems between modems. Fortunately, most modem vendors provided a mechanism for v.Fast products to be upgraded to V.34 compliance. Thus, in the remainder of this section we will focus our attention on the operational characteristics of the V.34 standard that superseded v.Fast products.

V.34 modem

The ITU V.34 standard represents a quantum leap in both complexity and operating rate capability in comparison to the V.32 bis standard. From an operational perspective, the V.34 standard's complexity can be judged by its documentation, which is approximately twice as thick as the V.32 bis standard. The doubling of modem documentation results from the addition of several new features such as an asymmetrical transmission capability, an auxiliary channel, non-linear encoding, precoding, and a significant increase in the trellis-code state to support trellis coding. When such previously developed features as error correction, compression, and different baud rate operations are considered, the V.34 standard offers well over 100 combinations of modulation schemes, baud rates, and other operating parameters. Unfortunately, many of the new features, while standardized for use in the transmitter section of the modem, are specified as options for implementation in the modem's receiver. Thus, interoperability can be a problem as the standard allows chip manufacturers to pick and choose from an extensive array of options.

Modulation

The original V.34 standard specified three mandatory baud rates—2400, 3000, and 3200. Under ideal conditions, a V.34 modem maps nine bits into each symbol, resulting in an operating rate of 3200 baud $\times$ 9 bits/baud, or 28.8 kbps. At this operating rate there are 960 points in the modem's signal constellation. To significantly reduce the probability of errors due to impairments causing a small shift in a point in the signal constellation, the V.34 modem includes non-linear coding, precoding, and a 16-state trellis-coding capability as well as optionally specified 32- and 64-state trellis coding. Each of these features will be discussed later in this section.

In addition to the mandatory baud rates, the V.34 standard specifies optional rates of 2743, 2800, and 3429. The 2743 and 2800 baud rates can be an important consideration when transmission occurs via an infrastructure where voice is digitized using Adaptive Differential Pulse Code Modulation (ADPCM), which uses prediction to reduce the voice digitization rate from PCM's 64 kbps to 32 kbps. You typically encounter ADPCM when communicating via satellite, on terrestrial circuits that are communicating via satellite, or on terrestrial circuits that are routed overseas or to Hawaii, Alaska, and other non-contiguous US locations. Since ADPCM fails when baud rates exceed 3000, the optional 2743 and 2800 baud rates permit relatively high-speed communications to occur over an ADPCM infrastructure.

The V.34 standard specifies two carrier frequencies for both mandatory and optional baud rates. Table 4.18 summarizes the carrier frequencies, bandwidth requirements, and maximum bit rate for the six baud rates included in the V.34 standard.

Also included in Table 4.18 is a column labeled 'Modem protocol', which indicates the difference in the use of carrier frequencies between a V.34 and a V.FC modem.

Table 4.18 V.34 Bandwidth Requirements, Maximum Operating Rates, and Carrier Frequencies

Symbol rate	Modem protocol	Carrier frequency	Bandwidth requirements	Maximum bit rate
2400	V.34	1600 Hz	400–2800 Hz	21 600
	V.34/VFC	1800 Hz	600–3000 Hz	21 600
2473	V.34	1646 Hz	274–3018 Hz	24 000
	VFC V.34	1829 Hz	457–3200 Hz	24 000
2800	V.34	1680 Hz	280–3080 Hz	24 000
	VFC/V.34	1867 Hz	467–3267 Hz	24 000
3000	V.34	1800 Hz	300–3300 Hz	26 400
	V.34/VFC	2000 Hz	500–3500 Hz	26 400
	VFC	1875 Hz	375–3376 Hz	26 400
3200	V.34	1829 Hz	229–3429 Hz	28 800
	VFC	1920 Hz	320–3250 Hz	28 800
3429	V.34	1959 Hz	244–3674 Hz	28 800

In examining the entries in Table 4.18, you will note that the variety of symbol (baud) rates provides the possibility of having a non-integral number of bits per symbol. When this occurs, a shell-mapping algorithm is used to generate the constellation points.

Options

In addition to three baud rates being specified as options, the V.34 standard allows asymmetrical transmission, an auxiliary channel, non-linear encoding, precoding, and two trellis-state codes to be optionally included in a modem's receiver. To appreciate the value of each option, as well as to understand why some V.34 modems implementing all options can cost considerably more than other V.34 modems that defer implementing one or more options, requires a discussion of each option. That discussion follows.

Asymmetrical transmission

The asymmetrical transmission capability of a V.34 modem enables it to send and receive at different operating rates. This feature removes the slowest common denominator acting as a break on the data exchange.

To appreciate the value of asymmetrical transmission, assume two users wish to communicate over the PSTN. Suppose one user's office is very close to a serving central office while the second user is located on a very long local copper loop that limits maximum data transfer capability. Without an asymmetrical transmission capability, communications would occur at the lower rate in both directions.

Auxiliary channel

The V.34 standard specifies a separate transmitted signal for the exchange of information on line quality and modem operating conditions. Unless the auxiliary channel option is included in the receiver, a pair of V.34 modems will not support the use of this option.

The auxiliary channel operates at 200 bps and is multiplexed by the modem over the regular channel, being demultiplexed at the other end of the connection. Thus, this option provides an inband management channel that could be used to transmit or receive management data from a bridge or router during a communications session.

Non-linear encoding

Non-linear encoding represents a technique used to increase the immunity of a modem to interference resulting from the pulse code modulation (PCM) of analog lines by the communications carrier for transport over the carrier's backbone digital infrastructure. Under non-linear coding, constellation points are spaced unequally so that the points in areas most susceptible to noise are farther apart. This technique reduces the probability that adjacent points will be confused with each other and can reduce the error rate when operating at 28.8 kbps by up to 50%.

Precoding

Precoding represents a form of equalization that reduces the amount of high-frequency noise on a line, also reducing the amount of intersymbol interference. This in turn permits the modem to obtain a high baud rate through the use of more bandwidth than might be obtainable if no precoding occurred. One difference between a V.FC and a V.34 modem is in the area of precoding. The V.34 standard, while similar to that of the V.FC, has a change that minimizes the dithering noise of the precoder.

Trellis coding

The use of trellis coding enables a modem to maintain a high baud rate on a noisy line. Under the V.34 specification, 16-, 32-, and 64-state trellis codes are specified for the transmitter; however, only one of the three must be implemented in the receiver.

A 64-state trellis code provides a superior level of performance but is very complex to implement. While the difference between a 16-state and a 64-state trellis code is minimal when transmission occurs on a good quality line, it could prevent a modem from having to drop down to a lower operating rate when line quality deteriorates.

Another difference between a V.FC and a V.34 modem is in their implementation of trellis coding. A V.FC modem uses a two-dimensional trellis code while a V.34 modem uses a four-dimensional code. The V.FC uses the two-dimensional trellis code to produce a 32-state trellis encoder. Although that encoder is more 'powerful' than the 16-state encoder in a V.34 modem, the latter also supports 32- and 64-state encoders.

Operation

A V.34 modem will exchange a test tone with another V.34 modem to profile the channel from 150 Hz to 3750 Hz. This channel profile operation determines the effective bandwidth available and is used to make such configuration decisions as the symbol rate, carrier frequency, constellation shape, encoding scheme, transmission power, and type of trellis encoder to use. Under ideal conditions, a V.34 modem will operate at 3200 baud, mapping nine bits per symbol to obtain a 28.8 kbps operating rate. The modem can drop back in steps of 2.4 kbps to 2.4 kbps and bump up in steps of 2.4 kbps if line quality improves.

Another reason for the incompatibility between V.FC and V.34 modems is a result of their channel-probing methods. A V.FC modem probes at 68 frequencies and uses 50 Hz spacing. In comparison, a V.34 modem uses 21 frequencies with 150 Hz spacing.

Operating limitations

One of the major problems associated with the use of the V.34 modem is the fact that many users will not be able to achieve its maximum operating rate capability of 28.8 kbps. There are two reasons for this, with the first directly related to the second condition, which is associated with a phone line problem.

One reason a V.34 modem fails to achieve a 28.8 kbps transmission rate results from one or both pairs of modems not incorporating the optional features in their receivers. When telephone line quality deteriorates, the lack of previously described optional receiver features makes it difficult, if not impossible, to maintain a 28.8 kbps operating rate.

A second reason why a V.34 modem may not achieve a 28.8 kbps operating rate is due to the bandwidth required to achieve that operating rate. As indicated in Table 4.18, 28.8 kbps requires a bandwidth of 3200 Hz (320 to 3250 Hz). Due to telephone company trunk components and switches, as well as faulty premises wiring and long local loops, the ability to obtain a 3200 Hz bandwidth on an end-to-end basis can become very difficult, resulting in a V.34 modem operating at 26.4 kbps, which requires 3000 Hz bandwidth, or at 24.0 kbps, which requires 2800 Hz bandwidth.

Upgrades

One of the features you will wish to consider when purchasing a V.34 modem is its upgrade capability. This capability is important if you purchase a modem that does not implement all V.34 optional features or if you wish to upgrade to the revised recommendation that provides a maximum data transmission capability of 33.6 kbps. Some vendors also permit a V.34 modem to support one-way 56 kbps transmission which is described later in this section.

Many modem manufacturers offer one of three methods for upgrading a V.34 modem—EPROM, replacement chips, or return to factory.

The use of Erasable Programmable Read Only Memory (EPROM) enables new code to be delivered via a software download. If a modem uses instructions stored in Programmable Read Only Memory (PROM), upgrades are shipped on replacement chips for installation by the user. The third method of upgrade results from the vendor burning code into application-specific integrated circuits (ASICs). The only way to upgrade this type of modem is by shipping it back to the vendor who will swap chip components.

33 600 bps

Not long after the V.34 recommendation was adopted, Rockwell Semiconductor Systems proposed a revision to the standard to provide a maximum operating rate capability of 33 600 bps. The Rockwell proposal was given the name V.34+, with other vendors and writers using the terms V.34 bis and V.34 Plus to describe the expected revision. However, Rockwell withdrew its proposal and the ITU decided to retain the original standard terminology as opposed to the creation of a new standard. Thus, the older V.34 recommendation defines a maximum operating rate of 28.8 kbps, while the revised recommendation defines a maximum operating rate of 33.6 kbps.

The new version of the V.34 recommendation was finalized in the fall of 1996 and added data rates of 33.6 kbps and 31.2 kbps to the original standard. Under the revised V.34 recommendation, the two new operating rates are optional, meaning that you must carefully examine the operating rates supported by a V.34 modem to determine its capability.

When a V.34 modem operates at a connect speed of 33.6 kbps with compression enabled, a theoretical maximum compression ratio of 4:1 means that the serial port from a terminal device must operate at 134.4 kbps. Recognizing the fact that the UART in most personal computers have a maximum transfer rate of 115.2 kbps, some vendors use a parallel port connector on their modems. This is because data transfers via a parallel port can be supported up to approximately 230 kbps. However, since the average compression ratio actually obtained by most users will be below 4:1, you can more than likely use a serial port interfaced V.34 modem without adversely affecting your data transfer capability. In comparison, if you anticipate using a recently introduced 56 kbps modem that is described next, you will more than likely require the use of a parallel interface or an enhanced serial adapter when employing data compression.

56 kbps

The rapid growth in remote access requirements, such as telecommunicating to an organization's LAN-based server or Internet surfing, resulted in the development of a unidirectional so-called '56 kbps' modem technology which was being considered by the ITU for standardization as the V.90 recommendation when this book was prepared. Unlike previously developed standardized modems that have a similar bidirectional operational capability, the 56 kbps modem provides its high-speed operational capability downstream from a source directly connected to a communications carrier's digital network infrastructure. In addition, only one analog-to-digital conversion can occur, otherwise the maximum rate of the modem is that of a V.34. In fact, the 56 kbps modem uses V.34 technology to provide an upstream 33.6 kbps transmission capability as well as a 33.6 kbps downstream capability when it cannot achieve a 56 kbps operating rate.

Operation

A 56 kbps modem achieves its downstream capability by 'intercepting' data in its digital form generated by an access server or similar device connected by a digital line to the central office of a communications carrier. Since digital transmission lines use repeaters instead of amplifiers that build up distortion, a better signal arrives at the central office. This concept is illustrated in Figure 4.36. Note that to achieve a downstream 56 kbps transmission capability, only one analog to digital conversion is permitted. In addition, the access server or similar device at one end must be directly connected to the digital network of the serving carrier, bypassing the normal analog local loop. This means that two 56 kbps modems cannot communicate at 56 kbps when communicating as an analog modem to analog modem since this type of connection requires the use of two analog conversions.

At the time this book was prepared there were two competing 56 kbps technologies that are both proprietary and incompatible with each other. One technology, referred to as X2, was developed by U.S. Robotics which was merged into 3Com Corporation. The second technology, referred to as V.flex2, was developed by Lucent Technologies and Rockwell Semiconductor Systems. Although both technologies support V.34 which enables one modem to interoperate with another at up to 33.6 kbps, they cannot interoperate at 56 kbps. In

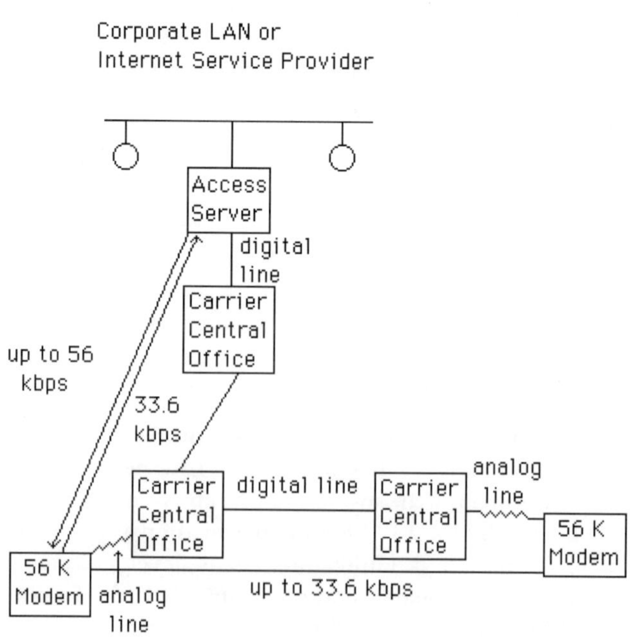

Figure 4.36 56 kbps modem communications require the elimination of one analog to digital conversion from the pair of conversions other modems operate upon. This is accomplished by locating one 56 kbps modem in a carrier's central office

addition, a Federal Communications Commission regulation precludes power levels that result in transmission above 53.3 kbps to protect telephone lines from crosstalk that can occur when the signal rate on an adjacent wire is too high. This means that unless this 20-year-old regulation is lifted, 56 kbps modems can only legally operate at a maximum rate of 53.3 kbps.

Work on the development of a 56 kbps modem standard referred to as the V.90 recommendation and activity to lift or revise the FCC regulation were in progress when this book was written. A summary of the operational characteristics of Bell System and ITU V-series type modems, as well as the previously discussed 56 kbps modem, is provided in Table 4.19.

Table 4.19 Modem operational characteristics

Modem type	Maximum data rate	Transmission technique	Modulation technique	Transmission mode	Line use
Bell System					
103A,E	300	Asynchronous	FSK	Half, full	Switched
103F	300	Asynchronous	FSK	Half, full	Leased
201B	2400	Synchronous	PSK	Half, full	Leased
201C	2400	Synchronous	PSK	Half, full	Switched
202C	1200	Asynchronous	FSK	Half	Switched
202S	1200	Asynchronous	FSK	Half	Switched
202D/R	1800	Asynchronous	FSK	Half, full	Leased
202T	1800	Asynchronous	FSK	Half, full	Leased
208A	4800	Synchronous	PSK	Half, full	Leased
208B	4800	Synchronous	PSK	Half	Switched
209A	9600	Synchronous	QAM	Full	Leased
212	0–300	Asynchronous	FSK	Half, full	Switched
	1200	Asynchronous/ synchronous	FSK	Half, full	Switched
ITU					
V.21	300	Asynchronous	FSK	Half, full	Switched
V.22	600	Asynchronous	PSK	Half, full	Switched/leased
	1200	Asynchronous/ synchronous	PSK	Half, full	Switched/leased
V.22 bis	2400	Asynchronous	QAM	Half, full	Switched
V.23	600	Asynchronous/ synchronous	FSK	Half, full	Switched
	1200	Asynchronous/ synchronous	FSK	Half, full	Switched
V.26	2400	Synchronous	PSK	Half, full	Leased
	1200	Synchronous	PSK	Half	Switched
V.26 bis	2400	Synchronous	PSK	Half	Switched
V.26 ter	2400	Synchronous	PSK	Half, full	Switched
V.27	4800	Synchronous	PSK		Leased
V.29	9600	Synchronous	QAM	Half, full	
V.32	9600	Asynchronous	TCM/QAM	Half, full	Switched/leased
V.32 bis	14 400	Asynchronous	TCM/QAM	Half, full	Switched/leased
V.33	14 400	Synchronous	TCM	Half, full	Leased
v.32 terbo	19 200	Asynchronous	TCM	Half, full	Leased
V.34	28 800	Asynchronous	TCM	Half, full	Switched/leased
V.90	56 000*	Asynchronous	TCM	Half, full	Switched/leased

* 56 kbps is unidirectional and is limited to operating with one analog to digital conversion.

Non-standard modems

There are three types of non-standard modems that deserve mention in this section even though they are obsolete when used as analog modems on conventional voice grade lines. This is because some of the concepts and a technology used by each type of modem are being incorporated into cable and digital subscriber line modems that are described later in this chapter.

Packetized ensemble protocol

One type of non-standard modem that reached the marketplace in the mid-1980s resulted in the development of a technology concept that is being applied to digital subscriber lines during the late 1990s. More formally known as a packetized ensemble protocol modem, this modem incorporated a revolutionary advance in technology due to the incorporation of a high-speed microprocessor and approximately 70 000 lines of instructions built into read-only memory (ROM) chips on the modem board.

Under the packetized ensemble protocol, the originating modem simultaneously transmits 512 tones onto the line. The receiving modem evaluates the tones and the effect of noise on the entire voice bandwidth, reporting back to the originating device the frequencies that are unusable. The originating modem then selects a transmission format most suitable to the useful tones, employing 2-bit, 4-bit or 6-bit quadrature amplitude modulation (QAM), and packetizes the data prior to its transmission. Figure 4.37 illustrates the carrier utilization of a packetized ensemble modem.

As an example of the efficiency of this type of modem, let us assume that 400 tones are available for a 6-bit QAM scheme. This would result in a packet size of 400×6 or 2400 bits. If each of the 400 tones is varied four times per second, a data rate of approximately 10 000 bps is obtained. It should be noted that the modem automatically generates a 16-bit cyclic redundancy check (CRC) for error detec-

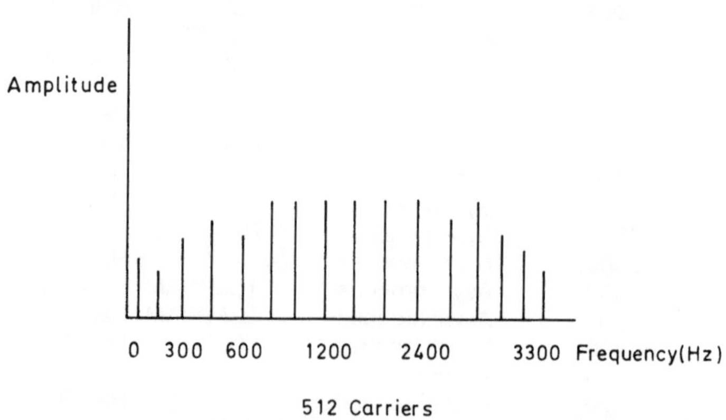

Figure 4.37 Packetized ensemble modem. (Reprinted with permission from *Data Communications Management*, © 1988 Auerbach Publishers, New York)

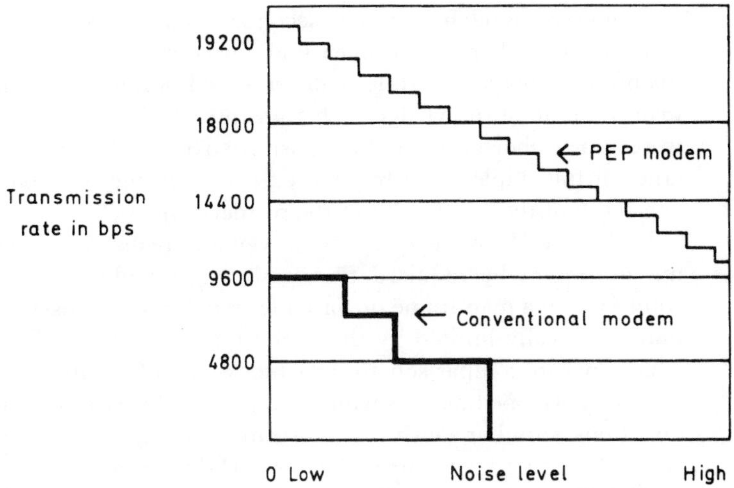

Figure 4.38 Transmission rate versus noise level. (Reprinted with permission from *Data Communications Management*, © 1988 Auerbach Publishers, New York)

tion, which is added to each transmitted packet. At the receiving modem, a similar CRC check is performed. If the transmitted and locally generated CRC characters do not match, the receiving modem will then request the transmitting modem to retransmit the packet, resulting in error correction by retransmission.

Two of the key advantages of a packetized ensemble protocol modem are its ability to automatically adjust to usable frequencies, which greatly increases the use of the line bandwidth, and its ability to lower its fallback rate in small increments. The latter is illustrated in Figure 4.38, which shows how this type of modem loses the ability to transmit on one or a few tones as the noise level on a circuit increases, resulting in a slight decrease in the data rate of the modem. In comparison, a conventional modem, such as a 9600 bps device, is designed to fall back to a predefined fraction of its main data rate, typically 7200 or 4800 bps.

The original Packetized Ensemble Modem was designed by Telebit Corporation and was marketed as the Trailblazer. Several vendors marketed similar modems under license to include a modem card for insertion into the system unit of an IBM PC or compatible personal computer and a stand-alone unit that can be attached to any computer with a standard RS-232 communications port. In addition to being compatible with other packetized ensemble protocol modems, several models of these devices are compatible with V.32, V.22 bis, V.22, 212A, and 103 type modems. This compatibility permits the personal computer user to use the device to access information utilities, other personal computers and mainframes that are connected to industry-standard modems.

Asymmetrical modems

Borrowing an old modem design concept, several vendors introduced asymmetrical modems. These modems in essence contain two channels which, in the early days of modem developments, were known as the primary and secondary channel.

Originally, modems with a secondary channel were used for remote batch transmission, where the primary high-speed channel was used to transmit data to a mainframe computer while a lower speed secondary channel was used by the mainframe to acknowledge each transmitted block. Since the acknowledgements were much shorter than the transmitted data blocks it was possible to obtain efficient full-duplex transmission even though the secondary channel might have one-tenth of the bandwidth of the primary channel.

In the late 1980s, several modem vendors realized that while high-speed transmission might be required to refresh a terminal's screen when the device was connected to a mainframe or for a file transfer, transmission in the opposite direction is typically limited by the user's typing speed or the shortness of acknowledgements in comparison to data blocks of information. Realizing this, modem vendors developed devices which use wide and narrow channels to transmit in two directions simultaneously as illustrated in Figure 4.39. The wide bandwidth channel permits a data rate of 9600 bps while the narrow bandwidth channel is used to support a data rate of 300 bps. Where these asymmetrical modems differ from older modems with secondary channels is in the incorporation of logic to monitor the output of attached devices and to then reverse the channels, permitting an attached terminal device to access the higher speed (wider bandwidth) channel when necessary. Although no standards exist for asymmetrical modems, several manufacturers attempted to formulate the use of common frequency assignments for channels. Because a 9600 bps asymmetrical modem initially had a retail price approximately two-thirds of a V.32 modem, at one time it was thought that asymmetrical technology would find a viable market. However, significant price reductions in V.32 and V.34 modems due to economies of scale have reduced the price of standardized modems below that of asymmetrical technology. Although asymmetrical modems designed for use on regular analog voice circuits failed to gain acceptance, the concept behind this technology was adapted for use with cable and digital subscriber line modems described later in this chapter.

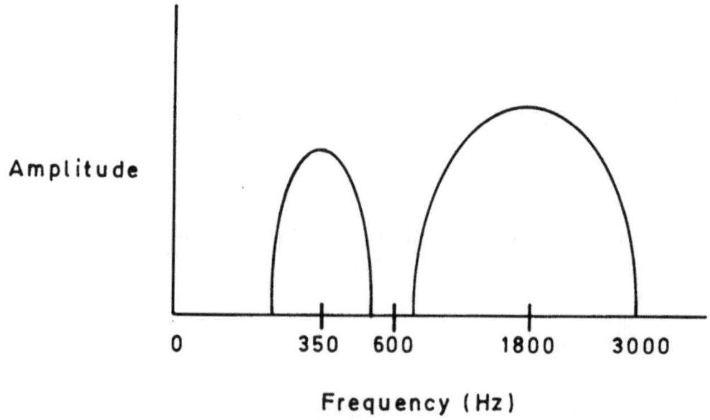

Figure 4.39 Asymmetrical modem channel assignment. (Reprinted with permission from *Data Communications Management*, © 1988 Auerbach Publishers, New York)

Ping-pong modems

Another type of modem operation that simulates full-duplex transmission is the 'ping-pong' or fast-turnaround modem. With this transmission method a modem sends data in one direction and then signals the remote modem when its transmission is completed. This signal informs the remote modem that it can now transmit data, because the originating modem has placed itself in the receive mode.

The faster the two modems are able to turn off their transmitter and turn on their receiver, the closer it appears that full-duplex transmission is occurring. Modems that employ a ping-pong transmission scheme include RAM buffers to hold data as the direction of transmission changes. This permits terminal devices connected to ping-pong operating modems to appear to continuously transmit data to the modem for modulation, although the modem is actually operating as a half-duplex device.

Modem handshaking

Modem handshaking is the exchange of control signals necessary to establish a connection between two data sets. These signals are required to set up and terminate calls, and the type of signaling used is predetermined according to one of three major standards, such as the Electronics Industries Association (EIA) RS-232 or RS-449 standard or the ITU V.24 recommendation. RS-232 and ITU V.24 standards are practically identical and are used by over 95% of all modems currently manufactured. To better understand modem handshaking, let us examine the control signals used by 103-type modems. The handshaking signals of 103-type modems and their functions are listed in Table 4.20, while the handshaking sequence is illustrated in Figure 4.40.

The handshaking routine commences when an operator at a remote terminal dials the telephone number of the computer or uses an intelligent modem to dial a predefined telephone number. At the computer site, a ring indicator (RI) signal at the answering modem is set and passed to the computer. The computer then sends a data terminal ready (DTR) signal to its modem, which then transmits a tone

Table 4.20 Modem handshaking signals and their functions

Control signal	Function
Transmit data	Serial data sent from device to modem
Receive data	Serial data received by device
Request to send	Set by device when user program wishes to transmit
Clear to send	Set by modem when transmission may commence
Data set ready	Set by modem when it is powered on and ready to transfer data; set in response to data terminal ready
Carrier detect	Set by modem when signal present
Data terminal ready	Set by device to enable modem to answer an incoming call on a switched line; reset by adaptor disconnecting call
Ring indicator	Set by modem when telephone rings

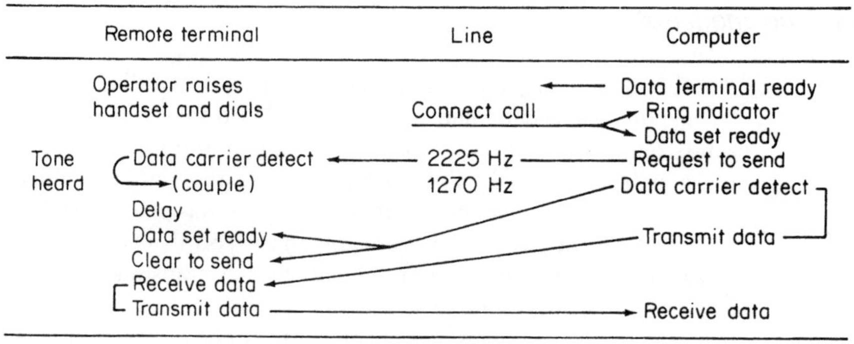

Figure 4.40 Handshake sequence

signal to the modem connected to the terminal. Upon hearing this tone when manually dialing, the terminal operator presses the data pushbutton on the modem. Upon depression of the data button for manually operated modems, the originating modem sends a data set ready (DSR) signal to the terminal, and the answering modem sends the same signal to the computer. At this point in time both modems are placed in the data mode of operation. When a call is dialed by an originating modem, the previously described process occurs automatically without operator intervention.

When accessing a remote computer, the distant device normally transmits a request for identification to the terminal. To do this the computer sets request to send (RTS) which informs its modem that it wishes to transmit data. The modem will respond with the clear to send (CTS) signal and will transmit a carrier signal. The computer's port detects the clear to send and carrier ON signals and begins its data transmission to the terminal. When the computer completes its transmission it drops the DTR signal, and the computer's modem then terminates its carrier signal. Depending upon the type of circuit on which transmission occurs, some of these signals may not be required. For example, on a switched two-wire telephone line, the RTS signal determines whether a terminal is to send or receive data, whereas on a leased four-wire circuit RTS can be permanently raised. For further information the reader should refer to specific vendor literature or appropriate technical reference publications.

Modem testing and problem resolution

Although most modems are extremely reliable devices that can be expected to provide users with years of communications capability, their complexity can result in a variety of problems. Some problems can result from the improper use or setting of one or more of the features built into modems, while other problems can be caused by equipment used with modems. In addition, modems are similar to other electronic devices in that over a period of time an increasing percentage of devices will fail. Regardless of the cause of the problem, modem users have a common goal to identify and correct communications problems.

To assist modem users in resolving problems most devices contain a series of built-in testing capabilities. In addition, all external modems have indicators that can provide users with the ability to resolve many types of communications problems.

Using modem indicators

If you examine the front panel of many external modems you will usually notice a series of mnemonic labels. These mnemonic labels are associated with light emitting diodes (LEDs) that are mounted under the modem's front panel. By understanding the meaning of the illumination or lack of illumination of the LEDs associated with a mnemonic label you can obtain an understanding of the cause of many communications problems as well as ascertain the status of a communications session you previously initiated.

For illustrative purposes we will first examine the front panel indicators of the U.S. Robotics Courier 2400 modem. This popular V.22 bis compatible modem has nine indicators as illustrated in Figure 4.41(a). Although we will primarily focus our attention upon the Courier 2400, you should note that there are no standards that define the inclusion, use, or labeling of modem status indicators. To indicate this fact, Figure 4.41(b) illustrates the front panel indicators for the Telebit T1000 modem. This modem is V.22 bis compatible and like the Courier 2400 is designed for use on the PSTN. As we examine the indicators on the Courier 2400 modem we will compare and contrast them to those on the T1000. This will show that although mnemonics may vary between modems, for most devices the meaning associated with indicators will be very similar. Although the Courier 2400 and Telebit T1000 modems can be considered as relics in an era of 28.8 and 33.6 kbps devices, their indicator panels are very similar in composition to most of the more modern modems. Thus, we can obtain an indication of the use of modem indicators for a wide range of products by reviewing the indicators on those modems.

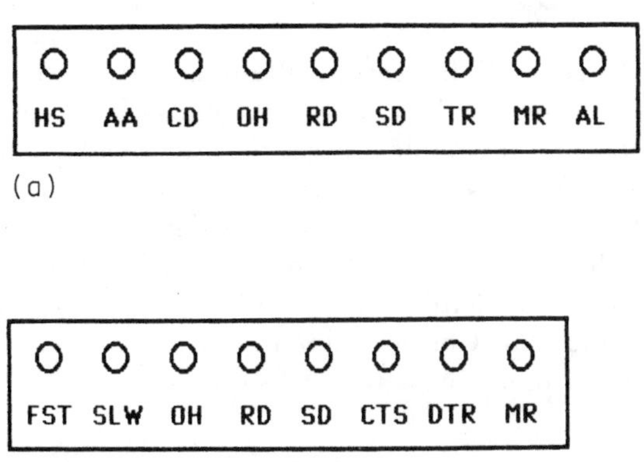

Figure 4.41 Modem indicators: (a) U.S. Robotics Courier 2400; (b) Telebit T1000

HS Indicator

The HS indicator illuminates when the Courier 2400 is communicating with another modem at 2400 bps, which is the highest speed at which that U.S. Robotics modem can operate. Although the illumination of the HS indicator indicates an operating rate of 2400 bps on a Courier 2400, its illumination on a different modem can indicate a different operating rate. For example, some 33.6 kbps V.34 modems have an HS indicator which, when lit, indicates an operating rate of 33.6 kbps. One common problem whose cause can be isolated through the use of the HS indicator is 'slow response'.

Upon occasion, a prior personal computer user may have changed communications software to operate a modem at a lower operating rate, such as at 1200 bps for a particular application. When the next person uses the computer to dial a different modem the prior operating speed selection may force the modem into 1200 bps operation. Other causes of the 'slow response' problem can be a defective answering modem that operates only at 1200 bps, or the user dialing a 1200 bps rotary instead of a 2400 bps rotary. Regardless of the cause, by examining the HS indicator you will note that the reason for slow response is not an overloaded computer system but the local or remote modem. Then, you can examine the data rate setting of your communications program and the telephone number dialed to further isolate the cause of the problem.

In examining the mnemonic status indicators for the T1000 modem, note that this device has two similar indicators to the Courier 2400's HS indicator—FST and SLW. The FST indicator illuminates when the T1000 is operating in its fast transmission mode using its proprietary Packetized Ensemble Protocol. When this indicator is illuminated, it informs you that the modem is communicating with another device that is also using the Packetized Ensemble Protocol. The SLW indicator indicates that the modem is operating in one of several slow modes—Bell System 103, 212A, or ITU V.22 or V.22 bis. Thus, you can use the FST and SLW indicators of the T1000 in a manner similar to that described for the Courier 2400 HS indicator.

AA Indicator

The Courier AA indicator is illuminated when the modem is powered on, on-line with a calling modem, and in its auto-answer mode of operation. This indicator denotes that the modem is set to receive calls instead of originate calls. Since the Courier auto-answer capability is set by placing DIP switch element 5 in an upward position, the failure of the modem to answer calls and the lack of illumination of the AA indicator would denote an improper DIP switch element setting. You can also change the modem's mode to allow it to answer calls by issuing an appropriate command. The Courier 2400, like all Hayes compatible modems, can be placed in an answer mode by setting its S0 register to a non-zero value.

CD indicator

The carrier detect (CD) indicator is illuminated when the local Courier 2400 has received a carrier signal from a distant modem that it recognizes. The iluminator of this indicator tells you that a valid carrier signal exists between the local and remote modem and that data transmission is possible.

If the CD indicator should go off during a communications session its lack of illumination informs you that data transmission is no longer possible. The cause of a lack of continuity between modems can range from noise on a line to the remote modem losing power and you should simply redial whenever you notice that the CD indicator has become extinguished.

CTS Indicator

In comparison to the Courier 2400 note that the T1000 does not have a CD indicator. Instead, you can turn on the modem's speaker to hear the presence or absence of a carrier tone or observe the CTS indicator status.

The clear to send (CTS) indicator is illuminated when the modem is ready to accept data from the attached terminal device for modulation. If no carrier signal is present the modem will not modulate data and its CTS indicator will not be illuminated.

OH Indicator

The off hook (OH) indicator is illuminated when the Courier 2400 has taken control of the telephone line. The illumination of this indicator denotes that the modem has successfully performed a function similar to a person picking up a telephone handset. The lack of illumination of the OH indicator denotes the failure of the modem to take control of the line, a problem usually resulting from someone previously disconnecting the modem from the telephone line. If this indicator does not illuminate when you issue a dialing command you should check the cable from the modem to the telephone wall jack to verify it is correctly connected.

RD Indicator

The receive data (RD) indicator of the Courier modem will flash when a data bit is received from the telephone line or when the modem is sending a result code to an attached terminal device. You can use the RD indicator to isolate several types of communications problems. As an example, assume the RD indicator is flashing but no data is being displayed on an attached personal computer. This would indicate that either the cable from the modem to the computer is defective or the personal computer's serial port has failed. Although the T1000 modem has the same RD indicator, readers should note that on some modems this indicator is labeled RX.

SD Indicator

The send data (SD) indicator functions in a reverse manner to the RD indicator, that is, the SD indicator flashes each time a bit is sent to the Courier 2400 modem from an attached terminal device.

You can use the SD indicator to verify that data is reaching the modem from an attached terminal device once a carrier detect signal is present. This can be accomplished by pressing characters on the terminal keyboard and observing the

SD indicator. If the SD indicator flashes, this indicates that data from the attached terminal is reaching the modem and it also verifies that there is continuity on pin 2 of the cable connecting the terminal and modem.

TR indicator

The terminal ready (TR) indicator on the Courier 2400 modem reflects the state of the terminal device's data terminal ready (DTR) control signal. The TR indicator is thus labeled DTR on the T1000 modem as well as on many other modems.

When the TR indicator is illuminated this denotes that the terminal or computer port is ready to establish communications. Similarly, a lack of illumination indicates that the attached terminal or computer port is not ready to establish communications. You can easily observe the latter state by powering on your modem prior to powering on an attached personal computer. Only after the computer is powered on and you have loaded communications software and prepared your program to go on-line will the TR indicator illuminate.

MR indicator

The modem ready (MR) indicator illuminates when the Courier 2400 modem is powered on. When you first power on your modem the MR indicator should illuminate to inform you that the device is operational.

AL indicator

The analog loopback (AL) indicator is illuminated when your modem is in an analog loopback mode of operation. When your modem is placed in this mode of operation data transmitted from the attached terminal device is sent to the modem's transmitter where it is modulated and looped back to the modem's receiver. The receiver demodulates the data and sends it back to the attached terminal device where you can visually verify the operation of the modem.

When the AL indicator is illuminated the modem is not connected to the telephone line and cannot transmit data. Thus, you must enter the appropriate command to place the modem back into its data mode to communicate with another modem.

Other indicators

Three other common modem indicators that warrant discussion are CS or CTS, ARQ, and SYN.

The CS or CTS indicator denotes the state of the RS-232 clear to send control signal. When the indicator is illuminated, this fact tells you that the modem has received a valid carrier tone and is ready to modulate data. The ARQ (automatic repeat request) indicator illuminates when the modem is in its MNP error detection and correction mode, having successfully made an MNP connection with a remote modem. The SYN (synchronous) indicator is applicable to modems that can operate synchronously. This indicator illuminates when the modem is placed into its synchronous mode of operation.

Modem testing

To assist users in isolating problems, most modems include a built-in diagnostic testing capability. This diagnostic testing capability ranges in scope from a simple self-test of the modem's circuitry to several more sophisticated types of loopback tests.

Self-test

Most modems include a self-test capability that is either initiated as a stand-alone test or is used in conjunction with a loopback test. The basic modem self-test is designed to verify the operation of the modem's internal circuitry and uses a pattern generator to produce a known sequence of data which will be used by the transmitter.

Figure 4.42 illustrates the operation of a basic modem self-test. Once initiated, the modem's transmitter is tied to its receiver breaking any prior connection to a communications line. Data modulated by the transmitter is demodulated by the receiver and passed to a pattern comparator which consists of circuitry as well as a section of ROM that contains the same data sequence which is generated by the pattern generator. The pattern comparator thus allows the demodulated data to be compared to the sequence generated when the self-test is initiated. If the received data does not match the contents of the pattern comparator the modem's circuitry will cause a status indicator, normally labeled ER for error, to illuminate. This action informs you that the modem is defective and should be returned to the manufacturer.

The basic self-test previously described was originally limited to inclusion in leased line modems. This was because most switched network modems can be easily attached to terminals or personal computers which can be used as a pattern generator. The operator's vision can then be used as a pattern comparator. Today, a growing percentage of switched network modems include a self-testing capability that can be used with one or more loopback modes of operation to test a local or remote modem or the communications path between modems.

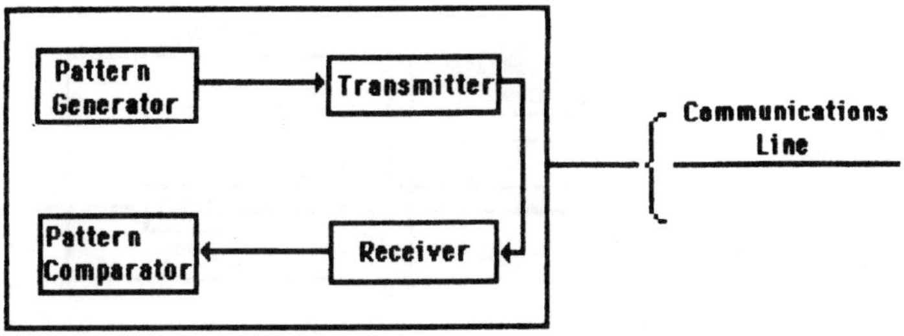

Figure 4.42 Modem self-test

Loopback tests

There are four types of loopback tests modems may be able to perform. Each of these tests is illustrated in Figure 4.43. Readers should note that the loopback testing capability of modems varies both by manufacturer as well as by a manufacturer's product line. Although most leased line modems will be capable of initiating all four loopback tests illustrated in Figure 4.43, many switched network modems are limited to performing a lesser number of loopback tests.

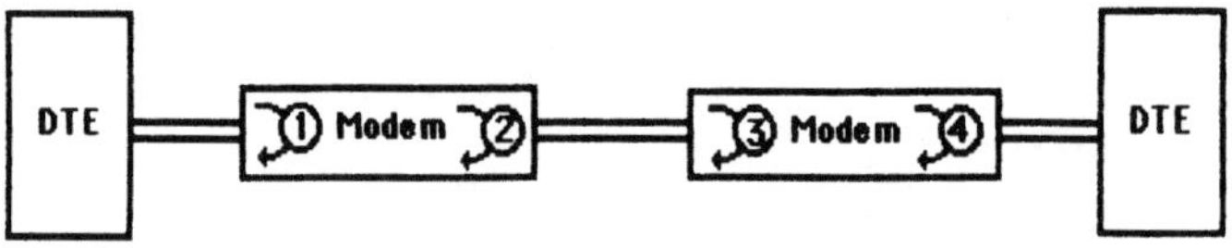

Figure 4.43 Modem loopbacks: 1 local digital loop, 2 local analog loop, 3 remote analog loop, 4 remote digital loop

The local digital loop test illustrated in Figure 4.43 is employed to test the operation of the data terminal equiment (DTE) to include determining if data is leaving the terminal or computer port. In comparison, the local analog loop tests the digital and analog circuits of the modem to determine if they are in working condition. The remote analog loop is used for testing the line to the remote modem, while the remote digital loop is used to check the operating condition of the line and the local and remote modems. Now that we have an overview of the use of each loopback test, let us examine them in more detail.

Local digital loopback

The initiation of a digital loopback test causes the local modem to tie its transmitter and receiver sections together, in effect, bypassing its modulator and demodulator. The input data received from a connected DTE on pin 2 of an RS-232 conductor are thus routed to pin 3 which is the receive data conductor. In effect, a local digital loop results in the establishment of two loops as illustrated in Figure 4.44.

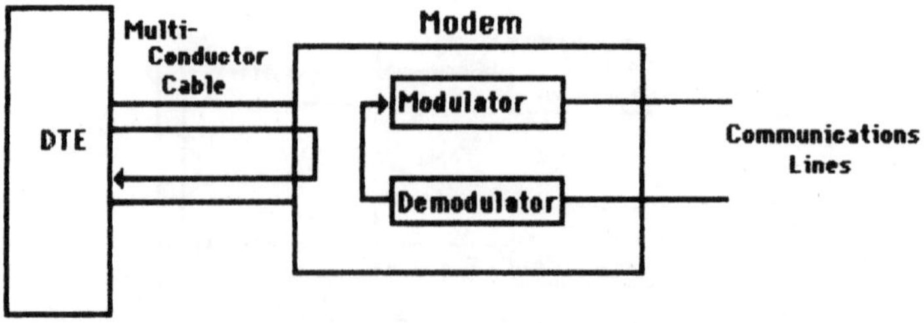

Figure 4.44 Local digital loop test

A local digital loop can be used to test the operation of the DTE as well as verify the multiconductor cable connecting the DTE to the modem. To accomplish those functions you would first initiate a local digital loop.

Once this is accomplished, you can type on the keyboard of your terminal or personal computer and visually observe the characters echoed back to your terminal or PC. After you are satisfied that your DTE and cable are performing correctly you can terminate the loopback.

Local analog loopback

The local analog loopback test is designed to verify the operation of the analog circuits of a modem, including its modulator and demodulator. During this test two loops are established, resulting in the modem's modulator being looped to its receiver, while the send and receive paths of the communications line are tied together. The local analog loop test is illustrated in Figure 4.45.

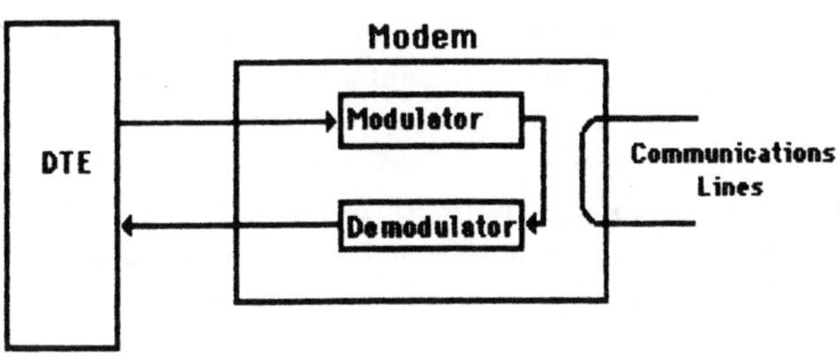

Figure 4.45 Local analog loop test

Remote analog loopback

The remote analog loopback results in the remote modem bridging the transmit and receive wire pairs of a leased line together. This allows you to use test equipment at the local site to determine the quality of the line. Since modems designed for use on the PSTN operate on a two-wire system this loopback test is not included in switched network modems.

Remote digital loopback

The remote digital loopback test results in the remote modem demodulating received data and then passing it to its transmitter for remodulation back onto the line. This test is similar to the local digital loopback test previously illustrated in Figure 4.44. Once the remote modem is placed into its digital loopback mode of operation data transmitted from the local site is looped at the remote modem back to the data originator, in effect, providing a mechanism for checking both the quality of the circuit as well as the operation of the local and remote modems.

While analog and digital loopbacks are the main tests built into modems, several vendors offer additional diagnostic capabilities that may warrant attention. A few of these diagnostic tests that deserve mention include bit error rate testing and alarm threshold monitoring. Normally these two diagnostic functions are implemented by the user obtaining a network control system (NCS) designed to operate with modems manufactured by the company that produced the NCS. Typically, the NCS is a microcomputer-based system that monitors the status of the modems in an organization's network, generates alarms when certain predefined conditions occur, and in some cases it actually performs corrective action prior to a disruption in network operation occurring.

The NCS is normally installed at the central computer site and on a periodic basis queries the status of the remote modems on both point-to-point and multidrop lines. Through the 1980s most NCS systems worked in conjunction with modems that perform testing and status queries via the use of a secondary channel. This permits data transmission to continue on the primary channel, with testing and responses to the testing flowing concurrently at a lower data rate on the secondary channel. Since then NCS systems were developed to work in conjunction with intelligent modems designed to recognize certain datapatterns as a request to perform testing and/or issue status responses. When used with this type of modem, one's actual information flow of data is interrupted by the testing.

To allow users to implement network control functions in a data communications network without having to replace existing modems, some vendors offer what is commonly called a 'modem wraparound unit' or modem network adapter. Such devices are cabled to both ends of the modem and in schematic diagrams appear to wraparound the modem, hence the term 'wraparound unit'. Figure 4.46 illustrates the utilization of an NCS manufactured by one vendor with local and remotely located modems manufactured by a second vendor by the employment of two modem wraparound units. Through the use of the wraparound units, the NCS can conduct modem and line tests, even when the modems are manufactured by different vendors.

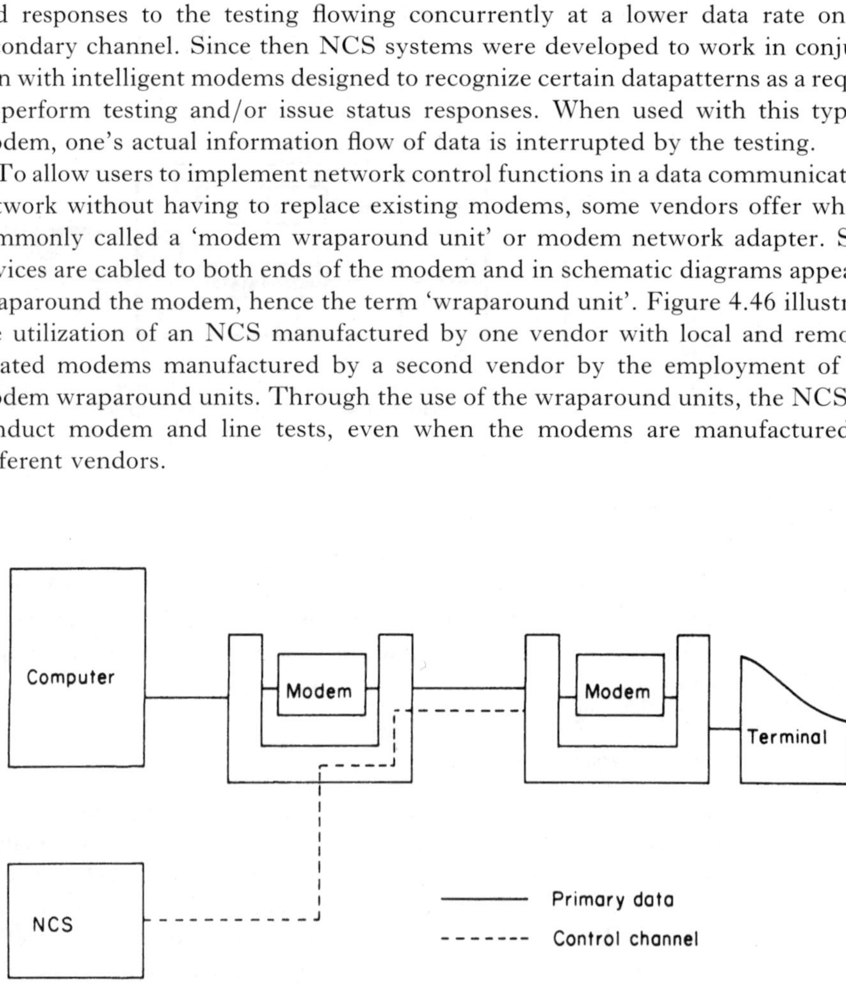

Figure 4.46 Modem wraparound unit utilization. The modem wraparound unit permits the use of a network control station produced by one vendor with modems manufactured by other vendors

4.3 INTELLIGENT MODEMS

In this section we will examine how microprocessors and read only memory (ROM), random access memory (RAM), and erasable programmable read only memory (EPROM) are used to add intelligence to modems. Although the ability to add intelligence to modems can provide an almost infinite number of features and functions, we will focus our attention upon three key areas that are normally associated with the term 'intelligent modem'—command sets, error detection and correction, and data compression.

The command set used by a modem governs its operational capability, enabling automatic dialing and redialing of telephone numbers, selection of an operating transmission rate, and other functions to be performed via communications software or by a person typing commands from his or her terminal or personal computer to the modem.

An error detection and correction capability results in the transmitting modem adding a special 'trailer' to each block of data which is formed according to a predefined algorithm. The receiving modem performs the same algorithm on each received data block and compares its computation to the transmitted 'trailer'. If the transmitted and computed 'trailers' match, the data block is considered to be received correctly. If they do not match, the receiving modem will transmit this fact to the transmitting modem, which will then retransmit the block. The correction of data thus occurs by retransmission.

When an intelligent modem performs data compression it examines data for redundancy, replacing, as an example, strings of characters by a special code and count which the receiving modem recognises and uses to regenerate the data back into its original form. Through the use of data compression a modem's effective information transfer capability is increased since each character transmitted may represent more than one data character.

Command sets

During the 1970s, a small communications company called Bizcom developed the concept of adding intelligence to modems through the use of codes that the modem would recognize and act upon. The series of codes that a modem is designed to interpret and act upon is referred to as its command set.

In addition to the pioneering work of Bizcom, other modem manufacturers developed similar performing modems. Unfortunately, during the early days of intelligent modem developments the command sets built into modems manufactured by different vendors were more than likely to be incompatible with one another. Many communications software products were thus originally restricted to operating with a specific modem or a few modems based upon the command sets the software supported. Due to the popularity of the Hayes Microcomputer Products series of Smartmodems™ which at one time captured over 50% of the personal computer modem market, that firm's command set emerged as a *de facto* industry standard. During the 1980s many modem manufacturers incorporated a similar command set into their products, resulting in a high level of compatibility between communications software that supports the command set used by Hayes Smartmodems and modems manufactured by other vendors.

Today, many modem manufacturers advertise their products as 'Hayes compatible'; however, this term can be misleading as it may not indicate whether or not 'Hayes compatible' software will support all of the features of a particular modem. In addition, communications software advertised as supporting the Hayes command set may not support all of the features of a specific Hayes Microcomputer Products Smartmodem. To understand why these incompatibilities can occur and how they can be alleviated, let us first examine the Hayes command set and then focus our attention upon several modem products and the command sets they support.

The Hayes command set

The Hayes command set actually consists of a basic set of commands and command extensions. The basic commands, such as placing the modem off-hook, dialing a number and performing similar operations are common to all Hayes modems. The command extensions, such as placing a modem into a specific operating speed are only applicable to modems built to transmit and receive data at that speed.

The commands in the Hayes command set are initiated by transmitting an attention code to the modem, followed by the appropriate command or set of commands one desires the modem to implement. The attention code is the character sequence AT, which must be specified as all uppercase or all lowercase letters. The requirement to prefix all command lines with the code AT has resulted in many modem manufacturers denoting their modems as Hayes AT compatible.

The command buffer in a Hayes Smartmodem holds 40 characters, permitting a sequence of commands to be transmitted to the modem on one command line. This 40-character limit does not include the attention code, nor does it include spaces included in a command line to make the line more readable. Table 4.21 lists the major commands included in the basic Hayes command set. Other modems, such as Telebit's Trailblazer employ a command set that can be considered to be a superset of the Hayes command set. While communications software that uses the Hayes command set will operate with this modem, such software cannot utilize the full potential of the modem. This is because the Hayes command set does not support several Telebit Trailblazer features, such as its proprietary method of error detection.

The basic format required to transmit commands to a Hayes compatible intelligent modem is shown below.

AT Command[Parameter(s)]Command[Parameter(s)] ... Return

Each command line includes the prefix AT, followed by the appropriate command and the command's parameters. The command parameters are usually the digits 0 or 1, which serve to define a specific command state. As an example, H0 is the command that tells the modem to hang up or disconnect a call, while H1 is the command that results in the modem going off-hook, which is the term used to define the action that occurs when the telephone handset is lifted. Since

Table 4.21 Basic Hayes command set (major commands)

Command	Description
AT	Attention
A	Answer call
A/	Repeat last command
B	Select the method of modem modulation
C*	Turn modem's carrier on or off
D	Dial a telephone number
E	Enable or inhibit echo of characters to the screen
F*	Switch between half- and full-duplex modem operation
H	Hang up telephone (on hook) or pick up telephone (off hook)
I	Request identification code or request check sum
L	Select the speaker volume
M	Turn speaker off or on
O	Place modem on-line
P	Pulse dial
Q	Request modem to send or inhibit sending of result code
R*	Change modem mode to 'originate-only'
S	Set modem register values
T	Touch-tone dial
V	Send result codes as digits or words
X	Use basic or extended result code set
Y	Enable or inhibit long space disconnect
Z	Reset the modem
+ + +	Escape command

Commands followed by an asterisk (*) are not applicable to the Hayes V-series Smartmodem 2400 but are recognized by earlier Hayes modems and modems manufactured by many other vendors.

many commands do not have parameters, those terms are enclosed in brackets to illustrate that they are optional. A number of commands can be included in one command line as long as the number of characters does not exceed 40, which is the size of the modem's command buffer. Finally, each command line must be terminated by a carriage return character.

Command use

To illustrate the utilization of the Hayes command set let us assume we desire to automatically dial New York City information. First, we must tell the modem to go off-hook, which is similar to one manually picking up the telephone handset. Then we must tell the modem the type of telephone system we are using, pulse or touch-tone, and the telephone number to dial. Thus, if we have a terminal or personal computer connected to a Hayes compatible modem, we would send the following commands to the modem:

AT H1
AT DT1,212-555-1212

In the first command, the 1 parameter used with the H command places the modem off-hook. In the second command, DT tells the modem to dial (D) a telephone number using touch-tone (T) dialing. The digit 1 was included in the telephone number since it was assumed we have to dial long-distance, while the comma between the long-distance access number (1) and the area code (212) causes the modem to pause for 2 s prior to dialing the area code. This 2 s pause is usually of sufficient duration to permit the long-distance dial tone to be received prior to dialing the area code number.

Since a Smartmodem automatically goes off-hook when dialing a number, the first command line is not actually required and is normally used for receiving calls. In the second command line, the type of dialing does not have to be specified if a previous call was made, since the modem will then use the last type specified. Although users with only pulse dialing availability must specify P in the dialing command when using a Hayes Smartmodem, several vendors now offer modems that can automatically determine the type of dialing facility the modem is connected to and then use the appropriate dialing method without requiring the user to specify the type of dialing. For other non-Hayes modems, when the method of dialing is unspecified, such modems will automatically attempt to perform a touch-tone dial and, if unsuccessful, then redial using pulse dialing.

To obtain an appreciation of the versatility of operations that the Hayes command set provides, assume two personal computer users are communicating with one another. If the users wish to switch from modem to voice operations without hanging up or redialing, one user would send a message via the communications program he or she is using to the other user indicating that voice communications is desired. Then, both users would lift their telephone handsets and type + + + (Return) ATH (Return) to switch from on-line operations to command mode (hang-up). This will cause the modems to hang-up, turning off the modem carrier signals and permitting the users to converse.

Result codes

The response of the Smartmodem to commands is known as result codes. The Q command with a parameter of 1 is used to enable result codes to be sent from the modem in response to the execution of command lines while a parameter of 0 inhibits the modem from responding to the execution of each command line.

If the result codes are enabled, the V command can be used to determine the format of the result codes. When the V command is used with a parameter of 0, the result codes will be transmitted as digits, while the use of a parameter of 1 will cause the modem to transmit the result codes as words. Table 4.22 lists the Basic Results Codes set of the Hayes Smartmodem 2400. As an example of the use of these result codes, let us assume the following commands were sent to the modem:

AT Q0
AT V1

The first command, AT Q0, would cause the modem to respond to commands by transmitting result codes after each command line is executed. The second command, AT V1, would cause the modem to transmit each result code as a word code. Returning to Table 4.22 this would cause the modem to generate the word

Table 4.22 Smartmodem 2400 result codes

Digit code	Word code
0	OK
1	CONNECT
2	RING
3	NO CARRIER
4	ERROR
5	CONNECT 1200
6	NO DIALTONE
7	BUSY
8	NO ANSWER (replaces NO CARRIER if the @ is included in the dial string)
9	Reserved for future use
10	CONNECT 2400

code 'CONNECT' when a carrier signal is detected. If the command AT V0 was sent to the modem, a result code of 1 would be transmitted by the modem, since the 0 parameter would cause the modem to transmit result codes as digits.

By combining an examination of the result codes issued by a Smartmodem with the generation of appropriate commands, software can be developed to perform such operations as redialing a previously dialed telephone number to resume transmission in the event a communications session is interrupted and automatically answering incoming calls when a ring signal is detected.

Extended AT commands

A large number of extended AT commands were developed by Hayes Microcomputer Products to control the operation and utilization of special modem features. Table 3.21 lists the extended AT commands supported by the Hayes V-series Smartmodem 2400. Note that each of these commands uses the ampersand (&) character to indicate it is an extension to the basic command set.

Hayes Smartmodems operating above 2400 bps support most, if not all, of the extended commands listed in Table 4.23 while adding certain extended commands that reflect the addition of certain features beyond those included in the V-series Smartmodem 2400. For example, a Hayes Smartmodem 9600 that supports the V.32 standard includes the extended commands &U0 and &U1. The &U0 extended command enables trellis coding when the modem operates at 9600 bps, while the &U1 extended command disables trellis coding.

All extended AT commands are initiated in the same manner as basic commands and have the same rules and constraints with respect to their use. For example, the &C command must be prefixed by 'AT' or 'at' to be recognized by the modem. Similarly, the modem must be in an off-line or an on-line command mode to be able to recognize and act upon the command. You should not confuse the Hayes V-series Smartmodem 2400 extended AT commands listed in Table 4.23 with the term 'extended AT commands' many modem vendors advertise. The latter is

Table 4.23 Enhanced AT command set (Hayes V-series Smartmodem 2400)

Command	Description
&C0	Assume data carrier always present
&C1	Track presence of data carrier
&D0	Ignore DTR signal
&D1	Assume command state when an on-to-off transition of DTR occurs
&D2	Hang up and assume command state when an on-to-off transition of DTR occurs
&D3	Reset when an on-to-off transition of DTR occurs
&D4	Reset when enter low power mode when DTR is low
&G0	No guard tone
&G1	550 Hz guard tone
&G2	1800 Hz guard tone
&J0	RJ-11/RJ-41S/RJ-45S telco jack
&J1	RJ-12/RJ-13 telco jack
&K0	Disable local flow control
&K3	Enable RTS/CTS local flow control
&K4	Enable XON/XOFF local flow control
&K5	Enable transparent XON/XOFF local flow control
&M0	Asynchronous mode
&M1	Synchronous mode 1
&M2	Synchronous mode 2
&M3	Synchronous mode 3
&R0	Track CTS according to RTS
&R1	Ignore RTS; always assume presence of CTS
&S0	Assume presence of DSR signal
&S1	Track presence of DSR signal
&T0	Terminate test in progress
&T1	Initiate local analog loopback
&T3	Initiate local digital loopback
&T4	Grant request from remote modem for RDL
&T5	Deny request from remote modem for RDL
&T6	Initiate remote digital loopback
&T7	Initiate remote digital loopback with self test
&T8	Initiate local analog loopback with self test
&W0	Save storable parameters of active configuration as profile 0
&W1	Save storable parameters of active configuration as profile 1
&X0	Modem provides transmit clock signal
&X1	Data terminal provides transmit clock signal
&X2	Receive carrier provides transmit clock signal
&Y0	Specify stored user profile 0 as power-up configuration
&Y1	Specify stored user profile 1 as power-up configuration
&Z	Store telephone number

usually a term used by vendors other than Hayes Microcomputer Products to define the commands used by their products to access and control such features as MNP, V.42, V.42 bis, and proprietary methods of error control and data compression. Microcom has defined a series of AT command extensions to control and access error control, data compression, and flow control. Used by the Microcom AX series of modems, the extensions have been adopted by Intel, Okidata, Practical Peripherals, Prometheus, and Ven-Tel. Modems manufactured by other vendors normally use unique commands and S register settings to access and control those previously mentioned features.

Modem registers

A third key to the degree of compatibility between non-Hayes and Hayes Smartmodems is the number, use and programmability of registers contained in the modem. Hayes Smartmodems contain a series of programmable registers that govern the function of the modem and the operation of some of the commands in the modem's command set. Table 4.24 lists the functions of the first 13 registers built into the Hayes Smartmodem 2400, to include the default value of each register and the range of settings permitted. These registers are known as S registers, since they are set with the S command in the Hayes command set. In addition, the current value of each register can be read under program control, permitting software developers to market communications programs that permit the user to easily modify the default values of the modem's S registers.

Table 4.24 S register control parameters

Register	Function	Default value	Range
S0	Ring to answer on	*	0–255
S1	Counts number of rings	0	0–255
S2	Escape code character	ASC11 43	ASC11 0–127
S3	Carriage return character	ASC11 13	ASC11 0–127
S4	Line feed character	ASC11 10	ASC11 0–127
S5	Backspace character	ASC11 8	ASC11 0–127
S6	Dial tone wait time (s)	2	2–255
S7	Carrier wait time (s)	30	1–255
S8	Pause time caused by comma (s)	2	0–255
S9	Carrier detect response time (1/10s)	6	1–255
S10	Time delay between loss of carrier and hang-up (1/10s)	7	1–255
S11	Touch-tone duration and spacing time (ms)	70	50–255
S12	Escape sequence guard time in units of 20 ms	50	0–255

*Usually depends on a switch setting

To understand the utility of the ability to read and reset the values of the modem's S registers, consider the time period a Smartmodem waits for a dial tone prior to going off-hook and dialing a telephone number. Since the dial tone wait time is controlled by the S6 register, a program offering the user the ability to change this wait time might first read and display the setting of this register during the program's initialization. The reading of the S6 register would be accomplished by the program sending the following command to the modem:

AT S6?

The modem's response to this command would be a value between 2 and 255, indicating the time period in seconds that the modem will wait for a dial tone. Assuming the user desires to change the waiting period, the communications program would then transmit the following command to the modem, where n would be a value between 2 and 255.

AT S6 $= n$

Compatibility

For a non-Hayes modem to be fully compatible with a Hayes modem, command-set compatibility, result-codes compatibility and modem-register compatibility is required. Of the three, the modem-register compatibility is usually the least important and many users may prefer to consider only command-set and response-codes compatibility when acquiring intelligent modems.

The rationale for omitting register compatibility from consideration is the fact that many non-Hayes modem vendors manufacture 'compatible' modems using the default values of the Hayes Smartmodem registers. This enables those manufacturers to avoid building the S registers into their modems, reducing the size, complexity and often also reducing the price of their modem. Thus, if the default values of the S registers are sufficient for the user and the modem under consideration is both command-set and result-code compatible the issue of register compatibility can normally be eliminated as an acquisition factor to consider.

A second rationale for omitting register compatibility from consideration is the fact that other non-Hayes modem vendors incorporate 100 or more registers into their products, usually with their first 13 being compatible to the first 13 Hayes S registers. The remaining registers are used mainly to control such advanced modem features as flow control, protocol spoofing, data compression, and error control.

Of the first 13 S registers perhaps the most critical to understand is the S7 register. This register is used by an originating modem to set the time that it will wait to receive a carrier tone from a distant modem prior to hanging up. For example, assuming you access a 9600 bps modem with a 1200 or 2400 bps modem, the setting of the S7 register will normally govern whether or not you will be able to establish communications. To understand the criticality of the S7 register it is important to understand how the remote modem operates upon receiving a call.

When a remote 9600 bps modem receives a call it first attempts to negotiate an operating rate. In response to the incoming call it responds with a carrier tone at a predefined frequency. If it does not receive a recognized response within approximately 15 s and supports multiple modulation schemes that are enabled the remote modem then transmits a second carrier tone which represents a lower operating rate. This process continues until the remote modem either receives a recognized carrier tone from the originating modem or the seting of the remote modem's S7 register is reached. If the latter occurs, the remote modem will disconnect the incoming call. While most data center and packet network dial-in ports have their modem S7 registers set to a minimum of 60 s, most modem users fail to note that when their modem is used to originate calls it may not receive a recognizable carrier tone for 45 or more seconds when accessing a higher-speed modem. Thus, you should normally change the factory default setting of 30 s of the S7 register to a higher value when dialing a high-speed modem with a Bell System 212A or V.22 bis modem. The preceding is also true if you use a V.32 modem set to operate at 9600 bps to access a V.34 modem set to operate at 28 800 bps, or whenever there are more than three operating rate differences between calling and answering modem.

Error detection and correction

Error detection and correction, also known as error control and error correction, has been implemented in a large number of switched network modems using a variety of techniques. Each technique follows a common methodology in that data transmitted by an attached terminal device formally known as data terminal equipment or DTE is first gathered into a block of characters. Next, an algorithm is applied to the block to generate one or more checksum characters that are appended to the block for transmission. The receiving modem performs the same algorithm on the block it receives, less its checksum character or characters. The checksum computed by the receiving modem is referred to as the locally generated checksum, which is compared to the transmitted checksum. If both the locally generated and transmitted checksums are equal, the data block is assumed to have been received error-free. Otherwise, the data block is assumed to have one or more bits in error, and the receiving modem then requests the transmitting modem to retransmit the data block. Thus, error correction is accomplished by retransmission.

Since error correction is accomplished by retransmission, all modems that perform error detection and correction must have buffers to temporarily store data until an acknowledgement from the receiving modem occurs, at which time the transmitting modem can discard the stored data block that was received error-free by the modem at the opposite end of the transmission path. Since modem data buffers are finite in size, a mechanism is required to control the flow of data from the attached terminal to the modem. This mechanism is known as flow control. As all modems that perform error detection and correction must perform flow control, we will first examine this subject area prior to investigating the methods by which error detection and correction are performed by modems.

Flow control

Flow control is a mechanism used to compensate for the difference between the rate data that reaches a device and the rate at which a device processes and transmits data. To illustrate the concept behind flow control, consider Figure 4.47 which illustrates its rationale. Assume this modem has an operating rate of 28 800 bps. If the compression ratio of the modem is 2:1, on the average, every two characters of data entering the modem will be compressed into one character. Thus, the terminal device can be configured to transfer data to the modem at 57 600 bps which is twice the data rate at which the modem transfers data onto the PSTN.

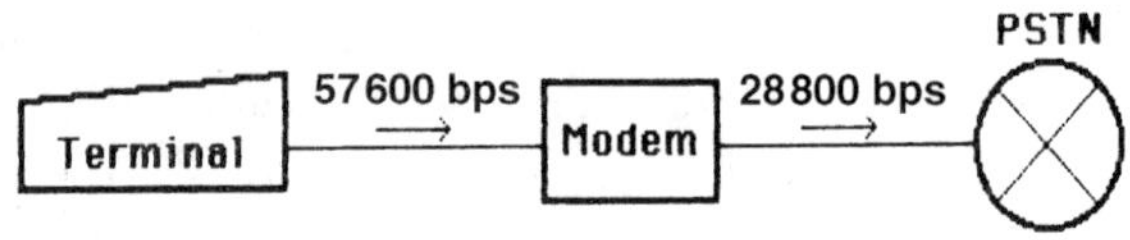

Figure 4.47 Rationale for flow control. When the modem achieves a 2:1 compression ratio, data entering the device at 57 600 bps is immediately placed on the line at 28 800 bps. When the compression ratio is under 2:1 data cannot exit the modem as fast as it enters the modem and is placed into a buffer area. To prevent the buffer from overflowing, the modem requires a mechanism to stop the terminal from sending additional data—a process known as flow control

As long as the modem can compress two characters into one, the terminal device can keep on transferring data to the modem at 57 600 bps. Suppose, however, a portion of the data entering the modem cannot be compressed or is compressible at a ratio less than 2:1. Either situation would cause data flowing into the modem to be lost. To prevent a potential data loss, modems that compress data or perform other functions, including error detection and correction, will include data buffers as temporary storage locations to compensate for the difference between the data flow into the modem and the rate at which the modem can process and transmit data.

Since data buffers represent a finite amount of storage, a modem that accepts data at one rate and transfers it at a different rate must be able to control the flow of data into its buffers. Known as flow control, this technique is used to prevent modem buffers from overflowing and losing data.

To illustrate how modem flow control operates, consider Figure 4.48 which illustrates a modem's buffer storage area. As data enters the modem from an attached terminal device at a faster rate than it can be placed on the line, the buffer begins to fill. When the occupancy of the buffer reaches a predefined high level (H_L) the modem initiates flow control to inhibit additional data from flowing into its buffer. Otherwise, the buffer could continue to fill until data eventually cannot be stored and is then lost. As data flows out of the buffer and is modulated and placed on the line, the point of low occupancy is reached (L_L). When this level of occupancy is reached the modem uses flow control to enable transmission to resume to the modem. Now that we have reviewed why flow control is used in modems, let us examine the methods by which it is implemented.

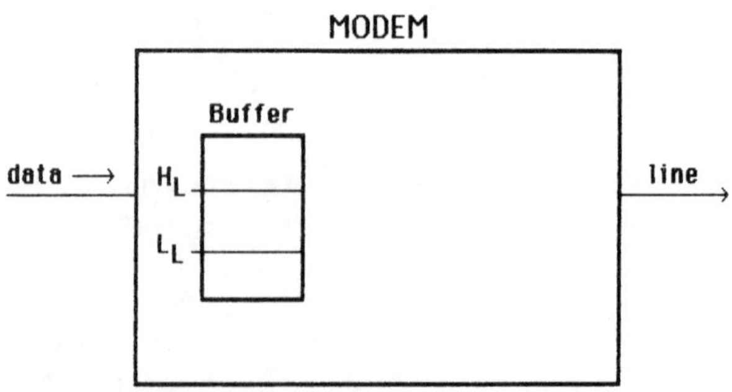

Figure 4.48 Flow control and buffer utilization. H_L = high level to disable data flow, L_L = low level to enable data flow

Methods of flow control

There are three primary methods by which flow control can be implemented, including the use of RTS/CTS control signals and the transmission of the character pairs XON/XOFF and ENQ/ACK. In addition, some devices support the use of a mixture of flow control methods.

RTS/CTS signaling

When a modem receives a RTS (request to send) signal from a terminal device and is ready to receive data, it will respond by raising its CTS (clear to send) signal. One method to control the flow of data into the modem is thus for the modem to turn off its CTS signal whenever it wants to stop the flow of data into its buffers. Whenever the modem wants the flow of data to resume it will then raise or turn on its CTS signal. This method of flow control is referred to as hardware flow control as the controlling signals at the interface govern the operation of the hardware.

XON/XOFF

The use of XON and XOFF characters for flow control is referred to as in-band signaling as the characters flow over pin 3 (receive data) to the attached terminal device. In comparison, the use of the CTS control signal which is not a data signal is referred to as out-band signaling.

Many asynchronous terminal devices recognize an ASCII code of 19, which represents the CTRL-S character, as a signal to suspend transmission. This XOFF signal is also known as a DC3 (device control number 3) character and is issued by the modem when it wants an attached DTE to suspend the flow of data to the modem. Once the modem has emptied its buffer to a low level it will transmit an XON character to the DTE, which serves as a signal for the DTE to resume transmission. The XON character has an ASCII code of 17 and represents the CTRL-Q character. This character is also known as the DC1 (device control number 1) character.

One of the problems associated with the use of XON/XOFF flow control is the use of those characters in a data file by an application program which, when transmitted, can cause a DTE to be infinitely suspended. For example, suppose you are transmitting a WordStar file to another computer system. That word processing program uses DC3s generated by a CTRL-S as markers for the beginning and end of underscoring. If your modem is set to echo characters and you have DC3s in a file, the first DC3 modulated and transmitted to the distant system is also echoed back to the attached terminal device. The DTE detects the DC3 as an XOFF request and correctly responds by suspending transmission and waiting for an XON from the modem which will never come. Here the obvious solution to the problem is to turn echo off for the file transfer. Unfortunately, this obvious solution is not obvious to many modem users, especially when they are transmitting and receiving binary files. For this reason it is highly recommended that you use hardware flow control if you intend to transfer binary files.

ENQ/ACK

The enquire/acknowledge method of flow control is used with certain Hewlett Packard computers and terminal devices. This method of flow control is based upon the DTE sending an enquire (ENQ) and then receiving an acknowledgement (ACK) prior to actually transmitting data. Under the Hewlett-Packard ENQ/ACK protocol the DTE that receives an ACK in response to its ENQ transmits a block of data of variable size that can be up to approximately 2000 characters in length.

When a modem is configured to use ENQ/ACK flow control it will respond to the DTE's ENQ with an ACK when it can accept data. At that time it will release the ENQ to flow to the remote modem.

DTE flow control

The previously described methods of flow control are also applicable in the reverse direction, with DTEs controlling the flow of data.

DTEs are similar to modems that contain buffers, in that when operating communications software, a finite area of storage is reserved for receiving data to be processed. If the received data is transferred to a peripheral device, such as a printer operating slower than the communications line, this situation could result in a loss of data. This is because the modem would be passing demodulated data to the DTE faster than it could empty its buffer area, eventually resulting in a loss of data when the buffer is full. Like modems, DTEs capable of performing flow control may be able to support three methods to accomplish this task: RTS/CTS signaling and the use of the character pairs XON/XOFF and ENQ/ACK.

Under RTS/CTS signaling the DTE will drop its RTS signal as an indicator that it wants to suspend receiving data. In a full-duplex operation the modem will stop sending data to the DTE when RTS drops and resume transmission when the RTS signal is raised. Under half-duplex operations the modem will transmit the contents of its buffer to the DTE and drop its CTS signal in response to the DTE lowering its RTS signal.

When XON/XOFF flow control is used by the DTE the local modem will suspend sending data to that device when it receives an XOFF. Upon receipt of an XON the modem will resume sending data to the DTE.

As previously mentioned, some Hewlett-Packard computers and terminals support the ENQ/ACK protocol. When both modems are configured to use the ENQ/ACK protocol the receipt of an ENQ from a remote DTE is passed through both modems to the local DTE. Until the local DTE responds with an ACK the local modem will not send data to that device.

Now that we have examined the methods by which modems perform flow control let us focus our attention upon the methods by which error detection and correction can be accomplished.

Methods of error detection and correction

Until 1989 error detection and correction was conspicuous by the absence of a *de jure* standard. Several error detection and correction techniques were developed based upon the use of cyclic or polynomial code error detection schemes previously described in Chapter 1. After a modem computes a CRC it is appended to the block of data to be transmitted. The receiving modem uses the same predefined generating polynomial to generate its own CRC based upon the received block and then compares the locally generated CRC with the transmitted CRC. If the two match, the receiving modem transmits a positive acknowledgement to the transmitting modem, which not only informs the distant modem that the data was received correctly but can also serve to inform the remote modem that if additional blocks of date remain to be transmitted the next block can be sent. If an error has occurred, the locally generated CRC will not match the transmitted CRC and the receiving modem will transmit a negative acknowledgement which informs the remote modem to retransmit the previously transmitted data block.

Although most modem manufacturers used the CRC-16 polynomial, $X^{16}+X^{15}+X^{5}+1$, which has the bit composition 11000000000010001, to operate against each data block, incompatibilities between the methods used to block data and transmit negative and positive acknowledgements resulted in the method of error detection and correction employed by one vendor being incompatible with the method used by another modem manufacturer. The one major exception to this incompatibility between vendor error detection and correction methods is the Microcom Networking Protocol (MNP), which has been licensed by Microcom to a large number of modem manufacturers. Until 1989, MNP was considered as a *de facto* standard due to a base of approximately one million modems supporting one or more MNP classes. In 1989 the MNP method of error detection and correction was recognized by the ITU as one of two methods for performing this function when the V.42 recommendation was promulgated.

Rationale

One of the issues that confuse many modem users is the rationale for using a modem's error detection and correction feature. After all, file transfer protocols such as XMODEM, YMODEM, ZMODEM, and their derivatives also provide an error detection and correction capability.

When a modem's error control feature is enabled and operates successfully in conjunction with a distant modem, error detection and correction is operating during the entire communications session. This means that regardless of the function you are performing, whether reading an electronic mail message, transferring a file, or sending a message to SYSOP, your transmission is protected. In comparison, the error detection and correction function embedded into a file transfer protocol operates only during the file transfer. Thus, a logical question you may have is why you should use a file transfer protocol with error detection and correction when your modem performs that function.

Although you need to use a file transfer protocol to transfer a file in an orderly manner, the error detection and correction function of the file transfer protocol may be redundant in most situations, because the file transfer protocol protects data from computer to computer, while the modem's error detection and correction feature protects data from modem to modem. Thus, the former protocol protects data leaving the serial port and flowing to the modem. However, since data errors due to cabling are extremely rare, in most cases the double protection from a file transfer protocol and error correcting modems between modems can actually cause transmission delays and is not necessary. Some modems incorporate a protocol spoofing feature, the use of which eliminates the transmission delays associated with the use of error detection and correction file transfer protocol with error correction modems. If the modem you are using does not support a suitable form of protocol spoofing, you should consider the use of a 'streaming file transfer protocol' to transfer files when using an error correcting modem. This term refers to a protocol that does not perform any error checking and whose use improves the transfer speed of information. An example of a streaming file transfer protocol is YMODEM-G. Since streaming file transfer protocols do not preform error detection and correction, they should only be used with modems operating in their error correction mode of operation.

MNP

The Microcom Networking Protocol (MNP) was developed by the modem manufacturer Microcom, Inc., to provide a sophisticated level of error detection and correction as well as to enhance the data file transfer of intelligent modems. Microcom has licensed their MNP for use by other modem vendors, resulting in a large number of manufacturers incorporating this protocol into their products. Today, the MNP error-correcting protocol is considered a *de facto* industry standard, with over ten million modems having this feature when this book was published.

The MNP protocol was designed in a layered fashion like the OSI reference model developed by the International Organization for Standardization. MNP contains three layers instead of the seven layers in the OSI reference model. Figure 4.49 illustrates the correspondence between the OSI reference model and the MNP protocol.

The MNP link layer is responsible for establishing a connection between two devices. Included in the link layer is a set of negotiations that are conducted between devices to enable them to agree upon such factors as the transmission mode (full-or half-duplex), how many data messages can be transmitted prior to requiring a confirmation and how much data can be obtained in a single message. After these values are established the link layer initiates the data transfer process as well as performing error detection and correction through the use of a frame checking scheme.

Figure 4.50 illustrates the format of an MNP frame of information which has similarities to both bisynchronous and HDLC communications. Each frame contains three bytes which act as a 'start flag'. The SYN character tells the receiver that a message is about to arrive, the combination of data link escape (DLE) and start of text (STX) informs the receiver that everything following is part of the message. The first header describes the user data, such as the duplex setting, number of data messages before confirmation, etc. The session header defines additional information about the transmitted data which enables the automatic negotiation of the level of service that can be used between devices communicating with one another. Currently there are ten versions or classes of the MNP protocol

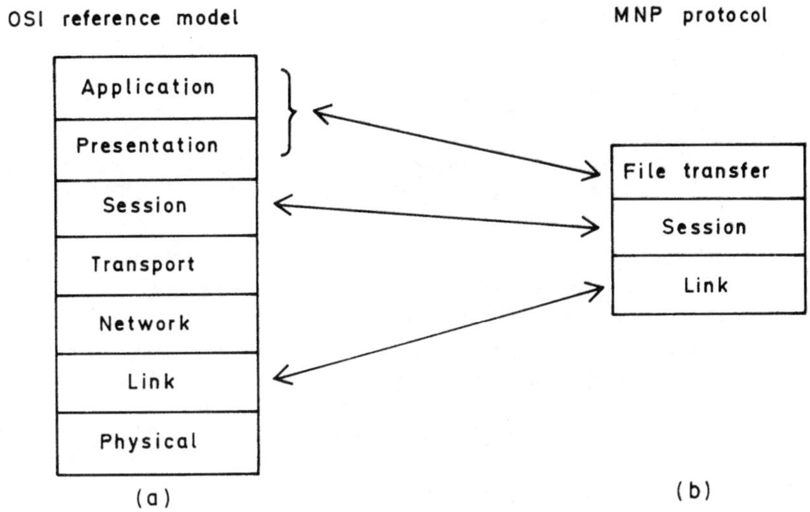

Figure 4.49 OSI reference model and MNP protocol: comparison

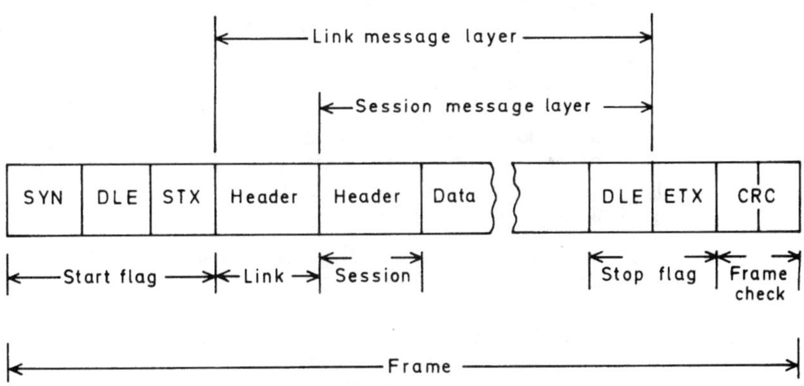

Figure 4.50 MNP block format

(see Table 4.25), with each higher level adding more sophistication and efficiency. When an MNP link is established the protocol assumes that the devices on both sides can only operate at the lowest level. Then, the devices negotiate with each other to determine the highest mutually supported class of MNP services they can support. If a non-MNP device is encountered the MNP device reverts to a 'dumb' operating mode, providing an MNP modem with the ability to be used with non-MNP devices.

The error correction capability of MNP actually occurs under its Class 4 operation. Modems that are MNP error control compatible will thus be advertised as MNP Class 4 compliant. Under Class 4, the actual framing of data depends upon whether the data is asynchronous or synchronous. If the attached terminal device transmits asynchronous data, the MNP frame format is as indicated in

Table 4.25 MNP protocol classes

Protocol class	Description
Class 1	The lowest performance level, uses an asynchronous byte-oriented half-duplex method of exchanging data. The protocol efficiency of a Class 1 implementation is about 70% (a 2400 bps modem using MNP Class 1 will have a 1690 bps throughput).
Class 2	Uses asynchronous byte-oriented full-duplex data exchange. The protocol efficiency of a Class 2 modem is about 84% (a 2400 bps modem will realize 2000 bps throughput).
Class 3	Uses synchronous bit-oriented full-duplex data exchange. This approach is more efficient than the asynchronous, byte-oriented approach, which takes 10 bits to represent 8 data bits because of the 'start' and 'stop' framing bits. The synchronous data format eliminates the need for start and stop bits. Users still send data asynchronously to a Class 3 modem, but the modems communicate with each other synchronously. The protocol efficiency of a Class 3 implementation is about 108% (a 2400 bps modem will actually run at a 2600 bps throughput).
Class 4	Adds two techniques—Adaptive Packet Assembly and Data Phase Optimization. In the former technique, if the data channel is relatively error-free, MNP assembles larger data packets to increase throughput. If the data channel is introducing many errors, then MNP assembles smaller data packets for transmission. Although smaller data packets increase protocol overhead, they concurrently decrease the throughput penalty of data retransmissions—more data are successfully transmitted on the first try. Data phase optimization is a technique for eliminating some of the administrative information in the data packets, which further reduces protocol overhead. The protocol efficiency of a Class 4 implementation is about 120% (a 2400 bps modem will effectively yield a throughput of 2900 bps).
Class 5	This class adds data compression, which uses a real-time adaptive algorithm to compress data. The real-time capabilities of the algorithm allow the data compression to operate on interactive terminal data as well as file transfer data. The adaptive nature of the algorithm refers to its ability to continuously analyze user data and adjust the compression parameters to maximize data throughput. The effectiveness of data compression algorithms depends on the data pattern being processed. Most data patterns will benefit from data compression, with performance advantages typically ranging from 1.3 to 1.0 and 2.0 to 1.0, although some files may be compressed at an even higher ratio. Based on a 1.6 to 1 compression ratio, Microcom gives Class 5 MNP a 200% protocol efficiency, or 4800 bps throughput in a 2400 bps modem installation
Class 6	This class adds 9600 bps V.29 modulation, universal link negotiation, and statistical duplexing to MNP Class 5 features. Universal link negotiation allows two unlike MNP Class 6 modems to find the highest operating speed (between 300 and 9600 bps) at which both can operate. The modems begin to talk at a common lower speed and automatically 'negotiate' the use of progressively higher speeds. Statistical duplexing is a technique for simulating full-duplex service over half-duplex high-speed carriers. Once the modem link has been established using full-duplex V.22 modulation, user data streams move via the carrier's faster half-duplex mode. The modems, however, monitor the data streams and allocate each modem's use of the line to best approximate a full-duplex exchange. Microcom claims that a 9600 bps V.29 modem using MNP Class 6 (and Class 5 data compression) can achieve 19.2 kbps throughput over dial circuits.

Table 4.25 (*continued*)

Protocol class	Description
Class 7	Uses an advanced form of Huffman encoding in conjunction with a predictive Markov algorithm to represent user data in the shortest possible Huffman codes. This compression technique is called *Enhanced Data Compression*. Enhanced Data Compression has all of the characteristics of Class 5 compression, but, in addition, predicts the probability of repetitive characters in the data stream. Class 7 compression, on the average, reduces data by 42%. In addition to Class 5 and Class 7 data compression, modern implementation of MNP now supports the ITU-T V.42 bis data compression standard
Class 8	Adds ITU V.29 fast-train modem technology to Class 7 Enhanced Data Compression, enabling half-duplex devices to emulate full-duplex transmission.
Class 9	Combines V.32 modem modulation technology with Class 7 Enhanced Data Compression, resulting in full-duplex throughput that can exceed that obtainable with a V.32 modem by 300%. Class 9 also employs *selective retransmission*, in which error packets are retransmitted, and *piggybacking*, in which acknowledgement information is added to the data. Retransmission is facilitated by indicating in the error or Negative Acknowledgement (NAK) packet the order sequence of each of the failed messages. Instead of having to retransmit all messages from the point of error to include the good messages, only the failed messages are resent.
Class 10	Adds Adverse Channel Enhancement (ACE), which optimizes modem performance in environments with poor or varying conditions, such as cellular communications, rural telephone service, and some international connections. Adverse Channel Enhancements fall into five categories: • Negotiated Speed Upshift: modem handshake begins at lowest possible modulation speed, and when line conditions permit, the modem upshifts to the highest possible speed. • Robust Auto-Reliable Mode: enables MNP10 modems to establish a reliable connection during noisy call setups by making multiple attempts to overcome circuit interference. In comparison, other MNP classes make only one call setup attempt. • Dynamic Speed Shift: causes an MNP10 modem to adjust its operating rate continuously throughout a session in response to current line conditions. This adjustment includes both downshifting and upshifting across modulation schemes, which enables modems to use the highest possible operating rate a given line connection can support. • Aggressive Adaptive Packet Assembly: results in packet sizes varying from 8 to 256 bytes in length. Small data packets are used during the establishment of a link, and there is an aggressive increase in the size of packets as conditions permit. The computations required under Aggressive Adaptive Packet Assembly to determine the optimum packet size are performed during idle time. • Dynamic Transmit Level Adjustment (DTLA): designed for cellular operations, DTLA results in the sampling of the modem's transmit level and its automatic adjustment to optimize data throughput. Once a cellular connection is established, transmit-level statistics are continually sampled using link management packets. When necessary, transmit levels are adjusted based on the existing conditions to optimize data throughput.

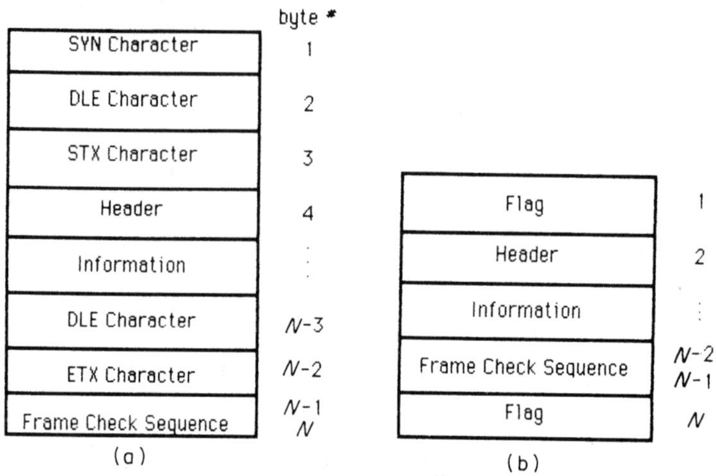

Figure 4.51 MNP frame format (DLE data link escape, STX start of text, ETX end of text): (a) Asynchronous data frame format; (b) Synchronous data frame format

Figure 4.51(a). If the attached terminal device transmits synchronous data, the frame format is as indicated in Figure 4.51(b). For both frame formats the frame check sequence (FCS) characters are generated by a CRC-16 polynomial.

Under MNP a special frame called a link request (LR) that has a predefined format is used to establish an error-corrected connection between two modems. This frame is also used to negotiate a variety of parameters between two MNP modems, including the maximum number of frames that can be outstanding and not acknowledged at a particular point in time, the maximum information field length and the framing mode to be used. The link request frame is developed by using a value of one in the header field in the synchronous frame format illustrated in Figure 4.51(b). If during negotiation the framing mode is negotiated for the transmission of asynchronous data the frame format illustrated in Figure 4.51(a) will be used.

One of the more interesting aspects of MNP is its Class 10, which is designed as an enabling technology architecture. MNP Class 10 can be viewed as residing on top of different modulation standards, such as V.22 bis, V.32, V.32 bis, V.34, and other standards as well as the *de jure* and *de facto* standards for data compression and error control. This enables an MNP Class 10 modem to provide users with maximum performance while assuring compatibility with all recognized international standards. Figure 4.52 illustrates the MNP Class 10 architecture.

MNP PDUs

From a technical perspective, an MNP frame's header and information field are called a Protocol Data Unit (PDU). The PDU provides the information required to define and negotiate facilities, manage communications, and transfer user data between modems. Eight types of PDUs are currently defined. Each has the general

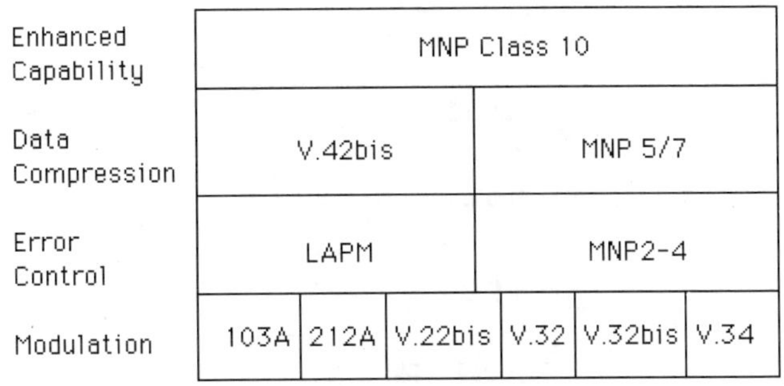

Figure 4.52 MNP Class 10 architecture

format:

LEN	TYPE	PARAMETERS...

where LEN represents the length of the PDU in bytes, TYPE is a numeric code of 1 through 9 that indicates the type of PDU, while PARAMETERS represent a string of octets that have special meaning. Both the position of the parameter in the PDU and the code in the TYPE octet are used to define the role of the parameter. With the ability to include up to 256 parameters in a PDU and define up to 256 different PDUs, MNP can be considerably expanded. In fact, it is very easy to enhance MNP by adding new parameters to PDUs or adding new PDUs since compatibility with existing MNP implementations is maintained as the use of new enhancements must be mutually agreed on.

Figure 4.53 illustrates the basic formats of the eight MNP PDUs which are transported within an asynchronous or synchronous frame. Both the length and type fields in each PDU are one octet in length. Up to 256 different parameters can be carried within a PDU and user data is attached to the LT PDU as a variable length field up to 256 bytes in length.

The Link Request PDU is used to initiate or acknowledge a connection request. This PDU can transport such parameters as the MNP service class, the number of messages that must be acknowledged, and the maximum amount of user data that will be carried in a message.

The Link Disconnect PDU is used to request that the link should be disconnected. Information in the parameters field indicates the reason for the disconnection request.

The Link Transfer PDU is used to move user data across an established connection. User data is attached to this PDU as a variable length field of up to 256 bytes. The only parameter that is included in this PDU is the Send Sequence Number (SSN) which identifies the location of the current PDU within a stream of PDUs being transmitted.

The Link Acknowledgement PDU is used to acknowledge the receipt of a message. Parameters within the PDU indicate how many additional Link Transfer

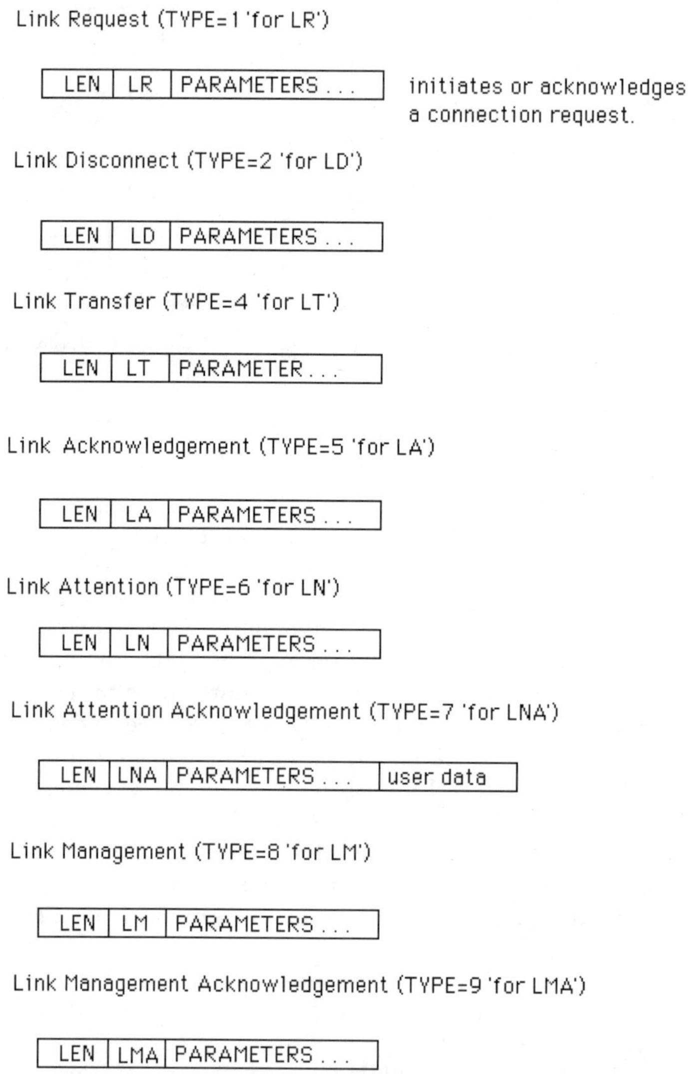

Figure 4.53 MNP ADUs

(LT) PDUs the receiver can accept prior to it becoming overwhelmed. Thus, this PDU can be viewed as a mechanism which controls a sliding window of PDUs.

The Link Attention PDU is used to interrupt communications between the transmitting and receiving device, such as when a PC user initiates a break signal. The parameters within this PDU indicate whether the transmitted data that was sent but not yet delivered should be delivered prior to the message interrupt, after the interrupt, or discarded.

The Link Attention Acknowledgement PDU is used by a receiving device to acknowledge the receipt of a previously transmitted Link Attention (LN) PDU. Among the parameters included in this PDU is the SSN of the LN PDU being acknowledged.

The Link Management PDU is used to implement MNP Class 6 and Class 10 functions. Thus, this PDU would be used to direct a change in the operating rate of a modem during a communications session.

The last PDU presently defined, Link Management Acknowledgement, acknowledges receipt of a Link Management PDU. Thus the LMA PDU would include the SSN of the LMA being acknowledged.

MNP session

An MNP session involves the exchange of PDUs as illustrated by the time chart in Figure 4.54. Communication is initiated when one modem transmits a Link Request PDU. The receiving modem responds with its own Link Request PDU permitting the two to agree upon a set of operating parameters when the originator returns its Link Acknowledgement PDU. This three-way handshaking process is referred to as the MNP call establishment process.

Once a call is established the originator transmits data using Link Transfer PDUs. The receiver acknowledges a specific number of Link Transfer PDUs with a Link Acknowledgement PDU, with the number of PDUs based upon negotiations

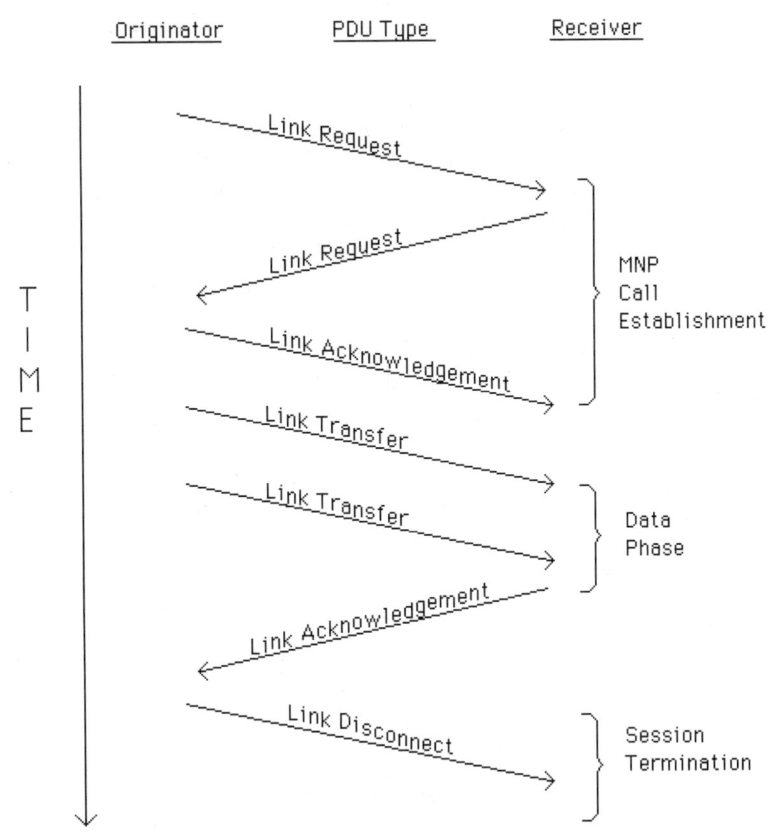

Figure 4.54 Simple MNP session

that previously occurred in the exchange of Link Requests during the call establishment process. The exchange of user data and acknowledgements is known as the data phase.

At any time either modem can issue a Link Disconnect PDU. Once issued, the sender assumes the link is terminated. Similarly, the receiver of the LD PDU also considers the connection terminated upon receipt of the sender's request to disconnect.

If an error should occur, a modem will transmit a Link Acknowledgement PDU. That PDU will tell the originator to transmit LT PDUs following the one whose SSN is specified in the Link Acknowledgement parameter. Thus, error control, while not indicated in Figure 4.54, represents a relatively simple process under MNP.

Cellular enhancements

One major problem that occurs on cellular transmission that is much more pronounced than when transmission occurs on land lines is noise. This is especially true when a user is really mobile, transmitting from a moving vehicle. As the vehicle moves, the laptop or notebook modem constantly changes its physical (spatial) relationship with cellular base towers. What might represent bothersome noise on a voice call may result in an excess number of retransmissions that result in the communications protocol terminating a session.

Although MNP-10 provides a reasonable level of cellular transmission capability, when used in a mobile environment transmission at data rates above 9.6 kbps are very difficult to obtain. In addition, MNP requires two compatible Class 10 modems to enhance cellular transmission capability.

Recognizing the limitations of MNP-10, Rockwell Semiconductor developed an extension to MNP-10 known as MNP-10EC, with the suffix initials standing for Enhanced Cellular. MNP-10EC can be considered as a new and improved version of MNP-10 for which Rockwell adapted the Physical Layer so that it is 'cellular aware', in which a signal blank condition occurring when moving between cell sites or when a signal is obscured by a building is recognized as a temporary condition. This enables the modem to recover the signal when it reappears without terminating the connection.

The actual operation of MNP-10EC is at the Protocol layer, with that layer combined with the cellular-aware Physical layer in Rockwell cellular modems. The Protocol layer is responsible for an adjustment of the modem's operating rate and packet size based upon signal quality and error rates. Thus, both signal blanking and signal quality impairments are compensated for by MNP-10EC. According to Rockwell, operating rates up to 14.4 kbps are achievable through the use of its MNP extension.

Enhanced Throughput Cellular

A competitive cellular-specific protocol to MNP-10EC is known as Enhanced Throughput Cellular (ETC). Developed by AT&T's formally owned Paradyne unit, this protocol represents a proprietary extension to V.32 bis and V.42

technologies. An ETC-capable modem can change its operating rate every five seconds to accommodate fluctuating levels of line noise, signal strength, and jitter.

The key to performance of an ETC modem is the manner by which it transmits frames. The modem uses 32-byte frames and permits up to 15 frames to be outstanding prior to requiring an acknowledgement.

While both MNP-10EC and ETC represent important developments for achieving higher operating rates on cellular facilities, another often overlooked method for improving transmission is the connection of a modem to a cellular telephone. Most cellular modems are cabled to a cellular telephone which results in a small amount of signal loss. One exception to this is a cellular direct-connect feature recently introduced by Megahertz Corporation. This modem was manufactured to connect directly to the cellular telephone without requiring the use of a cable. At the time this book was written Megahertz Corporation's Modem PC Card with a cellular direct-connect capability was capable of being used with certain Motorola and Nokia cellular phones.

LAP-M

As previously mentioned, the MNP error detection and correction method is one of two methods recognized under the ITU V.42 recommendation. In actuality, the MNP method was recognized as an alternative procedure under V.42, with the primary method known as link access protocol-modem or LAP-M.

Under V.42 the originating modem transmits an originator detection pattern (ODP). The ODP is defined as the bit sequence

01000 10001 11 . . . 11 01000 10011 11 . . . 11

which represents the DC1 character with even parity, followed by 8 to 16 ones, followed by DC1 with odd parity, followed by 8 to 16 ones. A V.42 compatible modem will respond to the ODP with an answer detection pattern (ADP), whose bit format will indicate that V.42 is supported or that no error-correcting protocol is desired. If the ADP is not observed within a predefined period of time, for which V.42 modems use a default value of 750 ms, it is assumed that the distant modem does not possess V.42 error-correcting capability. In this situation the originating modem may fall back to a non-error-correcting mode of operation, or if it incorporates MNP, it can then attempt to negotiate an MNP error detection and correction mode of operation.

Under the LAP-M protocol a different frame structure from MNP is used. In addition, the frame check sequence characters are generated by a different polynomial. Thus, MNP and LAP-M are completely incompatible with each other.

Compatibility issues

Since V.42 specifies MNP as an alternative procedure it is essentially an option that modem manufacturers may or may not include in a V.42 modem. Thus, if you wish to communicate using the MNP method of error detection and correction you can either consider purchasing a modem that is MNP Class 4 compatible or which is V.42 compatible and supports the alternative error-correction procedure. Fortunately, most modem vendors that have adopted V.42 support also support the alternative error-correction procedure more commonly referred to as MNP Class 4.

Data compression

Similar to error detection and correction, modem manufacturers have incorporated a variety of data compression algorithms into their products. Until 1989, most compression methods, while representing a proprietary scheme of one vendor, were licensed by that vendor to other manufacturers. This resulted in several *de facto* data compression standards being incorporated into modems manufactured by different vendors.

In 1990 the CCITT promulgated the V.42 bis recommendation which defines a new data compression method known as Lempel-Ziv as an international standard. Unlike V.42, which concerns error detection and correction and specifies MNP Class 4 as an alternative, V.42 bis does not specify an alternative method of data compression.

Rationale

A data compression performing modem results in each modulated bit conveying more information than an equivalent modem that does not compress data. To illustrate this concept assume that over a period of time the ratio of characters flowing into a modem's compression circuitry to the number of compressed characters output and modulated by the modem is 2:1. This 2:1 compression ratio indicates that, on average, a compression performing modem can transfer information twice as fast as a similar modem that does not compress data.

Modem vendors have incorporated data compression into their products to provide users with two key advantages over non-compression performing modems —a reduction in transmission time and an alternative to purchasing higher-speed and higher-cost modems. The reduction in transmission time is probably the more obvious of the two advantages as compacted data takes less time to be transmitted. Concerning the latter advantage, consider a data compression modem that operates at 14 400 bps and has a 2:1 compression ratio. This modem would provide you with an effective throughput of 28 800 bps and would probably be less expensive than a 28 800 bps modem since the modulation scheme required to operate at the higher data rate is more complex and more costly than the modulation method used by 14 400 bps modems.

MNP Class 5 compression

Of the two methods of data compression supported by the Microcom Networking Protocol, Class 5 is the more popular as it predates Class 7. Since each MNP Class is downward negotiable for compatibility, a modem that supports MNP Class 7 will communicate with a modem that is MNP Class 5 compatible using MNP Class 5 data compression.

MNP Class 5 specifies that the sending modem apply two modifications to the transmitted data stream in an attempt to reduce the number of bits actually sent. The first manipulation or data compression method is referred to as run-length encoding, while the second method of compression is known as adaptive frequency encoding.

MNP Class 5 uses run-length encoding to avoid sending long sequences of repeated data octets, where each octet represents eight bits that can define a character in a particular character set or any binary value from 0 to 255 represented by the individual bit settings within the octet. The MNP Class 5 version of run-length encoding results in a repetition count being inserted into the data stream to represent the number of repeated data octets which follow the first three occurrences of a sequence. The first three repeated data octets, which are actually sent, signal the beginning of a run-length encoded sequence. The next octet is always a repetition count which has a maximum value of 250. If the repeated sequence is only three octets in length, a repetition count of 0 is used. Four octets consisting of three repeated data octets and a count octet are thus used to compress any repeating sequence from 3 to 250 octets. Figure 4.55 illustrates the format of MNP Class 5 run-length encoding as well as a few examples of its operation.

MNP Class 5 run-length encoding can be considered as the first level of a two-level data compression scheme. The second level of compression is known as adaptive frequency encoding and is applied to the data stream after any repeated data octets are removed by the use of run-length encoding.

In adaptive frequency encoding a compression token is substituted in the data stream for the actually-occurring data octet in an attempt to transmit fewer than eight bits for each data octet. The token used changes with the frequency of

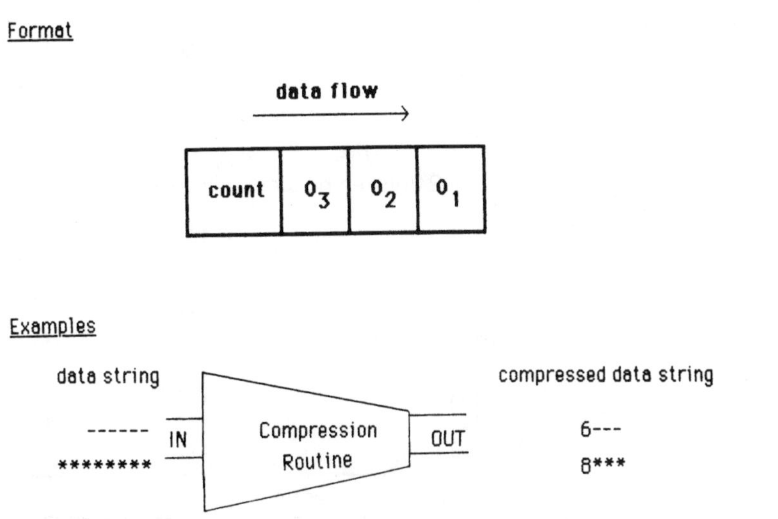

Figure 4.55 MNP class 5 run length encoding ($O_1 = O_2 = O_3$ = repeating octets)

Table 4.26 MNP Class 5 octet/token mapping at compression initialization

Data octet (decimal value)	Header (MSB → LSB)	Body (MSB → LSB)
0	000	0
1	000	1
2	001	0
3	001	1
4	010	00
5	010	01
6	010	10
7	010	11
8	011	000
9	011	001
10	011	010
11	011	011
12	011	100
13	011	101
14	011	110
15	011	111
16	100	0000
17	100	0001
18	100	0010
19	100	0011
20	100	0100
21	100	0101
22	100	0110
23	100	0111
24	100	1000
25	100	1001
26	100	1010
27	100	1011
28	100	1100
29	100	1101
30	100	1110
31	100	1111
32	101	00000
33	101	00001
34	101	00010
[35–246-token header/body continue in same pattern]		
247	111	1110111
248	111	1111000
249	111	1111001
250	111	1111010
251	111	1111011
252	111	1111100
253	111	1111101
254	111	1111110
255	111	11111110

occurrence of the actual data octet so that shorter tokens are substituted for more frequently occurring data octets.

The compression token used to represent a data octet is composed of two parts: a fixed-length header and a variable-length body. The header is three bits long and, in general, indicates the length of the body portion of the token. There are three special cases, however. First, there are two tokens with a header of 0. In these two cases, the true length of the body is 1. Finally, when the header indicates a length of 7 and the body is seven 1-bits, then the actual length is 8.

At initiation of data compression, the relative frequency of occurrence of each data octet is 0. For purposes of octet/token mapping, however, data octet 0 (00000000 binary) is assumed to be the most frequently occurring octet and is represented by the first of the shortest tokens, data octet 1 (00000001 binary) is taken to be the next most frequently occurring and is represented by the next token, etc., to data octet 255 (11111111 binary), initially assumed to be the most infrequently occurring data octet. Data octet/token mapping at compression initialization is shown in Table 4.26

In order to encode a data octet, the token to which it is currently mapped is substituted for the actual data octet in the data stream. After this substitution, the frequency of occurrence of the current data octet is increased by one. If the frequency of this data octet is greater after incrementing than the frequency of the next most frequently occurring data octet, then the compression tokens of the current data octet and the next most frequently occurring data octet are exchanged. The frequency of the current data octet is then compared to the frequency of the data octet which is now the next most frequently occurring data octet. If the frequency of the current data octet is greater, then the compressed tokens are once again swapped. This cycle continues until no more swaps are needed at which time the mapping of data octets and compression tokens is correctly adapted based on the relative frequency of the data octets.

Once the data octet/compression token mappings are sorted by frequency, the frequency count of the current character is compared to the fixed limit value of 255 (decimal). If this limit has been reached, then the frequency of occurrence of each data octet is scaled downward by dividing each frequency by 2. Integer division is used, thus any remainder after division is discarded (e.g., $3/2 = 1$).

As previously noted, the repetition count, the fourth octet in a run-length encoded sequence of repeated data octets, is also mapped to a compression token. The token used is that which is mapped to the count of the most frequently occurring data octet. The token used for a count of 5 would thus be 01001. Note also, that this count octet does not increase the frequency of occurrence of the data octet to which the token is mapped. Further, the repetitions of the run-length encoded data octet represented by the count do not contribute to the frequency of occurrence of the repeated data octet.

MNP Class 7 enhanced data compression

MNP Class 7 enhanced data compression builds upon the concept of combining run-length encoding with the use of an adaptive encoding table that contains a single column representing each character ordered by their frequency of occurrence. Under MNP Class 7, run-length encoding is combined with the use of a first-order Markov model. This model is used to predict the probability of the occurrence of a character based upon the value of the previous character, with an adaptive table of 256 character columns used to represent the ordered frequency of occurrence of each 'succedding' character.

Markov model

To encode a character the compressor selects a code which depends on the immediately preceding encoded character. The selected code is based on the frequency with which a character follows the previous character. For example, the probability of a 'U' following a 'Q' is very high and generally the 'U' will be encoded as 1 bit. Likewise, an 'H' following a 'C' has a different probability than an 'R' following a 'C', and will be coded according to its frequency of occurrence.

As noted in Table 4.27, the compressor keeps up to 256 coding tables, one for each possible 8-bit character (or pattern). To code a character for transmission, it uses the previous character to select the appropriate coding table. For example, when an 'A' is transmitted the model looks under the 'A' pointer for the next character, which is ordered according to its frequency of occurrence. If a 'C' is the next character, it is compressed based upon its location in the table. The model looks next under the 'C' pointer to find the following character, and so on. Each table contains the codes for characters following the previous character, and is organized according to the rules of Huffman coding.

Table 4.27 MNP Class 7 enhanced data compression

Pointers	A	B	C	D	E		
Characters coded	T	L	H	O	D	••	Up to
according to their	H	E	O	A	R		maximum
frequency of following	C	U	R	E	S		of 256
the previous character,	M	.		.	N		characters
i.e., the pointer	B	.		.	P		each

Huffman coding

In Table 4.27, each column of characters under the pointer character is compressed according to the rules of Huffman coding. Huffman coding changes the number of bits representing a character when the character's frequency of occurrence changes sufficiently. Huffman can adapt to various alphabets (e.g., ASCII, EBCDIC, and all uppercase) and languages (natural language, compiler code, and spreadsheets) without being pre-informed of the data used.

Unlike Class 5, Huffman coding can represent a character with only one bit, if it occurs often enough. In general, if one character occurs twice as often as another, its code is half as long.

MNP Class 7 is adaptive, meaning that it changes the coding of the data when the frequency of character occurrence changes. The compressor starts off with no assumptions about the data coding, and even learns that it is ASCII/English based on the data itself. The compression tables are empty at the start of each connection and are built as data is pased. When the data divides naturally into characters and the 'working set' of frequent characters is not too large (natural language is good here), then adaptation will create a coding structure which compresses the data well.

Run-length encoding

Multiple consecutive copies of the same character (or 8-bit pattern) are compressed for transmission using run-length encoding. If the encoder has sent the same character three consecutive times, the encoder sends the count of the remaining identical consecutive characters as a single 4-bit nibble. For example, a series of 'A's would be sent as 'AAA' with the remaining number of 'A's sent as a 4-bit nibble, while, a series of five 'A's would be sent as three Huffman encodings of 'A' and a nibble of binary two.

Decoding

To decode an encoded data stream, the receiving modem assumes it has an exact copy of the sending modem's compression data structures because:

(i) both modems reset their structures at the start of each connection;
(ii) communication has been error-free (MNP assures this);
(iii) decoding the data stream causes identical data structure updates to match the encoder's.

Since the receiving modem has the same compression tables as the sender, it knows which set of Huffman data decodes the incoming bit stream. The receiving modem compares the incoming bits against the Huffman codes in the table until there is a complete match. At this time, the character is delivered to the DTE and the table is updated.

V.42 bis

V.42 bis data compression is a modified version of the Lempel-Ziv method of compression that was developed approximately 20 years prior to the promulgation of the recommendation. The V.42 bis compression method uses an algorithm in which a string of information received from an attached DTE is encoded as a variable length codeword. To facilitate the development of codewords strings are stored in dictionaries at the encoding and decoding device and are dynamically updated to reflect changes in the composition of data.

The key to the efficiency of the V.42 bis compression process is its dictionary which is dynamically built and modified. The dictionary can be considered to represent a set of trees in which each root corresponds to a character in the alphabet, as illustrated in Figure 4.56. Each tree represents a set of known strings beginning with one specific character, and each node or point in a tree represents

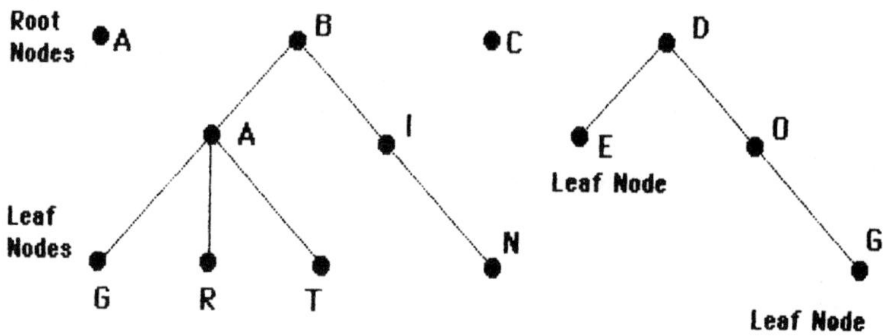

Figure 4.56 V.42 tree-based dictionary representation

one set of strings. The trees in Figure 4.56 thus represent the strings A, B, BA, BAG, BAR, BAT, BI, BIN, C, D, DE, DO and DOG.

Each of the leaf nodes shown in Figure 4.56 represents a node that has no other dependent nodes, in effect, representing the last character in a string. Conversely, a node that has no parent is known as a root node and represents the first character in a string.

Initially, each tree in the dictionary consists of a root node, with a unique codeword assigned to each node. As data is received from an attached DTE a string matching process occurs in which a sequence of DTE originated characters is matched against the dictionary. This process begins with a single character. If the string matches a dictionary entry and the entry was not created since the last invocation of the string matching procedure, the next DTE originated character is read and appended to the string. This process is repeated under the previously described conditions until the maximum string length is reached or the string does not match a dictionary entry or matches an entry created since the last invocation of the string matching procedure. At that time the last character appended to the string is removed and represents an 'unmatched' character while the characters in the string are encoded as a codeword.

Under V.42 bis the maximum string length can range from 6 to 250 and is negotiated between modems. The number of codewords has a default value of 512, which is its minimum value. A maximum value is not, however, specified and any value above the default value can be negotiated between two modems.

During the compression process the dictionary is dynamically modified by the addition of new strings based upon the compostion of the data. New strings are formed by appending a single character that was not matched from a string matching operation to an existing string, which results in the addition of a new node to a tree. As a result of the replacement of strings by codewords V.42 bis data compression is approximately 20 to 30% more efficient than MNP Class 5 compression and probably 5 to 10% more efficient than MNP Class 7 enhanced data compression. While the increased efficiency of V.42 bis can be expected to result in many modem manufacturers providing this capability in future products, for many modem users the difference between V.42 bis and either MNP Class 5 or Class 7 operations may not be perceptible. This is because interactive access to mainframe computers normally involves queries based upon short strings of information that are not very susceptible to data compression. In comparison, users with extensive file transfer requirements will find a more perceptible difference favouring V.42 bis. Thus, you should carefully consider their transmission requirements prior to purchasing a modem that peforms data compression as MNP Class 5 or Class 7 may be efficient as V.42 bis for certain applications.

Compatibility issues

The adoption of the V.42 bis recommendation resulted in some vendors providing V.42 bis chip set upgrades for their products. Other vendors announced that all future products will incorporate V.42 bis support. Although V.42 bis is the recommended standard, unlike V.42 it does not specify MNP as an alternative method of data compression. Recognizing that over 15 million modems have been sold that support MNP Class 5 or Class 7 data compression, most modem manufacturers have incorporated both MNP Class 5 and V.42 bis compression into their products. Microcom offers both MNP Class 7 and V.42 bis support. Vendors other than Microcom may be limited to Class 5, because at the time this book was prepared, MNP Classes 1 through 4 were in the public domain and different licensing arrangements were applicable to the higher levels of the MNP protocol.

Modems that support both V.42 bis and MNP Class 5 data compression provide you with the door to the future as well as a gateway to the past. In the future you can expect that most, if not all, data compression-performing modems will be V.42 bis compatible. In the interim, MNP Class 5 compatibility provides modems with the ability to operate in a data-compressed mode with over 15 million currently installed modems.

Throughput issues

No doubt many readers are familiar with the automotive section of the Saturday or Sunday newspaper in which the performance of a particular car is reviewed. Such reviews describe engine RPMs, torque, and other performance parameters that, for the most part, are irrelevant to the average driver. A similar case can be made for many reviews of data compression-performing modems in trade publications: the author discusses how a file was uploaded or downloaded in so many seconds due to the high operating rate and compression efficiency of the modem. What many modem reviews fail to mention is that it is extremely rare for you to obtain that level of performance for any sustained period of time during a typical communications session. In fact, the use of the compression feature of a modem may actually result in an increase in the time required to transfer certain types of files.

Although a 2:1 modem compression ratio means that you can expect to achieve twice the modem's operating rate for file transfer operations, it is assumed that you are transferring a standard ASCII text file. In actuality, most information utilities and bulletin board systems store files in a compressed format, using such programs as ARC, ARJ, PAK, and PKZIP to economize on file storage space. Attempting to use modem compression to transfer a previously compressed file can actually result in an expansion of the file during data transfer between modems if you are using

MNP5, or a marginal improvement in data transfer time if you are using the V.42 bis data compression method. Thus, a good rule of thumb is to transfer compressed files without using MNP5 data compression if that is the only compression method supported by your modem.

Negotiation problems

As previously indicated in this section, both error detection and correction and data compression are negotiated by the flow of data between modems once the basic modulation handshaking process is completed. This data flow occurs through the transmission of frames between modems. These frames describe the settings and capabilities of each modem. Although MNP, V.42, and V.42 bis all explicitly define how data frames are encoded and exchanged, significant problems can occur if one modem that has negotiation features enabled communicates with another modem that either does not have such capabilities or has those capabilities disabled. When either situation occurs, the negotiation frame will be passed onward by the modem that does not have enabled negotiation capabilities. If the modem is connected to an encryption device, data PBX, or another device that is data sensitive, the negotiation frame can cause havoc.

Figure 4.57 illustrates one example of *modem negotiation havoc*. In this example a user with a notebook containing a built-in datacryptor uses the switched telephone network to communicate in a secure manner with a mainframe computer. If the modem attached to the notebook computer has its negotiation capabilities enabled,

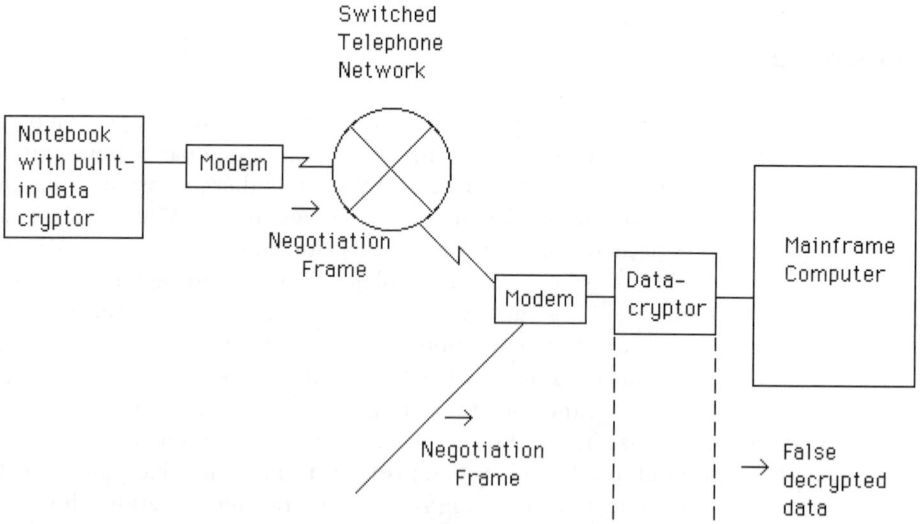

Figure 4.57 Modem negotiation havoc. When one modem transmits negotiation frames and the other either has its negotiation capability disabled or lacks a negotiation capability, the frames are passed through the modem as data. In this network configuration, the datacryptor performs its intended operation, resulting in false decrypted data sent to the mainframe, which results in a loss of synchronization between datacryptors

upon establishment of a carrier signal it will initiate handshaking by transmitting a negotiation frame. If the distant modem either lacks a negotiation capability or has that feature disabled, the negotiation frame passes through the modem as it is assumed to represent demodulated data. Unfortunately, the datacryptor also assumes it is demodulated data and initiates its decryption operation. Not only does this result in false decrypted data being passed to the mainframe but, in addition, it results in a loss of synchronization between the pair of datacryptors. Thus, when the notebook user attempts to transmit a request to the mainframe, data is encoded based upon one position in a pseudo-random key sequence generated by the datacryptor built into the notebook, and decoded at the mainframe by the datacryptor using a different position in a pseudo-random key sequence since the negotiation frame preceded the data. Needless to say, the mainframe receives what appears to be 'gobbledegook'.

Another common example of modem negotiation havoc can occur in a data PBX environment. When one modem transmits negotiation frames and the other either has its negotiation capability disabled or lacks a negotiation capability, the frames are passed through the modem as data. In this network configuration the datacryptor performs its intended operation, resulting in false decrypted data being sent to the mainframe which results in a loss of synchronization between datacryptors. If a data PBX is behind a modem, the negotiation frame could be misinterpreted as a routing code, and could cause a misrouting before a user could enter a correct routing code. To correct such situations, you can enable error detection and correction and data compression features on modems at the other end of the communications link, if those features were previously disabled, or replace modems that don't support those features with modems that do. A third option is to disable those features on the modem transmitting the negotiation frame. However, doing so eliminates the possibility of using those features.

Simultaneous voice and data operations

The growth in 'telecommuting' has resulted in many people working from their homes, accessing LAN servers and corporate mainframes to perform work that might otherwise require a long commute into the office. Accompanying the growth in telecommuting is an increase in the requirement to communicate verbally with other telecommuters or with staff at the office while maintaining a data transmission session. Although the Integrated Services Digital Network (ISDN) provides the ability to maintain two separate conversations on one pair of telephone wires, enabling a voice and a data version to coexist simultaneously, the technology is relatively expensive in comparison to the cost of using the analog PSTN and is far from ubiquitous. Recognizing those limitations, as well as the requirements of telecommuters to converse simultaneously with a distant party while transmitting data, resulted in the development of simultaneous voice and data (SVD) capability into certain modem products.

Currently two methods are used to provide a simultaneous voice and data capability. One method, referred to as *digital simultaneous voice and data* (DSVD), uses speech compression technology to digitize voice into a low bit rate which enables users to carry on a normal conversation over the PSTN while transmitting data at rates up to 33.6 kbps. The second method results in the transmission of voice in its native, analog format, while reducing the data rate to provide separate channels for voice and control information. This method is commonly referred to as analog simultaneous voice and data (ASVD). Although ASVD can be more aesthetically pleasing to a listener's ear in certain situations, most implementations of a SVD capability work digitally to obtain a higher data transfer operating rate capability. Two competitive and incompatible techniques for obtaining a simultaneous voice and data transmission capability were being offered by Multi-Tech Systems and Rockwell Semiconductor Systems when this book was published.

Multi-Tech Systems markets a series of DSVD modems using codecs (coders-decoders) to digitize the full spectrum of audio bandwidth from 0 to 3300 Hz.

Using what is referred to as the Multi-Tech Supervisory Protocol (MSP) for communications between modems, voice, data, and supervisory information are separately packetized for transmission. All three types of packets are multiplexed onto a single stream of data which enables a common circuit to support voice and data simultaneously; however, to eliminate potential voice distortion voice packets receive priority over data packets. Figure 4.58 illustrates the MSP packet structure. Note that each packet uses one 8-bit SYNC byte header for synchronization.

The ID/LI Information byte includes an ID (bit number 7) and a Packet Length Indicator (LI) field which ranges in value from 01h to 07Fh for data and qualified packets. Voice packets are preset to 24 or 39 bytes in length based upon the data operating rate. Thus, the ID/LI Information byte is set to a value of 00h in voice packets.

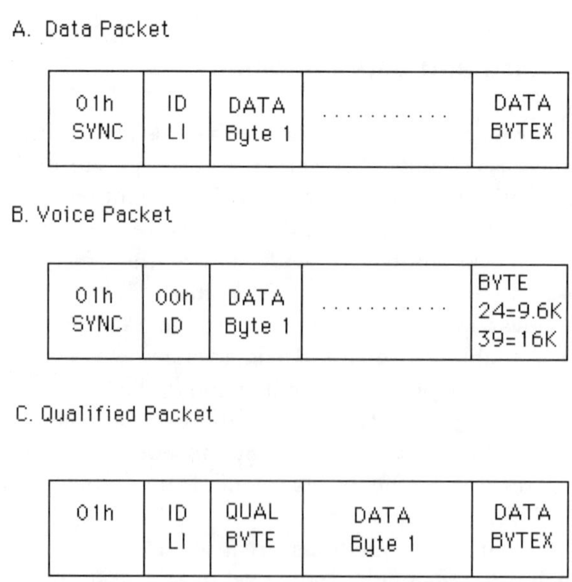

Figure 4.58 Multi-Tech Supervisory Prtotocol packet structure

The Qualifier (QUAL) byte, which is present in Qualified packets, identifies the nature of the command that follows that byte. Those commands fall into three broad categories—'immediate action' commands, 'engage stream mode for' commands, and 'interpret the following as' commands. Examples of 'immediate action' commands include ACK, NAK, XON, XOFF, and BREAK. Examples of 'engage stream mode for' commands include COMMANDS, VOICE and DATA, while 'interpret the following as' commands include commands, responses and status. Thus, the MSP functions in a manner similar to the Higher-Level Data Link Control (HDLC) protocol in that it supports command/response transmission sequences.

Although Multi-Tech's MSP provides simultaneous voice with data at an operating rate up to 16 kbps, the protocol does not attempt to detect and take advantage of periods of voice silence. A competitive simultaneous voice and data product from Rockwell known as DigiTalk and incorporated into that vendor's RC288ACi/SVD chip set both digitizes voice into a low bit rate and detects periods of silence, enabling modems using the chip set to shift to higher data operating rates during periods of voice silence. This technique permits a V.34 modem using a DigiTalk chip set to obtain a non-compressed data transfer rate of 28.8 kbps during periods of voice silence.

Synchronous dialing language

In comparison to the Hayes command set that is a *de facto* standard for asynchronous modem operations there are several methods used to support synchronous modem operations. The earliest method of automating synchronous dialing was accomplished by the use of an automatic calling unit. In this method of establishing calls over the PSTN a computer or another DTE device is connected to an automatic calling unit via an RS-366 (parallel) or an RS-232 (serial) interface. The DTE controls the calling unit, which dials the telephone number and upon completion passes the call to the modem. Figure 4.59 illustrates this procedure.

Other methods that have been developed to obtain synchronous dialing capability include the incorporation of an RS-366 interface into some modems, the use of an integrated external keypad through which the operator stores dialing sequences which the modem dials when the attached DTE raises the Data Terminal Ready signal and the use of a synchronous dialing language which

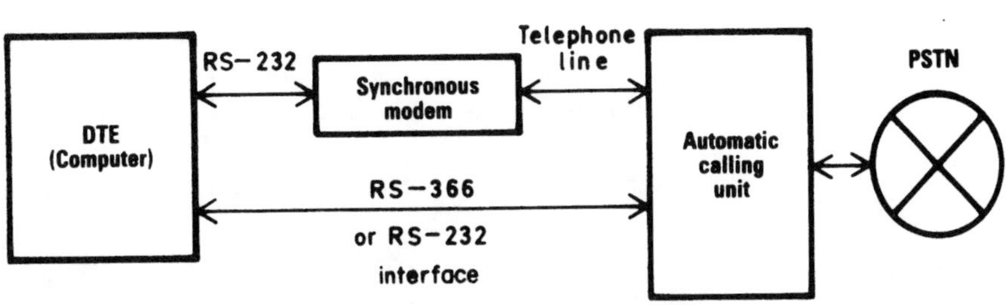

Figure 4.59 Using an automatic calling unit. Automatic calling units support both RS-232 (serial) and RS-366 (parallel) interfaces

combines dial control and data over a common RS-232 interface. Concerning the latter, there are currently several methods competing to become *de facto* industry standards.

Hayes Microcomputer Products' proposed standard is the Hayes Synchronous Interface (HSI). This interface consists of a hardware independent connection between a user's application program and a synchronous communications driver which directly controls the hardware. Under the HSI method, dialing commands to a synchronous modem are issued asynchronously, with the modem directed to change into a synchronous mode of operation once a conncetion with a distant device is made. Obviously, this method is only applicable for use with modems that can work both synchronously and asynchronously; however, it does permit the ubiquitous Hayes AT command set to be used.

A second method being used to perform automatic dialing operations with synchronous modems is based upon the use of the Racal-Vadic Synchronous Auto-Dial Language (SADL). Under SADL special modem controlling messages in both bisynchronous and SDLC protocols are predefined. This enables software developers to modify application programs to support the use of SADL compatible modems for dialing on the PSTN.

The third method being used for synchronous dialing is similar to the previously described SADL technique. This method is called 'SyncUp', which is sponsored by Universal Data Systems and is restricted to sending dialing commands under the bisynchronous protocol.

4.4 MULTIPORT MODEMS

The integration of modems and limited-function multiplexers into a device known as a multiport modem offers significant benefits to data communications users who require the multiplexing of only a few channels of data. Users who desired to multiplex a few high-speed data channels prior to the introduction of multiport modems were required to obtain both multiplexers and modems as individual units which were then connected to each other to provide the multiplexer and data transmission requirements of the user. Since multiplexers are normally designed to support both asynchronous and synchronous data channels, the cost of the extra circuitry and the additional equipment capacity was an excess burden for many user applications.

The recognition by users and vendors that a more cost-effective, less wasteful method of multiplexing and transmitting a small number of synchronous channels for particular applications led to the development of multiport modems. By the combination of the functions of a time division multiplexer (TDM) with the functions of a synchronous modem, substantial economies over the past data transmission methods can be achieved for certain applications.

Operation

A multiport modem is basically a high-speed synchronous modem with a built-in TDM that uses the modem's clock for data synchronization, rather than requiring one of its own, as would be necessary when separate modems and multiplexers are combined. In contrast with most traditional TDMs, a multiport modem multiplexes only synchronous data streams, instead of both synchronous and asynchronous data streams. An advantage of the built-in, limited-function multiplexer is that it is less complex and expensive, containing only the logic necessary to combine into one data stream information transmitted from as few as two synchronous data channels rather than the minimum capacity of four or eight channels associated with most separate multiplexers. The data channels in a multiport modem normally comprise a number of 2400 bps data streams, with the number of channels available being a function of the composite transmission speed.

Selection criteria

When investigating the potential use of multiport modems for a particular application, you should determine the speed combinations and the number of selectable channels available, as well as the ability to control the carrier function (mode of operation) independently for each of the channels. For example, one 9600 bps multiport modem can have as many as six different modes of operation; however, only one mode can be functioning at any given time. As illustrated in Figure 4.60, operating speeds can range in combination from a single channel at 9600 bps through four 2400 bps channels.

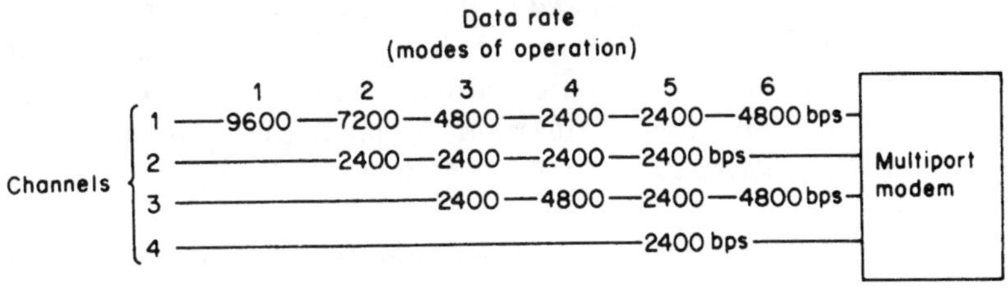

Figure 4.60 9600 multiport modem schematic utilization diagram. Multiport modem with six modes of operation is schematized here to show all possible data rate combinations for networking flexibility

Application example

A typical application of a 28 800 bps multiport modem is illustrated in Figure 4.61. This example shows a pair of four-channel 28 800 bps multiport modems servicing two CRT terminal devices, a synchronous printer operating at up to 1500 characters per second or 12 000 bps, and three low-speed synchronous terminals connected by a traditional TDM. The output of the 3-channel TDM is a 7200 bps synchronous data stream, which is in turn multiplexed by the multiport modem. Here the multiport modem's multiplexer combines the three asynchronous

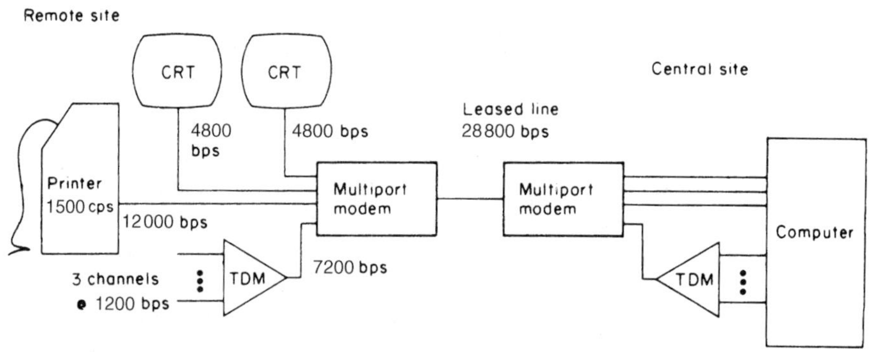

Figure 4.61 Multiport modem application example. A pair of four-channel multiport modems services two CRTs, a 1500 cps printer, and three low speed terminal devices over a single transmission line

multiplexed 1200 bps channels with the three synchronous unmultiplexed channels into a single multiplexed synchronous data stream. At the central site where the computer is located, the multiport modem at that end splits the 28 800 bps stream into four data streams; one data stream is then channeled through another 3-channel, traditional TDM, whose output data streams in turn are connected to the computer. The 3-channel TDM takes the 7200 bps synchronous data stream from the multiport modem and demultiplexes it into three 1200 bps asynchronous data streams, which are passed to the appropriate computer ports. The remaining data streams produced by the demultiplexer in the multiport modem are connected to three additional computer ports. As this example demonstrates, the high-speed multiport modem's utilization in conjunction with other communications components permits a wide degree of flexibility in the design of a data communications network.

A few examples of multiport modem and channel combinations available are listed in Table 4.28. It should be noted that not all multiport modem channel combinations listed in Table 4.28 may be available for a particular vendor's modem that operates at the indicated aggregate throughput, since some vendors offer only a 4-port multiplexer with their modem, while other vendors may offer an additional number of ports.

Although most manufacturers of multiport modems produce equipment that appears to be functionally equivalent, you should exercise care in selecting equipment because of the differences that exist between modems but are hard to ascertain from vendor literature. Although the modes of operation available for utilization are quite similar regardless of manufacturer, the number of ports or data channels supported by a multiplexer can differ significantly. Another variance concerns the use of an independently controlled carrier signal in some multiport modems. By using multiport modems that permit an independently controlled carrier signal for each channel, you can combine several polled circuits and further reduce leased line charges.

Table 4.28 Examples of multiport modem channel combinations. The wide degree of flexibility that can be provided by multiport modems in a network configuration is a function of several factors: throughput, available modes and channels, and data rates

Modem aggregate throughput	Operating mode	Multiport speed combinations							
		1	2	3	4	5	6	7	8
19 200	1	19200							
	2	16800	2400						
	3	14400	4800						
	4	14400	2400	2400					
	5	12000	7200						
	6	12000	4800	2400					
	7	12000	2400	2400	2400				
	8	9600	9600						
	9	9600	4800	4800					
	10	9600	4800	2400	2400				
	11	7200	7200	4800					
	12	7200	4800	2400	2400	2400			
	13	7200	2400	2400	2400	2400	2400	2400	
	14	4800	4800	4800	4800				
	15	4800	4800	4800	2400	2400			
	16	4800	4800	2400	2400	2400	2400		
	17	4800	2400	2400	2400	2400	2400	2400	
	18	2400	2400	2400	2400	2400	2400	2400	2400
14 400	1	14400							
	2	12000	2400						
	3	9600	4800						
	4	9600	2400	2400					
	5	7200	7200						
	6	7200	4800	2400					
	7	7200	2400	2400	2400				
	8	4800	4800	4800					
	9	4800	4800	2400	2400				
	10	2400	2400	2400	2400	2400	2400		
9600	1	9600							
	2	7200	2400						
	3	4800	2400	2400					
	4	2400	2400	4800					
	5	2400	2400	2400	2400				
	6	4800	4800						
7200	1	7200							
	2	4800	2400						
	3	2400	2400	2400					
4800	1	4800							
	2	2400	2400						

Standard and optional features

A range of standard and optional data communications features are available for users of multiport modems, including almost all of the features available in regular non-multiplexing modems; also available are unique multiport modem features such as multiport configuration selection, individual port testing, individual port display, and a data communications equipment (DCE) interface.

The multiport selector feature permits you to alter the multiport configuration simply by throwing a switch into a new position. This feature can be especially useful for an installation such as the one shown in Figure 4.62 (top) where communications requirements differ between shifts. For example, in the figure, during daytime operations three low-speed asynchronous, 4800 bps, terminal devices with a composite speed of 19 200 bps are serviced at this installation by one channel of

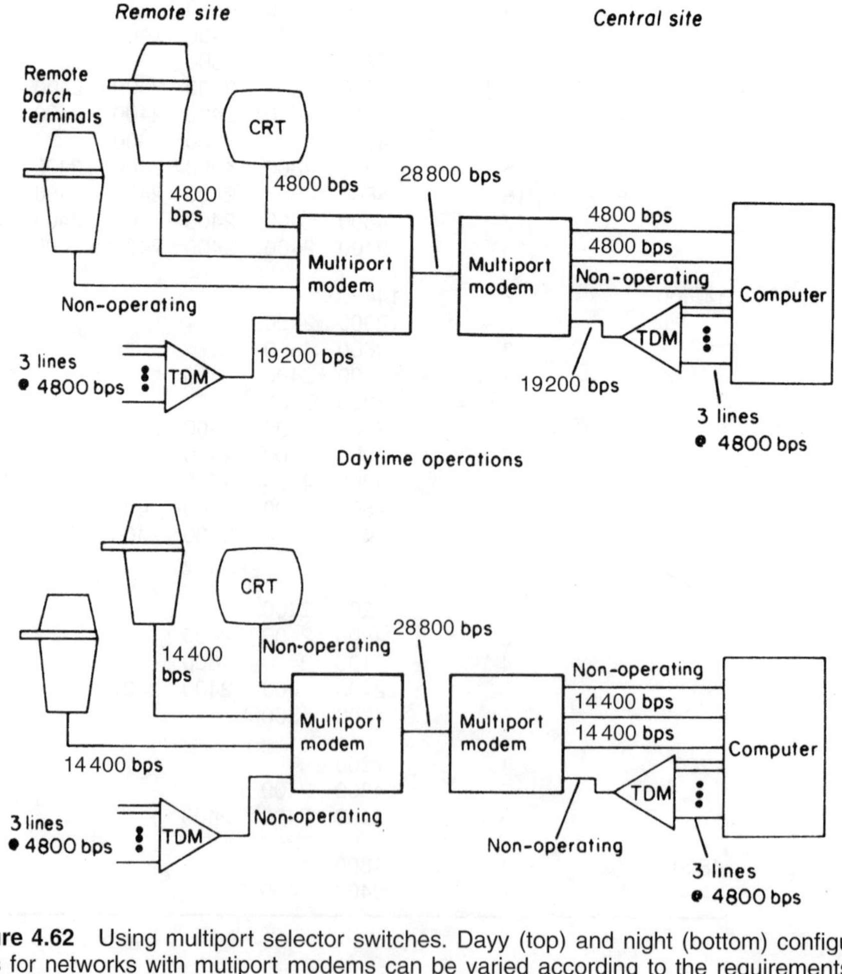

Figure 4.62 Using multiport selector switches. Dayy (top) and night (bottom) configurations for networks with mutiport modems can be varied according to the requirements of different operations

the multiport modem. One remote batch terminal and a CRT are serviced by two additional channels, each of which operates at 4800 bps. Because of daytime load requirements, the second batch terminal cannot be operated since the modem's maximum aggregate speed of 28 800 bps has been reached. On the assumption that the installation does not require the servicing of interactive users at night, one possible reconfiguration is shown in Figure 4.62 (bottom). The multiport selector permits both remote batch terminals to be serviced until the start of the next business day by two 14 400 bps channels while everything else in this network is shut down.

Numerous multiport modems contain a built-in test pattern generator and an error detector which permit users of such modems to determine if the device is faulty without the need of an external bit error rate tester. The use of this feature normally permits the individual ports of the modem to be tested.

Another option offered by some multiport modem manufacturers is a data communications equipment (DCE) interface. This option can be used to integrate remotely located terminals into a multiport modem network. Whereas the standard data terminal equipment (DTE) interface may require data sources to be collocated and within a 50-foot radius of the multiport modem, the DCE interface permits one or more data sources to be remotely located from the multiport modem. Installation of a multiport modem with a DCE option on one port permits that modem's port to be interfaced with another modem. This low-speed conventional modem can be used to provide a new link between the multiport modem's location and terminals located at different sites. As shown in Figure 4.63, the installation of a multiport modem with a DCE option on port 3 permits the port on that modem to be interfaced with another modem. This new modem can then be used to provide a new link between the multiport modem at location 1 and an additional remote batch terminal which is located at a second site.

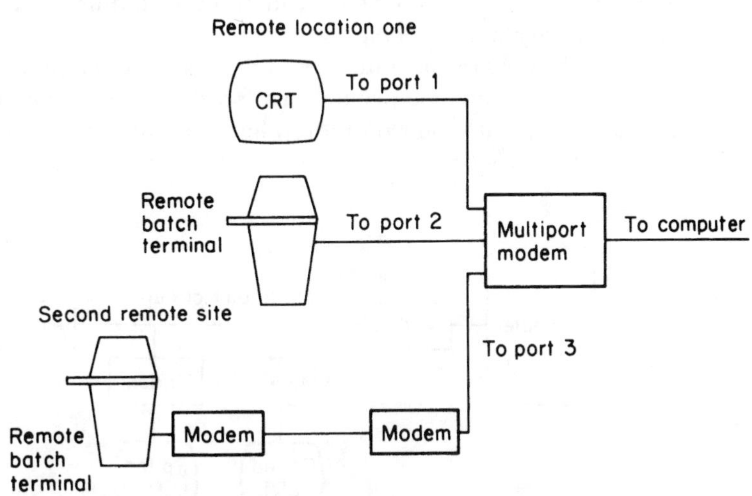

Figure 4.63 Multiport modem data communications equipment option. Using a data communications equipment (DCE) interface on port 3 of the multiport modem permits a second remote site to share the communications line from the first site to the computer

4.5 MULTIPOINT MODEMS

To alleviate substantial confusion, it should be noted that a multiport modem contains a built-in multiplexer which enables two or more separate data streams to be combined for transmission over a single circuit. In contrast, a multipoint modem is basically a modem designed to achieve fast polling acquisition times on multipoint lines.

Multipoint lines, or multidrop lines as they are also referred to, are usually installed for applications that require interactive terminal access from a number of geographically dispersed locations into a central computer facility.

An airline reservation system in which dispersed terminals randomly access the computer's database to determine flight information and seat availability is a representative application that could use multipoint lines. Another example would be an inventory control system where terminals are located at many warehouses and are used to report shippings and arrivals so that company inventories are continuously updated. For either application, the key item of interest is that each terminal only uses a small fraction of the total time available to all terminals connected to the line to complete a transaction, and the terminal is addressable and can recognize messages for which it is a recipient.

Although most terminals connected to multipoint circuits contain a buffer, this buffer primarily serves to enhance data throughput and is not a necessity. For example, consider the situation where an operator is entering data on the screen of a CRT through an attached keyboard. If the CRT does not have a buffer, the time it takes to transmit the data depends upon the operator's speed in typing it. During that time communications with all other terminals on the line are suspended. If the terminal has a buffer, the transmission speed from the terminal to the computer can be at a much higher rate than the operator's typing rate. Thus, once the operator has filled the CRT screen with data, the depression of a transmit key will permit the computer to select and receive the data from the terminal at a higher transfer speed and in a shorter time interval.

During the time the operator is entering data the computer is free to select other terminals. The wait time per terminal on the common line is therefore reduced. Since a slow operator on a multidrop line without buffered terminals could obtain an unjustified proportion of the total transmission time, some multidrop systems

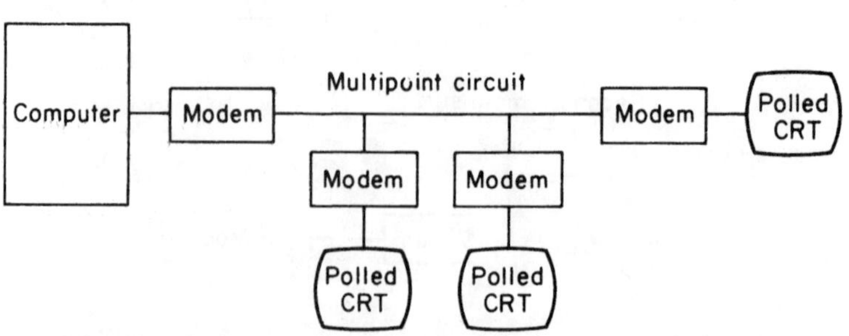

Figure 4.64 Typical multipoint circuit. On a multipoint circuit many polled terminals share the use of common communications facilities

incorporating unbuffered terminals have a built-in time-out feature. This time-out feature permits another operator to gain control of the line if the first operator pauses for a time greater than the time-out feature permits. When either type of terminal is used in the previously described working environment, then many terminals can share the same communications circuit on an interleaved basis. The polling and selecting protocol used will make it appear to each terminal operator as if a private connection existed for his or her exclusive use for transmitting and receiving data from the computer. A typical multipoint circuit used to connect three terminals is illustrated in Figure 4.64.

Factors affecting multipoint circuits

When the applicability of a multipoint circuit is being investigated, several parameters warrant careful investigation; two such parameters are the response time and the transaction rate of the terminals.

Response time

From a broad viewpoint, response time is the time interval from when an operator presses a transmit key at the terminal until the first character of the response appears back at that terminal. This response time consists of the many delay times associated with the components on the circuit, the time required for the message to travel down the circuit to the computer and back to the terminal, as well as the processing time required by the computer.

Transaction rate

The transaction rate is a term used to denote the volume of inquiries and responses that must be carried by the circuit during a specific period of time. This rate is normally expressed in terms of a daily average and as a peak for a specific period of the day.

Additional factors that affect the data transfer rates on multipoint circuits include the line protocol used and its efficiency, the transmission rate of the modems, and the turnaround time of the line. While there are many factors that contribute to multipoint line efficiency utilization, the focus of this section will be on multipoint modem characteristics which should be investigated to obtain a more efficient transmission process on such circuits.

Delay factors

When a terminal connected to a computer via a multipoint circuit is polled, several factors, which by themselves may appear insignificant, accumulate to degrade response time.

In the transmission of a message or poll from the computer to a terminal on the multipoint line, the first delay encountered is caused by the modem's internal delay time (D_m). This is the delay that would be seen if the time between the first bit entering the modem and the first modulated tone put on the circuit were measured. For a poll transmitted from the computer to a terminal or a response transmitted from the terminal to the computer, total modem delay time is equal to two times the modem's internal delay time ($2D_m$), since the poll or message is transmitted through two modems. In this case, the internal delay time of the modem is equivalent to the delay one would measure if the modems were placed back to back and the time beween the first bit entering the first modem and the first bit demodulated by the second modem were compared. This is shown in Figure 4.65. Depending upon the type of modem used, this internal delay time can vary from a few milliseconds to 20 ms or more.

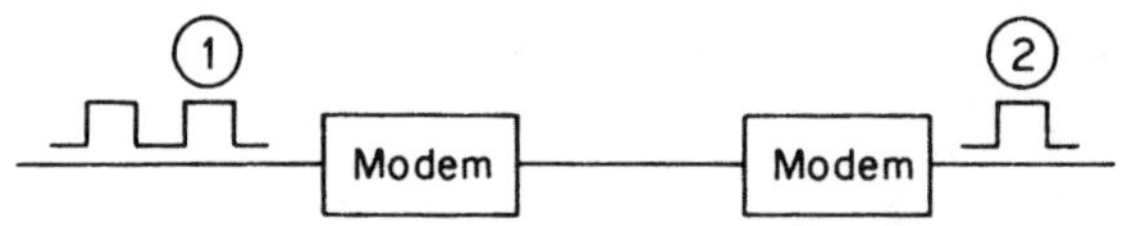

Figure 4.65 Internal modem delay time. Time difference from first bit transmitted (1) to first bit received (2) is denoted as total internal modem delay time ($2D_m$)

The second delay time on a multipoint circuit is a function of the distance between the computer and the terminal with which it is communicating. This delay time is called transmission or propagation delay (D_p) and represents the time it takes for the signal to propagate down the line to the receiving location. Although this delay time is insignificant for short transmission distances, coast-to-coast transmission can result in a transmission delay of approximately 30 ms on a terrestrial circuit or 250 ms if transmission occurs via a satellite.

When a response to a transmitted poll or message is returned from the remote terminal, several additional delay factors are encountered. First, the terminal itself causes a delay, since some time is required for the terminal to recognize the poll and initiate a response. The terminal delay (D_t) is usually a few milliseconds, but can be considerably longer, depending upon the design of the terminal and the software protocol used.

Since every modem on the multipoint line shares the use of this common circuit, only one modem at any point in time may have its carrier on. The carrier of each modem is turned on in response to the connected terminal or computer raising its request to send (RTS) signal, which indicates that a message is to be transmitted. The modem at the opposite end of the transmission path will then require some time interval to recognize this signal and adjust its internal timing. If the modems have automatic and adaptive equalization, additional time is spent adapting the modem to the incoming signal's characteristics.

Once these functions are completed, the modem located at the opposite end of the transmission path will raise its clear to send (CTS) signal which is required in response to the RTS signal if a return message is to be transmitted. This delay time is referred to as the request to send/clear to send delay time (RTS/CTS) and is usually denoted as $D_{R/C}$. Since the RTS/CTS delay time occurs twice in responding to a poll, the total RTS/CTS turnaround delay is $2D_{R/C}$. Disregarding terminal, propagation, and processing delays, the total turnaround delay attributable to the modems is M_D, where

$$M_D = 2(D_{R/C} + D_m)$$

Although the RTS/CTS delay varies by the type of modem used, delays can range from about 5 to 100 ms or more and normally cause most of the line's turnaround delay.

When multipoint networks were initially implemented, modem operating rates were at speeds of 2400 bps or lower. These modems were manually equalized and were adjusted and set at installation time. For this category of modems, the RTS/CTS delay ranges from about 10 to 20 ms, depending upon manufacturer. Owing to the advancement in modem transmission rates as well as new applications which require higher traffic rates and shorter response times, multipoint networks with 4800 bps modems were implemented. At this data rate, two types of modem equalization are used, the manual type previously discussed and that which is automatically equalized. In comparison to the static nature of manual equalization, automatic equalization permits the modem to continuously monitor and compensate for changing line conditions. An initial period of time is, however, required for the modem to 'train' on the signal each time the transmission direction reverses. For manually equalized modems, the RTS/CTS delay time is normally between 10 and 20 ms, since they are static in nature. For automatically equalized modems, the training time necessary for the modem to adapt to the incoming signal adds significantly to the RTS/CTS delay time, with such delays increased to 50 ms or more.

Throughput problems

To illustrate the effect of the increased delay on data throughput, let us examine the transmission of a 50-character data block. If each character consists of eight data bits, then the data block would contain 400 bits of information. If 2400 bps modems are used for transmission, the actual time required to transmit the block would be 400/2400 or 166 ms.

Let us assume that the RTS/CTS delay time of the 2400 bps modems was 10 ms and their internal delay time was 5 ms. The total delay time associated with the transmission and acknowledgement of the data block would then be 2 X (10 + 5) or 30 ms. Since the time to transmit the data block is 166 ms and the delay time associated with the block is 30 ms, the overhead attributable to the delay is 30/(166 +30) or 15.3%. Given a modem operating rate of 2400 bps, the effective data transfer rate becomes:

$$2400 - (2400 \times 0.153) = 2032 \text{ bps}$$

Now let us examine the effect of doubling the modems' operating rate to 4800 bps. At 4800 bps, modems employing automatic equalization originally had an

RTS/CTS delay time of 50 ms. Assuming the internal delay time remained constant at 5 ms, the total delay time to transmit the data block and receive an acknowledgement would become

$$2 \times (50 + 5) = 110 \text{ ms}$$

When the modems operate at 4800 bps, the time to transmit a 400-bit block is 400/4800 or 83.3 ms. Then the overhead attributable to the delay times is 110/(83.3 + 110) or 56.9%. Given the modem operating rate of 4800 bps, the effective data transfer rate becomes

$$4800 - (4800 \times 0.569) = 2069 \text{ bps}$$

Based upon the preceding analysis, it is obvious that the doubling of a multipoint modem's operating rate only resulted in an insignificant increase in the effective data transfer rate. If this situation were allowed to persist, most data communications users would never benefit from upgrading the operating rate of modems used on multipoint circuits. Fortunately, modem developments resulted in an increase in the effective data transfer rate that has a high correspondence to the increase in a modem's operating rate.

Multipoint modem developments

Until the early 1970s, users were forced to trade off the benefits derived from automatic equalization with the longer RTS/CTS delay time that was obtainable through the use of a modem equipped with this feature. Fortunately, many modem manufacturers incorporated techniques that reduce the RTS/CTS delay time while permitting automatic equalization.

One manufacturer implemented a so-called 'gearshift' technique where data transmission begins at a rate of 2400 bps, using a modulation technique that does not require extensive equalization. This reduces the RTS/CTS delay to a level of about 9 ms, which is the delay time normally associated with manually equalized modems transmitting at 2400 bps. Next, as transmission proceeds at 2400 bps, the receiving modem automatically equalizes on the incoming signal. After an initial transmission of 64 bits of data is received at the 2400 bps data rate, the training cycle is completed and both the sending and receiving modems 'gearshift' up to the faster 4800 bps data rate to continue transmission. While this technique will reduce the RTS/CTS delay time, the actual number of data bits transmitted during an interval of time will depend upon the size of the message transmitted. This is because the first 64 bits of each message are transmitted at 2400 bps prior to the modem gearshifting to the 4800 bps data rate.

Another technique used to increase the number of bits of information transmitted was obtained by incorporating a microprocessor into a modem. This microprocessor is used to perform equalization and provides a very fast polling feature which increases the data traffic transmitted during a period of time when compared to the standard Bell System equivalent 4800 bps modems or the gearshift-type modem. This comparison is illustrated in Table 4.29.

Table 4.29 Comparison of data traffic transmitted by 4800 bps multipoint modems

	Data bits		
Time (ms)	Modem with microprocessor	Modem using 'gearshift' technique	Bell System equivalent
9	0	0	0
20	24	24	0
26	58	41	0
36	106	65	48
50	173	133	115

Although the use of multipoint modems has significantly decreased since their peak period of usage during the mid-1980s, they can still be used effectively and economically to interconnect locations with relatively low volume transmission requirements. However, if you anticipate using multipoint modems you should select products with a minimal RTS/CTS delay to obtain a higher effective data transfer capability.

Remote multipoint testing

Since it is much more difficult to ascertain the cause of problems on a multipoint line than on a point-to-point circuit, modem manufacturers developed several testing features that are either incorporated into multipoint modems or offered as an option for end-user selection. One of the more common testing features offered with multipoint modems is remote testing. This feature results in each remote multipoint modem containing a unique address, permitting a central site modem to send an address code and command signal to the remote modems on the multipoint circuit, which will then place each remote site modem into an analog or digital loop-back mode of operation. By incorporating an address, each remote modem can then be tested from the remote site.

4.6 SECURITY MODEMS

The goal in the development of the security modem is to identify authorized users based upon the telephone number from which they originate calls on the PSTN. To accomplish this goal the security modem is designed to receive calls originated on the PSTN, interrogate the user's identity based upon a code or password, disconnect the user and then initiate a call on the PSTN to a predefined telephone number stored in memory in the modem. Since the security modem initiates a callback to the user attempting access, another name commonly associated with this device is a callback modem.

Operation

A callback or security modem contains a battery powered buffer area into which the network administrator enters an access code and telephone number for users authorized to dial a PSTN telephone number that is connected to the modem. Depending upon the security modem used, between four and twelve characters are commonly available for the authorization code.

Once a network user receives the authorization code and dials the number assigned to the security modem, that device first prompts them to enter their code. Assuming the code is entered correctly, the modem then disconnects the user from the line and originates a call to the telephone number associated with the access code.

Since the security modem simply verifies the telephone number of the originator of the call it should not be confused with data encryption devices that scramble transmission based upon a predefined algorithm. Thus, the protection of the security modem is relegated to verifying that the data call originator has an authorized access code and is located at an authorized telephone number.

Memory capacity and device access

The memory capacity and network administrator access to a security modem differ considerably between devices. Most security modems have a buffer area that only permits the storage of three to five access codes and associated telephone numbers. A few security modems have the capacity to store up to approximately 20 access codes and associated telephone numbers, representing the maximum capacity of devices in this equipment category. Concerning network administrator access, most modems require the network administrator to first cable a terminal to the device to enter the access codes and telephone numbers assigned to each access code. Then, once this is accomplished, the security modem can be connected to the computer or multiplexer port it is designed to service. This type of modem is difficult to modify, since it requires the network administrator to place it out of service and recable a terminal to the device to alter previously entered data. Then, after modifications are made to the access codes and telephone numbers to add new users or delete previously authorized users, the network administrator can disconnect the terminal and recable the security modem to the device it services.

A second method of updating access codes and associated telephone numbers is available on other security modems. This method requires the system administrator to first enter a master access code when they initially attach a terminal to the device. Thereafter, the network administrator can use any terminal attached to a compatible modem and dial the security modem via the PSTN. Once the master access code is entered, the network administrator can remotely enter or modify access codes and their associated telephone numbers.

Device limitations

Since it is a natural tendency of many persons to post access codes on terminals, the use of a security modem is only as good as the policy and enforcement of the policy concerning the use and distribution of access codes. In addition, since the security modem simply verifies the location of a terminal by a callback over the PSTN, data calls are still susceptible to illicit monitoring. Other limitations associated with security modems include the fixed telephone numbers stored in the modem's memory, the requirement and cost of a second call and the requirement of call originators to manually place some modems into an answer mode to receive the callback from the security modem.

Since a security modem requires the network administrator to associated fixed telephone numbers to access codes, this device is impractical for the traveling businessman to use. As an example of this, consider a salesperson who visits several customers during the day. Since a maximum of 20 access codes and associated telephone numbers can be entered into a security modem, even if the salesperson contacts the network administrator on a regularly scheduled basis it would be a demanding task to continuously adjust the security modem to correspond to one's sales calls.

The requirement of the security modem to initiate callbacks can considerably add to the cost of the organization's telephone bill. This is because the telephone companies in many metropolitan areas base the cost of a call upon message units. Since the initial call to the security modem results in a minimum of one message unit being used, even if the call only required 30 s for the user to enter his or her access code, message units can rapidly build up. For long-distance calls the cost associated with the requirement for an initial call and a callback may be more pronounced. This is due to the tariff structure of long-distance calls, where the cost of the first minute can be considerably higher than the cost per minute for succeeding minutes. As the initial call to the security modem is normally billed at the higher first-minute rate, this can considerably expand the communications cost of an organization over a period of time.

Another key limitation of security modems is the storage capacity of the device. If the modem, as an example, is limited to storing five access codes and associated telephone numbers a significant number of devices may be required to support a large number of users. As an example of this consider a geographical area that contains 50 terminals that will dial a multiplexer. If only six terminals are expected to be active at one time, six dial-in lines and conventional modems connected to the multiplexer would be sufficient. If security modems are required, 10 devices would be necessary, since each modem is assumed to store a maximum of five access codes and telephone numbers. This in turn would require 10 dial-in lines and 10 ports on the multiplexer, increasing the cost of communications facilities and equipment necessary when security modems are used. In addition, although a common rotary could be used with conventional modems, users dialing security modems would be restricted to accessing specific telephone numbers to which the modem with their access code is connected. This in turn further limits the flexibility associated with using security modems.

4.7 LINE DRIVERS

There are four basic methods used to connect a terminal device to a computer. These methods are listed in Table 4.30.

Table 4.30 Terminal-to-computer circuit connections

Direct connection of terminals through the use of wire conductors
Connection of terminals through the utilization of line drivers
Connection of terminals through the utilization of limited distance modems
Connection of terminals through the utilization of modems or digital service units

It is interesting to observe that each of the methods listed in Table 4.30 provides for progressively greater distances of data transmission while incurring progressively greater costs to the users. In this section the limitations and cost advantages of the first two methods will be examined in detail.

Direct connection

The first and most economical method of providing a data circuit is to connect a terminal device directly to a computer through the utilization of a wire conductor. Surprisingly, many installations limit such direct connections to 50 feet in accordance with terminal and computer manufacturers' specifications. These specifications are based upon EIA RS-232 and ITU V.24 standards. If the maximum 50-feet standard is exceeded, manufacturers may not support the interface, yet terminals have been operated in a reliable manner at distances in excess of 1000 feet from a computer over standard data cables. This contradiction between operational demonstrations and usage and standard limitations is easily explained.

If you examine the RS-232 and V.24 standards, such standards limit direct connections to 50 feet of cable for data rates up to 20 kbps. Since the data rate is inversely proportional to the width of the data pulses transmitted, taking capacitance and resistance into account, it stands to reason that slower terminals can be located further away than 50 feet from a computer without incurring any appreciable loss in signal quality. Simply stated, the longer the cable length the weaker the transmitted signal at its reception point and the slower the pulse rise time. As transmission speed is increased, the time beween pulses is shortened until the original pulse may no longer be recognized at its destination. This becomes more obvious when you consider that a set amount of distortion will affect a smaller (less wide) pulse than a wider pulse.

In Figure 4.66, the relationship between transmission speed and cable length is illustrated for distances up to approximately 3400 feet and speeds up to 40 000 bps. This figure portrays the theoretical limits of data transfer speeds over an unloaded length of 22 American wire gauge (AWG) cable. Many factors can have an effect on the relationship between transmission speed and distance including noise, distortion introduced owing to the routing of the cable, and the temperature of the surrounding area where the cable is installed. The ballast of a fluorescent fixture, for instance, can cause considerable distortion of a signal transmitted over a relatively short distance.

The diameter of the wire itself will affect total signal loss. If the cross-sectional area of a given length of wire is increased, the resistance of the wire to current

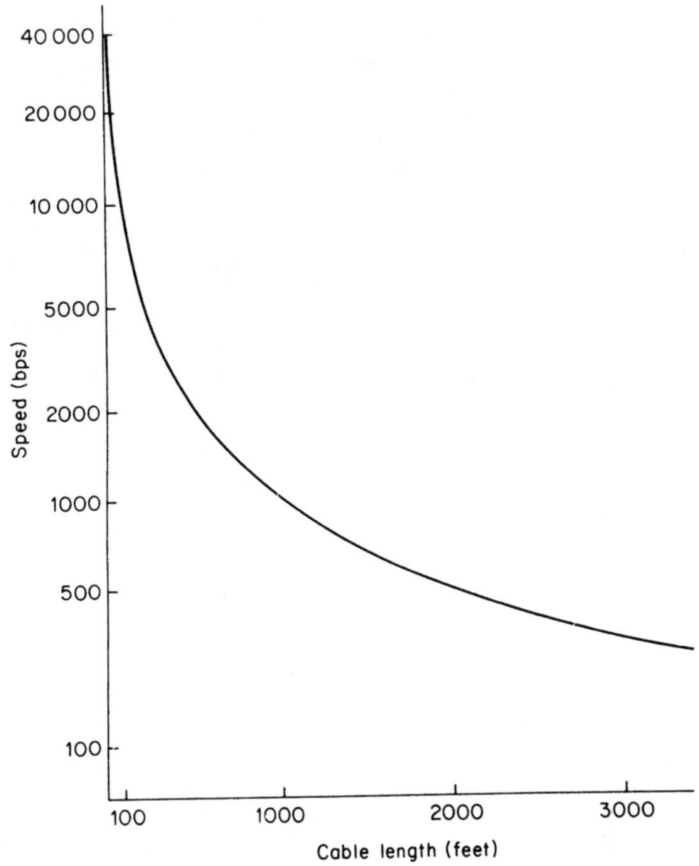

Figure 4.66 Speed and cable length relationship

flow is reduced. Table 4.31 shows the relationship between the dimensions and resistances of several types of commercially available copper wire denoted by gauge numbers. By increasing the gauge from 22 to 19, the resistance of the wire is reduced by approximately one-half.

Another method which can be utilized to extend the length of a direct wire connection is limiting the number of signals transmitted over the data link. After the connect sequence or handshaking is accomplished, only two signal leads are required for data transfer: the transmitted data and receive data leads. With some minor engineering at both ends of the data link and an available dc voltage source, the remaining signals can be held continuously high, permitting the use of a simple paired cable to complete the data link.

Cable length can be further extended by the use of commercially available low-capacitance shielded cable. The shield consists of a thin wrapping of lead foil around the insulated wires and is quite effective in reducing the overall capacitance of the data cable with a very modest increase in price over standard unshielded cables. The use of low-capacitance shielded cables is strongly recommended when several cables must be routed through the same limited diameter conduit. Once the

Table 4.31 Relationship between wire diameter and resistance

Gauge number	Diameter (inch)	Ohms/1000 feet at 7°F
10	0.102	1.02
11	0.091	1.29
12	0.081	1.62
13	0.072	2.04
14	0.064	2.57
15	0.057	3.24
16	0.051	4.10
17	0.045	5.15
18	0.040	6.51
19	0.036	8.21
20	0.032	10.30
21	0.028	13.00
22	0.025	16.50
23	0.024	20.70
24	0.020	26.20
25	0.018	33.00
26	0.016	41.80
27	0.014	52.40
28	0.013	66.60
29	0.011	82.80

practical limitation of cable length has been reached, signal attenuation and line distortion can become significant and either reduce the quality of data transmission or prevent its occurrence. One method of further extending the direct interface distance between terminals and a computer is by incorporating a line driver into the cable connection.

Using line drivers

As the name implies, a line driver is a device which performs the function of extending the distance a signal can be transmitted down a line. A single line driver, depending upon manufacturer and transmission speed, can adequately drive signals over distances ranging from hundreds of feet up to a mile. One manufacturer introduced a line driver capable of transferring signals at a speed of 100 kbps at a distance of 5000 feet and a 1 Mbps signal over a distance of 500 feet using a typical multipair cable.

A multitude of names have been given to the various brands of line drivers to include local data distribution units and modem eliminator drivers. For the purpose of this discussion a line driver is a stand-alone device inserted into a digital transmission line in order to extend the signal distance.

In Figure 4.67 the distinction between single and multiple line drivers is illustrated and contrasted with limited distance modems which will be explored in Section 4.8. The primary distinction between line drivers and limited distance

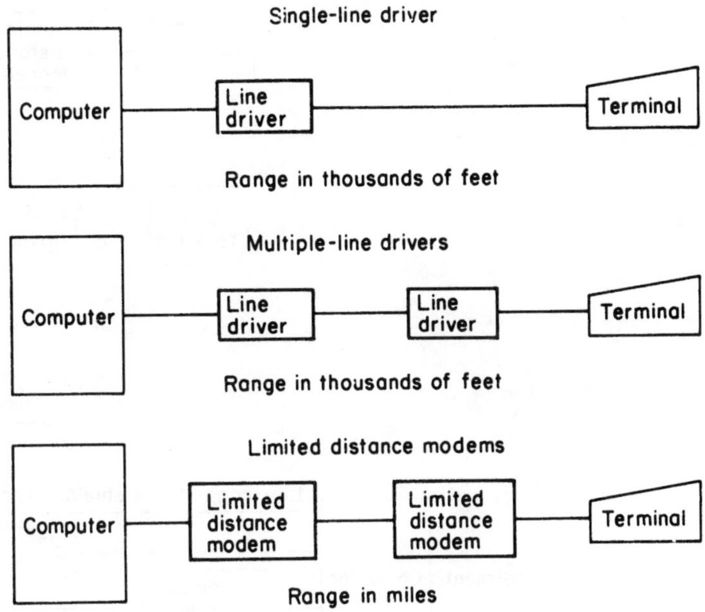

Figure 4.67 Line drivers and limited distance modems. When line drivers are used, the signal remains in its digital form for the entire transmission. Distances can be extended by the addition of one or more line drivers which serve as digital repeaters. For longer distances, limited distance modems can be utilized where the transmitted data is converted into an analog signal and then reconverted back into its original signal by the modem

modems is that two identical units must be used as limited distance modems to pass data in analog form over a conductor, whereas a line driver serves as a repeater to amplify and reshape digital signals.

Although some models of line drivers can theoretically have an infinite number of repeaters installed along a digital path, the cost of the additional units as well as the extra cabling and power requirements must be considered. Generally, the use of more than two line drivers in a single digital circuit makes the use of limited distance modems a more attractive alternative.

Applications

The characteristics of line drivers become important when one considers their incorporation into a data link. If the EIA RS-232 signals are accepted, amplified, regenerated, and passed over the same leads, they can be used as repeaters. If, however, the line driver also serves as a modem eliminator by providing a synchronous clock, inserting RTS/CTS delays and reversing transmit and receive signals, care should be taken when attempting to use them as repeaters. If this type of line driver is used and strapping options are provided for the RTS/CTS delay and such desirable features as internal/external clock, it is a simple matter to

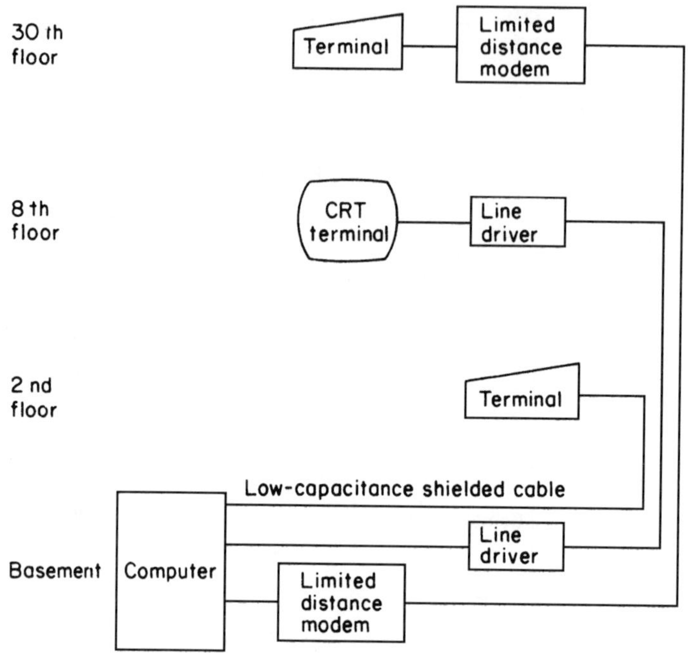

Figure 4.68 A variety of methods can be used to connect a terminal to a computer located in the same building

convert it to a repeater by setting the delay to zero, setting the clock to external, and using a short pigtail cable to reverse the signals.

Figure 4.68 illustrates a typical application where line drivers would be installed. In this office building the computer system is located in the basement. The three terminals to be connected to the computer are located on different floors of the building. A remote terminal located on the second floor of the building is only 100 feet from the computer and is directly connected by the use of a low-capacitance shielded cable. The second terminal is located on the eighth floor, approximately 500 feet from the computer and is connected by the use of line drivers to extend the signal transmission range. A third terminal, located on the 30th floor, uses a pair of limited-distance modems for transmission since a large number of line drivers would be cost-prohibitive.

4.8 LIMITED-DISTANCE MODEMS

As the name implies, limited-distance modems are designed for data transmission over relatively short distances when compared to traditional modems. The utilization of such devices can result in dramatic savings in comparison to the cost of using conventional modems for the transmission of data over short distances. These devices have operating rates ranging from 110 bps to over 1 million bps for distances ranging from 1 to 20 miles or more. Names given to these devices include not only limited-distance modems but such descriptive terms as modem bypass units, short-range data sets, short-haul modems, and wire line modems.

Rationale and status

The rationale for the use of limited distance modems is one of economics. Since they are designed to transmit over relatively short distances, they do not require equalization circuitry and other features associated with more expensive analog leased line modems. Although the migration to the use of digital transmission facilities, coupled with a significant decrease in the cost of conventional modems, considerably reduced demand for limited distance modems, they still represent an economical method to interconnect devices within a building or campus area.

Contrasting devices

In contrast to line drivers where one or more such devices are used to regenerate digital data and extend transmission ranges, limited distance modems require two matching components, one at each end of the circuit. In most cases, limited distance modems convert digital data into analog signals for transmission. Some devices on the market, however, convert a terminal's serial binary bits into bipolar return to zero signals to maintain transmission entirely in digital form. The utilization of this type of device can provide you with a direct-limited distance digital extension to a DATAPHONE® digital service/channel service unit (DSU/CSU).

Transmission media

Some of the distances that can be achieved with limited-distance modems are illustrated in Figure 4.69 when transmission is over unloaded, twisted-pair metallic wire cables. The representative transmission distances illustrated represent a refined composite derived from manufacturers' data sheets and should be used as a guide to the variations in methods of signaling and sensitivity in receiving level between devices. As discussed in Section 4.7, the smaller the wire gauge the greater the diameter of the wire and the lower the resistance of the wire to passage of current. The four-wire gauges selected for illustration in Figure 4.69 are used because they correspond to the common wire sizes used in the US telephone systems, and many users prefer to utilize the existing common carrier cables in lieu of routing their own private cables. Of course, there is no gauge restriction in routing a private cable, and in light of the differences shown in Figure 4.69 between 19- and 26-gauge cable, many users will install a lower gauge cable even when transmitting at a normal rate to alleviate the necessity of recabling if the data transmission rate should increase at a later data.

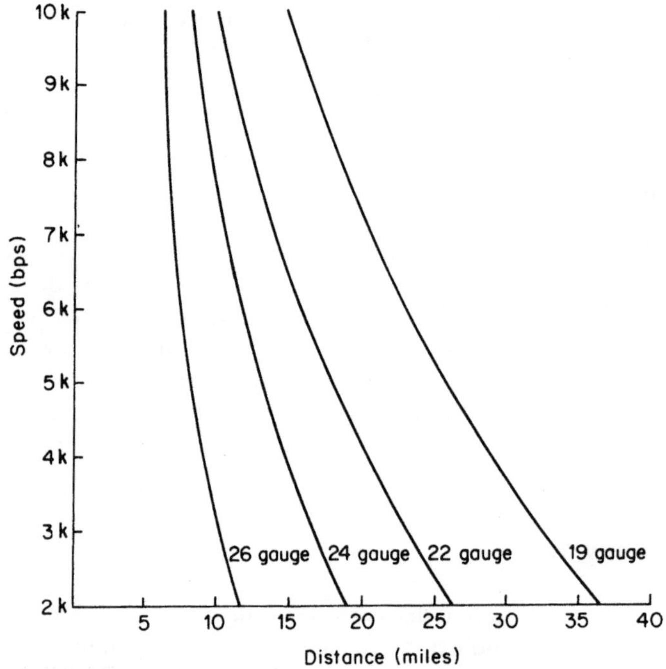

Figure 4.69 Representative transmission distance—miles per twin pair wire gauges (unloaded)

Table 4.32 Telephone wire sizes and resistances

Circuit type	Gauge (AWG)	Diameter (mm)	Resistance	
			1000 ft	Loop mile
Station wire	22	1.00	16	
Station wire	24	0.79	25	
Station wire	26	0.625	40	
Toll wire	19	1.42	8	
Interoffice wire	19	1.42	8	
Open wire lines				
Copper	10	4.00		6.7
Copper	12	3.00		10.2
Copper-clad steel	12	3.00		25
Copper-clad steel	14	2.00		44

Common telephone wire sizes and associated measured resistances are listed in Table 4.32. In this table, a loop mile is a term which is used to describe two wires connecting two points which are physically located 1 mile apart, an important measurement when one considers using carrier facilities for supporting their limited-distance transmission requirements. If the use of existing carrier facilities is being contemplated, a close liaison with the local telephone company should be established. Common carriers and their operating companies in some areas may not be completely familiar with this particular type of hardware or the tariff structure for the service to support limited-distance transmission. It is also important that the proposed limited-distance modem conforms to the specifications set forth in the Bell System publication 43401 entitled 'Transmission Specifications for Private Line Metallic Circuits'. This publication describes the signal level criteria objectives for private line metallic circuits (cable pairs without signal battery or amplification devices). In addition, the publication notes that the telephone companies have no obligation to provide private line channels on a metallic basis. Most manufacturers of limited-distance modems clearly specify that their equipment operate in accordance with the previously mentioned publication. If it is not explicitly stated, you may encounter delays and additional cost to ensure that the transmitter of the device is modified to comply with the specifications.

Operational features

Most limited-distance modems utilize a differential diphase modulation scheme and permit internal transmit timing or external derived timing from the associated data terminal. These devices act similarly to a line driver, and most will accommodate four-wire half/full duplex and two-wire half-duplex/simplex data transmission. Data rate switches on some asynchronous units provide selectable data rates ranging from 110 to 28 800 bps, while the selectable rates of most synchronous units inlude various increments of 2400 bps. Users often select a transmission speed only to realize that by the time the equipment is installed changing requirements may indicate a different speed; therefore, the ease of adjustment of the unit should be investigated.

Some manufacturers state that their limited-distance modems can be inserted into a transmission line to serve as a repeater, as illustrated in Figure 4.70, to

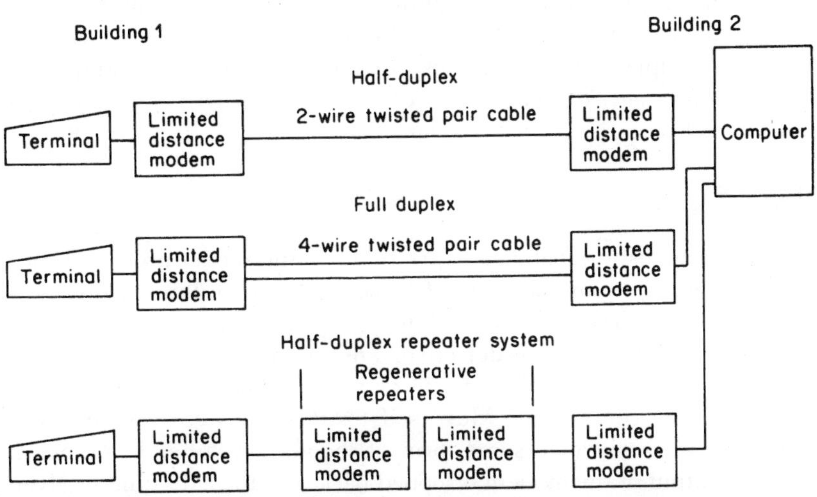

Figure 4.70 Typical point-to-point applications. If a limited-distance modem is planned to be used as a repeater, care should be taken to ensure that shelter, a power source, and access for diagnostic testing and maintenance is available

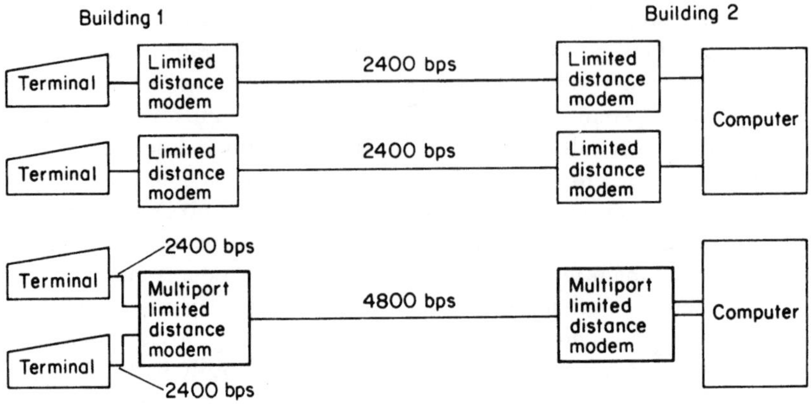

Figure 4.71 Multiport operation reduces cabling requirements. Using a limited distance modem with a multiport feature, one cable may be utilized to provide access to a computer from several terminals

further extend the transmission distance. An obvious limitation to inserting such devices to serve as repeaters or data regenerators is the fact that they must be sheltered, have an available power source, and be readily accessible for diagnostics and maintenance. Since these combinations are difficult to achieve at locations between buildings, normal utilization of limited-distance modems has their locations fixed at each end of a transmission medium.

A very desirable feature that is offered on some units is a multiport or split stream feature. This feature permits several collocated terminals to utilize a single limited-distance transmission link at a considerable cost saving over the less expensive but more limited capability single-channel device. The key advantage to the employment of a multiport limited-distance modem is that only one cable instead of many cables can be used to service the transmission requirements of multiple terminals, as illustrated in Figure 4.71. For additional information on the advantages of multiport operations, the reader is referred to Section 4.4.

Diagnostics

Diagnostic capabilities vary both by the model produced by a manufacturer and between manufacturers of these units. Some limited-distance modems have self-testing circuitry which permits the user to easily determine if the unit is operating correctly; or, if it is defective, it notifies the user of the operational status of the unit through the display of one or more lights which indicate equipment status and alarm conditions. Most self-testing features available with limited-distance modems involve some form of loopback testing. In such a test, the transmitter output of the modem is looped back or returned to the receiver of the same unit so that the transmitter signal can be checked for errors. Other tests which are available on some models include dc busback, in which the received data and clocking is transmitted back to the limited distance modem at the opposite end of the line to provide an end-to-end test and a remote loopback test which can be used to trigger a dc busback at a remote location.

4.9 BROADBAND MODEMS

In this section we turn our attention to two types of modems that provide a high data transmission operating rate approximately 1000 times or more that provided by the fastest modem designed for use on the public switched telephone network. Earlier in this chapter I referred to such devices as cable and digital subscriber line (DSL) modems. That reference, which is still commonly used, denotes the primary application or intended application of each type of modem. That is, a cable modem is designed to provide data transmission capability on cable TV systems, while a so-called digital subscriber line modem is designed to support high-speed data transmission on telephone twisted-pair wiring into the home or office. Since the term 'broadband' is commonly used to represent a data transmission operating rate at or above 1.5 Mbps, which most cable and DSL modems support, I have taken author liberties and classified them collectively as *broadband modems*.

Both cable and DSL modems provide the potential to revolutionize communications technology as well as substantially to alter the manner by which we commonly perform daily communications-related activities at work and at home. Within a short period of time you may be able to surf the Internet's World Wide Web via the twisted-pair telephone line or a coaxial TV cable at operating rates 1000 or more times greater than those obtainable through the use of analog voice-grade modems, order video-on-demand movies via either type of media, and use either transmission system to communicate by voice, picture phone or perhaps even a technology requiring megabyte operating rates that will be developed tomorrow.

Currently, the dropping of regulatory barriers as well as the availability and affordability of megabyte transmission technology is fueling a race between cable TV operators and local telephone companies as they develop plans to enter each other's industry. Although a telecommunications reform bill was finally passed into law after years of debate, the actual implementation of various aspects of that law raises some questions concerning the degree of competition that will occur. However, it is with a degree of certainty that competition will occur through the use of cable and DSL modems that are the subject of this section. Thus, a good place to commence a discussion of broadband modems is by first reviewing the basic infrastructure of telephone and cable TV transmission, placing special emphasis on the methods used to connect subscribers to their backbone network via different local loop technologies.

Telephone and cable TV infrastructure

The cabling infrastructure developed for telephone and cable TV (CATV) operations represents two diverse but to a degree potentially merging technologies. Telephone company plant is based upon the routing of twisted-pair wire from a serving central office to individual subscribers to provide two-way communications. In comparison, cable TV was based upon the use of coaxial cable to provide a one-way transmission path for video distribution to subscribers.

Telephone

Although the routing of wire-pairs directly to individual subscribers was long ago replaced in many areas by the multiplexing of several subscriber wire-pairs from a common location near groups of homes or offices to a telephone company central office as an economy measure, multiplexing maintains a direct path from the subscriber to the central office by either time or frequency, depending upon the technique employed. In addition, that path is bidirectional, enabling two-way voice and data to be transported from one subscriber to another via the telephone company wiring infrastructure.

Figure 4.72(a) provides a general schematic which illustrates the present telephone company wiring infrastructure. Although almost all long-distance transmission is carried via fiber optic cable, almost all local loop wiring is currently carried via metallic twisted-pair. Since the present twisted-pair local loop is limited by filters to a bandwidth under 3300 Hz, this also limits the maximum operating rate of voice-grade modems to approximately 33.6 kbps.

A second limitation involves the use of multiplexers that are referred to as a Digital Loop Carrier (DLC) system by a communications carrier. Although the use of a DLC system minimizes the length of local loops to subscribers, it places a limit on the actual bandwidth obtainable from a subscriber to a telephone company central office. This bandwidth limitation, as we will note later in this section, adversely affects the potential use of DSL modems and must be bypassed for the technology to operate.

Based upon several field trials conducted during the mid-1990s, telephone companies are beginning to expand the use of fiber by installing so-called 'fiber to the neighborhood'. Under this concept, fiber is routed from a central office to a common location in a housing subdivision, building complex, or similar site. Then, using electrical/optical converters the fiber trunk is connected to relatively short existing metallic twisted-pair wiring routed into homes and offices. Under this concept the high bandwidth of the fiber enables telephone companies to eliminate the filtering of twisted-pair local loops, and expand the bandwidth available per loop via multiplexing, which considerably expands the bandwidth available for use by modems. Thus, DSL modems can be used on the local loop to provide a much higher operating rate than that obtainable by the use of conventional twisted-pair wiring. Figure 4.72(b) illustrates the emerging telephone company wiring infrastructure.

Due to the considerable time and expense associated with recabling the existing telephone local loop infrastructure, several communications carriers are examining alternative methods to provide subscribers with high-speed transmission over existing local loop twisted-pair wire. Two promising techniques are known as High-Bit-Rate Digital Subscriber Line (HDSL) and Asymmetric Digital Subscriber Line (ADSL) which are described in detail later in this section. Although

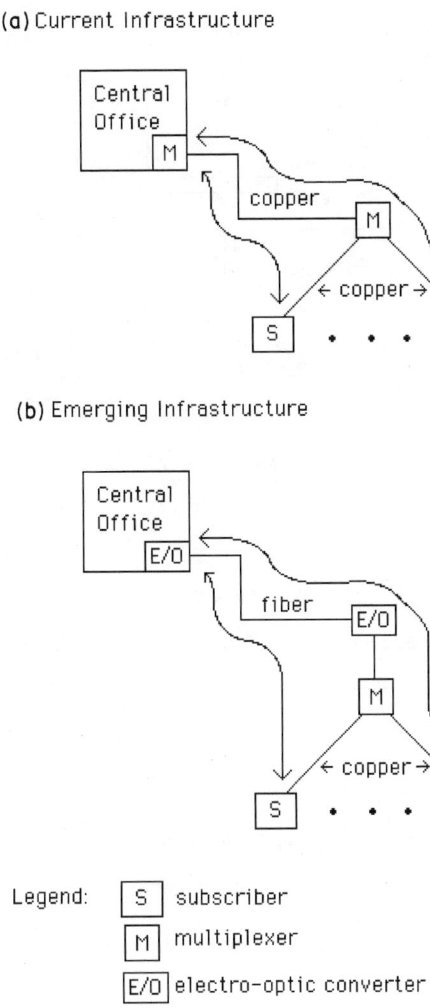

Figure 4.72 Current and emerging telephone company wiring infrastructure

HDSL and ADSL uses the existing twisted-pair local loop cable infrastructure, both require special equipment at the subscriber's premises and at the central office, the operation of which will be described later in this section.

Cable TV

The cable television (CATV) infrastructure was originally developed to provide one-way video signals to subscribers. To do so, its wiring infrastructure was designed using a tree structure, with signals transmitted from a headend located at the beginning of the tree onto branches for distribution to subscribers. Signals on main branches were further split onto additional branches, with amplifiers installed

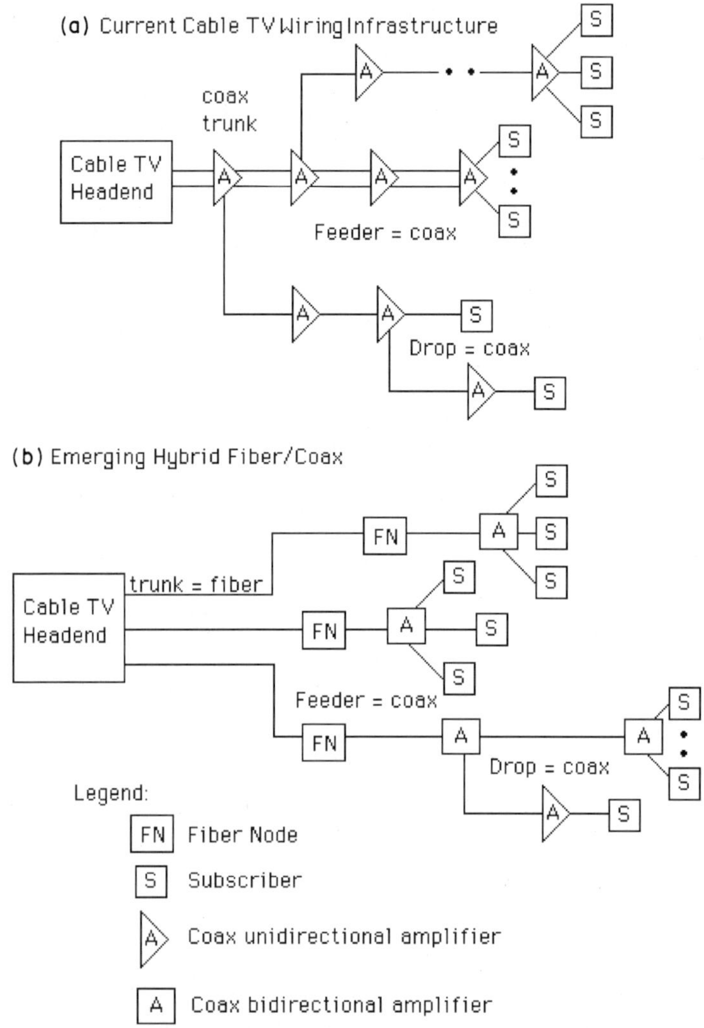

Figure 4.73 Current and emerging cable TV wiring infrastructure

to boost signal power since signal loss occurs due to cable distance as well as from the splitters encountered in the signal path.

The original bandwidth used on CATV systems ranged from 10 to 550 MHz, using 6 MHz per TV channel. Multiple signals are placed on a coaxial cable through the use of frequency division multiplexing equipment at the headend, with a tuner in subscriber set-top boxes used to extract the appropriate channel. This cabling and multiplexing scheme resulted in most CATV systems providing up to 83 channels to subscribers. The entire cable infrastructure of most pre-1995 CATV systems was based upon the use of coaxial cable as illustrated in Figure 4.73(a).

The CATV industry has had its share of trials to test different types of cabling and the support of two-way transmission by several competing techniques. One technique involves the use of a second cable to provide a return path, while the second technique uses a different portion of the coaxial cable bandwidth to obtain a return path.

Both techniques involve a considerable change to the existing CATV cabling infrastructure. For example, using a different portion of the frequency spectrum for the return path requires the installation of bidirectional amplifiers, while the use of a second cable requires both a new set of amplifiers and cabling.

Recognizing that the use of the CATV cabling infrastructure could not accommodate the requirements of a large subscriber base for video-on-demand, high-speed data transmission, and even digitized telephone service, cable operators began to install hybrid fiber/coax (HFC) systems during the mid-1990s. Under HFC a star cabling infrastructure is used, with fiber cable routed from a cable TV switch to an optical distribution node commonly located in a subdivision or building complex. Then, coax is routed from the distribution node to individual subscribers. Under the HFC architecture downstream frequencies of 50 MHz to 750 MHz can be subdivided to provide a variety of analog and video channels to include data services and telephony. Upstream operations are limited to the 5 MHz to 40 MHz frequency band for telephony, data services, and control channels used by interactive video set-top boxes.

Figure 4.73(b) provides a general schematic of the evolving CATV HFC cabling infrastructure. Note that the amplifiers shown in Figure 4.73(b) must bidirectional to enable upstream transmission from subscribers to be supported. Table 4.33 compares the current and evolving operating characteristics of CATV and telephone compares transmission facilities. The evolving CATV transmission method in Table 4.33 requires a degree of elaboration. Currently one-way CATV systems broadcast frequency-multiplexed signals. While the evolving HFC infrastructure will still use a broadcast transmission method for delivery of basic cable video services, it will also employ switching technology to provide routing and delivery of voice, data, and video-on-demand in a manner similar to that by

Table 4.33 Cable TV (CATV) versus telephone operating characteristics

Operating characteristic	CATV	Telephone
Bandwidth	6 MHz/TV channel	4 kHz/line
Use:		
Current	Video	Voice and data
Evolving	Video, data, voice	Voice, data, Video
Directionality:		
Current	One-way	Two-way
Evolving	Two-way	Two-way
Transmission method:		
Current	Broadcast	Switched
volving	Broadcast/switched	Switched
Media:		
Current	Coaxial cable	Twisted-pair
Evolving	Fiber + coaxial cable	Fiber + twisted-pair

which the telephone company uses switching to establish a connection between a calling and a called party for a voice or data communications session. Now that we have an appreciation for the current and evolving cabling infrastructures used for CATV and telephone local loop operations, let's turn our attention to the broadband modems that will provide high-speed transmission on each cabling system.

Cable modems

As an evolving technology there were several different engineering approaches being used to develop cable modems when this book was prepared. Although each approach results in an incompatibility between dissimilar vendor products, since it appears that each CATV operator will more than likely standardize their offering based upon a common vendor product, the inability of different cable systems to interoperate should not present a problem to subscribers. In fact, even if two cable systems use incompatible modems a subscriber on one system should be able to exchange electronic mail and operate other applications that require communications with subscribers on another system. This is true as long as the two systems are interconnected by a gateway and software on each system supports the application being used by each subscriber. This interoperability can be viewed as being similar to the Internet, where a person whose PC is attached to a LAN that in turn is connected to the Internet can communicate with a person who uses a modem to dial CompuServe, Prodigy, or America Online. In comparison, modem compatibility becomes an issue in the use of the switched telephone network where modem users are able to communicate with anyone connected to the PSTN as long as the called party has a compatible modem. Since CATV data operations can be expected to result in cable operators connecting the data portion of their network to commercial on-line services such as CompuServe and America Online as well as to the Internet, compatibility between cable modems manufactured by different vendors is not as important an issue as for modems developed for use on the PSTN. However, since standardization will make competition more meaningful for cable operators and vendors, a considerable amount of effort is presently being spent to standardize cable modem technology. One such effort is being made by the IEEE 802.14 Working Group under the title 'Standard Protocol for Cable-TV Based Broadband Communications Network'. Since the task of the IEEE 802.14 Working Group may not be completed for a considerable period of time, we will focus our attention on one announced product to obtain an appreciation for the operational capability of a cable modem.

LANcity LCP

One of the first modem manufacturers to provide a city-wide data communications capability through the development of a series of products to include a cable modem was LANcity of Andover, Massachusetts, which was acquired by Bay Networks. Founded in 1990, by the summer of 1994 LANcity had introduced equipment to provide cable TV connectivity to the Internet. In April 1995,

LANcity introduced the world's fastest and least expensive personal cable TV modem, named LCP (LANcity Personal Cable TV Modem). This modem can be set to operate in any 6 MHz TV channel in the transmit range from 5 to 42 MHz and the receive frequency range of 54 to 750 MHz.

This modem uses Quadrature Phase Shift Keying (QPSK) modulation to obtain a symmetrical 10 Mbps operating rate, requiring approximately 5 MHz of RF bandwidth in a 6 MHz channel. Under QPSK the phase of the carrier signal is varied based upon the composition of the digital data to be transmitted. For example, a digital '1' could be transmitted by generating a 180° phase shift in the carrier, while a '0' might be represented by a 0° phase shift. The 'quadrature' aspect of the modulation scheme results in the carrier being capable of being shifted to one of four possible phases (0°, 90°, 180°, 270°) based upon the dibit value of the data to be transmitted. Figure 4.74 illustrates the four-point QPSK constellation pattern.

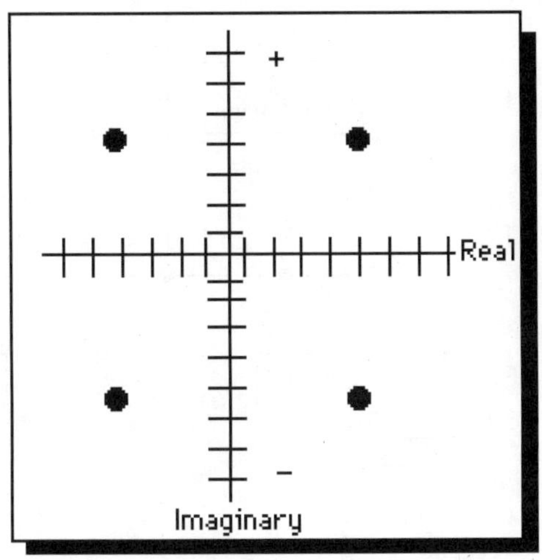

Figure 4.74 Four-point QPSK constellation

Access protocol

Based on the method by which CATV channels are routed on the main trunk and feeder cables through drop cables to the subscriber, bandwidth is shared among subscribers. In comparison, a local loop twisted-pair connection from the telephone company central office to a subscriber represents dedicated bandwidth. This means that access to the 6 MHz channel the cable modem uses must be shared among many subscribers, requiring an access protocol to govern the orderly flow of data onto the cable. LANcity and most cable modem vendors support the Ethernet Carrier Sense Multiple Access/Collision Detection (CSMA/CD) protocol to provide a mechanism for cable subscribers to gain orderly access to the 6 MHz TV channel. To accomplish this the LANcity LCP cable modem is cabled to an Ethernet 10 Mbps adapter card installed in a personal computer. The PC uses software to establish a TCP/IP protocol stack which results in the computer becoming in effect a workstation on a LAN that can represent up to 200 miles of CATV cable infrastructure.

In addition to requiring an Ethernet adapter card in each PC that will be connected to the cable modem, the upgrade of CATV plant to support data transmission via cable modems requires several additional components. These include a frequency converter, configuration server and router. Figure 4.75 illustrates their use on a CATV cabling infrastructure to provide subscribers with a bidirectional 10 Mbps operating capability using the LANcity LCP.

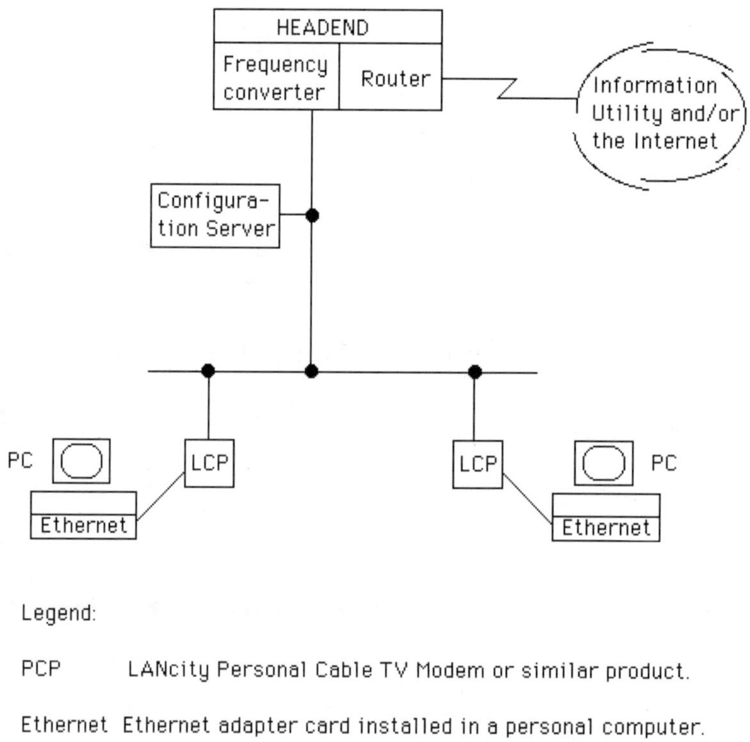

Figure 4.75 CATV plant components supporting data transmission

Frequency converter

The use of a frequency converter is required to change the reverse direction frequency to the forward direction frequency and vice versa. This conversion is required to enable one user to communicate with another as one channel provides a transmission path while a second channel is used for data reception. This conversion occurs at the headend and is limited to the 6 MHz channels used for data, ignoring other cable services. The LANcity product that performs frequency conversion is marketed as the LANcity TransMaster (LCT).

Configuration server

The configuration server supports the TCP/IP network formed on the CATV cable infrastructure. In doing so the server performs address resolution services as well as managing IP addresses for the client computers using cable modems.

Router

The function of the router is to encapsulate data into standardized frames for transport to external networks as well as to receive frames originating outside the CATV LAN.

Operating rate comparison

To obtain an appreciation of the potential capability of a 10 Mbps modem, Table 4.34 provides a comparison of the file transfer time for a 10 Mbyte file at operating rates from 9600 bps (V.32 modem) to the LCP cable modem's 10 Mbps rate. Although the data in Table 4.34 indicates that the 10 Mbps operating rate of the LANcity LCP modem provides a significant reduction in file transfer time, it should be noted that the transfer time of 8 seconds is a mathematically derived time that will not be obtainable even under the best of circumstances. In actuality, a cable modem user will share the 10 Mbps transmission path to the headend with other users using the CSMA/CD access protocol. This means your actual throughput will be governed by the number of other subscribers transmitting data. In addition, the occurrence of Ethernet frame collisions results in each Ethernet adapter using a random exponential backoff algorithm to generate a waiting time prior to retransmitting a frame. Although this might increase file transfer time to 10 to 20 seconds, or even 30 or more on a heavily utilized LAN, it still represents a considerable improvement over the V.34 28 800 bps transfer time of 46 minutes.

Table 4.34 10 Mbyte file transfer time comparison

Operating rate	Transfer time
9600 bps (V.32 modem)	138 minutes
14 400 bps (V.32 bis modem)	93 minutes
28 800 bps (V.34 modem)	46 minutes
64 000 bps (ISDN B channel)	20 minutes
1544 000 bps (T1 DSU)	52 seconds
4000 000 bps (Token-Ring)	20 seconds
10 000 000 bps (Ethernet)	8 seconds

IEEE 802.14 proposal

In a proposal to the IEEE 802.14 Working Group for the use of QAM for HFC downstream transmission, Scientific Atlanta noted that QAM is non-proprietary and was previously selected as the European Telecommunications Standard. In the firm's proposal to the IEEE two levels of modulation based upon 64 QAM and 256 QAM were defined to permit implementation flexibility. Based upon the proposed modulation specifications, the standardization of QAM for downstream transmission would result in a signaling rate of 5 MHz using a carrier frequency between 151 MHz and 749 MHz spaced 6 MHz apart to correspond to TV channel assignments.

The use of a 5 MHz signaling rate and 64 QAM which enables six bits to be encoded in one signal change permits a transmission rate of 6 bits/symbol × 5 MHz, or 30 Mbps. In comparison, the use of 256 QAM results in the packing of eight bits per signal change, resulting in a transmission rate of 8 bits/signal change × 5 MHz, or 40 Mbps. Through the use of forward error coding, which is briefly described later in this section, the data rate throughput is slightly reduced from the modem's operating rate. This reduction results from extra parity bits becoming injected into the data stream to provide the forward error detection and correction capability.

Constellation patterns

Figure 4.76 illustrates the signal constellation pattern for 64 QAM modulation while Figure 4.77 illustrates the pattern for one quadrant when 256 QAM modulation is used. Note that the signal element coding is differential quadrant coding using Gray coding within a quadrant. Under the Gray code the difference between two successive binary numbers is limited to one bit changing its state. Through the use of Gray code encoding the most likely error during demodulation in which an incorrect adjacent code is selected will result in a one-bit error when decoded at the receiver. Table 4.35 contains the binary and Gray code equivalent for 3-bit encoding.

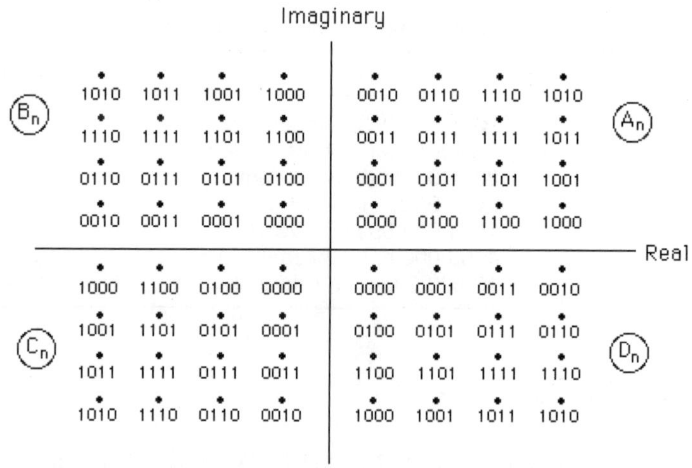

Figure 4.76 Signal constellation pattern for 64 QAM modulation. The binary numbers denote $b3_n$, $b2_n$, $b1_n$, $b0_n$, and the letters A_n, B_n, C_n, and D_n denote the four quadrants

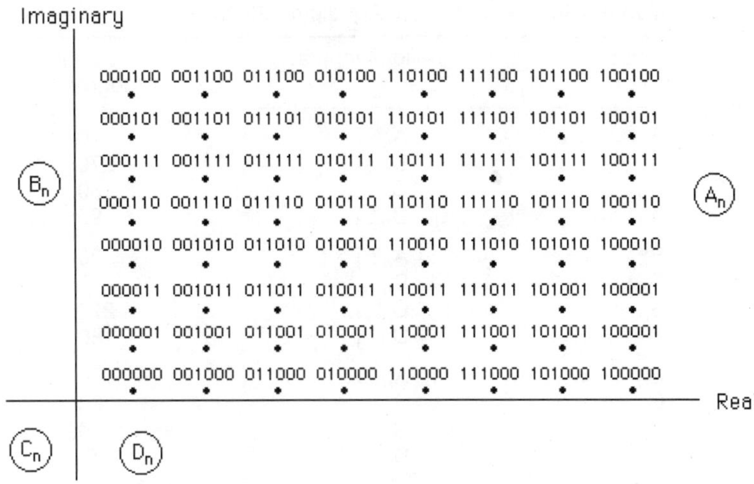

Figure 4.77 Constellation pattern for 256 QAM modulation (one quadrant illustrated)

Table 4.35 Binary and Gray code equivalence

Decimal	Binary	Gray Code
0	000	000
1	001	001
2	010	011
3	011	010
4	100	110
5	101	111
6	110	101
7	111	100

For both 64 and 256 QAM element coding the outputs are encoded using a phase change based upon the inputs and previous inputs. Table 4.36 lists the proposed 64 and 256 QAM element coding.

In addition to proposing the use of 64 and 256 QAM for downstream operations, Scientific Atlanta recognized the need for error correction and detection due to the noise levels on broadband systems. In the Scientific Atlanta proposal a Reed Solomon Parity code was proposed using interleaving to provide protection against random as well as burst errors. Thus, the demodulated data rates of 30 Mbps for 64 QAM and 40 Mbps for 256 QAM would result in actual corrected data rates of 27.378 and 36.504 Mbps, respectively.

DSL modems

As mentioned earlier in this section, a DSL modem derives its name from the fact that it was developed to support the transmission of data via copper twisted-pair wire on the line linking a subscriber to a central office. In actuality there is a family

Table 4.36 64 and 256 QAM element coding

Inputs ($b5_n b4_n$	Previous inputs	Phase change (degrees)	Outputs
A_n	A_{n-1}	0	00
A_n	B_{n-1}	270	10
A_n	C_{n-1}	180	11
A_n	D_{n-1}	90	01
B_n	A_{n-1}	90	01
B_n	B_{n-1}	0	00
B_n	C_{n-1}	270	10
B_n	D_{n-1}	180	11
C_n	A_{n-1}	180	11
C_n	B_{n-1}	90	01
C_n	C_{n-1}	0	00
C_n	D_{n-1}	270	10
D_n	A_{n-1}	270	10
D_n	B_{n-1}	180	11
D_n	C_{n-1}	90	01
D_n	D_{n-1}	0	00

of DSL modems being developed, each having features and operating characteristics developed to support different transmission rates over copper twisted-pair wire for different transmission distances. Two of the more popular types of DSL modems are HDSL and ADSL, both of which we will examine in this section.

In our previous examination of modems we noted that their operating rate is limited to approximately 33.6 kbps due to the voice channel bandwidth of approximately 3300 Hz limiting modem transmission capacity. One method recognized many years ago to transmit at a higher operating rate was to remove loading coils on a subscriber line, in effect increasing the bandwidth on a twisted-pair wire to approximately 1 MHz. Doing so enables a modem that packs two bits per baud to obtain a data transmission rate of 2 Mbps, while one that packs four bits per baud theoretically has the potential to operate at 4 Mbps when a 1 MHz bandwidth is available for use. Unfortunately, we do not live in a perfect world and the use of Digital Loop Carrier (DLC) multiplexers, a bridge tap on a subscriber line as well as the attenuation of high frequencies served as constraints which limited development of DSL modems. Improvements in line coding techniques resulted in the development of High-bit-rate Digital Subscriber Line (HDSL) technology during the early 1990s as a substitute for T1 and E1 transmission facilities, while Asymmetrical Digital Subscriber Line (ADSL) technology was developed to provide a high-speed transmission capability into homes and offices.

HDSL

HDSL represents a transmission technique which uses a dibit coding scheme referred to as 2 Binary, 1 Quaternary (2B1Q) that was originally developed for use with ISDN. Instead of transmitting on a single wire pair, HDSL uses two pairs (four wires) to split the T1 transmission rate of 1.544 Mbps so that each wire pair operates at 784 kbps. Since transmission on each wire pair occurs at 784 kbps, using dibit coding results in the signaling (baud) rate being reduced to 392K, enabling transmission distances up to 12 000 feet to be obtained on 24 or 26 gauge wire without repeaters.

In the European area the first series of HDSL modems developed for E1 transmission at 2.048 Mbps required three wire pairs for a total of six wires to achieve a targeted subscriber line distance. More modern E1 HDSL implementations operate on two-pair copper wire, each providing an operating rate of 1.168 Mbps to achieve an aggregate transmission rate of 2.048 Mbps. Figure 4.78 illustrates the use of HDSL to provide an extended T1 or E1 transmission capacity. Note that the HTU-C is the mnemonic for an HDSL Transceiver Unit located in a carrier's central office while HTU-R represents the unit located at the subscriber's remote site. Both actually represent broadband transmission devices that operate over two-pair copper wire. Since conventional T1 and E1 transmission lines require the use of repeaters at approximately 6000 feet, the use of HDSL enables relatively long transmission distances to be obtained without having to construct shelter and provide power for repeaters on the loop.

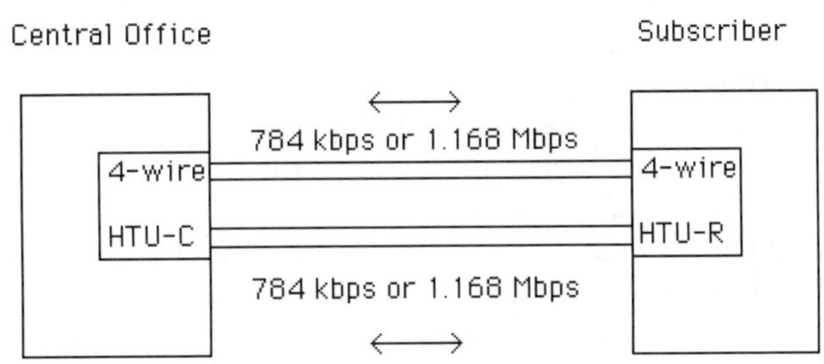

Figure 4.78 Using HDSL

ADSL

Asymmetric Digital Subscriber Line (ADSL) represents a technology developed in recognition of the physical characteristics of frequency. That is, since high frequencies attenuate more rapidly than low frequencies, a larger band of frequencies is allocated to downstream transmission than to upstream transmission. While this reduces the data rate at which a subscriber can transmit towards a central office, it results in the received signal being relatively high in signal strength which reduces the effect of crosstalk at the central office. This in turn enables a greater subscriber line transmission distance to be achieved than if a symmetrical transmission method was employed.

A typical ADSL modem provides two 'data' channels, with the return channel having a maximum operating rate of 576 kbps, and results in the asymmetric operation of the modem. In actuality, an ADSL circuit with ADSL modems connected to each end of a twisted-pair line has three channels—a high-speed downstream channel, a medium-speed upstream channel, and a standard voice

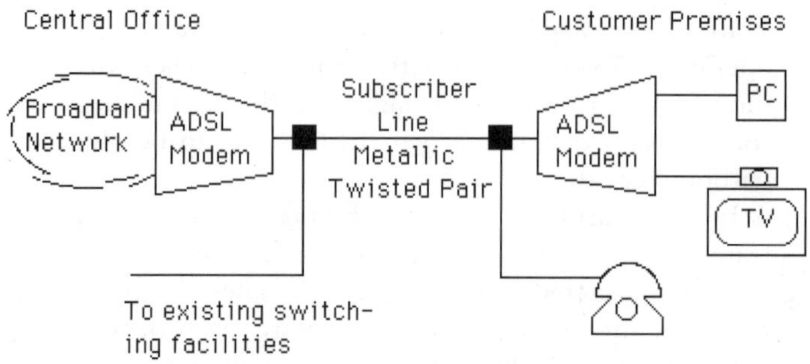

Figure 4.79 Asymmetric Digital Subscriber Line. An ADSL supports three channels formed by frequency—a high-speed downstream channel, a medium-speed upstream channel, and a conventional telephone channel. The latter is formed through the use of filters

telephone channel. The latter is split off from the digital modems by filters, which ensures subscribers can continue to obtain the use of a voice telephone channel on the existing twisted-pair connection even if one or both ADSL modems fail.

Figure 4.79 illustrates the basic operation of an ADSL circuit. The downstream operating rate depends upon several factors, including the length of the subscriber line, its wire gauge, the presence or absence of bridged taps, and the level of interference on the line. Based upon the fact that line attenuation increases with line length and frequency, while it decreases as the wire diameter increases, we can note ADSL performance in terms of the wire gauge and subscriber line distance. Ignoring bridged taps which represent sections of unterminated twisted-pair cable connected in parallel across the cable under consideration, various tests of ADSL lines provided a general indication of their operating rate capability. That capability is summarized in Table 4.37.

Table 4.37 ADSL performance

Operating rate	Wire gauge	Subscriber line distance
1.5/2.0 Mbps	24 AWG	18 000 feet
1.5/2.0 Mbps	26 AWG	15 000 feet
6.1 Mbps	24 AWG	12 000 feet
6.1 Mbps	26 AWG	9 000 feet

Operation

ADSL operations are based upon advanced digital signal processing and the employment of specialized algorithms to obtain high data rates on twisted-pair telephone wire. Currently there are two competing technologies used to provide ADSL capabilities—Discrete Multitone (DMT) modulation and Carrierless Amplitude/Phase (CAP) modulation. The first technology, DMT, represents an American National Standards Institute (ANSI) standard. In comparison, CAP represents a proprietary technology developed by Paradyne Corporation, now an independent company which was previously part of AT&T. At the time this book was prepared CAP had been licensed to a number of communications carriers throughout the world. Both DMT and CAP permit the transmission of high-speed data using Frequency Division Multiplexing (FDM) to create multiple channels on twisted-pair. Through the use of FDM the copper twisted-pair subscriber line is partitioned into three parts by frequency as illustrated in Figure 4.80. FDM assigns one channel for downstream data and a second channel for upstream data, while the third channel from 0 to 4 kHz is used for normal telephone operations. The downstream path can be subdivided through time division multiplexing to derive several high and low speed subchannels by time. In a similar manner the upstream channel can be subdivided.

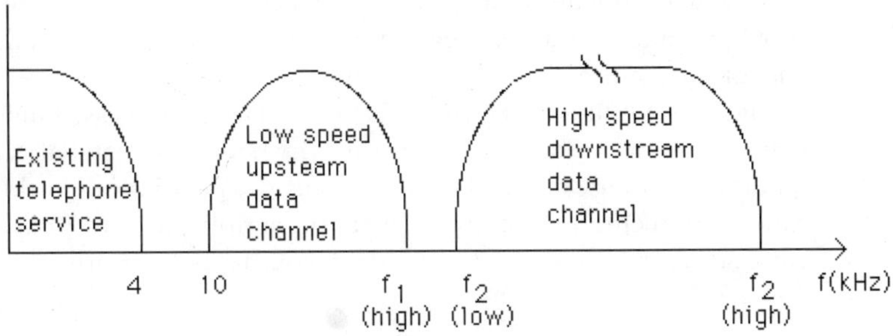

Figure 4.80 ADSL frequency spectrum

Discrete Multitone modulation

The concept behind Discrete Multitone (DMT) modulation is similar to that used in the Telebit Packetized Ensemble Protocol (PEP) modem described earlier in this chapter. That is, under DMT modulation available bandwidth is split or subdivided into a large number of independent subchannels. Since the amount of attenuation at high frequencies depends upon the length of the subscriber line and wire gauge, a DMT modem at the central office must determine which subchannels are usable. To do so that modem sends tones to the remote modem where they are analyzed. The remote modem responds to the central office modem subchannel scan at a relatively low speed which significantly reduces the possibility of the signal analysis performed by the remote modem being misinterpreted. Based upon the returned signal analysis, the central office modem will use up to 256 4-kHz wide subchannels for downstream transmission. Through a reverse measurement process the remote modem will use up to 32 4-kHz wide subchannels for upstream transmission.

One of the key advantages of DMT is its ability to take advantage of the characteristics of twisted-pair wire which can vary from one local loop to another. This makes DMT modulation well suited for obtaining a higher data throughput than is obtainable through the use of a single carrier transmission technique.

Channel encoding

Through the use of Quadrature Amplitude Modulation (QAM) up to 256 signal constellations are formed at 4 kHz intervals. Data to be transmitted is thus subdivided for encoding on separate channels and summed after demodulation at the receiver.

Carrierless Amplitude/Phase (CAP) modulation

Carrierless Amplitude/Phase (CAP) modulation is a derivative of QAM that was developed by Paradyne. Unlike DMT which subdivides the bandwidth of the wire into 4 kHz segments, CAP uses the entire bandwidth in the upstream and downstream channels. Under CAP serial data is encoded by mapping a group of bits into a signal constellation point using two-dimensional eight-state Trellis coding with Reed–Solomon forward error correction. The latter automatically protects transmitted data against impairments due to crosstalk, impulse noise, and background noise.

Figure 4.81 illustrates the CAP modulation process. Once a group of bits are mapped to a predefined point in the signal constellation, the in-phase and quadrature filters are used to implement the positioning in the signal constellation. Since this technique simply adjusts the amplitude and phase without requiring a constant carrier, the technique is referred to as 'carrierless'.

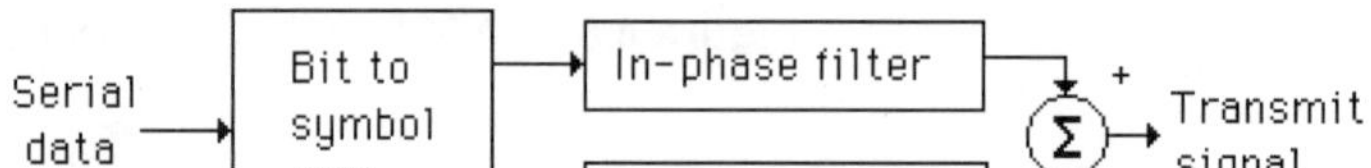

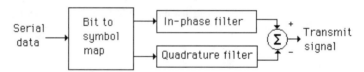

Figure 4.81 Carrierless Amplitude/Phase (CAP) modulation

The ADSL unit developed by Paradyne uses a CAP-256 line code (256-point signal constellation) for downstream operations, using bandwidth from 120 kHz to 1224 kHz. The composite signaling rate is 960 kbaud and seven bits are packed into each signal change to provide a downstream operating rate of 6.72 Mbps. However, the use of Reed–Solomon forward error correction reduces the actual payload to 6.312 Mbps plus a 64 kbps control channel. In the upstream direction the Paradyne device uses a CAP-16 line code in the 35 kHz to 72 kHz frequency band to obtain a composite signaling rate of 24 kbaud across 16 subchannels. Packing three bits per signal change, an upstream line rate of 72 kbps is obtained, of which 64 kbps is available for data.

Although cable and DSL modems are in their infancy when compared with more mature technologies, both types of devices hold the potential to revolutionize the manner by which we communicate. When we look to the past and think about how the video recorder and personal computer impacted our lives, we can also think about the future and plan for the coming revolution in communications that will occur through the use of broadband modems.

4.10 DIGITAL SERVICE UNITS

At the beginning of the 1970s, communications carriers began offering communication systems designed exclusively for the transmission of digital data. Specialized carriers, including the now-defunct DATRAN, performed a considerable service to the information-processing community through their pioneering efforts in developing digital networks. Without their advancements, major communications carriers may have delayed the introduction of an all-digital service.

In December 1974, the FCC approved the Bell System's DATAPHONE® digital service (DDS), which was shortly thereafter established between five major cities. Since then the service has been rapidly expanded to the point where, by 1998, more than 100 cities had been added to the DDS network.

Since the initial offering of DDS by AT&T, most communications carriers in the US and Europe have tariffed equivalent facilities. Today you can use digital transmission facilities from MCI, Sprint, and British Telecom, as well as from many other carriers.

Comparison of facilities

With the voice-grade type of analog transmission, the data travels in a continuous manner; although it is easily amplified, any noise or distortion along the link is also amplified. In addition, the data signals become highly attenuated or weakened by the telephone characteristics originally geared to voice transmission. For the analog transmission of data, expensive and complex modems must be employed at both ends of the link to shape (modulate) and reconstruct (demodulate) the digital signals.

When digital transmission facilities are used, the data travels from end to end in digital form with the digital pulses regenerated at regular intervals as simple values of one and zero. Inexpensive digital service units are employed at both ends of the link to condition the digital signals for digital transmission.

AT&T's DDS is strictly a synchronous facility providing full-duplex, point-to-point, and multipoint service limited to speeds of 2.4, 4.8, 9.6, 19.2 and 56 kbps. In addition to these leased-line services, AT&T introduced a switched 56 kbps digital service in 1985 as well as switched 384 kbps and 1.544 Mbps services that were in limited use when this book was prepared. Access to AT&T's Switched 56 service is obtained by dialing a '700' number, which was available in over 100 US cities in early 1998.

Rates for leased-line digital services are based primarily upon distance and transmission speeds. This type of digital service is normally cost effective for high-volume users that could justify the expense associated with a dedicated communications facility. In comparison, pricing for switched 56 kbps service is usage sensitive, based on both connection duration and distance between calling and called parties. Due to the cost of this service essentially corresponding to usage, it is attractive for such applications as the backup of critical DDS and T1 lines, peak-time overload usage to eliminate the necessity of installing additional leased digital circuits, and infrequent activities that may require a high data rate, such as still-frame and full-motion videoconferencing and facsimile transmission.

Terminal access to the DDS network is accomplished by means of a digital service unit which alters serial unipolar signals into a form of modified bipolar signals for transmission and returns them to serial unipolar signals at the receiving end. The various types of service units will be discussed in detail later in this section.

Digital signaling

One of the most critical issues to be addressed in the design of digital transmission facilities is the method by which binary data will be encoded as signal elements for transmission. The selection of one encoding method over another affects both the cost of constructing transmission facilities and the resulting quality of transmission obtained from the use of such facilities.

As previously indicated in Section 1.3, there are a number of digital encoding techniques communications carriers considered prior to selecting Alternate Mark Inversion (AMI) with a 50% duty cycle. Its selection eliminated high frequency components of a signal that could interfere with other transmissions, and concentrated power in the middle of the transmission bandwidth which minimized distortion. In addition, as a bipolar signal it prevents dc voltage buildup, enabling transformer coupling which in turn permits both power and signal to be carried over a common cable, and separated to power repeaters on a span line. Although AMI was well suited for its intended purpose, it is a very simple modulation technique which uses unibit encoding, where one bit is packed into each signal change. This means that a T1 circuit operating at 1.544 Mbps requires the use of the frequency spectrum from 0 to 1.544 MHz, resulting in a maximum transmission distance of 6000 feet prior to requiring a repeater to rebuild the signal. Although AMI is still widely deployed, its transmission limitation resulted in the development of HDSL previously described in Section 4.9.

When using a bipolar transmission method such as AMI, a bit error occurs in the form of a bipolar violation. Thus, let's turn our attention to this topic.

Bipolar violations

Bipolar transmission requires that each data pulse representing a logical one is transmitted with alternating polarity. A violation of this rule is defined as two successive pulses that have the same polarity and are separated by a zero level.

A bipolar violation indicates that a bit is missing or miscoded. Some bipolar violations are intentional and are included to replace a long string of zeros that could cause a loss of timing and receiver synchronization or to transmit control information. Figure 4.82 illustrates one example of a bipolar violation. Figure 4.82(a) shows the correct encoding of the bit sequence 010101010 using bipolar return to zero signaling. In Figure 4.82(b), the third 1 bit is encoded as a negative pulse and represents a violation of the bipolar return to zero signaling technique

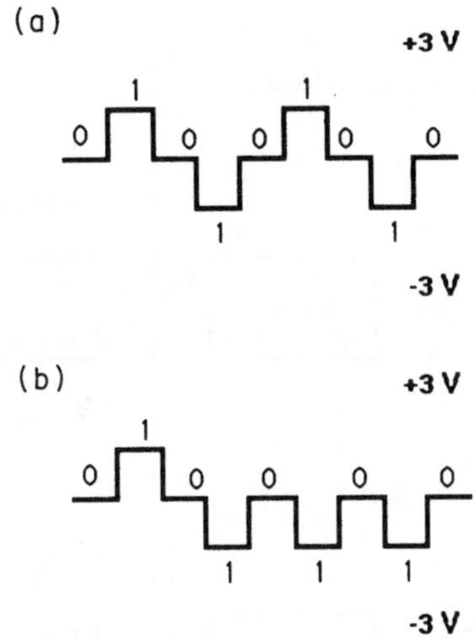

Figure 4.82 Bipolar violations. Two successive negative or positive pulses represents a bipolar violation of a bipolar return-to-zero signaling technique: (a) bipolar coding of data; (b) bipolar violation

where 1s are alternately encoded as positive and negative voltages for defined periods. In Section 4.11, we will examine several methods used to develop bipolar violations that are used to maintain synchronization when a string of consecutive zeros is encountered. These methods are commonly referred to as zero suppression codes.

DDS structure

DDS facilities are routed from a subscriber's location to an Office Channel Unit (OCU) located in the carrier's serving central office. Since there is a variety of multiplexing methods that can be employed by a communications carrier to combine DDS facilities onto a T-carrier, let us focus our attention upon two methods that will illustrate the relationship between DDS and a T1 circuit. Figure 4.83 illustrates the multiplexing arrangement within an AT&T serving central office that supports DDS transmission at 9.6 kbps and 56 kbps.

The data service units (DSUs) at the subscriber's location can be viewed as 'digital modems' since they modulate the unipolar signal received from data terminal equipment, including computer ports, multiplexer ports and terminal ports, into a modified bipolar signal suitable for transmission on the DDS network. Originally, separate channel service units (CSUs) and data service units were required to interface equipment to the DDS network. Today, most vendors manufacture DSUs that, in effect, combine the functions performed by separate DSUs and CSUs.

The user TDMs shown in Figure 4.83 illustrate two methods by which end-users can transmit data to maximize the data handling capacity of different DDS facilities. The user TDM shown in the upper left corner of Figure 4.83 illustrates

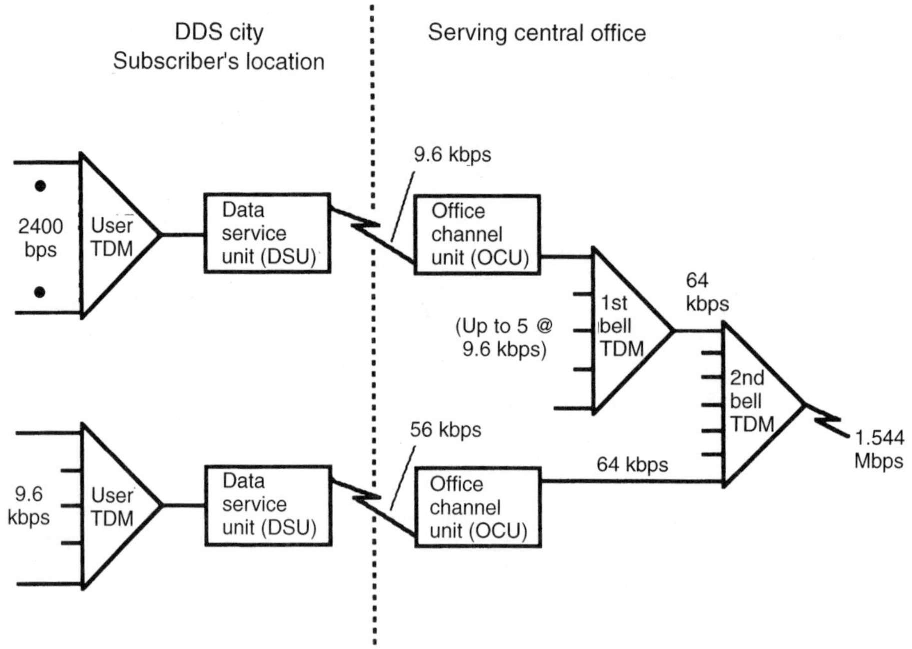

Figure 4.83 DDS multiplexing arrangement

how asynchronous transmission can be supported on DDS which is an all-synchronous transmission facility. In this example, 2400 bps asynchronous data sources are multiplexed into a 9.6 kbps synchronous data source for transmission onto DDS via the use of a DSU operating at 9.6 kbps. In the lower left portion of Figure 4.83, five 9.6 kbps asynchronous or synchronous data sources are multiplexed to obtain a 56 kbps synchronous data rate suitable for transmission on DDS. In both examples, one physical DDS circuit is used to transmit multiple logical channels of data.

The signals from the DSUs are terminated into a complementary office channel unit in the serving central office. From there, they enter into a multiplexing hierarchy which may carry voice as well as data.

Framing formats

One of the more interesting aspects of DDS is the constraints upon its transmission rate resulting from the formats used to encode user data. User data transmitted at 56 kbps is increased to a DS0 64 kbps data rate at the OCU, and that device inserts groups consisting of seven bits of customer data into an 8-bit byte as illustrated in Figure 4.84(a). In this encoding format the control bit (C) added to every seven bits of customer data is set to a '1' if the byte contains customer data, while a value of '0' indicates that the byte contains network control data, such as idle or maintenance codes or control information. Since a DS0 signal results in the transmission of an 8-

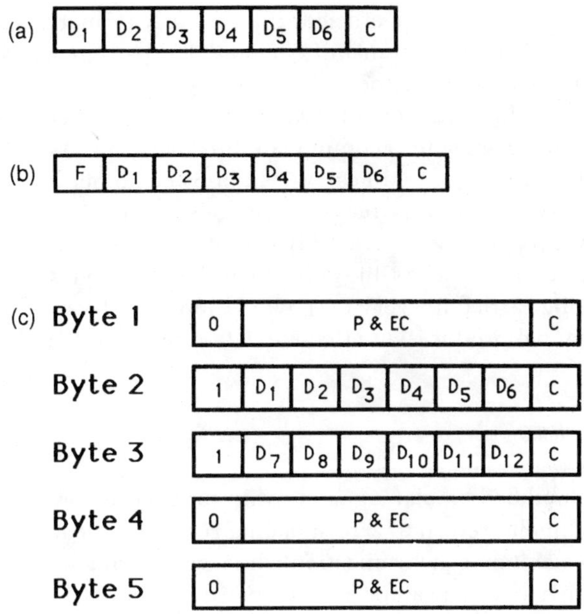

Figure 4.84 DDS framing formats: (a) 56 kbps; (b) 2.4, 4.8, 9.6 kbps; (c) 19.2 kbps (F frame bit, D data bits, C control bit, P&EC parity and error correction bits)

bit byte 8000 times per second, this framing format results in 8 kbps of control bits being added to the 56 kbps customer data rate.

The construction of DS0 signals from the 2.4, 4.8 and 9.6 kbps DDS subrates is illustrated in Figure 4.84(b). As indicated, customer data is inserted into 8-bit bytes with six bits of user data framed by a frame bit (F) and a control bit (C). Once 2.4, 4.8 or 9.6 kbps DDS data streams are framed, one of two methods is used to place the framed data onto a DS0 channel. When 'byte stuffing' is used, the frame bit is set to '1' and the customer data are repeated the required number of times to create a 64 kbps DS0 signal. Thus, the 8-bit byte containing six bits of user data is repeated at 5, 10, and 20 times to enable 9.6, 4.8 and 2.4 kbps DDS data to be placed on a 64 kbps channel. When the F bit is set to '1' the frame format illustrated in Figure 4.84(b), is referred to as a DS0-A format. Thus, DS0-A data can bypass the first level TDM illustrated in Figure 4.83 and be fed directly into the second level TDM.

The second method of placing 2.4, 4.8 or 9.6 kbps DDS data onto a DS0 channel involves the use of the first level TDM illustrated in Figure 4.83. When this occurs, five 9.6, ten 4.8, or twenty 2.4 kbps formatted signals are multiplexed onto a single DS0 channel. To distinguish between the repeating of the same data resulting from byte stuffing and the multiplexing of different DDS signals, the framing bit is altered from all ones in byte stuffing to a subrate framing pattern to indicate multiplexing of different DDS data sources. When this framing pattern occurs, the resulting framing format is referred to as a DS0-B format. Obviously, DS0-B formatting is more efficient than DS0-A formatting as the latter would require 20 DS0 channels to transmit twenty 2.4 kbps signals, while the former would require only one DS0 channel.

The introduction of 19.2 kbps DDS service required a substantial framing format change to accommodate this data rate. As illustrated in Figure 4.84(c), five bytes are required to carry 19.2 kbps customer data since three bytes are used for parity and error correction functions. In this framing format the frame bit in each byte (bit 1) results in a '01100' repeating pattern. Twelve bits of customer data are placed into two six-bit groups contained in bytes 2 and 3, resulting in 12 data bits being carried in every five-byte group of 40 bits. Thus, the use of the 64 kbps DS0 channel produces an effective data rate of $64 \times 12/40$, or 19.2 kbps.

Signaling structure

A modified bipolar signaling structure is used on DDS facilities. The modification to bipolar return to zero signaling results in the insertion of zero suppression codes to maintain synchronization whenever a string of six or more zeros is encountered. Otherwise, repeaters on the span line routed between the carrier office and the customer may not be able to obtain clocking from the signal and could then lose synchronization with the signal.

To ensure a minimum ones density, at 2.4, 4.8, 9.6, and 19.2 kbps any sequence of six consecutive zeros is encoded as $000X0V$, where

0 denotes 0 V transmitted (binary 0)
X denotes a zero or + or $-A$ V, with the polarity determined by conventional bipolar coding
V denotes + or $-A$ V, with the polarity in violation of the bipolar rule.

Figure 4.85 illustrates the zero suppression sequence used to suppress a string of six consecutive zeros. For transmission at 56 kbps, any sequence of seven consecutive zeros is encoded as $0000X0V$.

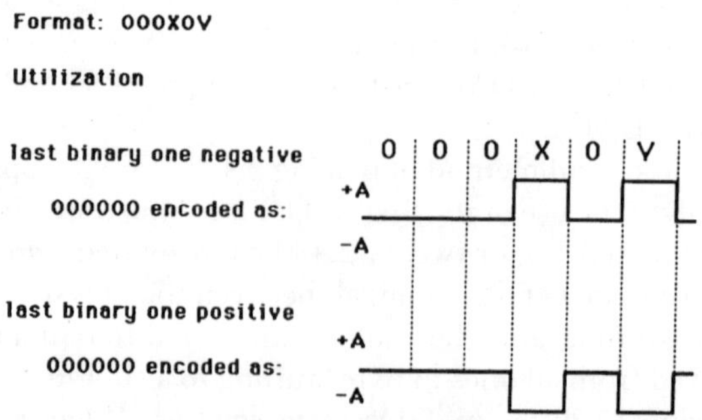

Figure 4.85 DDS zero suppression sequence

Timing

Precise synchronization is the key to the success of an all-digital network. Timing ensures that data bits are generated at precise intervals, interleaved in time and read out at the receiving end at the same interval to prevent the loss or garbling of data.

To accomplish the necessary clock synchronization on the AT&T digital network, a master clock is used to supply a hierarchy of timing in the network. Should a link to the master clock fail, the nodal timing supplies can operate independently for up to two weeks without excessive slippage during outages. In Figure 4.86, the hierarchy of timing supplies as linked to AT&T's master reference clock is illustrated. As shown, the subsystem is a treelike network containing no closed loops.

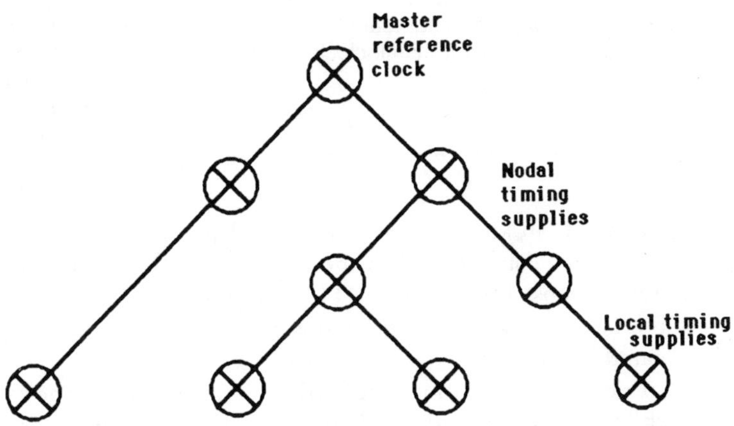

Figure 4.86 DDS timing subsystem

Service units

When DDS was introduced, both a channel service unit (CSU) and a data service unit (DSU) were required to terminate a DDS line.

The DSU converts the signal from data terminal equipment into the bipolar format used with DDS and T1 facilities.

The DDS master reference clock is an atomic clock that is accurate to 0.01 part per million (PPM). This clock was installed by AT&T at Hillsboro, Missouri, which is the geographic center of the United States and whose location ensures a minimal variance in propagation delay time between DDS nodes connected to the master reference clock. The master reference clock oscillates at a rate known as the basic system reference frequency and is the most accurate of three timing sources used by digital facilities. The other two sources of timing include channel banks and loop timing where clocking is obtained from a high speed circuit.

The DSU interface to the DTE is accomplished by the use of a standard 25 pin EIA RS-232/V.24 female connector on the 2.4 kbps to 19.2 kbps units. The

Table 4.38 DSU interchange circuits

RS-232 interface (DCE)		V.35 interface (DCE)	
Pin	Signal	Pin	Signal
1	Chassis ground	P	Transmit data (A)
2	Transmit data	S	Transmit data (B)
3	Receive data	R	Receive data (A)
4	Request-to-send	T	Receive data (B)
5	Clear-to-send	C	Request-to-send
6	Data set ready	D	Clear-to-send
7	Signal ground	H	Data terminal ready
8	Carrier detect	E	Data set ready
9	Positive voltage	B	Signal ground
10	Negative voltage	F	Receive line signal detect
15	Transmit clock	Y	Transmit timing (A)
17	Receive clock	AA	Transmit timing (B)
20	Data terminal ready	V	Receive timing (A)
24	External transmit clock	X	Receive timing (B)
		U	External transmit timing
		W	External transmit timing
		L	Local loopback

wideband, 56 kbps device utilizes a 34 pin ITU, V.35 (Winchester) female type connector. Table 4.38 lists the RS-232 and V.35 interchange circuits commonly used by most DSUs. Since DTEs are normally attached to the DSU, the latter's interface is normally configured as data communications equipment (DCE) by the manufacturer.

Prior to deregulation, the CSU was provided by the communications carrier, while the DSU could be obtained from the carrier or from third-party sources. This resulted in an end-user connection to the DDS network similar to that illustrated in the top portion of Figure 4.87 where the CSU terminated the carrier's four-wire loop and the DSU was cabled to the CSU. In this configuration the CSU terminates the carrier's circuit. In addition, a separate CSU was designed to perform signal regeneration, monitor incoming signals to detect bipolar violations and perform remote loopback testing. Interfacing between the DSU and CSU is accomplished by the use of a 15-pin female D-type connector which utilizes the

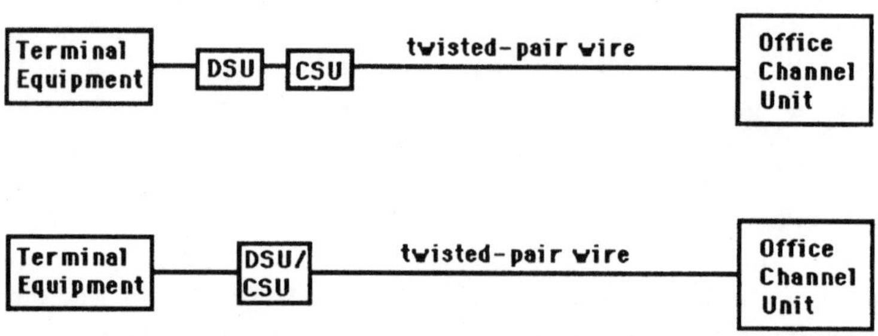

Figure 4.87 DSU/CSU connection

first six pins, where pin 1 is signal ground, pin 2 is status indicator, pins 3 and 4 are the receive signal pair, while pins 5 and 6 are the transmit signal pair. Today most communications carriers and third-party vendors manufacture combined DSU/CSU devices, integrating the functions of both devices into a common housing which is powered by a common power supply. The lower portion of Figure 4.87 illustrates the connection of end-user terminal equipment to DDS using a combined DSU/CSU unit.

DSU/CSU tests and indicators

Through the use of intentional bipolar violations, the DSU/CSU can generate a request to the OCU for the loop-back of the received signal onto the transmit circuit or it can interpret DDS network codes and illuminate relevant indicators on the device. When the loop-back button on the DSU/CSU is pressed, the device will generate four successive repetitions of the sequence $0B0X0V$ when operating at data rates up to 19.2 kbps or $N0B0X0V$ at 56 kbps, where

B denotes + or $-A$ V, with the polarity determined by bipolar coding for a binary 1
X denotes a 0 V for coding of binary 0 or B, depending upon the required polarity of a bipolar violation
V denotes + or $-A$ V, with the polarity determined by the coding of a bipolar violation
N denotes a do not care condition where the coding for a binary 0 or binary 1 is acceptable

In Figure 4.88 the DDS loop-back 6-bit sequence is illustrated for data rates at or under 19.2 kbps. Note that the sequence transmitted is dependent upon whether the previous binary 1 was transmitted as a positive or negative voltage.

Other bipolar violation sequences used by DDS include an idle sequence which indicates that a DTE does not have data to transmit, an out-of-service sequence and an out-of-frame sequence. The idle sequence is generated by the DSU while the out-of-service and out-of-frame sequences indicate a problem in the DDS network

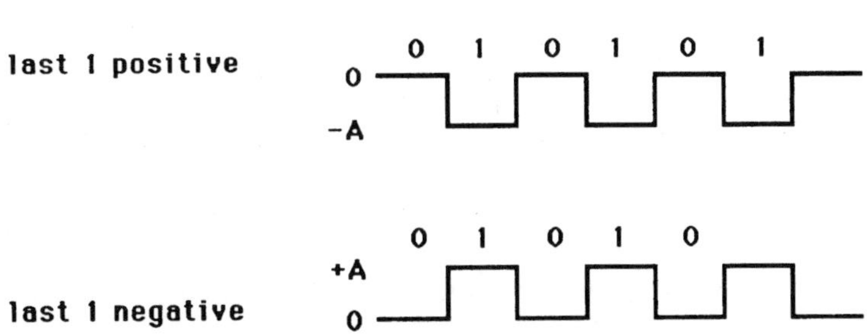

Figure 4.88 DDS loop-back sequence (data rates up to 19.2 kbps). DDS loop-back codes are intentional bipolar violations

and are generated by the network and used by the DSU to illuminate an appropriate indicator on the device.

DDS II

During 1988, AT&T introduced a new version of its DDS facility commonly referred to as DDS II. One of the key advantages of DDS II is its capability to provide a diagnostic channel along with the primary subrate channel. This diagnostic channel is obtained through a modification to the framing used by DDS and requires the use of special DSU/CSU that support the new channel.

Through the use of DDS II, end-users can perform non-disruptive testing or use the channel for network management purposes. The key to obtaining the ability to derive a secondary channel on DDS is the use of the network control or C bit.

In conventional DDS the C bit, which is bit eight in each DDS 8-bit byte, is transmitted as a binary one whenever a DTE requests access to a channel by turning its request to send (RTS) signal on. With the C bit continuously set to a one the DTE can transmit an unrestricted stream of data to include continuous zeros since every eighth bit will be automatically set to a one. By robbing this bit once every third byte, AT&T established a virtual path for diagnostic use.

The diagnostic channel data rate for 56 kbps DDS II is obtained by multiplying the full DS0 rate of 64 kbps by 1/8 which represents the C bit's portion of the DS0 rate to obtain 8 kbps. Next, since the bit robbing occurs every third byte, the resulting data rate becomes $8000 \times \frac{2}{3}$, or $2666\ \frac{2}{3}$ bps. Similarly, dividing 2666 2/3 bps by the number of 19.2, 9.6, 4.8 or 2.4 kbps channels multiplexed onto a DS0 channel results in the diagnostic data rate for DDS II at those data rates.

AT&T added diagnostic channel capability throughout the DDS network. Although older DSU/CSU devices can support transmission on DDS II, those devices cannot support the use of the secondary channel capability provided by this modification to DDS. To do so requires the use of newer DSU/CSU devices that support the multiplexing of diagnostic data onto every third C bit position.

Analog extensions to DDS

AT&T provides an 831A data auxiliary set which allows analog access to DDS for customers located outside the DDS servicing areas. The 831A connects the EIA RS-232 interfaces between a data service unit (500A-type) and a voice-band data set or modem. The 831A contains an 8-bit elastic store, control, timing, and test circuits which allow loop-back tests toward the digital network. The elastic store is a data buffer that is required by the DSU to receive data from the modem in time with the modem's receive clock. The data is then held in the elastic store until the DSU's transmit clock requests it. Thus, the buffer serves as a mechanism to overcome the timing differences between the clocks of the two devices. In the reverse direction, no buffer is required when the DSU's receive clock is used as the modem's external transmit clock. When the modem cannot be externally clocked or when one DSU is connected to a second DSU or a DTE that cannot accept an

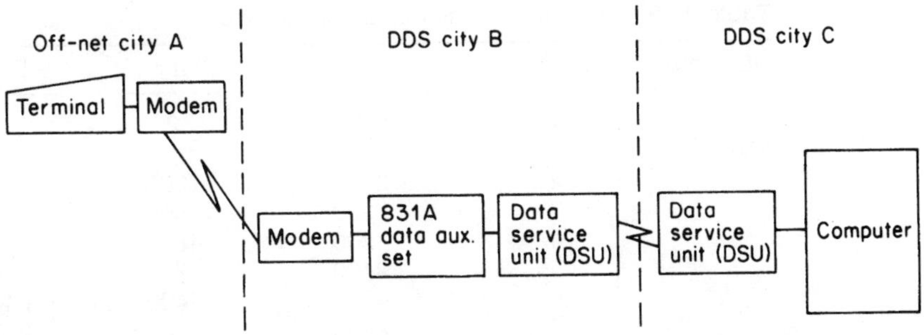

Figure 4.89 Analog extension to DDS. In order to obtain an analog extension to a digital network, a device known as a data auxiliary set, which provides an interface between a modem and a service unit, must be installed

external clock, a second elastic store will be required. Figure 4.89 illustrates a typical analog extension to a DDS servicing area.

Applications

As discussed in previous sections, the requirement for modems in an off-net analog extension could negate any real savings, gained in utilizing DDS. A network arrangement in the form of Figure 4.90, on the other hand, could easily achieve the high-performance characteristics inherent in DDS while reducing the overall costs of creating two independent data links.

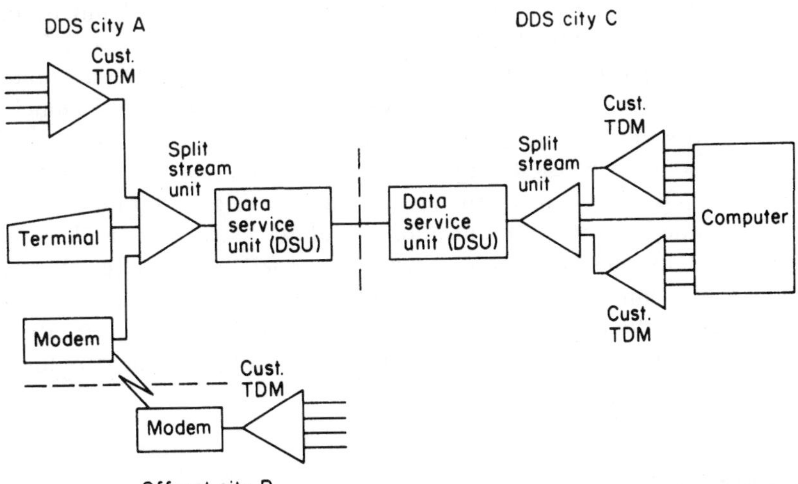

Figure 4.90 Multiplexing over DDS utilizing split stream units. An inexpensive split stream unit, or limited-function synchronous multiplexer, can offer considerable flexibility when interfacing into the DDS network

Table 4.39 Split stream unit modes of operation

DDS rate	Mode	SSU	Channel	Data	Rate
	1	1	2	3	4
9600	1	9600			
	2	2400	7200		
	3	7200	2400		
	4	4800	4800		
	5	2400	4800	2400	
	6	4800	2400	2400	
	7	2400	2400	2400	2400
4800	1	4800			
	2	1200	3600		
	3	3600	1200		
	4	2400	2400		
	5	1200	2400	1200	
	6	2400	1200	1200	
	7	1200	1200	1200	1200
2400	1	2400			
	2	600	1800		
	3	1800	600		
	4	1200	1200		
	5	600	1200	600	
	6	1200	600	600	
	7	600	600	600	600

This example shows the use of a device marketed by AT&T which is called a split stream unit (SSU). The SSU is similar to the multiplexer incorporated into now obsolete Bell System 209 data sets, in that the user can select various combinations of data transfer rates on up to four individual channels, up to the maximum capability of the DDS line. The SSU plugs directly into the DSU, with operational settings at one-half or one-quarter the specified DDS data rate. The unit provides local loop and remote loop testing of each individual channel and very effectively lifts the four speed restrictions of DDS service. Table 4.39 lists the operational modes of SSUs designed for use on 2400, 4800, and 9600 bps DDS circuits.

Of course, you can also obtain a variety of multiplexers manufactured by independent communications vendors that can be used in place of SSUs. Since DDS is a synchronous network that operates at a series of fixed data rates, you can also employ multiplexers as a means to concentrate data for communications over a DDS line. You can also use multiplexers to convert asynchronous data into synchronous data stream that is compatible with the data format for which DSUs are designed.

KiloStream service

Although British Telecom's KiloStream service is similar to DDS, there are several significant differences that warrant discussion.

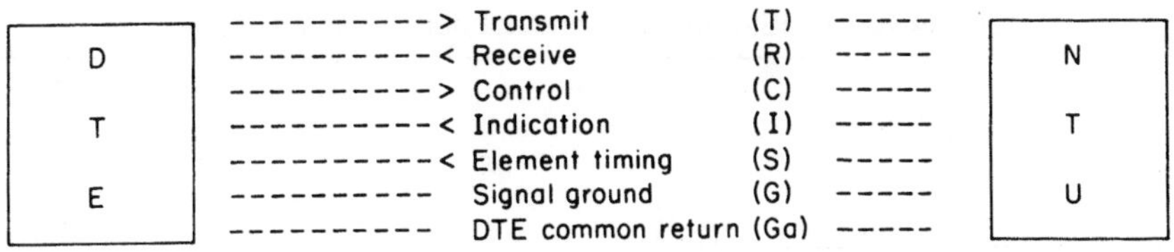

Figure 4.91 ITU X.21 interface circuits

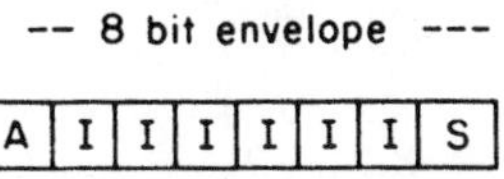

Figure 4.92 KiloStream envelope encoding. (A Alignment bit which alternates between '1' and '0' in successive envelopes to indicate the start and stop of each 8-bit envelope. S Status bit which is set or reset by the control circuit and checked by the indicator circuit. I Information bits)

The British Telecom customer is provided with an interface device which is called a network terminating unit (NTU), which is similar to a DSU. The NTU provides an ITU interface for customer data at 2.4, 4.8, 9.6 or 48 kbps to include performing data control and supervision, which is known as structured data. At 64 kbps, the NTU provides an ITU interface for customer data without performing data control and supervision, which is known as unstructured data.

The NTU controls the interface via ITU recommendation X.21, which is the standard interface for synchronous operation on public data networks. An optional V.24 interface is available at 2.4, 4.8 and 9.6 bps while an optional V.35 interface can be obtained at 48 kbps. The X.21 interface is illustrated in Figure 4.91. Here the control circuit (C) indicates the status of the transmitted information—data or signaling, while the indication circuit (I) signals the status of information received from the line. The control and indication circuits control or check the status bit of an 8-bit envelope used to frame six information bits.

Customer data is placed into a $6+2$ format to provide the signaling and control information required. This is known as envelope encoding and is illustrated in Figure 4.92.

The NTU performs signal conversion, changing unipolar non-return to zero signals from the V.21 interface into a diphase WAL 2 encoding format. This ensures that there is no dc content in the signal transmitted to the line, provides isolation of the electronic circuitry from the line, and provides transitions in the line signal to enable timing to be recovered at the distant end. Table 4.40 lists the NTU operational characteristics of KiloStream.

The KiloStream network

In the KiloStream network, the NTUs on a customer's premises are routed via a digital local line to a multiplexer operating at 2.048 Mbps. This data rate is the European equivalent of the T1 line in the United States that operates at 1.544

Table 4.40 KiloStream NTU operational characteristics

Customer data rate (kbps)	DTE/NTU interface	Line data rate (kbps)	NTU operation
2.4	X.21	12.8	6+2 envelope encoding
4.8	X.21	12.8	6+2 envelope encoding
9.6	X.21	12.8	6+2 envelope encoding
48	X.21	64	6+2 envelope encoding
64	X.21	64	No envelope encoding
48	X.21 bis/V.35	64	6+2 envelope encoding
2.4	X.21/V.24	12.8	6+2 envelope encoding
4.8	X.21/V.24	12.8	6+2 envelope encoding
9.6	X.21/V.24	12.8	6+2 envelope encoding

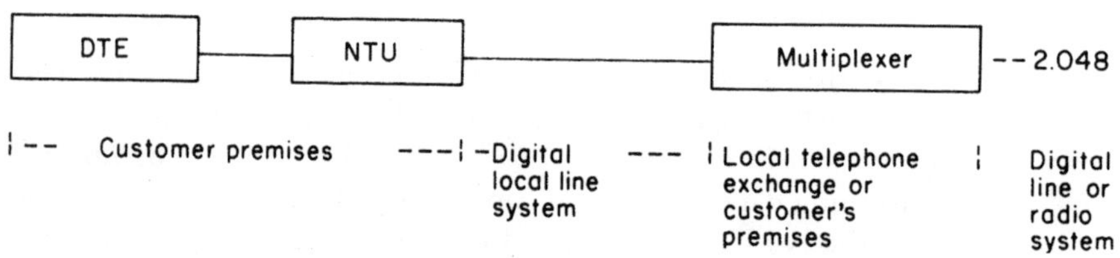

Figure 4.93 The KiloStream connection

Mbps. The multiplexer can support up to 31 data sources and may be located at the local telephone exchange or on the customer's premises if traffic justifies. It is connected via a digital line or a radio system into the British Telecom KiloStream network as illustrated in Figure 4.93.

Unlike true CEPT, the 2.048 Mbps T-carrier used for KiloStream uses 31 DS0 channels. Normal, CEPT uses one channel for synchronization (framing) and a second channel for signaling. Since there is no direct voice signaling on KiloStream, time slot 16, which normally would carry that information, can be used for data. From an examination of Figure 4.94, the reader will note that a DTE operating rate of 2.4, 4.8 or 9.6 kbps results in a line rate of 12.8 kbps, which is precisely one-fifth of the DS0 64 kbps data rate.

4.11 CHANNEL SERVICE UNITS

A channel service unit (CSU) is a communications device that was developed to terminate a T1 line in North America and an E1 line in Europe. Since a data service unit (DSU) commonly includes a built-in CSU, it is natural for many readers to be confused with respect to the use and operating characteristics of CSUs. To alleviate this potential confusion, let us distinguish between modern standalone CSUs and CSUs built into DSUs.

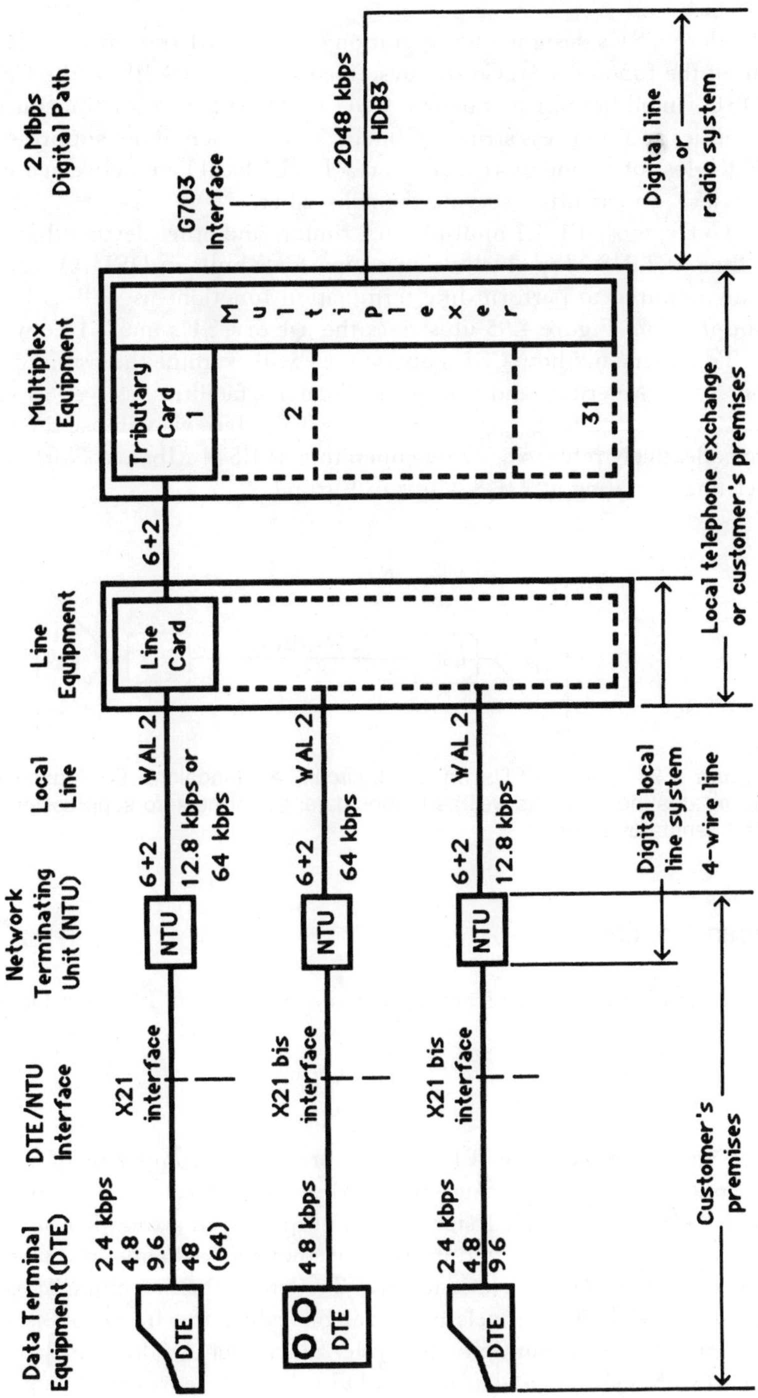

Figure 4.94 KiloStream structure

Comparison to DSU/CSU

Modern CSUs designed for operation on T1 or E1 transmission facilities perform all of the functions previously described in Section 4.10 for the CSU portion of a DSU. In addition, a standalone CSU will frame data for transmission on T1/E1 facilities and suppress strings of binary zeros, which, if not suppressed, could result in the loss of timing by repeaters on a T1/E1 local loop between a subscriber and a carrier's central office serving the subscriber.

Today, most T1/E1 multiplexers, routers and other devices that operate at 1.544 Mbps or 2.048 Mbps include an option for a built-in DSU. A standalone CSU is thus required to perform line termination functions as well as framing and zero suppression. Figure 4.95 illustrates the use of CSUs in a T1 network segment.

To understand how CSUs operate, we will examine the framing structures used on North American and European T-carrier facilities, as well as several methods commonly used to suppress strings of zeros. Here we will use the term 'T-carrier' to collectively refer to a T1 line operating at 1.544 Mbps in North America and an E1 line operating at 2.048 Mbps in Europe.

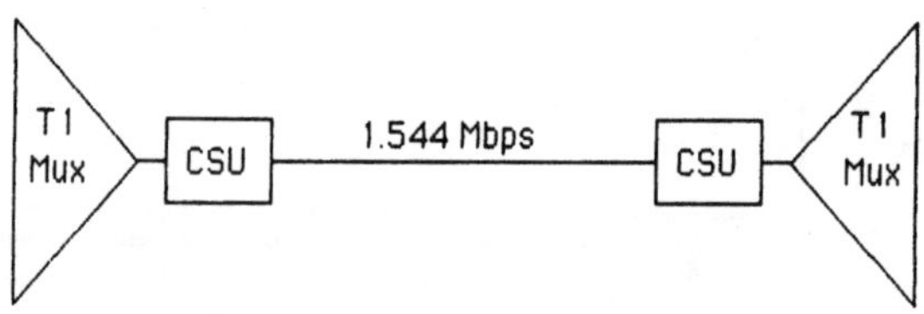

Figure 4.95 Using CSUs on a T1 circuit. A standalone CSU used on a T1 circuit terminates the circuit as well as frames data, performs zero suppression, and recognizes and generates alarms

North American framing

In North America, two framing formats are used on T1 lines: D4 and ESF.

D4 framing

In North America the T1 signal represents a composite of 24 separate DS0 channels, each representing one PCM encoded voice signal digitized at 64 kbps. The 24 channels in a T1 signal are multiplexed in a round-robin order to ensure each channel is transmitted in turn and that every channel receives a turn prior to any channel receiving a second turn. To denote the beginning of each sequence of 24 digitized DS0 channels a special bit called the frame bit is prefixed to the beginning of each multiplexing cycle. Since each DS0 channel is encoded into a PCM word using eight data bits, 24 DS0 channels represents a sequence of 192 data bits. The full pattern of one frame bit and 192 data bits is called the DS1 (digital signal level 1) frame and represents a total of 193 bits. Since sampling occurs 8000

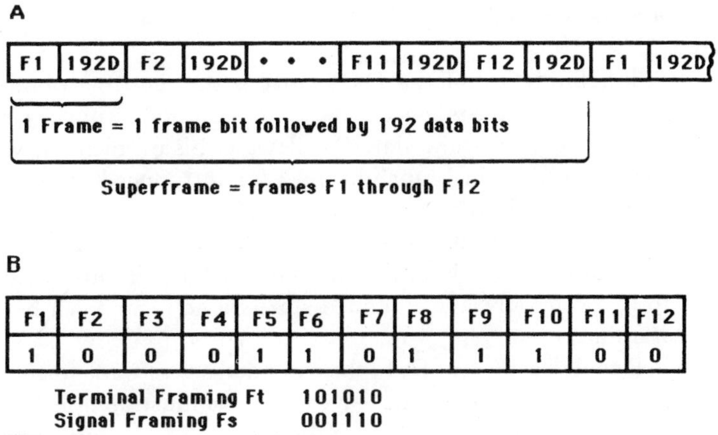

Figure 4.96 D4 framing structure and framing pattern. The D4 framing pattern represents the hex characters 8CD which are continuously repeated. (a) D4 Framing Structure, (b) D4 Framing Pattern

times per second, 193 bits × 8000 samples per second results in the 1.544 Mbps operating rate of a T1 circuit. As this operating rate include 8000 frame bits, only the remaining 1.536 Mbps is actually available to the user.

One of the most popular methods of framing the DS1 signal is called D4 framing. This framing technique takes its name from the AT&T D4 channel bank used in that communications carrier's network. Under D4 framing the frame bits in 12 consecutive frames are grouped together to form a superframe whose frame bits are used to form a repeating pattern. Figure 4.96 illustrates the D4 framing structure and framing pattern. Under D4 framing, the 1.544 Mbps data stream must meet the following requirements.

(i) It must be encoded as a bipolar, AMI, non-return to zero signal to ensure that the signal has no dc component and can be transformer coupled, permitting the circuit to carry power for the repeaters.
(ii) Each pulse must have a 50% duty cycle with a nominal voltage of 3.0 V.
(iii) There can be no more than 15 consecutive '0's' present in the data stream, which defines the minimum 1's density of the circuit.
(iv) The D4 framing pattern is embedded in the data stream.

The framing bits in the D4 superframe consist of six F_t (terminal framing) bits that are used to synchronize the bit stream and six F_s (signal framing) bits that are used to define multiframe boundaries.

The F_t bit conveys a pattern of alternating 0's and 1's (101010), which is used to define frame boundaries, enabling one slot to be distinguished from another. Due to this, it is also known as a frame alignment signal. The F_s bit conveys a pattern of 001110, which is used to define multiframe boundaries. This enables one frame to be distinguished from another, permitting frames 6 and 12 to be identified for the extraction of their signaling bits. Note that the composite D4 framing pattern represents the hex characters 8CD which are continuously repeated.

Extended superframe format

A key limitation of the D4 format is the fact that obtaining a communications capability between devices on a T1 circuit for testing or control purposes required the use of a DS0 time slot. To alleviate this problem, as well as to provide the T1 user with additional capability, AT&T introduced an extended superframe format in early 1985. Although this new framing format require the installation of equipment that supports the frame format and has consequently been slowly introduced, by the late 1990s a majority of T1 circuits in North America conformed to it. Eventually, the extended superframe format can be expected to replace D4 framing.

Denoted as F_e and ESF, the extended superframe format extends D4 framing to 24 consecutive frame bits—F1 to F24 as illustrated in Figure 4.97.

Unlike D4 framing in which the 12 framing bits form a specific repeating pattern, the ESF pattern can vary. ESF consists of three types of frame bits.

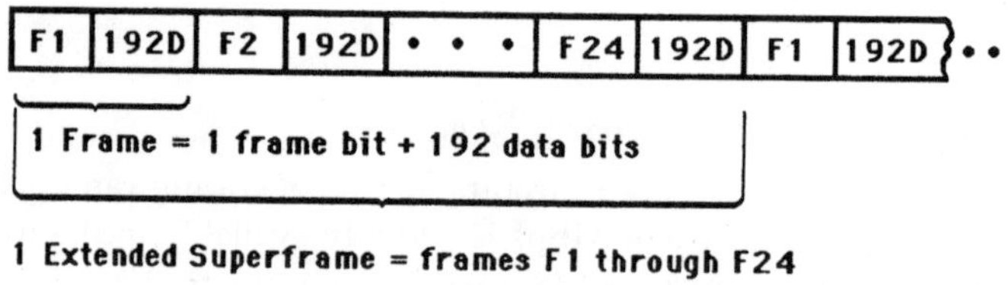

Figure 4.97 The extended superframe

Derived data link

The ESF 'd' bits, which represent a derived data link, are used by the telephone company to perform such functions as network monitoring to include error performance, alarm generation and reconfiguration to be passed over a T1 link. The 'd' bits appear in the odd frame positions, e.g., 1, 3, ..., 21, 23. Since they are used by 12 of the 24 framing bits, the 'd' bits represent a 4 kbps data link.

The data link formed by the 12 'd' frame bits is coded in to higher level data link control (HDLC) protocol format known as BX.25. Figure 4.98 illustrates the data link format carried by the 'd' bits.

One of the primary goals in the development of the BX.25 protocol was to provide a mechanism to extract performance information from ESF compatible CSUs. Doing so allows circuit quality monitoring without taking the circuit out of service and is a major advantage of ESF over D4 framing. Standard maintenance messages that are defined in AT&T's publication 54016 include messages that can return performance data concerning the number of errored seconds (ES), severely errored seconds (SES), and failed seconds (FS). An errored second is a second that contains one or more bit errors, while a severely errored second is considered to be a second with 320 or more bits in error. If ten consecutive severely errored seconds occur, this condition is considered as a failed signal state. Then, each signal in a failed signal state is considered to be a failed second. Table 4.41 summarizes the standard maintenance messages transmitted on the ESF data link.

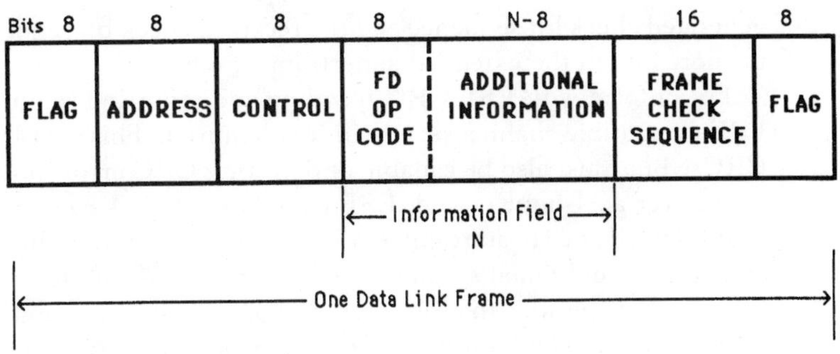

Figure 4.98 ESF data link format

Table 4.41 ESF data link maintenance messages

Send One Hour Performance Data
Upon receiving this command the ESF CSU will supply the following:
current status, elapsed time of current interval, ES and FS in the current 15-minute interval, number of valid intervals, count of ES and FS in 24-hour register, ES and FS during the previous four 15-minute intervals.

Send 24-Hour 'ES' Performance Data
Upon receiving this command, the ESF CSU will supply the following:
current status, elapsed time of current interval, ESs and FSs in current 15-minute interval, number of valid intervals, count of ESs and FSs in 24-hour register, and ESs during previous 96 15-minute intervals.

Send 24-Hour 'FS' Performance Data
Upon receiving this command, the ESF CSU will supply the following:
current status, elapsed time of current interval, ESs and FSs in the current 15-minute interval, number of valid intervals, count of ESs and FSs in 24-hour register, and FSs during previous 96 15-minute intervals.

Reset Performance Monitoring Counters
Upon receiving this command, the ESF CSU will reset all interval times and ES and FS registers and supply the current status.

Send Errored ESF Data
Upon receiving this command, the ESF CSU will supply current data present in ESF error event registers. Each count represents one error event (65535 max.).

Reset ESF Register
Upon receiving this command, the ESF CSU will reset the ESF error event register and supply the current status.

Maintenance Loop-Back (DLB)
Energizes upon receiving the proper code embedded in the 4 kbps data link. This loop-back loops through the entire CSU.

Error check link

Frame bits 2, 6, 10, 14, 18 and 22 are used for a CRC-6 code. The 6-bit cyclic redundancy check sum is used by the receiving equipment to measure the circuit's bit error rate and represents 2 kbps of the 8 kbps framing rate. The CRC-6 code yields an accuracy of 98.4% and the occurrence of a mismatch between the locally generated check bit sequence and the received check bit sequence indicates that one or more bits in the extended superframe are in error.

To conform with ESF CRC-6 coding and reporting requirements, the use of an ESF compatible channel service unit is required. This CSU not only generates the CRC-6 but must also be capable of detecting CRC errors and storing a CRC error count over a 24-hour period. ESF compatible CSUs contain buffer storage which enables the device to store current line status information, including all error events and errored and failed seconds for the current 15-minute period and the previous 96 15-minute periods that represent the prior 24-hour period. To enable the carrier to retrieve this data, as well as to reset any or all counters and activate or deactivate loop-back testing on the local span line, the CSU must also have the ability to respond to network commands. An ESF compatible CSU must thus have the capability to send and receive data based upon the BX.25 formation via the 4 kbps data link.

A competing standard from the T1E1 committee requires the CSU to broadcast this information all of the time instead of upon request from the carrier. This standard is designed to overcome the potential problem resulting from an Interexchange Carrier (IEC) resetting CSU registers prior to a Local Exchange Carrier (LEC) reading the CSU registers.

The previously mentioned competing standard is officially known as ANSI T1.403. Under the T1.403 standard performance reports are transmitted every second and only three seconds of historical data are maintained. A T1.403 compliant CSU transmits a performance report of 112 bits about the current second and three previous seconds every second. This report contains a range of CRC-6 errors, frame bit errors, bipolar violations and other error event information for each second.

The key difference between AT&T Publication 54016 and the ANSI T1.403 ESF method is in the philosophy behind each ESF method since both use the 4 kbps data link for transmitting performance reports. AT&T's philosophy is one of in-line query—response under which parameters are organized into performance

Table 4.42 ESF framing pattern

Frame bit	Composition	Frame bit	Composition
1	d	13	d
2	C1	14	C4
3	d	15	d
4	0	16	0
5	d	17	d
6	C2	18	C5
7	d	19	d
8	0	20	1
9	d	21	d
10	C3	22	C6
11	d	23	d
12	1	24	1

d = data link
C*x* = CRC-6 bit *x*

registers for 15-minute intervals, and can be stored and retrieved for up to a 24-hour period. In comparison, under ANSI T1.403, parameters are broadcast every second and store only three seconds of data. Although AT&T and several communications carriers support both methods, AT&T appears to be migrating towards the ANSI method even though that method does not maintain historical data, nor does it allow user access to registers.

Framing pattern

The third type of frame bits are used to generate the framing pattern. Here, frame bits 4, 8, 12, 16, 20 and 24 are used to generate the F_e framing pattern whose composition is 001011. These six bits result in a 2 kbps framing pattern. In Table 4.42 you will find a summary of the ESF framing pattern.

CEPT PCM-30 format

CEPT PCM-30 is a PCM format used for time division multiplexing of 30 voice or data circuits onto a single twisted pair cable using digital repeaters and is primarily used in Europe.

The standard CEPT frame is 32 channels × 8 bits/channel or 256 bits. With 8000 samples per second, the CEPT data rate becomes 8000 × 256, or 2.048 Mbps. Note that under this format there are no framing bits added to a frame as done under the North American T1 format. This is because the framing bits are carried within specific time slots. Thus, a European CSU must examine bits within a specific time slot.

Frame composition

Each CEPT PCM-30 frame consists of 32 time slots to include 30 voice, one alignment and one signaling, with each time slot represented by eight bits. Since each PCM channel is sampled 8000 times per second, the standard CEPT-30 data rate is 32 × 8 × 8000, or 2.048 Mbps.

Alignment signal

An alignment signal (0011011) is transmitted in bit positions 2 to 8 of time slot 0 in alternating frames. This signal is used to enable each channel to be distinguished at the receiver. Bit position 1 in time slot zero carries the International bit, while frames not containing the frame alignment signal are used to carry National and International signaling and alarm indication for loss of frame alignment. Figure 4.99 illustrates the composition of the CEPT-30 frame and multiframe, where the multiframe consists of 16 frames, numbered from frame 0 to frame 15.

To avoid imitation of the frame alignment signal, alternating frames fix bit 2 to a 1 in time slot zero which is the reason why a 1 is entered into that bit position for odd time slot 0 frames.

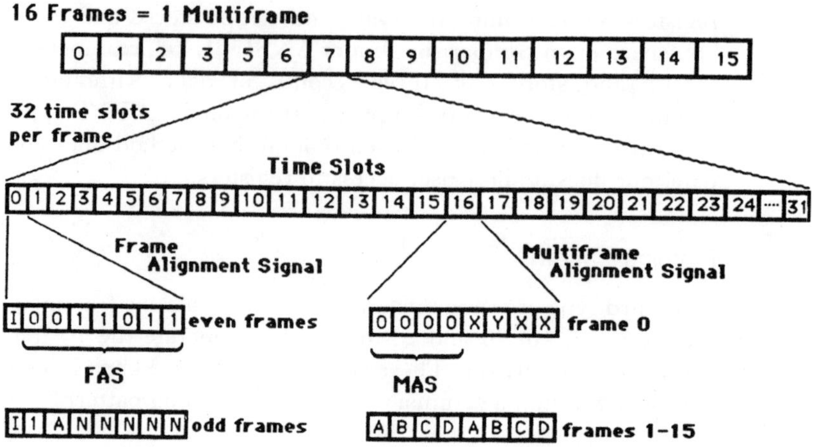

Figure 4.99 CEPT PCM-30 frame and multiframe composition (i international bit, N national bit, A alarm indication signal, FAS frame alignment signal, ABCD ABCD signaling bits, X extra bit for signaling, Y loss of multiframe alignment, MAS multiframe alignment signal)

CEPT CRC option

For enhanced error monitoring capability, CEPT PCM-30 includes a CRC-4 option. Under this option, a group of eight frames known as a submultiframe (SMF) is treated as a long binary number. This number is multiplied by X^4 (10000) and divided by X^4+X+1 (10011). The four-bit remainder is transmitted in bit position 1 (1 bit in Figure 4.99) in time slot zero in even frames which contain the frame alignment signal. After the receiver computes its own CRC-4 check it uses bit position 1 in time slot zero of frames 13 and 15 for CRC error performance reporting. Table 4.43 summarizes how these bits are used for CRC error performance reporting purposes.

Table 4.43 CEPT PCM-30 CRC error performance reporting

Bit 1 Frame 13	Bit 1 Frame 15	
1	1	CRC for SMF I, II error free
1	0	SMF II in error, SMF I error free
0	1	SMF II error free, SMF I in error
0	0	Both SMF I and II in error

T-carrier signal characteristics

As previously discussed, there are several advantages to transmitting T-carrier signals in a bipolar alternate mark inversion (AMI) format, including the absence of a dc component to the signal and the ability to obtain clock recovery from the signal in an all-ones condition. Unfortunately, the disadvantage associated with this signaling method is that a sequence of spaces is encoded as a period of zero voltage or no signal and repeaters on a span line cannot recover clocking without a signal occurring every so often.

For repeaters to properly recover clocking, a certain number of binary ones must be contained within a transmitted signal. The process used to ensure a minimum number of ones flows on a T1 or E1 line is called zero suppression and is performed by CSUs. Since North American and European approaches to this problem differ, we will examine the encoding methods used to ensure an appropriate signal contains a minimum number of binary ones for each location.

North America

In North America, AT&T publication 62411 sets the ones density requirement to be 'n' ones in each window of $8\ X\ (n+1)$ bits, where n varies from 1 to 23. This means that a T1 carrier cannot have more than 15 consecutive zeros ($n = 1$) and there must be approximately three ones in every 24 consecutive bits ($n = 2$ to 23). Two methods used to provide this minimum ones density include binary 7 zero code suppression and binary 8 zero substitution.

Binary 7 zero code suppression

Under the binary 7 zero code suppression method a binary one is substituted in bit position 7 of each time slot if all eight positions are zeros. An example of this method of ensuring a minimum ones density is illustrated at the top of Figure 4.100(a). Although it might appear wiser to select the least significant bit for inversion, this cannot be done since the setting of a frame bit to zero when bit positions 2 to 8 in the previous time slot were set to zero would result in a string of 16 consecutive zeros if the bits in the time slot following the frame bit were zero and bit position 8 was used for substitution. Figure 4.100(b) illustrates this worst-case scenario which explains why bit position 7 in each time slot is used for bit value inversion to ensure a minimum ones density.

If a data channel contains all 0s, the data can be corrupted due to B7 zero suppression. Therefore a data channel normally is restricted to seven usable data bits, with one bit in the data channel set to a 1. This prevents the data channel from being corrupted, but also limits its data rates to 56 kbps.

When one bit is set to a 1 on a DS0 channel, the channel is known as a non-clear channel. The 56 kbps on a non-clear channel is also known as a DS-A channel.

A T1 clear channel is one in which all 64 kbps in each DS0 are usable. On private microwave systems, B7 zero code suppression is normally not required, permitting clear channel capability.

Binary 8 zero substitution

The binary 8 zero substitution (B8ZS) technique was developed by Bell Laboratories and is now sanctioned by the ITU for use in North America. This

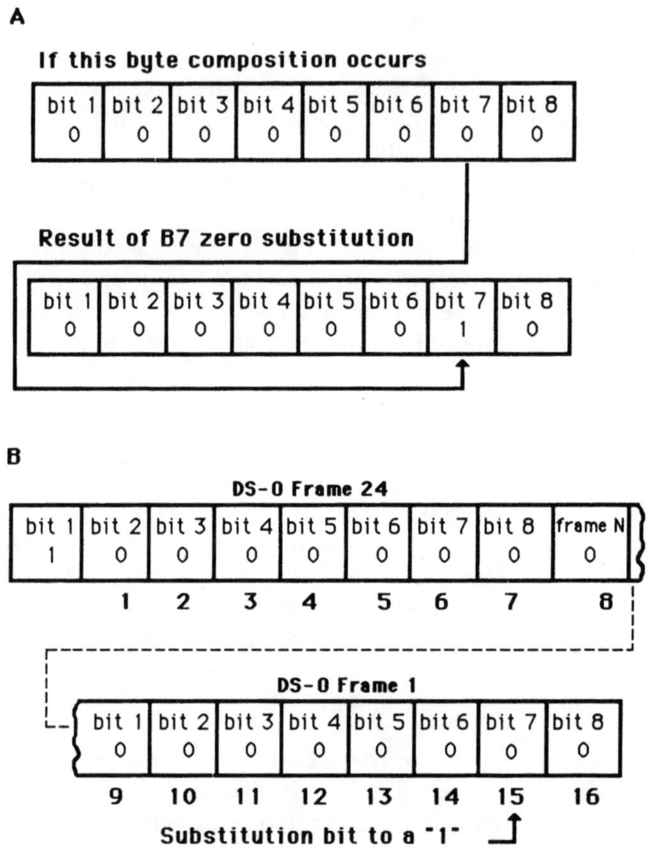

Figure 4.100 B7 zero code suppression: (a) B7 zero code suppression example; (b) worst-case scenario

method of ensuring a minimum ones density was placed into operation during the mid 1980s and offers a significant improvement over binary 7 zero code suppression as it both maintains a minimum ones density and also provides a clear channel capability, permitting each DS0 channel to carry data at 64 kbps.

Under B8ZS coding, each eight consecutive 0s in a byte are removed and replaced by a B8ZS code. If the pulse preceding an all-zero byte is positive, the inserted code is $000 + - 0 - +$. If the pulse preceding an all-zero byte is negative, the inserted code is $000 - + 0 + -$. Figure 4.101 illustrates the use of B8ZS coding in which an all-zeros byte is replaced by one of two binary codes, with the actual code used based upon whether the pulse preceding the all-zeros byte was positive or negative.

Both examples result in bipolar violations occurring in the fourth and seventh bit positions. Both carrier and customer equipment must recognize these codes as legitimate signals and not as bipolar violations or errors for B8ZS to work to enable a receiver to recognize the code and restore the original eight zeros.

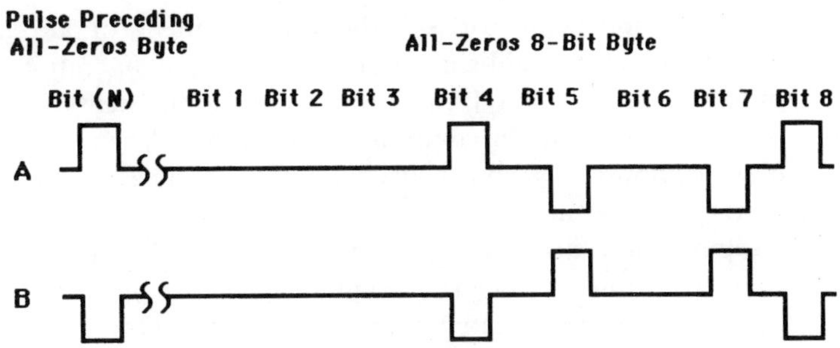

Figure 4.101 B8ZS coding

Europe

In Europe the high density bipolar 3-zero maximum (HDB3) coding is used by CEPT PCM-30 to obtain a minimum ones density for clock recovery from received data. Under HDB3, the data stream to be transmitted is monitored for any group of four consecutive zeros. A four-zero group is then replaced with an HDB3 code. Two different HDB3 codes are used to ensure that the bipolar violation pulses from adjacent four-zero groups are of opposite polarity as indicated in Figure 4.102. The selection of the HDB3 code is based upon whether there was an odd or even number of ones since the last bipolar violation (BV) occurred. If an odd number of ones occurred since the previous bipolar violation, the coding method in Figure 4.102 (a), is used to replace a sequence of four zeros. If an even number of ones occurred since the previous bipolar violation the coding method in Figure 4.102(b), is used to replace a sequence of four zeros.

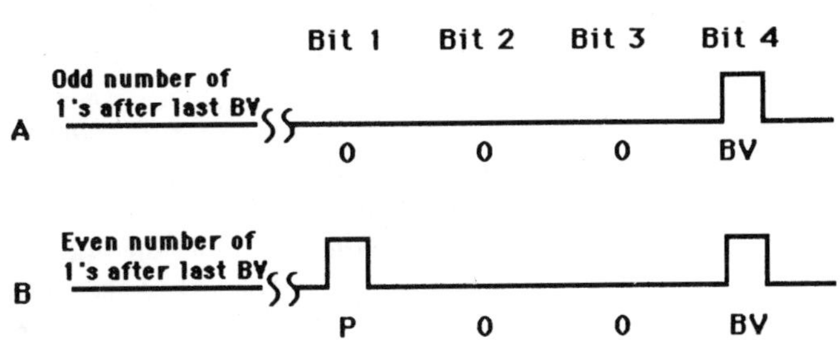

Figure 4.102 HDB3 coding (P polarity bit, BV bipolar violation)

4.12 PARALLEL INTERFACE EXTENDERS

When an application arises that will require batch processing at a site remotely located from a mainframe computer, several approaches can be considered to satisfy this requirement.

A traditional approach is the establishment of a remote batch processing operation. The establishment of this type of operation normally requires the procurement of several communications components in addition to the remote batch terminal. First, a communications controller must be installed and interfaced to the computer if such a device does not already exist at the computational facility. This controller, in conjunction with the computer, performs such tasks as character assembly and disassembly, transmission error checking by generating check characters from the received data blocks, and comparing the check character to the check character generated by the remote batch terminal, as well as performing numerous traffic management functions. Next, a teleprocessing software module will be added and integrated to the computer's operating system to perform and control the transmission discipline. This software may not only be costly but may affect computer performance since it typically requires hundreds of thousands to millions of or more memory locations. Last, a transmission medium and either a high-speed modem or DSU to translate the signals into an acceptable form for transmission must be installed. This traditional remote batch processing operation is illustrated in Figure 4.103.

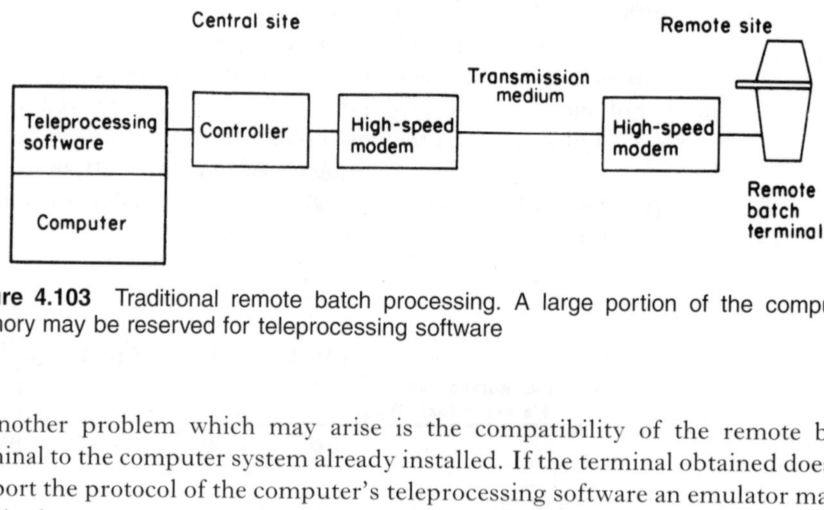

Figure 4.103 Traditional remote batch processing. A large portion of the computer's memory may be reserved for teleprocessing software

Another problem which may arise is the compatibility of the remote batch terminal to the computer system already installed. If the terminal obtained does not support the protocol of the computer's teleprocessing software an emulator may be required to provide an acceptable transmission link. In any event, transmission from the computer to the terminal will most likely require code conversion, since most terminals cannot accept the computer's native code. Realizing these problems, a device was introduced which permits transmission from selected computers to a variety of computer peripherals without the necessity of the addition of special teleprocessing software or a communications controller. This device is called a parallel interface extender.

Another application that can require the extension of the parallel interface of a mainframe computer channel involves the connection of remote LANs and high-speed peripheral devices to a centrally located mainframe. Since the device in effect extends the parallel channel of the mainframe, another name for this device is a channel extender.

Extender operation

A parallel interface or channel extender is a device which translates the parallel protocol of an input/output or channel (I/O) channel from such devices as a computer or selected computer peripheral units into a serial protocol which is suitable for transmission over a normal serial communications link. At the other end of the link, another parallel interface extender or similar operating device translates the serial protocol formed by the first parallel interface extender back into the original parallel protocol transmitted by the I/O channel for reception by devices similar to the standard peripherals used for local data processing at a computer center. From a broad viewpoint, a parallel interface extender can be compared as being similar in performance to a multiplexer, combining the data from a number of leads of the parallel I/O channel into a single bit stream for transmission over a serial communications link, as shown in Figure 4.104.

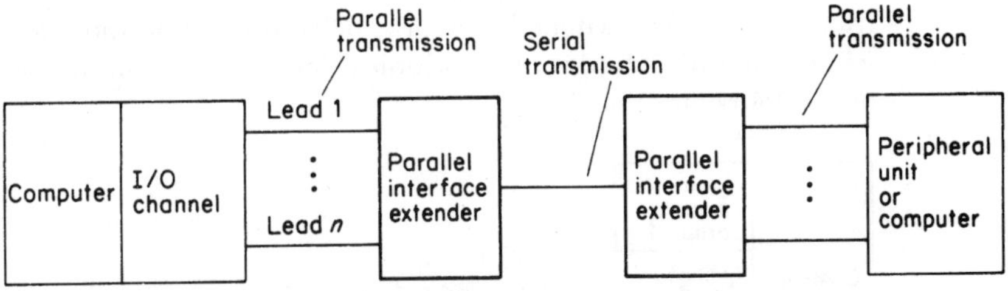

Figure 4.104 Parallel interface or channel extender operation. Operation is similar to a multiplexer, combining the data from a number of leads from a computer or peripheral I/O channel into a single bit stream for transmission

By permitting a computer to utilize its regular I/O channel when communicating with a remotely located peripheral device, the necessity of obtaining specialized communications software is alleviated, and communications with remote peripheral units can then take place with the same software which is regularly used for communicating with the local peripherals at the computer center. Depending upon the computer configuration in use, a parallel interface extender may reduce operating system software requirements from 10 to 30% or more, when compared with communicating to remote devices by using a line controller and the required teleprocessing software modules. Another advantage obtained through the use of parallel interface extenders is the ability to program remote applications in Fortran, Cobol and other higher level languages, as well as in assembly language, using the READ, WRITE statements of Fortran and Cobol or the GET and PUT macros of assembly language to perform input and output remote peripheral functions. In addition to the translation of parallel to serial and serial to parallel data, to perform remote peripheral functions the device encodes commands and status information into a serial bit stream at the transmission end of the link while the device at the receiving end of the link performs a decoding function to reconstruct the original commands, status, and data before passing such information to the remote peripheral unit.

Extender components

A parallel interface extender consists of a control unit and one or more line module groups, as shown in Figure 4.105. The control unit of the extender connects to the multiplexer channel of a computer and emulates the functions of several peripheral control units, such as card readers, card punches, magnetic tapes, and line printers, thus supporting the computer's standard software, as previously mentioned. Each line module group is connected to the control unit of the extender on one side and provides an interface for the connection of dedicated, switched, or leased lines on the opposite side. The line module contains all necessary line control and error control components, as well as a built-in modem which alleviates the necessity of having special communications software such as IBM's virtual telecommunications access method (VTAM) or synchronous data link control (SDLC), a communications controller, line adapter, and a separate modem. For alternate high-speed data transfer in either direction, half-duplex line modules can be used while full-duplex line modules provide data transfer in both directions at the same time. For data transfer over the switched network, half-duplex line modules are normally used while either half-duplex or full-duplex line modules can be used with dedicated or leased four-wire lines.

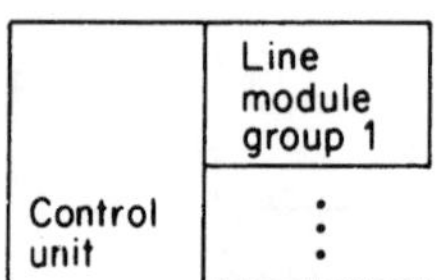

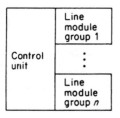

Figure 4.105 Parallel interface extender components. A parallel interface extender consists of a control unit and one or more line module groups. In addition, some manufacturers provide modems or DSUs/CSUs built in to the device

Application examples

One example of the use of a Parallel interface extender is a product that connects to an IBM system central processing unit via the system's byte multiplexer channel and can be used for such diverse applications as permitting two computers to communicate with each other in their native code, as in Figure 4.106; or it can be used to connect a computer to a variety of local or remotely located peripheral units,

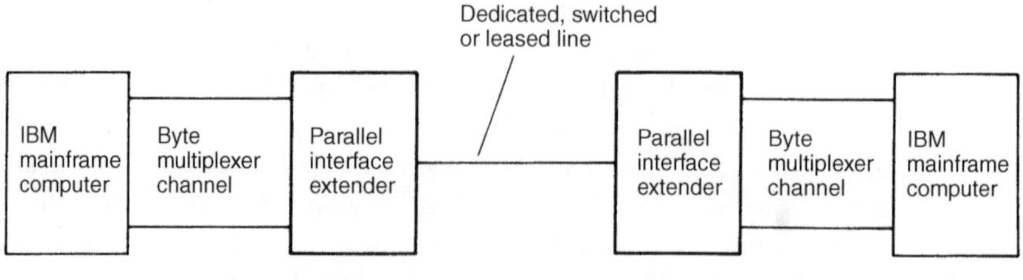

Figure 4.106 Intercomputer communications using a parallel intenace extender. Using a parallel interface extender, the parallel transmission of the byte multiplexer channel of a computer is converted into a serial data stream for transmission

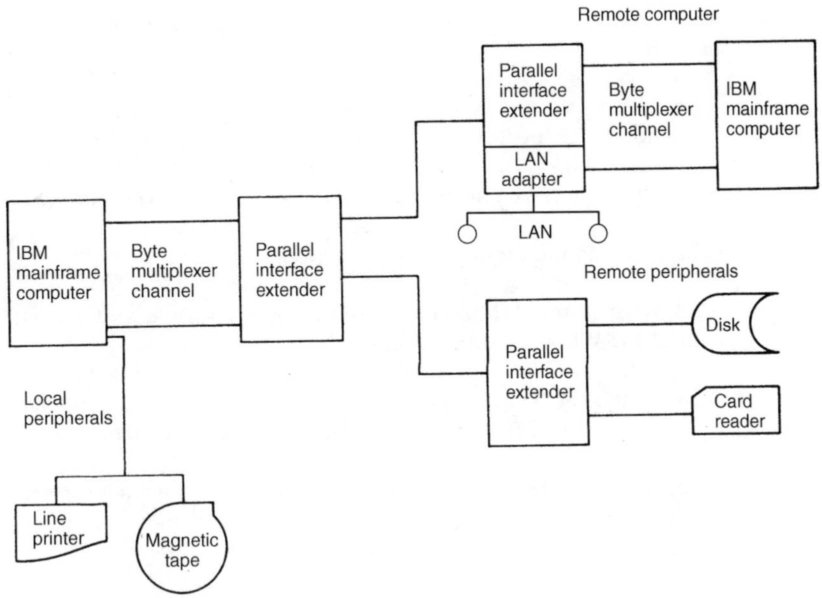

Figure 4.107 Local and remote peripherals as well as computers can be serviced. Through the utilization of a parallel interface extender, remote peripherals as well as remote computers are referenced by the central computer as if they were local devices

as shown in Figure 4.107. In examining Figure 4.107, note the integration of a LAN adapter into one of the remote parallel interface extenders (channel extender). Some extenders support the addition of Token-Ring or Ethernet LAN modules, permitting remote LANs as well as remote peripherals and remote mainframes to be integrated into a common network infrastructure.

Supporting data transmission rates from 4800 bps to 2.048 mbps, the parallel interface extender permits a level of data transfer that can be matched to the operating speed of most peripheral devices. Parallel interface extenders permit peripherals to be transferred to remote locations for other applications than those originally required; hence, the use of such equipment warrants further examination.

REVIEW QUESTIONS

4.1.1 What is the difference between an acoustic coupler and a modem with respect to their line connection?

4.1.2 What does the term 'Bell System' compatibility mean when discussing the operational characteristics of a modem?

4.1.3 If an originate mode acoustic coupler transmits a mark at f_1 and a space at f_2 and receives a mark at f_3 and a space at f_4, what would be the corresponding frequencies of an answer mode coupler to ensure communications compatibility?

4.1.4 Discuss the conventional utilization of originate and answer mode couplers and modems.

4.1.5 Why is it more likely than not that an American using his or her portable personal computer in Europe would not be able to communicate with a computer located in Europe?

4.2.1 What is a carrier signal? By itself, does it convey any information? Why?

4.2.2 How can the characteristics of a carrier signal be altered?

4.2.3 What is the difference between a bit per second and a baud? When can they be equivalent? When are they not equivalent?

4.2.4 What is the Nyquist relationship and why does it require modem designers to develop multilevel phase-shift keying modulation schemes for modems to operate at high data rates?

4.2.5 What does the signal constellation pattern of a modem represent? What is the normal relationship between the density of the signal constellation pattern and the susceptibility of a modem to transmission impairments?

4.2.6 Discuss the difference between Trellis-coded modulation and conventional quadrature amplitude modulation with respect to the density of the signal constellation and the susceptibility of a modem employing each modulation technique to transmission impairments.

4.2.7 How could you use a reverse channel? What is the difference between a reverse channel and a secondary channel?

4.2.8 What is a multiport modem and under what circumstances should you consider using this device?

4.2.9 Why are Bell System 212-type modems operating at 1200 bps sometimes compatible with V.22 modems while at other times they are incompatible?

4.2.10 Why are Bell System 202 type modems incompatible with ITU V.23 modems?

4.2.11 Discuss the compatibility of an ITU V.26 modem employing a pattern A phase change with a similar modem using the pattern B phase change.

4.2.12 Explain why the V.29 signal constellation pattern forms a mirror image.

4.2.13 What are the key differences between the ITU V.32 and V.32 bis modems with respect to their operating rates and fallback/fall-forward capability?

4.2.14 Does the V.32 terbo represent an industry standard modem?

4.2.15 Why do some V.34 modems have a maximum operating rate of 28.8 kbps while other V.34 modems can operate at 33.6 kbps?

4.2.16 What is the maximum operating rate of a 56 kbps modem when a transmission path flows over two analog–digital conversion points?

4.2.17 Assume two modems connected to a leased line have both local and digital loop-back self-testing features. If communications were disabled, discuss the tests you would perform to determine if the line or one or both modems caused the communications failure.

4.2.18 What is a wraparound unit? When would it be used?

4.3.1 What are the three functions primarily associated with intelligent modems?

4.3.2 Why can the term 'Hayes compatible' be misleading?

4.3.3 What are result codes?

4.3.4 What command line entry would you use to set the carrier wait time of an intelligent modem to 120 s?

4.3.5 How does an intelligent modem correct detected errors?

4.3.6 What is the purpose of flow control?

4.3.7 Discuss three methods that are used for flow control.

4.3.8 At what level does the MNP protocol perform error detection and correction?

4.3.9 What is the difference between V.42 and V.42 bis?

4.3.10 How does run-length encoding function under MNP class 5?

4.3.11 What is the major problem affecting transmission on a cellular facility?

4.4.1 What are the advantages and disadvantages in using multiport modems instead of separate modems and multiplexers?

4.4.2 What function does a DCE option perform? How would you obtain a DCE option through the use of a cable?

4.5.1 What is the key difference between multipoint modems and conventional modems? Discuss the differences in throughput obtained on a multipoint circuit as the average block size transmitted increases.

4.5.2 Why is the effect of multipoint modems on satellite circuits minimal?

4.6.1 How does a security modem operate?

4.6.2 What two data elements would a network administrator enter into a security modem?

4.6.3 Discuss two limitations associated with the use of security modems.

4.7.1 Discuss the relationship between the data rate, wire gauge, and transmission distance.

4.7.2 What is the difference between a line driver and a limited-distance modem?

4.8.1 Discuss the difference between line drivers and limited-distance modems with respect to matching components.

4.8.2 If you are using 26 AWG cable, what resistance would your transmission encounter if the cable distance was 2000 feet?

4.8.3 What is the primary benefit from the use of a multiport limited distance modem?

4.8.4 Assume your organization has eight terminals that require a connection to the computer system located in the same building 600 feet distant. If the terminals operate at 2400 bps and can only transmit pulses 400 feet before the pulses become distorted, determine the most economic method to connect the terminals given the following cost for equipment and cable:

cable cost per foot	$ 0.50
line driver	$100.00
multiport limited-distance modem (4 port)	$250.00

4.9.1 Describe two limitations that can restrict transmission on a subscriber line to 33.6 kbps.

4.9.2 Why are many cable TV systems limited to a unidirectional transmission capability?

4.9.3 Why is the actual transmission obtained on a 6 MHz cable modem channel significantly less than the capacity of the channel?

4.9.4 What is the purpose of using 2B1Q coding on an HDLS transmission facility?

4.9.5 What is the purpose of using an asymmetrical transmission method that employs a small band of frequencies for transmission from a subscriber to a central office?

4.10.1 Why is the probability of a transmission error occurring on a digital transmission facility less than that on an analog transmission facility?

4.10.2 Why do digital transmission facilities employ bipolar signaling?

4.10.3 Define a bipolar violation.

4.10.4 Why is a string of six zeros converted into a zero suppression sequence on DDS?

4.10.5 What is an intentional bipolar violation? How is it used on DDS?

4.10.6 How can you transmit asynchronous data on DDS?

4.11.1 What is the primary difference between a CSU designed to operate on a T1 or E1 circuit and the CSU of a DSU/CSU designed to operate on a DDS transmission facility?

4.11.2 What is the actual data rate available for use on a T1 line?

4.11.3 How many bits are included in a superframe?

4.11.4 Discuss the advantages of the use of the extended superframe format over D4 framing.

4.11.5 How many bits are in an extended superframe?

4.11.6 Discuss the use of each of the three types of frame bits in the extended superframe.

4.11.7 How many bits are in a CEPT frame? Where are the framing bits in a CEPT frame?

4.11.8 How many bits are contained in the CEPT multiframe?

4.11.9 What is the advantage of binary 8 zero substitution over binary 7 zero code suppression?

4.12.1 What is the advantage of using a parallel interface extender with respect to teleprocessing software?

4.12.2 How could you use a central computer facility to distribute reports to several remotely located printers connected to the central computer via parallel interface extenders?

5

LAN INTERNETWORKING DEVICES

In the first four chapters of this book we became familiar with networking fundamentals, obtaining an overview of the characteristics of different types of networks and a general appreciation for the operation and utilization of several networking devices. Using that knowledge we will now begin to examine networking devices in detail. In this chapter we will focus our attention upon a core series of devices used to interconnect LANs that can be separated by a short cabling distance or that are located thousands of miles from one another . . . thus, the title of this chapter.

Devices we will examine in this chapter include what many persons refer to as the Three Musketeers of LAN internetworking—bridges, routers, and gateways. Although these devices play an important role in LAN internetworking, they are not the only products that provide users with the ability to access a local area network. Thus, we will also examine the operation and utilization of several additional internetworking devices in this chapter. Those additional devices include access servers, LAN switches, and several types of communications servers. While a Frame Relay Access Device (FRAD) can be considered to represent a LAN internetworking device, it also represents a communications product that can concentrate data from several network devices. Therefore we will defer a discussion of the operation and utilization of FRADs until Chapter 6 when we cover data communications concentration devices.

5.1 BRIDGES

A bridge is a LAN internetworking device originally developed to interconnect two networks within close proximity of one another. This type of bridge, which is often referred to as a local bridge, represents one of two types of bridges developed to interconnect local area networks. The second type of bridge is a remote bridge and converts LAN traffic into a wide area protocol which allows LANs to be interconnected via a WAN.

Although bridges were originally developed to connect two or more networks into a common network, they are also used as a mechanism to support network separation. That is, as the level of network utilization increases to a point where transmission delays adversely affect the operation and productivity of network users, such networks are typically segmented into two or more independent entities. Then, bridges are used to interconnect those entities together.

From a physical perspective, the key difference between local and remote bridges is in their area of network interfaces. A local bridge has two or more LAN ports, enabling it to interconnect two or more local area networks by becoming a direct participant on each network. In comparison, a remote bridge includes a serial port which enables it to be connected to a serial transmission facility. Thus, a remote bridge is a LAN participant and has the ability to transmit and receive frames via a WAN connection to and from a distant network.

Basic operation

Bridges operate at the data link layer of the OSI Reference Model. This means that they use Media Access Control (MAC) addresses to make their frame forwarding decisions. Figure 5.1 illustrates the use of a local bridge to interconnect two local area networks.

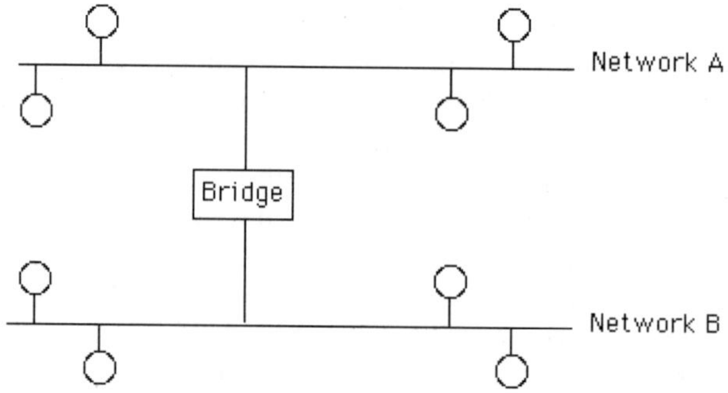

Figure 5.1 Using a local bridge to connect two LANs

When a bridge begins to operate it examines each frame transmitted on each network. By reading the source address included in each frame it obtains the ability to assemble a table of local addresses for each network. Since a destination address is included in each frame, it uses that address as a decision criterion for performing one of three operations—forwarding, filtering, or flooding the frame. If the destination address is in the local address table and represents a port other than the port the frame was received on, the bridge uses that address to forward the frame. If the frame's destination address is the port it was received on, the frame is already on the correct network. Thus, the bridge discards or filters the frame. If the destination address is not in the local address table, the bridge transmits the frame onto all ports other than the port the frame was received on, a technique referred to as flooding.

We can summarize the operation of the bridge shown in Figure 5.1 as follows:

- Bridge reads all frames transmitted on network A.
- Frames with destination address on network A are filtered by the bridge.
- Frames with destination address on network B are forwarded onto network B.
- The above process is reversed for traffic on network B.

Flooding

To better understand the concept of flooding requires the use of a bridge with three or more ports. Thus, let's examine Figure 5.2 which illustrates the operation of a four-port bridge that will serve as a mechanism to discuss the effect of flooding. Instead of 48-bit MAC addresses we will use addresses A through H for simplicity to indicate source addresses for each station on four networks connected to a common bridge.

Initially, when the bridge is powered on, its address/port table is empty. This situation is illustrated in the first address/port table entry at the bottom of Figure 5.2. Next, let's assume the device with source address A transmits a frame to the device whose address is E. Since a search of the address/port table provides no match for the destination address, the bridge transmits or floods the frame onto all ports other than the port it was received on. This means that the frame is forwarded

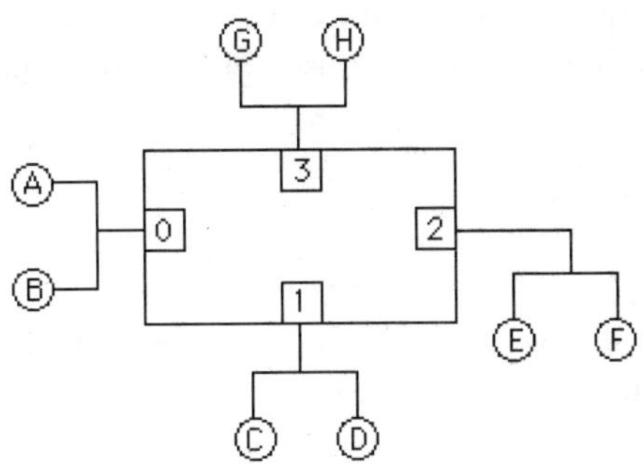

Address/port table entries.

1. Power-on state

Address	Port
-	0
-	1
-	2
-	3

2. Frame from A transmitted to E

Address	Port
A	0
-	1
-	2
-	3

3. Frame from G transmitted to A

Address	Port
A	0
-	1
-	2
G	3

Figure 5.2 Multiport bridge operation

onto the networks attached to ports 1, 2, and 3. Thus, the frame adversely affects the performance on the networks connected to ports 1 and 3 as it precludes other transmissions on those networks for the duration of the frame that will be discarded, since its intended recipient is not on either network. In addition to flooding the frame, the bridge also updates its address/port table as it noted that the station with source address A is on port 0. This update is shown in the second entry at the bottom of Figure 5.2.

Next, let's assume the workstation with address G transmits a frame to the station whose address is A. The bridge searches its address/port table, noting that address A is associated with port 0. Thus, the bridge forwards the frame onto the network connected to port 0. Note that the forwarding process depends upon the ability of a bridge to first learn the addresses of network devices. Thus, the forwarding process is sometimes referred to as a backward or reverse learning process. Returning to the bridge operation, once it notes address A is on port 0 and forwards the frame, it also notes that address G is not in the address/port table and updates the table. This update is indicated in the third entry at the bottom of Figure 5.2.

In addition to maintaining MAC addresses and their associated ports, a bridge time-stamps each entry. The time stamp is used to purge aged entries and enables the finite memory of the bridge to hold the most recently noted addresses. Since many workstations have significant periods of network inactivity, it is quite common for entries to be purged from a bridge's address/port table. However, once purged, the first frame with a destination address no longer in the address/port table will result in the bridge flooding the frame. Thus, flooding can be considered as a process associated with the use of bridges that will continue to occur periodically long after a bridge is powered on. This means that in addition to providing a mechanism to transmit frames to locations whose destinations are not presently known, flooding places an additional level of utilization on networks.

Filtering and forwarding

The process of examining each frame is known as filtering. The filtering rate of a bridge is directly related to its level of performance. That is, the higher the filtering rate of a bridge the lower the probability it will become a bottleneck to network performance. A second performance measurement associated with bridges is their forwarding rate. The forwarding rate is expressed in frames per second and denotes the maximum capability of a bridge to transmit traffic from one network to another.

Types of bridges

There are two general types of bridges—transparent and translating. Each type of bridge can be obtained as a local or remote device, with a remote device including a wide area network interface as well as the ability to convert frames into a WAN transmission protocol.

Transparent bridge

A transparent bridge provides a connection between two local area networks that employ the same data link protocol. Thus, the bridge shown in Figure 5.1 can be considered to be a transparent bridge. This type of bridge is used to connect two or more local area networks that employ identical protocols at the data link layer. At the physical layer, some transparent bridges have multiple ports that support different media. Thus, a transparent bridge does not have to be transparent at the physical level, although the majority of such bridges are.

Although a transparent bridge provides a high level of performance for a small number of network interconnections, its level of performance decreases as the number of interconnected networks increases. The rationale for this loss in performance is based upon the method used by transparent bridges to develop a route between LANs.

Translating bridge

A translating bridge provides a connection capability between two local area networks that employ different protocols at the data link layer. Since networks using different data link layer protocols normally use different media, a translating bridge will also provide support for different physical layer connections.

Figure 5.3 illustrates the use of a translating bridge to interconnect a Token-Ring and an Ethernet local area network. In this example, the bridge functions as an Ethernet node on the Ethernet and as a Token-Ring node on the Token-Ring. When a frame from one network has a destination on the other network, the bridge will perform a series of operations, including frame and transmission rate conversion. For example, consider an Ethernet frame destined to the Token-Ring network. The bridge will strip the frame's preamble and FCS, then it will convert the frame into a Token-Ring frame format. Once the bridge receives a free token the new frame will be transmitted onto the Token-Ring; however, the transmission rate will be at the Token-Ring network rate and not at the Ethernet rate. For frames going from the Token-Ring to the Ethernet the process would be reversed.

One of the problems associated with the use of a translating bridge is the conversion of frames from their format on one network to the format required for use on another network. As previously indicated in Chapter 4, the information field of an Ethernet frame can vary from 64 to 1500 bytes, while a Token-Ring can have a maximum information field size of 4500 bytes when the ring operates at 4 Mbps and 18 000 bytes when the ring operates at 16 Mbps. If a station on a Token-Ring network has a frame whose information field exceeds 1500 bytes in length, the bridging of that frame onto an Ethernet network cannot occur. This is because there is no provision within either data link protocol to inform a station that a frame flowing from one network to another was fragmented and requires reassembly. To effectively use a bridge in this situation requires software on each workstation on each network to be configured to use the smallest maximum frame size of any network to be connected together. In this example, Token-Ring

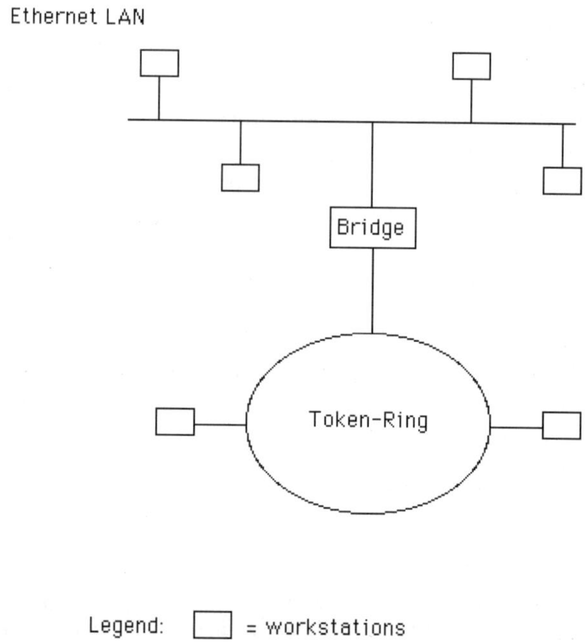

Figure 5.3 Translating bridge operation. A translating gridge connects local area networks that employ different protocols at the data link layer. In this example the translating bridge is used to connect an Ethernet local area network to a Token-Ring network

workstations would not be allowed to transmit information fields greater than 1500 bytes.

Features

The functionality of a bridge is based upon the features incorporated into this device. Table 5.1 lists 11 major bridge features which define both the functionality and performance level of a bridge.

Table 5.1 Bridge features

- Filtering and forwarding rate
- Selective forwarding capability
- Multiple port support
- Wide area network interface support
- Local area network media interface support
- Transparent operation at the data link layer
- Translating operation to link dissimilar networks
- Encapsulation operation to support wide area network usage
- Standalone and adapter-based fabrication
- Self-learning (transparent) routing
- Source routing

Filtering and forwarding

The filtering and forwarding rate indicates the ability of the bridge to accept, examine, and regenerate frames on the same network (filtering) and transfer frames onto a different network (forwarding). A higher filtering and forwarding rate indicates a higher performing bridge.

Selective forwarding

Some bridges have a selective forwarding capability. Bridges with this feature can be configured to selectively forward frames based upon predefined source and destination addresses. Through the use of a selective forwarding capability you can develop predefined routes for frames to take when flowing between networks as well as enable or inhibit the transfer of information between predefined workstations.

Figure 5.4 illustrates the use of the selective forwarding capability of two bridges to provide two independent routes for data transfer between an Ethernet and a Token-Ring network. In this example, you might enable all workstations with source address 1 and 2 to have data destined to the Token-Ring flow over bridge 1, while workstations with a source address of 3 and 4 that are transmitting data to the Token-Ring are configured to use bridge 2.

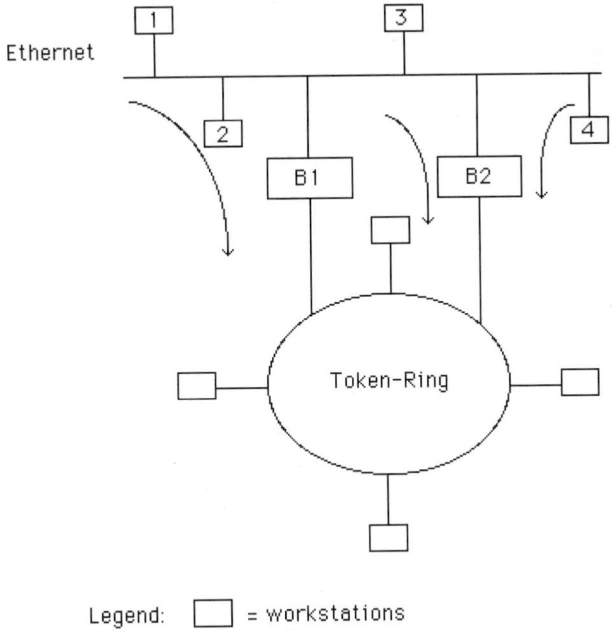

Figure 5.4 Using bridge selective forwarding capability. Using the selective forwarding capability of bridges enables the data flow between networks to be distributed based upon source or destination addresses

For readers familiar with Ethernet bridging, you are probably aware of a constraint referred to as the spanning tree path which precludes the ability to form a closed loop when bridging. The reason for the spanning tree is to prevent the continuous forwarding of frames in a circular manner that would adversely affect network performance. While the spanning tree algorithm would require one of the two bridges shown in Figure 5.4 to be placed into a standby state of operation, since the two bridges are considered to be selectively forwarding by address, this is not required. That is, by forwarding only over single paths the integrity of the spanning tree is maintained on the Ethernet network. Later in this section we will discuss the operation of the spanning tree algorithm in detail.

Multiple port support

The multiple port support capability of a bridge is related to its local and wide area network media interface support. Some bridges support additional ports beyond the two that make up a basic bridge. Doing so enables a bridge to provide connectivity between three or more local area networks.

Figure 5.5 illustrates one potential use of a multiple port bridge to link an Ethernet network to two Token-Ring networks.

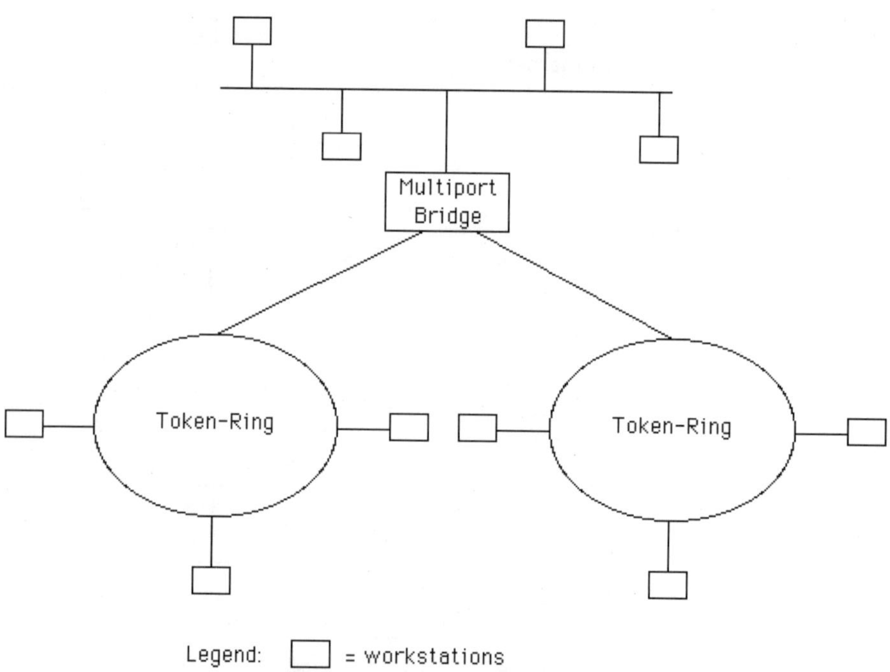

Figure 5.5 Using a multiport bridge. Through the use of a multiport bridge you can connect three or more local area networks

Local and wide area interface support

Local area media interfaces supported by bridges can include thin and thick Ethernet coaxial cable, IEEE 10BASE-T, 100BASE-T, and other types of twisted-pair cable. Wide area network interfaces are incorporated into remote bridges that are designed to provide an internetworking capability between two or more geographically dispersed LANs linked by a WAN. Common WAN media interfaces can include RS-232 for data rates at or below 19.2 kbps, X.21 for packet network access at data rates up to 128 kbps, and V.35 for data rates between 48 kbps and 6 Mbps.

Transparent operation

Although bridges are thought of as transparent to data, this is not always true. For interconnecting different networks located in the same geographical area bridges are normally transparent to data. However, some remote bridges use data compression algorithms to reduce the quantity of data transmitted between bridges connected via a wide area network. Such compression-performing bridges are not transparent to data, although they restore data to its original form.

Frame translation

For interconnecting different types of local area networks, bridges must perform a translation of frames. For example, an Ethernet frame must be changed into a Token-Ring frame when the frame is routed from an Ethernet to a Token-Ring network. As previously mentioned, since frames cannot be fragmented at the data link layer you must set the workstations on the Token-Ring network to the smallest maximum frame size of the Ethernet, or 1500 bytes.

When data is transferred between colocated local area networks the frame format on one network is suitable for transfer onto the other network or is modified for transfer when the MAC layers differ. When a bridge is used to connect two local area networks via a wide area network facility a WAN protocol is employed to control data transfer. The wide area network protocol is better suited for transmission over the WAN as it will normally incorporate error detection and correction, enable a large number of unacknowledged 'WAN' frames to exist to speed information transfer, support full-duplex data transfers, and is standardized. Examples of such wide area network protocols include IBM's SDLC, Digital Equipment Corporation's DDCMP, and the ITU-T's HDLC and X.25.

Frame encapsulation

Figure 5.6 illustrates the operation of a pair of remote bridges connecting two local area networks via a wide area network. For transmission from network A to network B, user data from a network A station is first converted into LLC and

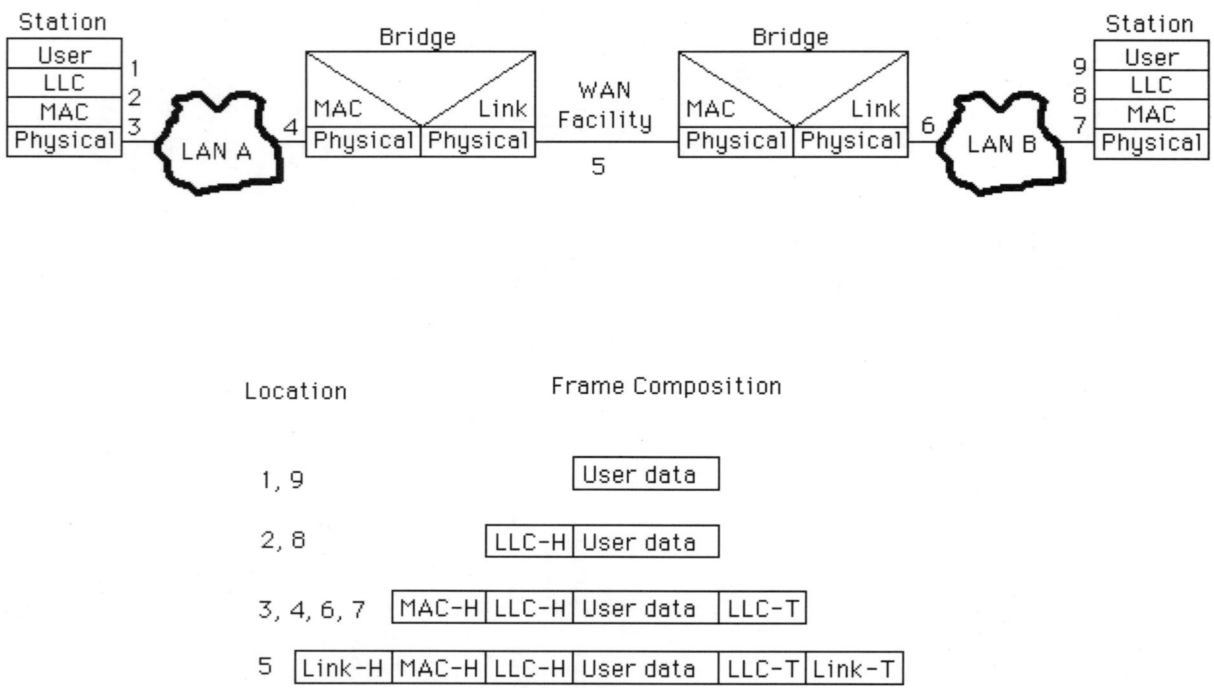

Figure 5.6 Remote bridge operation. A remote bridge wraps the Logical Link Control (LLC) and Media Access Control (MAC) frames in another protocol for transmission over a wide area network. H represents a header and T represents a trailing field

MAC frames. The bridge then encapsulates one or more LAN frames into the bridging protocol frame used for communications over the wide area network. Since the LAN frame is wrapped in another protocol, we say the LAN frame is tunneled within the WAN protocol. At the opposite end of the wide area network the distant remote bridge performs a reverse operation, removing the WAN protocol header and trailer from each frame.

Fabrication

Some bridges are manufactured as standalone products. Such devices can be considered as 'plug and play', as you simply connect the bridge to the media and power it on. Other bridges are manufactured as adapter cards for insertion into the system unit of a personal computer, workstation, or reduced instruction set computer (RISC). Through the use of software developed in conjunction with hardware you may obtain more flexibility in the use of this type of bridge than a standalone device whose software is fixed in ROM.

Routing methods

The routing capability of a bridge governs its abilty to interconnect local area networks. Two types of routing methods used by bridges are transparent and source routing. Transparent bridging was originally developed to support the interconnection of Ethernet networks and is based upon the use of the spanning tree algorithm to prevent closed loops from allowing frames to continuously circulate a group of connected networks. Source routing was developed to support bridging of Token-Ring networks and uses the routing information field (RIF) in a Token-Ring frame to create and use a route between source and destination. This enables source routing bridging to support a closed loop network structure. A third type of bridge is a source route-transparent bridge which enables Ethernet and Token-Ring networks to be connected while supporting the appropriate routing technique on each side of the bridge commensurate with the type of network the bridge is connected to.

To illustrate the problem associated with a closed loop architecture and the need for a spanning tree algorithm, consider Figure 5.7 which illustrates the connection of three Ethernet local area networks by the use of three two-port bridges.

In this example the interconnected networks form a circular or loop topology. Since a transparent bridge views workstations as being connected to either port 1 or port 2, a circular or loop topology will create problems. Those problems can result in an unnecessary duplication of frames which will not only degrade the overall level of performance of the interconnected network but quite possibly confuse end stations. For example, consider a frame whose source address is A and whose destination address is F. Both bridge 1 and bridge 2 will forward the frame. Although bridge 1 will forward the frame to its appropriate network using the most direct route, the frame will also be forwarded by bridge 3 to Ethernet 2, resulting in a duplicate frame arriving at workstation F. At workstation F a mechanism would be required to reject duplicate frames. Even if such a mechanism is available, the

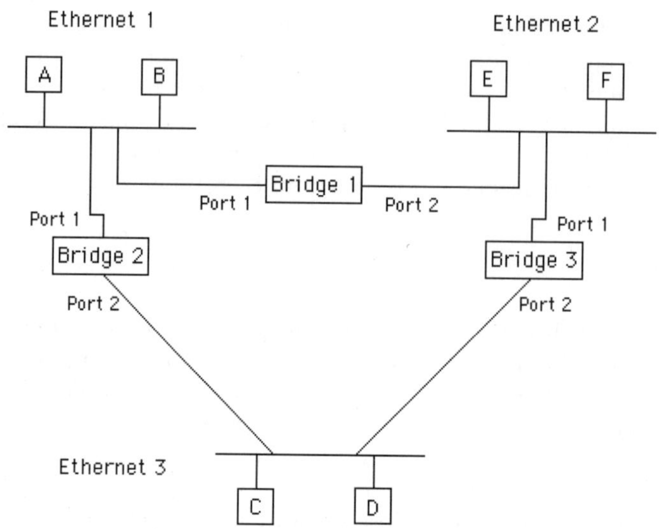

Figure 5.7 Transparent bridges do not support network loops. The construction of a circular or loop topology through the use of transparent bridge can result in an unnecessary duplication of frames as well as confuse end stations. To avoid those problems the Spanning Tree Protocol (STP) will open a loop by placing one bridge in a standby mode of operation

additional traffic flowing across multiple internet paths would result in an increase in network utilization. This in turn would saturate some networks, while significantly reducing the level of performance of other networks. For those reasons transparent bridging is prohibited from creating a loop or circular topology.

Spanning tree protocol

The problem of active loops was addressed by the IEEE Committee 802 in the 802.1D standard with an intelligent algorithm known as the Spanning Tree Protocol (STP). The STP is based upon graph theory and converts a loop into a tree topology by disabling a link. This action ensures there is a unique path from any node in an internet to every other node. Disabled nodes are then kept in a standby mode of operation until a network failure occurs. At that time, the spanning tree protocol will attempt to construct a new tree using any of the previously disabled links.

Operation

To illustrate the operation of the spanning tree protocol, we must first become familiar with the difference between the physical and active topology of bridged networks. In addition, we should familiarize ourselves with a number of terms associated with the spanning tree algorithm defined by the protocol. Thus, we will also review those terms prior to discussing the operation of the algorithm.

Physical versus active topology

In transparent bridging, a distinction is made between the physical and active topology resulting from bridged local area networks. This distinction enables the construction of a network topology in which inactive but physically constructed routes can be placed into operation if a primary route should fail and in which the inactive and active routes would form an illegal circular path violating the spanning tree algorithm if both routes were active at the same time.

Figure 5.8 (a) illustrates one possible physical topology of bridged networks. The cost (C) assigned to each bridge will be discussed later in this section. Figure 5.8 (b) illustrates a possible active topology for the physical configuration shown in Figure 5.8 (a).

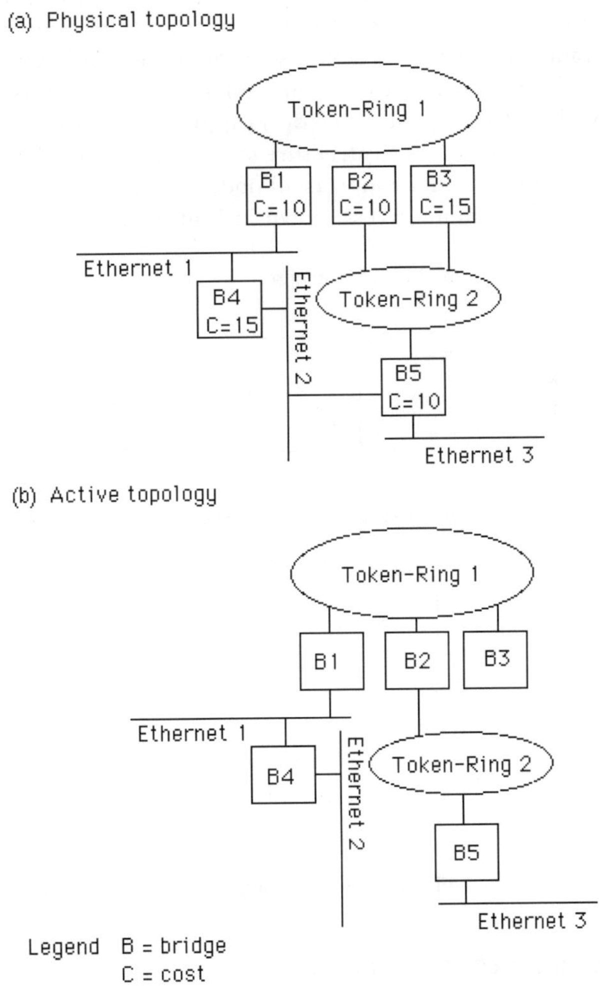

Figure 5.8 Physical versus active topology. When transparent bridges are used, the active topology cannot form a closed loop in the Internet

When a bridge is used to construct an active path, it will forward frames through those ports used to form active paths. The ports through which frames are forwarded are said to be in a forwarding state of operation. Ports that cannot forward frames due to their operation forming a loop are said to be in a blocking state of operation.

Under the spanning tree algorithm, a port in a blocking state can be placed into a forwarding state and provides a path that becomes part of the active network topology. This new path must not form a closed loop and usually occurs due to the failure of another path, or a bridge component, or the reconfiguration of interconnected networks.

Spanning tree algorithm

The basis for the spanning tree algorithm is a tree structure since a tree forms a pattern of connections that has no loops. The term 'spanning' is used because the branches of a tree structure span or connect subnetworks.

As a review for readers unfamiliar with graph theory, let's examine the concept behind spanning trees. To do so appropriately we need a point of reference, so let's begin with the graph structure shown in Figure 5.9 (a). A spanning tree of a graph is a subgraph that connects all nodes and represents a tree. Figure 5.9 (a) has eight distinct spanning trees, shown in Figure 5.9 (b).

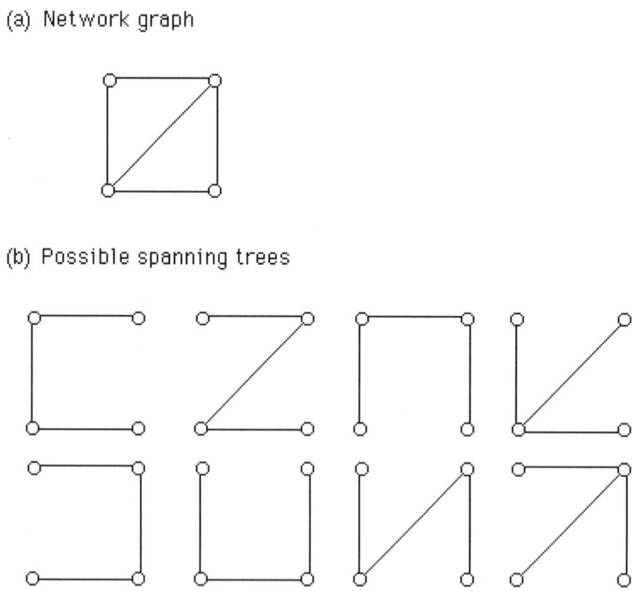

Figure 5.9 Forming spanning trees from a network graph

Minimum spanning tree

Suppose the links connecting each node are assigned a length or weight. Then, the weight of a tree represents the sum of the weights of its links or edges. If the weights or lengths of the links or tree edges differ, then different tree structures will have different weights. Thus, the identification of the minimum spanning tree requires us to examine each of the spanning trees supported by a graph and identify the structure that has the minimum length or weight.

The identification of the minimum spanning tree can be accomplished by listing all spanning trees and finding the minimum weight or length associated with the list. This is a brute force method that always works but is not exactly efficient, especially when a graph becomes complex and can contain a significant number of trees. A far better method is obtained by the use of an appropriate algorithm.

Kruskal's algorithm

There are several popular algorithms developed for solving the minimum spanning tree of a graph. One of those algorithms is Kruskal's algorithm which is relatively easy to understand and will be used to illustrate the computation of a minimum spanning tree. Since we need weights or lengths assigned to each edge or link in a graph, let's revise the network graph previously shown in Figure 5.9 (a) and add some weights. Figure 5.10 illustrates the weighted graph.

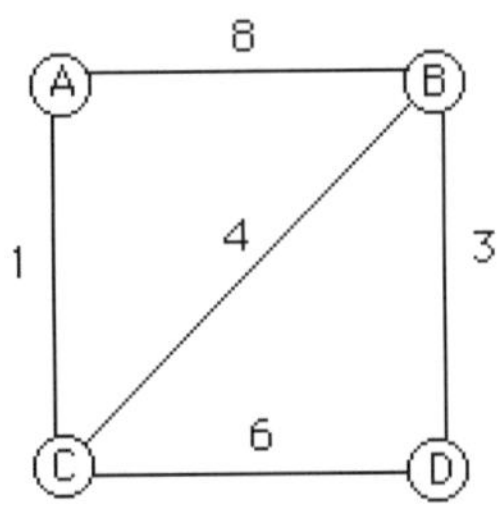

Figure 5.10 A weighted network graph

Kruskal's algorithm can be expressed as follows:

1. Sort the edges of the graph (G) into increasing order by weight or length.
2. Construct a subgraph (S) of G and initially set it to the empty state.
3. For each edge (e) in sorted order, if the endpoints of the edges (e) are disconnected in S, add them to S.

Using the graph shown in Figure 5.10, let's apply Kruskal's algorithm as follows:

1. Sorting the edges of the graph into increasing order by weight or length produces the following table:

Edge	Weight/length
A–C	1
B–D	3
C–B	4
C–D	6
A–B	8

2. Set the subgraph of G to the empty state. Thus, S = null.
3. For each edge add to S as long as the endpoints are disconnected. Thus, the first operation produces:

$$S = A, C \quad \text{or}$$

The next operation produces:

$$S = (A, C) + (B, D) \quad \text{or}$$

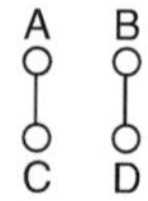

The third operation produces:

$$S = (A, C) + (B, D) + (C, B) \quad \text{or}$$

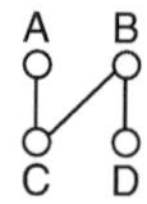

Note that we cannot continue as the endpoints in S are now all connected. Thus, the minimum spanning tree consists of the edges or links (A,C) + (B,D) + (C,B) and has the weight 1 + 3 + 4, or 8.

Now that we have an appreciation for the method by which a minimum spanning tree is formed, let's turn our attention to its applicability in transparent bridge-based networks.

Root bridge and bridge identifiers

Similar to the root of a tree, one bridge in a spanning tree network will be assigned to a unique position in the network. Known as the root bridge, this bridge is assigned as the top of the spanning tree and has the potential to carry the largest amount of network traffic due to its position.

Since bridges and bridge ports can be active or inactive a mechanism is required to identify bridges and bridge ports. Each bridge in a spanning tree network is assigned a unique bridge identifier. This identifier is the MAC address on the bridge's lowest port number and a two-byte bridge priority level. The priority level is defined when a bridge is installed and functions as a bridge number. Similar to the bridge priority level, each adapter on a bridge which functions as a port has a two-byte port identifier. Thus, the unique bridge identifier and port identifier enable each port on a bridge to be uniquely identified.

Path cost

Under the spanning tree algorithm, the difference in physical routes between bridges is recognized and a mechanism is provided to indicate the preference for one route over another. That mechanism is accomplished by the ability to assign a path cost to each path. Thus, you could assign a low cost to a preferred route and a high cost to a route you only want to be used in a backup situation.

Once path costs are assigned to each path in a network, each bridge will have one or more costs associated with different paths to the root bridge. One of those costs is lower than all other path costs. That cost is known as the bridge's root path cost and the port used to provide the least path cost towards the root bridge is known as the root port.

Designated bridge

As previously discussed, the spanning tree algorithm does not permit active loops in an interconnected network. To prevent this situation from occurring, only one bridge linking two networks can be in a forwarding state at any particular time. That bridge is known as the designated bridge, while all other bridges linking two networks will not forward frames and will be in a blocking state of operation.

Constructing the spanning tree

The spanning tree algorithm employs a three-step process to develop an active topology. First, the root bridge is identified. In Figure 5.8(b) we will assume bridge 1 (B1) was selected as the root bridge. Next, the path cost from each bridge to the root bridge is determined and the minimum cost from each bridge becomes the root path cost. The port in the direction of the least path cost to the root bridge, known as the root port, is then determined for each bridge. If the root path cost is the same for two or more bridges linking LANs, then the bridge with the highest priority will be selected to furnish the minimum path cost. Once the paths are selected, the designated ports are activated.

In examining Figure 5.8(a), let us now use the cost entries assigned to each bridge. Let us assume bridge 1 was selected as the root bridge as we expect a large amount of traffic to flow between Token-Ring 1 and Ethernet 1 networks. Therefore, bridge 1 will become the designated bridge between Token-Ring 1 and Ethernet 1 networks.

In examining the path costs to the root bridge, note that the path through bridge 2 was assigned a cost of 10, while the path through bridge 3 was assigned a cost of 15. Thus, the path from Token-Ring 2 via bridge 2 to Token-Ring 1 becomes the designated bridge between those two networks. Hence, Figure 5.8(b) shows bridge 3 inactive by the omission of a connection to the Token-Ring 2 network. Similarly, the path cost for connecting the Ethernet 3 network to the root bridge is lower by routing through the Token-Ring 2 and Token-Ring 1 networks. Thus, bridge 5 becomes the designated bridge for the Ethernet 3 and Token-Ring 2 networks.

Bridge protocol data unit

One question that is probably in readers' minds by now is how does each bridge know whether or not to participate in a spanned tree topology? Bridges obtain topology information by the use of Bridge Protocol Data Unit (BPDU) frames.

The root bridge is responsible for periodically transmitting a 'HELLO' BPDU frame to all networks to which it is connected. According to the spanning tree protocol, HELLO frames must be transmitted every 1 to 10 seconds. The BPDU has the group MAC address 800143000000 which is recognized by each bridge. A designated bridge will then update the path cost and timing information and forward the frame. A standby bridge will monitor the BPDUs but does not update nor forward them.

When a standby bridge is required to assume the role of the root or designated bridge as the operational states of other bridges change, the HELLO BPDU will indicate that a standby bridge should become a designated bridge. The process by which bridges determine their role in a spanning tree network is an iterative process. As new bridges enter a network they assume a listening state to determine their role in the network. Similarly, when a bridge is removed, another iterative process occurs to reconfigure the remaining bridges.

Although the STP algorithm procedure eliminates duplicate frame and degraded network performance, it can be a hindrance for situations where multiple paths between networks are desired. Another disadvantage of the spanning tree protocol is when it is used in remote bridges connecting geographically dispersed networks. For example, suppose Ethernet 1 was located in Los Angeles, Ethernet 2 in New York and Ethernet 3 in Atlanta. If the link between Los Angeles and New York was placed in a standby mode of operation, all frames from Ethernet 2 routed to Ethernet 1 would be routed through Atlanta. Depending upon the traffic between networks, this situation may require an upgrade in the bandwidth of the links connecting each network to accommodate the extra traffic flowing through Atlanta. Since the yearly cost of upgrading a 56 or 64 kbps circuit to a 128 kbps fractional T1 link can easily exceed the cost of a bridge or router, you may wish to consider the use of routers to accommodate this networking situation. In comparison, when using local bridges, their higher operating rate in interconnecting local area networks will normally allow an acceptable level of performance to occur when LAN traffic is routed through an intermediate bridge.

Source routing

Source routing is a bridging technique developed by IBM for connecting Token-Ring networks. The key to the implementation of source routing is the use of a portion of the information field in the Token-Ring frame to carry routing information and the transmission of 'discovery' packets to determine the best route between two networks.

The presence of source routing is indicated by the setting of the first bit position in the source address field to a binary 1. When set, this indicates that the information field is preceded by a route information field (RIF) which contains both control and routing information.

Figure 5.11 illustrates the composition of a Token-Ring route information field (RIF). This field is variable in length and is developed during a discovery process which is described later in this section.

The control field contains information which defines how information will be transferred and interpreted as well as the size of the remainder of the RIF. The length bits identify the length of the RIF in bytes, while the D bit indicates how the field is scanned, left to right or right to left. Since vendors have incorporated

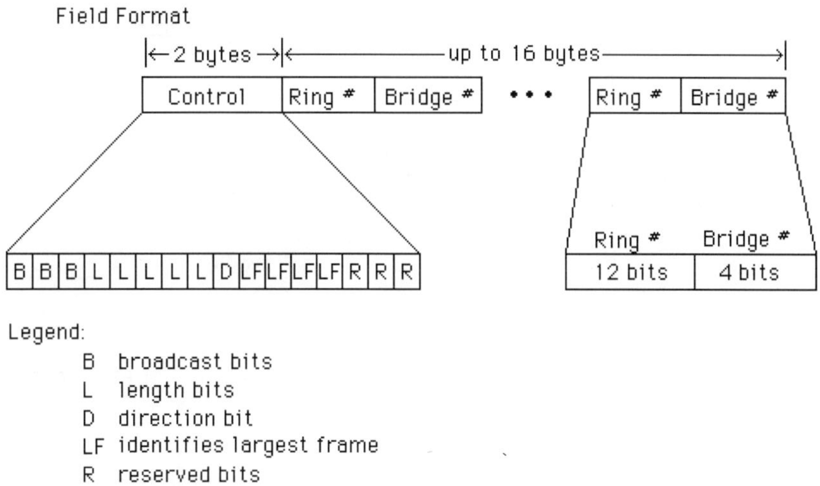

Figure 5.11 Token-Ring route information field. The field is variable in length

different memory in bridges which may limit frame sizes, the LF bits enable different devices to negotiate the size of the frame. Normally a default setting indicates a frame size of 512 bytes. Each bridge can select a number and, if supported by other bridges, that number is then used to represent the negotiated frame size. Otherwise, a smaller number used to represent a smaller frame size is selected and the negotiation process is repeated.

Up to eight route number subfields, each consisting of a 12-bit ring number and a 4-bit bridge number, can be contained in the routing information field. Both ring numbers and bridge numbers are expressed as hexadecimal characters, with three hex characters used to denote the ring number and one hex character used to identify the bridge number.

Operation

To illustrate the concept behind source routing, consider the network illustrated in Figure 5.12. In this example let us assume two Token-Ring networks are located in Atlanta and one network is located in New York.

Each Token-Ring and every bridge are assigned ring and bridge numbers. For simplicity ring numbers R1, R2 and R3 were used, although, as previously explained, those numbers are actually represented in hexadecimal. Similarly, for simplicity bridge numbers are shown as B1, B2, B3, B4 and B5 instead of a hexadecimal character.

When a workstation wants to originate communications it is responsible for finding the destination by transmitting a discovery packet to network bridges and other network workstations whenever it has a message to transmit to a new destination address. Assuming workstation A wishes to transmit to station C, it sends a route discovery packet which contains an empty route information field and its source address as indicated in the upper left portion of Figure 5.12. This packet is

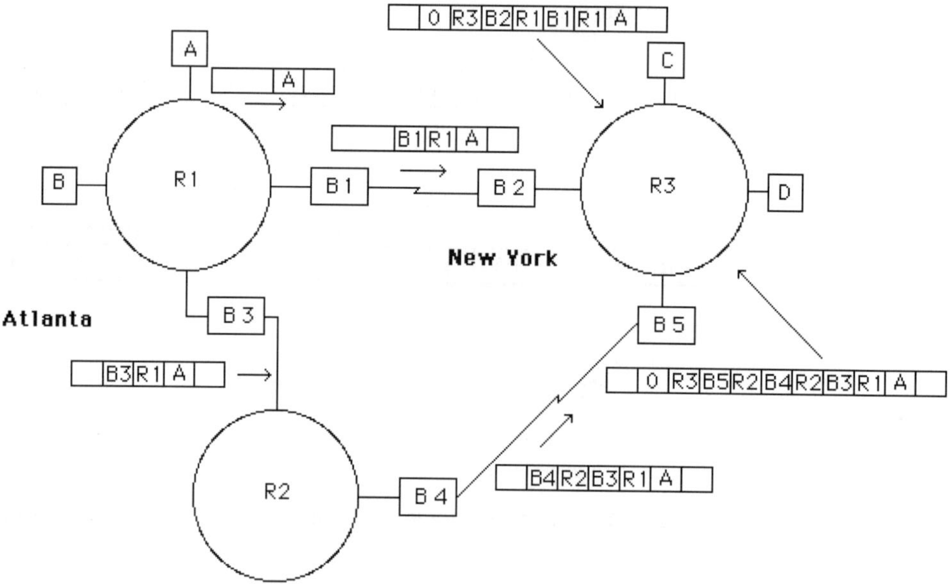

Figure 5.12 Source routing discovery operation. The route discovery process results in each bridge entering the originating ring number and its bridge number into the route information field

recognized by each source routing bridge in the network. When received by a source routing bridge the bridge enters the ring number from which the packet was received and its own bridge identifier in the packet's routing information field. The bridge then transmits the packet to all its connections with the exception of the connection on which the packet was received, a process known as flooding. Depending upon the topology of the interconnected networks, multiple copies of the discovery packet will reach the recipient. This is illustrated in the upper right corner of Figure 5.12 in which two discovery packets reach station C. Here one packet contains the sequence R1B1R1B2R30 where the zero indicates there is no bridging in the last ring. The second packet contains the route sequence R1B3R2B4R2B5R30. Station C then picks the best route based upon either the most direct path or the earliest arriving packet and transmits a response to the discover packet originator. The response indicates the specific route to use and workstation A then enters that route into memory for the duration of the transmission session.

Advantages

There are several advantages associated with the use of source routing. One advantage is the ability to construct mesh networks with loops for a fault-tolerant design which cannot be accomplished with the use of transparent bridges. Another advantage is the inclusion of routing information in the information frames. Several vendors have developed network management software products which use that information to provide statistical information concerning network activity. Those products may assist you in determining wide area network link utilization, the need to modify the capacity of those links, or whether one or more workstations are hogging communications between networks.

Disadvantages

Although the preceding advantages are considerable, they are not without a price. That price includes a requirement to specifically identify bridges and links, higher bursts of network activity and an incompatibility between Token-Ring and Ethernet networks. In addition, due to the structure of the route information field which supports a maximum of seven entries, routing of frames is restricted to crossing a maximum of seven bridges.

When using source routing bridges to connect Token-Ring networks you must configure each bridge with a unique bridge/ring number. In addition, unless you wish to accept the default method by which workstations select a packet during the route discovery process, you will have to reconfigure your LAN software. Thus, source routing creates an administrative burden not present when using transparent bridges.

Due to the route discovery process the flooding of packets occurs in bursts when stations are powered on or after a power outage. Depending upon the complexity of an interconnected series of networks, the discovery process can degrade network performance. Perhaps the biggest problem is for organizations that require the interconnection of Ethernet and Token-Ring networks.

A source routing bridge can only be used to interconnect Token-Ring networks since it operates on route information field data which is not included in an Ethernet frame. Although transparent bridges can operate in Ethernet, Token-Ring and mixed environments, their use precludes the ability to construct loop or mesh topologies and inhibits the ability to establish operational redundant paths for load sharing.

Different LAN operating systems use the RIF data in different ways. Thus, the use of a transparent bridge to interconnect Ethernet and Token-Ring networks may require the same LAN operating system on each network. To alleviate these problems several vendors introduced source routing transparent (SRT) bridges.

Source routing transparent bridges

A source routing transparent bridge supports both IBM's source routing and IEEE 802.1D transparent spanning tree protocol operations. This type of bridge can be considered as two bridges in one and has been standardized by the IEEE 802.1 committee.

Operation

Under source routing the MAC packets contain a status bit in the source field which identifies whether or not source routing is to be used for a message. If source routing is indicated the bridge forwards the frame as a source routing frame. If source routing is not indicated, the bridge determines the destination address and processes the packet using a transparent mode of operation, using routing tables generated by a spanning tree algorithm.

Advantages

There are several advantages associated with the use of SRT bridges. First and perhaps foremost, their use enables different networks to use different LAN operating systems and protocols. This capability enables you to interconnect networks developed independently of one another and allows organization departments and branches to use LAN operating systems without restriction. Secondly, and also very important, SRT bridges can connect Ethernet and Token-Ring networks while preserving the ability to mesh or loop Token-Ring networks. Thus, their use provides an additional level of flexibility for network construction.

Network utilization

In concluding our examination of the use of bridges, we will turn our attention to the topologies they support. In doing so we will examine serial and parallel bridging and how network topologies resulting from the use of bridges can provide mechanisms to enhance performance of subdivided networks as well as to provide alternate communications paths between connected networks.

Serial and sequential bridging

Figure 5.13(a) illustrates the basic use of a bridge to interconnect two networks serially. Suppose monitoring of each network indicates a high level of intranetwork utilization. One possible configuration to reduce intra-LAN traffic on each network can be obtained by moving some stations off each of the two existing networks to form a third network. The three networks would then be interconnected through the use of an additional bridge as illustrated in Figure 5.13(b). This extension results in sequential bridging and is appropriate when intra-LAN traffic is necessary but minimal. Both serial and sequential bridging are applicable to transparent, source routing and SRT bridges which do not provide redundancy nor the ability to balance traffic flowing between networks. Each of these deficiencies can be alleviated through the use of parallel bridging. However, this bridging technique creates a loop and is only applicable to SRT bridges.

Parallel bridging

Figure 5.13(c) illustrates the use of parallel bridges to interconnect two Token-Ring networks. This bridging configuration permits one bridge to back up the other, providing a level of redundancy for linking the two networks as well as a significant increase in the availability of one network to communicate with another. For example, assume the availability of each bridge used in Figure 5.13(a) (serial)

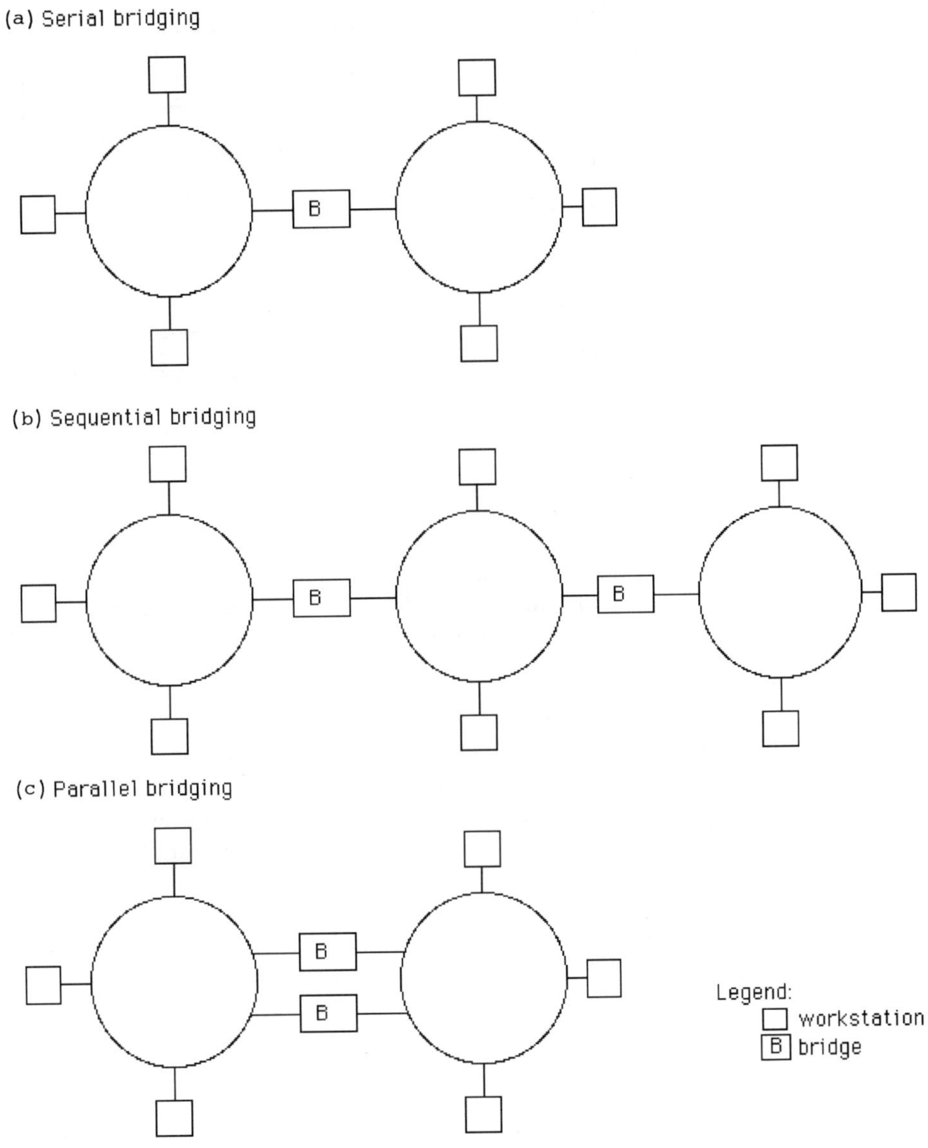

Figure 5.13 Serial, sequential, and parallel bridging

and Figure 5.13(c) (parallel) is 90%. The availability through two serially connected bridges would be 90% × 90%, or 81%. In comparison, the availability through parallel bridges would be 100%−(unavailability of bridge 1 × unavailability of bridge 2) = 100%−(10% × 10%) = 99%

The dual paths between networks also improve inter-LAN communications performance as communications between stations on each network can be load balanced. Thus, the use of parallel bridges can be expected to provide a higher level of inter-LAN communications than the use of serial or sequential bridges.

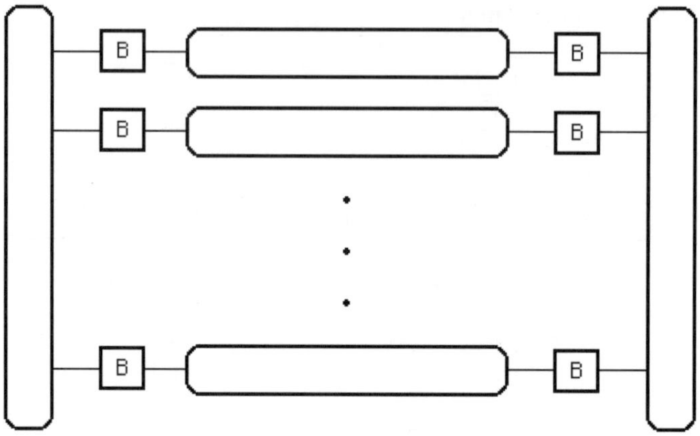

Figure 5.14 Constructing redundant backbone rings. Parallel bridging is often used in building to construct redundant backbone rings. Backbone rings are normally routed vertically within a building and connected via parallel bridge to 'floor' networks

One of the more common uses of parallel bridging is to construct redundant backbone rings within a building. Typically, organizations will establish independent networks on each floor in a building and require a method both to interconnect those networks as well as to obtain a load sharing and redundancy capability. Figure 5.14 illustrates how those goals can be accomplished through the use of parallel bridging to connect each 'floor' network to two backbone networks routed up the vertical shafts common in most buildings. Depending upon internetwork communications requirements, the backbone networks could be a different type of network than the floor networks. For example, each floor network might be a 4 Mbps Token-Ring, while the backbone networks could be 16 Mbps Token-Ring or even 100 Mbps FDDI networks.

From a performance perspective, a workstation on one network requires at most the crossing of only two bridges to reach any other workstation in the building. This provides a much lower routing delay than connecting networks sequentially. In addition, the use of one or more backbone networks simplifies the addition of more floor networks to the building internetwork at a later date. For example, if the organization expands and leases a new floor it can connect a floor network to the building internetwork by the use of one or two bridges.

5.2 ROUTERS

A router is a device that operates at the network layer of the ISO OSI Reference Model. What this basically means is that a router examines network addresses and makes decisions about whether or not data on a local area network should remain on the network or should be transmitted to a different network. Although this level of operation may appear to be insignificant in comparison with a bridge which operates at the data link layer, in actuality there is a considerable difference in the routing capability of bridges and routers.

Comparison to bridges

A bridge uses 48-bit MAC addresses it associates with ports as a mechanism to determine whether a frame received on one port is ignored, forwarded onto another port, or flooded onto all ports other than the port it was received on. At the data link layer there is no mechanism to distinguish one network from another. Thus, the delivery of frames between networks requires bridges to learn MAC addresses and base their forwarding decisions on such addresses. Although this is an acceptable technique for linking a few networks together, as the number of networks increases and the number of workstations on interconnected networks increases, bridges would spend a considerable amount of time purging old entries and flooding frames, causing unacceptable performance problems. This resulted in the development of routers that prevent the flooding of frames between networks and operate by using network addresses.

To understand the concept behind routing, consider Figure 5.15 which illustrates two networks connected by a pair of routers. In this example, assume address A, B, C and D represent MAC addresses of workstations, and E and F represent the MAC address of each router connected to each network. Assuming workstation A has data to transmit to workstation C, an application program executing on workstation A assigns network 20 as the destination address in a packet it prepares at the network layer. That packet is transported via one or more MAC frames to MAC address E, which is the data link address of the network adapter card installed in router 1.

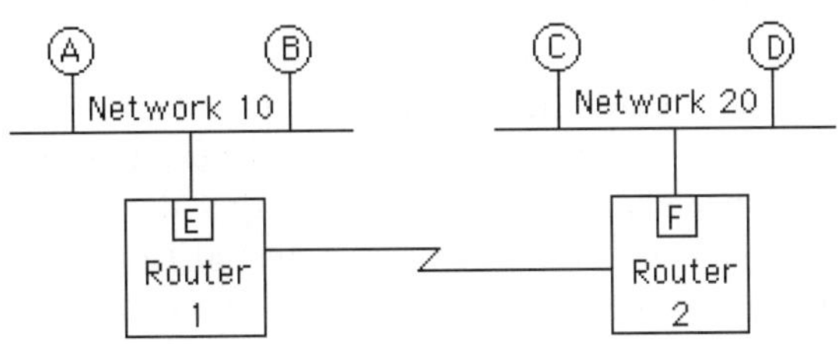

Figure 5.15 Routing between two networks. When a workstation on one network transmits data to a device on another network, an application program provides the destination network address which is used by routers as a basic for their routing decisions

In actuality, each device on a network can have a unique network address, with a portion of the address denoting the network and the remainder of the address denoting each host or router interface. This enables the application program on workstation A to transmit a packet destined to workstation C to router 1, with the data link layer passing the passing the packet to MAC address E as a frame or sequence of frames. The router receives the frame or sequence of frames explicitly addressed to it and notes that the packet has the destination address for network 20. The router then searches its routing tables and determines that it should relay the packet to router 2. When the packet is received at router 2, it recognizes the fact that network 20 is connected to the router via its connection to the LAN by an adapter card using MAC address F. Thus, the router will transport the packet to its destination using one or more frames at the data link layer with a MAC source address of F.

This represents a simplified description of the routing process. A more detailed description will be presented later in this section.

Network layer operations

The ability to operate at the network layer enables a router to extend internetworking across multiple data links in an orderly and predefined manner. This means that a router can determine the best path through a series of data links from a source network to a destination network. To accomplish this the router operates using a network protocol, such as the Internet Protocol (IP), Digital Equipment Corporation's DECnet Phase V, and Novell's IPX. This networking protocol must operate on both the source and destination networks when protocol-dependent routers are used. If protocol-independent routers are used you can interconnect networks using different protocols. The protocol-independent router can be considered as a sophisticated transparent bridge. Its operation and utilization are described later in this section. In comparison, since a bridge operates at the data link layer, it can always be used to transfer information between networks operating different network protocols. This makes a bridge more efficient for linking networks that have only one or a few paths, while a router is more efficient for interconnecting multiple network links via multiple paths.

Network address utilization

Unlike a bridge which must monitor all frames at the MAC layer, a router is specifically addressed at the network layer. This means that a router has to examine only frames explicitly addressed to that device. In communications terminology, the monitoring of all frames is referred to as a promiscuous mode of operation, while the selective examination of frames is referred to as a non-promiscuous mode of operation.

Another difference between the operation of bridges and routers is the structure of the addresses they operate upon. Bridges operate at the data link layer, which means that they typically examine physical addresses that are contained in read-only memory on adapter cards and used in the generation of frames. In comparison, routers operate at the network layer where addresses are normally assigned by a network administrator to a group of stations having a common characteristic, such as being connected on an Ethernet in one area of a building. This type of address is known as a logical address and can be assigned and modified by the network administrator.

Table operation

Similar to bridges, routers make forwarding decisions using tables. Unlike a bridge that may employ a simple table look-up procedure to determine if a destination address is on a particular network, a router may employ a much more sophisticated forwarding decision criterion. For example, a router may be configured to analyze several paths based upon an algorithm and to dynamically select a path based upon the results of the algorithm.

Advantages of use

The use of routers provides a number of significant advantages in comparison to the use of bridges. To illustrate those advantages we will examine the use of routers shown in Figure 5.16 in which four corporate offices containing seven local area networks are interconnected through the use of four routers. In this example networks A and B are located in a building in Los Angeles, networks C and D are located in New York, network E is located in Washington, DC, and networks F and G are located in Miami.

Multiple path transmission and routing control

Suppose a station on network A in Los Angeles requires transmission to a station on network G in Miami. Initially, router R1 might use the path R1–R4 to transmit data between networks. If the path should fail or if an excessive amount of traffic flows between Los Angeles and Miami using that path, router R1 can seek to establish other paths, such as R1–R3–R4 or even R1–R2–R3–R4. In fact, many routers will consider each packet as a separate entity, routing the packet to its destination over the best available path at the time of transmission. Although this could conceivably result in packets arriving at R4 out of sequence, routers have the ability to resequence packets into their original order prior to passing data onto the destination network.

Flow control

As data flows through multiple paths towards its destination it becomes possible for a link to become congested. For example, data from a station on network C and network E routed to network G might build up to the point where the path R3–R4 becomes congested. To eliminate the possibility of packet loss, routers will use flow control. That is, they will inhibit transmission onto a link as well as notify other routers to inhibit data flow until there is an available level of bandwidth for traffic.

Frame fragmentation

As previously mentioned, bridges cannot break a frame into a series of frames when transmission occurs between networks with different frame sizes. This situation requires workstations to be configured to use the smallest maximum frame size of

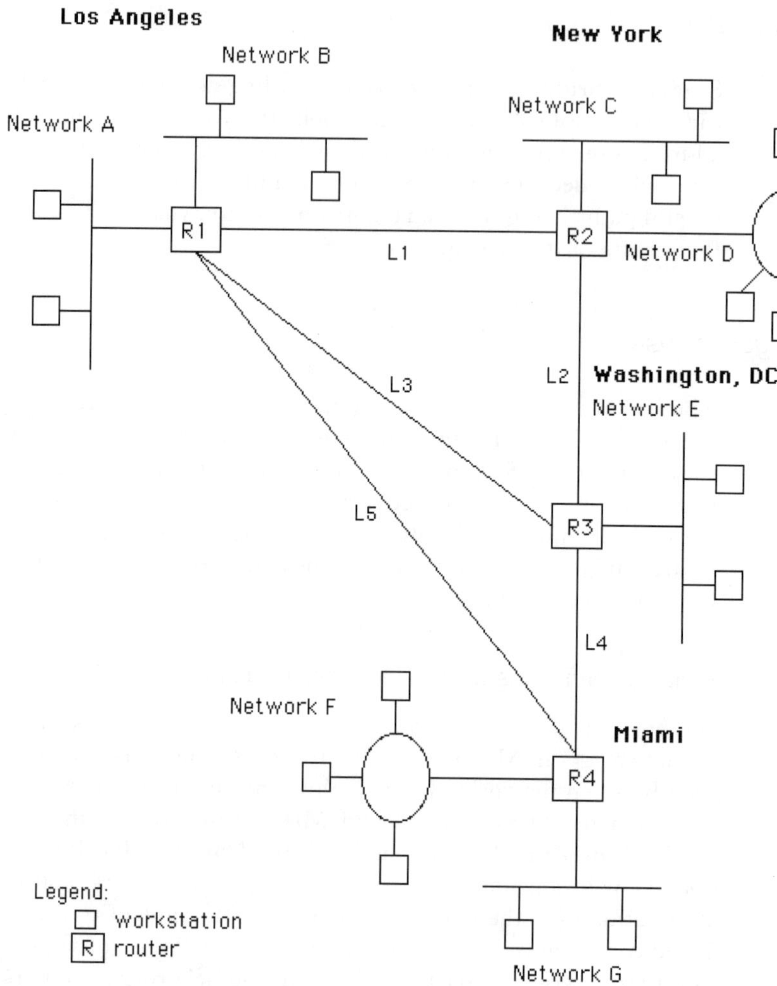

Figure 5.16 Building an internet using routers. Routers can be used to establish complex networks in which traffic is varied over network facilities based upon the operational status and utilization of different network paths

any network to be connected together. In comparison, most network protocols supported by routers include a provision for fragmentation of packets and their reassembly.

The higher level of functionality of routers over bridges is not without a price. That price is in terms of packet processing, software complexity, and cost. Since routers provide a more complex series of functions that bridges, their ability to process packets is typically one-half to two-thirds of the processing capability of bridges. In addition, the development time required to program a more complex series of functions adds to the cost of routers. Thus, routers are generally more expensive than bridges. Table 5.2 summarizes the major differences between bridges and routers in terms of their operation, functionality, complexity, and cost.

Table 5.2 Bridge/router comparison

Characteristic	Bridge	Router
Routing based upon an algorithm or protocol	Normally no	Yes
Protocol transparency	Yes	Only for protocol-independent routers
Uses network addresses	No	Yes
Promiscuous mode of operation	Yes	No
Forwarding decision	Elementary	Can be complex
Multiple path transmission	Limited	High
Routing control	Limited	High
Flow control	No	Yes
Frame fragmentation	No	Yes
Packet processing rate	High	Moderate
Cost	Less expensive	More expensive

Localizing broadcast storms

Two additional differences between bridges and routers concern their ability to localize broadcast storms and hardware configuration. Bridges will pass broadcast MAC layer frames from one network to another. In comparison, routers can be used to limit broadcasts to individual networks.

Hardware configuration

Concerning their hardware configuration, most bridges are PC based, resulting in their configuration flexibility being constrained by the number of system expansion slots in the system unit of the computer. In comparison, the design of many routers is based upon the use of a base chassis that can be significantly expanded to accommodate the addition of a large number of LAN and WAN modules. Thus, many routers can support a much wider range of LAN and WAN interfaces. For example, both Bay Networks and Cisco Systems market routers that can be configured to support 10BASE-T, 100BASE-T, Token-Ring, and FDDI LAN connections as well as T1, E1, ISDN, and the High Speed Serial Interface (HSSI) for WAN connections.

Based upon the preceding we can generalize the use of bridges for linking relatively homogeneous protocol networks where interconnected segments form one larger network. In comparison, the use of routers should be considered when we need to establish a hierarchical network structure, limit broadcast traffic between networks, and obtain an alternate routing capability that requires multiple WAN connectors.

IP support overview

The most popular network layer protocol supported by routers is the Network Protocol (IP) whose packet format was described in Chapter 4. Each IP network has a distinct network address and each interface on the network has a unique host address that represents the host portion of a 32-bit address. Since the IP address occurs at the network layer while frames that move data on a LAN use MAC addresses associated with the data link layer, a translation process is required to enable IP-compatible devices to use the transport services of a local area network. Thus, any discussion of how routers support IP requires an overview of the manner by which hosts use the services of a router.

When a host has a packet to transmit, it will first determine if the destination IP address is on the local network or a distant network, with the latter requiring the services of a router. To accomplish this, the host will use the subnet mask bits set in its configuration to determine if the destination is on the local network. For example, assume the subnet mask is 255.255.255.128. This means the mask extends the network portion of an IP address to 11111111.11111111.11111111.1, or 25 bit positions, resulting in seven (32 − 25) bit positions being available for the host address. This also means you can have two subnets, with subnet 0 containing host addresses 0 to 127 and subnet 1 having host addresses 128 to 255, the subnet being defined by the value of the 25th bit position in the IP address.

If we assume the base network IP address is 193.56.45.0, then the base network, the two subnets, and the subnet mask are as follows:

Base network:	11000001.00111000.00101101.00000000 = 193.56.45.0
Subnet 0:	11000001.00111000.00101101.00000000 = 193.56.45.0
Subnet 1:	11000001.00111000.00101101.10000000 = 193.56.45.128
Subnet mask:	11111111.11111111.11111111.10000000 = 193.56.45.128

Now suppose a host with the IP address 193.56.45.21 needs to send a packet to the host whose address is 193.56.45.131. By using the subnet mask, the transmitting host notes that the destination, while on the same network, is on a different subnet. Thus, the transmitting host will require the use of a router in the same manner as if the destination host was on a completely separate network.

Figure 5.17 illustrates the internal and external network view of the subnetted network. Note that from locations exterior to the network, routers forward packets

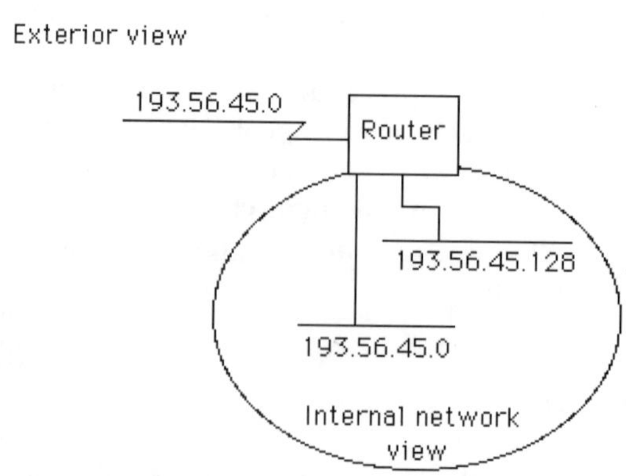

Figure 5.17 Using subnet masks to subdivide a common IP network address

to the router connecting the two subnets as if no subnetting existed. The corporate router is configured via the use of subnet masks to differentiate hosts on one subnet from those on the other subnet. From an interior view, packets originating on one subnet must use the resources of the router to reach hosts on the other subnet as well as hosts on other networks.

Once the transmitting host notes that the destination IP address is on either a different network or a different subnet, it must use the services of a router. Although each host will be configured with the IP address of the router, the host will transport packets via the data link layer, which requires knowledge of the 48-bit MAC address of the router port connected to the segment the transmitting host resides on.

ARP

The translation between IP and MAC addresses is accomplished by the use of the Address Resolution Protocol (ARP). To obtain the MAC address of the router's LAN interface the host will broadcast an ARP request. This request will be received by all stations on the segment, with the router recognizing its IP address and responding by transmitting an ARP response.

Since a continuous use of ARP would rapidly consume network bandwidth, hosts normally maintain the results of ARP requests in cache memory. Thus, once the relationship between an IP address and a MAC address is learned, subsequent requests to transmit additional packets to the same destination can be accomplished by the host checking its cache memory.

When packets arrive at the router destined for a host on one of the subnets, a similar process occurs. That is, the router must obtain the MAC addresses associated with the IP address to enable the packet to be transported by data link layer frames to its appropriate destination. Thus, in addition to being able to correctly support the transmission of packets from one interface to another, an IP-compatible router must also support the ARP protocol. Later in this section we will discuss and describe additional protocols routers can support.

Communications and routing protocols

For routers to be able to operate in a network they must normally be able to speak the same language at both the data link and network layers. This means that the routers must communicate with one another using the same route protocol as well as the same routing control protocol.

Routing protocols

Examples of route protocols include Transmission Control Protocol and the Network Protocol (TCP/IP), Xerox Network Services Network Transport Protocol (XNS), Digital Equipment Corporation's DECnet, Novell's IPX, and Apple

Computer's AppleTalk. Examples of routing protocols include the Routing Information Protocol (RIP), Open Shortest Path First (OSPF), Intermediate System to Intermediate System (IS–IS), and the Routing Table Maintenance Protocol (RTMP).

Handling non-routable protocols

Although many mainframe users consider IBM's System Network Architecture (SNA) as a router protocol, in actuality it is non-routable in the traditional sense of having network addresses. This means that for a router to support SNA or another non-routable protocol, such as NetBIOS, the router cannot compare a destination network address against the current network address as there are no network addresses to work with. Instead, the router must be capable of performing one or more special operations to handle non-routable protocols. For example, some routers may be configurable such that SNA addresses in terms of Physical Units (PUs) and Logical Units (LUs) can be associated with pseudo-network numbers, enabling the router to route an unroutable protocol. Another common method employed by some routers is to incorporate a non-routable protocol within a routable protocol, a technique referred to as tunneling. A third method and one considered by many to be the 'old reliable' mechanism is to use bridging. Later in this section, when we cover protocol-independent routers, we will describe methods that can be used to route non-routable protocols, to include SNA traffic.

Communications protocols

There are a wide variety of communications protocols in use today. Some of these protocols were designed specifically to operate on local area networks, such as Apple Computer's AppleTalk. Other protocols, such as X.25 and frame relay, were developed as wide area network protocols.

Table 5.3 lists 16 popular communications protocols. Readers are cautioned that many routers support only a subset of the protocols listed in Table 5.3.

Depending upon their support of communications protocols, routers can be classified into two classes—protocol dependent and protocol independent.

Table 5.3 Popular communications protocols

AppleTalk	ISO CLNS
Applo Domain VINES	HDLC
Banyan	Novell IPX
CHAOSnet	SDLC
DECnet Phase IV	TCP/IP
DECnet Phase V	Xerox XNS
DDN X.25	X.25
Frame Relay	Ungermann-Bass Net/One

Protocol-dependent routers

To understand the characteristics of a protocol-dependent router consider the network previously illustrated in Figure 5.16. If a workstation on network B wishes to transmit data to a second workstation on network F, router R1 must know that the second workstation resides on network F and the best path to reach that network. The method used to determine the network where the destination workstation resides determines the protocol dependency of the router.

If the workstation on network B tells router R1 the destination location, it must supply a network address in every LAN packet it transmits. This means that all routers in the network must support the protocol used on network B. Otherwise, workstations on network B could not communicate with workstations residing on other networks and vice versa.

NetWare IPX example

To illustrate the operation of a protocol-dependent router let us assume networks B and C use Novell's NetWare as their LAN operating system. The routing protocol used at the network layer between a workstation and a server is known as IPX. This protocol can also be used between servers as well as other protocols.

In comparison with IP where network addresses are assigned by a centralized administrative group (InterNIC), IPX network addresses are administrator assigned. A four-byte destination network and a six-byte destination node which represents the MAC address of the destination host function in a manner similar to an IP network and host address. However, since an IPX header includes the physical address of the destination network adapter card, there is no need to perform an ARP operation when a packet reaches the destination network. Instead, the router can use the transported destination MAC address to form an appropriate frame at the data link layer that will transport the IPX packet to its destination.

Figure 5.18(a) illustrates in simplified format the IPX packet composition for workstation 1 on network B transmitting data to workstation 2 on network C under Novell's NetWare IPX protocol.

After router R1 receives and examines the packet it notes that the destination address C requires the routing of the packet to router R2. Thus, it converts the first packet into a router (R1) to router (R2) packet as illustrated in Figure 5.18(b). At router R2 the packet is again examined. Router R2 notes that the destination network address C is connected to that router. Thus, router R2 reconverts the packet for delivery onto network C by converting the destination router address to a source router address and transmitting the packet onto network C. This is illustrated in Figure 5.18(c).

Addressing differences

In the preceding example note that each router uses the destination workstation and network addresses to transfer packets. If all protocols used the same format and addressing structure, routers would be protocol insensitive at the network layer. Unfortunately this is not true. For example, under TCP/IP addressing conventions

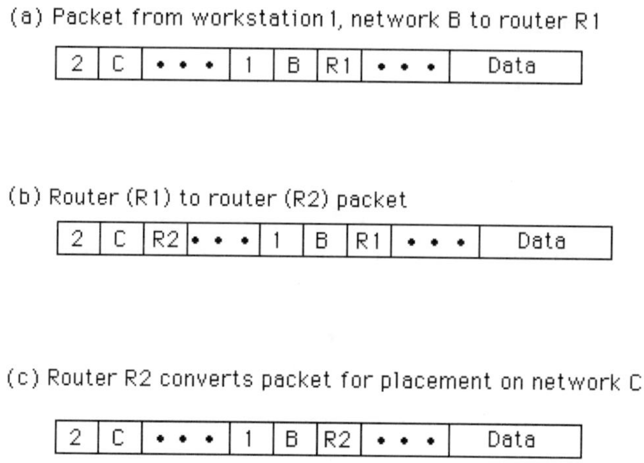

Figure 5.18 NetWare IPX routing

are very different from that used by NetWare. This means that networks using different operating systems require the use of multiprotocol routers that are configured to maintain multiple routing tables, examine each packet received to determine the network layer protocol, and use that information in conjunction with the destination address contained in the packet and its appropriate routing table to correctly forward the packet.

Non-routable protocol support

Another problem associated with protocol-dependent routers is the fact that some LAN protocols cannot be routed using that type of device. This is because some LAN protocols, such as NetBIOS and IBM's LAN Server, unlike NetWare, DECnet and TCP/IP, do not include routing information within a packet. Such protocols are restricted to using the physical addresses of adapter cards, such as Token-Ring source and destination addresses. Since a protocol-dependent router must know the network on which a destination address is located, it cannot, in a conventional sense, route such protocols. Thus, a logical question is how does a router interconnect networks using an IBM LAN protocol or similar non-routable protocol?

The answer to this question will depend upon the method used by the router manufacturer to support non-routable protocols. As previously discussed, such methods can include bridging, tunneling, or the configuration of a router that enables pseudo-network addresses to be assigned to each device.

Protocol-independent routers

A protocol-independent router can be considered to function as a sophisticated transparent bridge. That is, it addresses the problem of network protocols that do not have network addresses, by examining the source addresses on connected networks to automatically learn what devices are on each network. The protocol-independent router assigns network identifiers to each network whose operating system does not include network addresses in its network protocol. This activity enables both routable and non-routable protocols to be serviced by a protocol-dependent router.

In addition to automatically building address tables like a transparent bridge, a protocol-independent router exchanges information concerning its routing directions with other network routers. This enables each router to build a map of the interconnected networks. The method used to build the network map falls under the category of a link state routing protocol which is described later in this section.

Advantages

There are two key advantages associated with the use of protocol-independent routers: their ability to automatically learn network topology, and their ability to service non-routable protocols. The first of these can considerably simplify the administration of a network. For example, in a TCP/IP network each workstation has an IP address and must know the IP addresses of other LAN devices it desires to communicate with.

IP addresses are commonly assigned by a network administrator and must be changed if a workstation is moved to a different network or a network is segmented due to a high level of traffic or another reason. In such situations all LAN users must be notified about the new IP address or they will not be able to locate the moved workstation. Obviously, the movement of workstations within a building between different LANs could become a considerable administrative burden. In comparison, the ability of a protocol-dependent router to automatically learn addresses removes the administrative burden of notifying users of changes in network addresses.

An exception to the preceding occurs through the use of the Dynamic Host Configuration Protocol (DHCP). Through the use of a DHCP server and appropriate client software, stations are dynamically assigned IP addresses for a relatively short period of time. Once they complete an application the server can reassign the address to a new station. Although the use of the DHCP can ease the administrative burden of configuring and reconfiguring IP workstations, it requires the use of a server and client software. Thus, there continues to be no free lunch in networking.

The ability to route non-routable protocols can be of considerable assistance in integrating IBM System Network Architecture (SNA) networks into an enterprise network. Otherwise, without the use of protocol-independent routers organizations may have to maintain separate transmission facilities for SNA and LAN traffic.

Supporting SNA traffic

Figure 5.19 illustrates an example of the use of protocol-independent routers to support both inter-LAN and SNA traffic. In this example an IBM SNA network in which a 3174 control unit with a Token-Ring adapter (TRA) resides at a remote site

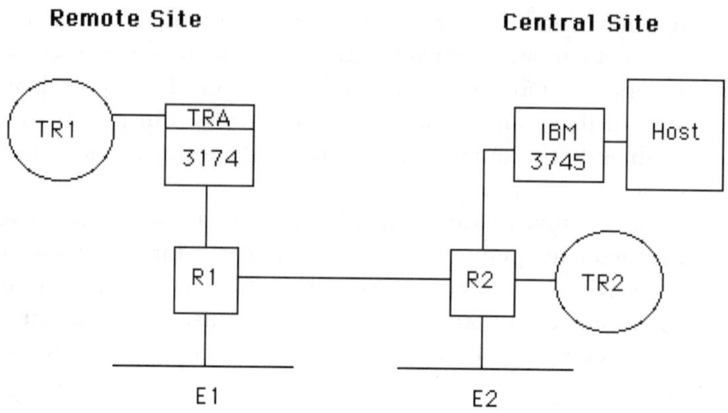

Figure 5.19 Supporting SNA traffic. A protocol-independent router can support SNA traffic as well as other LAN traffic over a common transmission facility

provides communications connectivity to an IBM 3745 front-end processor at a central site. Thus, routers must be capable of routing both SNA and LAN traffic to enable the use of a common transmission facility between the central and remote sites.

Methods to consider

There are essentially three methods by which SNA and LAN traffic can be combined for routing over a common network—encapsulation, conversion, or through protocol-independent routing. Under the encapsulation method, SNA packets are modified so that another protocol's header, addressing and trailer fields surround each SNA packet. For example, a TCP/IP protocol-dependent router would encapsulate SNA into TCP/IP packets for routing through a TCP/IP network. Since a TCP/IP packet has over 60 bytes of overhead while the average SNA packet is 30 bytes in length, encapsulation can considerably reduce performance when transmission occurs over low-speed links. A second disadvantage of encapsulation is that it requires the existence of a corporate network using the encapsulation protocol. Otherwise, you would have to build this network to obtain an encapsulation capability.

The second method used for integrating SNA traffic with LAN traffic occurs through the use of protocol conversion. This technique eliminates the need for adding network headers and enhances the efficiency of the protocol integration efforts.

The third method by which an SNA network can be integrated with LAN traffic is through protocol-independent routing. Protocol-independent routers would assign a LAN device address to each SNA control unit and front-end processor. Then, SNA packets would be prefixed with source and destination addresses to permit their routing through the network. At the destination router the addressing information is removed and SNA packets are delivered to their destination in their original form. Since the addition of source and destination addresses adds significantly fever bytes than an encapsulation process, overhead is reduced in comparison to encapsulation. This in turn enables lower-speed links to be used to interconnect locations. Readers are referred to Section 5.3 for additional information concerning the use of gateways to integrate SNA and other protocols into a common network infrastructure.

Types of routing protocols

The routing protocol is the key element which enables the transfer of information across a network in an orderly manner. The protocol is responsible for developing paths between routers which is accomplished by a predefined mechanism by which routers exchange routing information. There are two basic types of routing protocols—vector distance and link state.

Vector distance protocol

A vector distance protocol constructs a routing table in each router and periodically broadcasts the contents of the routing table across the network. When the routing table is received at another router, that device examines the set of reported network destinations and the distance to each destination. The receiving router then determines if it knows a shorter route to a network destination, finds a destination it does not have in its routing table, or finds a route to a destination through the sending router where the distance to the destination changed. If any one of these situations occurs, the receiving router will change its routing tables.

The term vector distance relates to the information transmitted by routers. Each router message contains a list of pairs known as vector and distance. The vector identifies a network destination, while the distance is the distance in hops from the router to that destination.

Figure 5.20 illustrates the initial distance vector routing table for routers R1 and R2 previously illustrated in Figure 5.16. Each table contains an entry for each directly connected network and is broadcast periodically throughout the network. Here the distance column indicates the distance to each network from the router in hops.

(a) Router R1

Destination	Distance
E1	0
E2	0

(b) Router R2

Destination	Distance
E3	0
TR1	0

Figure 5.20 Initial distance vector routing tables

At the same time router R1 is constructing its initial distance vector table other routers are performing a similar operation. Figure 5.20(b) illustrates the composition of the initial distance vector table for router R2.

As previously mentioned, under a distance vector protocol the contents of each router's routing table are periodically broadcast. Assuming routers R1 and R2 broadcast their initial distance vector routing tables, each router uses the received routing table to update its initial routing table. Figure 5.21 illustrates the result of this initial routing table update process for routers R1 and R2.

(a) Router R1

Destination	Distance	Route
E1	0	direct
E2	0	direct
E3	1	R2
TR1	1	R2

(b) Router R2

Destination	Distance	Route
E1	1	R2
E2	1	R2
E3	0	direct
TR1	0	direct

Figure 5.21 Initial routing table update

As additional routing tables are exchanged the routing table in each router will converge with respect to the internetwork topology. However, to ensure each router knows the state of all links, routing tables are periodically broadcast by each router. Although this process has a minimal effect upon small networks, its use with large networks can significantly reduce available bandwidth for actual data transfer. This is because the transmission of lengthy router tables will require additional transmission time in which data cannot flow between routers.

Routing Information Protocol

One example of a popular vector distance routing protocol is the TCP/IP Routing Information Protocol (RIP). Under RIP participants are either active or passive. Active participants are normally routers which transmit their routing tables, while passive devices listen and update their routing tables based upon information supplied by other devices. Normally host computers operate as passive participants, while routers operate as active participants.

Under RIP an active router broadcasts its routing table every 30 seconds. Each routing table entry contains a network address and the hop count to the network.

To illustrate an example of the operation of RIP, let's draw a network in terms of its links and nodes, replacing the four routers by the letters A, B, C and D for

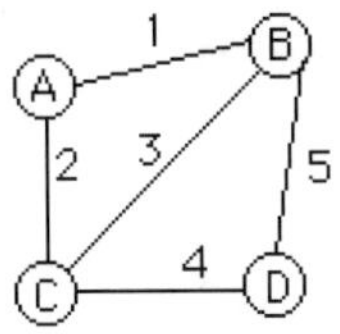

Figure 5.22 Redrawing the network in Figure 5.10 in terms of its links and nodes

simplicity of illustration. Figure 5.22 contains the revised network consisting of four nodes and five links.

When the routers are powered up they only have knowledge of their local conditions. Thus, each routing table would contain a single entry. For example, the table of router n would have the following value:

From *n* to	Link	Hop count
n	local	0

For the router represented by node A, its table would then become:

From A to	Link	Hop count
A	local	0

Thirty seconds after being turned on, node A will broadcast its distance vector (A = 0) to all its neighbors, which in Figure 5.22 are nodes B and C. Node B receives on link 1 the distance vector A = 0. Upon receipt of this message, it updates its routing table as follows, adding one to the hop count associated with the distance vector supplied by node 1:

From B to	Link	Hop count
B	Local	0
A	1	1

Node B can now prepare its own distance vector (B = 0, A = 1) and transmit that information on its connections (links 1, 3 and 5).

During the preceding period node C would have received the initial distance vector transmission from node A. Thus, node C would have updated its routing table as follows:

From C to	Link	Hop count
C	Local	0
A	2	1

Once it updates its routing table, node C will then transmit its distance vector (C = 0, A = 1) on links 2, 3 and 4.

Assuming the distance vector from node B is now received at nodes A and C, each will update their routing tables. Thus, their routing tables would appear as follows:

From A to	Link	Hop count
A	Local	0
B	1	1

From C to	Link	Hop count
C	Local	0
A	2	1
B	3	1

At node D, the initial state is first modified when it receives the distance vector (B = 0, A = 1) from node B. Since D received that information on link 5, it updates its routing table as follows, adding one to each received hop count:

From D to	Link	Hop count
D	Local	0
B	5	1
A	5	2

Now, let's assume node D receives the update of node C's recent update (C = 0, A = 1, B = 1) on link 4. As it does not have an entry for node C, it will add it to its routing table by entering C = 1 for link 4. When it adds 1 to the hop count for A received on link 4, it notes that the value is equal to the current hop count for A in its routing table. Thus, it discards the information about node A received from node C. The exception to this would be if the router maintained alternate routing entries to use in the event of a link failure. Next, node D would operate upon the vector B = 1 received on link 4, adding one to the hop count to obtain B = 2. Since that is more hops than the current entry, it would discard the received distance vector. Thus, D's routing table would appear as follows:

From D to	Link	Hop count
D	Local	0
C	4	1
B	5	1
A	5	2

The preceding example provides a general indication of how RIP enables nodes to learn the topology of a network. In addition, if a link should fail, the condition can be easily compensated for as, similar to bridge table entries, those of routers are also time-stamped and the periodic transmission of distance vector information would result in a new route replacing the previously computed one.

Link state protocol

A link state routing protocol addresses the traffic problem associated with large networks that use a vector distance routing protocol. It does this by transmitting routing information only when there is a change in one of its links. A second difference between vector distance and link state protocols concerns the manner in which a route is selected when multiple routes are available between destinations. For a vector distance protocol the best path is the one that has fewest intermediate routers or hops between destinations. In comparison, a link state protocol can use multiple paths to provide traffic balancing between locations. In addition, a link state protocol permits routing to occur based upon link delay, capacity and reliability. This provides the network manager with the ability to specify a variety of route development situations.

SPF algorithms

Link state routing protocols are implemented through the use of a class of algorithms known as Shortest Path First (SPF). Unfortunately, the name associated with this class of algorithms is a misnomer as routing is not based upon the shortest path.

The use of SPF algorithms requires each participating router to have complete knowledge of the network topology. Each router participating in an SPF algorithm then performs two tasks—status testing of neighboring routers and periodically transmitting link status information to other routers.

To test neighboring routers a short message is periodically transmitted. If the neighbor replies, the link is considered to be up. Otherwise, the absence of a reply after a predefined period of time indicates that the link is down.

To provide link status information each router will periodically broadcast a message which indicates the status of each of its links. Unlike the vector distance protocol in which routes are specified, an SPF link status message simply indicates whether or not communications are possible between pairs of routers. Using information in the link status message, routers are able to update their network map.

In comparison to vector distance protocols in which tables are required to be exchanged, link state protocols such as SPF algorithms exchange a much lower volume of information in the form of link status queries and replies. Then, SPF participating routers simply broadcast a status of each of its links that other routers use to update their network map. This routing technique permits each router to compute routes independently of other routers and eliminates the potential for table flooding that can occur when a vector state protocol is used to interconnect a large number of networks.

Operation example

To illustrate the operation of a link state routing protocol let us return to the internetwork configuration previously illustrated in Figure 5.16. Figure 5.23 indicates the initial network map for router R1. This map lists the destination of all networks on the internet from router R1, their distance and route. Note that if multiple routes exist to a destination each route is listed, as doing so defines a complete network topology as well as allowing alternate routes to be selected if link status information indicates that one or more routes cannot be used.

Let us assume that at a particular point in time the link status messages generated by the routers in the internet are as indicated in Figure 5.24. As indicated in Figure 5.16, L1 represents the link connecting routers R1 and R2, L2 connects routers R2 and R3, L3 connects routers R1 and R3, L4 connects routers R3 and R4, while L5 connects routers R1 and R4. Note that both routers R2 and R3 determined that link L2 is down. Using this information Router R1 would then update the status column for its network map. Since link L2 is down, all routes that require a data

Destination	Distance	Route	Status
A	0	direct	up
B	0	direct	up
C	1	R2	up
C	2	R3-R2	up
C	3	R4-R3-R2	up
D	1	R2	up
D	2	R3-R2	up
D	3	R4-R3-R2	up
E	1	R3	up
E	2	R2-R3	up
E	2	R4-R3	up
F	1	R4	up
F	2	R3-R4	up
F	3	R2-R3-R4	up
G	1	R4	up
G	2	R3-R4	up
G	3	R2-R3-R4	up

Figure 5.23 Router R1 initial link state network map

R1 link status

Link	Status
L1	Up
L3	Up
L5	Up

R3 link status

Link	Status
L2	Down
L3	Up
L4	Up

R2 link status

Link	Status
L1	Up
L2	Down

R4 link status

Link	Status
L4	Up
L5	Up

Figure 5.24 Link status messages

flow between R2 and R3 would have their status changed to down. For example, destination C via route R3, R2 would have its status changed to down. Since the minimum distance to network C is 1 hop via router R2, the failure of link L2 would not affect data flow from router R1 to network C. Now consider the effect of link L1 becoming inoperative. That would affect route R2 which has the minimum distance to network C. Of course, when a new link status message indicates that a previously declared down link is up, each router's network map would be updated accordingly.

5.3 GATEWAYS

The well-known phrase 'one person's passion is another person's poison' can in many ways apply to gateways. The term gateway was originally coined to reference a device which provides a communications path between two local area networks. That term is still commonly used to reference a device interconnecting networks, and many programs used to configure workstation and server network operations request the entry of a 'gateway' network address, even though it is more appropriate to use the term router. A second popular use of the term gateway is to reference a device which enables users to access applications. Since the first type of application gateway developed was designed to facilitate access to a mainframe, many persons equate gateways with mainframe access although this is not necessarily always true. For example, you can have a gateway that provides access to an electronic mail application residing on a server. In this case some vendors market such products as email gateways.

In this section we will focus our attention upon application gateways and not routers, even though the latter are still referenced as gateways. To better distinguish between the two, the gateways we will examine operate at all seven layers of the OSI Reference Model, and perform any protocol conversion required to enable an application to be accessed.

To illustrate the functionality of an application gateway, consider one of the most common types of such devices, an electronic mail gateway. An electronic mail (email) gateway converts documents from one email format into another. For example, an internal corporate network might be operating Lotus Development Corporation's cc:MAIL. If they require connectivity with the Internet, the gateway must convert internal cc:MAIL documents so that they can be transported via the Simple Mail Transport Protocol (SMTP) which is a TCP/IP electronic mail application. Similarly, SMTP mail delivered via the Internet to the organization's gateway must be converted by the gateway into the format used by cc:MAIL.

Overview

Gateways are protocol specific in function, typically used to provide access to a mainframe computer. Some vendors manufacture multiprotocol gateways. Such products are normally manufactured as adapter cards containing separate processors that are installed in the system unit of a personal computer or a specially designed vendor hardware platform. When used in conjunction with appropriate

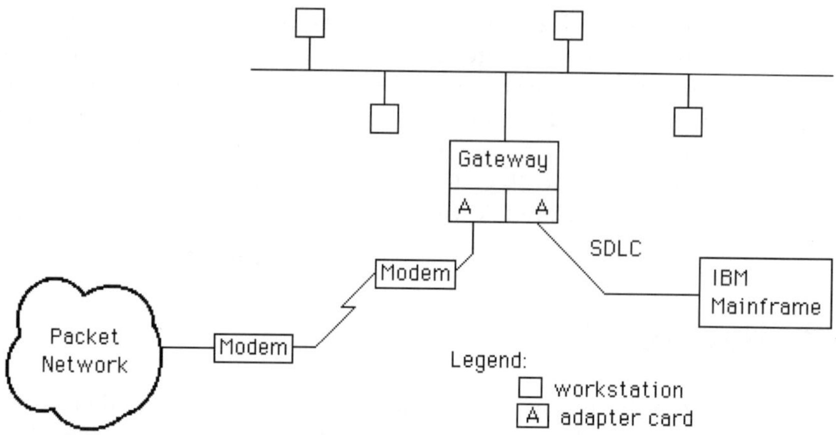

Figure 5.25 Multiprotocol gateway operation. A multiprotocol gateway can be used to provide local area network stations access to different computational facilities, either directly or through the use of a packet network

vendor software, this type of gateway is actually an N-in-1 gateway, where N references the number of protocol conversions and separate connections the gateway can perform.

Figure 5.25 illustrates the use of a multiprotocol gateway to link LAN stations to an IBM mainframe via an SDLC link and via an X.25 connection to a packet switching network. Once connected to the packet switching network, LAN traffic may be further converted by gateway facilities built into that network, or traffic may be routed to a packet network node and transmitted from that node to its final destination in its X.25 packet format.

Gateways are primarily designed and used for LAN–WAN connections and not for inter-LAN communications. Due to the more sophisticated functions performed by gateways, they are slower than routers in providing network throughput. In addition, due to the large number of protocol options that may require consideration when configuring a gateway, its installation can be considerably more difficult than the setup of a router.

Mainframe access

Since the mainframe market is dominated by IBM which has a market share in excess of 70%, any discussion of the use of gateways to access mainframes will be oriented towards IBM mainframe access. Thus, our discussion of gateway methods to access mainframes will be similarly oriented.

Control unit connectivity

One of the earliest methods of providing LAN users with access to an IBM mainframe was through the use of the IBM 3174 Subsystem Control Unit.

Introduced in late 1986, the 3174 can be used to connect dumb terminals and PCs using emulation boards via coaxial cable, shielded twisted-pair wire, and telephone-type or unshielded twisted-pair wire. One of the more interesting features of the 3174 is its ability to be connected to IBM's Token-Ring local area network. This is accomplished through the use of a Token-Ring adapter card installed in a 3174 expansion slot, converting the 3174 into an active participant on the Token-Ring network.

When used with a Token-Ring adapter (TRA) the Token-Ring in a 3174 is limited to supporting up to 140 downstream Physical Units (PUs). Here the PU can be considered as a gateway PC, with the software on each gateway determining the number of LUs supported on the Token-Ring. For example, Novell NetWare gateway software can be obtained to provide support for 16, 32 or 97 LUs. In this type of networking configuration, which is illustrated in Figure 5.26, the communications controller polls the control unit, the control unit polls individual PU gateways, and each gateway is responsible for polling the LUs serviced by the gateway. In this example the gateway PC is known as a downstream PU, which polls downstream LUs.

Since SNA's architecture can be considered as a polling structure it is wise to limit the number of LUs on a gateway, otherwise the polling time from the gateway PC to the LUs representing different sessions on each workstation or the screen and printer of a workstation can result in excessive delays. Thus, a good rule of

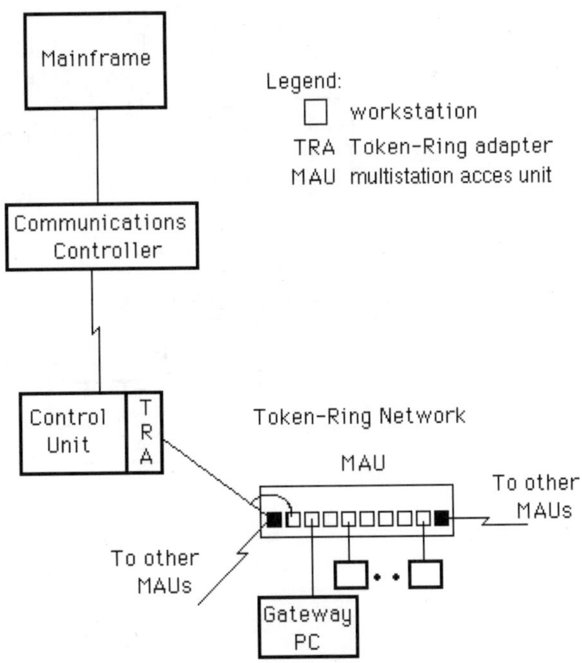

Figure 5.26 Using a control unit Token-Ring Adaptor (TRA) to interconnect a Token-Ring network. Through the installation of a TRA, both local and remote control units provide a connection capability to a Token-Ring network

thumb is to add another downstream PU in the form of an additional gateway for every 32 workstations on a local area network.

Ethernet connectivity

In Figure 5.26 we illustrated how a Token-Ring network could be connected to an SNA network through a 3174 control unit Token-Ring adapter. A similar mixture of hardware and software can be used to connect an Ethernet/IEEE 802.3 network to an SNA network. By examining how this is accomplished we can note some of the numerous vendor gateway construction options available for connecting different types of LANs into an SNA network.

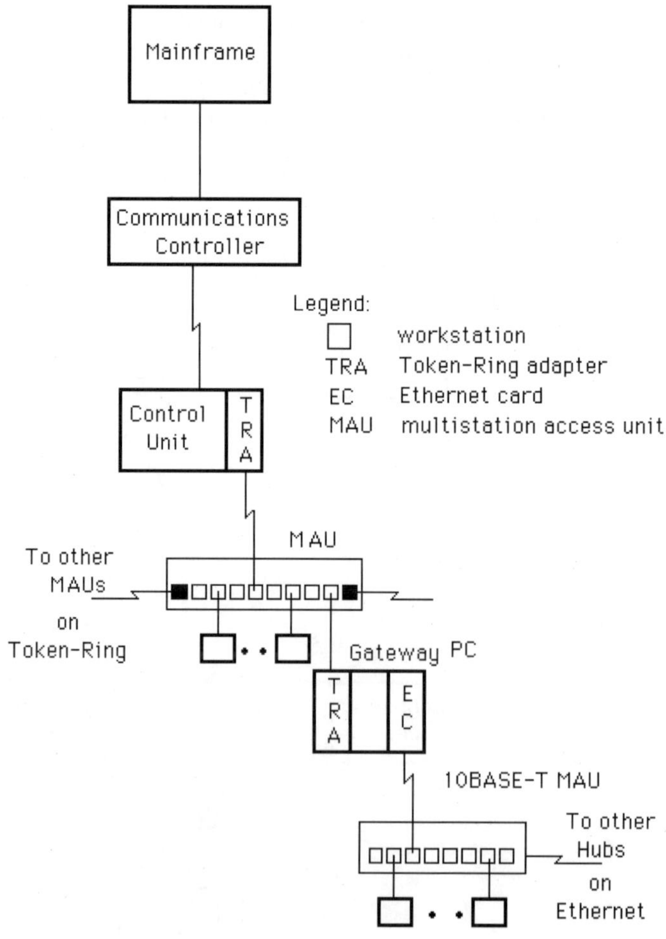

Figure 5.27 Using a control unit TRA to interconnect an Ethernet network. Through the installation of a Token-Ring adapter and an Ethernet card and appropriate software, the gateway PC provides a connection from a 10BASE-T network to the mainframe via a Token-Ring network

Figure 5.27 illustrates the use of a gateway PC to connect an Ethernet 10BASE-T network to an SNA network. In this example the gateway PC contains a Token-Ring adapter card for connection to a Multistation Access Unit (MAU) and an Ethernet card for connection to a 10BASE-T wire hub. Although it may appear that the gateway functions as a bridge, it operates at a much higher level, although it does transfer information between the Token-Ring and the Ethernet network. If the Ethernet is operating Novell NetWare, the gateway PC translates LUs into IPX addresses and vice versa. This address conversion is based upon configuring the gateway PC software when the gateway is installed. Since LU assignments are defined in the NCP in the communications controller, this means that the gateway installation process must occur in close coordination with NCP programmers that encode the NCP to recognize the gateway as a downstream PU with LUs assigned to that PU. In addition, to enable workstations on the Ethernet to gain full screen access to mainframe applications, each workstation desiring such access must execute an appropriate terminal emulation program.

In the example illustrated in Figure 5.27 one vendor markets a gateway that can be configured to operate as one to eight separate PUs, with each PU capable of supporting up to 32 LUs. Since two LUs on one PU are reserved, that vendor's product provides the ability to support up to 254 LUs. This provides the servicing of up to 127 workstations that use separate LUs for screen and printer, or for a lesser number of workstations that require the ability to execute multiple SNA sessions in the form of additional LUs.

Alternative gateway methods

The Token-Ring and Ethernet gateways discussed above use the services of a TRA on a control unit, requiring the physical presence of a control unit to interconnect to an SNA network. Recognizing that the additional hardware cost of a control unit can limit the effectiveness of that gateway method, several vendors, including IBM and third-party vendors such as Eicon Technology of Montreal, Canada, market alternative solutions which enable different types of LANs to access SNA networks over WAN facilities. Thus, let's look at a few of those alternative solutions.

SDLC connectivity

An SDLC gateway consists of a pair of adapter cards installed in a personal computer and appropriate software that performs the required conversion from the packet format used on the local area network to an SNA data stream. One card is an SDLC adapter which provides the framing for the bit-oriented protocol used by SNA. In actuality, most vendor SDLC gateways include SNA functions in Read Only Memory (ROM) on the adapter card, which makes the card function as if it was a series of 3174 or 3274 control units (multiple PUs) with each PU associated with a group of LUs. This second adapter is typically a Token-Ring or Ethernet adapter, used to connect the gateway to either a Token-Ring or an Ethernet local area network.

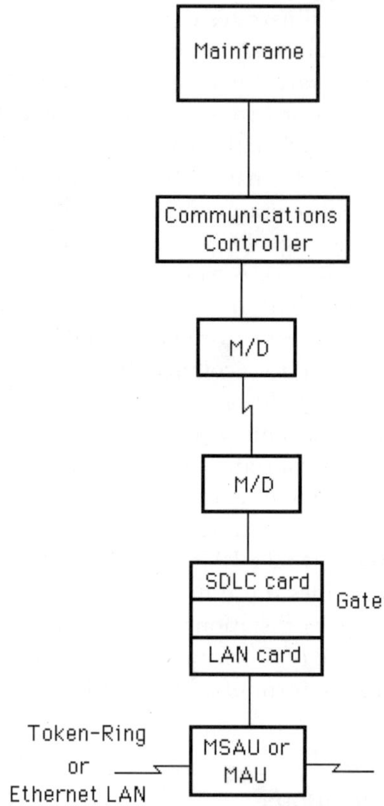

Legend:

M/D modem or data service unit
MSAU multistation access unit
MAU media access unit

Figure 5.28 Using an SDLC gateway. An SDLC gateway provides access to an SNA network by providing communications between a LAN and a communications controller

Figure 5.28 illustrates the use of an SDLC gateway to obtain access to a communications controller via a wide area network or an extended distance cable. Concerning the latter, the use of an SDLC gateway may permit the connection of a LAN to a communications controller via a lengthy cable within a building. This may be more attractive than the use of the gateway via a control unit TRA whose cable distance is limited to LAN cabling restrictions.

Both RS-232 and V.35 connectors can be obtained with most SDLC adapters. The use of an RS-232 connector limits the SDLC transmission rate to 19.2 kbps while the use of a V.35 connector enables digital transmission facilities at 56 kbps to be used to connect the gateway to the communications controller.

Some SDLC gateways are limited to supporting one PU and 32 LUs. Other SDLC gateways considerably expand upon that basic level of support. For example, an Eicon SNA gateway product which uses SDLC connectivity to a communications controller supports 32 PUs and up to 254 sessions. In addition, an Eicon gateway can be configured using up to four cards which results in a total of 128 PUs and 1016 sessions that can be supported by one gateway PC. Of course, limiting the transmission to either 19.2 kbps or 56 kbps per card may severely restrict LAN performance when accessing the mainframe.

X.25 connectivity

The ability of an IBM mainframe to support X.25 requires an IBM software program known as NCP Packet Switching Interface (NPSI) to be obtained and loaded as an NCP module in a communications controller. Through the use of NPSI an IBM communications controller, such as the 3745, can be directly connected to a packet switching network. This in turn enables any terminal device capable of supporting the X.25 protocol to communicate with the communications controller via the use of a packet network's transmission facilities.

To take advantage of NPSI and the use of a packet switching network as a data transport mechanism, several vendors introduced X.25 gateways. This type of gateway is also constructed through the use of adapter cards installed in the system unit of a personal computer. Typically, two or three adapter cards may be required, with the actual number dependent upon a vendor's use of ROM versus loadable software. One adapter card provides the connection to the local area network while a second adapter card packetizes data for transmission on the packet network. Either a third card or perhaps loadable software makes the gateway PC function as a downstream PU with LUs, in effect functioning as a control unit. Data from the local area network is converted into a 3270 format and encapsulated into an X.25 data stream by the gateway personal computer. That data is routed through the packet network to the IBM communications controller, where the NPSI port converts the datastream back into the 3270 format created by the gateway PC. Figure 5.29 illustrates the operation of an X.25 gateway.

Similar to an SDLC gateway, the major constraint of an X.25 gateway concerns its operating rate, throughput, and PU and LU support. The operating rate of an X.25 card is limited to 56 kbps, though its, throughput may be considerably less than that of an SDLC gateway operating at the same data rate. This is because the X.25 gateway not only packetizes data adding additional overhead but in addition has its data checked as it flows through the packet network. The additional delays due to the error checking performed at packet switches can add between 0.25 and 0.5 seconds to the response time of network users in comparison to the use of an SDLC gateway. However, one NPSI port can provide support for more than one X.25 gateway connected via a packet network to the mainframe location. In comparison, each SDLC gateway requires the use of a separate SDLC port on the communications controller, making the decision between the use of SDLC and X.25 gateways consider the cost of public network usage charges versus building a network as well as a trade-off between performance and hardware cost in the form of additional communications controller ports. Since an X.25 gateway can be considered to encapsulate the functions of an SDLC gateway into X.25 packets, it

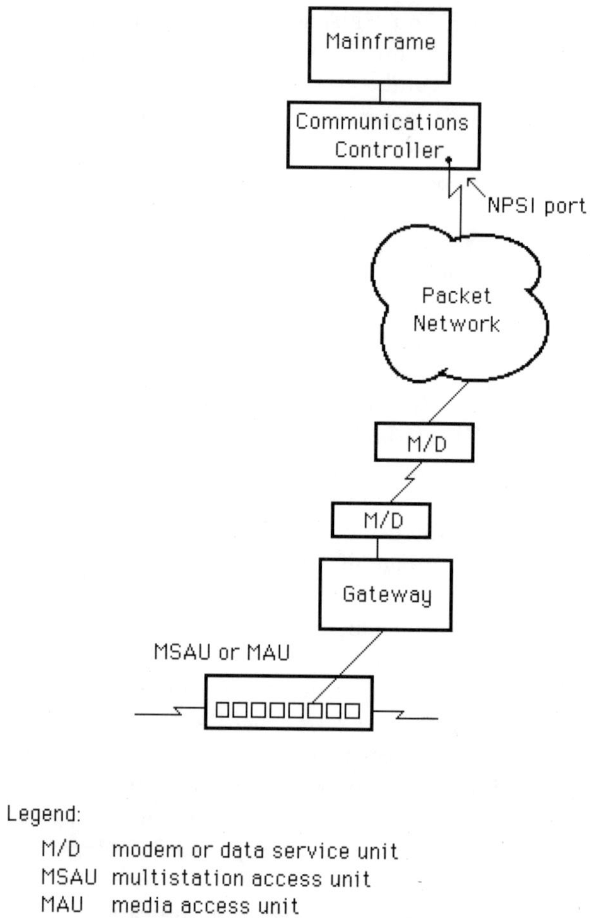

Figure 5.29 X.25 gateway operation. An X.25 gateway converts LAN packets into a 3270 data stream and encapsulates the data into X.25 packets for transmission through a packet network

should come as no surprise that their PU and LU support should be the same. This appears to be true for all vendor products examined by the author.

The TIC connection

The use of a Token Interface Coupler (TIC) installed in a communications controller provides a gateway access method very similar to a control unit with a TRA. That is, the TIC is cabled to a MAU port and one or more gateway PCs are also cabled to an MAU port.

The primary differences between the use of a communications controller TIC and a control unit TRA are in the areas of cost, network interconnection distance, operating rate and PU and LU support.

A TIC can cost well over \$5000. In comparison a TRA costs under \$1000. Concerning network interconnection distance, the use of a TIC restricts access to the LAN cabling distance, since the communications controller must be cabled to an MAU under lobe length distance restrictions. In comparison, a local control unit can be connected via coaxial cable to a communications controller, which enables the gateway distance to be extended. If a remote control unit is used, that device functions similar to a remote communications controller since both the TRA and the TIC cabling distances to a MAU are governed by lobe distance restrictions.

In a local environment the communications controller and control unit are both channel-attached to a host computer, and their data transfer capabilities are similar. In a remote environment the transmission rate of a control unit is restricted to a maximum of 56 kbps. In comparison, a remote communications controller can operate at T1/E1 data rates, providing over 20 times the data transfer capability of a control unit. Thus, the TIC can provide a higher level of throughput when used at a remote location.

The biggest difference between the use of a TRA and a TIC is in the area of PU and LU support. The NCP on a communications controller can support up to 9999 PUs per TIC. Then, each gateway PC functioning as a PU will support a grouping of LUs based upon the gateway software used. Another key difference between the use of a TRA and a TIC concerns the method of gateway communications.

When a control unit TRA provides a connection to the Token-Ring network the communications controller polls the control unit and the control unit polls each downstream PU, with each gateway polling its LUs. When a TIC is used each downstream PU requests service from the TIC by using a 'dial-up' service when it requires service. Thus, this can considerably reduce the polls flowing on an attached local area network and results in the TIC being able to theoretically support up to 9999 PUs.

3278/9 coaxial connection

A rather outdated and limited function gateway is based upon the use of a 3278/9 coaxial adapter card. Instead of emulating a 3X74 control unit like SDLC and X.25 gateways, the coaxial adapter permits a gateway PC to be connected to a port on a 3X74 control unit. That port can be configured as a distributed function terminal (DFT) port. When used in this manner the DFT port provides access to five sessions as it represents five LUs. Gateway software then divides the five SNA mainframe sessions among contending workstations on the local area network. This means that a coaxial adapter-based gateway is limited to providing a maximum of five simultaneous host sessions. Similar to the other gateways described in this section, a Token-Ring or Ethernet adapter card would be installed in the gateway to provide a connection to the local area network. Figure 5.30 illustrates the hardware used to provide a 3278/9 coaxial cable gateway.

Although a coaxially connected gateway is limited in its session support, it operates at coaxial cable data transfer rates to the control unit. If a local control unit is used the operating rate of a coaxially connected gateway can approach 2 Mbps. In comparison, SDLC and X.25 type gateways are limited to a 56 kbps data transfer rate. Thus, coaxially connected gateways can provide a high level of SNA access performance for small local area networks when such networks are at the mainframe location. In addition, this method eliminates the necessity to obtain a TIC or

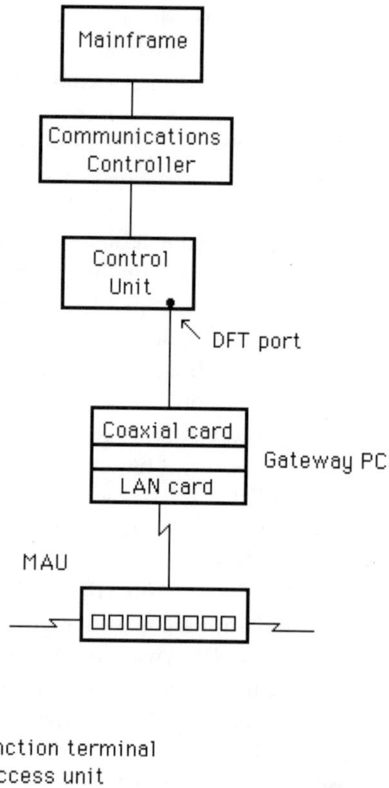

Figure 5.30 3278/9 coaxial connectivity. Through the use of a 3278/9 coaxial adapter card, a LAN card, and a DFT port on a control unit, up to five LAN workstations can simultaneously access an SNA network

TRA and can be used with older 3274 control units that cannot support the installation of a Token-Ring adapter. Thus, the coaxially connected gateway also represents the lowest cost gateway.

Using the 3172 Interconnect Controller

The IBM 3172 Interconnect Controller can be considered to represent the 'Swiss Army knife' equivalent of gateway computers. The 3172 can be directly channel-attached to S/370 and S/390 computers via either a parallel or an ESCON channel, and supports the connection of a variety of LAN and WAN connections as well as a mixture of protocols that can run over those connections. Those protocols include TCP/IP and IPX/SPX which enable the 3172 to provide an interconnection to the Internet and Novell-based LANs.

Figure 5.31 illustrates how a 3172 Interconnect Controller can be used as a gateway between an IBM host and the Internet. In this example, the 3172 Interconnect Controller enables TCP/IP communications to flow on the Token-

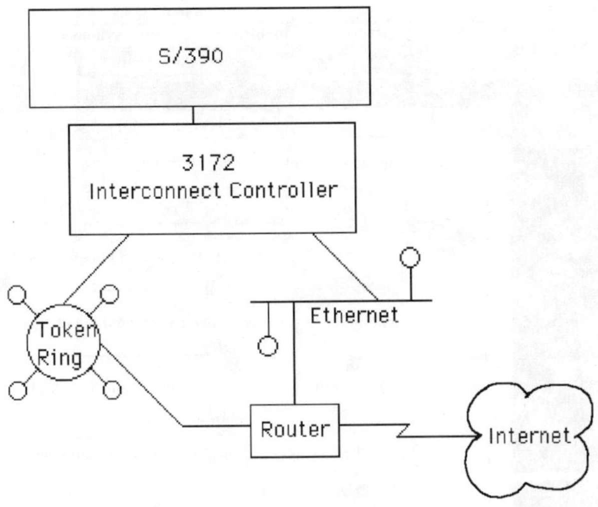

Figure 5.31 Using the 3172 Interconnect Controller

Ring and Ethernet networks shown connected to the 3172. This in turn enables the router- based connections to the Internet to reach the mainframe via the 3172.

Software considerations

Although the 3172 and other gateway products provide users with the ability to route data between devices that communicate using different protocols, you must also consider the software on the end station accessing the host to obtain a full gateway capability. IBM's 3270 Information Display system uses special codes to enable and disable a variety of terminal features associated with that vendor's original series of fixed logic terminal devices. As PCs began to replace dumb terminals, emulation programs were required to enable PC-based LAN workstations to obtain keyboard and screen operation compatibility with IBM hosts. One of the most popular IBM 3270 display station emulation programs is the PCOM/3270, an acronym for IBM's Personal Communications/3270 series of emulation programs.

If the 3172 can be considered as the Swiss Army knife of gateways, the PCOM/3270 can be considered to represent the Swiss Army knife of emulation programs. PCOM/3270 supports a variety of mainframe attachment methods, ranging from LAN adapters and coaxial cable to asynchronous and synchronous serial communications. Figure 5.32 illustrates the PCOM/3270 Customize Communication display screen which shows four options for communicating from a coaxial-based PC to a S/390 host. Note that the entry 'COAX' in the column labeled 'Adapter:' is shown selected, which results in the display of four attachment methods shown in the middle of the illustration. In comparison, the use of a LAN adapter card provides a terminal emulation capability for a larger number of attachment options. Figure 5.33 illustrates the PCOM/3270 configuration display after the LAN option was selected for the adapter card. In this example, the highlighted bar is shown

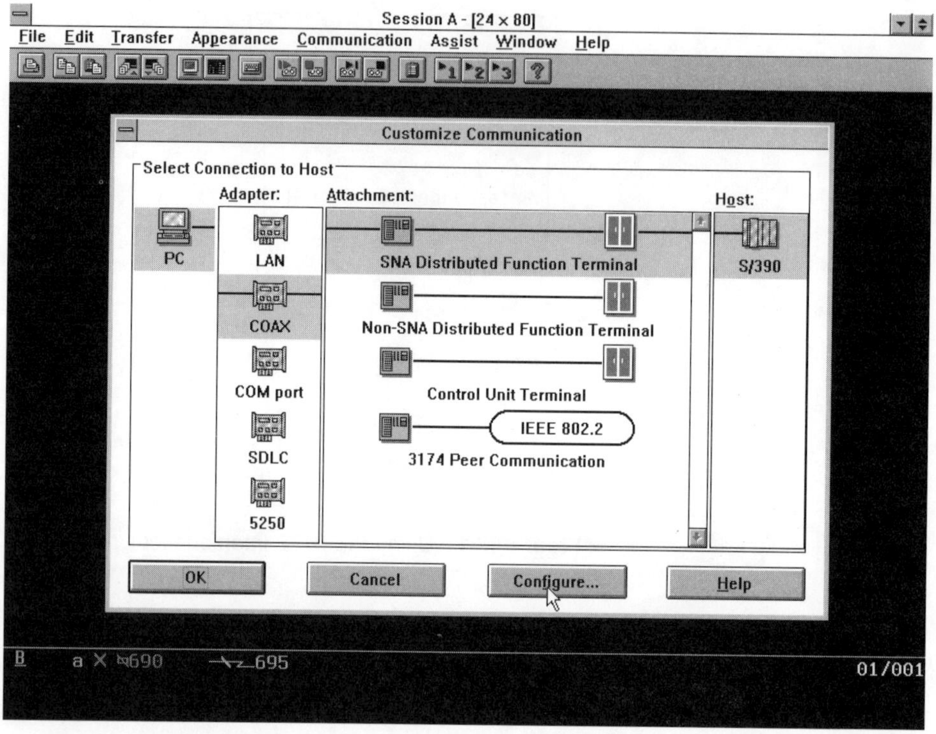

Figure 5.32 Using PCOM/3270 to select an appropriate attachment method when using a coaxial cable adapter for host communications

placed over the IEEE 802.2 attachment method. Thus, if we were configuring a LAN station or an Ethernet or Token-Ring network connected to a S/390 via a TIC or a similar gateway method, we would select that attachment method.

Once you select your attachment method, you can customize your workstation's session parameters. Figure 5.34 illustrates the customization screen for an SDLC connection from a PC to a S/390 host. Although the default screen display of 24 lines by 80 columns is shown, PCOM/3270 supports a number of screen size options that work relatively well with SVGA-based workstations. For example, if your requirements include a 132-column display you can easily adjust the screen size by a simple point and click operation.

TN3270 operations

One of the more interesting methods for accessing IBM mainframes is via a special version of Telnet known as TN3270. Through the use of a TN3270 program you can access IBM mainframes connected to the Internet via the Internet, as well as via workstations on a LAN operating a TCP/IP protocol stack. This capability results from the fact that TN3270 represents a TCP/IP program that operates on top of a TCP/IP protocol stack.

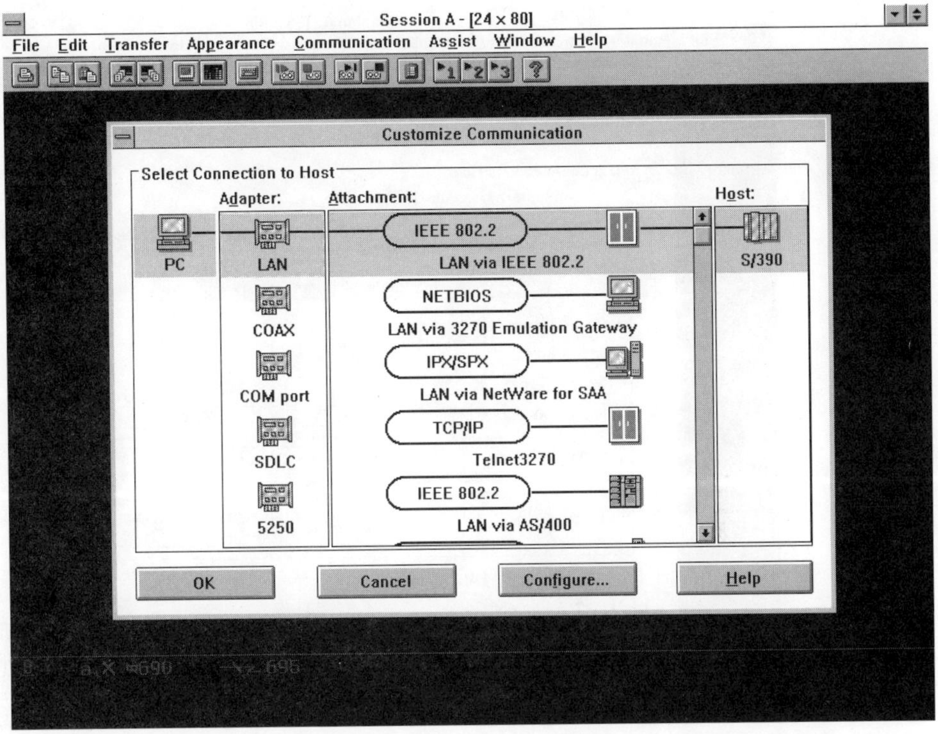

Figure 5.33 Using PCOM/3270 to select a LAN attachment via an IEEE 802.2 network

Figure 5.35 illustrates the connection configuration of the NetManage Chameleon TN3270 terminal emulation program to initiate a session with a mainframe whose IP address is 198.78.46.1. Note that the session request will occur on TCP port 23 and the PC running TN3270 will operate as a 3270 model 2 display which uses a 24 by 80 character display. Figure 5.36 illustrates a TCP/IP connection to an IBM mainframe using TN3270 via an Internet connection to a 3172 which in turn is connected to the mainframe. Although the TN3270 program used by the author uses a GUI interface, the actual display generated by the host is text based. This is illustrated in Figure 5.36 which illustrates the use of TN3270 to access the author's calendar stored on a mainframe calendar system.

Now that we have an appreciation for APPN and SNA gateways and the use of emulation software, we can turn our attention to the integration of IBM traffic into a TCP/IP network. Since this capability can be considered to represent an SNA to TCP/IP gateway performed by Data Link Switching, we will next focus our attention upon this gateway technique.

Data Link Switching

One of the major problems associated with SNA- and APPN-based networks is the fact that both are essentially proprietary. As the use of TCP/IP expanded for

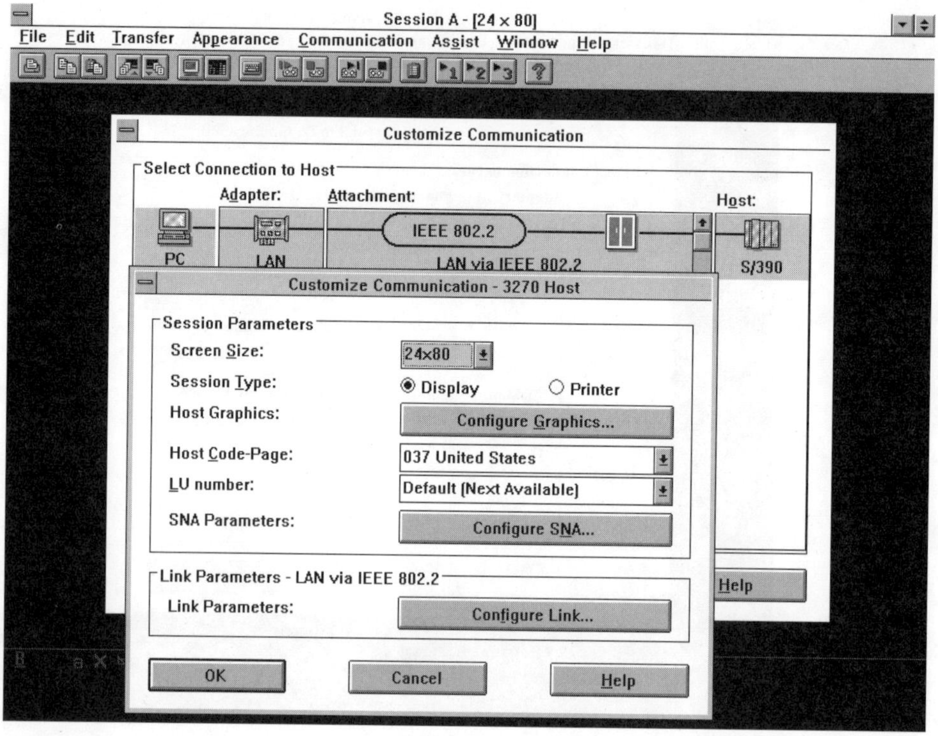

Figure 5.34 Using PCOM/3270 to adjust session parameters

interconnecting LANs on private intranets as well as for communications with the Internet, many organizations were forced to maintain two separate networks, one for SNA and one for a second network protocol which is commonly TCP/IP.

In an effort to better support customer requirements for improving connectivity and lowering the cost associated with multiprotocol networking, IBM developed a tunneling mechanism referred to as Data Link Switching (DLSw). DLSw was first introduced by IBM in 1992 as a feature of its 6611 bridge–router series of products referred to as the Nway 6611 Network Processor. IBM submitted its effort to the Internet Engineering Task Force (IETF) which resulted in DLSw being standardized as RFC 1434. It was also standardized by an IBM-sponsored, multivendor forum known as the APPN Implementors' Workshop.

Overview

Data Link Switching was developed as a mechanism to support SNA and NetBIOS data traffic via both bridged and router-based networks in a multiprotocol environment. Although DLSw is primarily used to tunnel SNA and NetBIOS under IP, it can also be used to tunnel other protocols. Since DLSw enables organizations to merge their SNA and APPN networks with their IP-based networks, many redundant communications links become candidates for removal,

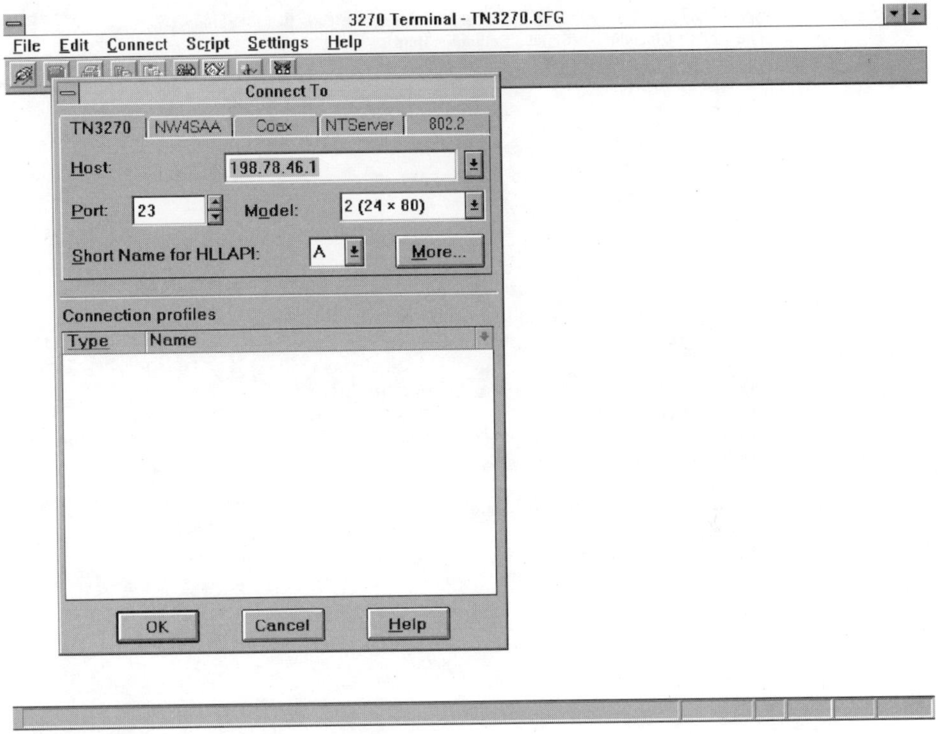

Figure 5.35 Using the NetManage Chameleon TN3270 terminal emulation program to obtain TCP/IP access to an IBM SNA host

enabling organizations to operate a more efficient and less costly network infrastructure.

Operation

Under DLSw the connection-oriented protocols of SNA and NetBIOS in the form of Logical Link Control Type 2 (LLC2) and Synchronous Data Link Control (SDLC) packets are encapsulated into IP packets. Figure 5.37 illustrates an example of the manner by which point-to-point SDLC traffic and LAN-based LLC2 traffic are integrated into an IP router-based network. In examining Figure 5.37, note that the tunneling effort involves wrapping the IP header around SDLC or LLC2 data. Since such data then becomes encapsulated, as you might expect, another term used to reference the transport of SDLC and LLC2 data in IP packets is encapsulation.

Although the actual tunneling or encapsulation process appears to be relatively simple to accomplish, in actuality it is a complicated process. The complication results from the fact that SNA uses connection-oriented protocols, LLC2 and SDLC, which are based upon positive acknowledgment with retransmission (PAR). As such, if an ACK is not received within a period of time after a sequence of frames are transmitted, a timer will expire, resulting in the sending station

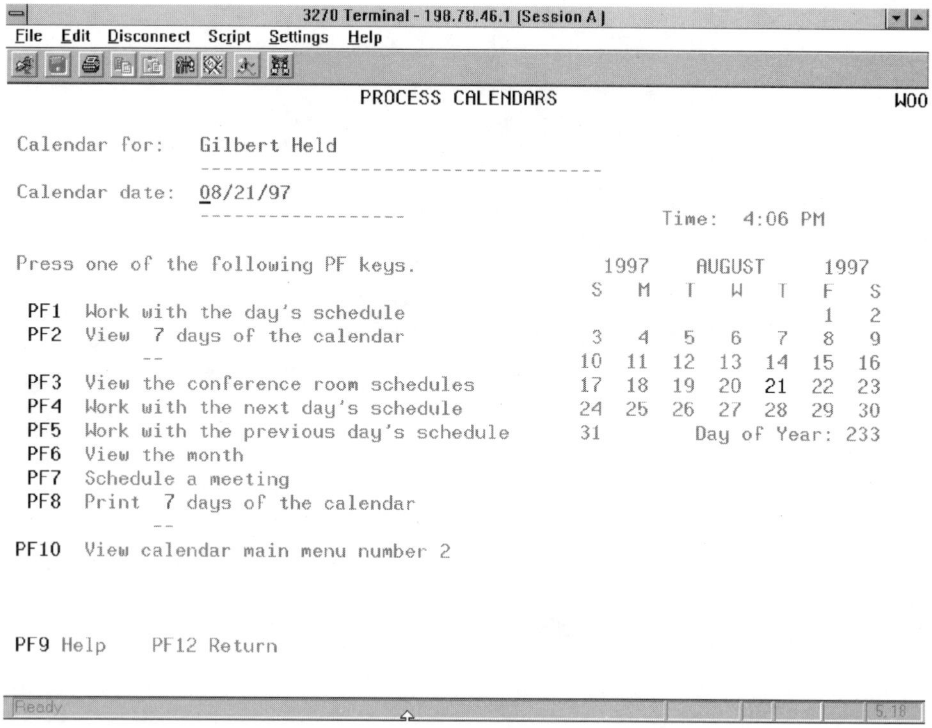

Figure 5.36 Using a TN3270 session to access a mainframe-based calendar system

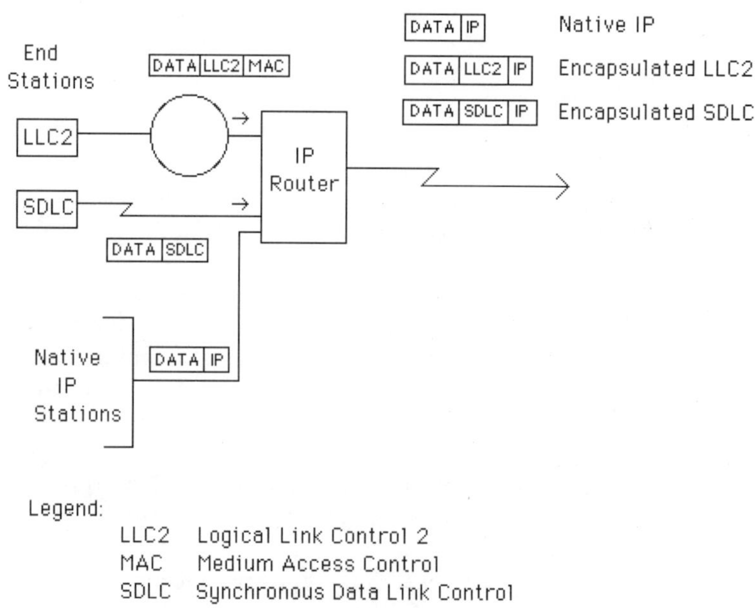

Figure 5.37 The DLSw tunneling process

retransmitting the data. If, due to network congestion, a circuit failure or other impairments repeat, after a predefined number of repeats the connection will be terminated, resulting in the loss of any SNA and NetBIOS sessions in progress. Since IP is a connectionless protocol, the transport of SDLC and LLC2 data within IP would very likely result in periodic session timeouts as traffic density varies during the day. To prevent this situation from occurring, DLSw relies on spoofing, with the sending router acknowledging frames as they are received. The routers then use a reliable transport protocol, such as TCP, to ensure that data arrives at its intended destination. A slightly different procedure is used for SDLC which relies on the constant polling between primary and secondary SDLC stations. DLSw-capable routers perform proxy polling. That is, the sending router intercepts polls from the SDLC primary station while another router polls the SDLC secondary station as if it was the SDLC primary station. Since polls are not passed onto the IP network, transmission efficiency is enhanced.

Although any encapsulation or tunneling method adds overhead in the form of an additional header, the use of spoofing and proxy polling can be considered as a significant counterbalance. Thus, in most cases the overhead associated with the use of additional headers will have a negligible effect upon the overall performance of DLSw.

Communications servers

A communications server is a relatively new term being used for a multifunction protocol gateway. The idea behind the communications server is to provide network managers and LAN administrators with protocol independence, enabling network design and restructuring decisions to be made independent of existing network topology constraints associated with network protocols currently being used.

Although Data Link Switching can be considered to represent a mechanism that provides a multiprotocol gateway capability, it is based upon the use of tunneling of LLC2 and SDLC in IP. Thus, it is restricted to enabling specific protocols to be transported under IP which, while an acceptable solution for the networking requirements of many organizations, may not be sufficient for use by other organizations. Another problem associated with the use of tunneling or encapsulation is the fact that application data must first be structured through one protocol stack prior to being encapsulated and processed through a second protocol stack. This means that in addition to an additional load being placed on the network in the form of dual headers, there is also an additional processing time at each device that performs the encapsulation function.

MPTN

Recognizing the previously described problems, IBM developed a protocol conversion facility which works on several types of hardware platforms operating its AIX and OS/2 operating systems as well as under Microsoft's Windows NT. This protocol conversion facility is called Multiprotocol Transport Networking

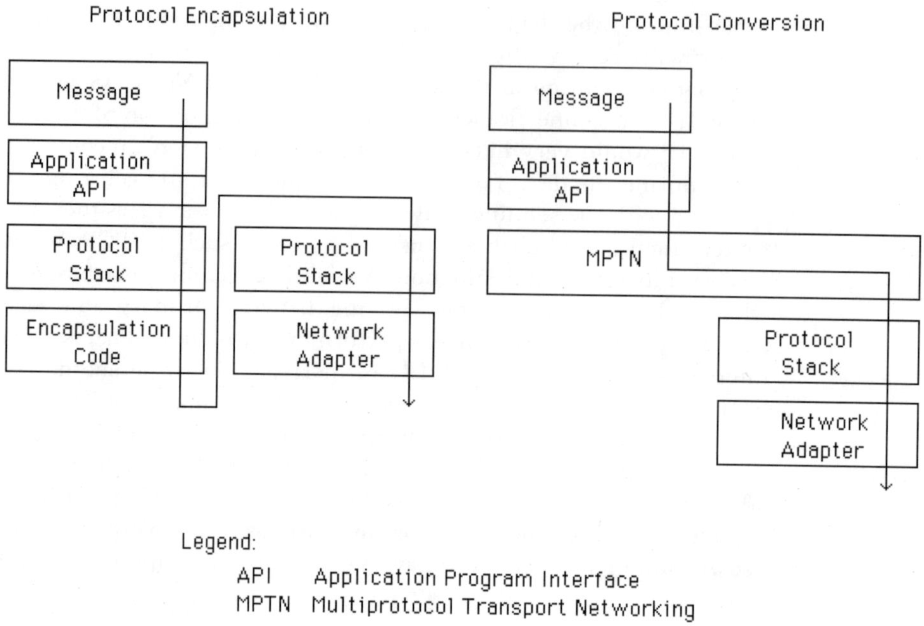

Figure 5.38 Comparing protocol encapsulation and conversion

(MPTN) and employs protocol conversion instead of encapsulation whenever possible.

MPTN operates at the Application Programming Interface (API) layer. Operating at the lower portion of the Application Layer in the protocol stack, MPTN would, for example, convert the sockets interface to use SNA protocols instead of TCP/IP, or the APPC interface to use TCP/IP protocols instead of SNA. Through the use of MPTN, an application invokes its preferred API without knowledge of the actual network protocol that will be used. MPTN then converts the API calls to use the protocol of the desired transport network. Figure 5.38 provides a comparison between protocol encapsulation and MPTN's protocol conversion.

Through the use of protocol conversion the transport of information is not tied to a specific protocol. This means that from a theoretical basis a communications server could be developed to provide a large number of protocol conversions that could provide organizations with a protocol-independent capability. However, from a practical standpoint the development of conversion software is a time-consuming effort and most communications servers are limited to providing a gateway between SNA and TCP/IP networks, although certain IBM products also provide a conversion capability for Novell IPX as well as IBM's LAN Manager's NetBIOS.

Figure 5.39 illustrates the use of two communications servers to integrate IPX- and TCP/IP-based LANs with an SNA network. In this example one communications server provides conversion from the TCP/IP-based networks to SNA while the second server performs a similar function for an IPX network.

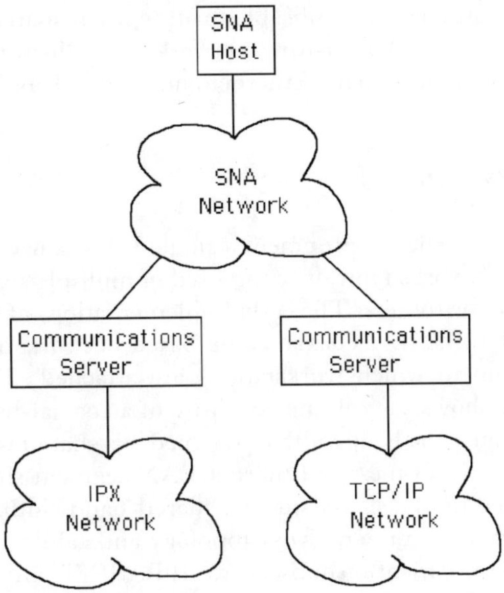

Figure 5.39 Using communications servers

5.4 LAN SWITCHES

Until the early 1990s the primary method employed to overcome the effect of network congestion was segmentation, subdividing a network into two or more entities interconnected by a bridge or router. Today, network managers and administrators have several options, including the use of a higher operating rate network such as Fast Ethernet or ATM or the use of LAN switches. In this section we will focus our attention on the latter, which are also referred to as switching hubs.

To obtain an appreciation for the role of LAN switches, we will first review the operation of conventional hubs, including the bandwidth constraints associated with their use. We will then describe the operational methods supported by various types of LAN switches. Using this information, we will explore the use of both Ethernet and Token-Ring switches to gain an understanding of the key features built into many products, as well as why the presence of some features and the absence of others can result in degraded performance instead of an expected improvement. Lastly, we will examine how the use of certain switch features can result in network problems and how those problems can be alleviated through the use of other device features.

Conventional hub bottlenecks

Conventional hubs, which were developed to facilitate the cabling of network devices, also function as a bottleneck with respect to the use of network bandwidth.

This bottleneck is applicable to all types of shared media networks, including both Ethernet and Token-Ring networks. To illustrate why this occurs, let's examine the operation of both Ethernet and Token-Ring hubs.

Ethernet hub operation

In an Ethernet environment a single LAN is usually referred to as a segment, with large networks typically composed of multiple segments connected by one or more bridges or routers. The early implementations of Ethernet in the form of 10BASE-5 and 10BASE-2 coaxial cable-based networks resulted in the use of a common medium to which workstations are attached. This is illustrated in Figure 5.40 which shows the cabling structure of a coaxial-based Ethernet network.

Because the bandwidth of the media is shared with only one user able to transmit at any given time, the Ethernet LAN segment shown in Figure 5.40 is commonly referred to as a shared-media, shared-bandwidth network.

A change in the network topology and cabling structure of Ethernet resulted in the development of hub-centric 10BASE-T and 100BASE-T networks, in which cabling from individual network devices to dedicated ports on the hub resulted in a star-wiring configuration. When two or more hubs are interconnected to form a common network, the wiring topology resembles a star–bus structure as illustrated in Figure 5.41. Although the wiring topology changed, the use of hubs did not alter the fact that the network remained a shared-media, shared-bandwidth network.

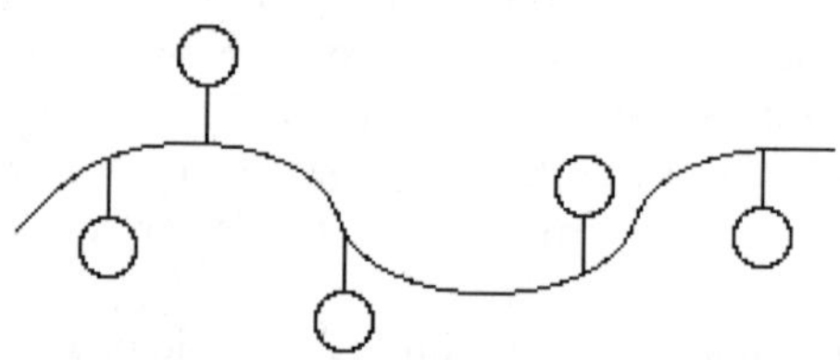

Figure 5.40 A shared-media, shared-bandwidth Ethernet LAN segment. 10BASE-5 and 10BASE-2 Ethernet networks consist of a coaxial run to which network devices are attached

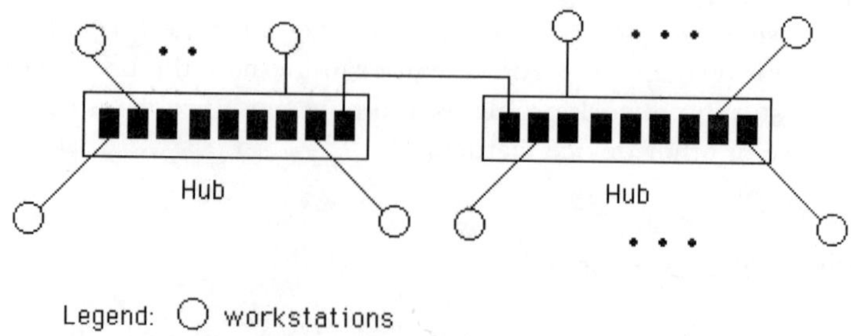

Figure 5.41 A two-hub 10BASE-T Ethernet network

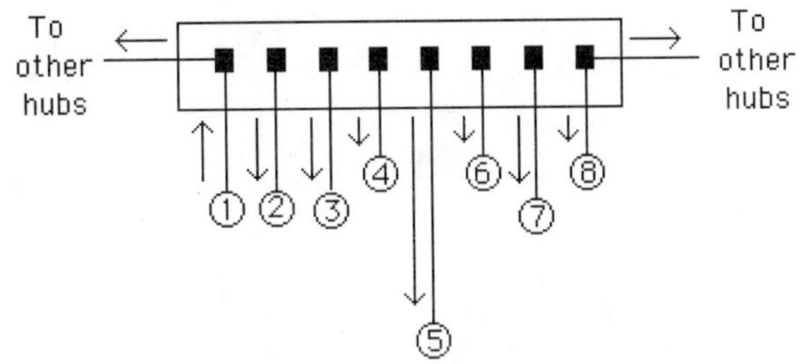

Figure 5.42 Conventional hub dataflow. A conventional hub functions as a data regenerator, outputting an incoming frame received on one port onto all other ports

To illustrate the problem associated with the use of a shared-media, shared-bandwidth network, let's examine the operation of a conventional Ethernet hub. Figure 5.42 illustrates the data flow when one workstation (node 1) transmits a frame to another workstation, file server, gateway or other network device which is connected either to the same hub or to another hub connected to the data originator's hub. Since the hub functions as a data regenerator, the frame is repeated onto each connection to the hub to include interconnections to other hubs. This restricts data flow to one workstation at a time, since collisions occur when two or more attempt to gain access to the media at the same time.

Token-Ring hub operation

Although data flow on a Token-Ring network is circular, this type of network is also a shared-media, shared-bandwidth network. In a Token-Ring network environment, hubs, referred to Multistation Access Units or MAUs, are connected via their Ring In and Ring out ports to form a star–ring topology similar to that shown in Figure 5.43. The actual data flow of a frame is from one device to the next and includes flowing down the cable, called a lobe, connecting the device to the MAU port, to an attached device and back to the port prior to flowing to the next port. Since only one frame can flow on the network at any point in time, access to the bandwidth is also shared. Thus, a Token-Ring network also represents a shared-media, shared-bandwidth network.

Bottleneck creation

Conventional Ethernet hubs create network bottlenecks because all traffic flows through a shared backplane in the hub. Thus, every device connected to an Ethernet hub competes for a slice of the bandwidth of the backplane. In a Token-Ring environment devices compete to acquire a token, resulting in the sharing of network bandwidth in a similar manner. The end result of this bandwidth sharing is an average transmission rate per device that is many times below the operating rate of the network. For example, consider a departmental 10BASE-T network operating at 10 Mbps consisting of 12 interconnected eight port hubs that supports

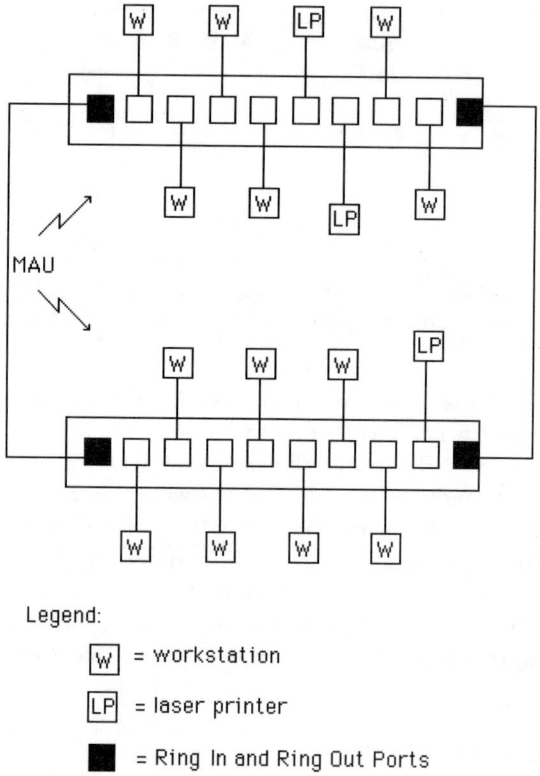

Figure 5.43 The connection of Token-Ring MAUs forms a star–ring topology

a total of 96 devices. The average slice of bandwidth available for each device is therefore 10 Mbps/96 or approximately 104 kbps. Note that although each device transits and receives data at the LAN operating rate of 10 Mbps, their average data transfer capability is approximately 104 kbps since each device must compete with 95 other devices to obtain access to the network. Similarly, a 96-node Token-Ring network would result in each device attached to that network having an average data transfer capability of 4 Mbps or 16 Mbps divided by 96, depending upon the operating rate of the network. This means that over a period of time the addition of network users, the introduction of one or more graphic-based applications, or growth in the use of current applications can result in a severely taxed network. When this type of situation occurs, you can consider a variety of techniques to enhance network performance, including network segmentation through the use of a bridge or router, migrating your existing infrastructure to a different and higher operating rate technology, or employing LAN switches.

Switching operations

The development of LAN switches has its foundation, like many other areas of modern communications, in telephone technology. Shortly after the telephone was invented the switchboard was developed to enable multiple simultaneous conversations to occur without requiring telephone wires to be installed in a complex matrix between subscribers. Later, telephone office switches were developed to route calls based upon the telephone number dialed, followed in a similar manner by the development of bridges in a LAN environment. Bridges can be considered to represent an elementary type of switch due to their limited number of ports and simplistic switching operation. That switching operation is based upon whether or not the destination address in a frame 'read' on one port is known to reside on that port.

A key limitation of a bridge is its common inability to perform more than one frame-forwarding operation at any point in time. Recognizing this performance bottleneck, equipment designers developed a product that can perform multiple simultaneous frame-forwarding operations. That device, which has evolved into a multi-billion dollar business, is the LAN switch. The LAN switch is based upon matrix switches which for decades were successfully employed in telecommunications operations. By adding buffer memory to store address tables, frames flowing on LANs connected to different ports could be simultaneously read and forwarded via the switch fabric to ports connected to other networks.

Basic components

Figure 5.44 illustrates the basic components of a four-port intelligent switch. Like a bridge that reads frames flowing on a network to construct a table of source

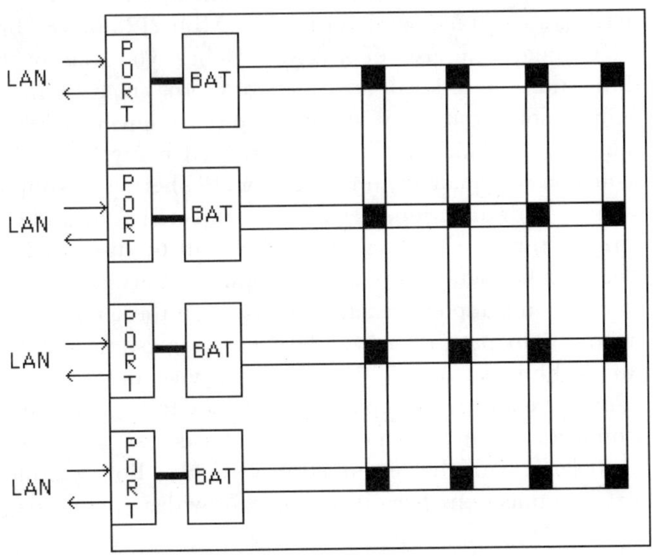

Figure 5.44 Basic components of an intelligent switch. An intelligent switch consists of buffers and address tables (BAT), logic and a switching fabric which permits frames entering one port to be routed to any port in the switch. The destination address in a frame is used to determine the associated port with that address via a search of the address table, with the port address used by the switching fabric for establishing the cross-connection

addresses, the tables in a LAN switch can be created through a self-learning process. This allows the destination address in frames flowing on LANs connected to a switch port to be compared to a table of destination addresses and associated port numbers. When a match occurs between the destination address of a frame flowing on a network connected to a port and the address in the port's address table, the frame is copied into the switch and routed through the switch fabric to the destination port, where it is placed onto the network connected to that port. If the destination port is in use due to a previously established cross-connection between ports, the frame is maintained in a buffer until it can be switched to its destination.

Key advantages of use

A key advantage associated with the use of LAN switches results from their ability to support parallel switching, permitting multiple cross-connections between source and destination to occur simultaneously. For example, if four 10BASE-T networks were connected to the four-port switch shown in Figure 5.44, two simultaneous cross-connections, each at 10 Mbps, could occur, resulting in an increase in bandwidth to 20 Mbps. Here each cross-connection represents a dedicated 10 Mbps bandwidth for the duration of a frame. Thus, from a theoretical perspective, an N-port LAN switch supporting a 10 Mbps operating rate on each port provides a throughput up to $(N/2) \times 10$ Mbps. For example, a 128-port LAN switch would support a throughput up to $(128/2) \times 10$ Mbps or 640 Mbps, while a network constructed using a series of conventional hubs connected to one another would be limited to an operating rate of 10 Mbps, with each workstation on that network having an average bandwidth of 10 Mbps/128 or 78 kbps.

Through the use of LAN switches you can overcome the operating rate limitation of a local area network. In an Ethernet environment, the cross-connection through a switching hub represents a dedicated connection so there will never be a collision. This fact enabled many switching hub vendors to use the collision wire-pair from conventional Ethernet to support simultaneous transmission in both directions between a connected node and hub port, resulting in a full-duplex transmission capability that will be discussed in more detail later in this section. In fact, a similar development permits Token-Ring switching hubs to provide full-duplex transmission, since if there is only one station on a port there is no need to pass tokens and repeat frames, raising the maximum bidirectional throughput between a Token-Ring device and a switching hub port to 32 Mbps. Thus, the ability to support parallel switching as well as initiate dedicated cross-connections on a frame-by-frame basis can be considered the key advantages associated with the use of LAN switches. Both parallel switching and dedicated cross-connections permit higher bandwidth operations.

Delay times

Switching occurs on a frame-by-frame basis, with the cross-connection torn down after being established for routing one frame. Thus, frames can be interleaved from two or more ports to a common destination port with a minimum of delay. For example, consider a maximum length Ethernet frame of 1526 bytes to include a 1500 byte data field and 26 overhead bytes. At a 10 Mbps operating rate each bit time is $1/10^7$ seconds or 100 ns. For a 1526 byte frame the minimum delay time if one frame precedes it in attempting to be routed to a common destination becomes:

$$1526 \text{ bytes} \times \frac{8 \text{ bits}}{\text{byte}} \times \frac{100 \text{ ns}}{\text{bit}} = 1.22 \text{ ms}$$

This delay time represents blocking resulting from frames on two service ports having a common destination and should not be confused with another delay time referred to as latency. Latency represents the delay associated with the physical transfer of a frame from one port via the switch to another port and is fixed based upon the architecture of the switch. In comparison, blocking delay depends upon the number of frames from different ports attempting to access a common destination port and the method by which the switch is designed to respond to blocking. Some switches simply have large buffers for each port and service ports in a round-robin fashion when frames on two or more ports attempt to access a common destination port. This method of service is not similar to politics as it does not show favoritism; however, it also does not consider the fact that some attached networks may have operating rates different from other attached networks. Other switch designs recognize that port buffers are filled based upon both the number of frames having a destination address of a different network and the operating rate of the network. Such switch designs use a priority service scheme based upon the occupancy of the port buffers in the switch.

Switching techniques

There are three switching techniques used by intelligent switching hubs: cross-point, also referred to as cut-through or 'on the fly', store-and-forward, and a hybrid method which alternates between the first two methods based upon the frame error rate. As we will soon note, each technique has one or more advantages and disadvantages associated with its operation.

Cross-point switching

The operation of a cross-point switch is based upon an examination of the destination of frames as they enter a port on the switching hub. The switch uses the destination address as a decision criterion to obtain a port destination from a look-up table as soon as the destination address in the frame is read. This technique enables the switch to initiate its table look-up and cross-connection operation without having to read the entire frame, providing a very fast switching operation that minimizes delay.

Figure 5.45 illustrates the basic operation of cross-point or cut-through switching. Under this technique the destination address in a frame is read prior

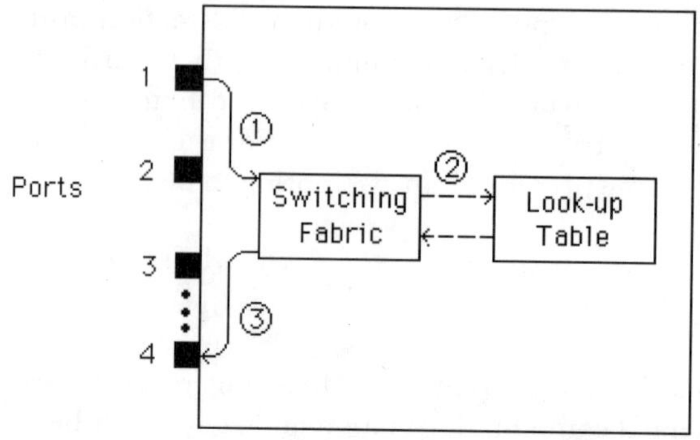

Figure 5.45 Cross-point/cut-through switching. A cross-point or cut-through operating switch reads the destination address in a frame prior to storing the entire frame (1). It forwards that address to a look-up table (2) to determine the port destination address which is used by the switching fabric to provide a cross-connection to the destination port (3)

to the frame being stored (1). That address is forwarded to a look-up table (2) to determine the port destination address which is used by the switching fabric to initiate a cross-connection to the destination port (3). Since this switching method only requires the storage of a small portion of a frame until it is able to read the destination address and perform its table look-up operation to initiate switching to an appropriate output port, latency through the switch is minimized.

Latency functions as a brake on two-way frame exchanges. For example, in a client–server environment the transmission of a frame by a workstation results in a server response. Thus, the minimum wait time is twice the latency for each client–server exchange, lowering the effective throughput of the switch. Since a cross-point switching technique results in a minimal amount of latency, the effect upon throughput of the delay attributable to a LAN switch using this switching technique is minimal.

Store-and-forward

In comparison to cut-through switching, a store-and-forward LAN switch first stores an entire frame in memory prior to operating on the data fields within the frame. Once the frame is stored, the switching hub checks the frame's integrity by performing a cyclic redundancy check (CRC) upon the contents of the frame, comparing its computed CRC against the CRC contained in the frame's Frame Check Sequence (FCS) field. If the two match, the frame is considered to be error-free and additional processing and switching will occur. Otherwise, the frame is considered to have one or more bits in error and will be discarded.

In addition to CRC checking, the storage of a frame permits filtering against various frame fields to occur. Although a few manufacturers of store-and-forward LAN switches support different types of filtering, the primary advantage advertised by such manufacturers is data integrity. Whether or not this is actually an advantage depends upon how you view the additional latency introduced by the storage of a full frame in memory as well as the necessity for error checking. Concerning the latter, switches should operate error-free, so a store-and-forward switch only removes network errors which should be negligible to start with.

When a switch removes an errored frame, the originator will retransmit the frame after a period of time. Since an errored frame arriving at its destination network address is also discarded, many persons question the necessity of error checking by a store-and-forward LAN switch. However, filtering capability, if offered, may be far more useful as you could use this capability, for example, to route protocols carried in frames to destination ports far more easily than by frame destination address. This is especially true if you have hundreds or thousands of devices connected to a large switch. You might set up two or three filters instead of entering a large number of destination addresses into the switch.

Figure 5.46 illustrates the operation of a store-and-forward LAN switch. Note that a common switch design is to use shared buffer memory to store entire frames, which increases the latency associated with this type of switch. Since the minimum length of an Ethernet frame is 72 bytes, the minimum one-way delay or latency, not counting the switch overhead associated with the look-up table and switching fabric operation, becomes:

$$\begin{aligned} &96\ \mu s + (72 \text{ bytes} \times 8 \text{ bits/byte} \times 100 \text{ ns/bit}) \\ &= (9.6 \times 10^{-6}) + (576 \times 100 \times 10^{-9}) \text{ seconds} \\ &= 67.2 \times 10^{-6} \text{ seconds} \end{aligned}$$

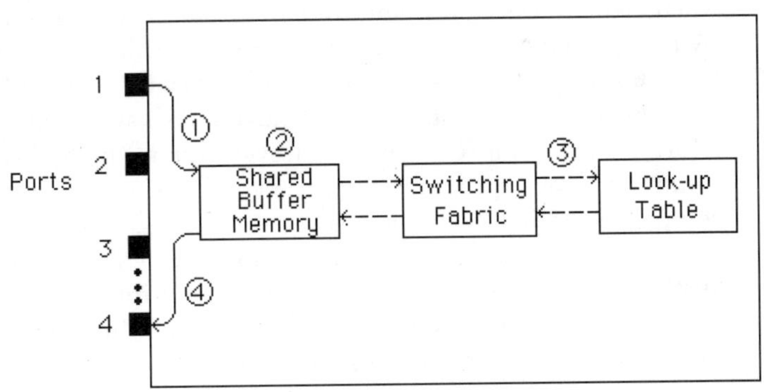

Figure 5.46 Store and forward switching. A store and forward LAN switch reads the frame destination address (1) as it is placed in buffer memory (2). As the entire frame is being read into memory, a look-up operation (3) is performed to obtain a destination port address. Once the entire frame is in memory, a CRC check is performed and one or more filtering operations may be performed. If the CRC check indicates the frame is error-free, it is forwarded from memory to its destination address (4), otherwise it is disregarded

Here 9.6 μs represents the Ethernet interframe gap, while 100 ns/bit is the bit duration of a 10 Mbps Ethernet LAN. Thus, the minimum one-way latency of a store-and-forward Ethernet switch is 0.000 0672 seconds, while a round-trip minimum latency is twice that duration. For a maximum length Ethernet frame with a data field of 1500 bytes, the frame length becomes 1526 bytes. Thus, the one-way maximum latency becomes:

$$96\,\mu\text{s} + (1526 \text{ bytes} \times 8 \text{ bits/byte} \times 100 \text{ ns/bit})$$
$$= (9.6 \times 10^{-6}) + (12\,208 \times 100 \times 10^{-9}) \text{ seconds}$$
$$= 0.001\,2304 \times 10^{-9} \text{ seconds}$$

Hybrid

A hybrid switch supports both cut-through and store-and-forward switching, selecting the switching method based upon monitoring the error rate encountered by reading the CRC at the end of each frame and comparing its value to a computed CRC performed 'on the fly' on the fields protected by the CRC. Initially the switch might set each port to a cut-through mode of operation. If too many bad frames are noted on a port, the switch will automatically set the frame processing mode to store-and-forward, permitting the CRC comparison to be performed prior to the frame being forwarded. This permits frames in error to be discarded without having them pass through the switch. Since the 'switch', no pun intended, between cut-through and store-and-forward modes of operation occurs adaptively, another term used to reference the operation of this type of switch is adaptive.

The major advantages of a hybrid switch are that it provides minimal latency when error rates are low and discards frames by adapting to a store-and-forward switching method so it can discard errored frames when the frame error rate rises. From an economic perspective, the hybrid switch can logically be expected to cost more than a cut-through or store-and-forward switch as its software development effort is more comprehensive. However, due to the highly competitive market for communications products, upon occasion its price may be reduced below competitive switch technologies.

In addition to being categorized by their switching technique, LAN switches can be classified by their support of single or multiple addresses per port. The former method is referred to as port-based switching, while the latter switching method is referred to as segment-based switching.

Port-based switching

A LAN switch which performs port-based switching supports only a single address per port. This restricts switching to one device per port; however, it results in a minimum amount of memory in the switch as well as provides for a relatively fast table look up when the switch uses a destination address in a frame to obtain the port for initiating a cross-connect.

Figure 5.47 illustrates an example of the use of a port-based switch. In this example M user workstations use the switch to contend for the resources of N servers. If $M > N$, then a switch connected to Ethernet 10 Mbps LANs can

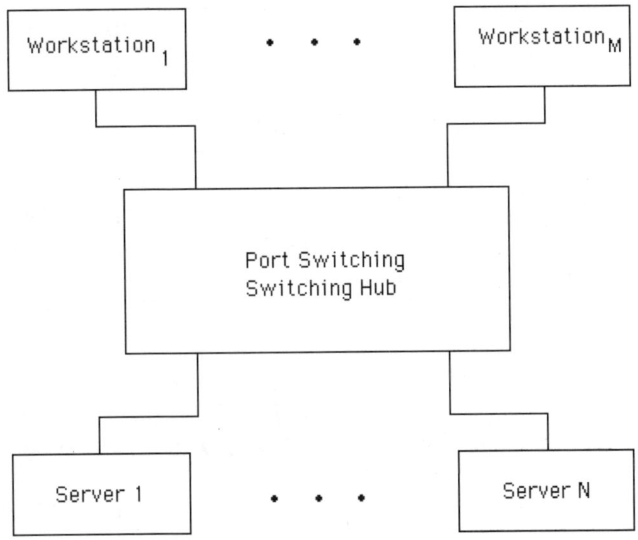

Figure 5.47 Port-based switching. A port-based switch associates one address with each port, minimizing the time required to match the destination address of a frame against a table of destination addresses and associated port numbers

support a maximum throughput of $(N/2) \times 10\,\text{Mbps}$, since up to $N/2$ simultaneous client–server frame flows can occur through the switch.

It is important to compare the maximum potential throughput through a switch to its rated backplane speed. If the maximum potential throughput is less than the rated backplane speed, the switch will not cause delays based upon the traffic being routed through the device. For example, consider a 64-port switch that has a backplane speed of 400 Mbps. If the maximum port rate is 10 Mbps, then the maximum throughput (assuming 32 active cross-connections were simultaneously established) becomes 320 Mbps. In this example the switch has a backplane transfer capability sufficient to handle the worst-case data transfer scenario. Now let's assume that the maximum backplane data transfer capability was 200 Mbps. This would reduce the maximum number of simultaneous cross-connections capable of being serviced to 20 instead of 32 and adversely affect switch performance under certain operational conditions.

Since a port-based switch has to store only one address per port, search times are minimized. When combined with a pass-through or cut-through switching technique, this type of switch results in a minimal latency to include the overhead of the switch in determining the destination port of a frame.

Segment-based switching

A segment-based switching technique requires a LAN switch to support multiple addresses per port. Through the use of this type of switch, you achieve additional

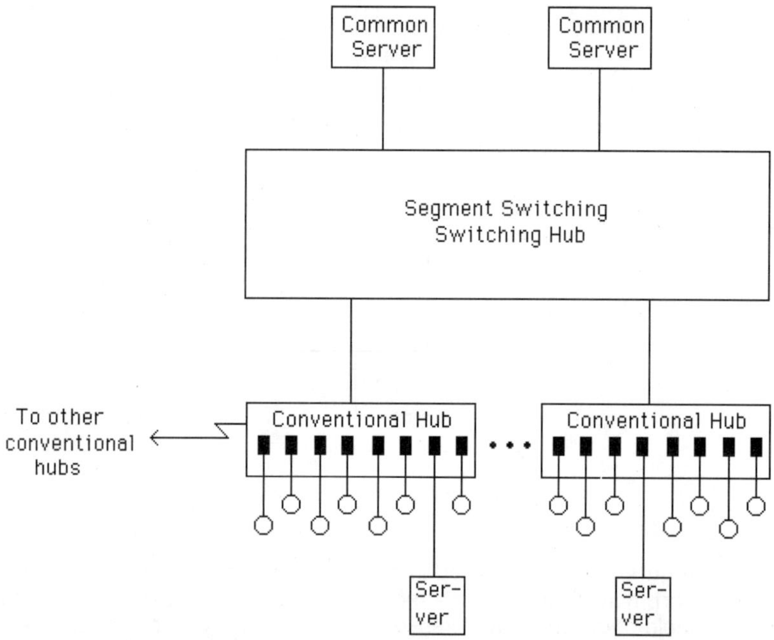

Figure 5.48 Segment-based switching. Through the use of a segment-based switching hub, you can maintain servers for use by workstations on a common network segment as well as provide access by all workstations to common servers

networking flexibility since you can connect other hubs to a single segment-based switch port.

Figure 5.48 illustrates an example of the use of a segment-based switch in an Ethernet environment. Although two segments in the form of conventional hubs with multiple devices connected to each hub are shown in the lower portion of Figure 5.48, note that a segment can consist of a single device, resulting in the connection of one device to a port on a segment switch being similar to a connection on a port switch. However, unlike a port switch that is limited to supporting one address per port, the segment switch can, if necessary, support multiple devices connected to a port. Thus, the two servers connected to the switch at the top of Figure 5.48 could, if desired, be placed on a conventional hub or a high-speed hub, such as a 100BASE-T hub, which in turn would be connected to a single port on a segment switch.

In Figure 5.48 each conventional hub acts as a repeater and forwards every frame transmitted on that hub to the switch, regardless of whether or not the frame requires the resources of the switch. The segment switch examines the destination address of each frame against addresses in its look-up table, forwarding only those frames that warrant being forwarded. Otherwise, frames are discarded as they are local to the conventional hub. Through the use of a segment-based switching hub, you can maintain the use of local servers with respect to existing LAN segments as well as install servers whose access is common to all network segments. The latter is illustrated in Figure 5.48 by the connection of two common servers shown at the top of the LAN switch. If you obtain a store-and-forward segment switch which supports filtering, you could control access to common servers from individual workstations or by workstations on a particular segment. In addition, you can also use the filtering capability of a store-and-forward segment-based switch to control access from workstations located on one segment to workstations or servers located on another segment.

Using LAN switches

Today you can acquire network-specific switches that operate with a single type of LAN or switches with a translating capability that enable dissimilar networks to be serviced. In addition, some LAN switches incorporate a built-in routing feature that enables the switch to perform basic bridging and routing. Due to the almost infinite number of methods by which LAN switches can be employed, we will focus our attention upon several common generic methods; however, we will have to use specific types of LANs to illustrate and describe the operation of different switching methods.

Network redistribution

Network redistribution involves the movement of bandwidth-intensive workstations off conventional hubs, connecting them directly to ports on a switch. Figure 5.49 illustrates an example of network redistribution through the use of a LAN switch.

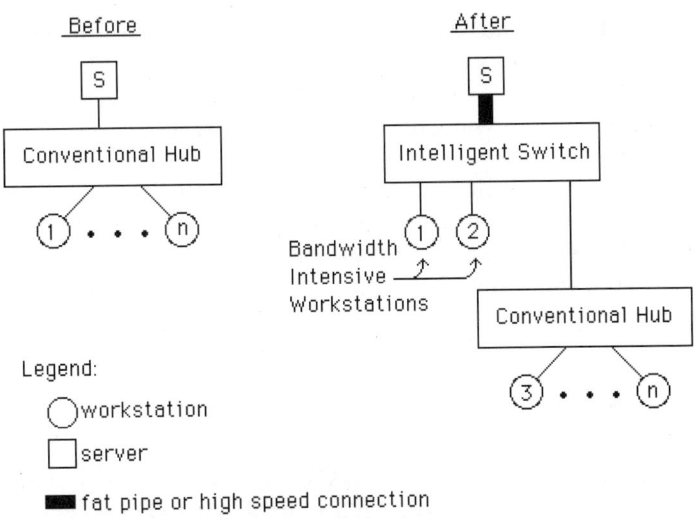

Figure 5.49 Network redistribution. Through the movement of bandwidth-intensive workstations off conventional hubs, you can minimize the effect of bottlenecks they cause on other workstations remaining connected to conventional hubs

In the left half of Figure 5.49 ('before'), a conventional hub is shown providing support for n nodes to a common server. Assuming two workstations require access to a visual database, transmit or receive large files, or perform other bandwidth-intensive applications, those workstations are redistributed onto a switch as shown in the right half of Figure 5.49 ('after'). Note that only one server is shown in the 'after' network schematic, with the server relocated from the conventional hub to the switch and connected to the switch by either a fat pipe or a high-speed connection. Here the term 'fat pipe' is used to reference a group of switch ports that function as a single entity, providing a higher bandwidth than is obtainable by a single connection of a switch to a network device. If a conventional connection was used the redistribution of workstations would have a negligible effect upon performance, as access to the server would not increase, only enhancing any peer-to-peer communications that may occur.

Server segmentation

Since access to data on servers is normally the reason why network performance degrades, another technique commonly used is to segment servers. Figure 5.50 illustrates an example of server segmentation obtained through the use of a LAN switch.

In the top portion of Figure 5.50 ('before'), two servers are shown on a network consisting of interconnected conventional hubs. In this example access to either server is constrained by the operating rate of the network. For example, if the top portion of Figure 5.50 represented a 10BASE-T network the maximum bandwidth to a server would be 10 Mbps, which is the operating rate of the LAN, while the average bandwidth would be 10 Mbps/n, where n represents the total number of nodes on the network.

Through the use of a LAN switch, servers can be placed on their own network segment as illustrated in the lower portion of Figure 5.50 ('after'). In addition, through the use of a fat pipe or high-speed connection between each server and the switch you can enhance access to each server.

If you simply moved each conventional hub onto a port on the switch, your ability to enhance network access would be limited. This limitation would result from the fact that the simultaneous access of workstations on different conventional hubs to different servers only provides the ability to double bandwidth. Thus, you would more than likely want to consider connecting high-activity workstations directly to the switch, as shown in the lower portion of Figure 5.50.

Backbone operation

A third common use of LAN switches is as a backbone for connecting lower operating rate LANs or LAN switches. In doing so, you can create a tiered network hierarchy.

Figure 5.51 illustrates the use of one 100BASE-T switch to interconnect a group of 10BASE-T switches. Although individual stations are shown connected to each

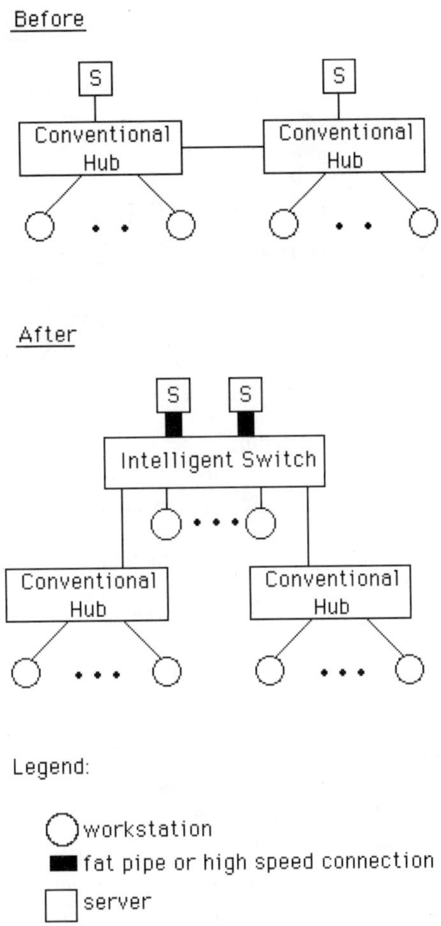

Figure 5.50 Server segmentation

10 Mbps switch port, this discussion is also applicable to the use of segment-based switches including those that are used to connect individual workstations on some ports and network segments on other ports. In this example note that workgroup servers remain local to each 10 Mbps switch, resulting in a majority of 10 Mbps switch traffic remaining local to each switch. Only when communications are required to a departmental server or a workgroup server connected to another switch will traffic be routed to the 100 Mbps backbone switch.

An alternative to the use of a tiered structure backbone switch is the use of a conventional 100 Mbps LAN for backbone operations, such as a conventional 100BASE-T non-switching hub. Although the cost of a non-switching hub is considerably less than that of a switch, it permits only one frame at a time to flow through the hub between interconnected devices. In comparison, the use of a 100 Mbps switch with n ports would support a throughput of $100 \times n/2$ as it permits $n/2$ simultaneous cross-connections.

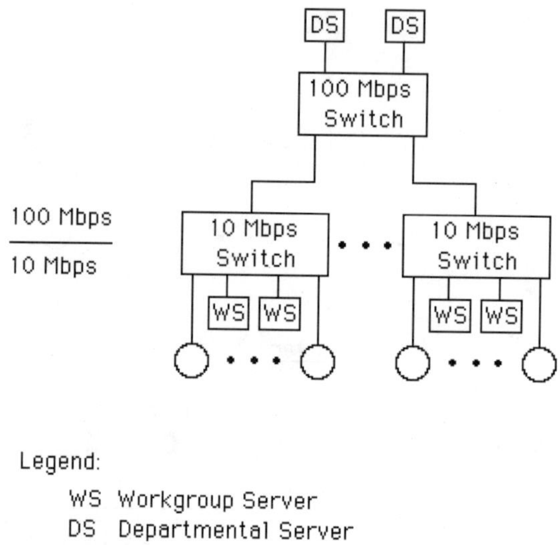

Figure 5.51 Using a high-speed switch as a network backbone

Handling speed incompatibilities

One of the main problems associated with the use of LAN switches occurs when a device connected on a high-speed port requires communications with a device connected on a lower operating rate port. For example, assume a server is connected to a switch at 100 Mbps while a workstation is connected via a 10 Mbps operating rate. Since buffer storage on each port is finite, it becomes possible for a client query followed by a long server response to result in the loss of frames once a buffer is filled. A similar problem can occur in the reverse direction when the cumulative operating rate of many clients attempting to access a common device exceeds the operating rate of the connection between the switch and that device. Recognizing this problem, vendors developed several methods to regulate the transmission of information, a process referred to as flow control. Flow control is primarily applicable to Ethernet switches, since the rotating token can be used to delay transmission on Token-Ring switches. Two popular methods used by switch vendors to regulate the transmission of information include backpressure and server software modules.

Backpressure

Backpressure is a term used to represent the generation of a false collision signal. Because a collision signal causes an Ethernet workstation or server to delay further transmission based on an exponential backoff algorithm, it provides a mechanism for implementing flow control. That is, once buffer storage in the switch has reached a predefined level of occupancy, the switch will generate a false collision signal. As the transmitting device delays further transmission, the switch's destination port has the opportunity to empty the contents of its buffer, precluding the occurrence of data loss.

Although backpressure is an effective flow control mechanism, its use requires a second wire pair. This makes it mutually exclusive with full-duplex transmission (FDX) since workstations and servers directly cabled to switch ports never encounter a collision and can operate in a FDX mode by using the collision wire pair for transmission in the opposite direction. If you require full-duplex transmission and want to preclude the loss of frames via flow control, you must turn to the use of a server software module.

Server software module

Several switch vendors developed software that operates on Windows NT and NetWare servers that regulate the flow of data between switch ports and those servers. To accomplish flow control the switch transmits a predefined signal to the module operating on the server, while a second signal is used to inform the module to resume transmission. The major difference between the use of backpressure and server software modules is that the latter can provide support for full-duplex transmission.

ATM considerations

Although ATM equipment has been gaining considerable momentum for use for wide area networks, its use in a LAN environment is primarily as a backbone for interconnecting what many persons refer to as legacy LANs, such as Token-Ring and Ethernet, instead of as a replacement of those networks. The reason for the lack of ATM to the desktop is more than likely due to its cost, lack of multimedia desktop applications, and the fact that both Ethernet and Token-Ring LANs can continue to provide a reasonable level of performance by segmentation and connection of segments via a high-speed backbone.

The use of ATM switches as a backbone for connecting legacy LAN traffic results in a series of incompatibilities that must be overcome. The major incompatibilities between ATM and legacy LANs include their connection method, address, and transport mechanism. An ATM network uses a connection-oriented switching mechanism to route traffic in the form of 53-byte cells from source to destination using Virtual Path Identifiers (VPIs) and Virtual Channel Identifiers (VCIs). In comparison, legacy LANs are connectionless, transmitting variable-length frames containing 48-bit MAC addresses to all stations on a network, with the addressed station then reading the frame. To overcome these incompatibilities, the ATM Forum developed a specification referred to as LAN Emulation (LANE). LAN Emulation represents a method whereby ATM's connection-oriented infrastructure provides a service which enables stations on legacy LANs to connect to other legacy stations as well as devices connected to ATM switches by mimicking or emulating the connectionless operation of legacy networks.

LAN Emulation Services

ATM provides LAN Emulation through a client–server model. The client is commonly implemented on an ATM adapter installed in a legacy LAN switch. The server portion of the model which provides the emulation services is commonly implemented in an ATM switch; however, some vendors implement the emulation services in a router connected to an ATM switch.

LAN Emulation Services are performed by a LAN Emulation Server (LES), a Broadcast and Unknown Server (BUS), and a LAN Emulation Configuration Server (LECS). These three components can be provided either as a single centralized service or distributed and configured to provide redundancy. The LES is responsible for registering MAC addresses as well as for converting between MAC and ATM addresses.

Unlike legacy LANs that include a broadcast capability, ATM is a connection-oriented point-to-point network. Thus, a mechanism is required to facilitate obtaining unknown addresses by querying each station on an ATM network as well as for converting a legacy broadcast into appropriate ATM connections. The Broadcast and Unknown Server (BUS) assumes this responsibility by handling all of the broadcasting and multicasting functions.

The third component of LAN Emulation Services is the LAN Emulation Configuration Server (LECS). The LECS is responsible for providing configuration information about the ATM network and assigns individual LANE clients to emulated LANs by directing them to a LES.

LAN Emulation operation

To illustrate the operation of LAN Emulation, consider Figure 5.52 which illustrates the use of an ATM backbone switch to provide communications between stations on two legacy LAN switches as well as from those stations to servers and a router connected to the ATM backbone. When a station on a legacy LAN first generates a frame that requires transportation via an ATM backbone, the LEC automatically locates the LECS via the use of a well-known ATM VPI, VCI address. This is denoted by the line numbered 1 in Figure 5.52.

The LECS returns the address of the LES and the BUS. This is indicated by the line numbered 2 in Figure 5.52. Once the address of the LES is known, the LEC can transmit an address resolution message to the LES. This is indicated by the line numbered 3. If the LES previously resolved a MAC address into an ATM address, it can directly provide the ATM address from its cache memory, as indicated by the line numbered 4. The LEC then uses the ATM address to establish a connection through the ATM switch to the destination legacy LAN switch, as indicated by the line numbered 5.

If the LES did not have the required ATM address, the LEC would send a request to the BUS. The BUS would broadcast the address resolution message to all end stations, with the LEC associated with the destination station returning its ATM address to both the LES and the originating LEC. This enables the LEC to initiate a connection to the appropriate ATM endpoint. In addition, the response is placed in the cache memory of both the LECs and the LES so that subsequent requests via the same or different LECs can be expediently serviced.

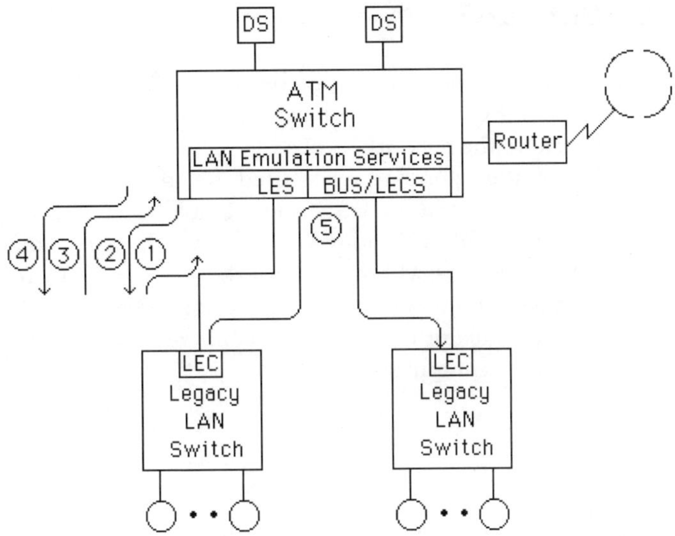

1. LEC locates the LECS.
2. LECS provides addresses of LES and BUS.
3. LEC requests ATM address of destination via an address resolution message to the LES.
4. LES provides ATM address.
5. LEC initiates connection.

Legend:

DS Departmental Server
○ Workstations on legacy LANS
BUS Broadcast and Unknown Server
LES LAN Emulation Server
LECS LAN Emulation Configuration Server

Figure 5.52 ATM Emulation enables legacy LANs to operate via an ATM backbone

LAN Emulation constraints

One of the major problems associated with the use of ATM as a backbone for legacy LANs involves the LAN Emulation process. If the backbone switch performing emulation services should fail, all communications between legacy LANs also fail. Another problem is the temporary failure of an ATM switch due to a bad power supply causing previously learned addresses to be purged from cache memory. If a network using an ATM backbone has thousands of stations on legacy LANs, it can take 15 to 20 minutes or more until the MAC to ATM addresses are again resolved, adversely affecting performance between legacy LANs during that learning period. For these reasons, many organizations using an ATM backbone continue to place workgroup servers on local switches instead of moving them to a 'server farm' on the backbone. This infrastructure enables local switches to continue to provide access to workgroup servers even if the backbone should fail.

5.5 ACCESS SERVERS

An access server is a conversion device which enables remote terminal devices, such as a remote PC, to access a local area network as if the remote device was directly connected to the network. Since the access server enables remote access to a distant LAN, another popular term used for this product is 'remote access server'; however, the acronym RAS should not be confused with a specific Microsoft product included in Windows NT 4.0 which enables dial-in and Internet access to an NT 4.0 server. Microsoft uses the term RAS to reference a specific product while other vendors and trade publications may use the term to reference either a different product or the generic capability to access a distant LAN remotely located from another computer.

Overview

The best way to become familiar with the basic capability and features of a generic remote access server is to examine its typical hardware configuration. Figure 5.53 illustrates the general components of a remote access server in the form of a hardware schematic of the backplane of this product. Examining the figure from left to right, the terminal connector provides a mechanism to control the server, enabling a PC or conventional terminal to be connected to the server as a console. The next two connectors shown, a DB-15 AUI port and an RJ-11 jack, enable the server to be connected to a coaxial cable or twisted-pair-based Ethernet LAN. The two jacks on the right end of the server's backplane typically enable a variety of different types of WAN connections, such as 56 kbps digital dial-up, T1 or E1, and ISDN.

Although Ethernet connectors are shown in Figure 5.53, different types of remote access servers support different types of LANs, ranging from Token-Ring and FDDI to Fast Ethernet and, by the time you read this book, Gigabit Ethernet. Through the use of a T1 WAN connection, a remote access server can support up to 24 simultaneous analog or digital connections, while the use of an E1 connection enables the support of up to 30 sessions over a common WAN facility.

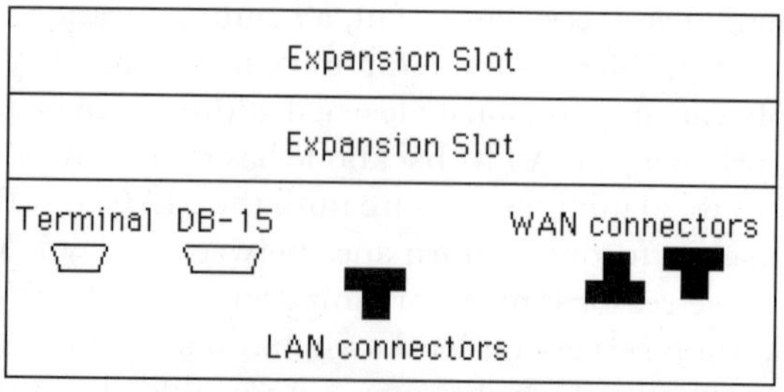

Figure 5.53 Basic hardware components of an access server

Utilization

The most common use of a remote access server is for providing dial-up access to the Internet. To do so an Internet Service Provider (ISP) installs a remote access server at their facility and connects the server to their internal LAN. Dial-in modem subscribers and subscribers accessing the ISP via ISDN can use their communications carrier to reach the remote access server which then converts the incoming access request into a LAN session, with access verified via an access manager server. Once access is verified, the session is accepted and the ISP enables subsequent requests to be routed onto the Internet.

Figure 5.54 illustrates the use of a remote access server to provide dial-up access to the Internet via an ISP. In examining Figure 5.54 it should be noted that some remote access servers include a bank of modems which results in demodulation and remodulation occurring at the ISP's facility. In other situations the ISP may use its remote access server to support ISDN and switched 56 kbps access, using built-in DSUs instead of modems to support digital access over a wide area network connection. Although not shown in Figure 5.54, most ISPs have a mixture of different types of servers on their LAN. These can include mail servers, Web servers, and different types of security or access manager servers.

In addition to Internet access, remote access servers are commonly used to provide remote access to academic, government, and corporate LANs. If such LANs are connected to a backbone organizational network, then the use of a remote access server can provide access to a broader network. Figure 5.55 illustrates how a remote access server could be used to provide organizational employees with access to a corporate network via a variety of wide area network connections. In this

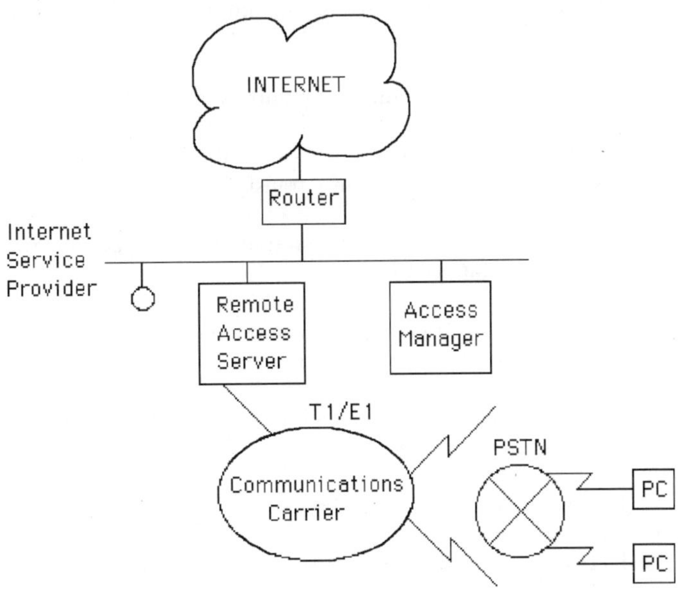

Figure 5.54 Using a remote access server to obtain dial-up access to the Internet

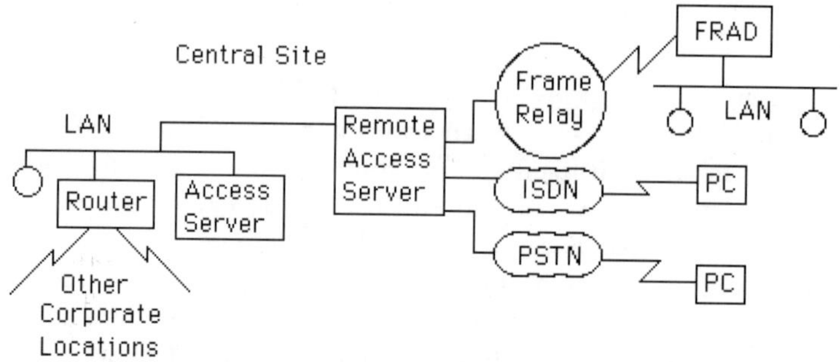

Figure 5.55 Using an access server to access a corporate

example, access to the corporate central site is shown occurring via frame relay, ISDN, and dial-up.

REVIEW QUESTIONS

5.1.1 Where does a bridge operate with respect to a layer of the OSI Reference Model?

5.1.2 From a physical perspective what is the difference between a local and a remote bridge?

5.1.3 What type of address is used by a bridge for making frame forwarding decisions?

5.1.4 Describe the terms forwarding, filtering, and flooding with respect to the operation of a bridge.

5.1.5 Why do bridges time-stamp entries in their address/port tables?

5.1.6 What is the relationship between the type of data line protocols employed by two LANs connected by the use of a translating bridge?

5.1.7 What do you have to do to a translating bridge to permit Ethernet and Token-Ring networks to be interconnected?

5.1.8 What is the difference between the physical and active topology of a network formed through the use of transparent bridges?

5.1.9 What is a minimum spanning tree? How can it be found given a graph containing weights assigned to each link in the graph?

5.1.10 What is a root bridge?

5.1.11 What is the difference between a designated bridge and a standby bridge?

5.1.12 What is routing information placed under source routing?

5.1.13 How does a Token-Ring workstation locate a path to its destination via a data flow over bridges and other rings?

5.1.14 Discuss two advantages and two disadvantages associated with the use of source routing.

5.1.15 How does a source routing transparent bridge operate?

5.1.16 Assume the availability level of a bridge is 99.9%. If two bridges are placed in series to interconnect three LANs, what is the probability of data flowing from the first to the third LAN, ignoring the availability level of each network?

5.2.1 Where does a router operate with respect to the seven-layer OSI Reference Model?

5.2.2 What is the difference between bridges and routers with respect to flooding?

5.2.3 Discuss three features of routers that are not included in bridges.

5.2.4 What is the function of the Address Resolution Protocol?

5.2.5 What is the relationship between the number of address table entries in a router and its support for multiple protocols?

5.2.6 Name three techniques used by a router to handle non-routable protocols.

5.2.7 What are two key advantages associated with a protocol-independent router?

5.2.8 What do the terms 'vector' and 'distance' reference under a vector distance routing protocol?

5.2.9 What values are broadcast between routers that use the routing information protocol?

5.2.10 What is the advantage of a link state protocol over a vector state protocol with respect to network traffic?

5.2.11 Discuss the difference between vector distance and link state protocols with respect to traffic balancing.

5.3.1 Name two types of gateways and discuss their general functionality.

5.3.2 Where does an application gateway operate with respect to the ISO Reference Model?

5.3.3 Describe how a multiprotocol gateway can be used.

5.3.4 Describe the hardware and physical connection required to enable a user on a Token-Ring network to access an IBM mainframe via a 3174 control unit.

5.3.5 Describe the hardware and physical connections required to enable a user on an Ethernet network to access an IBM mainframe via a 3174 control unit.

5.3.6 Describe a key disadvantage associated with the use of an SDLC gateway.

5.3.7 Why is the throughput of an X.25 gateway less than that of an SDLC gateway operating at the same data rate?

5.3.8 What are two key differences between the use of a Token Interface Coupler (TIC) in a communications controller and a Token-Ring adapter in a control unit with respect to their gateway capability?

5.3.9 What is the key limitation associated with the use of a 3278/9 coaxial connection gateway?

5.3.10 Describe three connection options supported by the IBM PCOM/3270 program.

5.3.11 What is the purpose of Data Link Switching spoofing?

5.3.12 How does Data Link Switching enhance the transmission of SDLC data?

5.4.1 Name three methods you can use to overcome the effect of network segmentation.

5.4.2 What is the average transmission rate per station on a 20-device LAN that operates at 10 Mbps?

5.4.3 Why or why not is a bridge a LAN switch?

5.4.4 Describe the advantage of parallel switching over serial switching.

5.4.5 What would be the theoretical maximum throughput of a 64-port 10BASE-T LAN switch?

5.4.6 Define LAN switch latency.

5.4.7 Describe three LAN switching techniques.

5.4.8 Which LAN switching technique has a minimum amount of latency?

5.4.9 Which LAN switching technique results in a maximum amount of latency?

5.4.10 Which LAN switching technique adapts to the error rate of frames flowing through the switch?

5.4.11 What are the advantages associated with the use of a hybrid switch?

5.4.12 What is the difference, with respect to addresses recognized per port, between port- and segment-based LAN switches?

5.4.13 How many cross-connections can a LAN switch with a backplane speed of 600 mbps that supports 24 Fast Ethernet 100BASE-T ports handle?

5.4.14 Discuss three common methods by which the use of LAN switches facilitates enhancing network performance.

5.4.15 What is a fat pipe?

5.4.16 Describe two methods used for flow control by LAN switches.

5.4.17 Describe three incompatibilities between ATM and legacy LANs.

5.4.18 What technique is used to enable frames generated by legacy LANs to be routed over an ATM backbone?

5.5.1 What is the primary function of an access server?

5.5.2 Discuss two communications applications facilitated by the use of access servers.

6

WIDE AREA NETWORK DATA CONCENTRATION EQUIPMENT

The communications devices included in this chapter perform a variety of functions which govern their utilization for selected application areas, with their inclusion here based upon their common function of concentrating data. In this chapter, the operation and utilization of devices designed primarily to accomplish data concentration will be covered. Specific devices to be investigated in this chapter include a variety of multiplexing equipment, concentrators, and front-end processors, as well as equipment which permits modems, service units, lines, and the ports of the computers and other devices to operate on a shared use basis. In addition, a device which splits a data stream into two streams for transmission to take advantage of the difference in the tariff between high-speed and lower operating rate leased lines will also be covered.

Although the title of this chapter includes the term 'wide area network', this term is included only because the primary function of each device is to concentrate two or more data sources for transmission over a wide area network transmission facility. Thus, the title does not imply that these devices cannot be used to interconnect LANs. In fact, several devices covered in this chapter, including different types of multiplexers, packet assembler/disassemblers, Frame Relay Access Devices, control units, and front-end processors either directly or indirectly enable LANs to be interconnected over a wide area network.

Equipment sizing

One of the more frequent problems in data communications involves determining the appropriate capacity of a communications device. This usually involves selecting the number of ports or channels to be installed in the communications device to satisfy your communications requirements in an efficient manner. If too few ports are installed, users will encounter busy signals or may be placed in queues, resulting in a loss of productivity, if the users are within your organization, or the possible loss of customers, if users are from outside the organization. If too many ports are installed, many ports will be underutilized, probably resulting in the expenditure of funds for unnecessary hardware.

Owing to the importance of equipment sizing, a review of the mathematics associated with such problems is presented in Appendix A. The material presented in that appendix can be used for sizing such devices as multiplexers, concentrators, and front-end processors as well as for determining an appropriate configuration for a remote access server.

6.1 MULTIPLEXERS

With the establishment of distributed computing, the cost of providing required communications facilities became a major focus of concern to users. Numerous network structures were examined to determine the possibilities of using specialized equipment to reduce communications costs. For many networks where geographically distributed users accessed a common computational facility, a central location could be found which would serve as a hub to link those users to the computer. Even when transmission traffic was low and the cost of leased lines could not be justified on an individual basis, quite often the cumulative cost of providing communications to a group of users could be reduced if a mechanism was available to enable many communications devices to share common transmission facilities. This mechanism was provided by the utilization of multiplexers whose primary function is to provide the user with a reduction of communications costs. This device enables one high-speed line to be used to carry the formerly separate transmissions of a group of lower speed lines. The use of multiplexers should be considered when a number of devices communicate from within a similar geographical area or when a number of leased lines run in parallel for any distance.

Evolution

From the historical perspective, multiplexing technology can trace its origination to the early development of telephone networks. Then, as today, multiplexing was the employment of appropriate technology to permit a communications circuit to carry more than one signal at a time.

In 1902, 26 years after the world's first successful telephone conversation, an attempt to overcome the existing ratio of one channel to one circuit occurred. Using specifically developed electrical network terminations, three channels were derived from two circuits by telephone companies.

The third channel was denoted as the phantom channel, hence the name 'phantom' was applied to this early version of multiplexing. Although this technology permitted two pairs of wires to effectively carry the load of three, the requirement to keep the electric network finely balanced to prevent crosstalk limited its practicality.

Comparison with other devices

Multiplexers were originally prewired, fixed logic devices; they produced a composite output transmission by sharing frequency bands (frequency division multiplexing) or time slots (time division multiplexing) on a predetermined basis, with the result that the total transmitted output was equal to the sum of the individual data inputs. Multiplexers were also originally transparent to the communicator, so that data sent from a terminal device through a multiplexer to a computer were received in the same format and code by the computer as its original form.

Today many modern multiplexers are based upon programming concepts developed for a now obsolete data communications device referred to as a concentrator. Concentrators were minicomputer-based communications devices that obtained a high degree of functionality through specialized software. A key function performed by concentrators was a dynamic sharing technique. If a terminal device was not active, the concentrator would not reserve a space in its composite high-speed output as a traditional multiplexer would.

This scheme, commonly known as dynamic bandwidth allocation, enabled a larger number of data sources to share the use of a high-speed line through the use of a concentrator than when such devices were connected to a multiplexer, since the traditional multiplexer allocates a time slot or frequency band for each terminal, regardless of whether the terminal is active. For this reason, statistics and queuing theory played an important role in the planning and utilization of concentrators. Due to the stored program capacity of concentrators, those devices could be programmed to perform a number of additional functions. Such functions as the preprocessing of sign-on information and code conversion were used to reduce the burden of effort required by the host computer system.

The advent of statistical and intelligent multiplexers, which are discussed later in this section, resulted in their gradual substitution for concentrators. Through the use of built-in microprocessors, these multiplexers can now be programmed to perform numerous functions previously available only through the use of concentrators.

Device support

In general, any device that transmits or receives a serial data stream can be considered a candidate for multiplexing. Data streams produced by the devices listed in Table 6.1 are among those that can be multiplexed. The intermix of devices as well as the number of any one device whose data stream is considered for multiplexing is a function of the multiplexer's capacity and capabilities, the economics of the application, and cost of other devices which could be employed in that role, as well as the types and costs of high-speed lines being considered.

Multiplexing techniques

Two basic techniques were developed for multiplexing: frequency division multiplexing (FDM) and time division multiplexing (TDM). Within the time

Table 6.1 Candidates for data stream multiplexing

Analog network private line modems
Analog switched network modems
Digital network data service units
Digital network channel service units
Data terminals
Data terminal controllers
Minicomputers
Routers
Computer ports
Computer–computer links
Other multiplexers

division technique, two versions are available—fixed time slots which are employed by traditional TDMs and most modern T1 multiplexers and variable use of time slots which are used by statistical and intelligent TDMs.

Frequency division multiplexing

Frequency division multiplexing (FDM) represents the earliest use of multiplexing by communications carriers. Although FDM was quite common when the backbone infrastructure of communications carriers was analog, its use has considerably diminished in tandem with the conversion of carrier backbone infrastructures to digital. However, as we approach the new century, there are still a few applications that continue to use FDM equipment and justify an overview of the technology. In addition, as previously noted in this book, digital subscriber lines which are emerging as a new high-speed local loop technology are based upon FDM, further justifying an examination of a technology many persons previously considered dated.

In frequency division multiplexing, the available bandwidth of a line is split into smaller segments called data bands or derived channels. Each data band in turn is separated from another data band by a guard band which is used to prevent signal interference between channels, as shown in Figure 6.1. Typically, frequency drift is the main cause of signal interference, and the size of the guard bands is structured to prevent data in one channel drifting into another channel.

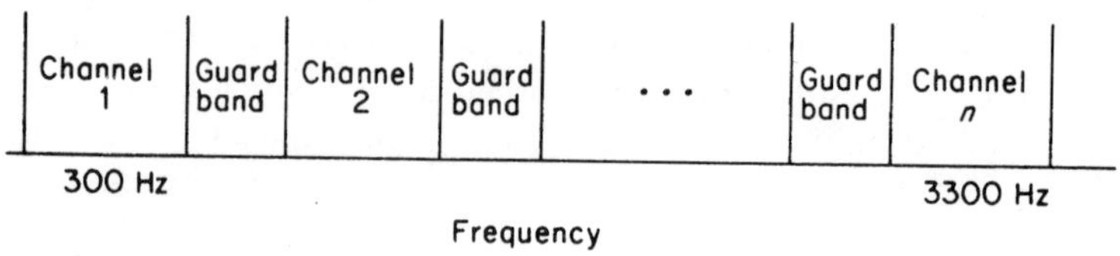

Figure 6.1 FDM channel separations. In frequency division multiplexing the 3 kHz bandwidth of a voice-grade line is split into channels or data bands separated from each other by guard bands

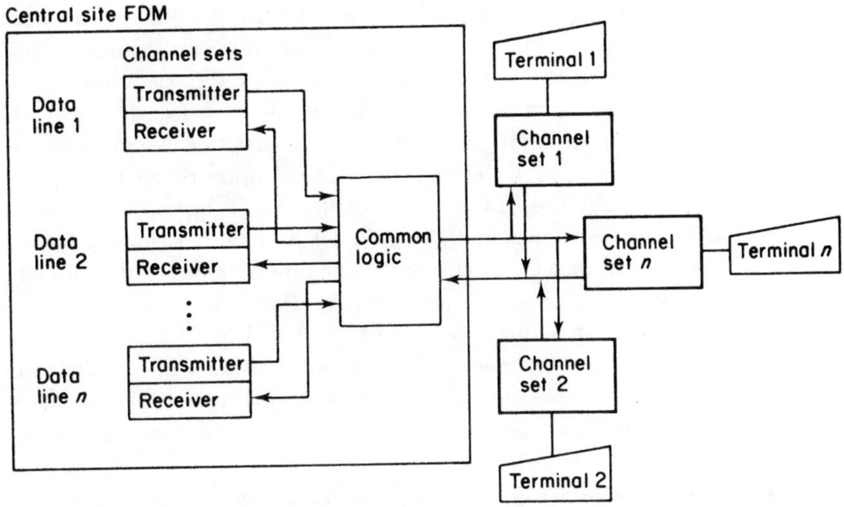

Figure 6.2 Frequency division multiplexing. Since the channel sets modulate the line at specified frequencies, no modems are required at remote locations

Physically, an FDM contains a channel set for each data channel as well as common logic, as shown in Figure 6.2. Each channel set contains a transmitter and receiver tuned to a specific frequency, with bits being indicated by the presence or absence of signals at each of the channel's assigned frequencies. In FDM, the width of each frequency band determines the transmission rate capacity of the channel, and the total bandwidth of the line is a limiting factor in determining the total number or mix of channels that can be serviced. Although a multipoint operation is illustrated in Figure 6.2, FDM equipment can also be utilized for the multiplexing of data between two locations on a point-to-point circuit. Data rates up to 1200 bps can be multiplexed by an FDM connected to a voice grade analog line. Typical FDM channel spacings required at different data rates for operation on a voice-grade line are listed in Table 6.2.

Table 6.2 FDM voice-grade leased line channel spacings

Speed (bps)	Spacing (Hz)
75	120
110	170
150	240
300	480
450	720
600	960
1200	1800

An FDM's aggregate data handling capacity depends upon the mixture of subchannels as well as on the type of line conditioning added to a circuit. A chart of FDM subchannel allocations for a voice-grade leased line is illustrated in Figure 6.3. This chart can be used to compute the mixture of data subchannels that can be transmitted via a single voice-grade leased line when frequency division multiplexing is employed. The figure is based upon data subchannel spacing standards formulated by the ITU. As illustrated, 17 75-bps subchannels can be multiplexed on an unconditioned (C0) circuit, 19 subchannels on a C1 conditional circuit, 22 subchannels on a C2 circuit and 24 channels on a C4 conditioned circuit. For higher data rates ITU standards allocate a fractional proportion of the bandwidth allocated to the previously discussed data rate of 75 bps.

With the development of terminals operating at speeds that were not multiples of ITU frequencies, such as 134.5 bps teleprinters, a number of vendors developed FDM equipment tailored to make more efficient use of voice-grade circuits than permitted by ITU standards.

Since the physical bandwidth of the line limits the number of devices which may be multiplexed, FDM is mainly used for multiplexing low-speed asynchronous terminals. An advantage obtained through the use of such equipment is its code transparency. Once a data band is set, any terminal operating at that speed or less can be used on that channel without concern for the code of the terminal. For example, a channel set to carry 300 bps transmission could also be used to service an IBM 2741 terminal transmitting at 134.5 bps or a Teletype 110 bps terminal. Another advantage of FDM equipment is that no modems are required since the channel sets modulate the line at specified frequencies, as shown in Figure 6.2. At the main computer site, the FDM multiplexer interfaces computer ports through channel sets. The common logic acts as a summer, connecting the multiplexer channel sets to the leased line.

At each remote location, a channel set provides the necessary interface between the terminal device at that location and the leased line. When using FDM equipment, individual data channels can be picked up or dropped off at any point on a telephone circuit. This characteristic permits the utilization of multipoint lines and can result in considerable line charge reductions based upon developing a single circuit which can interface multiple terminals. Each remote terminal device to be serviced only needs to be connected to an FDM channel set which contains band-pass filters that separate the line signal into the individual frequencies designated for that terminal. Guard bands of unused frequencies are used between each channel frequency to permit the filters a degree of tolerance in separating out the individual signals.

FDM normally operates in a full-duplex transmission mode on a four-wire circuit by having all transmit tones sent on one pair of wires and all receive tones return on a second pair. FDM can also operate in the full-duplex mode on a two-wire line. This can be accomplished by having the transmitter and receiver of each channel set tuned to different frequencies. For example, with 16 channels available, one channel set could be tuned for channel 1 to transmit and channel 9 to receive, while another channel set would be tuned to channel 2 to transmit and channel 10 to receive. With this technique, the number of data channels is halved.

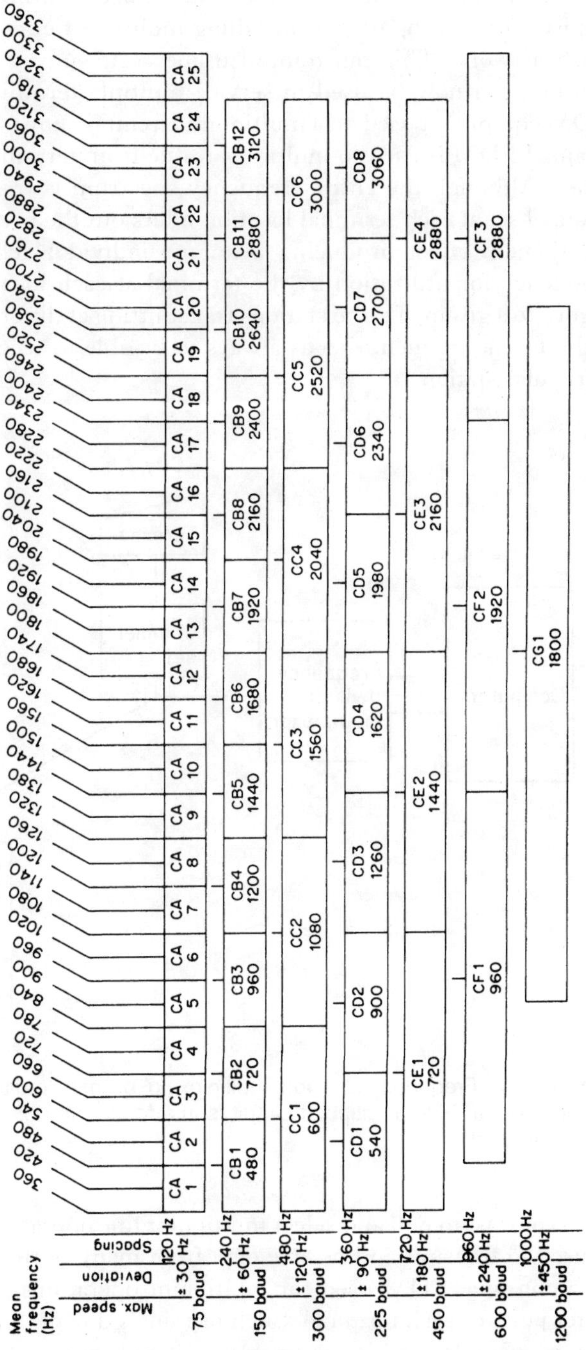

Figure 6.3 ITU FDM subchannel allocations for a voice-grade leased line

FDM utilization

As mentioned previously, one key advantage in utilizing FDM equipment is the ability afforded the user in installing multipoint circuits for use in a communications network. This can minimize line costs since a common line, optimized in routing, can now be used to service multiple terminal locations. An example of FDM equipment used on a multipoint circuit is shown in Figure 6.4, where a three-channel FDM is used to multiplex traffic from terminals located in three different cities. Although the entire frequency spectrum is transmitted on the circuit, the channel set at each terminal location filters out the preassigned bandwidth for that location, in effect producing a unique individual channel that is dedicated by frequency for utilization by the terminal at each location. This operation is analogous to a group of radio stations transmitting at different frequencies and setting a radio to one frequency so as always to be able to receive the transmission from a particular station.

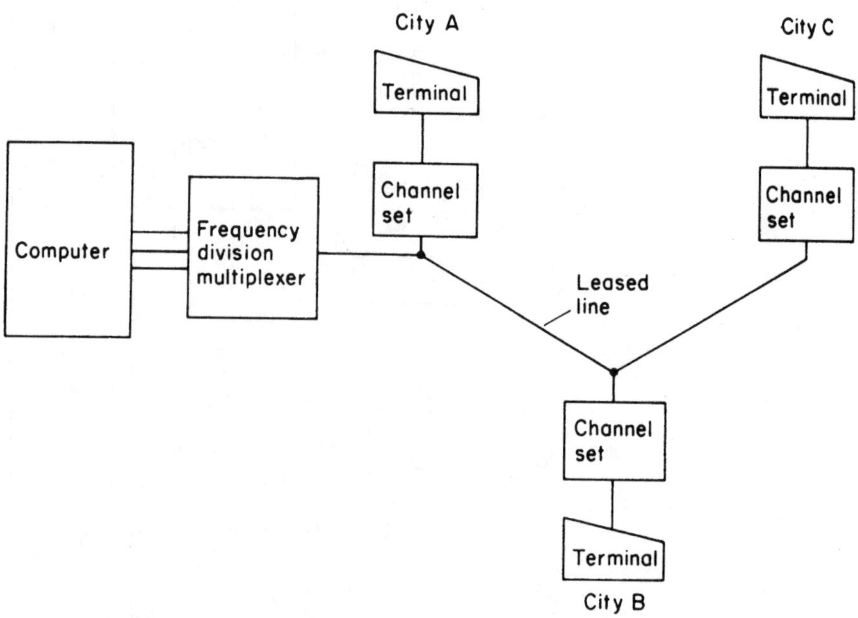

Figure 6.4 Frequency division multiplexing permits multipoint circuit operations. Each terminal on an FDM multipoint circuit is interfaced through the multiplexer to an individual computer port

In contrast to poll and select multipoint line operations where one computer port is used to transmit and receive data from many buffered terminals connected to a common line, FDM used for multipoint operations as shown requires one computer port for each terminal. Such terminals do not, however, require a buffer area to recognize their addresses, nor is poll and select software required to operate in the computer. When buffered terminals and poll and select software are available, polling by channel can take place, as illustrated in Figure 6.5. In this example,

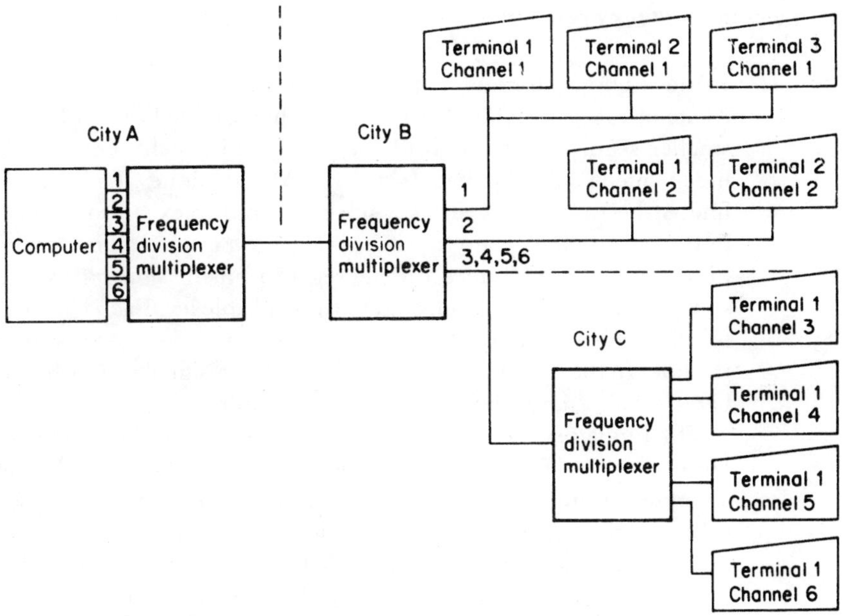

Figure 6.5 FDM can intermix polled and dedicated terminals in a network. Of the six channels used in this network, channels 1 and 2 service a number of polled terminals, while channels 3 to 6 are dedicated to service individual terminals

channels 1 and 2 are each connected to a number of relatively low-traffic terminals which are polled through the multiplexer system. Terminals 3 and 6 are presumed to be higher traffic stations and are thus connected to individual channels of the FDM or to individual channel sets.

Current status

Similar to FDM used by common carriers to multiplex many voice conversations, the use of FDM by organizations to multiplex data has greatly diminished. By the early 1990s, the use of FDM was essentially limited to niche markets, such as slow-speed terminals connected via multidrop lines to a common computer facility. In this type of application environment, terminals at different sites are typically used to transmit information concerning the arrival and departure of trucks. If it is only critical to know when a truck arrived and departed a specific depot and not its actual manifest, slow-speed terminals connected via multidrop lines using FDM are more than adequate. If large quantities of data must traverse the network, including manifest information, the speed limitations of FDM would probably make it unacceptable and result in the use of TDM technology.

In the late 1990s a new role for the use of FDM was evolving. That role involves the splitting of a telephone subscriber's local access line into two asymmetric channels by assigning more bandwidth in one direction than in the other direction. This technique, in conjunction with other communications technology described in Chapter 4, forms the basis for asymmetric digital subscriber lines (ADSL).

Time division multiplexing

In the FDM technique, the bandwidth of the communications line serves as the frame of reference. The total bandwidth is divided into subchannels consisting of smaller segments of the available bandwidth, each of which is used to form an independent data channel. In the TDM technique, the aggregate capacity of the line is the frame of reference, since the multiplexer provides a method of transmitting data from many devices over a common circuit by interleaving them in time. The TDM divides the aggregate transmission on the line for use by the slower-speed devices connected to the multiplexer. Each device is given a time slot for its exclusive use so that at any one point in time the signal from one device is on the line. In the FDM technique, in which each signal occupies a different frequency band, all signals are being transmitted simultaneously.

The fundamental operating characteristics of a TDM are shown in Figure 6.6. Here, each low- to medium-speed terminal is connected to the multiplexer through an input/output (I/O) channel adapter. The I/O adapter provides the buffering and control functions necessary to interface the transmission and reception of data from the terminals to the multiplexer. Within each adapter, a buffer or memory area exists which is used to compensate for the speed differential between terminal devices and the multiplexer's internal operating speed. Data is shifted from terminal devices to the I/O adapter at different rates (typically 110 to 19 200 bps and multiples of 56 or 64 kbps), depending upon the speed of the device; but when data is shifted from the I/O adapter to the central logic of the multiplexer, or from central logic to the composite adapter, it is at the much higher fixed rate of the TDM. On output from the multiplexer to each terminal device the reverse is true, since data is first transferred at a fixed rate from central logic to each adapter and then from the adapter to the terminal device at the data rate acceptable to the device.

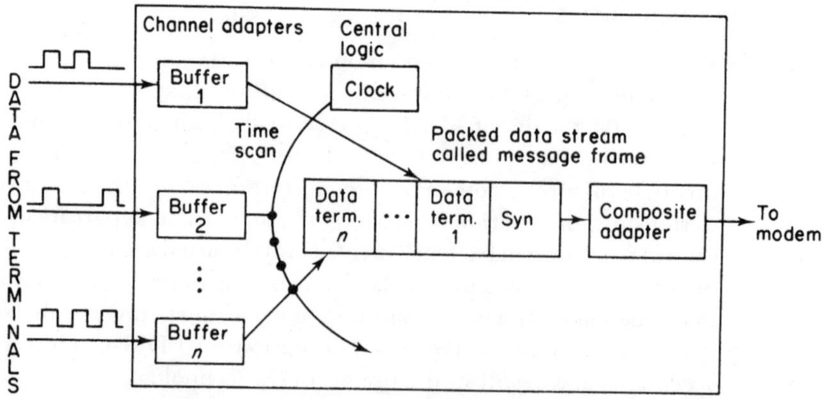

Figure 6.6 Time division multiplexing. In time division multiplexing, data are first entered into each channel adapter buffer area at a transfer rate equal to the device to which the adapter is connected. Next, data from the various buffers are transferred to the multiplexer's central logic at the higher rate of the device for packing into a message frame for transmission

Depending upon the type of TDM system, the buffer area in each adapter will accommodate either bits or characters.

The central logic of the TDM contains controlling, monitoring, and timing circuitry which facilitates the passage of individual terminal data to and from the high-speed transmission medium. The central logic will generate a synchronizing pattern which is used by a scanner circuit to interrogate each of the channel adapter buffer areas in a predetermined sequence, blocking the bits of characters from each buffer into a continuous, synchronous data stream which is then passed to a composite adapter. The composite adapter contains a buffer and functions similar to the I/O channel adapters. It now, however, compensates for the difference in speed between the high-speed transmission medium and the internal speed of the multiplexer.

The multiplexing interval

When operating, the multiplexer transmits and receives a continuous data stream known as a message train, regardless of the activity of connected terminal devices. The message train is formed from a continuous series of message frames which represents the packing of a series of input data streams. Each message frame contains one or more synchronization characters followed by a number of basic multiplexing intervals whose number is dependent upon the model and manufacturer of the device. The basic multiplexing interval can be viewed as the first level of time subdivision which is established by determining the number of equal sections per second required by a particular application. Then, the multiplexing interval is the time duration of one section of the message frame.

When TDMs were first introduced the section rate was established at 30 sections per second, which then produced a basic multiplexing interval of 0.033 seconds or 33 ms. Setting the interval to 33 ms made the multiplexer directly compatible with a 300-baud asynchronous channel which transmits data at up to 30 characters per second (cps). With this interval, the multiplexer was also compatible with 150-baud (15-cps) and 110-baud (10-cps) data channels, since the basic multiplexing interval was a multiple of those asynchronous data rates. Later TDMs had a section rate of 120 sections per second, which then made the multiplexer capable of servicing a range of asynchronous data streams up to 1200 bps. Modern TDMs have a section rate of at least 19 200 sections per second, providing support for asynchronous data streams up to 19 200 bps.

TDM techniques

The two TDM techniques available are bit interleaving and character interleaving. Bit interleaving is generally used in systems which service synchronous terminal devices, whereas character interleaving is generally used to service asynchronous terminal devices.

When interleaving is accomplished on a bit-by-bit basis, the multiplexer takes one bit from each channel adapter and then combines them as a word or frame for transmission. As shown in Figure 6.7(a), this technique produces a frame containing one data element from each channel adapter. When interleaving is accomplished

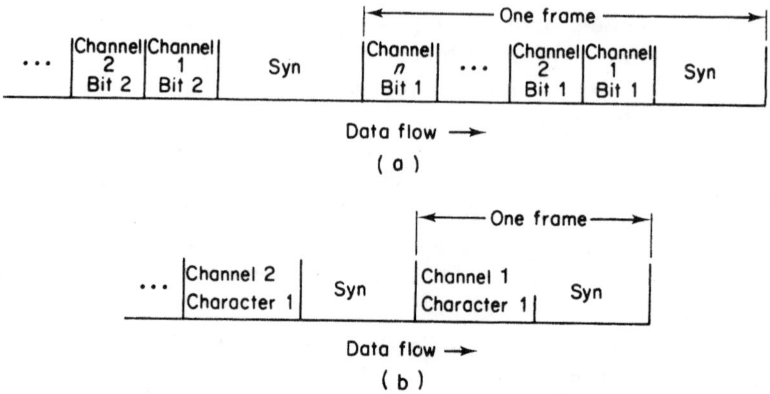

Figure 6.7 (a) Time division interleaving bit-by-bit. The first bit from each channel is packed into a frame for transmission. (b) Time division multiplexing character-by-character. One or more complete characters are grouped with a synchronization character into a frame for transmission

on a character-by-character basis, the multiplexer assembles a full character into one frame and then transmits the entire character, as shown in Figure 6.7(b). Although a frame containing only one character of information is illustrated in Figure 6.7, to increase transmission efficiency most multiplexers transmit long frames containing a large number of data characters to reduce the synchronization overhead associated with each frame. Thus, while a frame containing one character of information has a synchronization overhead of 50%, a frame containing four data characters has its overhead reduced to 20%, and a frame containing nine data characters has a synchronization overhead of only 10%, assuming constant slot sizes for all characters.

When TDMs were first developed, the larger buffer area required for character-by-character multiplexing made this technique more expensive. Today the primary difference between the two techniques is in the area of latency or delay. Bit multiplexing results in a minimal delay and is very suitable for supporting high-speed synchronous data streams. In comparison, character-by-character multiplexing enables the multiplexer to perform a variety of operations on data, such as stripping start, stop, and parity bits from asynchronous characters and compressing groups of characters first stored in a buffer. However, such operations add a degree of latency which may not be suitable for high-speed data sources.

TDM applications

The most commonly used TDM configuration is the point-to-point system, which is shown in Figure 6.8. This type of system, which is also called a two-point multiplex system, is shown linking a mixture of terminal devices to a centrally located multiplexer. As shown terminal devices can be connected to the multiplexer in a variety of ways: by a leased line running from the terminal's location to the multiplexer, by a direct connection if the user's terminal device is within the same building as the multiplexer and a cable can be laid to connect the two, or via the

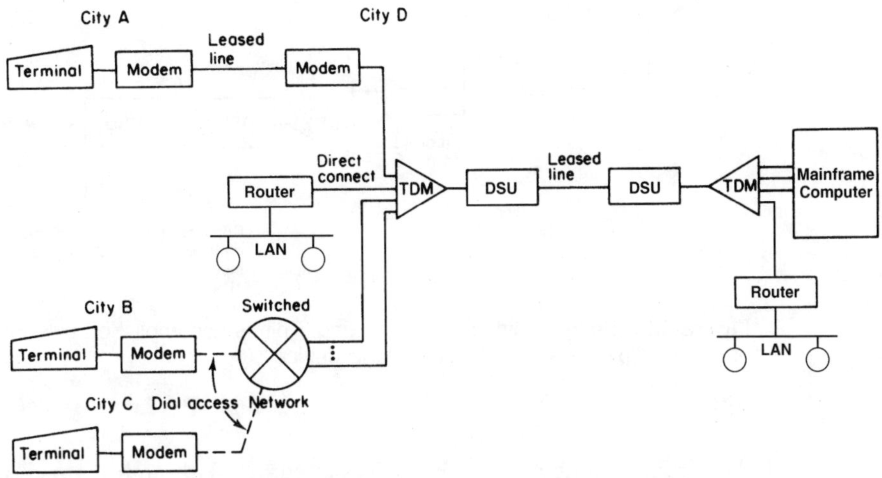

Figure 6.8 Time division multiplexing point-to-point. A point-to-point or two-point multiplexing system links a variety of data users at one or more remote locations to a central computer facility

switched network to call the multiplexer over the dial network. For the latter method, since the connection is not permanent, several terminal devices can share access to one or more multiplexer channels on a contention basis.

As shown in Figure 6.8, terminal devices in cities B and C use the dial network to contend for two multiplexer channels which are interfaced to an automatic answer unit on the dial network. This network structure enables a number of remote users to contend for access to the two dial-in lines. Also note in Figure 6.8 that a router which provides the capability to interconnect LANs is shown directly input to the TDM at city D. This illustrates how TDMs can be used to support a variety of different networking applications. Although many organizations replaced the use of dial-in lines to TDM ports with remote access servers that allow remote users to directly connect to a LAN, in some situations the use of TDM ports and a telephone company rotary can be considerably more economical while providing the same type of access. That is, if remote users only require access to a distant mainframe it may be more economical to avoid the use of an access server and use multiplexer ports to provide access to the distant computer as indicated in the lower portion of Figure 6.8.

Series multipoint multiplexing

A number of multiplexing systems can be developed by linking the output of one multiplexer into a second multiplexer. Commonly called series multipoint multiplexing, this technique is most effective when terminals are distributed at two or more locations and the user desires to alleviate the necessity of obtaining two long-distance leased lines from the closer location to the computer. As shown in Figure 6.9, four devices are multiplexed at city A onto one high-speed channel which is transmitted to city B where this line is in turn multiplexed along with the

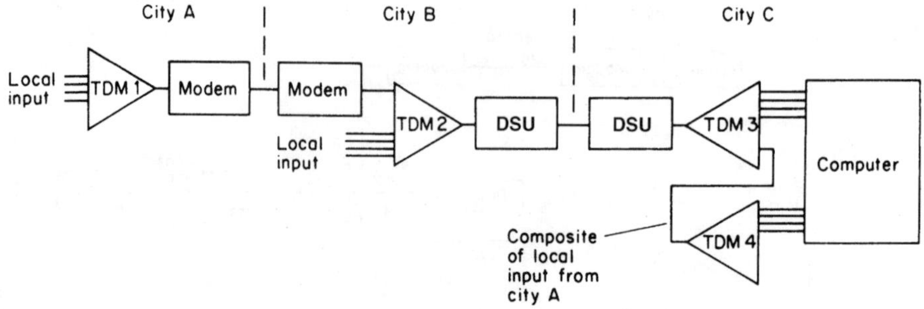

Figure 6.9 Series multipoint multiplexing. This is accomplished by connecting the output of one multiplexer as input to a second device

data from a number of other devices at city B. Although the user requires a leased line between city A and city B, only one line is now required to be installed for the remainder of the distance from city B to the computer at city C. If city A is located 50 miles from city B, and city B is 2000 miles from city C, 2000 miles of duplicate leased lines are avoided by using this multiplexing technique.

Multipoint multiplexing requires an additional pair of channel cards to be installed at multiplexers 2 and 3 and high-speed modems or DSUs to be interfaced to those multiplexers to handle the higher aggregate throughput when the traffic of multiplexer 1 is routed through multiplexer 2; but, in most cases the cost savings associated with reducing duplicated leased lines will more than offset the cost of the extra equipment. Since this is a series arrangement a failure of either TDM2 or TDM3 or a failure of the line between these two multiplexers will terminate service to all terminals connected to the system.

Hub-bypass multiplexing

A variation of series multipoint multiplexing is hub-bypass multiplexing. To be effectively used, hub-bypass multiplexing can occur when a number of remote locations have the requirement to transmit to two or more locations. To satisfy this requirement, remote terminal traffic is multiplexed to a central location which is the hub, and the terminal devices which must communicate with the second location are cabled into another multiplexer which transmits this traffic, bypassing the hub.

Figure 6.10 illustrates one application where hub bypassing might be utilized. In this example, eight devices at city 3 require a communications link with one of two computers; six devices always communicate with the computer at city 2, while two devices use the facilities of the computer at city 1. Data from all eight devices are multiplexed over a common line to city 2 where the two channels that correspond to the terminal devices which must access the computer at city 1 are cabled to a new multiplexer, which then remultiplexes the data to city 1. When many remote locations have dual location destinations, hub-bypassing can become very economical. Since the data flows in series, an equipment failure will, however, terminate access to one or more computational facilities, depending upon the location of the break in service.

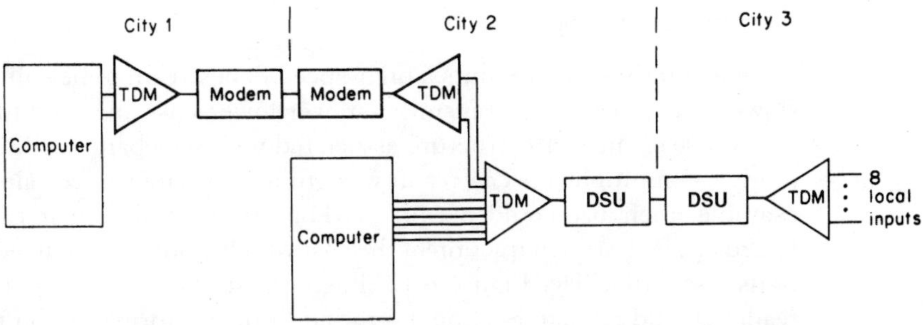

Figure 6.10 Hub-bypass multiplexing. This method should be considered when a number of terminals have the requirement to communicate with more than one location

Although hub-bypass multiplexing can be effectively used to connect collocated terminal devices to different destinations, if more than two destinations exist a more efficient switching arrangement can be obtained by the employment of a port selector or a multiplexer that has port selection capability. The reader is referred to Section 6.9 for information concerning port selection.

Front-end substitution

Although rarely utilized, a TDM may be installed as an inexpensive front end for a computer, as shown in Figure 6.11. When used as a front end, only one computer port is then required to service terminal devices which are connected to the computer through the TDM. The TDM can be connected at the computer center, or it can be located at a remote site and connected over a leased line and a pair of modems or DSUs. Since demultiplexing is conducted by the computer's software, only one multiplexer is necessary.

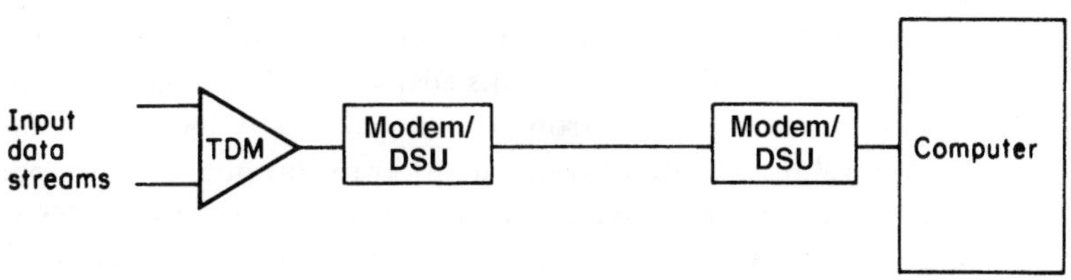

Figure 6.11 TDM system used as a front end. When a TDM is used as a front-end processor, the computer must be programmed to perform demultiplexing

Due to the wide variations in multiplexing techniques of each manufacturer, no standard software has been written for demultiplexing; and, unless multiple locations can use this technique, the software development costs may exceed the hardware savings associated with this technique. In addition, the software overhead associated with the computer performing the demultiplexing may degrade its performance to an appreciable degree and must be considered.

Inverse multiplexing

Inverse multiplexing permits a high-speed data stream to be split into two or more slower data streams for transmission over lower-cost lines and modems or DSUs.

Because of the tariff structure associated with wideband facilities, the utilization of inverse multiplexers can result in significant savings in certain situations. As an example, their use could permit 128 kbps transmission over two 56 kbps digital leased or dial lines at a fraction of the cost which would be incurred when using a T1 transmission facility. Figure 6.12 illustrates the use of inverse multiplexers. The reader should refer to Section 4.3 for additional information on these devices.

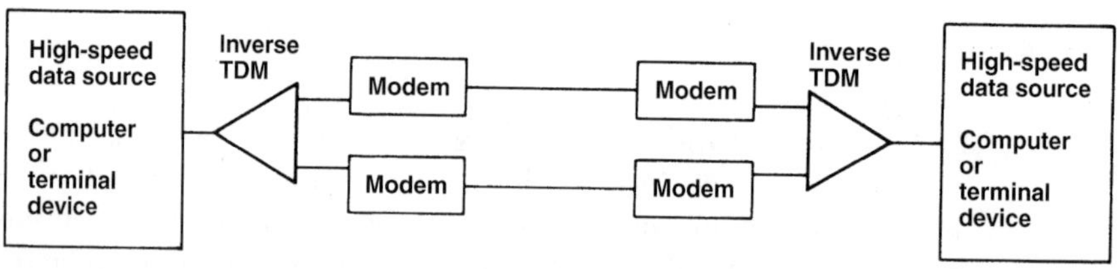

Figure 6.12 Inverse multiplexing. An inverse multiplexer splits a serial data stream into two or more individual data streams for transmission at lower data rates

Multiplexing economics

The primary motive for the use of multiplexers in a network is to reduce the cost of communications. In analyzing the potential of multiplexers, you should first survey users to determine the projected monthly connect time of each terminal device. Then, the most economical method of data transmission from each individual terminal device to the computer facility can be computed. To do this, direct dial costs should be compared with the cost of a leased line from each terminal location to the computer site.

Once the most economical method of transmission for each individual terminal to the computer is determined, this cost should be considered the 'cost to reduce'. The telephone mileage costs from each terminal city location to each other terminal city location should be determined in order to compute and compare the cost of utilizing various techniques, such as line dropping and the multiplexing of data by combining several low-to medium-speed data streams into one high-speed line for transmission to the central site.

In evaluating multiplexing costs, the cost of telephone lines from each terminal location to the 'multiplexer center' must be computed and added to the cost of the multiplexer equipment. Then, the cost of the high-speed line from the multiplexer center to the computer site must be added to produce the total multiplexing cost. If this cost exceeds the cumulative most economical method of transmission for individual terminal devices to the central site, then multiplexing is not cost-justified. This process should be reiterated by considering each city as a possible multiplexer center to optimize all possible network configurations. In repeating this process, terminals located in certain cities will not justify any calculations to prove or

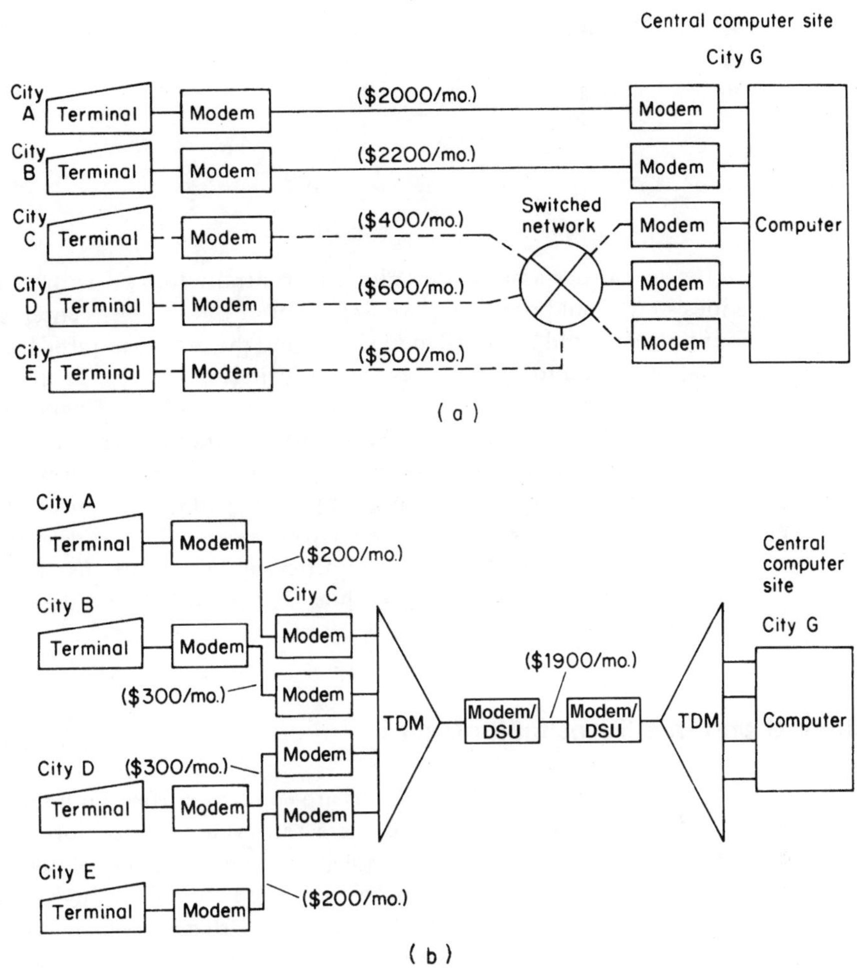

Figure 6.13 Multiplexing economics. On an individual basis, the cost of five terminals accessing a computer system (a) can be much more expensive than when a time division multiplexer is installed (b)

disprove their economic feasibility as multiplexer centers, because of their isolation from other cities in a network.

An example of the economics involved in multiplexing is illustrated in Figure 6.13. In this example, assume the volume of traffic from the devices located in cities A and B would result in a dial-up charge of $3000 per month if access to the computer in city G was over the switched network. The installation of leased lines from those cities to the computer at city G would cost $2000 and $2200 per month, respectively. Furthermore, let us assume that the terminals at cities C, D, and E only periodically communicate with the computer, and their dial-up costs of $400, $600 and $500 per month, respectively, are much less than the cost of leased lines between those cities and the computer. Then, without multiplexing, the network's most economical communications cost would be

Location	Cost per month
City A	\$2000
City B	\$2200
City C	\$ 400
City D	\$ 600
City E	\$ 500
Total cost	\$5700

Let us further assume that city C is centrally located with respect to the other cities so we could use it as a homing point or multiplexer center. In this manner, a multiplexer could be installed in city C, and the terminal traffic from the other cities could be routed to that city, as shown in Figure 6.13(b). Employing multiplexers would reduce the network communications cost to \$2900 per month which produces a potential savings of \$2800 per month, which should now be reduced by the multiplexer costs to determine net savings. If each multiplexer costs \$500 per month, then the network using multiplexers will save the user \$1800 each month. Exactly how much saving can be realized, if any, through the use of multiplexers depends not only on the types, quantities, and distributions of terminal devices to be serviced but also on the leased line tariff structure and the type of multiplexer employed.

Statistical and intelligent multiplexers

In a traditional TDM, data streams are combined from a number of devices into a single path so that each device has a time slot assigned for its use. While such TDMs are inexpensive and reliable, and can be effectively employed to reduce communications costs, they make inefficient use of the high-speed transmission medium. This inefficiency is due to the fact that a time slot is reserved for each connected device, whether or not the device is active. When the device is inactive, the TDM pads the slot with null characters and cannot use the slot for other purposes.

Null characters are inserted into the message frame since demultiplexing occurs by the position of characters in the frame. If these mulls are eliminated, a scheme must then be employed to indicate the origination port or channel of each character. Otherwise, there would be no way to correctly reconstruct data and route it to its correct computer port during the demultiplexing process.

A statistical multiplexer uses a microprocessor to combine signals from a number of active devices while ignoring inactive devices. This technique, which dynamically allocates the bandwidth of the high-speed transmission medium is, as you might expect, referred to as dynamic bandwidth allocation.

By dynamically allocating bandwidth as required, statistical multiplexers permit more efficient utilization of the high-speed transmission medium. This permits the multiplexer to service more terminal devices without an increase in the high-speed link as would a traditional multiplexer. The technique of allocating time slots on a demand basis is known as statistical multiplexing and means that data is transmitted by the multiplexer only from the devices that are actually active.

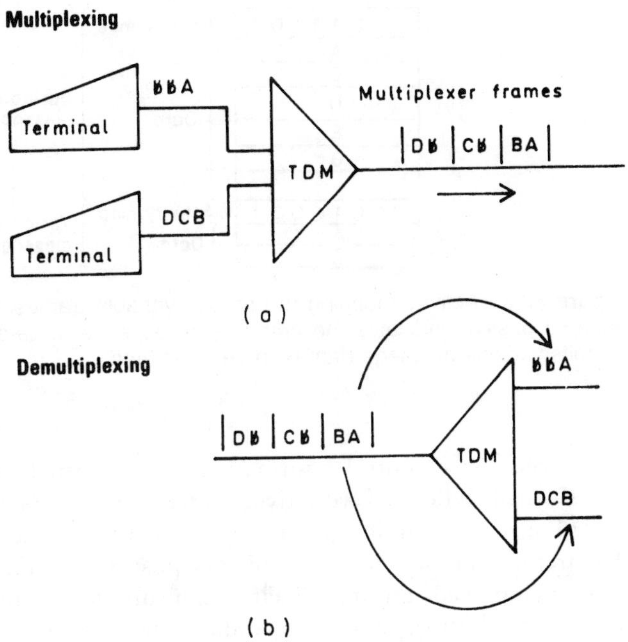

Figure 6.14 Multiplexing and demultiplexing by TDMs. (b̸ absence of line activity during multiplexer scan, null character inserted into message frame)

Depending upon the type of TDM, either synchronization characters or control frames are inserted into the stream of message frames. Synchronization characters are employed by conventional TDMs, while control frames are used by TDMs which employ a high-level data link control (HDLC) protocol between multiplexers to control the transmission of message frames.

The construction technique used to build the message frame also defines the type of TDM. Conventional TDMs employ a fixed frame approach as illustrated in Figure 6.14. Here, each frame consists of one character or bit for each input channel scanned at a particular period of time. As illustrated, even when a particular terminal device is inactive, the slot assigned to that device is included in the message frame transmitted since the presence of a pad or null character in the time slot is required to correctly demultiplex the data. In the lower portion of Figure 6.14 the demultiplexing process which is accomplished by time slot position is illustrated. Since a typical terminal device may be idle 90% of the time, this technique contains obvious inefficiencies.

Statistical frame construction

A statistical multiplexer employs a variable frame building technique which takes advantage of idle times to enable more devices to share access to a common circuit. The use of variable frame technology permits previously wasted time slots to be

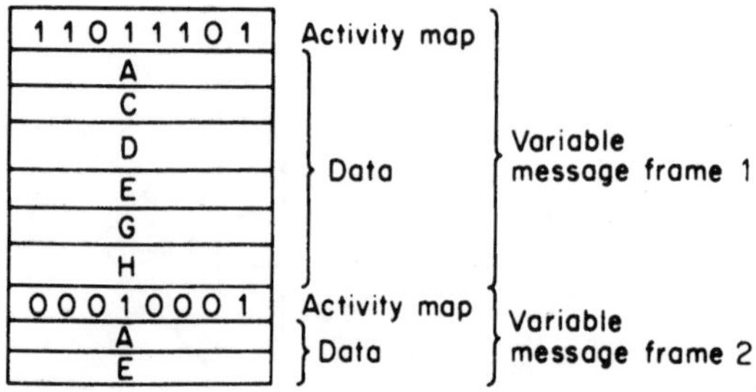

Figure 6.15 Activity mapping to produce variable frames. Using an activity map where each bit position indicates the presence or absence of data for a particular data source permits variable message frames to be generated

eliminated, since control information is transmitted with each frame to indicate which terminal devices are active and have data contained in the message frame.

One of many techniques that can be used to denote the presence or absence of data traffic is the activity map which is illustrated in Figure 6.15. When an activity map is employed, the map itself is transmitted before the actual data. Each bit position in the map is used to indicate the presence (bit position set to '1') or absence (bit position set to '0') of data from a particular multiplexer time slot scan. The two activity maps and data characters illustrated in Figure 6.15 represent a total of eight characters which would be transmitted in place of 16 characters that would be transmitted by a conventional multiplexer.

Another statistical multiplexing technique involves buffering data from each data source and then transmitting the data with an address and byte count. The address is used by the demultiplexer to route the data to the correct port, while the byte count indicates the quantity of data to be routed to that port. Figure 6.16 illustrates the message frame of a four-channel statistical multiplexer employing the address and byte count frame composition method during a certain time interval. Note that since channels 3 and 4 had no data traffic during the two particular time intervals, there was no address and byte count nor data from those channels transmitted on the common circuit. Also note that the data from each channel is of variable length. Typically, statistical multiplexers employing an address and byte count frame composition method wait until either 32 characters or a carriage return is encount-

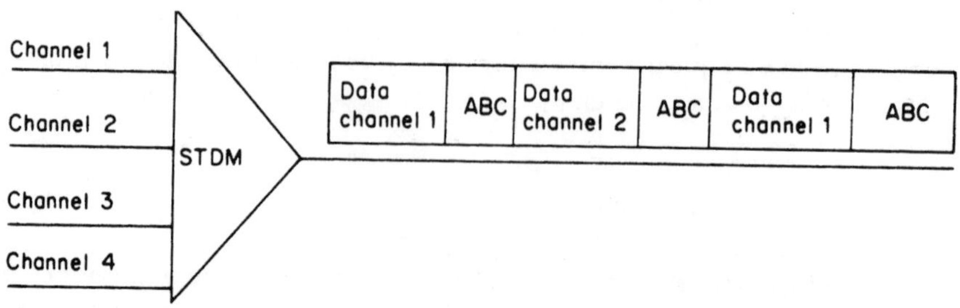

Figure 6.16 Address and byte count (ABC) frame composition

ered prior to forming the address and byte count and forwarding the buffered data. The reason 32 characters was selected as the decision criterion is that it represents the average line length of an interactive transmission session.

A few potential technical drawbacks of statistical multiplexers exist which users should note. These problems include the delays associated with data blocking and queuing when a large number of connected terminal devices become active or when a few terminals transmit large bursts of data. For either situation, the aggregate throughput of the multiplexer's input active data exceeds the capacity of the common high-speed line, causing data to be placed into buffer storage.

Another reason for delays is when a circuit error causes one or more retransmissions of message frame data to occur. Since the connected terminal devices may continue to send data during the multiplexer-to-multiplexer retransmission cycle, this can also fill up the multiplexer's buffer areas and cause time delays.

If the buffer area should overflow, data would be lost which would create an unacceptable situation. To prevent buffer overflow, all statistical multiplexers employ flow control as a mechanism to transmit a traffic control signal to attached terminals and/or computers when their buffers are filled to a certain level. Such control signals inhibit additional transmission through the multiplexer until the buffer has been emptied to another predefined level. Once this level has been reached, a second control signal is issued which permits transmission to the multiplexers to resume.

Flow control

Similar to intelligent modems discussed in Chapter 4, there are several flow control techniques employed by statistical multiplexers. Those methods supported by statistical multiplexers include inband signaling, outband signaling, and clock reduction. Inband signaling involves transmitting XOFF and XON characters to inhibit and enable the transmission of data from terminals and computer ports that recognize these flow control characters. Since many terminal devices and computer ports do not recognize these control characters, a second common flow control method involves raising and lowering the clear to send (CTS) control signal on the RS-232 or V.24 interface. Since this method of buffer control is outside the data path where data is transmitted on pin 2, it is known as outband signaling.

Both inband and outband signaling are used to control the data flow of asynchronous devices. Since synchronous devices transmit data formed into blocks or frames, the use of either inband or outband signaling would most likely break a block or frame. This would cause a portion of a block or frame to be received, which would result in a negative acknowledgement when the receiver performs its cyclic redundancy computation. Similarly, when the remainder of the block or frame is allowed to resume its flow to the receiver, a second negative acknowledgement would result.

To alleviate these potential causes of decrease of throughput, multiplexer vendors typically reduce the clocking speed furnished to synchronous devices. For example, a synchronous device operating at 4800 bps might first be reduced to 2400 bps by the multiplexer halving the clock. Then, if the buffer in the multiplexer continues to fill, the clock might be further reduced to 1200 bps.

Service ratio

The measurement used to denote the capability of a statistical multiplexer is called its service ratio, which compares its overall level of performance in comparison to a conventional TDM. Since synchronous transmission by definition denotes blocks of data with characters placed in sequence in each block, there are no gaps in this mode of transmission. In comparison, a terminal operator transmitting data asynchronously may pause between characters to think prior to pressing each key on the terminal. The service ratio of STDMs for asynchronous data is thus higher than the service ratio for synchronous data. Typically, STDM asynchronous service ratios range between 2:1 and 3.5:1, while synchronous service ratios range between 1.25:1 and 2:1, with the service ratio dependent upon the efficiency of the STDM as well as its built-in features, including the stripping of start and stop bits from asynchronous data sources. In Figure 6.17, the operational efficiency of both a statistical and a conventional TDM are compared. Here we have assumed that the STDM has an efficiency three times that of the TDM.

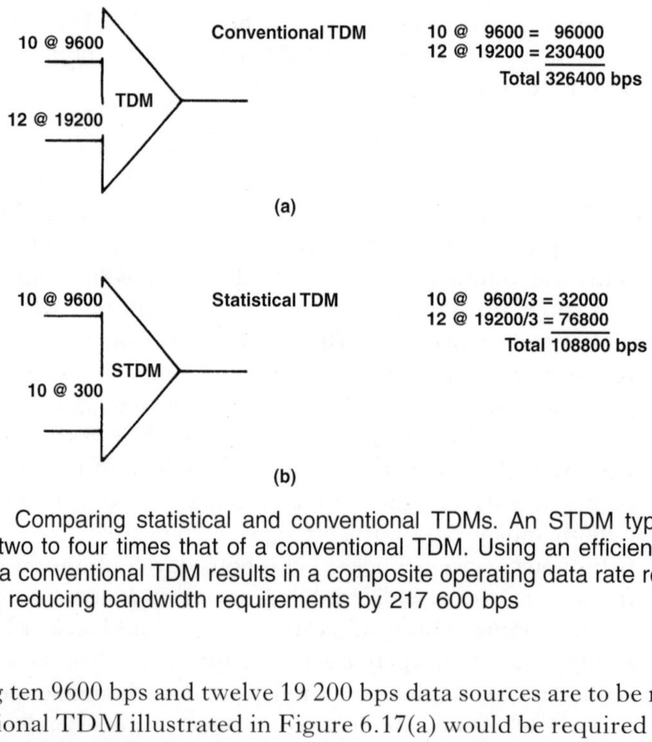

Figure 6.17 Comparing statistical and conventional TDMs. An STDM typically has an efficiency of two to four times that of a conventional TDM. Using an efficiency level three times that of a conventional TDM results in a composite operating data rate requirement of 108 800 bps, reducing bandwidth requirements by 217 600 bps

Assuming ten 9600 bps and twelve 19 200 bps data sources are to be multiplexed, the conventional TDM illustrated in Figure 6.17(a) would be required to operate at a data rate of at least 326 400 bps, requiring the use of a fractional T1 line operating at a minimum of 384 000 bps. For the STDM shown in Figure 6.17(b), assuming a three-fold increase in efficiency over the conventional TDM, the composite data rate required will be 108 800 bps. This allows the use of a 128 kbps fractional T1 line instead of a more expensive 384 kbps line that would be required if a conventional TDM was used.

Data source support

Some statistical multiplexers support only asynchronous data while other multiplexers support both asynchronous and synchronous data sources. When a statistical multiplexer supports synchronous data sources it is extremely important to determine the method used by the STDM vendor to implement this support.

Some statistical multiplexer vendors employ a band pass channel to support synchronous data sources. When this occurs, not only is the synchronous data not multiplexed statistically, but the data rate of the synchronous input limits the overall capability of the device to support asynchronous transmission. Figure 6.18 illustrates the effect of multiplexing synchronous data via the use of a band pass channel. When a band pass channel is employed, a fixed portion of each message frame is reserved for the exclusive multiplexing of synchronous data, with the portion of the frame reserved proportional to the data rate of the synchronous input to the STDM. This means that only the remainder of the message frame is then available for the multiplexing of all other data sources.

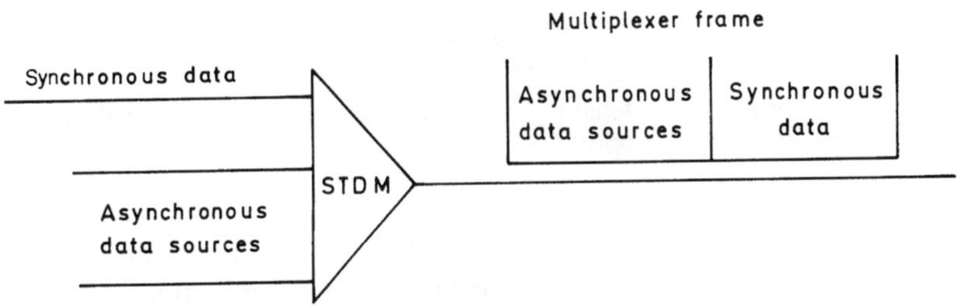

Figure 6.18 The use of a band pass channel to multiplex synchronous data. The synchronous data source is always placed into a fixed location on the multiplexer frame. In comparison, all asynchronous data sources contend for the remainder of the multiplexer frame

As an example of the limitations of band pass multiplexing, consider an STDM that is connected to a 56 kbps DSU and supports a synchronous terminal operating at 33 600 bps. If band pass multiplexing is employed, only 22 400 bps is then available in the multiplexer for the multiplexing of other data sources. In comparison, assume another STDM statistically multiplexes synchronous data. If this STDM has a service ratio of 1.5:1, then a 33 600 bps synchronous input to the STDM would on the average take up 22 400 bps of the 56 000 bps operating line. Since the synchronous data is statistically multiplexed, when that data source is not active other data sources serviced by the STDM will flow through the system more efficiently. In comparison, the band pass channel always requires a predefined portion of the high-speed line to be reserved for synchronous data, regardless of the activity of the data source.

Switching and port contention

Two features normally available with more sophisticated statistical multiplexers are switching and port contention. Switching capability is also referred to as alternate

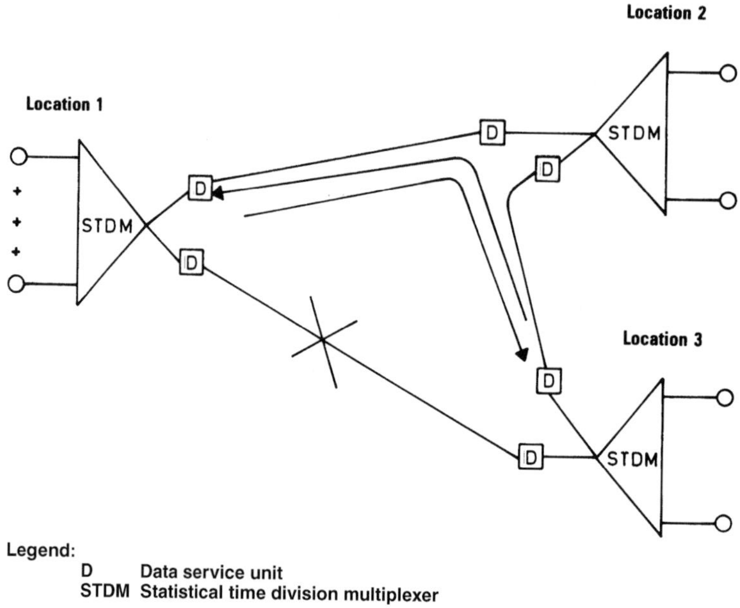

Figure 6.19 Switching permits load balancing and alternate routing if a high-speed line should become inoperative

routing and requires the multiplexer to support multiple high-speed lines whose connection to the multiplexer is known as a node. Switching capability thus normally refers to the ability of the multiplexer to support multiple nodes. Figure 6.19 illustrates how alternate routing can be used to compensate for a circuit outage. In the example shown, if the line connecting locations 1 and 3 should become inoperative an alternate route through location 2 could be established if the STDMs support data switching.

Port contention is normally incorporated into large capacity multinodal statistical multiplexers that are designed for installation at a central computer facility. This type of STDM may demultiplex data from hundreds of data channels, however, since many data channels are usually inactive at a given point in time, it is a waste of resources to provide a port at the central site for each data channel on the remote multiplexers. Thus, port contention results in the STDM at the central site containing a lesser number of ports than the number of channels of the distant multiplexers connected to that device. Then, the STDM at the central site contends the data sources entered through remote multiplexer channels to the available ports on a demand basis. If no ports are available, the STDM may issue a 'NO PORTS AVAILABLE' message and disconnect the user or places the user into a queue until a port becomes available.

ITDMs

One advance in statistical multiplexer technology resulted in the introduction of data compression into STDMs. Such devices intelligently examine data for certain characteristics and are known as intelligent time division multiplexers (ITDM). These devices take advantage of the fact that different characters occur with different frequencies and use this quality to reduce the average number of bits per character by assigning short codes to frequently occurring characters or sequences of characters and long codes to seldom-encountered characters or less frequently encountered sequences of characters.

The primary advantage of the intelligent multiplexer lies in its ability to make the most efficient use of a high-speed circuit in comparison to the other classes of TDMs. Through compression, synchronous data traffic which normally contains minimal idle times during active transmission periods can be boosted in efficiency. Intelligent multiplexers typically permit an efficiency four times that of conventional TDMs for asynchronous data traffic and twice that of conventional TDMs for synchronous data traffic.

STDM/ITDM statistics

Although the use of statistical and intelligent multiplexers can be considered on a purely economic basis to determine if the cost of such devices is offset by the reduction in line and modem or DSU costs, the statistics that are computed and made available to the user of such devices should also be considered. Although many times intangible, these statistics may warrant consideration even though an economic benefit may at first be hard to visualize. Some of the statistics normally available on statistical and intelligent multiplexers are listed in Table 6.3. Through a careful monitoring of these statistics, network expansion can be preplanned to cause a minimum amount of potential busy conditions to users. In addition,

Table 6.3 Intelligent multiplexer statistics

Multiplexer loading: % of time device not idle
Buffer utilization: % of buffer storage in use
Number of frames transmitted
Number of bits of idle code transmitted
Number of negative acknowledgements received

$$\text{Traffic density} = \frac{\text{non-idle bits}}{\text{total bits}}$$

$$\text{Error density} = \frac{\text{NAKs received}}{\text{frames transmitted}}$$

$$\text{Compression efficiency} = \frac{\text{total bits received}}{\text{total bits compressed}}$$

$$\text{Statistical loading} = \frac{\text{number of actual characters received}}{\text{maximum number which could be received}}$$

$$\text{Character error rate} = \frac{\text{characters with bad parity}}{\text{total characters received}}$$

frequent error conditions can be noted prior to user complaints and remedial action taken earlier than normal when conventional devices are used.

Features to consider

Table 6.4 lists the primary selection features to consider when evaluating statistical multiplexers. Although many of these features were previously discussed, a few features were purposely omitted from consideration until now. These features include auto baud detect, flyback delay and echoplex, and primarily govern the type of terminal devices that can be efficiently supported by the statistical multiplexer.

Table 6.4 Statistical multiplexer selection features

Feature	Parameters to consider
Auto baud detect	Data rates detected
Flyback delay	Settings available
Echoplex	Selectable by channel or device
Protocols supported	2780/3780, 3270, HDLC/SDLC, other
Data type supported	Asynchronous, synchronous
Service ratios	Asynchronous, synchronous
Flow control	XON-XOFF, CTS, clocking
Multinodal capability	Number of nodes
Switching	Automatic or manual
Port contention	Disconnect or queued when all ports in use
Data compression	Stripping bits or employs compression algorithm

Auto baud detect is the ability of a multiplexer to measure the pulse width of a data source. Since the data rate is proportional to the pulse width, this feature enables the multiplexer to recognize and adjust to different speed terminals accessing the device over the switched telephone network.

On older electromechanical printers, a delay time is required between sending a carriage return to the terminal device and then sending the first character of the next line to be printed. This delay time enables the print head of the terminal to be repositioned prior to the first character of the next line being printed. Many statistical multiplexers can be set to generate a series of fill characters after detecting a carriage return, enabling the print head of an electromechanical terminal to return to its proper position prior to receiving a character to be printed. This feature is called flyback delay and can be enabled or disabled by channel on many multiplexers.

Since some networks contain full-duplex computer systems that echo each character back to the originating terminal the delay from twice traversing through statistical multiplexers may result in the terminal operator obtaining the feeling that his or her terminal is non-responsive. When echoplexing is supported by an STDM, the multiplexer connected to the terminal immediately echoes each character to the terminal, while the multiplexer connected to the computer discards characters echoed by the computer. This enables data flow through the multiplexer system to be more responsive to the terminal operator. Since error detection and correction is built into all statistical multiplexers, a character echo from the computer is not necessary to provide visual transmission validation and is safety eliminated by echoplexing.

The other options listed in Table 6.4 should be self-explanatory, and the user should check vendor literature for specific options available for use on different devices.

Utilization considerations

Although networking requirements govern the type of multiplexer that will result in the best price-performance, several general comparisons can be made between devices. In comparison to FDM, the principal advantages of TDM include the ability to service high input data sources, the capacity for a greater number of individual inputs, the performance of data compression (intelligent multiplexers), the detection of errors, and the request for retransmission of data (statistical and intelligent multiplexers). The key differences between multiplexers are tabulated in Table 6.5.

Table 6.5 Multiplexer comparisons

	FDM	TDM	STDM	ITDM
Efficiency	poor	good	better	best
Channel capacity	poor	good	better	best
High-speed data	very poor	poor	better	best
Configuration change				
Data rate	good	fair	good	good
Number of channels	poor	good	better	better
Installation ease	poor	poor	good	good
Problem isolation	poor	poor	good	good
Error detection				
retransmission	n/a	n/a–good	automatic	automatic
Multidrop capability	good	n/a	possible	possible

6.2 T1/E1 MULTIPLEXERS

The primary utilization of T1 and E1 multiplexers is to concentrate both data and digitized voice on a 'T series' or 'E series' digital facility. To obtain an appreciation of the utilization of T1 and E1 multiplexers it is necessary to examine the process by which voice conversations can be digitized as well as the relationship between voice digitization and the data formats of the T1 and E1 carriers.

The T-carrier

T1 transmission popularity is a relatively recent phenomenon, with commercial services available only since the mid-1980s, although the technology for this transmission method dates back to the early 1960s. At that time, the T-carrier was used exclusively by telephone companies and was based upon the time division multiplexing of digitized voice.

Since the development of the T-carrier was based upon a voice digitization method referred to as Pulse Code Modulation (PCM), we will first examine that technique prior to turning our attention to the composition of the T-carrier and its European equivalent known as the E-carrier.

PCM

Pulse Code Modulation (PCM) represents one of the earliest techniques developed to digitize voice, and serves as a reference for comparing the quality of other methods against its 'toll quality' voice reconstruction. Under PCM an analog signal such as the human voice is digitized into a 64 kbps data stream based upon a three-step process. Those steps include sampling, quantization, and coding.

Sampling

The first step in the PCM process involves sampling an analog wave to extract information from that continuously varying signal. The sampling rate used by PCM is based upon the Nyquist theorem that requires a rate twice the bandwidth to faithfully reproduce a signal. Since the bandwidth of a telephone channel ranges in the frequency spectrum from 0 to 4 kHz, this resulted in the selection of a sampling rate of 8000 samples.

Figure 6.20 illustrates an example of the sampling of an analog wave. The resulting samples form what is referred to as a Pulse Amplitude Modulation (PAM) signal. That signal consists of a series of samples of the analog wave. Note that the height of each sample can have an infinite number of values.

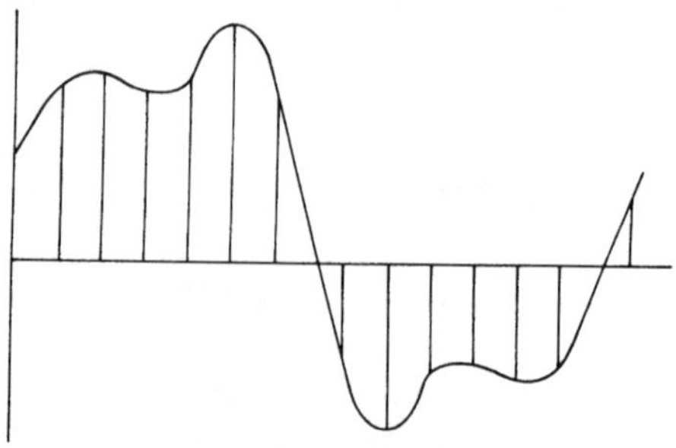

Figure 6.20 Pulse amplitude modulation

Quantization

Since it would be impossible to encode PAM samples into a fixed number of bit positions, quantization is used as a mechanism to convert each PAM sample into a discrete value. Thus the second step in the PCM process is quantization.

One of the problems engineers faced when they were developing quantization methods was the type of interval to use. To illustrate this problem, consider Figure 6.21 which shows a uniform series of quantization intervals used to provide a discrete value to each PAM sample. Note that when a PAM sample falls into a particular quantization interval, its value is based on the discrete value assigned to the center of the quantization interval. Also note that the difference between the value of the interval used to denote the sample and the actual height of the sample represents a quantization error referred to as quantization 'noise'. Although an obvious method to reduce quantization noise is to increase the number of quantization levels, this would also increase the number of bits required to encode each PAM signal. Thus, other methods were investigated as a mechanism to reduce quantization noise while minimizing the number of bits per sample.

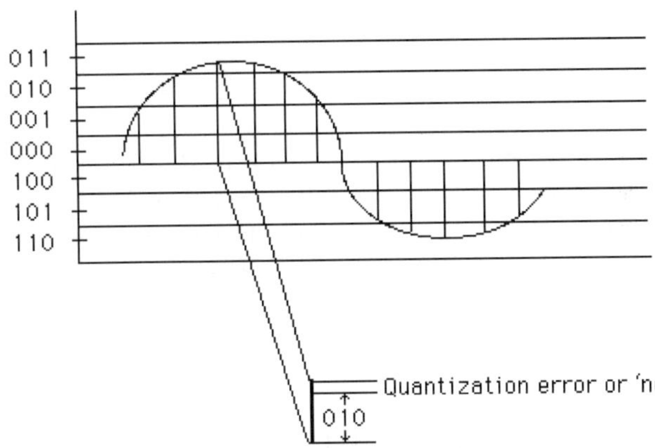

Figure 6.21 Using uniform quantization intervals to assign values to each PAM signal

Two techniques used to reduce the effect of quantization noise while minimizing the number of bits per sample are companding and non-linear quantization. Companding is a term used to represent the compression and expansion of speech energy. When a voice signal is compressed, its high-energy components are lowered while its low-energy components are raised in value, resulting in a lower dynamic range for the signal to be quantized and encoded.

The compression and latter expansion functions are logarithmic and follow one of two 'laws'—the A law which is used in European PCM systems and the μ (pronounced mu) law used in North America. The curve for the A law can be plotted from the formula

$$Y = \frac{AX}{1 + \log A} \qquad 0 \leq v \leq \frac{V}{A}$$

$$Y = \frac{1 + \log (AX)}{1 + \log A} \qquad \frac{V}{A} \leq v \leq V$$

where $A = 87.6$.

The curve for the μ law can be plotted from the following formula:

$$Y = \frac{\log (1 + \mu X)}{\log (1 + \mu)}$$

where μ had a value of 100 for the original North American T1 system employing PCM while latter systems use a value of 255. For the preceding formulas:

$$X = \frac{v}{V} \quad \text{and} \quad Y = \frac{i}{B}$$

where v represents the instantaneous input voltage, V is the maximum input voltage, i is the number of the quantization step, and B is the number of quantization steps on each side of the center of the range.

Coding

The third step in the PCM process is coding. This process is very closely related to quantization since it must convert the interval in which a sample falls into a binary value. In both North American and European systems eight bits are used to identify the quantization interval into which a sample falls, resulting in the data rate of PCM becoming 8000 samples/second X 8 bits/sample, or 64 kbps. However, the actual encoding method differs between each system.

In North America PCM systems subdivide the sample interval into 15 non-linear segments called chords (actually 16 since segments cutting the origin are co-linear and are counted as one) with 16 linear intervals per segment that are referred to as steps. Since the code level for 256 quantization values goes from 0 to 255 to represent a range from peak negative to peak positive, four bits are used to denote the chord and four bits the step in a chord. In comparison, European PCM uses a 13-segment approximation of the A-law curve which goes from negative to positive.

In the coding method developed for A-law quantization, the first bit indicates whether the quantization interval is above (1) or below (0) the origin. The following three bits identify the chord while the next four bits identify the step in the chord.

DS1 framing

In the United States, AT&T originally developed a 24-channel PCM system known as DS1, which resulted in the multiplexing of 24 voice conversations onto

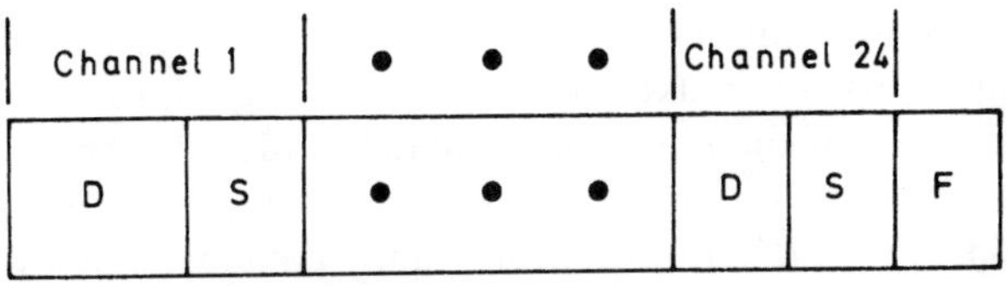

Figure 6.22 The DS1 frame (D seven data bits representing nearest digital value of the analog signal at the time of sampling, S signaling bit, F framing bit)

one 1.544 Mbps line. In the DS1 system, the analog signal of each voice channel was originally quantized through the use of a seven-level code, permitting 2^7 or 128 quantizing steps. To every seven bits that represent a coded quantum step an additional bit was added for signaling, resulting in a total of eight bits used to represent one sample of the analog signal. In addition, for each frame representing 24 analog signal samples, an additional bit was added. This bit is known as the framing bit and results in the composition of the DS1 frame illustrated in Figure 6.22. Since each voice signal is sampled 8000 times per second under PCM, 8000 frames per second are transmitted on a DS1 system. This results in a data rate of 193 X 8000 or 1.544 Mbps. Here the 193 bits per frame result from multiplying eight bits/channel times 24 channels and adding one framing bit. Today modern PCM systems use an eight-level code which results in the same data rate of 1.544 Mbps.

Digital signal levels

In the United States, the telephone network's digital hierarchy contains several digital signal (DS) levels. With the exception of the first level known as DS0 signaling, each succeeding level is made up of a number of lower level signals.

Table 6.6 lists the four most widely used DS levels, their data rate and use in a telephone network. In such networks, 24 DS0 signals form one DS1 signal, while four DS1s make up a DS2 and seven DS2s form one DS3 signal.

Table 6.6 Digital transmission hierarchy

Digital signal level	Data rate	Telephone company network use
DS0	64 kbps	Basic voice bandwidth data channel encoded via PCM; digital data service (DDS), analog/digital channel-bank inputs
DS1	1.544 Mbps	This is the well-known T1 carrier which consists of 24 DS0 signals. Used for point-to-point communications between telephone company offices
DS2	6.312 Mbps	Used between telephone company central offices as well as inter- and intra-building communications
DS3	44.736 Mbps	Used in high-capacity digital radio, coaxial cable and fiber optic transmission systems for communications between telephone company offices

If you multiply the data rate of the lower level signal by the number of lower level signals used to make up a higher level signal, the sum will be slightly less than the data rate of the higher level signal. This difference is used for framing the lower level signals onto the higher level signal and is not employed for actual data transmission. Thus, multiplying the data rate of a lower level signal by the number of signals that make up a higher level signal can be used to obtain the actual data carrying rate of a DS level, while the data rate in Table 6.6 is the signaling rate of data and frame information.

Framing changes

The name of the service represented by the original DS1 system has changed several times since the early 1960s based upon changes in the composition of the framing structure. Today communications carriers offer D4 and ESF framing. D4 framing is an older technique that is gradually being replaced by the ESF framing technique.

D4 framing

In D4 framing a sequence of 12 frame bits is used to develop a precise pattern employed by T1 multiplexers to keep the bit stream in synchronization. Figure 6.23 illustrates the D4 framing pattern, showing the value of the 193rd bit in each of the first 12 frames transmitted on a T1 circuit. This 12-bit frame pattern continuously repeats itself, providing the synchronization signal used by equipment attached to a T1 circuit. Readers are referred to Section 4.11 for detailed information concerning the D4 framing format.

Frame number	F1	F2	F3	F4	F5	F6	F7	F8	F9	F10	F11	F12
Frame bit value	1	0	0	0	1	1	0	1	1	1	0	0

Figure 6.23 D4 framing pattern

ESF framing

A more recently introduced T1 framing pattern called the Extended Super Frame (ESF) format contains 24 frame bits. Unlike the D4 pattern which repeats itself, the ESF consists of three types of frame bits, two of which can vary. These three types of frame bits include line control, cyclic redundancy checking, and a frame pattern for synchronization.

In the ESF frame the bits in the odd frames from 1 through 23 are used by the telephone company to perform network monitoring, set alarm conditions, and perform other control operations. Frame bits in frames 2, 6, 10, 14, 18, and 22 are used for cyclic redundancy checking. Finally, frame bits in frames 4, 8, 12, 16, 20, and 24 are used to form the repeating pattern 001011, which is used for synchronization.

The ESF framing format represents a considerable improvement over D4 framing as it supports both performance measurements as well as the transmission of those measurements. CSUs that support the ESF framing format store performance measurements based upon the use of a CRC-6 check to determine whether or not frames are received error free. CSUs contain registers that store data concerning frame errors, error-free seconds, and other performance information. Telephone company equipment uses the odd framing bits to send messages to CSUs to periodically dump their registers and retrieve performance information concerning the T1 circuit connected to the CSU. Readers are referred to Section 4.11 for detailed information concerning the ESF framing format and the operation and utilization of CSUs.

T1 signal characteristics

The 1.544 Mbps T1 bit stream is a bipolar signal which is also called alternate mark inversion (AMI). Under this signaling format, the one pulses have an alternating polarity, thus, if the nth one bit is represented by a positive pulse, the $(n + 1)$th one bit will be represented by a negative pulse. In comparison to alternating polarities for one bits, all zero bits are represented by a zero voltage. In addition to bipolar signaling a T1 data stream must have a minimum number of one bits in the data stream to provide an acceptable level of timing to repeaters when transmission occurs on metallic circuits. Techniques known as zero suppression which provide a minimum ones density are described in Section 4.11

European E1 facilities

In Europe, a 32-channel system was developed to encode and multiplex voice signals in comparison to the 24-channel system used in the United States and other North American countries. Under the 32-channel system, 30 channels transport digitized voice signals, while the remaining two channels are used to provide signaling and synchronization information. Since each channel operates at 64 kbps which represents eight bits used for each sample at a sample rate of 8000 samples per second, 64 kbps times 32 channels results in a composite data rate of 2.048 Mbps on a European E1 facility, which is technically referred to as a G703/732 channel.

Table 6.7 lists the composition of the 32 channels that are used to establish the G703/732 frame structure. In comparing the North American T1 frame structure to the European E1 G703/732 structure, there are two key differences between each system in addition to the different data rates. First, North American T1 systems derive the 1.544 Mbps data rate from the use of 24 channels while the European E1 G703/732 system uses 30 voice channels plus separate synchronization and signaling channels, with each channel operating at 64 kbps to produce a 2.048 Mbps data rate. Secondly, the North American T1 system uses the 193rd bit in each frame for

Table 6.7 European G703/732 frame structure

Time slot	Type of information
0	Synchronizing
1–15	PCM encoded speech
16	Signaling
17–31	PCM encoded speech

Table 6.8 Signaling characteristics comparison

	T1	E1 G703/732
Composite data rate (Mbps)	1.544	2.048
Number of channels	24	32
Channel data rate (kbps)	64	64
Synchronization	Frame bit	Channel 0
Signaling	Eighth bit in sixth frame	Channel 16

synchronization, whereas the G703/732 system provides a separate 64 kbps channel for this function. Table 6.8 compares the signaling characteristics of the T1 and G703/732 systems. Due to these major differences between systems, many T1 multiplexers designed for operation in North America will not perform properly if used in Europe, while other T1 multiplexers can operate correctly on both sides of the Atlantic. Readers are referred to Section 4.11 for additional information concerning the E1 system.

The T1 multiplexer

When T1 became available for subscriber use in 1984, it was initially relegated to use in voice networks. This limited utilization was based upon the lack of equipment to efficiently combine voice and data on one T1 facility as well as the availability of telephone company equipment designed exclusively for the combination of voice channels onto one T1 circuit. Due to the cost of less than 20 analog circuits used for transmitting data equaling the cost of one T1 facility while providing approximately 6% of the capacity of a T1 facility, many communications equipment manufacturers developed T1 multiplexers designed to economically multiplex both voice and data onto a composite T1 or E1 channel.

In addition to performing data multiplexing, most T1 multiplexer manufacturers offer users a variety of optional voice digitization modules that can be used to digitize voice signals at data rates ranging from the standard 64 kbps PCM rate to data rates as low as 2.4 kbps. Such modules can be employed to increase the number of voice channels that can be transmitted on a T1 line by a factor of two to four or more over the normal 24 or 30 channels obtainable when PCM digitization is employed. The use of these multiplexers enables organizations to effectively integrate voice, data and video information onto a common T1 facility. Prior to examining how T1 multiplexers can be employed in a network and their operational features and capabilities, a discussion of the various voice digitization techniques commonly available is warranted to obtain an understanding of the benefits and limitations of use of modules performing voice digitization in multiplexers.

Voice digitization techniques

The first series of voice digitization techniques developed were based upon different methods used to sample a speech waveform and are collectively referred to as waveform coding. Waveform coding voice digitization techniques include pulse code modulation (PCM), Adaptive Differential PCM (ADPCM) and continuously variable slope delta (CVSD) modulation. Although waveform coding techniques normally result in a relatively high bit rate for encoded analog speech, they also result in a very good quality of reconstructed speech.

The fact that more simultaneous voice conversations could be transported at low data rates than at higher data rates per digitized signal resulted in the development of additional techniques used to digitize voice. Those techniques can be classified into two additional classes—voice coders and hybrid coders. Voice coders examine the spectral components of speech such as pitch, tone, and energy level, coding values for key speech parameters instead of samples of actual speech. At the receiver, the values of those parameters are used to synthesize speech. Voice coding is performed by a device commonly referred to as a vocoder and results in the generation of a low bit rate for digitized speech. The resulting reconstructed speech, however, may sound at best 'hollow'. Popular examples of vocoding include a family of linear prediction coding products developed during the 1970s and 1980s.

Recognizing the limitations associated with waveform coding and vocoding resulted in a new series of voice digitization methods that combines the best features of both. This series of digitization methods is commonly classified as hybrid coding. Under hybrid coding speech is sampled and speech parameters are extracted in a similar way to the vocoding process. However, those parameters are then used in an iterative process to reconstruct the sample, compare it to the original sample, and adjust the speech parameters until there is a minimum difference between the two. Once this is accomplished, the revised speech parameters are used to represent a short period of speech. Examples of hybrid coders include a family of code-excited linear predictive coders (CLEP) that are being used in applications ranging in scope from voice over frame relay and Internet telephony to inclusion in T1 and E1 multiplexers to enable additional simultaneous voice conversations to be transported over T1 and E1 lines.

Waveform coding

The development of PCM was followed by a number of additional waveform coding techniques. Each of these new techniques was developed as a mechanism to reduce the data rate required to transport a digitized voice conversation, enabling an additional number of conversations to be transported on high-speed digital transmission facilities. Two commonly used waveform coding techniques that followed the development of PCM are adaptive differential PCM (ADPCM) and continuous variable slope delta (CVSD) modulation, both of which are available as digitization modules for use in many T1 and E1 multiplexers.

ADPCM

Adaptive differential pulse code modulation is based upon the premise that human speech does not dramatically change. This fact enables the use of a transcoder to predict the value of the next sample based upon the last few samples. Then, instead of quantizing speech, only the difference between the speech signal and the prediction is quantized. Since the variance between the difference between real and predicted speech is less than that of real speech samples, quantization can be performed with fewer bits. In fact, under ADPCM the use of four bits is employed instead of the eight bits used under PCM. Since ADPCM uses the same 8000 samples per second rate of PCM, the resulting digitization rate is reduced to 32 kbps, or one-half the PCM digitization data rate.

In the mid-1980s the ITU standardized a 32 kbps ADPCM as recommendation G.721. Later variations in the original standard resulted in the development of recommendation G.723 for 24 kbps and 40 kbps ADPCM.

The major problem associated with the use of ADPCM concerns its ability to support modem transmission. ADPCM is based upon the use of a predictor which is possible since human speech does not significantly change from one sample to the next. However, this is not true for high-speed modem modulation since the analog modulation can significantly change between samples. For this reason, ADPCM normally does not support modem transmission above 9600 bps. Although ADPCM is used on a few international circuits, it is primarily employed on internal organization voice networks developed through the use of T1 or E1 multiplexers incorporating ADPCM modules. This enables an organization to obtain the advantages associated with ADPCM, while retaining some PCM channels for both voice and modem transmission.

CVSD

In the continuously variable slope delta (CVSD) modulation digitization technique, the analog input voltage is compared to a reference voltage. If the input is greater than the reference a binary '1' is encoded, while a binary '0' is encoded if the input voltage is less than the reference level. This permits a one-bit data word to represent each sample.

At the receiver, the incoming bit stream represents changes to the reference voltage and is used to reconstruct the original analog signal. Each '1' bit causes the receiver to add height to the reconstructed analog signal, while each '0' bit causes the receiver to decrease the analog signal by a set amount. If the reconstructed signal was plotted, the incremental increases and decreases in the height of the signal will result in a series of changing slopes, resulting in the naming of this technique: continuously variable slope delta modulation.

Since only changes in the slope or steepness of the analog signal are transmitted, a sampling rate higher than the PCM sampling rate is required to recognize rapidly changing signals. Typically, CVSD samples the analog input at 16 000 or 32 000 times per second. With a one-bit word transmitted for each sample, the CVSD data rate normally is 16 kbps or 32 kbps. Other CVSD data rates are obtainable by varying the sampling rate. Some T1 multiplexer manufacturers offer a CVSD option which permits sampling rates from 9600 to 64 000 samples per second, resulting in a CVSD data rate ranging from 9.6 kbps to 64 kbps, with the lower sampling rates reducing the quality of the reconstructed voice signal. Normally, voice signals are well recognizable at 16 kbps and above, while a data rate of 9.6 kbps will result in a marginally recognizable reconstructed voice signal.

Vocoding

Growth in communications during the 1970s and 1980s resulted in research into methods to generate low bit rate digitized speech. One series of methods was based upon sampling human speech and extracting parameters that would enable speech to be synthesized. This series of speech digitization methods is referred to as vocoding, and its operation requires an overview of the basic properties of speech. Thus, prior to discussing the operation of a commonly used vocoder, let's first examine how speech is produced along with its components.

Speech is produced when air is forced from our lungs through our vocal cords along our vocal track. Our vocal cords vibrate as well as open and close, producing a periodic pitch referred to as voiced sounds. In addition to voiced sounds, as air is forced from our lungs while our vocal cords are open we generate a noise-like turbulence-type sound referred to as unvoiced sound. A third component of speech results from the closure of our vocal tract which allows air pressure from our lungs to build up. When that pressure is released, we generate what is referred to as a plosive sound.

The actual modeling of speech depends upon the type of model developed and how the initial analysis is performed. Concerning the latter, some sounds do not fall into a single class. Instead, they represent a mixture of classes and require a model to take this into consideration to provide a more natural sounding reproduction of modeled speech. Since linear predictive coding represents one of the most popular types of vocoders, let's turn our attention to this voice digitization method.

Linear predictive coding

In linear predictive coding, analog voice input is analyzed and converted into a set of digital parameters that represent the voice input. To reconstruct a digitized voice signal a synthesizer is used to develop an analog voice output based upon the received set of digitized parameters. The key to the utility of linear predictive coding is the limiting of the analysis of voice signals to four sets of voice parameters, permitting a very low data rate to be obtained for transmitting analog data in digital form.

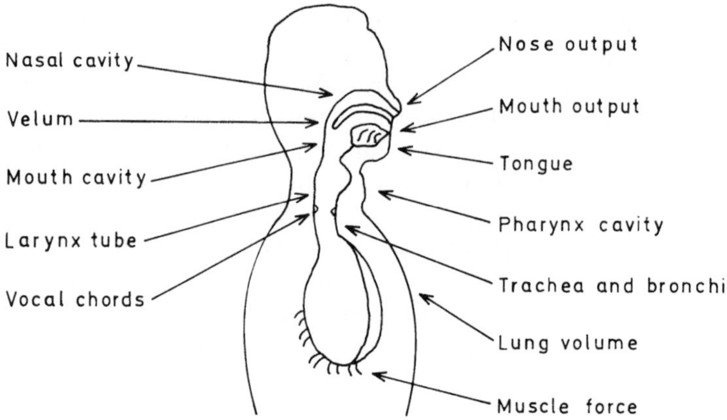

Figure 6.24 Speech-producing elements of the human vocal tract. (Reprinted with permission from *Data Communications Management*, © 1988 Auerbach Publishers, New York)

Under linear predictive coding speech is first sampled by a 12-bit analog to digital converter. Next, the output of the converter is used as input to four parametric detectors. These parametric detectors include a pitch detector, a voice/unvoiced detector, a power detector and a spectral data decoder.

To appreciate the employment of parametric detectors, consider Figure 6.24 which illustrates the speech-producing elements of the human vocal tract. The pitch detector analyzes data to obtain the fundamental pitch frequency at which human vocal cords vibrate. Next, the voice/unvoiced detector senses whether sound is caused by the vibration of vocal cords (voice) or by sounds such as 'shhh' (unvoiced), which do not vibrate. The power detector determines the amplitude or loudness of the sound, while the spectral data detector models the resonant cavity formed by the throat and the mouth.

Through the use of parametric detectors, a parametric model of human speech is constructed, resulting in the generation of a series of speech coefficients which represent human voice. These speech coefficients developed through the use of linear predictive coding are then transmitted instead of the actual amplitude of the signal which is used by PCM and other waveform coding techniques. As a result of the modeling of human speech instead of actual speech, a very low data rate is obtained to represent a voice conversation. The use of linear predictive coding allows digitized voice to be transmitted at data rates as low as 2400 or 4800 bps, enabling the communications manager to consider a variety of strategies to integrate voice into low-speed data networks or to obtain multiple voice channels on one physical line.

One of the most popular implementations of linear predictive coding is LPC10 which obtains its name as a result of a linear prediction of 10 speech components. Under LPC10 speech is sampled at an 8 kHz rate. Ten speech parameters are predicted and encoded into 42 bits while seven bits are used to represent pitch and voicing and five bits for gain. This results in 54 bits per frame using a framing rate that repeats 44.44 times per second. Thus, the resulting data rate becomes 54 bits/frame × 44.44 frames/second, or 2400 bps.

Although LPC10 results in a very low bit rate for digitized speech, it has several disadvantages. Those disadvantages include the fact that the resulting synthesized speech can sound 'metallic', voiced excitations appear unnatural, and, since only speech is modeled, background noise can interface with the model. The recognition of those problems resulted in the development of a new series of voice digitization techniques referred to as hybrid coding.

Hybrid coding

As previously mentioned in this section, hybrid coding represents a voice digitization technique which combines waveform and vocoding. The most commonly used hybrid coding methods are a family of code excited linear prediction (CELP) coders that represent state-of-the-art voice digitization methods used in a large number of applications from T1 and E1 multiplexer modules to Frame Relay Access Devices (FRAD) and Internet telephony.

CELP coding

A Code Excited Linear Prediction (CELP) coder uses speech samples to model both the vocal tract and pitch as well as to predict parameters for other speech parameters in a manner similar to a linear prediction vocoder. However, instead of using the model parameters 'as is', a CELP coder will compare the results of the model to the original speech sample. Then, the hybrid coder will either use an iterative process to adjust parameters to reduce the error between the model and the actual sample to an acceptable level, or transmit model parameters and a compressed representation of the errors. The latter technique is used by the family of CELP voice digitization methods, with the term 'code' referencing the automatic creation of a codebook. The codebook stores a series of vocal tract and pitch predictor values. Instead of transmitting the actual values, CELP coders use the index of the codeword that produces the best quality speech.

Until the mid-1990s the computationally intensive operations required to search codebook entries, perform a series of iterative speech synthesis operations, and compare generated speech against sampled speech to select a best-fit codebook entry were beyond the capability of most microprocessors. The development of high-speed digital signal processor (DSP) chips resulted in the development of low-cost hardware to perform CELP operations. In addition, the substantial increase in the capability of Pentium, Power PC, and other modern microprocessors enabled software developers to include versions of CELP as software modules which form the basis for many Internet telephone applications.

Several versions of CELP were standardized by the ITU. A 16 kbps version which has a minimal delay was standardized as Recommendation G.728, while another version that operates at 8 kbps was standardized as Recommendation G.729. A third ITU standard which was developed to support the transmission of voice over packet networks, such as the Internet and Frame Relay networks, supports selectable operating rates of 6.4 kbps and 7.3 kbps. Standardized as an annex to Recommendation G.729, this standard references a Conjugate Structure–Algebraic Code Excited Linear Prediction (CS–ACELP) voice digitization method.

T1 multiplexer employment

Modern T1 multiplexers are microprocessor-based time division multiplexers designed to combine the inputs from a variety of data, voice and video sources onto a single communications circuit that operates at 1.544 Mbps in North America and 2.048 Mbps in Europe. Table 6.9 lists the typical input channel rates accepted by most T1 multiplexers. It should be noted that, although digitized voice is treated as synchronous input, its digitized data rate can vary considerably based upon the type of optional voice digitization modules offered by the T1 multiplexer manufacturer.

Table 6.9 Typical T1 multiplexer channel rates

Type	Data rates (bps)
Asynchronous	110, 300, 600, 1200, 1800, 2400, 3600, 4800, 7200, 9600,19 200
Synchronous	2400, 4800, 7200, 9600, 14 400, 16 000, 19 200, 32 000, 38 400, 40 800, 48 000, 50 000, 56 000, 64 000, 112 000, 115 200, 128 000, 230 400, 256 000, 460 800, 700 000, 756 000
Voice	2400, 4800, 6400, 8000, 9600, 16 000, 32 000, 48 000, 64 000

In most applications, input to the T1 multiplexer's voice channels results from an interface to an organization's Private Branch Exchange (PBX), resulting in one or more tie lines being obtained through the use of two T1 multiplexers and a T1 carrier facility. Figure 6.25 illustrates a typical T1 multiplexer application where voice, video, and data are combined onto one T1 carrier facility. In this example, it was assumed that PCM digitization channel modules were selected for use in the T1 multiplexer, resulting in the 10 voice channels on the PBX interface using 640 kbps of the available 1.544 Mbps T1 operating rate.

Since digitized video normally requires a data rate of 384 kbps to be effective, it was assumed that the organization using the T1 multiplexer has a conference room to connect to a distant location for video conferencing. Thus, 384 kbps input to the T1 multiplexer in Figure 6.25 represents a digitized video conferencing signal. Similarly, it is assumed that the organization has two data centers, one at each location where there is also a PBX. This permits mainframe computer-to-mainframe computer transmission to occur at 128 kbps. To support LAN to LAN communications between distant locations, a router is shown connected at a data rate of 384 kbps to a port on the T1 multiplexer. Note that this results in a mixture of voice, video, conventional mainframe and LAN communications to be transported over a common digital transmission facility instead of separate facilities. Since the monthly cost of between four and six 56 kbps circuits is equivalent to the

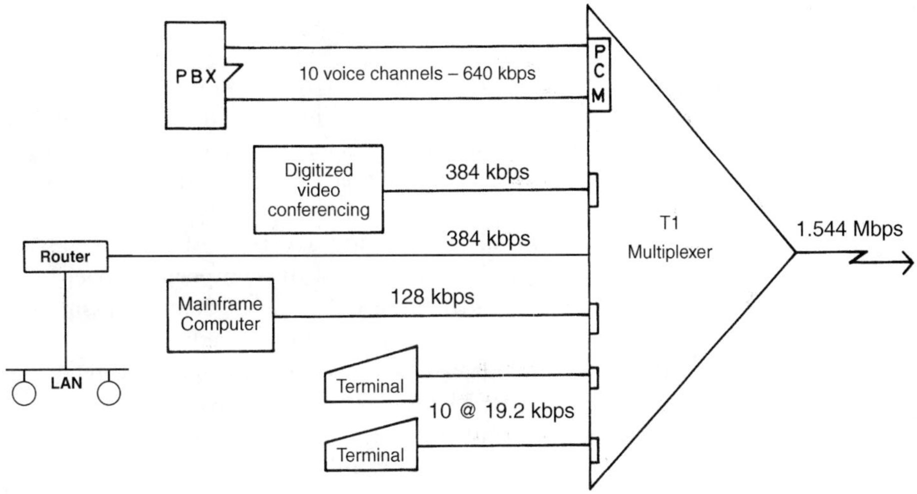

Figure 6.25 Typical T1 multiplexer application

cost of a T1 circuit, the potential to reduce network costs by integrating different applications onto high-speed digital circuits is one of the driving forces behind the use of T1 and E1 multiplexers. Finally, it was assumed that 10 data terminals, each operating at 19.2 kbps at one site, required access to the computer located at the other end of the T1 link.

Features to consider

Table 6.10 lists the major features of T1 and E1 multiplexers that warrant consideration during an acquisition process. While all of the listed features are important to consider, they may not be relevant to certain situations based upon the immediate and long-term requirements of a specific organization.

Table 6.10 T1 and E1 multiplexer features to consider

Bandwidth utilization method
Bandwidth allocation method
Voice interface support
Voice digitization support
Internodal trunk support
Subrate channel utilization
Digital access cross-connect capability
Gateway operation support
Alternate routing and route generation
Redundancy
Maximum number of hops and nodes supported
Diagnostics
Configuration rules

Bandwidth utilization

Inefficient T1 and E1 multiplexers assign data to the carrier facility using 64 kbps DS0 channels for each data source as illustated in Figure 6.26(a). In this example the assignment of input data sources is fixed to predefined channels, resulting in the inability of the multiplexer to take advantage of the inactivity of different data sources.

More efficient multiplexers employ a variety of demand assigned bandwidth techniques to make more efficient use of the composite carrier bandwidth. This is illustrated in Figure 6.26(b), in which a basic demand assignment feature of a multiplexer dynamically assigns bandwidth based upon the activity of the data sources. In this example it was assumed that several 9.6 kbps data sources became active along with two PCM digitized voice conversations and were dynamically assigned to the carrier bandwidth in their order of activation. Note that this method of bandwidth assignment normally results in an increase in available bandwidth since the probability of all inputs becoming active at one time is usually very low. In addition, having the capability to allocate bandwidth based upon the data rate of the

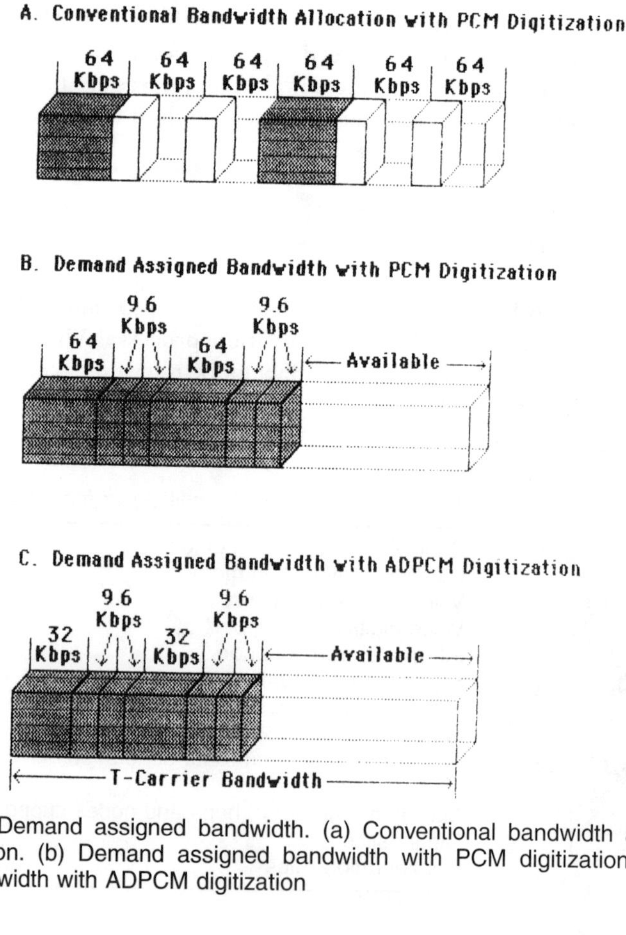

Figure 6.26 Demand assigned bandwidth. (a) Conventional bandwidth allocation with PCM digitization. (b) Demand assigned bandwidth with PCM digitization. (c) Demand assigned bandwidth with ADPCM digitization

data sources and not by a DS0 channel basis allows the 9.6 kbps data sources to occupy significantly less bandwidth. Thus, demand assignment with dynamic bandwidth allocation results in a considerable improvement in the use of a T or E-carrier's data transmission capacity in comparison to a conventional bandwidth allocation process.

Figure 6.26(c) illustrates the effect upon bandwidth allocation based upon the use of a more efficient voice digitization module in multiplexers. In this example it was assumed that ADPCM voice digitization modules were used in the multiplexer instead of PCM voice digitization modules. The use of ADPCM reduces the bandwidth required for carrying each voice conversation to 32 kbps, further increasing the available bandwidth of the carrier to support other data sources.

Bandwidth allocation

Most T1 and E1 multiplexers use time division multiplexing schemes to allocate bandwidth to each voice and data channel as well as portions of DS0 channels. Techniques used for bandwidth allocation can include the demand assignment of bandwidth previously illustrated in Figure 6.26(b) and 6.26(c), as well as non-contiguous resource allocation and the packetization of voice, data and video.

Figure 6.27 illustrates the advantage of non-contiguous resource allocation over the conventional method of allocating DS0 channels. In the top portion of that illustration a section of T-carrier bandwidth supporting three voice calls is shown. Here, each call is placed in a contiguous portion of the T-carrier bandwidth. In Figure 6.27(b), it was assumed that call B was completed and its bandwidth became available for use. Now suppose a data source (D) became active that required more bandwidth than that freed by the completion of call B. Under the non-contiguous resource allocation method the bandwidth required to accommodate the data source could be split into two or more non-contiguous sections of the T-carrier bandwidth as illustrated in Figure 6.27(c).

The third method of bandwidth assignment was pioneered by Stratacom with that firm's introduction of a T1 multiplexer that packetizes both voice and data. The Stratacom multiplexer uses 'Fast Packet' technology where the term fast packet refers to the fact that information is transmitted across the network in packet format without error checking being performed at nodes in the network instead of a time division multiplexed format. Although the external interface to voice and data is the same as a conventional T1 multiplexer, the internal operation of the Stratacom multiplexer is considerably different from other devices.

The Stratacom fast packet multiplexer generates packets only when data sources are active, using a packet length of 193 bits which corresponds to the North American T1 frame length. Figure 6.28 illustrates the Stratacom frame format. Although 20 bits, in effect, function as overhead to provide a destination address (16 bits), priority (two bits), and error correction to the header by the use of a Hamming code (six bits), the efficiency of packetized multiplexing can be considerable. This is because the technique takes advantage of the fact that voice conversations have periods of silence and are typically half-duplex in nature. This enables packet technology to provide an efficiency improvement of approximately 2:1 over

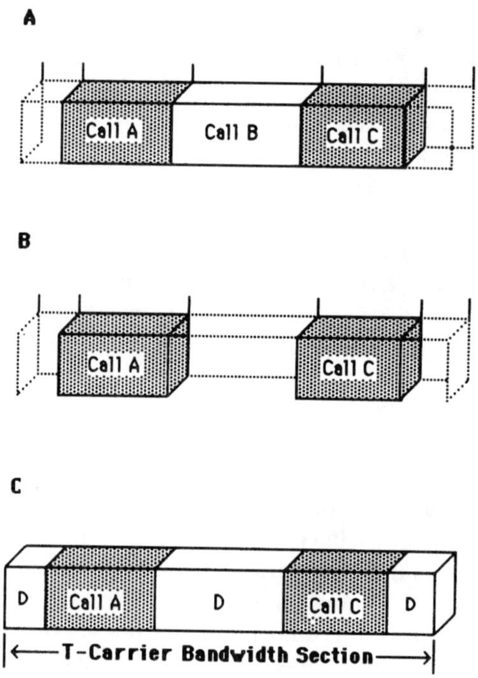

Figure 6.27 Bandwidth allocation methods. Non-contiguous resource allocation of bandwidth enables input to the T1 multiplexer to be split into portions of the available bandwidth. (a) T-carrier bandwidth section supporting three calls. (b) Call B is completed and its bandwidth becomes available. (c) Data Source D multiplexed into non-contiguous sections of bandwidth

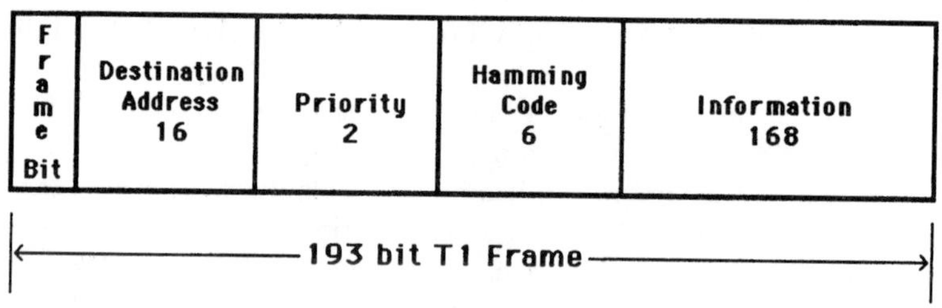

Figure 6.28 Stratacom packet format. The Stratacom T1 multiplexer packetizes voice and data sources into 193 bit frames containing 168 information bits

conventional time division multiplexing of voice. With the addition of ADPCM voice digitization modules the Stratacom fast packet multiplexer can support up to 96 voice conversations on a T1 circuit.

One of the problems associated with the use of packet technology to transport digitized voice is the fact that you cannot delay voice. Thus, unlike data packets that can be retransmitted if an error is detected, packetized voice cannot tolerate the delay of retransmission. In addition to not being able to retransmit voice, you must also consider the effect of a large number of voice channels becoming active. When too many channels become active the total bit rate of the digitized input channels can exceed the output bit rate of the T1 circuit. To avoid too much delay to specific channels, some channels will be skipped since a listener can tolerate a 125 ms delay. Another problem associated with packetized voice is the delay that can occur as the packets are routed from node to node in a complex network. To overcome this problem Stratacom incorporates a priority field in its packet which enables certain packets to be processed and routed before other types of packets.

Voice interface support

Since most T1 and E1 multiplexer applications include the concentration of voice signals, the type of voice interfaces supported for two-wire and four-wire applications is an important multiplexer feature to consider. Prior to examining the types of voice interfaces supported by T1 and E1 multiplexers, let us first review some of the more common types of voice signaling methods since it is the signaling method that is actually supported by a particular interface.

Two of the most common types of telephone signaling include loop signaling and E&M signaling. Loop signaling is a signaling method employed on two-wire circuits between a telephone and a PBX or between a telephone and a central office. E&M signaling is a signaling method employed on both two-wire and four-wire circuits routed between telephone company switches.

In loop signaling, the raising of the telephone handset results in the activation of a relay at the PBX or central office, causing current to flow in a circuit formed between the telephone set and the PBX or central office. The raising of the handset, referred to as an off-hook condition, results in the PBX or central office returning a dial tone to the telephone set. As the subscriber dials the telephone number of the called party, the dialed digits are received at a telephone company central office which then signals the called party by sending signaling information through the telephone company network. Once the call has been completed the replacement of the handset onto the telephone set, a condition known as onhook, causes the relay to be deactivated and the circuit previously formed to open.

A second type of telephone off-hook signaling that flows in a loop is ground start signaling. This method of signaling is also used on two-wire circuits between a telephone set and a PBX or central office. Unlike loop start signaling in which loop seizure is detected at the PBX or central office, ground start allows the detection of loop seizure to occur from either end of the line.

E&M signaling

E&M signaling is used on both two-wire and four-wire circuits connecting telephone company switches. Here the M lead is used to send ground or battery signals to the signaling circuits at a telephone company switch, while the E lead is used to receive an open or ground from the signaling circuit. In E&M signaling the local end asserts the M lead to seize control of the circuit. The remote end receives the signal on the E lead and toggles its M lead as a signal for the local end to proceed.

The local end then sends the address by toggling its M lead, in effect, placing dialing pulses on that lead which is used by the remote end to effect the desired connection. Once a call is completed, either party will drop its M lead, resulting in the other side responding by dropping its M lead. Currently, there are five types of E&M signaling: types I, II, III, IV and V. The major difference between E&M signaling types relates to the method by which an on-hook condition is established—ground or open.

Table 6.11 lists some of the more common types of voice interface cards supported by many T- and E-carrier multiplexer vendors. The two-wire and four-wire transmission only interfaces are designed to support permanent two-wire and four-wire connections between two points that do not require the passing of signaling information. Both types of interfaces are normally used to support a data modem connection through a T- or E-carrier multiplexer.

Table 6.11 Common types of voice interface cards

2-wire transmission only
4-wire transmission only
2-wire E&M
4-wire E&M
2-wire Foreign Exchange (FX)
4-wire Foreign Exchange (FX)

The two-wire and four-wire E&M interfaces usually support the connection of PBXs and telephone company equipment to a T- or E-carrier multiplexer. As previously mentioned, there are five types of E&M signaling, with each type applicable to both two-wire and four-wire operations.

The two-wire foreign exchange office interface is designed to support the attachment of a T- or E-carrier multiplexer to a PBX or central office switching equipment that provides an open or closed foreign exchange termination point.

Voice digitization support

In addition to the method of bandwidth assignment and allocation, a third major feature affecting the efficiency of T- and E-carrier multiplexers is the type of voice digitization modules the device supports. Although most T- and E-carrier multiplexers support the use of PCM and ADPCM, some vendors also support the use of adapter cards that contain proprietary voice digitization modules. One example of this is adaptive speech interpolation which changes the digitization rate of selected voice channels from 32 kbps to 24 kbps as available bandwidth becomes saturated. Although some proprietary techniques may offer advantages in both the fidelity of a reconstructed voice signal as well as in the bandwidth required to carry the signal, their use restricts an organization to one vendor's product.

Although PCM and ADPCM are supported by most T1 and E1 multiplexers, the type of ADPCM supported warrants consideration. While ITU standards govern digitization at 24, 32 and 40 kbps, some vendors also support proprietary rates of 14, 21 and 28 kbps. Similarly, although a low delay version of CELP was standardized by the ITU that operates at 16 kbps, several vendors announced products during 1997 that use what they refer to as Enhanced CELP (E-CELP) which results in speech being digitized at data rates of 4.8, 7.47 and 9.6 kbps. Other vendors, while supporting basic ITU standards, add silence suppression to reduce the bandwidth requirements of a particular coding method to half its normal rate. This becomes possible because a conversation is essentially half-duplex in nature with periodic pauses in speech.

When evaluating the capability of voice digitization modules, it is important to note that the clarity of reproduced speech can be as important as or more important than obtaining a low data rate. That clarity is usually expressed in terms of a value between 1 and 5 for a standardized voice quality test, with a value of 5 indicating a very high level of recognition of speech passed through a particular voice digitization method. Thus, you should examine both the data rate produced by a particular voice digitization technique as well as the quality of reconstructed voice, and use both in selecting an appropriate method to satisfy organizational requirements.

Figure 6.29 illustrates the effect of the use of several types of voice digitization modules upon the capacity of a T-carrier. If standard PCM modules are used, the T-carrier becomes capable of supporting either 24 or 30 voice calls depending upon whether a North American or European T-carrier facility is used Figure 6.29(a). When 32 kbps ADPCM modules are used to digitize voice, the voice-carrying capacity of the T-carrier is doubled as shown in Figure 6.29(b). The third example, Figure 6.29(c), which illustrates the use of 24 kbps ADPCM, shows the voice-carrying capacity of a T-carrier tripling, while the fourth example (Figure 6.29(d)) shows how the voice-carrying capacity of a T-carrier can be quadrupled through the use of ADPCM and DSI.

Internodal trunk support

The internodal trunk support feature of T- and E-carrier multiplexers references the ability of the device to connect to North American and European transmission facilities. To support North American T-carrier facility usage, the multiplexer must not only operate at 1.544 Mbps but, in addition, support the required communications carrier framing—D4 or ESF. To support European E-carrier facility usage the multiplexer must operate at 2.048 Mbps and support CEPT 30-PCM framing.

Subrate channel utilization

Channel utilization is a function of the subrate multiplexing capabilities of the T- and E-carrier multiplexer. Many T- and E-carrier multiplexers support asynchronous data rates from 50 bps to 19.2 kbps and low-speed synchronous data rates from 2.4 kbps to 19.2 kbps, permitting multiple data sources to be placed onto one DS0

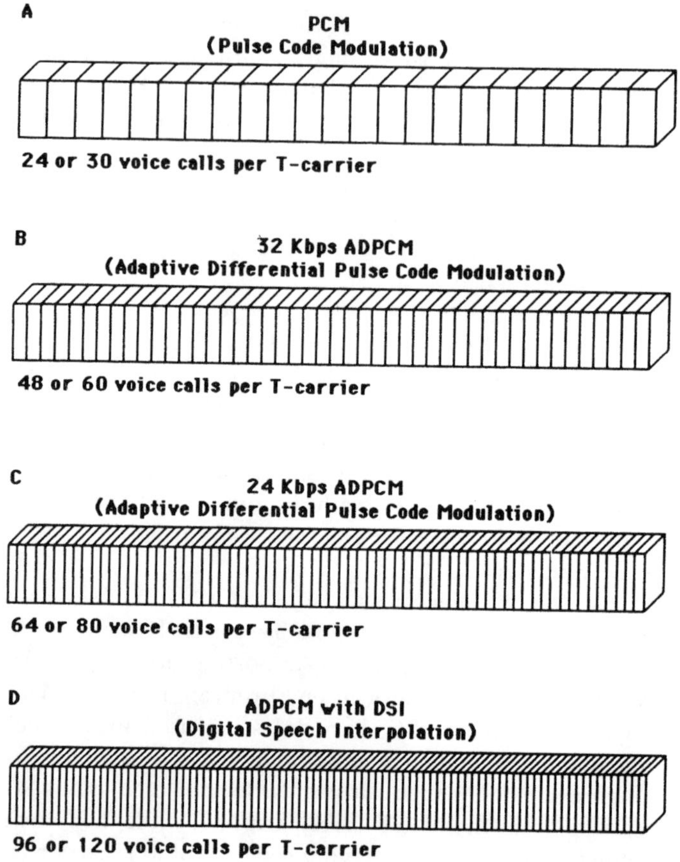

Figure 6.29 Trunk capacity as a function of digitization

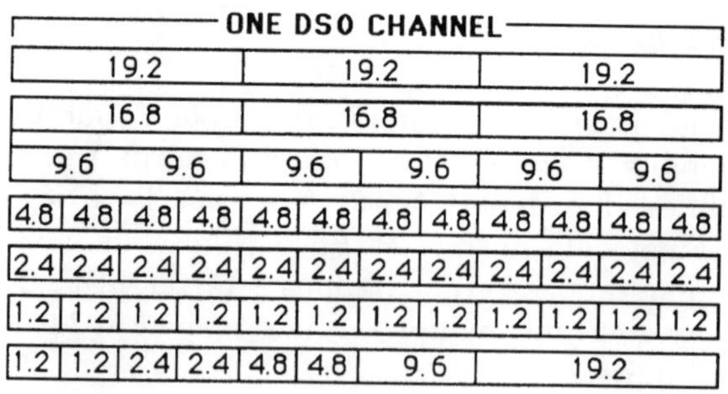

Figure 6.30 Typical subrate channel utilization

channel. Figure 6.30 illustrates one of the methods by which a T1 or E1 multiplexer vendor's equipment might multiplex subrate data channels onto one DS0 channel. Unfortunately, not all vendors provide a subrate multiplexing capbility in their equipment. When this occurs, subrate data sources are bit padded to operate at 64 kbps which can considerably reduce the ability of the multiplexer to maximize bandwidth utilization. In such situations users can obtain one or more subrate multiplexers to combine several data sources to a 64 kbps data rate; however, this may result in a higher cost than obtaining T- or E-carrier multiplexers that include a built-in subrate multiplexing capability.

Digital access cross-connect capability

The ability of a multiplexer to provide digital access cross-connect operations can be viewed as the next step up in terms of functionality from point-to-point multiplexer operations. Although communications carrier DACS are limited to switching DS0 channels, many multiplexer vendors include the capability to drop and insert/bypass subrate channels or digitized voice encoded at bit rates under 64 kbps, permitting sophisticated T-carrier networks to be constructed.

Figure 6.31 illustrates an example of the use of a digital access cross-connect feature used in three T1 multiplexers labeled A, B, and C. In this example, channel 8 on multiplexer A is routed to multiplexer C where it is dropped, freeing that DS0 channel for use as the T-carrier is then routed to multiplexer B's location. Thus, a data source at location C could be inserted into the T-carrier on channel 8, resulting in channel 8 being routed from C to B as shown in the figure. In this example it was assumed that all other DS0 channels were simply passed through or bypass multiplexer C and are then routed to multiplexer B. Thus, a digital access and cross-connect capability can be used to establish a virtual circuit through an intermediate multiplexer (bypass) without demultiplexing the data, allow intermediate nodes to add data to the data stream (insert), as well as permit an intermediate node to act as a terminating node (drop) for other multiplexers.

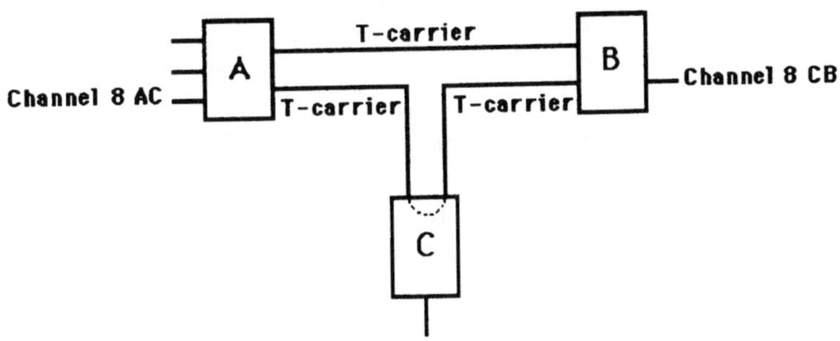

Figure 6.31 DS0 cross-connect. Although many multiplexers support the cross-connection of DS0 channels, some T1 multiplexers also permit the cross-connection of subchannels

Gateway operation support

To function as a gateway requires a T-carrier multiplexer to support a minimum of two high-speed circuits. In addition, the T-carrier multiplexer must perform several other operations that must be coordinated with the use of T- and E-carrier multiplexers connected to the gateway multiplexer.

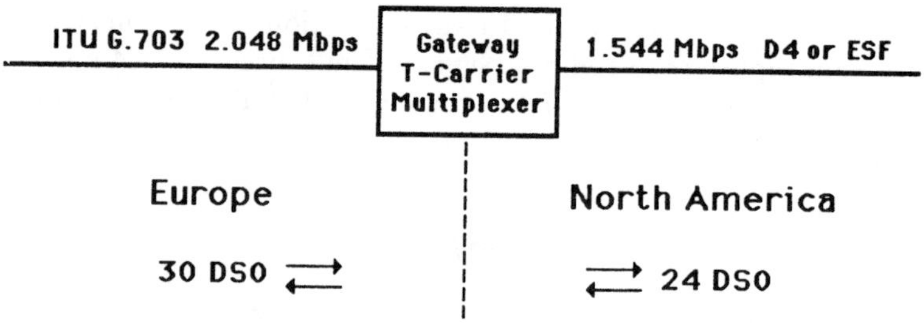

Figure 6.32 Gateway operation. The operation of a gateway results in the dropping of six DS0 channels when a European CEPT-30 facility is converted to a North American T1 link

Three of the main problems associated with connecting European and North American T- and E-carrier circuits through the use of a gateway multiplexer involve compensating for the differences between European E-carriers and North American T-carriers with respect to the number of DS0 channels each carrier supports, the method by which signaling is carried in each channel, and the method by which performance monitoring is accomplished.

Since a European E-carrier contains 30 DS0 channels, while a North American T1 link supports 24, the gateway multiplexer will map 30 DS0 channels to 24, resulting in the loss of 6 channels. This means that the end-user's organization is limited to the effective use of 24 channels on a European connection via a gateway multiplexer.

For signaling conversion the gateway multiplexer will move AB or ABCD signaling under D4 and ESF frame formats into channel 16 for North American to European conversion. For signaling conversion in the other direction, appropriate bits will be moved from channel 16 to the robbed bit positions used in D4 and ESF framing.

The area of performance monitoring is presently either ignored by gateway multiplexers or handled on an individual link basis to the gateway device. Since many European systems use a CRC-4 check while ESF employs a CRC-6 check no conversion is performed by multiplexers that support performance monitoring since the results would be meaningless for a link consisting of both North American and European facilities connected through a gateway. Instead, the gateway will provide statistics treating each connection as a separate T- and E-carrier facility. Figure 6.32 illustrates the placement of a gateway T1 multiplexer connecting a North American T1 facility to a European CEPT-30 facility.

Alternate routing and route generation

If a T-carrier network consists of three or more multiplexers interconnected by those carrier facilities, both the alternate routing capability and the method of route generation are important features to consider.

Basically, route generation falls into two broad areas: paths initiated by tables constructed by operators and dynamic paths automatically generated and maintained by the multiplexers. To illustrate both alternate routing and route

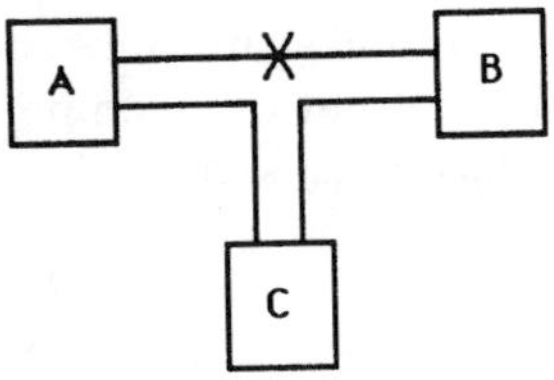

Figure 6.33 Alternate routing. T-carrier multiplexers provide a variety of methods to effect alternate routing to include using predefined tables and by the dynamic examination of current activity at the time of failure

generation, consider the T-carrier network illustrated in Figure 6.33. In this illustration the T-carrier circuit connecting multiplexers A and B has failed. With an alternate routing capability some or all DS0 channels previously carried by circuit AB must be routed from path AC to path CB to multiplexer B.

If the multiplexers employ alternate routing based upon predefined tables assigned by operations personnel, DS0 channels previously routed on path AB will be inserted into the T-carrier linking A to C and then routed onto path CB based upon the use of those tables. As the DS0 channels from path AB are inserted into the T-carrier linking A to C, DS0 channels on path AC must be dropped, a process referred to as bumping. If calls were in progress on path A to C and C to B when the failure between A and B occurred, those calls may be dropped depending upon whether or not the multiplexers employ a priority bumping feature or have the capability to downspeed voice digitized DS0 channels.

Priority bumping refers to the ability to override certain existing DS0 subchannels based upon the priority assigned to DS0 channels that were previously carried on the failed link and the priorities assigned to DS0 active channels on the operational links. Downspeed refers to the capability of multiplexers to shift to a different and more efficient voice digitization algorithm to obtain additional bandwidth to obtain the ability to carry DS0 channels from the failed link on the operational circuits. One example of downspeed would be switching from 32 kbps ADPCM to 24 kbps ADPCM, resulting in freeing up 12 kbps per operational DS0 channel.

When alternate routing and route generation is dynamically performed by multiplexers, those devices examine current DS0 activity and establish alternate routing based upon predefined priorities and the current activity of DS0 channels. If predefined tables are used, the multiplexers do not examine whether or not a particular DS0 channel is active prior to performing alternate routing, however, some multiplexers may have the ability to perform forced or transparent bumping regardless of whether they use fixed tables or dynamically generate alternate paths. Under forced bumping, DS0 channels are immediately reassigned, whereas under transparent bumping current voice or data sessions are allowed to complete prior to their bandwidth being reassigned.

Redundancy

Since the failure of a T1 or E1 multiplexer can result in a large number of voice and data circuits becoming inoperative, redundancy can be viewed as a necessity similar to business insurance. To minimize potential downtime you can consider dual power supplies as well as redundant common logic and spare voice and data adapter cards. Doing so may minimize downtime in the event of a component failure as many multiplexers are designed to enable technicians to easily replace failed components.

Maximum number of hops and nodes supported

As T1 and E1 multiplexers are interconnected to form a network each multiplexer can be considered as a network node. When a DS0 channel is routed through a multiplexer that multiplexer is known as a hop. Thus, the maximum number of hops refers to the maximum number of internodal devices a DS0 channel can traverse to complete an end-to-end connection.

In addition to the maximum number of hops users must also consider the maximum number of nodes that can be networked together. The maximum number of addressable nodes that can be managed as a single network is normally much greater than the maximum number of hops supported since the latter is constrained by the delay to voice as DS0 channels are switched and routed through hops.

Diagnostics

Most T1 and E1 multiplexers provide both local and remote channel loop-back capability to facilitate fault isolation. Some multiplexers have built-in test pattern generation capability which may alleviate the necessity of obtaining additional test equipment for isolating network faults.

Configuration rules

Figure 6.34 illustrates a typical T1 multiplexer cabinet layout which is similar to the manner in which a multiplexer would be installed in an industry standard 19-inch rack. In examining multiplexer configuration rules, a variety of constraints may

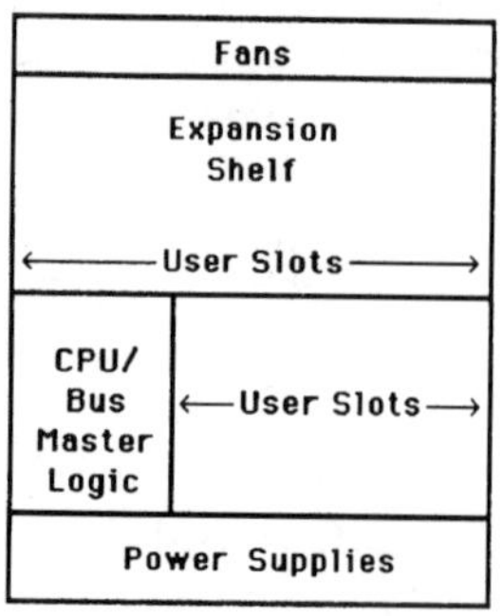

Figure 6.34 Typical T1 or E1 multiplexer cabinet

exist to include the number of trunk module cards, voice cards and data cards that can be installed. Other constraints will include the physical number of channels supported by each card and the type of voice digitization modules that can be obtained. Depending upon end-user requirements, additional expansion shelves may be required to support additional cards. When this occurs, additional power supplies may be required and their cost and space requirements must be considered.

6.3 SUBRATE VOICE/DATA MULTIPLEXERS

Until recently there was little advantage to be gained by attempting to multiplex voice and data onto a subrate digital circuit such as the popular and widely available 56 kbps and 64 kbps transmission facilities. The reason why little or no advantage was obtainable was based upon the fact that PCM and ADPCM, while providing toll quality voice, required a relatively high bandwidth that filled all or half of a 64 kbps circuit. Thus, the use of PCM resulted in no additional bandwidth that could be used for multiplexing other data sources, while the use of ADPCM would allow two voice conversations or one voice conversation and 32 kbps of data on a common 64 kbps transmission facility to be multiplexed.

The development of low bit rate voice digitization techniques such as LD-CELP and other versions of a family of similar hybrid coders made it possible to transport multiple voice conversations on relatively low cost 56 and 64 kbps digital circuits. The introduction of subrate voice/data multiplexers was targeted to branch office environments where only a few simultaneous calls between the branch and home office or between branch offices were required to be supported.

Operation

The subrate voice/data multiplexer is similar in many ways to T1 and E1 multiplexers. Although they support a composite 56/64 kbps output instead of 1.544 Mbps or 2.048 Mbps, they acquire their functionality through the addition of different types of card modules similar to T1 and E1 multiplexers. That is, adding a four-port voice card module that supports an 8 kbps version of CELP would enable four simultaneous calls while consuming 32 kbps of bandwidth.

Unlike voice that cannot tolerate delays, data card modules support flow control and their input is statistically multiplexed. This enables a higher level of data input bandwidth than the data portion of the composite output channel since the multiplexer will dynamically allocate bandwidth. In fact, most subrate voice/data multiplexers dynamically allocate all of the high-speed bandwidth; however, once a voice conversation commences they may then reserve 8 kbps of bandwidth for the conversation since the delay associated with bumping or with gracefully waiting to acquire bandwidth currently used for data could degrade the voice conversation in progress.

Utilization

Figure 6.35 illustrates an example of the use of an eight-port subrate voice/data multiplexer. In this example four ports are used to service voice while the remaining four data ports service three directly connected terminal devices and a computer.

If we assume the CELP modules operate at 8 kbps, then the occurrence of four voice calls would reduce the available 64 kbps composite bandwidth to 32 kbps for the four data sources. If we further assume that all four data sources are synchronous and the multiplexer has a 2:1 service ratio for synchronous data, then the four data inputs can equal 64 kbps prior to flow control becoming necessary to avoid buffer overflow. Depending upon the concurrent use of terminals and the computer shown in Figure 6.35, you could probably connect each terminal at 9.6 kbps and the computer at 19.2 kbps without experiencing any visually noticeable delays. Since both voice and data share a common transmission facility, the configuration shown in Figure 6.35 at a minimum can eliminate up to four simultaneous long-distance calls from a branch office to another branch office or corporate headquarters. By comparing your organization's long-distance telephone call economies against the cost of equipment, you can determine if the use of a subrate voice/data multiplexer is economically feasible. For many organizations, a full payback is commonly obtained within 12 months or less and is perhaps the key driving force behind the significant expansion in the use of subrate voice/data multiplexers during the mid to late 1990s.

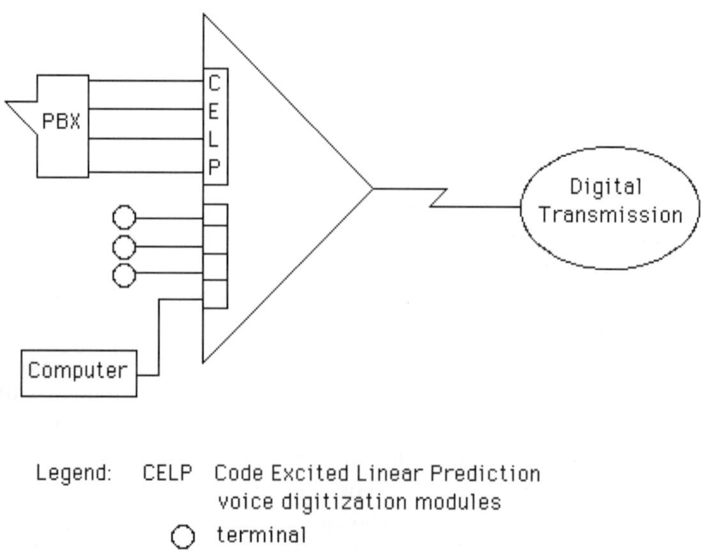

Figure 6.35 Using a subrate voice/data multiplexer

6.4 INVERSE MULTIPLEXERS

Through the introduction of a class of data communications equipment known as inverse multiplexers, network users can obtain transmission at high-speed data rates through the utilization of multiple lower operating rate lines. These devices also provide network configuration flexibility and provide reliable backup facilities during leased line outage situations. In addition, since high-speed transmission may not be available at some locations, these devices have the extra advantage of extending high-speed transmission to every location where more available lower operating rate leased lines and low speed DDS service can be obtained.

Initially in this section we will focus our discussion of inverse multiplexers to devices that split a data stream into two substreams and recombine those substreams at the other end of the communications link. Then, using the previously discussed material as a foundation, we will examine the use of inverse multiplexers that have four and six subchannel capacities. In concluding this section we will discuss a new role developed for the use of inverse multiplexers through their connection to the switched digital network. Called bandwidth-on-demand inverse multiplexers, these devices can be used for videoconferencing, T1 and E1 circuit backup and disaster recovery operations.

Operation

An inverse multiplexer splits a data stream at the transmitting station, enabling multiple substreams to travel down different paths to a receiving station. Such a data communications technique has several distinct advantages over single-channel wideband communications lines. Using leased lines increases network routing flexibility and permits the use of the analog direct distance dialing (DDD) or digital switched network as a backup in the event of the failure of one or more leased lines.

Similar in design and operation, devices produced by several manufacturers permit data transmission at speeds up to 1.544 or 2.048 Mbps by combining the transmission capacity of multiple circuits. Their operation can be viewed as reverse time division multiplexing. Input data streams are split into multiple paths by the unit's transmitting section.

The most basic inverse multiplexer produces two serial bit streams by dividing all incoming traffic into two paths. In its simplified form of operation, all odd bits are transmitted down one path and all even bits down the other. At the other end, the receiver section continuously and adaptively adjusts for differential delays caused by two-path transmission and recombines the dual bit streams into one output stream, as illustrated in Figure 6.36.

Each inverse multiplexer contains a circulating memory that permits an automatic training sequence, triggered by modem equalization, to align the memory to the differential delay between the two channels. This differential delay compensation allows, for example, the establishment of a 128 kbps circuit consisting of a 64 kbps satellite link, and a 64 kbps ground or undersea cable, as shown in Figure 6.37. Since the propagation delay on the satellite circuit would be 250 ms or more while the delay on the cable link would be less than 30 ms, without adjusting for the

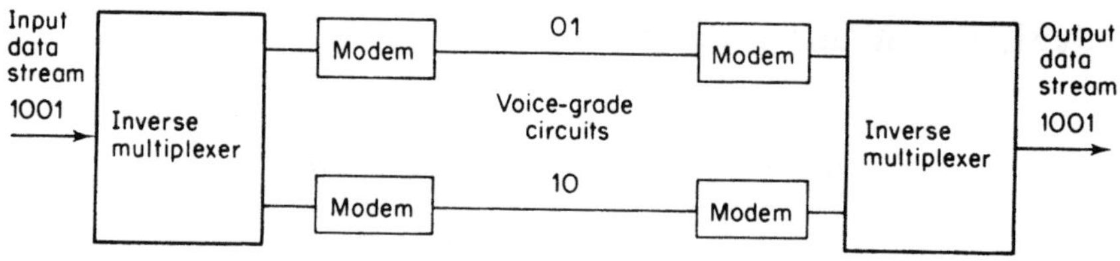

Figure 6.36 Inverse multiplexer operation. An inverse multiplexer splits a transmitted data stream into odd and even bits which are transmitted over two voice-grade circuits and recombined at the distant end

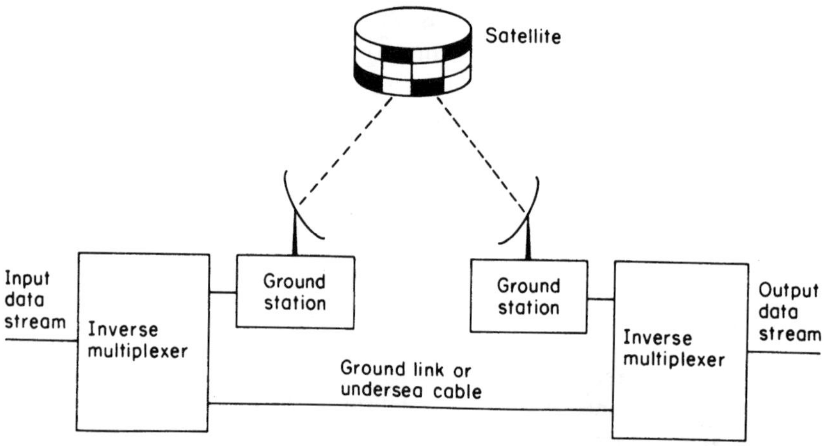

Figure 6.37 Inverse multiplexing using satellite–terrestrial circuits. One type of 'plexing' configuration could involve the use of a satellite link for one data stream with a ground or undersea cable link for the other

difference in transmission delays the bit stream could not be reassembled correctly. In this type of application, the failure of one circuit can be compensated by transmitting the entire data load over the remaining channel at one-half the normal rate.

Typical applications

As stated previously, one key advantage obtained by using inverse multiplexers is the cost savings associated with using a few multiple low-speed lines in place of a single higher operating rate transmission facility. Another advantage afforded by these devices gives the user the ability to configure and reconfigure their network based upon the range of distinct speeds available to meet changing requirements. For example, consider the use of four 33.6 kbps speed-selectable modems used with a pair of inverse multiplexers. You could use this hardware configuration with two

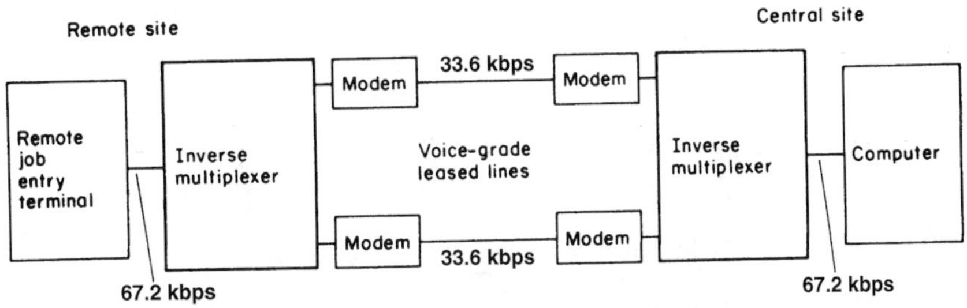

Figure 6.38 Computer-to-computer transmission. Inverse multiplexer at remote computer site splits data stream into odd and even bit streams that travel at data rates up to 67.2 kbps over conditioned voice-grade lines and recombine at the CPU located at the central site. In the example illustrated, four 33.6 kbps modems provide a 67.2 kbps transmission capacity

analog leased lines to obtain a 67.2 kbps transmission capacity between two locations. As an alternative, you could use a pair of Data Service Units and a 56 kbps digital transmission leased line with backup obtained via a pair of 33.6 kbps modems that could provide PSTN redundancy at 33.6 kbps or a lower operating rate based upon the quality of a dual backup connection.

Figure 6.38 illustrates the use of a pair of inverse multiplexers used with four 33.6 kbps modems and two analog leased lines to obtain a computer-to-computer transmission capability of 67.2 kbps. Note that this is without considering the effect of data compression. If the modems can achieve a 2:1 average compression ratio, the use of the network configuration shown in Figure 6.38 would provide a computer-to-computer transmission capability of 137.4 kbps. Since it is still easier in many locations to obtain an analog leased line than a digital leased line, the configuration shown in Figure 6.38 can represent a practical method to obtain an interim high-speed transmission capability usually associated with the use of a digital circuit. It is easier to obtain an analog circuit in an era when most communications carrier infrastructures are digital, because the vast majority of local loops are analog. In many locations the distance of a subscriber from a carrier's office may be beyond the capability of digital transmission without a major investment by the carrier to enable digital transmission. Thus, the use of inverse multiplexers also provides a mechanism to quickly obtain the transmission capability of a group of analog lines while waiting for a carrier to extend their digital network to your organizational location.

Another application for the use of inverse multiplexers could be to connect two geographically separated local area networks at a data rate of 112 kbps by aggregating the transmission capacity of two 56 kbps digital leased lines as illustrated in Figure 6.39. Upon occasion, locations to be interconnected may not have fractional T1 capability. In addition, depending upon the tariff structure and circuit availability of a communications carrier, it may be more economical and quicker to use two 56 kbps digital circuits than one fractional T1. This is because fractional T1 is provided by installing and provisioning a T1 line from a carrier's central office to the subscriber.

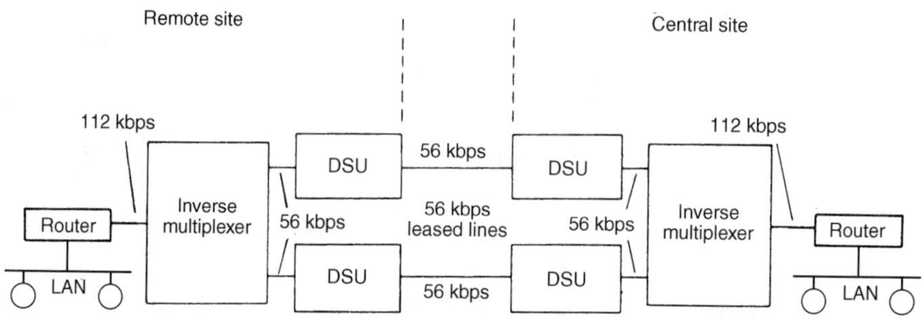

Figure 6.39 By aggregating the transmission of 56 kbps digital leased lines, an inverse multiplexer can provide a fractional T1 transmission capacity

Contingency operations

Through the use of inverse multiplexers you obtain the ability to consider a variety of network options not normally associated with the use of analog or digital leased lines. For example, many inverse multiplexers can be set to monitor the utilization of one circuit linking the two devices. When utilization reaches a predetermined threshold one inverse multiplexer could be programmed to use the switched analog or switched digital network as a mechanism to remove potential bottlenecks. This concept is illustrated in Figure 6.40(a) which shows how the switched analog

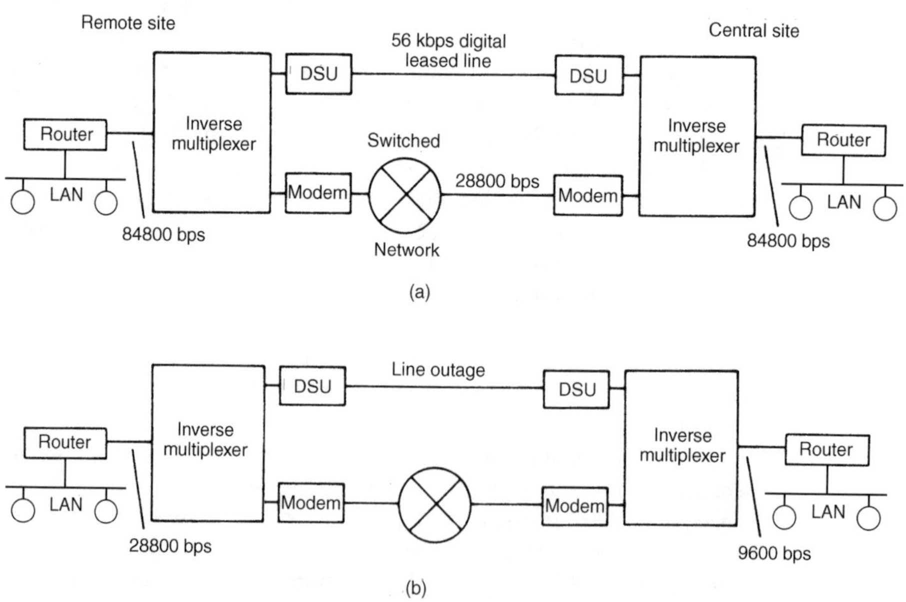

Figure 6.40 Methods of circuit compensation. (a) Dial restore on PSTN to compensate for failure of digital leased line. (b) Continued operations on one line at a data rate of 28 800 bps or less. Leased-line failure can be remedied directly over the switched analog or switched digital network

network is used as a mechanism to enhance network operations when utilization on the leased line reaches a predefined level. Note that if the leased line should fail, the use of the switched analog network enables communications at 28 800 bps via the use of dial modems. This is illustrated in Figure 6.40(b). Also note that you could consider the use of switched 56 kbps digital service as a method of circuit compensation.

Economics of use

One advantage in the use of an inverse multiplexer is the cost savings associated with using voice-grade analog or low-speed 56 kbps digital leased lines in comparison to a fractional or full T1 transmission facility. Although circuit tariffs follow a sliding scale of monthly per-mile fees based upon distance, the cost of a T1 circuit is often up to six times that of a similar-distance voice-grade analog or 56 kbps digital leased line. This means that the cost of four modems or DSUs, two inverse multiplexers, and two voice-grade analog or 56 kbps digital leased lines will normally be less than that of one T1 circuit and two channel service units.

Although a T1 circuit provides over 24 times the transmission capacity of a 56 kbps circuit, there are many network applications that require significantly less

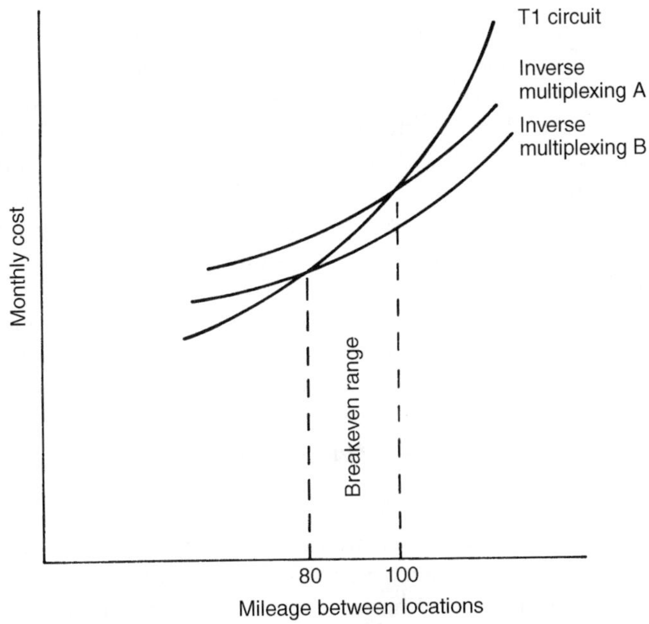

Figure 6.41 Economics of inverse multiplexing. In comparing the monthly cost of one T1 circuit and two channel service units to the cost of two voice-grade analog or 56 kbps digital leased lines, two inverse multiplexers, and four voice-grade modems or DSUs, the breakeven range will be between 80 and 100 miles

bandwidth. Thus, aggregating between two and six analog or 56 kbps digital circuits can provide a significant cost saving over the use of a full T1 and many types of fractional T1 circuits that require a full T1 access line to a carrier's central office.

Since tariffs follow a sliding scale, with the monthly cost per mile of a short circuit much more expensive than that of a long circuit, there are certain situations where the preceding does not hold true. This economic exception is usually encountered when connecting locations 80 to 100 miles or less distant from one another. Normally, if the two locations to be connected are over 80 to 100 miles apart, inverse multiplexing of two analog or 56 kbps digital leased lines is a more economic method of performing communications, while T1 is more economically advantageous at distances under that mileage range. The economics of inverse multiplexing are illustrated in Figure 6.41 that the 80- to 100-mile economic decision point range results from the fact that the monthly lease cost of inverse multiplexers varies between vendors.

A second advantage associated with using inverse multiplexing is the reduction in ordering time such devices may allow. Unfortunately, in many locations a notification period ranging up to several months in duration may be required for the installation of a T1 circuit. In comparison, the telephone company can usually install one or more voice-grade 56 kbps digital circuits within 30 days of receiving an order.

Extended subchannel support

As initially mentioned in this section, a number of vendors extended the capabilities of their inverse multiplexers to the support of four and six subchannels. Using four subchannel inverse multiplexers with eight 28 800 bps modems would permit a composite data transfer rate of 115 200 bps, while six subchannel inverse multiplexers with twelve 28 800 bps modems would permit a data rate of 172 800 bps to be obtained. When 56 kbps digital circuits are used instead of analog leased lines, a four-subchannel inverse multiplexer permits a composite data rate of 224 kbps, while a six-subchannel inverse multiplexer permits a composite data rate of 336 kbps.

Although the utilization of two subchannel inverse multiplexers is fairly common, the employment of four and six subchannel devices is far less frequently encountered, primarily owing to economics. This is because the monthly cost of the additional pairs of modems or DSUs and extra voice-grade analog or 56 kbps digital leased lines will rapidly approach and then exceed the cost of one fractional T1 circuit and two CSUs.

The use of four and six subchannel inverse multiplexers is thus usually governed by the unavailability of high-speed digital transmission facilities and not by the economics associated with their use. In addition, because of the ease of channel speed selection and line restoration capability, the use of inverse multiplexers can provide more efficient line utilization and better backup capabilities than are currently available on T1 transmission facilities. The utilization of inverse multiplexers can therefore be expected to continue, probably without regard to future tariff changes.

Bandwidth-on-demand

The extended availability of 56 kbps and 64 kbps switched digital service has resulted in a new use for inverse multiplexers by providing a bandwidth-on-demand capability. Such inverse multiplexers are designed to originate multiple calls over the switched digital network at 56 kbps or 64 kbps. By the use of multiple calls the inverse multiplexer provides users with the desired bandwidth they demand, hence the name for this communications device.

First introduced in late 1991, bandwidth-on-demand inverse multiplexers were initially marketed to satisfy end-user videoconferencing applications in an economical manner. Instead of organizations having to lease a fractional or full T1 circuit to interconnect videoconferencing sites, businesses could now use the switched digital network via the use of inverse multiplexers to initiate a series of calls only for the duration of the videoconference, in effect providing a mechanism for substantially reducing communications cost. Once the videoconference was concluded the calls would be terminated, a process referred to as circuit breakdown. Unless there are numerous videoconference sessions each day, the aggregate cost per minute per call will be significantly less than the cost of a leased fractional or full T1 circuit.

Modern bandwidth-on-demand inverse multiplexers support a variety of digital interfaces such as ISDN, fractional T1, and T1 lines. Concerning the latter, it is often more economical to install a T1 access line from a communication carrier's central office to a subscriber's premises even if the subscriber only requires the use of four or six 56 kbps or 64 kbps switched calls to be aggregated. In such situations four or six channels in the T1 circuit would be terminated at the carrier's central office and connected to the switched digital network.

Figure 6.42 illustrates the use of a bandwidth-on-demand inverse multiplexer interfaced to a T1 local loop access line to support videoconferencing at 384 kbps by aggregating six 64 kbps DSO channels on the T1 line. In actuality the inverse multiplexer is connected to six distinct digital telephone numbers at the carrier's central office and dials six calls to the distant location.

To illustrate the economics involved in the use of aggregated digital calls, note that in 1998 the average cost of a switched 64 kbps call was 6 cents/minute. Thus

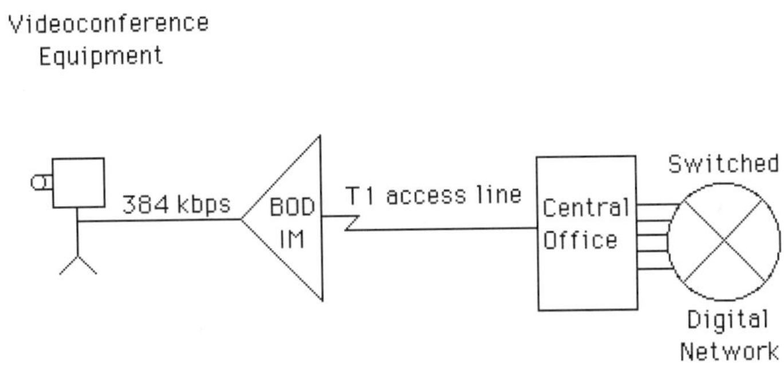

Figure 6.42 Using a bandwidth-on-demand inverse multiplexer (BODIM) with a T1 access line. By dialing six 64 kbps DSO channels from the T1 access line, an inverse multiplexer obtains a 384 kbps transmission capability over the switched digital network

aggregating six calls results in the cost of a 384 kbps videoconferencing becoming 36 cents/minute, or \$21.60 per hour. Since a fractional T1 line operating at 384 kbps can cost approximately \$2.00 per mile per month, you can easily compute a breakeven point between the use of the switched digital network and the use of a leased line. For example, assuming a requirement exists for a one-hour videoconference twice per week. Then, excluding the cost of inverse multiplexers, the cost of dial-up digital transmission becomes \$21.60 per call × 2 calls/week × 4 weeks/month, or \$172.80 per month. Since the cost of a 384 kbps leased line is approximately \$2.00/mile per month, this means that without considering the cost of the pair of inverse multiplexers the breakeven point would be \$172.80 per month ÷ \$2.00 per mile/month, or 86.4 miles. That is, if the two locations were greater than 86.4 miles from one another, it would be more economical to use the switched digital network.

Now let's assume each inverse multiplexer costs \$2500 and has a useful life of 3 years or 36 months. Then, the monthly cost associated with using the switched digital network becomes:

$$\frac{\$2500X2}{36} + \$172.80, \quad \text{or} \quad \$311.69$$

The breakeven point now becomes \$311.69/\$2.00, or 155.85 miles. Of course, another advantage associated with the use of the digital switched network is the fact that, unlike the use of a leased line that only enables videoconferencing with one distant location, you can use the digital switched network to communicate with any location connected to that network.

Since the failure of a T1 circuit can result in the inoperability of a large number of voice and data channels, many organizations, such as banks and insurance companies, back up such critical circuits with fractional or full T1 circuits. This can be very expensive and, in addition, results in circuits that are rarely used but which are paid for month after month. Recognizing this situation, several inverse multiplexer vendors have targeted the disaster backup and recovery market for their products. When used for disaster backup and recovery the user obtains the ability to use the switched digital network to backup fractional or full T1 lines. However, until single or multiple 56 kbps or 64 kbps calls are made the user only pays a small monthly fee for the digital access line, which can be considerably less than the cost of a leased digital line.

6.5 PACKET ASSEMBLER/DISASSEMBLER

The packet assembler/disassembler (PAD) is a specialized type of multiplexer originally developed to permit multiple terminals at one location to obtain access to a packet switching network via a common circuit. Originally PADs were designed to convert asynchronous start–stop protocols to an X.25 protocol. Later, many vendors expanded the functionality of PADs by adding additional protocol support to these devices.

Applications

The primary use of PADs is to enable non-X.25 networking devices to be connected to a packet switching network. There are many kinds of PADs, with their features and operation governing where they can be used to attach devices to an X.25 network. Some PADs support only asynchronous terminals, while other PADs support bisynchronous devices, SNA/SDLC devices or a mixture of terminals using different protocols.

PADs are used in both public and private X.25 networks. A public X.25 network is owned and operated by a communications carrier, which may sell or lease PADs to companies using the network. In addition, companies may purchase PADs for use on a public network once the device is certified by the communications carrier for use on their network. A private X.25 network can be constructed by an organization purchasing or leasing PADs and other network equipment and circuits

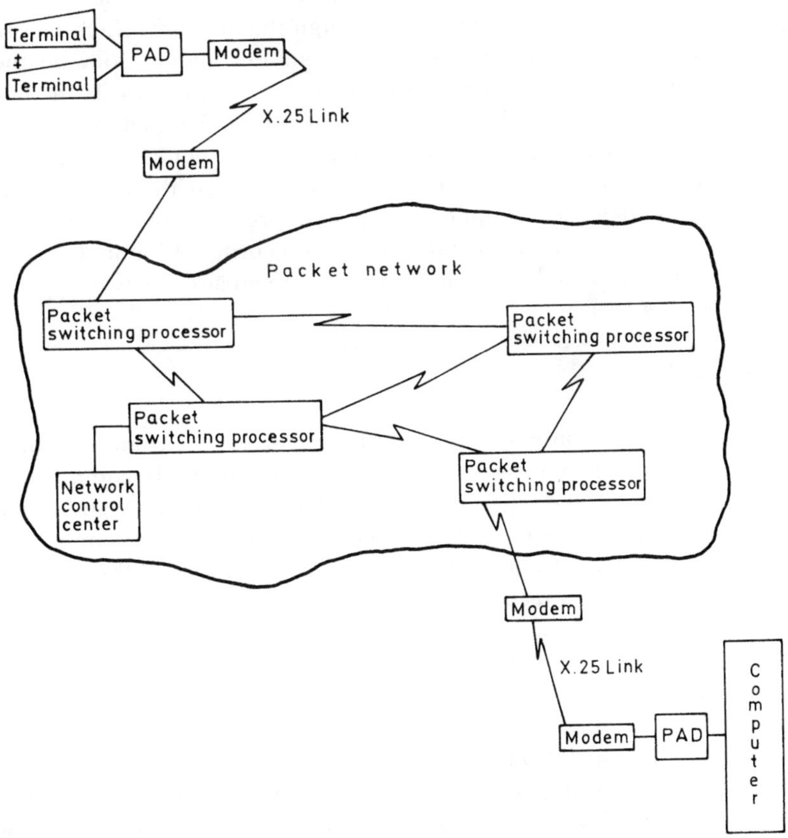

Figure 6.43 Using PADs in an X.25 network. By installing a PAD at a location where two or more terminals requiring access to an X.25 network are located the cost of communications to that network can be reduced. At a computer site, a PAD can be used to convert the X.25 packet network protocol to the protocol supported by a host computer

or it can be formed by a communications carrier subdividing a portion of their network for the exclusive use of an organization. Regardless of the method used, the company that has a private X.25 network can either purchase or lease PADs from communications carriers or third-party vendors.

Figure 6.43 illustrates two of the main uses of PADs for connecting non-X.25 devices to an X.25 network. In the top left corner of Figure 6.43 several non-X.25 terminals are assumed to require a connection to a packet network. Although the terminals could either dial a port on the packet network or be routed to the network via individual leased lines, it is normally more economical to install a PAD at the terminal location. Then, the terminals can be connected to the PAD and the PAD in turn can be connected to the packet network by the use of a single circuit over which data is transmitted according to the X.25 protocol.

Once the X.25 data enters the packet network it will flow through the network based upon its destination code and the activity on different trunks in the network. As indicated in Figure 6.43, most X.25 networks include a number of packet switching processors interconnected to one another by trunk lines which form a mesh structure. If any trunk should become overloaded or fail, the network control center can issue commands through the network which will result in the alternate routing of data. Once data reaches a destination node on the network, it can be transmitted to a host computer in one of two ways. If the host computer directly supports the X.25 protocol a leased line can be used to connect the computer to the packet network. If the host computer does not support the X.25 protocol, a PAD can be used as a conversion device between the multiplexed X.25 protocol and individual ports supporting specific protocols on the computer. This conversion process is illustrated at the lower right corner of Figure 6.43 where a PAD is used to interface a non-X.25 host computer to a packet network.

Types of PADs

The most common type of PAD supports asynchronous terminals and its operation is defined by a set of ITU recommendations. Recommendation X.3 defines the functions and operating parameters of the PAD and asynchronous terminals attached to that device. Included in the X.3 recommendation is a set of 22 parameters. Table 6.12 lists the meaning of each of the PAD parameters defined by the ITU X.3 recommendation.

X.3 parameters

Each of the PAD parameters has two or more possible values. Fourteen of the more commonly altered parameters and their possible values are listed below.

1:m—Escape from data transfer. This parameter defines whether or not you can escape from data transfer mode to command mode. In effect, this parameter determines if you can give the PAD X.28 command signals after data transfer is initiated. Readers should note that ITU Recommendation X.28 (described in Section 2.2) controls the PAD, while Recommendation X.3 specifies the PAD

Table 6.12 ITU X.3 PAD parameters

X.3 parameter number	X.3 meaning
1	Escape from data transfer
2	Echo
3	Selection of data forwarding signal
4	Selection of idle timer delay
5	Ancillary device control
6	Control of PAD service signals
7	Procedure on receipt of break signal
8	Discard output
9	Padding after carriage return
10	Line folding
11	Terminal speed
12	Flow control of the terminal pad
13	Line feed insertion after carriage return
14	Padding after line feed
15	Editing
16	Character delete
17	Line delete
18	Line display
19	Editing PAD service signals
20	Echo mask
21	Parity treatment
22	Page wait

parameters. Possible values are: 0—escape not possible; 1—escape possible with DLE (control P); 32-126—escape possible with a defined character.

2:m—Echo. This parameter controls the echoing of characters on the user's screen as well as their forwarding to the remote DTE. Possible values are: 0—no echo; 1—echo.

3:m—Selection of data forwarding signal. This parameter defines a set of characters that act as data forwarding signals when they are entered by the user. Coding of this parameter can be a single function or the sum of any combination of the functions listed below. As an example, a 126 code represents the functions 2 through 64, which results in any character to include control characters being forwarded. Possible values of parameter 3 are: 0—no data forwarding character; 1—alphanumeric characters; 2—character CR; 4—characters ESC, BEL, ENQ, ACK; 8—characters DEL, CAN, DC2; 16—characters ETX, EOT; 32—characters HT, LF, VT, FF; 64—all other characters: X'00' to X'1F'.

4:m—Selection of idle timer delay. This parameter is used to specify the value of an idle timer used for data forwarding. Possible values of this parameter include: 0—no data forwarding on time-out; 1—units of 1/20 s, maximum 255.

5:m—Ancillary device control. This parameter allows flow control using XON and XOFF (DC1 and DC3) from the PAD to the terminal. Possible values of this parameter are: 0—no use of XON and XOFF for flow control; 1—use of XON and XOFF for flow control.

7:m—Procedure on receipt of break signal. This parameter specifies the operation to be performed upon entry of a break character. Possible values of this parameter include: 0—nothing; 1—send an interrupt; 2—reset; 4—send an indication of break PAD; 8—escape from data transfer state; 16—discard output. Similar to parameter 3, parameter 7 can be coded as a single function or as the sum of a combination of functions.

9:m—Padding after carriage return. This parameter allows pad characters to be inserted after carriage returns to provide time for the print head of electromechanical printers to be positioned to column 1 after printing a line. Otherwise, the first few characters at the beginning of a new line could be lost. Possible values of this parameter are: 0—no padding; 1 to 7—the number of padding characters.

10:m—Line folding. This parameter determines how many characters are printed on a line. When a line contains more characters than the value in line folding output exceeding the value is printed on a subsequent line or series of lines. Possible values of this parameter are: 0—no line folding; 1 to 255—line folding with the number specifying the maximum number of characters per line.

12:m—Flow control of the terminal PAD. This parameter permits flow control of received data using X-On and X-OFF characters. Possible values are: 0—no flow control; 1—flow control.

13:m—Line feed insertion after carriage return. This parameter instructs the PAD to insert a line feed (LF) into the data stream following each carriage return (CR). Possible values are: 0—no LF insertion; 1—insert an LF after each CR in the received data stream; 2—insert an LF after each CR in the transmitted data stream; 4—insert an LF after each CR in the echo to the screen. The coding of this parameter can be as a single function or a combination of functions by summing the values of the desired options.

15:m—Editing. This parameter permits the user to edit data locally or at the host. If the local editing is enabled, the user can correct any data buffered locally, otherwise it must flow to the host for later correction. Possible values of this parameter include: 0—no editing in the data transfer state; 1—editing in the data transfer state.

16:m—Character delete. This parameter permits the user to specify which character in the ASCII (International Alphabet Number 5) character set will be used to indicate that the previously typed character should be deleted from the buffer. Possible values for this parameter include: 0—no character delete; 1-127—character delete character.

17:m—Line delete. This character is used to enable the user to specify which character in the character set denotes that the previously entered line should be deleted. Possible values are: 0—no line deleted; 1-127—line delete character.

18:m—Line display. This parameter enables the user to define the character which will cause a previously typed line to be redisplayed. Possible values are: 0—no line display; 1-127—line display character.

20:m—Echo mask. This parameter is only applicable when parameter 2 is set to 1. When this occurs parameter 20 permits the user to specify which characters will be echoed. Possible values include: 0—no echo mask (all characters echoed); 1—no echo of character CR; 2—no echo of character LF; 4—no echo of characters VT, HT, FF; 8—no echo of characters BEL, BS; 16—no echo of characters ESC,

ENQ; 32—no echo of characters ACK, NAK, STX, SOA, EOT, ETB, and ETX; 64—no echo of characters defined by parameters 16, 17, and 18; 128—no echo of all other characters in columns 0 and 1 of International Alphabet Number 5 and the character DEL.

In addition to supporting asynchronous terminals a variety of PADs are marketed to support vendor-specific protocols. Most PADs in this category support IBM bisynchronous or SNA/SDLC terminals, although a few vendors manufacture equipment to support other vendor-specific protocols.

The physical location of the PAD can be in one of three places. First, the PAD can reside on an X.25 network node, with terminals required to access the node via the PSTN or by leased lines. This is the most common PAD location, since the majority of persons accessing packet networks use asynchronous terminals and dial a node via the PSTN. Secondly, a stand-alone PAD can be installed at the end-user's terminal or host computer location as previously illustrated in Figure 6.43. The use of this type of PAD permits the multiplexed X.25 protocol to be routed on one circuit between the packet node and the end-user location, in effect serving to minimize the communications cost of the end-user. The third PAD location is inside a computer or terminal device. This type of PAD is normally a special adapter card that can be installed within a personal computer or on the channel adapter of a front-end processor. Most PADs built on an adapter card permit the personal computer user to establish simultaneous communications with two or more remote computers on one X.25 connection as illustrated in Figure 6.44. This capability permits the user to switch from one host session to another without requiring the log-off from one computer system prior to signing onto a second system.

As previously discussed when we examined packet switching networks earlier in this book, the use of X.25 networks results in error checking at each node in the network. This results in cumulative delays or latency when data is routed through multiple nodes, and makes X.25 performance detrimental for linking LANs. Although X.25 networks remain viable for transporting email, mainframe, and

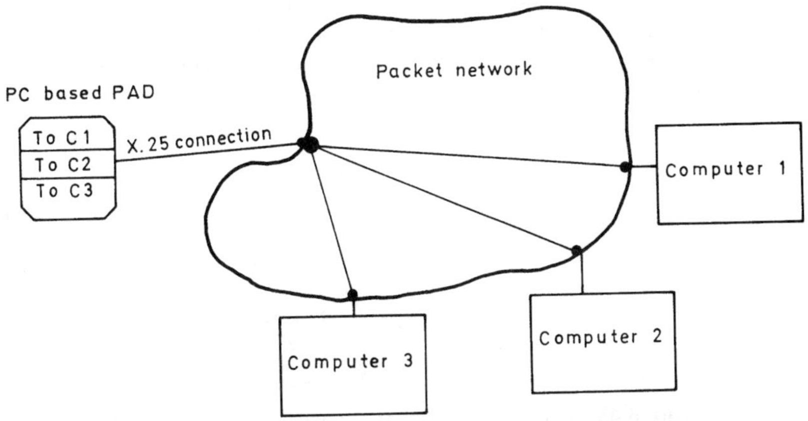

Figure 6.44 PC based PAD utilization

credit card data, their use for interconnecting geographically separated local area networks has been replaced to a large degree by the use of Frame Relay networks. Since the Frame Relay Access Device (FRAD) is the key device used to obtain access to a Frame Relay network, Section 6.6 is focused upon the operation and utilization of that networking device.

6.6 FRAME RELAY ACCESS DEVICE

A Frame Relay Access Device (FRAD) can be considered to represent a combination protocol conversion and encapsulation product, enabling a variety of LAN and WAN protocols to be transported over a Frame Relay network. Since the basic function of a FRAD is to provide access into a Frame Relay network, all FRADs transmit and receive data via the use of Frame Relay frames to and from a Frame Relay network. Depending upon the type of internetworking a FRAD is manufactured to support, it may encapsulate or convert LAN and WAN protocols in or into Frame Relay frames. In addition, based upon other functions included in the design of a FRAD, it may perform a variety of additional functions. One example of a relatively new feature added to FRADs is the ability to support voice over Frame Relay. To do so, most FRADs include frame fragmentation and priority queuing as well as modules that digitize voice at very low data rates.

Hardware overview

From a hardware perspective, a FRAD consists of a series of software modules operating in memory that control the data flow to at least one LAN via a network adapter card, and to a Frame Relay network via a serial port. Figure 6.45 illustrates the use of a pair of FRADs to interconnect two geographically separated Ethernet LANs via a Frame Relay network.

In examining Figure 6.45 note that some FRADs are modular devices, enabling the addition of LAN and WAN port modules that can be used to serve additional devices. For example, adding Token-Ring or ATM port modules enables the FRAD to provide connectivity to Token-Ring and ATM networks. Similarly, the use of serial port modules may permit a FRAD to support SDLC, bisynchronous, and asynchronous data streams.

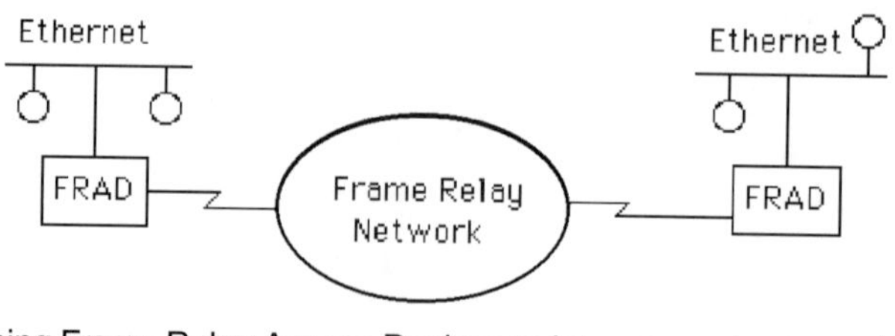

Figure 6.45 Using Frame Relay Access Devices to interconnect geographically separated LANs via a Frame Relay network

Comparison to routers

Routers and FRADs are very similar with respect to their fabrication and the distinction between the two can be blurred; however, there are some key differences between these two internetworking devices. A router is designed to move traffic between networks based upon network routing protocols, such as IP or IPX. Although some routers can also transport legacy traffic such as SNA SDLC, bisynchronous, and asynchronous transmission, they usually do so by encapsulating such traffic into IP packets, a process also referred to as tunneling. Although a router can be used to provide access to a Frame Relay network, some do so via transmission of IP packets to a network provider's switch which then provides a translation of IP into Frame Relay frames. Thus, this can result in encapsulation and conversion when legacy traffic is converted by a router. A router that can be configured to support Frame Relay directly on WAN links is commonly referred to as an Internetworking FRAD (I-FRAD).

The I-FRAD

An I-FRAD is a much more sophisticated product than a FRAD as it includes support for routing protocols, enabling it to be used to connect a corporate location to the Internet as well as to its own internal router-based network. One of the key advantages associated with the use of an I-FRAD is its ability to provide a multinetworking support capability. That is, a three-port I-FRAD consisting of one LAN port and two serial ports could be used to connect LAN users to another organizational network via a Frame Relay network provider as well as to the Internet. Figure 6.46 illustrates this potential use of an I-FRAD.

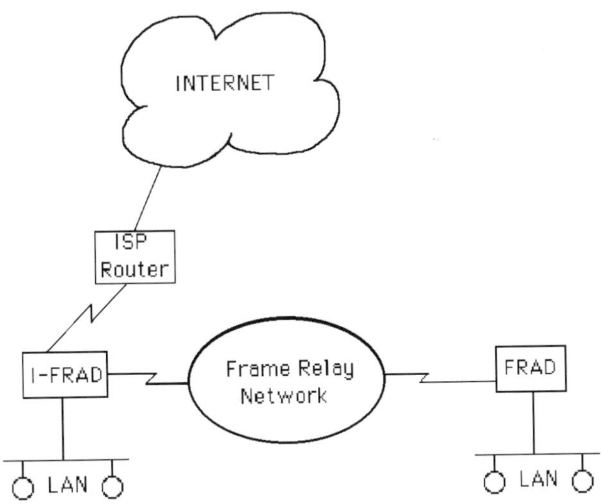

Figure 6.46 AN I-FRAD provides access to a Frame Relay network as well as routing to IP-baded networks such as the Internet

Protocol support

As previously mentioned, FRADs can support several legacy protocols such as IBM's SNA/SDLC and its older bisynchronous transmission protocol in addition to asynchronous, Ethernet and Token-Ring protocols. For many organizations, a key consideration when selecting a FRAD is its method of support for IBM's Synchronous Network Architecture (SNA). Although SNA represents a legacy mainframe-based communications architecture, there are over 50 000 SNA networks in existence. This means that organizations with mainframe networks and LANs that wish to take advantage of the capability of Frame Relay to replace or supplement expensive SNA leased line networks need to obtain FRADs capable of supporting Synchronous Data Link Control (SDLC), the protocol used on SNA networks.

There are at least four techniques you can consider for transmitting SNA over a Frame Relay network. Those methods are summarized in Table 6.13

Table 6.13 Techniques for transporting SNA via a Frame Relay Network

- SNA/SDLC encapsulation into TCP/IP
- SNA/SDLC conversion to SNA/LLC and encapsulation into TCP/IP
- Data Link Switching
- SNA/SDLC conversion to SNA/LLC2 and encapsulation into Frame Relay based on RFC 1490

SNA/SDLC encapsulation into TCP/IP

One of the oldest techniques developed to route SNA traffic via a Frame Relay network is to first encapsulate it into TCP/IP. Then, TCP/IP is encapsulated within the Frame Relay frame. This results in each frame carrying six bytes of Frame Relay header overhead, 20 bytes of TCP header overhead, and 20 bytes of IP header overhead in addition to the SNA header. Due to the high overhead associated with this technique, it is not widely used.

SNA/SDLC conversion to SNA/LLC2

Under this SNA transmission technique legacy SNA traffic is first converted into SNA/LLC2 and then encapsulated into TCP/IP. Although this technique results in the use of Frame Relay, TCP, and IP headers similar to the direct encapsulation of SNA/SDLC into TCP/IP, the conversion to SNA/LLC2 results in the elimination of a significant amount of the polling activity required by SNA. This reduces the risk of session timeouts due to the delays resulting from multiple encapsulations.

Data Link Switching

Data Link Switching (DLSw) was developed by IBM as a mechanism to enable SNA to be transported via IP routers, assuming that the network infrastructure is an IP backbone. Under DLSw, each SNA packet is encapsulated in IP for routing, and TCP for error recovery. Since the DLSw header is 16 bytes in length, this transmission technique results in the expansion of frame overhead to 56 bytes.

When DLSw is transported via Frame Relay, an additional six bytes of Frame Relay overhead is added which can result in the total overhead approaching the size of many SNA packets. This overhead is illustrated in Figure 6.47(a). On low-speed Frame Relay connections such as a 56 kbps link, the high amount of overhead can result in session timeouts. Thus, DLSw is only useful when LAN traffic predominates over legacy transmission or when time sensitivity is not a key issue. An example of the latter would be transporting files instead of interactive query–response SNA traffic via a Frame Relay network.

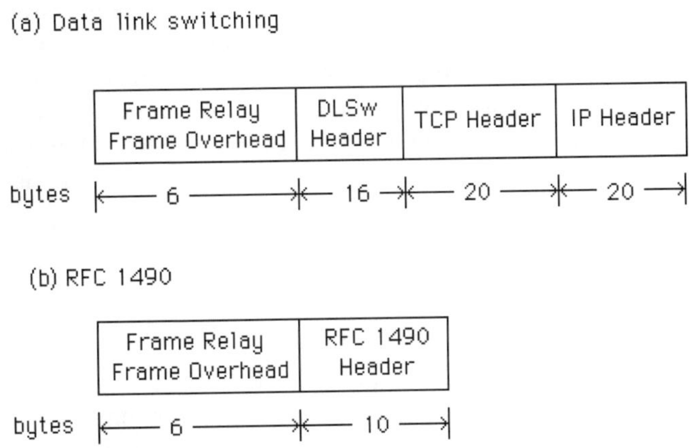

Figure 6.47 Comparing data link switching and RFC 1490 to move SNA traffic via a Frame Relay network

RFC 1490

Recognizing the problems associated with the three previously discussed methods developed to support the transmission of SNA/SDLC traffic over a Frame Relay network, the Internet Engineering Task Force (IETF) developed the RFC 1490 standard as a mechanism to transport SNA directly over a Frame Relay network with a minimum of overhead. Under RFC 1490 SNA/SDLC traffic is converted to LLC2 which provides an error recovery and flow control capability. As illustrated in Figure 6.47(b), the RFC 1490 header consists of only 10 bytes, reducing the total overhead when an RFC 1490 packet is transported by Frame Relay to 16 bytes.

Voice over Frame Relay

The transmission of voice over a Frame Relay network requires a FRAD to perform several functions. Those functions include fragmenting frames, prioritizing voice over data, and removing the variable delay associated with frames arriving at their destination. The latter is due to the variable delays occurring from random processing loads placed upon switches in a Frame Relay network. In addition, most FRADs designed to support voice also include digitization modules developed to convert voice into a low bit rate digital data stream. Thus, to obtain an appreciation of the functionality of a FRAD that supports voice we must discuss its ability to perform fragmentation, prioritization, buffering, and digitization.

Fragmentation

Without frame fragmentation it becomes possible for a relatively long frame transporting data to be serviced and placed on a circuit routed to a Frame Relay network between two relatively short frames transporting sequences of a digitized conversation. At the destination the lengthy data frame would result in one voice frame being significantly delayed from the next, causing an annoying delay that distorts the reconstructed voice conversation. Recognizing this problem, FRADs that support voice do so by fragmenting frames so that they do not exceed a certain length based upon the type of data transported. Doing so limits the potential delay distortion when a frame transporting data or fax is serviced between frames transporting voice.

Prioritization

A second technique employed by FRADs designed to support voice is the use of priority queues. Typically, frames transporting voice and fax are placed in high-priority queues while data which is more tolerant of a delay is placed in a lower-priority queue. The FRAD then services high-priority queues in preference to lower-priority queues. However, to ensure that frames transporting voice do not significantly delay frames transporting data, a FRAD will typically use a service ratio such as 5:3, meaning that it will extract frames five times from a high-priority queue for transmission in comparison to every three extracted from a low-priority queue for transmission.

Buffering

When voice is digitized and converted into a series of frames, there are uniform delays between each frame. As those frames flow through a network, random processing loads placed on network switches results in the arrival of frames at their destination with variable delays betwen frames. Since the reconstruction of voice from randomly delayed frames would sound awkward, FRADs that support voice

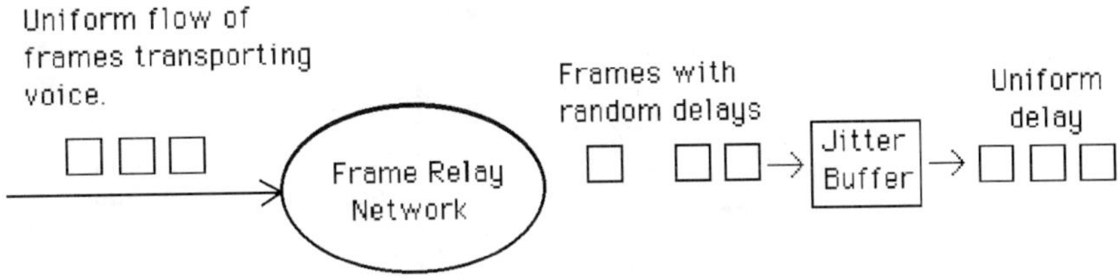

Figure 6.48 Using a jitter buffer. A jitter buffer in a FRAD enables frames transporting voice that arrive with random delays to be extracted at a uniform rate, enabling more natural sounding speech to be reconstructed

include a jitter buffer whose general operation is illustrated in Figure 6.48. The jitter buffer enables frames arriving with variable delays to be extracted with a uniform delay, enabling the received voice signal to appear more natural sounding when it is reconstructed.

Voice digitization

Most FRADs that support voice over Frame Relay are available with a variety of voice digitization modules. Some modules represent waveform coding methods such as PCM and ADPCM, and their use results in a digitized voice data stream operating at 64 kbps and 32 kbps, respectively. Other voice coding modules supported include several hybrid coding methods based upon a family of Code Excited Linear Predicting (CELP) algorithms that reduce a digitized voice conversation to a data rate of 16 kbps or less. Readers are referred to Section 6.3 in this chapter for information concerning different voice compression methods and the advantages and disadvantages associated with each method. In 1997, the Frame Relay Forum's technical committee developed an Implementation Agreement (IA) for Voice over Frame Relay (VoFR). This IA defines how FRADs from different manufacturers can communicate their settings to one another so that they can interoperate. This capability should eventually enable voice over Frame Relay to support interoperability between different organizations which could significantly enhance the use of this evolving technology.

6.7 FRONT-END PROCESSORS

A front-end processor is a minicomputer first developed to offload serial communications processing from a mainframe computer. Since its initial development the functionality and capability of front-end processors have been significantly enhanced, with some products now including routing, LAN connectivity, and other modern communications networking functions.

Front-end processors provide a large volume of network communications power in support of a particular computer system. Thus, they are considered to be 'closely coupled' to a particular host computer and may be specifically programmed to

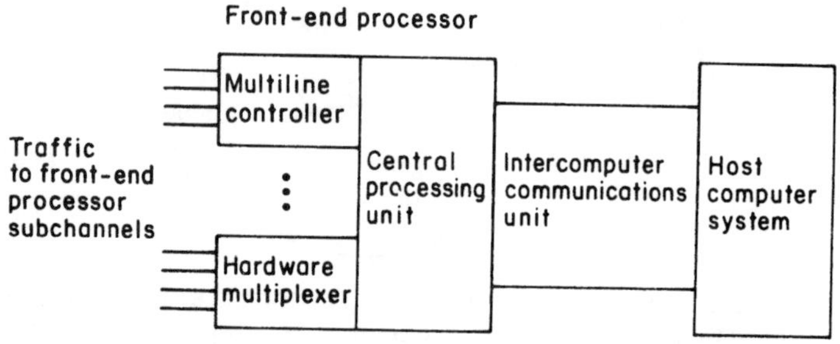

Figure 6.49 Fron-end processors. A front-end processor can be considered the heart of a host or main frame based computer network, relieving the host computer system of many software burdens, performing such functions as code conversion, character blocking, and character deblocking

operate with that computer and its operating system. A typical front-end processor configuration is illustrated in Figure 6.49. In examining the hardware configuration of a typical front-end processor shown in Figure 6.49, note that the multi-line controller (MLC) represents a grouping of single line controllers, and a single line controller represents a serial port, buffer area, and logic which enables received bits to be grouped into characters for parallel operations and parallel formed characters to be placed bit by bit onto a serial communications line. By performing serial to parallel and parallel to serial conversion, as well as adding appropriate communications protocol characters to a data stream, performing CRC computations and similar communications functions, the front-end processor offloads a considerable amount of communications processing from a host computer. This frees the host computer of a large portion of the burden of communications processing, enabling more of its CPU cycles to be devoted to processing application programs.

MLCs enable a mixture of asynchronous and synchronous lines to be serviced at data rates ranging from 50 bps to 2.048 Mbps. By configuring each port for a specific protocol the MLC also enables multiple protocols to be supported at different data rates over a grouping of line terminations.

Another device encountered on some front-end processors is a local communications multiplexer which provides for time division multiplexing by character, to and from the front-end processor, for a variety of low-speed terminals at transmission rates up to 9600 bps. These local multiplexers can handle terminals with differing communications speeds and code settings, with the character demultiplexing performed by software in the front-end processor. In addition to performing network and communications processing activities that are normally associated with front-end processors, owing to the larger memory and word size, quite often they can be used to perform message switching functions by the addition of modular software.

The operating system which supervises the overall control and operation of all system functions is the key element of a front-end processor. Although numerous software elements must be evaluated, major consideration should be given to determining supported line protocols as well as supported processor communications. In the first area, most vendors divide their supported line protocols into several categories or classes of support. Normally, category one refers to vendor-

Table 6.14 Front-end processor software modules

Module	Functional description
Character/message assembly/disassembly	Assemble the serial bit stream into characters and messages for transfer to the host computer and vice versa
Data formatting	Restructure incoming data to a more compact format to permit an increase in host computer efficiency
Code conversion	Convert the data codes of different terminals into the code employed by the host computer
Message switching	Route messages from one terminal to another without requiring the data to pass through the host computer
Polling	Query input/output devices to see if they have information to transfer or are available to accept information
Error checking	Check incoming data to insure transmission errors did not occur; reject data blocks and request retransmission if errors are computed to have occurred
Protocol support	Ability to communicate with different terminal devices according to a standard method of interface
Automatic operation	Handle automatic answering and/or outward calling on the public switched telephone network
Statistics	Compute such important parameters as traffic density and circuit availability as well as other statistics which may be essential for the effective management of a data communications network

developed and tested software to support certain line protocols. Category two usually refers to vendor-developed but non-qualified tested software; while category three refers to customer-developed interfaces designed to support certain terminal line protocols.

The functionality of front-end processors is obtained through hardware and software. Table 6.14 lists the major front-end processor software modules offered by most vendors of this device, followed by a brief functional description of the module. Later in this section we will examine the hardware functionality of specific IBM front-end processors.

Communications controllers

In place of the term front-end processor IBM labeled its equivalent hardware as a communication controller. Members of the IBM communication controller family include the IBM 3704, 3705, 3720, 3725, 3745 and 3746 which are designed to operate with IBM host computers ranging from system/360 through system/390 products as well as AS/400 and 4300 series computer systems. Although there are numerous hardware differences between members of the IBM communication controller family, the examination of the IBM 3725, 3745 and 3746 presented in this section can be used as a reference for reviewing the hardware functionality and operational capability of the earlier models in this family.

IBM 3725

The IBM 3725 is a modular, programmable communication controller that can be attached to an IBM computer system in one of two ways: channel-attached or link-attached. In a channel-attached mode of operation the 3725 must be physically located in close proximity to the host computer system as illustrated in the top portion of Figure 6.50. In this method of attachment the communication controller is physically connected to a data channel on the host computer system and data is transferred in parallel between the computer and the controller.

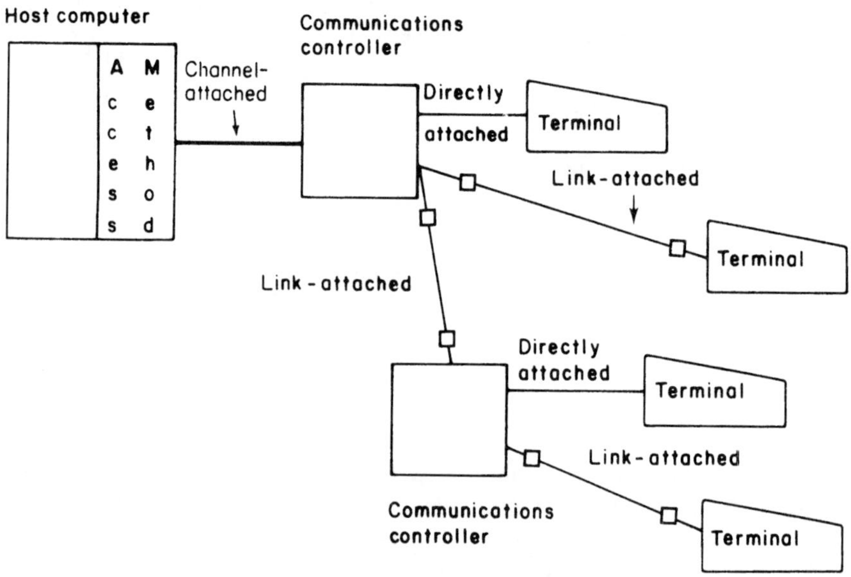

Figure 6.50 IBM communications controller attachments. (□ modem or DSU)

In a link-attached mode of operation the communication controller is connected via a telecommunications line through the use of modems or DSUs to another communication controller, which can be channel-attached to the host computer or may be another link-attached controller. In this attachment method, data is transferred serially. In the lower portion of Figure 6.50, a link-attached controller connected to the channel-attached controller previously discussed is illustrated.

Terminal devices can be connected both directly and remotely via a communications facility to either a link-attached or channel-attached communication controller.

The 3725 communication controller contains three logical subsystems which are illustrated in Figure 6.51.

The control subsystem consists of a central control unit, main storage, and channel adapters. The transmission subsystem consists of up to eight line attachment bases (LABs), each containing one or more communication scanners and line attachment hardware called line interface couplers (LICs). The maintenance and

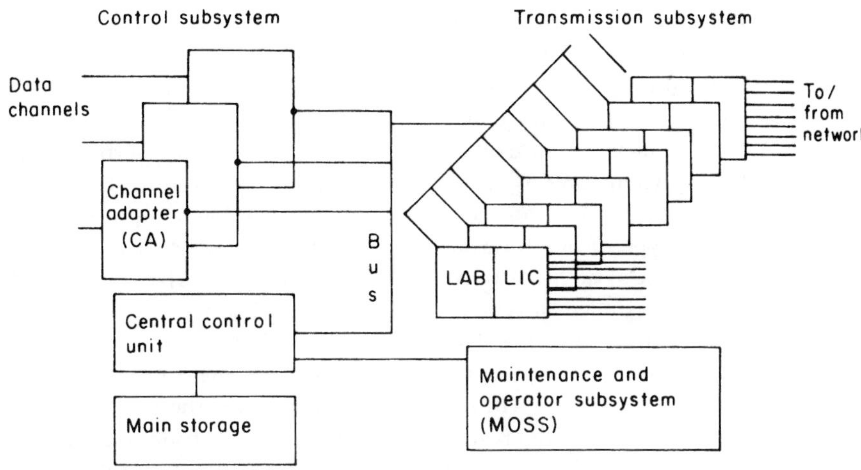

Figure 6.51 3275 communications controller subsystems (LAB line attachment base, LIC line interface coupler)

operator subsystem (MOSS) consists of a processor and storage, a diskette drive, control panel, and up to two externally connected operator consoles.

The channel adapters provide the physical interface between the communication controller and its direct attachment to a host computer system, permitting the controller to be attached to a byte-multiplexer, block multiplexer, or selector channel on the host computer. Up to two channel adapters can be installed in a 3725, while four additional channel adapters can be installed in the 3725's expansion unit, which is formally called the 3726 Communication Controller Expansion.

The line attachment bases house comunication scanners and line interface couplers. Two types of LABs can be selected by the end-user based upon the data transmission rate of the connected line. The type A line attachment base (LABA) contains one communication scanner and supports up to 32 low- and medium-speed lines, operating at or under 9600 bps. These lines can be connected to up to eight LICs, with each LIC supporting the physical connection of up to four lines.

The second type of LAB is the type B (LABB) device which contains two communication scanners and can support up to eight high-speed lines or 32 medium-speed lines, attached through a maximum of eight LICs. Up to three LABs can be installed in the basic 3725 controller, permitting it to service up to 96 lines. Five additional LABs can be installed in the 3726 Communication Controller Expansion, providing support for a total of 256 lines. The first two LABs must be type A devices, while the remaining six line attachment bases can be eight type A or type B, as required.

The scanner in each line attachment base is a hardware device that contains a microprocessor, storage for the scanner microcode, and a buffer area for servicing the lines. Each scanner is near universal in terms of support, as scanners can service a mixture of line speeds and communication protocols to include asynchronous, bisynchronous, and synchronous data link-control. Depending upon the type of line connection, the scanner will insert or delete control characters, perform error detection and correction, and perform such functions as bisynchronous code translation.

The line interface coupler provides the physical connection between the IBM communication controller and a directly connected terminal or communications facility terminated by a modem or digital service unit. Several types of LICs are available for selection, with their actual use based upon such factors as the line speed, type of line interface (RS-232/ITU V.24 or ITU V.35), and line protocol. Some LICs can be installed in either LABA or LABB bases, while other LICs can only be installed in type B line attachment bases.

The third major subsystem of the 3725 is the maintenance and operator subsystem (MOSS). This subsystem has its own microcode and is used to detect and isolate failures within the controller. Since the microcode in MOSS operates independently of the rest of the controller, it can function even when the controller is inoperative. Up to two external operator consoles can be connected to the 3725 maintenance and operator subsystem.

Due to the wide acceptance of local area networks within organizations, IBM developed a Token-Ring interface coupler (TIC) which allows 3725 and more recently developed communications controllers to be connected to Token-Ring networks. The TIC on a 3725 can be considered as a gateway device and represents one of four methods by which Token-Ring networks can be connected to IBM communications controllers. Readers are referred to Section 4.3 for additional information concerning Token-Ring gateways.

IBM 3745

The 3745 communications controller was introduced by IBM during the late 1980s. A series of 3745 models are marketed by IBM which extended the useful life of this communications controller as well as added a networking capability which was significantly more versatile than its 3725 predecessor.

The 3745 has the same functional structure as the 3725, consisting of a maintenance and operator subsystem, control subsystem, and communications subsystem. The key differences between the 3745 and 3725 are in the number and types of communications links supported by the newer front-end processor, which is also referred to by IBM as a communications controller.

Although the exact configuration of a 3745 determines the actual number of low, medium, and high speed lines and LAN connections it can support, its capability far surpasses the 3725. For example, a 3745 can support a maximum of 896 low and medium speed transmission lines at data rates up to 256 kbps per line, as well as up to 16 high speed T1 or E1 lines operating at 1.544 Mbps and 2.048 Mbps, respectively. In addition, a 3745 can be configured to support a maximum of eight Token Ring and four Ethernet LAN connections through the installation of appropriate LAN adapters in its communications subsystem.

Similar to a 3725, a 3745 can be channel-attached to host computers and link-attached to other communications controllers and control units. Unlike the 3725 that was restricted to supporting Token-Ring LANs, the 3745 can also support

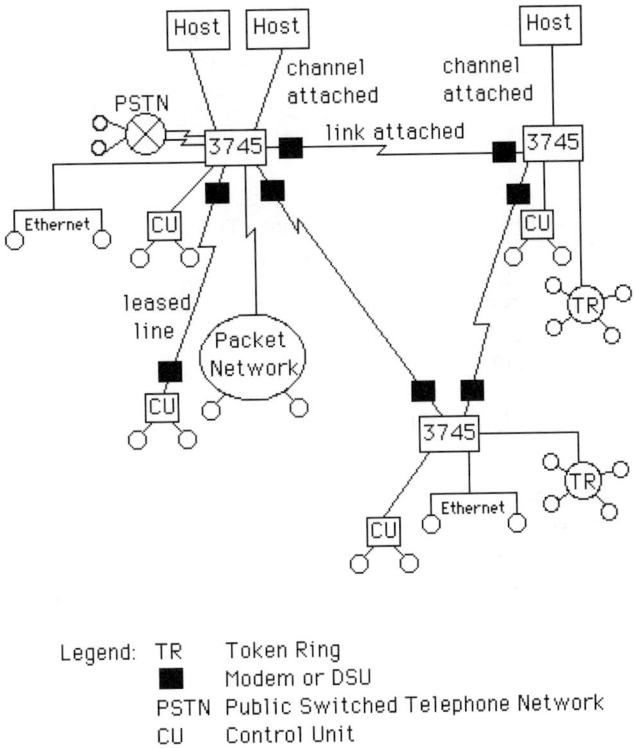

Figure 6.52 Using a 3745 communications controller permits Token-Ring and Ethernet connectivity while extending channel and link attachment capabilities

Ethernet LANs. When coupled with its increased serial line support capability, the 3745 becomes a true communications hub in an enterprise mainframe-centric based network. Figure 6.52 illustrates a possible network configuration based upon the use of several 3745 communications controllers to provide SNA and LAN communications access to multiple host computers at two corporate data centers.

In examining Figure 6.52, note that while two 3745s are channel-attached to host computers, a third communications controller operates as a remote controller, concentrating traffic from legacy terminal devices connected to a control unit as well as from workstations connected to Ethernet and Token-Ring networks for forwarding to other 3745s. These in turn provide access to host computers. Also note that the 3745s can run conventional SNA, which requires all sessions to be routed via a host or the more modern APPN in which sessions can be established between peers without requiring access to a host. Under APPN, it becomes possible, for example, for a station on an Ethernet LAN at one location to access a server at a different location.

Communications structure

The key to the communications capability of the 3745 is its communications subsystem structure. Figure 6.53 illustrates in schematic form the structure of the

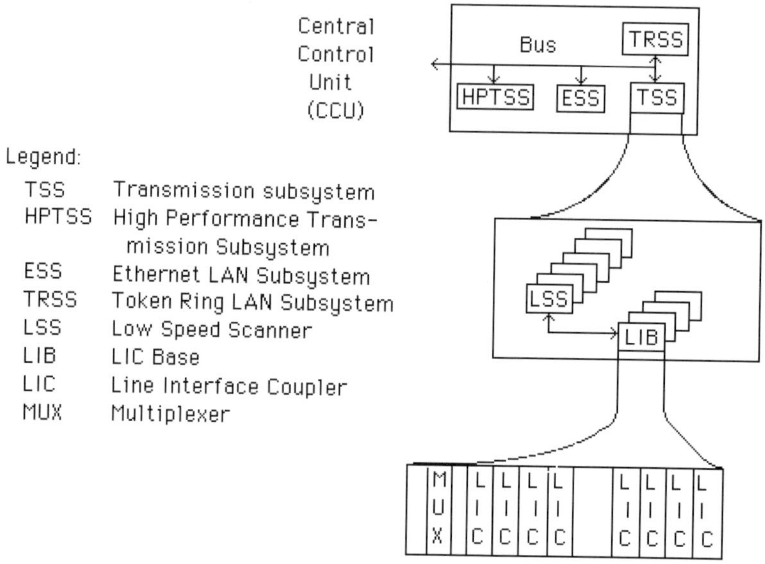

Figure 6.53 The 3745 communications controller communications subsystem

communications subsystem of a 3745 communications controller. That structure is based upon the use of a single central control unit (CCU) in a 3745; however, up to two CCUs can be installed in this communications controller.

In examining Figure 6.53 note that the serial line support capability of a 3745 occurs through the population of its transmission subsystem (TSS). That is, a TSS can have up to six low speed scanners (LSS), with each scanner supporting up to four Line Interface Coupler bases (LIBs). Each LIB in turn has one hardware multiplexer and up to eight line interface couplers (LICs). Since each LIC can support up to four ports, each LIB can support up to 32 lines. However, the actual number of lines that can be operated on a LIB or group of LIBs that make up an LSS depends upon the type of LIC. Each LIC has a 'weight' which is a value from 0.4 to 100 that represents the percentage of scanner occupation, with LIC weights based upon their operating rate, protocol, and character code. Thus, while it is physically possible for an LIC to support 32 lines, based upon the lowest LIC weight of 0.4, a maximum of 25 supported serial lines per LIC is actually supportable.

Returning to Figure 6.53, the ability to support Ethernet and Token-Ring LANs is obtained by the addition of adapters to the 3745's Token-Ring Subsystem (TRSS) and Ethernet Subsystem (ESS) modules. By adding appropriate low speed scanners, LIC bases, line interface couplers, and LAN adapters you can customize the hardware configuration of a 3745 to support a variety of networking requirements for SNA and APPN networks.

IBM 3746

A relatively new addition to IBM's communications product line is its 3746. Although the 3746 offloads communications functions from System/390 host computers, it is marketed as a 'multiprotocol controller' instead of a communications controller. The key reason for this name change is the fact that the 3746 operates as an Internet Protocol (IP) router as well as supports APPN, enabling separate SNA and IP networks to be consolidated over a single multiprotocol transport network.

A key advantage associated with the use of a 3746 is its ability to support communications without requiring the use of a Network Control Program (NCP) on the host and a communications controller. Another advantage associated with the use of the 3746 is its multiprotocol support capability which allows it to support APPN and IP over Token-Ring and Ethernet LANs, Frame Relay, SDLC, and point-to-point protocol (PPP) links, the latter a dial-up version of IP.

Hardware overview

A basic 3746 supports up to 16 adapters. Those adapters can consist of a mixture of communications line adapters, ESCON channel adapters, and Token-Ring adapters. The communications adapter consists of a communications line processor (CLP) which supports up to four line interface couplers. Each communications line adapter can support up to 120 lines operating in half- or full-duplex, each running SDLC, PPP, or Frame Relay at data rates from 600 bps to 2.048 Mbps. Via the use of a communications line adapter up to 500 active physical units can be supported, resulting from lines connected to control units which in turn support legacy terminals and PCs connected to the control unit.

The ESCON adapter permits a 3746 to be connected via optical fiber to the channel on a S/390. Up to two ESCON adapters can be used to enhance transmission to an attached S/390 while providing a duplicated path between the 3746 and a S/390.

The third type of adapter supported by the 3746 is a Token-Ring adapter. The Token-Ring adapter used by the 3746 consists of a Token-Ring processor and one or two Token-Ring couplers, permitting up to two rings to be directly connected to the controller. Although the 3746 can operate as a stand-alone communications networking device, it can also operate in conjunction with the NCP on a 3745 communications controller to expand the support of active devices. For example, a single Token-Ring adapter with either one or two TICs can support up to 500 active PUs. However, when a 3746 is connected to a 3745, PU support expands to 2000, with the additional 1500 being controlled by the NCT in the 3745. Figure 6.54 provides an example of the use of two 3746 multi-protocol controllers to interconnect geographically separated corporate locations via a mixture of network protocols. In examining Figure 6.54, note that since a 3746 has an IP routing capability, it provides the ability to connect a corporate network to the Internet. In addition, since it supports Frame Relay, it permits you to consider the use of that fast packet network as a substitute or supplement to leased lines. Thus, it provides a

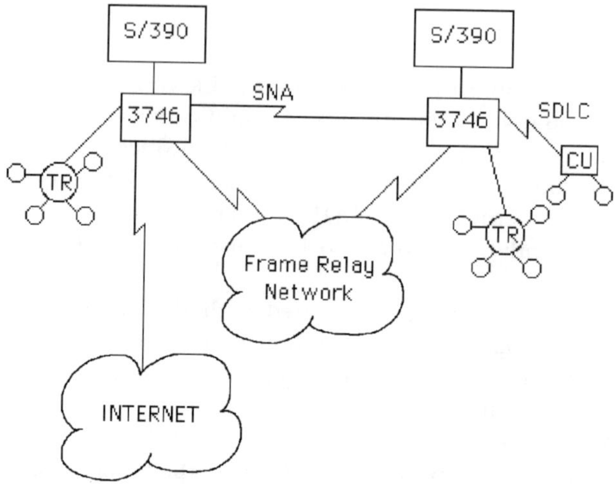

Figure 6.54 Using the 3746 multiprotocol controller to support a variety of communications requirements

degree of networking flexibility beyond that obtainable through the use of conventional front-end processors.

6.8 MODEM- AND LINE-SHARING UNITS

Cost-conscious persons are always happy to hear of ways to save money on the job. One of the things a data communications manager can best do to make his or her presence felt is to produce a realistic plan for reducing expenses. It may be evident that a single communication link is less costly than two or more. What is sometimes less obvious is the most economical and effective way to make use of even a single link.

Multiplexing is usually the first technique that comes to mind. But there are many situations where far less expensive, albeit somewhat slower, equipment is quite adequate. Here, terminals are polled one by one through a 'sharing device' that acts under the instructions of the host computer. Typically, the applications where this method would be most useful and practical would be those where messages are short and where most traffic between host computer and terminal devices moves in one direction during any one priod of time. The technique which can be called 'line-sharing' (as distinct from multiplexing) may work in some interactive situations, but only if the overall response time can be kept within tolerable limits. The technique is not as a rule useful for remote batching or RJE, unless messages can be carefully schedule so as not to get in each other's way because of the long run-time for any one job. Although line-sharing is inexpensive, it has some limits to its usefulness, particularly in situations where a multiplexer, most likely a TDM, can be used to produce additional economic leverage.

A similar device

Another device which functions similarly to modem- and line-sharing units, although its operation and usage is different, is a control unit. This device is built to work with a specific type of terminal device, usually manufactured by the vendor that produced the control unit. In addition, while modem- and line-sharing units operate basically transparent to the data flow, control units scan and interpret the data, looking for and operating upon orders received in the information flow. For additional information on control units the reader is referred to Section 6.10.

Operation

A TDM operates continuously to sample in turn each channel feeding it, either bit by bit or character by character; this produces an aggregate transmission at a speed equal to the sum of the speeds of all its terminals.

A conventional TDM operation is illustrated in Figure 6.55(a). For example, a multiplexer operating character by character assembles its first frame by taking the letter A from the first terminal, the letter E from the second, and the letter I from the third terminal. During the next cycle, the multiplexer takes the second

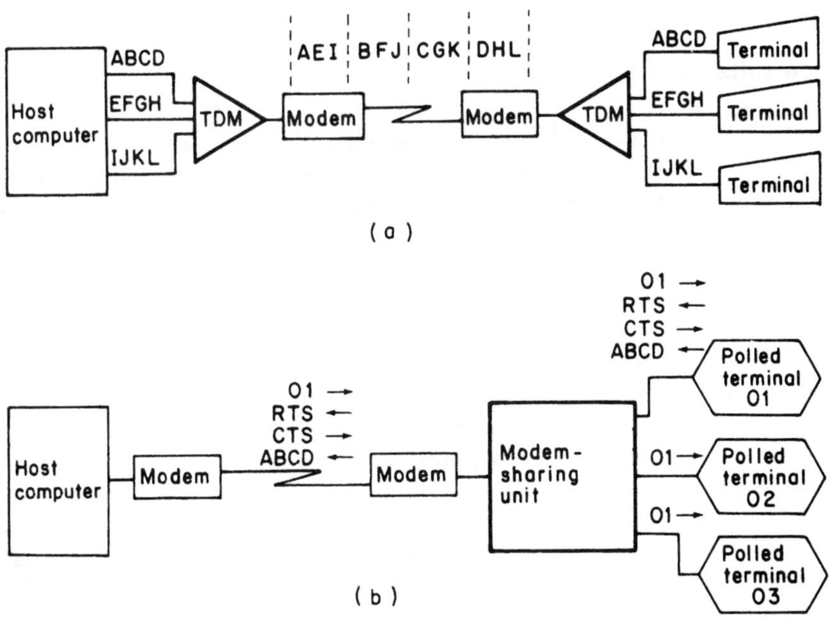

Figure 6.55 Multiplexing versus line-sharing. (a) Time division multiplex network. (b) Modem-sharing network. Multiplexer needs: a time or frequency division multiplex system (a) requires one computer port for each terminal and a multiplexer at each end. A sharing system (b) needs only one computer port. Because it requires terminals to be polled, a sharing system can be cost-effective for interactive operation, but may not be so for long messages such as are likely to move in remote job entry or remote batch types of applications

character of each message (B, F, and J, respectively) to make up its second frame. And the sampling continues in this way until traffic on the line is reversed to allow transmission from the computer to the terminals. The demultiplexing side of the TDM (operating on the receiving side of the network) assembles incoming messages and distributes them to their proper terminals or computer ports.

An FDM divides up the transmission link's total bandwidth into a number of distinct frequency bands, each of which is able to carry a low-speed channel. The FDM accepts and moves transmissions from all of its terminals and ports simultaneously and continuously.

A line-sharing network is connected to the host computer by a local link, through which the host polls the terminals one by one. The central site transmits the address of the terminal to be polled throughout the network by way of the sharing unit. This is illustrated in Figure 6.55(b). The terminal assigned this address (01 in the diagram) responds by transmitting a request to send (RTS) signal to the computer, which returns a clear to send (CTS), to prompt the terminal to begin transmitting its message (ABCD in diagram). When the message is completed, the terminal drops its RTS signal, and the computer polls the next terminal.

Throughout this sequence, the sharing device continuously routes the signals to and from the polled terminal and handles supporting tasks, such as making sure the carrier signal is on the line when the terminal is polled and inhibiting transmission from all terminals not connected to the computer.

Device differences

There are two types of devices that can be used to share a polled line: modem-sharing units and line-sharing units. They function in much the same way to perform much the same task—the only significant difference being that the line-sharing unit has an internal timing source, while a modem-sharing unit gets its timing signals from the modem it is servicing.

A line-sharing unit is mainly used at the central site to connect a cluster of terminals to a single computer port, as shown in Figure 6.56. It does, however,play a part in remote operation, when a data stream from a remote terminal cluster forms one of the inputs to a line-sharing unit at the central site to make it possible to run with a less expensive single-port computer.

In a modem-sharing unit, one set of inputs is connected to multiple terminals or processors, as shown in Figure 6.56. These lines are routed through the modem-sharing unit to a single modem. Besides needing only one remote modem, a modem-sharing network needs only a single two-wire (for half-duplex) or four-wire (for full-duplex) communications link. A single link between terminals and host computer allows all of them to connect with a single port on the host, a situation that results in still greater savings.

If multiplexing were used in this type of application the outlay would likely be greater, because of the cost of the hardware and the need for a dedicated host computer port for each remote device. A single modem-sharing unit, at the remote site, is all that is needed for a sharing system, but multiplexers come in pairs, one for each end of the link.

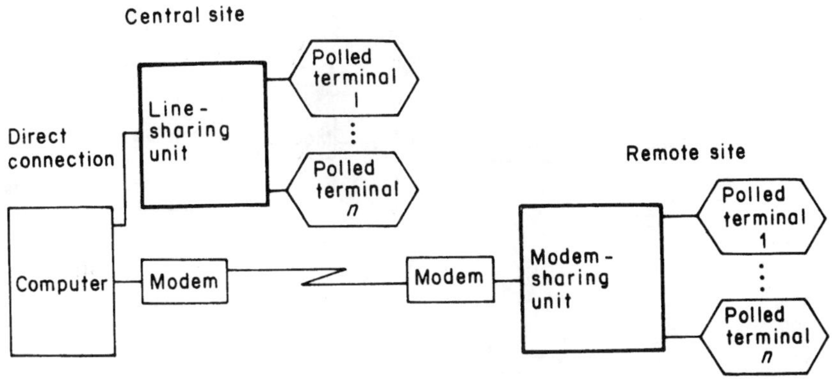

Figure 6.56 Line-sharing and modem-sharing use compared. Line-sharing units tie central site terminals to the computer, but modem-sharing units handle all the remote terminals. A line-sharing unit requires internal timing, whereas a modem-sharing unit gets its timing from the modem to which it is connected. In either case, access to the host is made through a single communications link—either a two-wire or four-wire—and a single port at the central site computer

The polling process makes sharing units less efficient than multiplexers. Throughput is cut back because of the time needed to poll each terminal and the line turnaround time on half-duplex links. Another problem is that terminals must wait their turn. If one terminal sends a long message others may have to wait an excessive amount of time, which may tie up operators if unbuffered terminals are used; but terminals with buffers to hold messages waiting for transmission will ease this situation.

Sharing unit constraints

Sharing units are generally transparent within a communications network. There are, however, four factors that should be taken into account when making use of these devices: the distance separating the data terminals and the sharing unit (generally set at no more than 50 feet under the RS-232 interface specification); the number of terminals that can be connected to the unit; the various types of modems with which the unit can be interfaced; and whether the terminals can accept external timing from a modem through a sharing unit. Then, too, the normal constraints of the polling process, such as delays arising from turnaround and response and the size of the transmitted blocks, must be considered in designing the network.

The 50-foot limit on the distance between terminal and sharing unit (RS-232/ITU V.24 standard) can cause problems if terminals cannot be clustered closely. A way to avoid this constraint is to obtain a sharing unit with a DCE option. This option permits a remote terminal to be connected to the sharing unit through a pair of modems, as illustrated in Figure 6.57. This in turn allows users the economic advantage of a through connection out to the farthest point. Since the advantage of modem-sharing units over a multipoint line is the reduction in the total number of

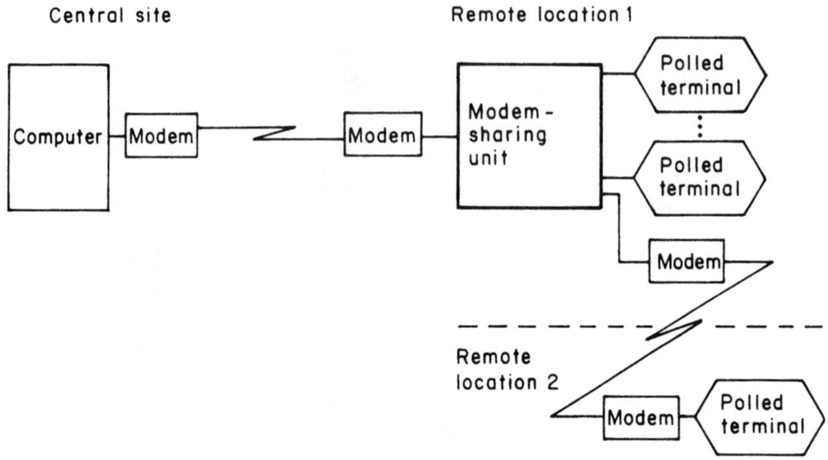

Figure 6.57 Extending the connection. Line- or modem-sharing units form a single link between a host computer and terminals. This system contains a modem-sharing unit with inputs from the terminals at its own site as well as from remote terminals. A line-sharing unit at the central site can handle either remote site devices or local devices more than 50 feet away from the host computer, which is the maximum cable length advisable under the RS-232/ITU V.24 standards

modems when terminals are clustered, only one or at most a few DCE options should be used with a modem-sharing unit, as it could defeat the economics of clustering the terminals to utilize a common modem.

It is advisable to check carefully into what types of modem can be supported by modem-sharing units, since some modems permit a great deal more flexibility of network design than others. For instance, if the sharing unit can work with a multiport modem, the extra modem ports can service remote batch terminals or dedicated terminals that frequently handle long messages. An example of this flexibility of design is shown in Figure 6.58. Some terminals that cannot accept external timing can be fitted with special circuitry through which the timing originates at the terminal itself instead of at the modem.

Other sharing devices

Sections 6.9 through 6.11 cover other sharing devices. Section 6.9 explores the use of port-sharing units, which are used in polled networks for the programmed selection of the computer port. Section 6.10 examines control units which although functioning similarly to sharing units are different in operation and usage. Section 6.11 discusses port selection (or port contention) units, where the unit provides random access to the computer ports under its control. The most significant differences in the various types of sharing units lies in their placement and function in the options available to them. Unlike line-, modem-, and port-sharing units, the port selection devices operate by time-sharing or contention. Access to any one port is provided on a first-come, first-served basis whenever a port is available. With

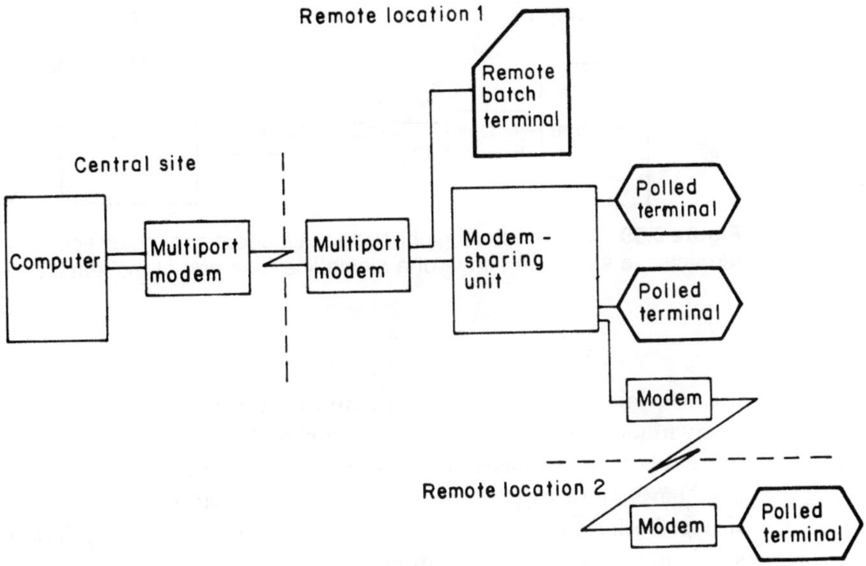

Figure 6.58 Multiple applications can share the line. Through the use of a modem-sharing unit with a data communications equipment interface, a terminal distant from the cluster (location 2) can share the same line segment (computer to location 1) that is used to transmit data to those terminals at location 1. With a second application that requires a remote batch terminal at location 1, additional line economies can be derived by installing multiport modems so both the polled terminals and the RBT can continue to share the use of one leased line from the computer to location 1

port selection, therefore, a large number of lines contend for a small number of ports. Users let go of a port by signing off in a way similar to that found in time-sharing and RJE applications. You should refer to these sections for additional information.

6.9 PORT-SHARING UNITS

An alternative to the utilization of modem- and line-sharing units in a communications network can be obtained through the employment of devices known as port-sharing units. In addition, the proper employment of such devices can be used to complement or supplement modem- and line-sharing units and in certain situations may result in large economies being realized.

When to consider

Port-sharing units are devices that are installed between a host computer and modem and that control access to and from the host for up to six terminals with the number of terminals limited by the capabilities of current hardware. In this way, port-sharing units are able to cut down on the number of computer ports needed for

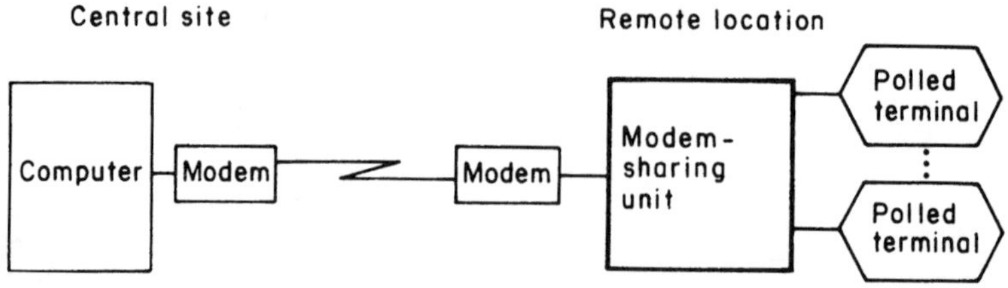

Figure 6.59 Modem-sharing unit usage. A modem-sharing unit permits a number of polled terminals to share the usage of a single line, one pair of modems, and a single computer port

these terminals. The port sharing unit is versatile, inexpensive, and available from many modem and terminal manufacturers. Its utilization can save the cost of one or more relatively expensive computer channels that does essentially the same job but may have more capabilities than are needed. Port-sharing units can be used to service both local and remote terminals and so expand the job that can be done by a single port of the host computer.

To put the concept of port-sharing into perspective, you should be aware of related devices designed to cut networking costs. Modem-sharing units and line-sharing units are available to minimize modem and line costs at remote locations, but they do not deal with the problem of overloading the host computer's ports. Modem- and line-sharing units are partial solutions to the cost of data communications networking, but they are limited to the types of modems they can handle; and they can, by themselves, complicate the life of the network designer.

A problem that surfaces when either modem- or line-sharing units are used by themselves is the distribution of polled terminals within the network. For either kind of sharing unit to be effective, terminals should be placed so that several are grouped close together. A typical modem-sharing unit employed to connect a number of terminals at a remote location for access to one computer port via a single pair of modems is illustrated in Figure 6.59. Some modem-sharing units can be obtained with a DCE interface option which is an RS-232 interface by which remote terminals at two or more locations can be connected to the modem-sharing unit through the installation of a pair of modems between the terminal and the modem-sharing unit. Such a configuration is shown in Figure 6.60. Although the use of the interface shown in Figure 6.60 permits the network to have a more flexible configuration, with a number of terminals remote from the sharing unit, the number of such terminals that can be served by any one unit is usually limited to one or two.

Another disadvantage of this arrangement is that it is rather like putting all your eggs in one basket. If either modem on the high-speed link between the computer and the modem-sharing unit should fail, or if the circuit itself goes down, all the remote terminals become inoperative.

As new applications develop and the number of remote terminals connected to the host computer increases, a situation can arise where the network designer runs out of ports to service the network. If no additional ports are available, a costly computer upgrade or the installation of a second computer system would represent

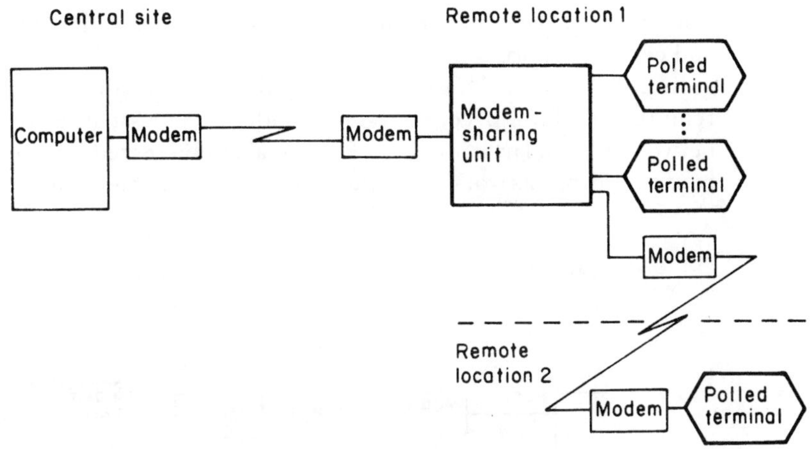

Figure 6.60 Network expansion using modem-sharing units with data communications equipment (DCE) interface. Remote terminals at two or more locations can share a common polled circuit when a DCE interface is present on the modem-sharing unit

a major economic burden. A method to obviate or postpone these types of equipment upgrades is through the utilization of port-sharing units.

Operation and usage

Port-sharing, then, is presented either as an alternative or as a supplement to modem- and line-sharing, in networks without multiplexers. A port-sharing unit is connected to a computer port and can transmit and receive data to and from two to six either synchronous or asynchronous modems, as shown in Figure 6.61.

Data from the computer port is broadcast by the port-sharing unit, which passes the broadcast data from the port to the first modem that raises a receiver-carrier detect (RCD) signal. Data for any other destination will be blocked by the unit until the first modem stops receiving. The port-sharing unit thus provides transmission by broadcast and reception by contention for the port connected to it. Like a modem-sharing unit, a port-sharing unit is transparent with respect to data

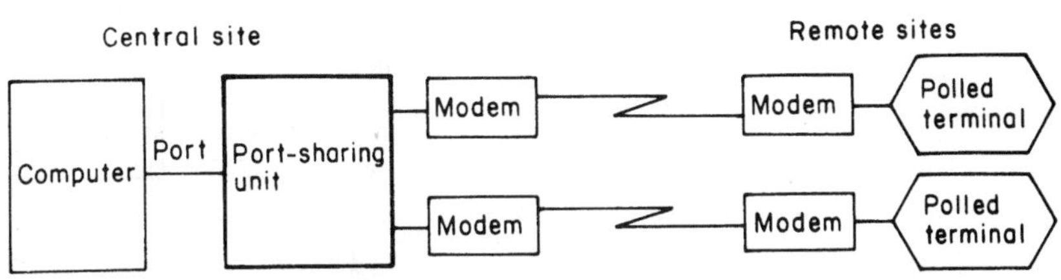

Figure 6.61 Using a port-sharing unit. The port-sharing unit lies between the host computer and the modems at the central site. One advantage is that if a failure occurs on the communications line or at the modem, only the terminal on that particular line goes down. On a polled, multidrop line, a line failure renders all terminals beyond it inaccessible

transmission. Data rates are limited only by the capabilities of the terminal, modem, and computer port.

In order to gain the same results without a port-sharing unit would, however, require a multidrop configuration. Both the port-sharing unit and a multidrop network allow a large number of terminals to be served by one computer port; but in a multidrop network the failure of any part of the circuit will put all terminals

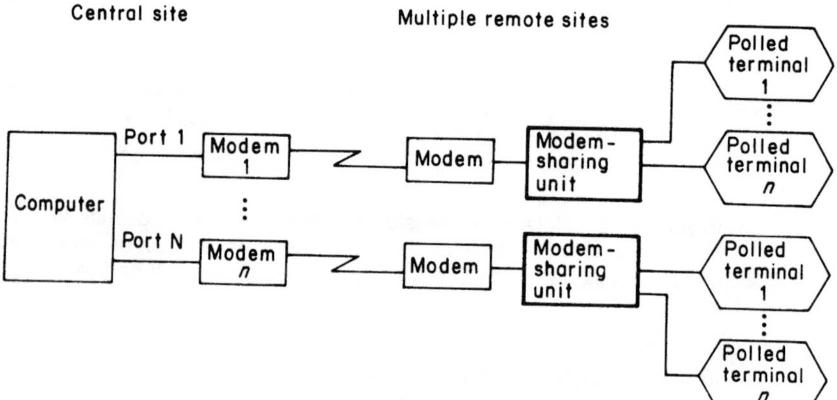

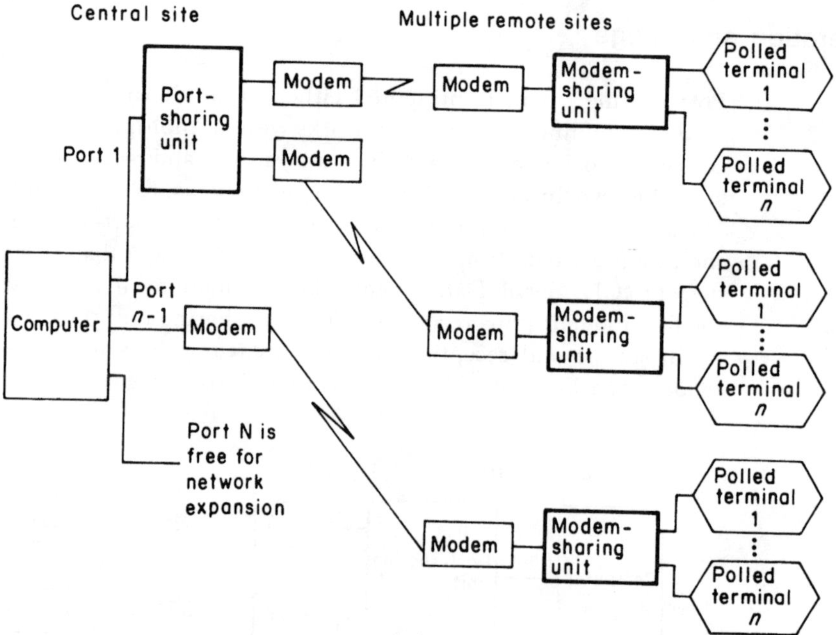

Figure 6.62 Two sharing techniques combined. When only modem-sharing units are used (a) a time may come when every port is in use and no further expansion on the network is possible. Rather than add another computer to serve a single new terminal, the user may prefer to invest in a port-sharing unit (b) that will solve the problem for about $500

beyond the failure out of action. In the configuration in Figure 6.61 however, failure of modem or outage on the line will only cut out a terminal on that segment. Failure of a computer port or of the port-sharing unit would of course bring down the entire network, but these devices are stable and such failures are fairly unusual.

Port-sharing as a supplement

Port-sharing units may also be used alongside modem-sharing units. If modem-sharing units alone are used, a situation can arise where there are not enough ports to serve the network, as in Figure 6.62(a). If each modem-sharing unit serves its full complement of terminals and all the computer ports are in use, expansion of the network, even by just one port, may require a second mainframe computer.

This problem can also be dealt with by the use of a port-sharing unit at the central site which by cutting down the number of ports currently needed allows a network to expand without additional computer ports. This is illustrated in Figure 6.62(b) which shows how one port-sharing unit with a two-modem interface can free a computer port from the configuration shown at the top of that illustration.

One versatile feature of port-sharing units is an option that allows the unit to accept a local interface instead of the normal RS-232 DCE interface, so that up to two local terminals may be operated without modems at the central site, as shown in Figure 6.63.

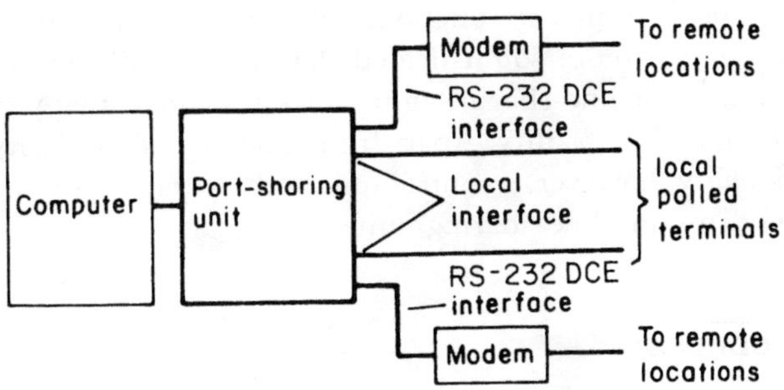

Figure 6.63 Connecting local peripherals. A local interface option to port-sharing lets local and remote sites be served by the same port. Peripherals are polled as if they were at remote sites

While both modem-sharing units and port-sharing units are similar in the way they are used, there is an important difference in the normal placing of their interfaces. Table 6.15 provides, a comparison of the characteristics of a port-sharing unit with those of modem- and line-sharing units.

A similar device

Another device which has a function similar to port-sharing units, although its operation and usage differ, is a port selector (or port contention) unit. This unit

Table 6.15 Features of sharing units

Feature	Modem-sharing/ line-sharing unit	Port-sharing unit
Transmit mode	Broadcast	Broadcast
Receive mode	Contention	Contention
Number of modems interfaced	2–32	2–6
Terminals supported	Polled	Polled
Options	RS-232 interface (MSU to modem)	Local interface (PSU to terminal)
Normal interface placement	Between modem terminal	Between computer port and modem

provides access to a computer port on a first-come, first-served basis. Instead of up to six lines contending for a port, a larger number of lines can reach the same port when it opens up. The availability of a port occurs when a current user of the port disconnects from the system. For additional information, the reader is referred to Section 6.11.

6.10 CONTROL UNITS

Although similar in functionality to modem- and line-sharing units, a control unit is a much more sophisticated device. Basically the control unit functions like a modem- or line-sharing unit, since it enables many devices to share the use of a common line facility. Apart from a degree of similarity with respect to their functionality, however, a control unit both operates and can be utilized differently from modem- and line-sharing units.

Control unit concept

The concept of the control unit originated in the 1960s with the introduction of the IBM 3270 Information Display System which consisted of three basic components- a control unit, display station, and printer. The control unit enabled many display stations and printers to share access to a common port on IBM's version of a front-end processor, marketed as a communication controller, or via a common channel on a host computer, when the control unit was directly attached to a host computer channel. Here the term display station represents a CRT display with a connected keyboard.

Since the introduction of the 3270 Information Display Unit, most computer vendors have manufactured similar products, designed to connect their terminals and printers to their computer systems. Control units vary considerably in configuration, ranging from a single control unit built into a display station to a configuration in which the control unit directs the operation of up to 64 attached display stations and printers. These attached display stations and printers are usually referred to by the term cluster; hence, another popular name for a control unit is cluster controller.

The major differences between IBM and other vendor control units is in the communications protocol employed for data transmission to and from the control unit, the types of terminals and printers that can be attached to the control unit, and the physical number of devices the control unit supports. In this section we will focus our attention upon IBM control units; however, when applicable, differences in the operation and utilization of IBM and other vendor devices will be discussed.

Attachment methods

IBM control units can be connected to a variety of IBM computer systems, ranging in scope from System/360 through System/390 processors to 43XX computers as well as certain System/3 models. The control unit can be connected either directly to the host computer via a channel attachment to a selector, multiplexer, or byte multiplexer channel on the host; or it can be link-attached to IBM's version of the front-end processor known as a communication controller as illustrated in Figure 6.64. Once connected, the control unit directs the operation of attached display stations and printers, with such devices connected via a coaxial cable to the control unit or via twisted-pair or newer models. Note that a generic type of control unit is indicated in Figure 6.64 by the use of the term 327X. Several control unit models are marketed by IBM, with numerous types of each model offered. The actual control unit installed depends upon the communications protocol (bisynchronous or synchronous data-link control) used for communications between the control unit and the host computer, the type of host system to which the control unit is to be

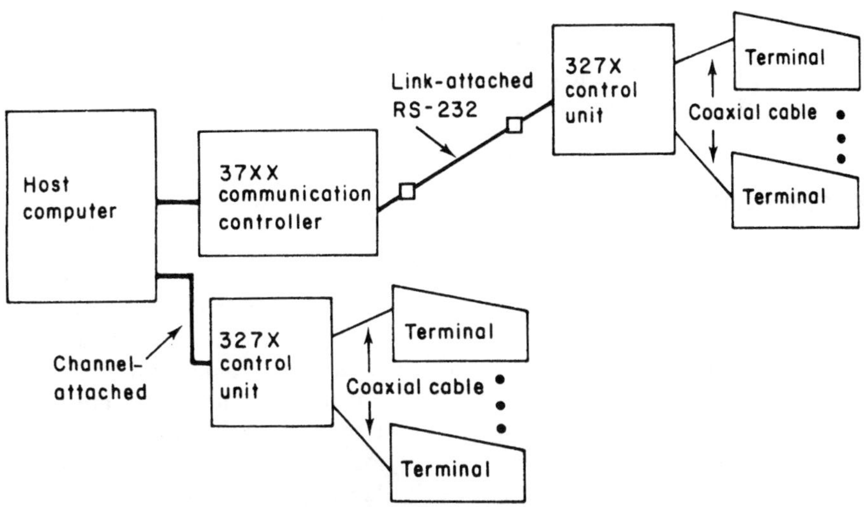

Figure 6.64 IBM 3270 Information Display System. Control units used in this system can be either link-attached to a communication controller or channel-attached to a host computer system. (□ = modem or DSU)

connected, and the method of attachment (channel- or link-attached), and the number of devices the control unit is to support.

Two of the most popular IBM control units throughout the mid-1980s were the 3274 and 3276 devices. The 3274 enables up to 32 devices to be connected while the 3276 is a table-top control unit with an integrated display station that permits up to seven additional devices to be connected to it. In 1987 IBM introduced its 3174 control unit which can be obtained in several different configurations and can be configured to support 64 devices eliminating VN like the 3274 which supports a maximum of 32 devices. Other major differences between the 3174 and 3274 are the availability of a Token-Ring option and a protocol conversion option for the newer 3174 control unit. The token-ring option permits the control unit to be connected to IBM's token-ring network, while the protocol conversion option permits asynchronous ASCII terminals to be connected to the control unit in place of normally more expensive EBCDIC devices.

Since the attachment of terminal devices to an IBM control unit requires constant polling from the control unit to the terminal, either dedicated or leased lines are normally used to connect the two devices together. Therefore the 3270 Information Display System is considered to be a 'closed system'. In addition, many IBM control units require terminals to be connected by the use of coaxial cable. This normally precludes the use of asynchronous terminals using the PSTN to access a 3270 Information Display System.

Devices marketed by other computer vendors vary considerably in comparison to the method of attachment employed by IBM. Most non-IBM control units permit terminal devices to be connected via an RS-232/ITU V.24 interface while this capability is only available on certain 3174 and 3274 models. Thus, terminal devices remotely located from the main cluster of terminals can be connected to such control units via tail circuits as illustrated in Figure 6.65. Another difference between IBM and other computer vendor control units concerns the method of attachment of the control unit to the host computer system. IBM control units can be channel-attached or link-attached, whereas most non-IBM control units are link-attached to the vendor's front-end processor, even when the control unit is located in close proximity to the computer room.

Unit operation

One of the major differences between a control unit and modem- and line-sharing units is the fact that the former is not a passive device. The 3270 data stream that flows between the control unit and the host computer system contains orders and control information in addition to the actual data. Both buffer control and printer format orders are sent to the control unit, where they are interpreted and acted upon. Buffer control orders are used by the control unit to perform such functions as positioning, defining, modifying, and assigning attributes and formatting data that is then written into a display character buffer that controls the display of information on the attached terminal. Thus, the terminal display information flows from the host computer to the control unit in an encoded, compressed format, where it is interpreted by the control unit and acted upon. Similarly, printer format

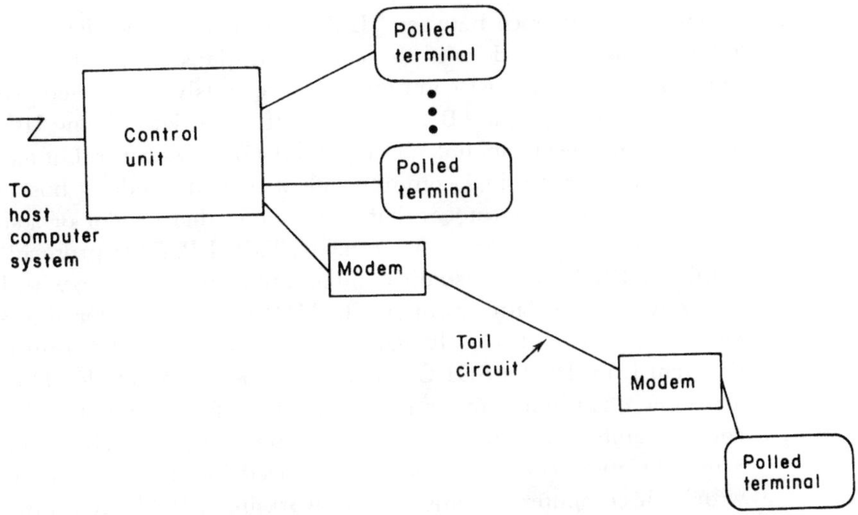

Figure 6.65 Non-IBM control units can connect terminals via tail circuits. Control units produced by many computer manufacturers other than IBM have RS-232/ITU V.24 interfaces for attaching terminal devices. This type of interface permits distant terminals to be connected to the controller via tail circuits

orders are stored in the printer character buffer in the control unit as data and are interpreted and executed by the printer's logic when encountered in the print operation. Since a network can consist of many control units, with each control unit having one or more attached terminal devices, both a control unit address and terminal address is required to address each terminal device.

In comparison to a sharing unit where the host polls each terminal, the control unit performs the polling operation. This enables a higher degree of data throughput, since the computer system can then transmit and receive data during what would normally be a polling interval if sharing units were employed.

Protocol support

Both binary synchronous communications and synchronous data link control (SDLC) communications are supported by most members of the IBM family of 3X7x control units. Some control units are soft-switch selectable, incorporating a built-in diskette drive that permits a communications module incorporating either communication protocol to be loaded into the control unit; other models may require a field modification to change the communications discipline.

By employing a communications discipline, end-to-end error detection and correction is accomplished between the control unit and the host computer system. In comparison, line- and modem-sharing units are essentially transparent to the data flow and do not perform error checking, which either is performed by the individual terminals connected to the sharing device or is not performed.

Another difference between IBM control units and devices manufactured by other computer vendors is the communications protocol employed to link the control unit to the host computer. As previously mentioned, IBM control units communicate employing BISYNC or SDLC, the latter being IBM's version of the ISO's higher-level data link control (HDLC). Most control units manufactured by other computer vendors communicate with that vendor's host computer system using a proprietory protocol that is unique to that vendor or a standard protocol. One example of the former is Honeywell's V.I.P. 7705 protocol which, although similar to BISYNC, is also different and thus incompatible with IBM's BISYNC. An example of the latter would be the HDLC protocol offered by several computer vendors which, while closely matching the standard HLDC protocol, is sufficiently different from IBM's SDLC version as to be incompatible. Thus, although there are numerous vendors that manufacture IBM plug-compatible control units, such control units normally will not operate with non-IBM computer systems. Similarly, most control units manufactured by other computer vendors for use with their computer systems will not work in an IBM environment.

Breaking the closed system

One of the key constraints of the IBM family of control units is the fact that it is a closed system, requiring terminal devices to be directly connected to the control unit. Without the introduction of a variety of third party products, this cabling restriction would preclude the utilization of personal computers and asynchronous terminals in a 3270 network. Fortunately, a variety of third party products reached the communications market that enable users operating 3270 networks to overcome the cable restriction of control unit connections. Foremost among such products are protocol converters and terminal interface units.

Protocol converters

When used in a 3270 network, a protocol converter will function as either a terminal emulator or a control unit/terminal emulator, depending upon the method employed to attach a non-3270 terminal into a 3270 network. When functioning as a terminal emulator, the protocol converter is, in essence, an asynchronous to synchronous converter that converts the line by line transmission display of an asynchronous terminal or personal computer into the screen-oriented display upon which the control unit operates. Similarly, the protocol converter changes the full screen-oriented display image transmitted by the control unit into the line-by-line transmission which the terminal or personal computer was built to recognize.

During this translation process cursor positioning, character attributes, and other control codes are mapped from the 3270 format into the format which the terminal or personal computer was built to recognize, and vice versa. A protocol converter functioning as a terminal emulator is illustrated at the top of Figure 6.66. It should be noted that some protocol converters are stand-alone devices while other protocol converters are manufactured as adapter boards that can be inserted into a system expansion slot within a personal computer's system unit.

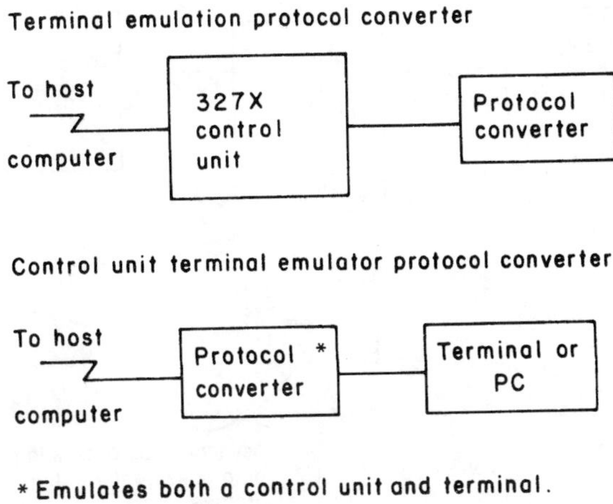

Figure 6.66 Protocol converters open the closed 3270 system. Through the use of protocol converters a variety of asynchronous terminals and personal computers can be connected to a 3270 network

A protocol converter functioning as a control unit/terminal emulator is illustrated in the lower portion of Figure 6.66. Since all terminals access a 3270 network through control units, stand-alone terminals or personal computers must also emulate the control unit function if they are to be directly attached to the host computer system. The reader is referred to Section 6.12 for detailed information concerning the operation and utilization of protocol converters.

Terminal interface unit

Another interesting device that warrants attention prior to concluding our discussion of control units is the terminal interface unit. This device is basically a coaxial cable to RS-232 converter, which enables coaxial cabled terminals to access asynchronous resources to include public databases and computer systems that support asynchronous transmission.

In an IBM 3270 network, terminal devices are connected directly to a control unit, precluding the use of the terminal from accessing other communications facilities. The terminal interface unit breaks the restriction, since it converts a coaxial interface used to connect many 3270 terminals to control units into an RS-232 interface at the flick of the switch on the unit. The use of this device is illustrated in Figure 6.67. When in a power-off state, the terminal interface unit is transparent to the data flow and the terminal operates directly connected to the control unit. When in an operational mode, the terminal interface unit converts the keyboard-entered data into ASCII characters for transmission and similarly converts received characters for display on the IBM terminal's screen. The terminal interface unit can thus be employed to make a coaxial cabled terminal multi-functional, permitting it to access the coaxial cable port on a control unit as well as an RS-232/ITU V.24 device, such as an asynchronous modem.

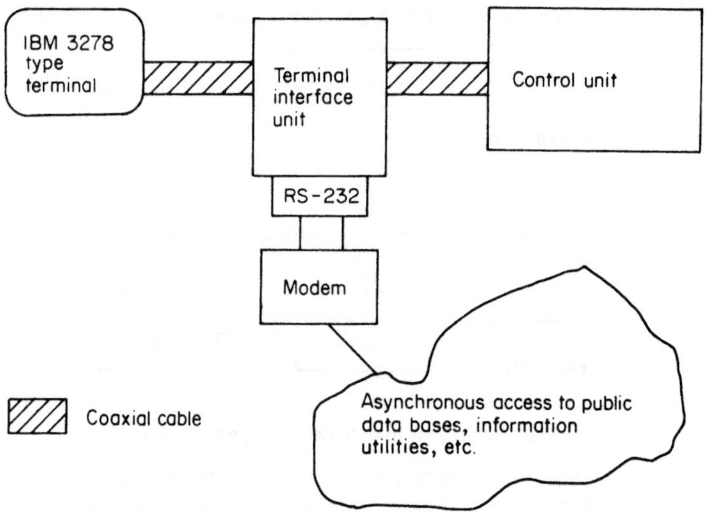

Figure 6.67 The terminal interface unit makes a coaxial cabled terminal multifunctional

6.11 PORT SELECTORS

A traditional method used to provide service to new terminals as they are added to a network is through the expansion of the number of front-end processor ports. A variety of approaches can be used to accommodate such terminals.The number of dial-in lines at the computer center can be expanded, or additional dedicated or direct connect lines can be installed to service terminals added at the computer facility. For remote locations, the addition or upgrading of multiplexers and the installation of additional leased lines may be required to provide new channels to enable the new terminals to connect and transmit to the computer.

Even when such a network expansion is completed, it is unlikely that all ports will be busy at the same time. Thus, while extra lines and additional communications components may be required to provide additional transmission paths to the computer, some front-end processor ports may be operating only a fraction of the day. Port selectors, through their ability to cross-connect incoming transmission to available ports, permit more efficient front-end processor utilization.

Types of devices

Several equipment manufacturers sell port selection devices. Most of these products are stand-alone devices, built to function as an interface between computer ports and lines which may emanate from multiplexers, direct dial lines, or dial-up lines. Some manufacturers call their stand-alone port selectors data PBXs as they are restricted to switching data. Other manufacturers, however, offer port selection as an optional or built-in feature in their statistical and intelligent multiplexers.

Some port selectors are designed only to contend for asynchronous, teletype terminal traffic; other devices can contend for both asynchronous and synchronous traffic within the same unit, with each type of traffic being contended for by one or more computer ports servicing that mode of transmission.

Operation

The utilization of port selectors permits terminals to be added to a network without a corresponding increase in the number of computer ports. In addition, the utilization of this device may permit a system contraction whereby a number of computer ports become unnecessary and can be returned to the manufacturer.

The basic function performed by a port selector can be viewed as a dynamic data switch similar to telephone rotaries (stepping switches that sequentially search for available telephone lines), except that the selector provides appropriate interfaces between computers and terminals to route a large calling terminal population to a lesser number of called computer ports. Some selectors have additional features specifically applicable to data networks. Although users tend to confuse port selectors with port-sharing units, their applications are for specific line environments that result from the utilization of different types of terminals for specific applications. Port-sharing units are used in polled networks where the computer controls the traffic flow, and terminals must have a buffer area to recognize polls to their address. Port selection units are used in contention networks, in which terminals transmit to the host on a random basis; and the access to any port is normally on a first-come, first-served basis.

Computer site operations

From a network viewpoint, a port selector is similar to a black box with N line side input connections and n port side output connections, with $N \geq n$ as shown in Figure 6.68. The port selector continuously scans all line side connections for incoming data from terminals connected to that side. At the same time, the selector maintains a status check of available ports so that when a terminal becomes operational and requests access, the selector connects the terminal to an available

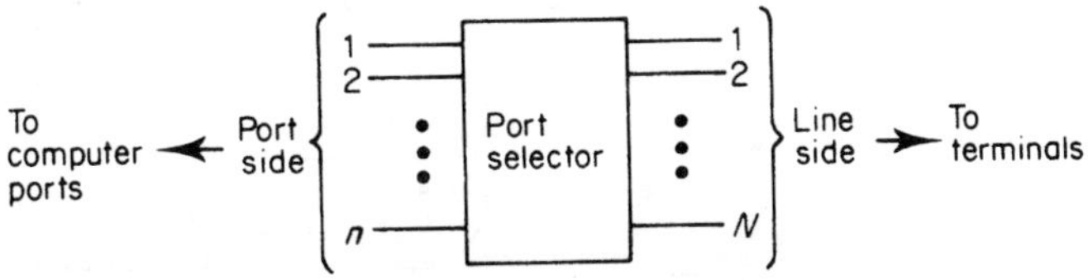

Figure 6.68 Port selector at computer site. At line side, port selectors can interface a variety of channels—from multiplexers, dial-in lines, leased lines, and direct connect lines. As terminal traffic becomes active, the port selector attaches the circuit to a predefined or randomly selected computer port for the duration of the transmission

port. Some port selectors can be arranged to form subgroups of contending terminals and ports, so that certain terminals can only be connected to one or more of an assigned group of ports. This capability is discussed more fully under 'Selector features' at the end of this section. If all of the designated ports are taken, the selector will continue to scan until a port becomes available or until the request for access ceases. Another option is a 'busy out' feature whereby the port selector signals new callers that all available ports are in use.

Other port selectors have the capability to place users into a queue when all ports are in use. Some port selectors offering a queuing feature inform the terminal user of their position in the queue only when their terminal access attempt is first initiated, while other port selectors continue to transmit the user's position in the queue each time the position changes or on a periodic basis.

Usage decisions

A typical dial-up computer network, as in Figure 6.69, indicates what must be taken into account in coming to a decision about port selectors. In this example, 48 computer ports handle messages from two multiplexers and a local rotary. Each multiplexer provides 16 terminal-to-computer connections, as well as 16 dial-in ports at the computer site. If the network is distributed over several time zones, peaks in utilization of the terminals will occur at different times in different places.

A typical profile of the number of users logged onto the network from each geographical area can be computed with statistical software packages provided by computer manufacturers. Profiles of the system's utilization are shown in Figure 6.70. While other networks may not reveal exactly the same patterns, they may be similar because normal working habits are a significant factor in the fluctuations of network utilization. At the start of working hours (Figure 6.70, point 1 on the third

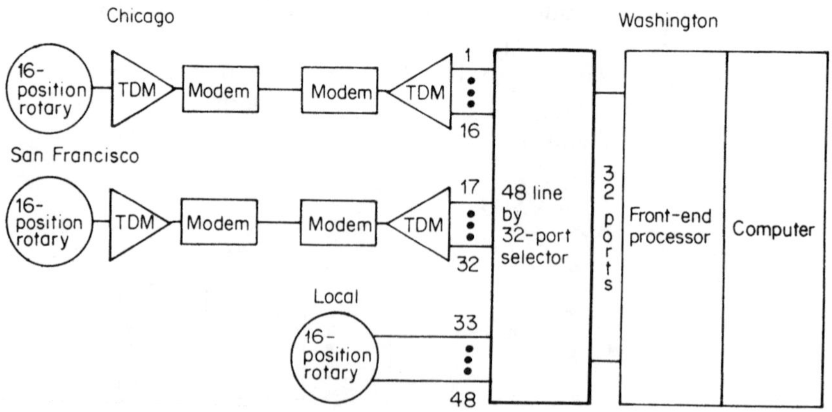

Figure 6.69 Traditional computer network. In a typical computer network a 48-port front-end processor would permit the 32 remote terminals and 16 local terminals to reach the computer simultaneously. This configuration ensures that no terminal user ever meets a busy signal—but it is expensive

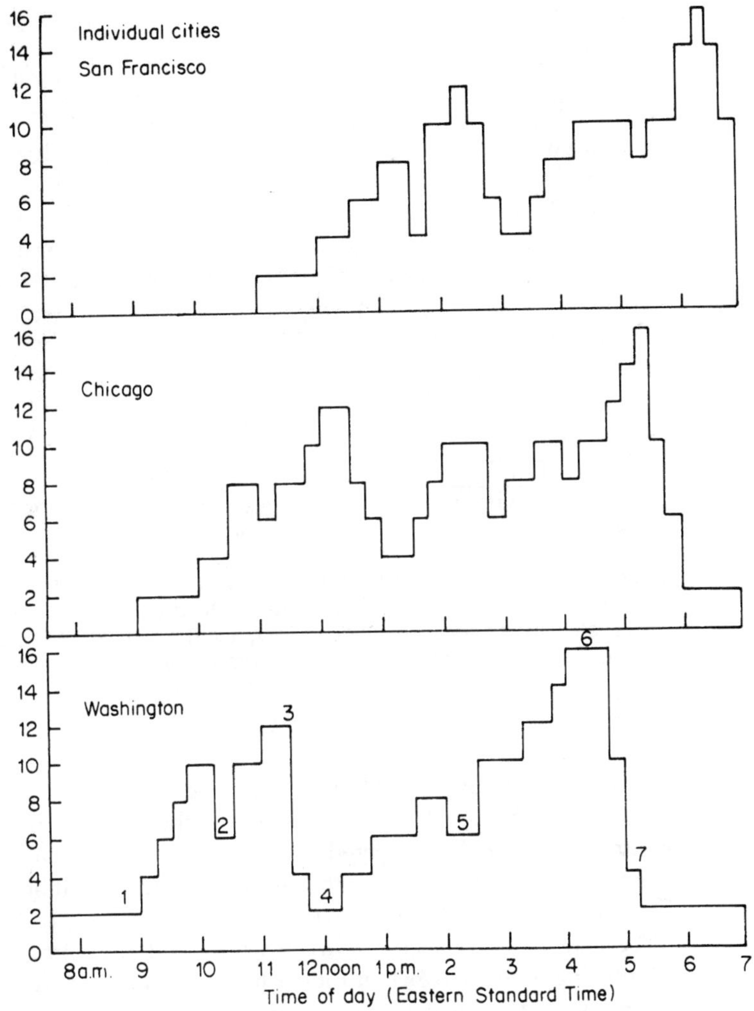

Figure 6.70 Utilization profiles. Users of a nationwide network are not likely all to be in their offices or doing the same amount of work at the same time. This situation works in favor of a port selection configuration; it will readily be seen that peak usage rarely coincides in the different, widely separated centers served by a large communications network

profile), use gradually builds as people arrive at the office and settle down to work. During the morning coffee break period (point 2), the number of users decreases temporarily, with the length of time and the degree of the drop-off varying from place to place. Morning peak use (point 3) is followed by a drop in activity during the lunch period (point 4). Use then builds up until the afternoon break (point 5) and peaks again as people rush to complete the day's work (point 6). As the close of business approaches (point 7), activity tapers off until only a few terminals remain on-line.

By combining the three local profiles, an overall network profile can be developed (Figure 6.71) which represents the number of terminals on-line as the day moves

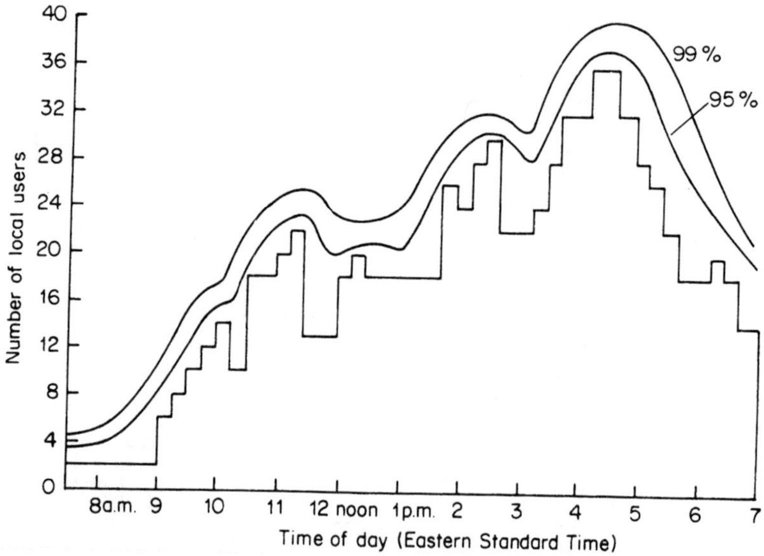

Figure 6.71 The total network profile. The curves represent the 95th and 99th percentiles of use for the entire network. They show that at peak periods only 37 or 40 terminals, respectively, will be either on-line or seeking access. The curves also show that during these same periods of peak use a substantial number of the remote terminals will inevitably be idle

forward (Eastern Standard Time). The smooth curves above this profile plots activity in the 95th and 99th percentiles, and they indicate the maximum number of terminals that can be expected to be on-line for 95 and 99 out of every 100 observations. Put another way, 95% of the time there will be up to 37 terminals on-line or seeking access between 4:15 p.m. and 4:45 p.m., and 99% of the time there will be as many as 40 users (between 4 p.m. and 5:30 p.m.).

Returning to the network in Figure 6.69, it is now apparent that 99% of the time, eight or more of the 48 computer ports are not in use, and that 95% of the time 11 or more ports will be idle. Thus, the use of a port selector becomes a question of economics against inconvenience. Is the cost of a number of mostly idle computer ports worth the seldom used advantage of being able to connect all terminals, simultaneously and without delay? A related question that the network designer must answer is whether the computer can process all messages rapidly enough when all terminals are on-line.

Let us assume that instead of installing the equipment illustrated in Figure 6.69, a port utilization study was conducted and that the 90th percentile of port usage was decided upon. This would mean that a 48-line by 32-port port selector would be required. After calculating the cost savings possible with such a port selector we can determine if the sacrifice of 16 continuously available ports is justifiable. Assuming such justification, the revised network which incorporates a port selector is illustrated in Figure 6.72. Other port selector configurations could result from an investigation into the savings possible with a 48-line by 37-port selector (95th percentile, Figure 6.71) or a 48-line by 40-port selector (99th percentile).

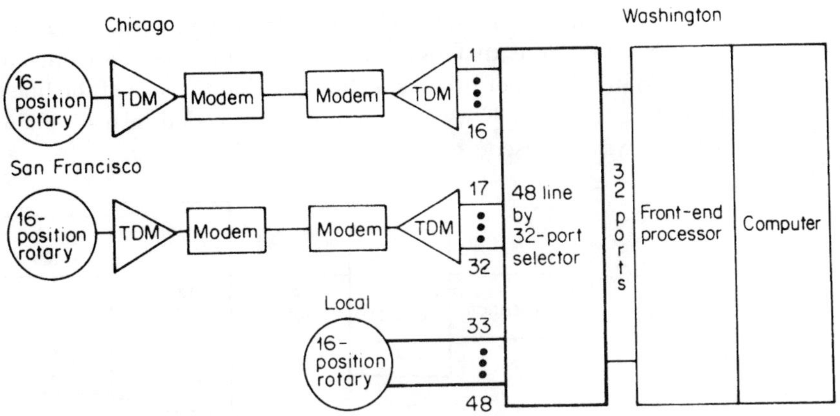

Figure 6.72 Network revision to incorporate a port selector. This network provides almost the same performance as the network shown in Figure 6.69. However, through the use of the port selector, only 32 computer ports are required. Because of the different time zones, it is unlikely that a remote terminal will get a busy signal

Port costs

Besides a savings in the circuit boards that make up the ports themselves, other computer costs are reduced because of the fewer ports. For example, using port selectors can result in additional savings for installations that have redundant computers, because every excess computer port that can be eliminated on one front-end processor can also be eliminated on the other. In addition, reductions may also be possible in the capital outlay for devices that switch between central processing units. One such redundant processor configuration is shown in Figure 6.73.

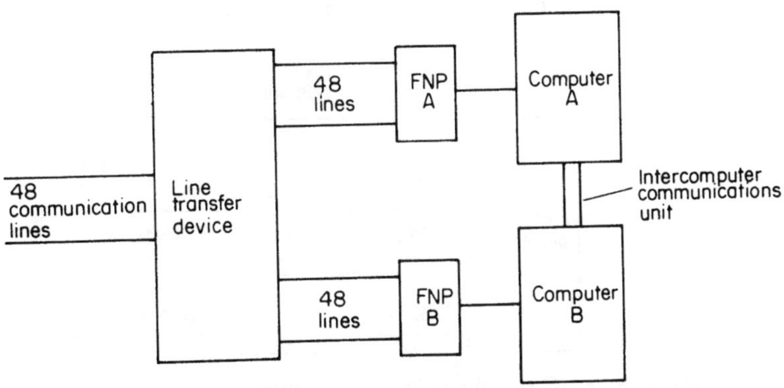

Figure 6.73 Redundant processor configuration. Savings in excess of twice the savings of a single processor may be obtained since a portion of the switching arrangement may be removed

The installation of a port selector to front-end a line transfer device is shown in Figure 6.74. Note that not only are the number of front-end processor ports reduced by 32 (16 per processor) but also that the size of the line transfer device can be reduced, since a 32 by 64 switch is now needed instead of a 48 by 96 switch if a selector was not used.

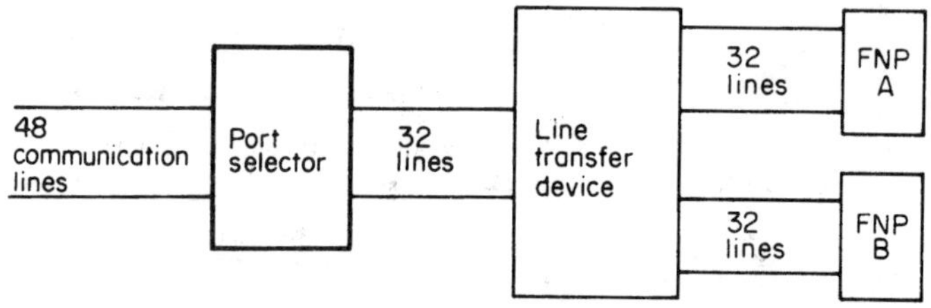

Figure 6.74 Port selector increases savings for redundant configurations. Using a port selector to front-end the line transfer device of a dual processor not only doubles the number of ports that can be eliminated but also reduces the size of the switch required

Load balancing

A key advantage obtained from the utilization of port selectors is the ability to balance the communications load when an installation has two or more front-end processors. This load balancing can be accomplished simply by wiring the cables from the port side of the selector in an alternate manner to each front-end processor. An installation with dual front-end processors would cable the leads from ports 1, 3, 5, 7, 9, ..., N-1 from the port selector to the first front-end processor and the leads from ports 2, 4, 6, 8, ..., N to the second front-end processor.

Selector features

A wide range of standard and optional port selector features should be considered. First, the basic size or capacity of the port selector and its expansion capability, if

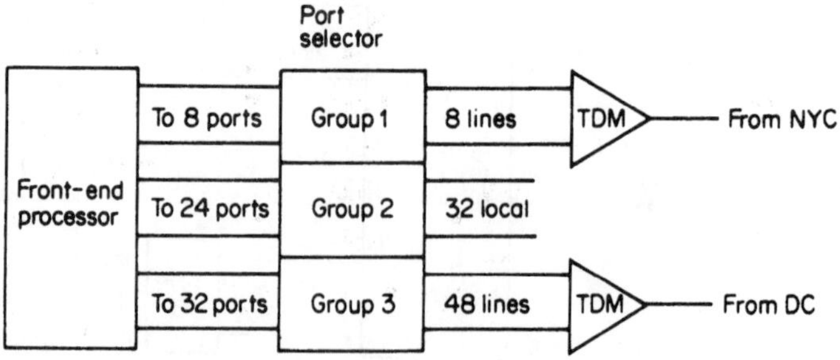

Figure 6.75 Grouping selector contention. Users from New York City always require immediate access so no contention is performed. Although 32 local terminals are connected to the selector, management has decided that at most 24 will be active at one time and has directed a 4:3 contention ratio be established for this group

any, should be determined. Selectors are offered as a base unit, with a predetermined number of calling channels and called ports available for interface. By the addition of line and port nests and adapters, the capacity of the selector can be increased.

Another feature to consider is the ability to route data sources to specific line groups as illustrated in Figure 6.75. Instead of being preconfigured to seek available ports within a predetermined group or groups of ports, automatic group selection is an option which permits the selector to determine which of the port groups a given user requires based upon the transmission of a control character from the user's terminal.

Line-switching network

One of the most useful features of a port selector is its ability to serve as the foundation for the construction of a line- or circuit-switched network. This can be accomplished by obtaining a port selector that has an automatic group selection feature.

To understand how a port selector can function as the foundation for a line-switching network, let us assume our organization has two data-processing centers —one located in Macon, Georgia, and second located in Miami, Florida. Suppose there are a large number of terminals located in Washington, DC, and New York City that require access to both data centers. One method that might be employed to provide communications support to the terminal users would be the installation of multiplexer systems to link each city to each data center. This would require, as an example, two multiplexers and high-speed modems to be installed in New York City. Then two circuits would also be required, with one circuit routed to Macon and the second to Miami.

At each data center, the circuit would be terminated with a high-speed modem and a multiplexer. In addition to installing dual multiplexers and high-speed modems at the terminal locations, there would be significant distances where the circuits would be routed in parallel. Thus, the utilization of a common multiplexer system at each terminal location could result in significant cost savings as well as operational efficiencies if a method were obtained to allow terminal users to access one multiplexer and control the routing of their terminal traffic to either data center. One method that will permit this to occur is by the installation of a port selector with an automatic group selection feature as illustrated in Figure 6.76. A port selector is installed in Macon, Georgia, which is the initial termination point of the multiplexed data traffic from New York and Washington, DC. The automatic group selection feature of the port selector is then configured to recognize two groups. Group 1 is assigned to the ports of the port selector that are cabled to the host computer system located in Macon. Group 2 is assigned to a group of ports that are routed from the port selector into a multiplexer that is then connected to a circuit that is routed to Miami.

When a terminal user in New York or Washington, DC reaches the port selector, he or she will be prompted by the port selector to enter the routing group number or similar parameter. Then, by entering a '1' the user would be routed to the computer

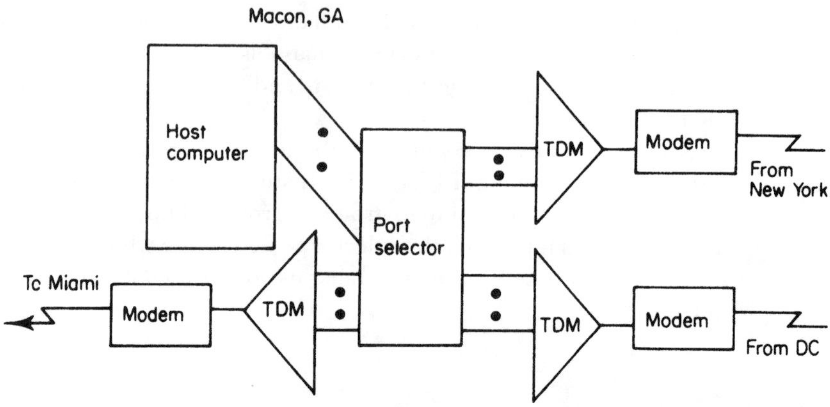

Figure 6.76 Constructing a circuit-switched network. A port selector with an automatic group select feature enables the utilization of a common network segment to access geographically separated computer systems

in Macon while the entry of a '2' would cause the user to be routed out of the port selector and into a multiplexer that would connect to the computer in Miami. Thus, in this example, the port selector enables the utilization of a common network segment for obtaining access to geographically distant computer systems.

Additional features available on selectors include diagnostic modules and busy-out switches. Selector diagnostic modules are used to determine which line is connected to which port and to display key diagnostic information for that particular cross-connect while busy-out switches permit user flexibility in changing existing contention ratios.

6.12 PROTOCOL CONVERTERS

A wide range of protocol converters are marketed, ranging in scope from devices that convert the physical and electrical interface from one standard to another, to devices that may perform five levels of conversion to include the physical and electrical interface. Table 6.16 lists the different levels of operational conversion that may be performed by a protocol converter.

Table 6.16 Operational conversion performed by protocol converters

Operational conversion	Level
Device functionality	1
Device operation	2
Protocol	3
Data code/speed	4
Physical and electrical interface	5

Operation

Note that to properly connect incompatible devices, a protocol converter may need to provide more than one level of operational conversion, with the number of levels of conversion based upon the incompatibility of the devices.

Physical/electrical conversion

At the lowest level of conversion, it may be necessary to convert the link-level connection from one interface standard to another. Thus, some protocol converters can be obtained which perform, as an example, RS-232 to RS-449 conversion or RS-232 to X.21 conversion.

Data code/speed conversion

Owing to the variation in operating codes and speeds of operation among devices, often code and speed conversion is required to obtain compatibility between equipment. At this level of conversion, the protocol converter performs as a speed and/or code converter. Although in many instances code conversion is obtained by the simple mapping of a bit string that represents a character in one code into a bit string that represents the same character in a different code, upon occasion code conversion can become quite complex.

One example of this would be a protocol converter employed to convert an asynchronous terminal's transmission into a bisynchronous communications format for connection to a host computer's BISYNC port. In this type of operation, the computer's port will probably be transmitting EBCDIC data in blocks, with each block containing a block check character at the end of the block. Similarly, the computer port will be expecting to receive EBCDIC data grouped into blocks, with a block check character appended to each block. If the asynchronous terminal device transmits ASCII data, not only must that data code be converted to EBCDIC, but in addition a computational operation must occur to generate the block check character. Although the previous illustration of code conversion involved developing the appropriate block check character from ASCII data, in actuality a protocol conversion is also being conducted.

Conversion categories

Protocol conversion can fall into three primary categories since most communications protocols fall into the categories of asynchronous, byte-oriented, and bit-oriented protocols. Common examples of protocols in each category include teletype (TTY) asynchronous transmission, IBM's various types of byte-oriented bisynchronous protocols, and the ISO's higher-level data link control (HDLC) bit-oriented protocol.

Converting a data stream from one protocol into another requires a considerable amount of work. As an example, in TTY transmission each character is framed by a start bit and one or more stop bits, with either parity used for simple error detection or no error detection and correction scheme employed. In bisynchronous transmission data is grouped into blocks, with a number of synchronization characters prefixed to the block to provide synchronization between the transmitting and receiving devices. In addition, special control characters consisting of unique patterns of bit configurations inform the receiver of such information as when the message starts, when the actual text starts, if a message consists of more than one block; it also informs the receiver about the end of the block containing the block check character, which is formed by a cyclic redundancy checking process that enables the receiver to ascertain if the block was received without a transmission error affecting the block.

A protocol converter, converting asynchronous data to bisynchronous data, thus must eliminate the start and stop bits from each asynchronous character (character stripping), block the characters into a bisynchronous block format, generate and appropriately place the transmission control characters within the data block, and compute a cyclic redundancy check character and append it to the block as the block check character.

Device operation conversion

Since the physical characteristics of many devices vary, one of the functions a protocol converter may have to perform is device operation conversion. Again returning to the previous asynchronous to bisynchronous example, one device operation conversion would involve the translation of bisynchronous printer spacing information into an equivalent number of carriage returns, line feed sequences recognized by the asynchronous device.

Device functionality conversion

At the highest level of protocol operation, device functionality conversion ensures that the data transmitted by one device is correctly interpreted by the other device. As an example, a computer's cursor positioning code sequences transmitted in one protocol would be mapped by the protocol converter into the equivalent cursor control codes of the device the converter supports. Other examples of device functionality conversion include the translation of Program Function (PF) keys on 3270 terminals to equivalent codes on other devices, translation of field attributes to include underline, blinking and high intensity display, and other terminal functions.

In an IBM 3270 networking environment, protocol converters, in essence, open the 3270 network to asynchronous terminal access. Both IBM and numerous third-party vendors manufacture standalone protocol converters. In addition, IBM markets an optional asynchronous emulation adapter (AEA) for use in its 3174 control unit which enables the 3174 to support asynchronous terminal access. Here the AEA can be considered to be a built-in control unit protocol converter.

Character versus block mode operation

Since an IBM 370 network is based upon the use of intelligent terminals, to be more effective protocol converters must support intelligent terminal operations. To understand how this can be accomplished, let us first examine how a synchronous IBM terminal operates when connected to a control unit which a protocol converter is designed to emulate. Then we can use this information to compare and contrast two different protocol converter operations to transmission efficiency.

In an IBM 3270 network, data typed by a terminal operator is transmitted to a control unit and echoed back to the terminal. The control unit stores data from each connected terminal until a PA, PF, or Enter key is pressed, at which time the contents of the control unit's buffer are transmitted. This blocking of data until a predefined key is pressed that could cause a screen change significantly reduces transmission between control units and a front-end processor, allowing up to 64 terminal devices to share access to a front-end processor via a control unit. A second advantage obtained from blocking data is the ability of the control unit to communicate with the front-end processor using a block checking algorithm which supports error detection and correction.

Figure 6.77 illustrates the typical data flow between a personal computer and a protocol converter. Some protocol converter manufacturers market software emulation programs which block data on a personal computer in a manner similar to the personal computer being an intelligent 3270 terminal connected to a control unit. That is, data is blocked and only transmitted to the protocol converter when a PA, PF, or Enter key is pressed.

A second and less sophisticated type of emulation program that works with protocol converters transmits a character each time a key is pressed. This type of emulator is a character mode emulator and can significantly increase transmission costs and result in the awkward display of characters on the personal computer

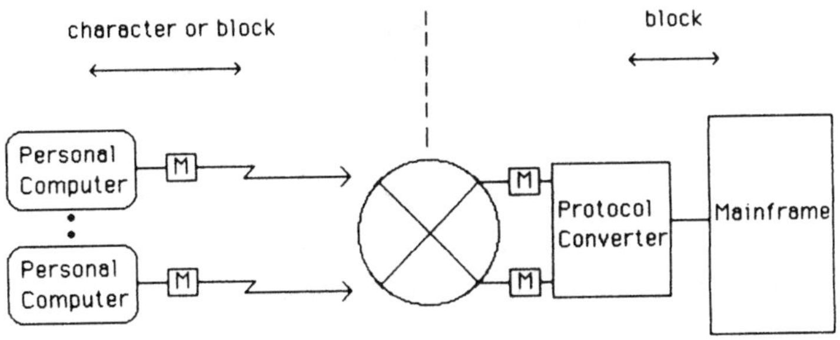

Figure 6.77 Protocol converter system data flow. Blocked transmission from a personal computer to a protocol converter allows error detection and correction between the PC and the protocol converter. In addition, block mode can significantly reduce the cost of communications and network delays when access occurs via a packet network

when access between the PC and the protocol converter is via a packet network. The additional cost results from the packet network forming packets containing single characters. Since most packet networks have a charge component based upon the number of packets carried, the use of a character mode emulator can become very expensive. Concerning delay time, since the character has to transmit the packet network to the protocol converter and then be echoed back to the personal computer, you can experience delays of up to 0.5 s from the time you type a character until it is displayed on your screen. This large delay results from each packet network node checking packets prior to forwarding them through the network, a condition that now occurs in each direction on single character packets. In comparison, a block mode emulator program used on your personal computer would immediately echo characters to your screen and only transmit data to the protocol converter when a PA, PF or Enter key was pressed. Therefore, you should always attempt to use a block mode emulator to access a protocol converter via a packet network.

Applications

The number of applications to which protocol converters can be employed is limited only by your imagination, such is the diversity of the devices. Figure 6.78 illustrates two popular application areas where protocol converters are typically employed.

In the top part of Figure 6.78, a protocol converter functioning as an asynchronous/synchronous converter is employed to permit an asynchronous terminal to be

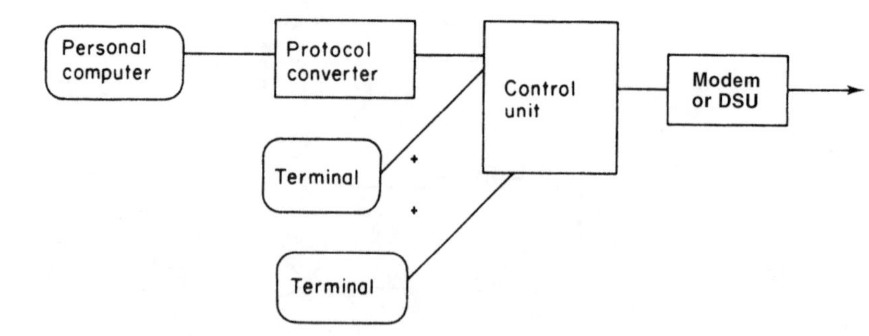

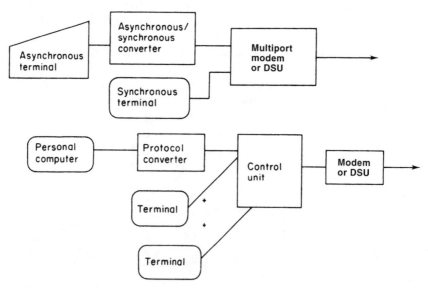

Figure 6.78 Typical protocol converter applications. Protocol converters can be used for asynchronous to synchronous conversion (top) or to make asynchronous personal computers that transmit data on a line-by-line format compatible with a screen-oriented display protocol (bottom)

interfaced to a multiport mode. Since the limited function multiplexer in the multiport modem multiplexes only synchronous data, the asynchronous data stream must be converted into a synchronous data stream prior to the modem or DSU multiplexing the data. The use of an asynchronous to synchronous protocol converter thus allows both synchronous and asynchronous devices to share the use of a common multiport modem or DSU.

A protocol converter functioning as a terminal emulator is illustrated in the lower part of Figure 6.78. In this mode of operation, the protocol converter changes the line by line transmission sequences of the personal computer into a screen-oriented format that matches the protocol supported by the control unit. In addition, the protocol converter will most likely perform data code, speed, device operation, and functionality conversion.

REVIEW QUESTIONS

6.1.1 What are the major differences between time division multiplexing and frequency division multiplexing?

6.1.2 How many 300-bps data sources could be multiplexed by a frequency division multiplexer and by a traditional time division multiplexer?

6.1.3 Why does frequency division multiplexing inherently provide multidrop transmission capability without requiring addressable terminals and poll and select software?

6.1.4 Describe a modern-day role for the use of frequency division multiplexing.

6.1.5 Assume your organization has two computer systems, one located in city A and another located in city B. Suppose your organization has eight terminals in city C and 10 terminals in city B, with the following computer access requirements:

Terminal location	Terminal destination	
	City A computer	City B computer
City B	6	4
City A	4	4

Draw a network schematic diagram illustrating the use of time division hub-bypass multiplexing to connect the terminals to the appropriate computers.

6.1.6 Why is a front-end processor substitution for a multiplexer at a computer site the exception rather than the rule?

6.1.7 Assume the cost of communications equipment and facilities is as follows:

24-channel TDM at $1000
T1 CSU/DSU at $800
Leased line city A to B $1000/month

If 24 64-kbps terminals are located in city A and the computer is located in city B, draw a schematic of the multiplexing network required to permit the terminals to access the computer, assuming each terminal can be directly connected to the multiplexer. What is the cost of this network for one year of operation?

6.1.8 Why must a statistical multiplexer add addressing information to the data it multiplexes?

6.1.9 Why are most statistical multiplexers ill suited for multiplexing multidrop circuits?

6.1.10 Assume the statistical multiplexer you are considering using has an efficiency 2.5 times that of a conventional TDM. If you anticipate connecting the multiplexer to a 56 kbps DSU, how many 9600 bps data sources should the multiplexer multiplex?

6.1.11 Assume the multiplexer discussed in Question 6.1.10 services synchronous data by the use of a bandpass channel. What would be the effect upon the number of 9600 bps asynchronous data sources supported in Question 6.1.10 if you must service a 19 200 bps synchronous data source by the use of a bandpass channel?

6.1.12 Assume you are considering the use of an intelligent multiplexer that has an efficiency three times that of a conventional multiplexer for asynchronous data traffic and 1.5 that of a conventional multiplexer for synchronous data traffic. If you have to multiplex eight 19 200-bps asynchronous terminals and four 2400-bps synchronous terminals on a line operating at 64 kbps, could the multiplexer support your requirement? Why?

6.1.13 How could you use the statistical loading data available from an intelligent multiplexer to determine if the vendor's literature concerning its efficiency in comparison to a conventional TDM is reasonable?

6.1.14 Draw a network schematic showing how the port contention option of a multiplexer could be used in a sub-bypass network.

6.2.1 Why do T1 carriers operate at different rates in the United States and Europe?

6.2.2 Why is the method a vendor uses in digitizing voice important to consider when evaluating a T1 multiplexer?

6.2.3 What are the three steps in Pulse Code Modulation?

6.2.4 What is the key difference between D4 and ESF framing on a T1 circuit with respect to performance measurements?

6.2.5 Describe the general operation of waveform, vocoding, and hybrid coding voice digitization techniques.

6.2.6 What is the key advantage obtained through the use of a Code Excited Linear Prediction (CELP) voice digitization method in comparison to PCM or ADPCM?

6.2.7 Draw a network diagram to illustrate how a T1 multiplexer could be used to support LAN to LAN communications, videoconferencing, and computer-to-computer transmission.

6.2.8 Illustrate the advantage obtained by the use of a non-contiguous resource allocation feature of a T1 multiplexer.

6.2.9 What are two problems associated with the use of applying packet technology to route calls through a series of T1 multiplexers?

6.2.10 Under what circumstances would you want to use a two-wire or four-wire transmission only voice interface on a T -carrier multiplexer?

6.2.11 What is the maximum number of voice channels that can be carried on T1 and CEPT-30 facilities if the T1 multiplexer uses PCM to digitize voice? How do those numbers change when 32 kbps and 24 kbps ADPCM are used?

6.2.12 How could you maximize the efficiency of a T1 multiplexer that does not support subrate multiplexing?

6.2.13 Discuss three functions a gateway T1 multiplexer must perform.

6.2.14 Describe two methods used by T-carriers to effect the alternate routing of DS0 channels. Which method would normally be more efficient?

6.2.15 What does the term 'downspeed' mean?

6.2.16 What does the term 'transparent bumping' refer to?

6.2.17 What is the difference between a multiplexer functioning as a hop and a multiplexer functioning as a node?

6.3.1 What is a subrate voice/data multiplexer?

6.3.2 To be effective, what type of voice compression should a subrate voice/data multiplexer support?

6.3.3 Between what type of organizational locations should subrate voice/data multiplexers be considered for use? Why?

6.4.1 You are comparing the cost of inverse multiplexing to using T1 facilities to transmit data at 128 kbps, and you determine the monthly cost of the following facilities and devices.

Facility/device	Monthly cost
T1 line	$9000
64 kbps line	$1250
Inverse multiplexer at	500
64 kbps DSU at	50
T1 CSU/DSU at	100

Would it be economical with these figures to use inverse multiplexers?

6.4.2 What are the primary advantages of inverse multiplexers in comparison to high-speed digital transmission at T1 or E1 data rates?

6.4.3 Why is the utilization of most inverse multiplexers limited to two channel devices?

6.5.1 What is the primary use of PADs?

6.5.2 What is the difference between ITU Recommendations X.28 and X.3?

6.5.3 What X.3 setting is required to enable XON and XOFF flow control from a PAD to a terminal?

6.5.4 Where are the three locations where PADs are installed?

6.6.1 What is the primary function of a FRAD?

6.6.2 Draw a network diagram which indicates how a FRAD could support the transmission of data from a Token-Ring and an Ethernet LAN via a Frame Relay network.

6.6.3 Discuss the key difference between a router and a FRAD.

6.6.4 What is an Internetworking-FRAD (I-FRAD)?

6.6.5 Discuss four techniques that can be used to transport SNA via a Frame Relay network.

6.6.6 Discuss the importance of frame fragmentation, prioritization, and buffering when voice is transmitted over a Frame Relay network.

6.7.1 What is the primary function of a front-end processor or communications controller?

6.7.2 What is the major difference between a channel-attached and a link-attached communications controller?

6.7.3 Describe the general function of five front-end processor software modules.

6.7.4 What are the major differences between IBM 3725 and 3745 front-end processors with respect to LAN support?

6.7.5 What is the purpose of a line interface coupler (LIC) weight?

6.8.1 On what type of circuits are modem- and line-sharing units used? What is the major difference between devices?

6.8.2 Discuss two constraints associated with the use of modem- and line-sharing units.

6.9.1 What is the primary constraint you should consider when employing port-sharing units in conjunction with line- or modem-sharing units?

6.9.2 Discuss two constraints associated with the use of port-sharing units.

6.10.1 Discuss the differences between sharing units and a control unit.

6.10.2 Why is it difficult to attach RS-232/ITU V.24 devices to an IBM 3270 Information Display System network?

6.10.3 Discuss some of the products one can obtain to connect RS-232/ITU V.24 devices into a 3270 network to include their operational features.

6.11.1 What is the major benefit derived from the use of a port selector?

6.11.2 What is the advantage of obtaining a port selector with queuing and queuing position display capability?

6.11.3 When should one consider the utilization of a port selector at a remote site?

6.11.4 Assume your primary computer system is located in New York and your organization has a second computer system in Boston. If your organization has a 16-channel multiplexer connection between Chicago and New York and a 32-channel multicplexer connection between Philadelphia and New York, draw a network schematic diagram to illustrate the use of a port selector to provide access for up to 16 simultaneous users to the Boston computer system using a minimum number of leased lines.

6.12.1 Describe five operational conversions performed by protocol converters.

6.12.2 Why is character mode emulation expensive when transmission occurs over a packet network?

7

SPECIALIZED DEVICES

In this concluding chapter we will turn our attention to a group of communications devices that do not gracefully fall into one of the categories covered in previous chapters. The first series of devices we will examine are communications switches which evolved from use in technical control centers to use on desks. A second group of products is designed to eliminate redundancies in data, which in turn enhances transmission. These products are referred to collectively as data compression performing devices. Although we will focus attention upon hardware, we will also note that the functionality and benefits that can be derived from their use can also be obtained from software incorporated into different products. The third group of products we will examine permits transmission via fiber optics within a building or campus, enabling us to consider a variety of additional networking strategies. In concluding this chapter, we will examine several devices to include access lists maintained by routers and firewalls.

7.1 DATA COMMUNICATIONS SWITCHES

Communication switches bring freedom and economy to network design and operations. Until recently, they were found mainly in network technical control centers, where they help in on-line monitoring, fault diagnosis, and digital and analog testing, but now they are also being used to reroute data quickly and efficiently and to replace several dedicated backup units with just one, enabling a single communications device to act as standby for several devices.

The kinds of switches available and how they may be combined or chained to fulfill different functions are described in the first part of this section. The second part will concentrate on applications, in particular on the use of switches to assure network uptime without a heavy investment in redundant equipment.

The four basic categories of switches are fallback, bypass, crossover, and matrix. Of these, two or more from the same or different categories may be chained to serve still other data communications requirements. Furthermore, within each category there are two types of switches: the so-called telco switches which transfer four-wire leased or two-wire dial-up telephone lines; and the Electronics Industry Association (EIA) or ITU switches referred to in this section as EIA switches, which transfer all 24 leads of an EIA RS-232 or ITU V.24 interface.

Fallback switches

The fallback switch is a rapid and reliable means of switching other network components from on-line to standby equipment. The EIA version selects either a pair of 24-pin-connected components, which, as shown in Figure 7.1, may be terminals, modems, DSUs or channels on a front-end processor.

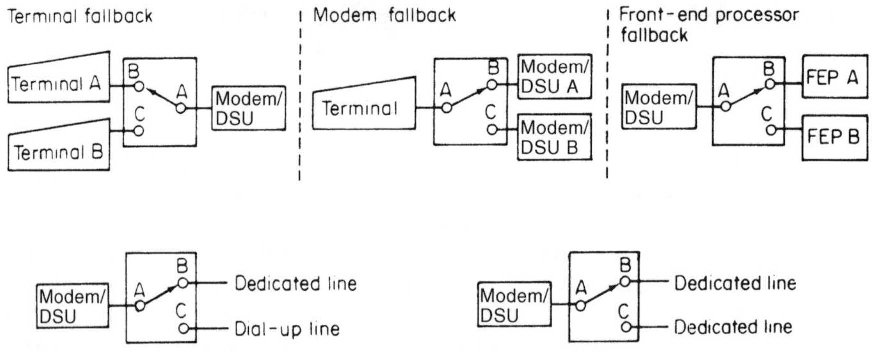

Figure 7.1 Fallback switches. Top: EIA (24-pin) fallback switch. Bottom: telco fallback switch. The EIA fallback switch transfers 24 leads at a time, while the telco switch transfers the two or four leads associated with telephone lines

In the first example, two terminals share a single modem. This configuration might be required, for example, when terminals have the same transmission speed but use different protocols, so that each communicates with a different group of remote terminals or computers.

In the second example, one terminal is provided with access to two modems or DSUs, one of which is redundant but needed for uptime reliability. Alternatively, the first modem or DSU might enable a terminal to transmit to another device at one data rate during one portion of the day, while the second lets it talk to a different device at a different data rate during other periods of the day. Then, depending on operational requirements, one terminal with a fallback switch for modem or DSU selection could be more practical than installing two terminals.

In a third application, an EIA fallback switch (Figure 7.1, top right) permits a modem or DSU to be transferred between front-end processors. Although called a line-transfer device by some manufacturers, in effect what you obtain is a device that selects which front-end processor will service the modem or DSU.

A telco fallback switch similarly allows the user to select one of two sets of telephone lines. As shown in Figure 7.1 (bottom), it can select one line from among various combinations of dedicated and dial-up lines that may have been installed to fit the needs of a particular application. Thus, for a large data transfer application which is of a critical nature, the telco fallback switch could be connected to a pair of leased lines, one of which is used as an alternative circuit in the event of an outage on the primary circuit.

Bypass switches

The EIA bypass switch connects several EIA interfaces of one type (for example, modems or DSUs to the same number plus a spare of another EIA interface type (for example, terminals) and can switch any member of the first group to the spare member of the second group. One application for bypass switches is at a computer installation (Figure 7.2, top left). Here, one front-end channel is reserved as a spare in case any of the existing channels, which normally service predetermined modems or DSUs should need to be connected quickly to a spare channel.

In another application (Figure 7.2, top right), the EIA bypass switch can substitute a standby spare terminal for a failed on-line terminal and do away with the need for a spare modem or DSU. Although seldom used for multiple terminal access, a bypass switch can also enable many terminals to share a single modem or DSU and a dial-up or leased line.

A telco bypass switch transfers any one of a group of two-wire or four-wire telco lines to a spare communications component. For example, as shown in Figure 7.2 (bottom left), if the first modem or DSU should fail, line 1 can be switched to the spare communications component (Figure 7.2, bottom right). Telco bypass

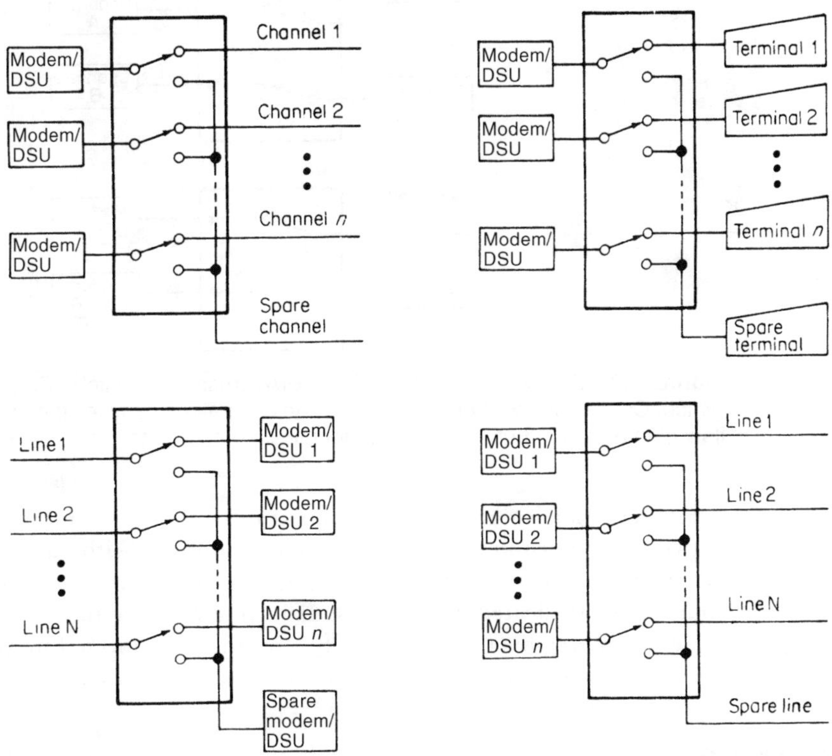

Figure 7.2 Bypass switches. Top: EIA bypass switch. Bottom: telco bypass switch. Bypass switches transfer either EIA or telco interfaces to spare components with a similar interface on the other side of the switch

switches can be used to switch leased or dial-up lines to modems, DSUs, automatic dialers, or even acoustic couplers.

Crossover switch

Crossover switches provide you with an easy method of interchanging the data flow between two pairs of communication components. Four connectors are associated with each switching module, one for each of the two pairs of communication components connected to that module.

As shown in Figure 7.3 (top), an EIA crossover switch permits the data flow to be reversed between two pairs of EIA interfaced components. In the example shown in Figure 7.3 (top), modem A (or DSU A), which is normally connected to the front-end processor channel A, and modem B (or DSU B), which is connected to the front-end processor channel B, are reversed when the switch is moved from the normal to the crossover mode of operation. Thus, modem A (or DSU A) then becomes connected to channel B and modem B (or DSU B) is connected to channel A upon crossover.

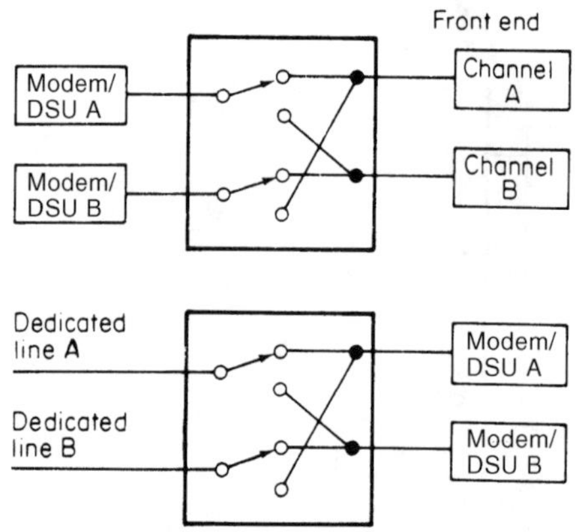

Figure 7.3 Crossover switches. Top: EIA crossover switch. Bottom: telco crossover switch. Crossover switches, either EIA or telco, make it easy for the operator to reroute the flow of information between pairs of identical components

Similarly, a telco crossover switch permits the user to interchange the data flow between two telco lines and two modems or DSUs. Although two dedicated lines are shown connected to the crossover switch in Figure 7.3 (bottom), you can also connect one dedicated line and one dial-up line or two dial-up lines to the switch. Here, upon crossover, line A, which is normally connected to modem A (or DSU A), becomes connected to modem B (or DSU B), and vice versa.

Matrix switch

With a matrix switch the user can interconnect any combination of a group of incoming interfaces to any combination of a group of outgoing interfaces. Matrix

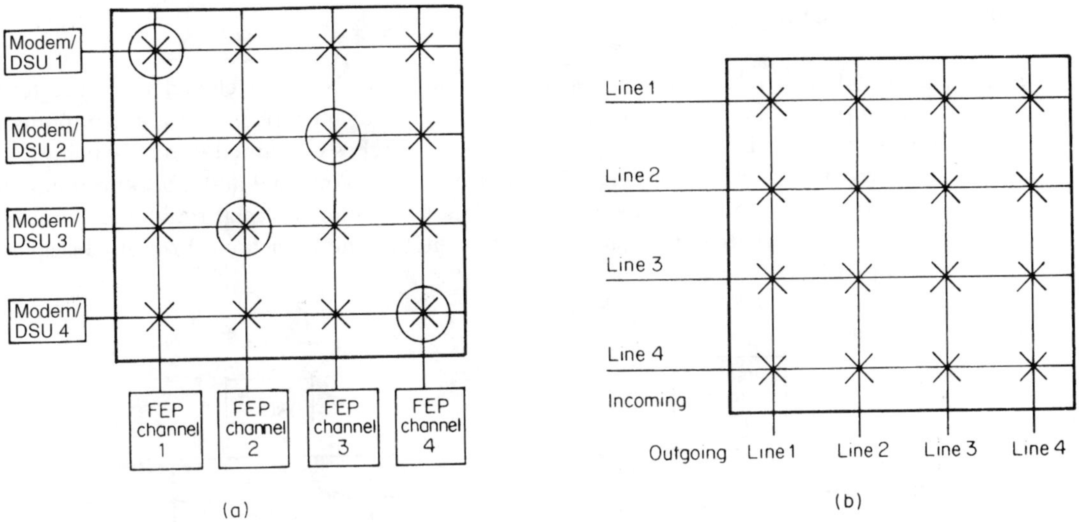

Figure 7.4 Matrix switches. (a) 4 by 4 EIA matrix switch. (b) 4 by 4 telco matrix switch. Matrix switches permit any combinations of a group of incoming interfaces to be rapidly connected to any combination of outgoing interfaces

switches are manufactured as an *n* by *n* matrix, with 4 by 4, 8 by 8, 16 by 16 combinations typically available. The user of a manual matrix switch makes an interconnection by depressing two pushbuttons on the switch simultaneously, one representing the incoming interface and the other representing the outgoing interface.

As shown in Figure 7.4 (a), an EIA 4 by 4 matrix switch is a quick and efficient way of connecting any combination of four modems to any combination of four front-end processor channels. The circles represent the depressed switch combinations, so that, in this case, modem 1 serves front-end processor channel 1, modem 2 serves front-end processor channel 3, modem 3 serves front-end processor channel 2, and modem 4 serves front-end processor channel 4. Further, with this configuration the user is free to designate one or more modem (or DSU) or front-end processor channels as spares or a combination of modems and channels as spares.

The telco 4 by 4 matrix shown in Figure 7.4 (b) similarly permits the transfer of any combination of four incoming lines to any combination of four outgoing lines.

A type of application warranting investigation of telco matrix switches arises when remote terminals require access to two or more adjacent computers. If the terminals are used heavily enough to justify installing leased lines from the remote sites to the central computers, the telco matrix switch enables the user to switch the incoming leased lines to outgoing cables which, via modems or DSUs, are connected to different computers.

Additional derivations

A number of additional switching functions have been developed from the four categories of switches previously discussed. For instance, a spare component backup switch is basically a pair of fallback switches contained in one housing. As shown in Figure 7.5 (top left), this switch permits a normal and a backup mode of operation. The normal mode permits data to be transferred through the primary component, whereas the backup mode switches the data flow through the spare components.

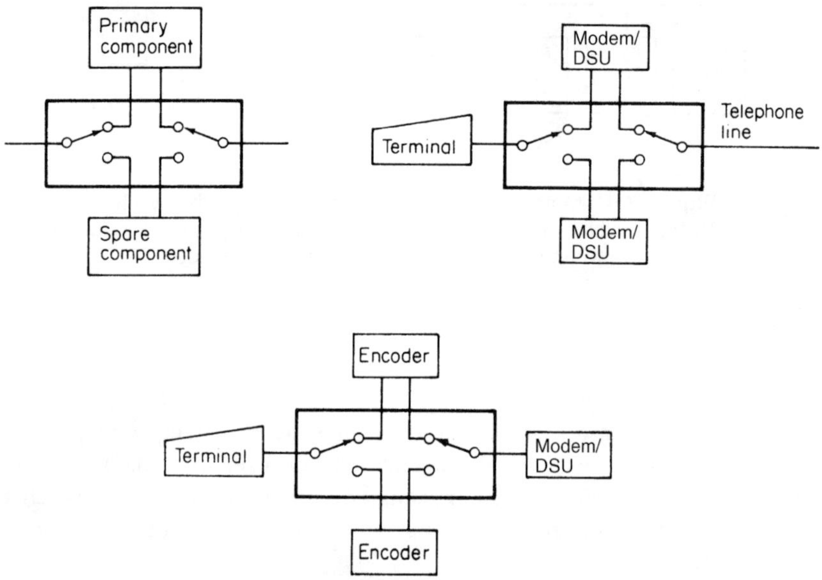

Figure 7.5 Backup switch variations. Top left: backup switch. Top right: backup switch, 3 of 4 EIA interface. Bottom: backup switch, 4 of 4 EIA interface. Paired EIA and telco fallback switches, which often come in one package, provide both a backup and a normal mode of operation

In another configuration (Figure 7.5, top right), a pair of modems or DSUs are the primary and spare components connected to one terminal, and the switch selects the modem or DSU to be used in transferring data between the terminal and the telco line. Because three EIA interfaces are involved, this configuration is called a 3 of 4 EIA interface bypass switch. In a 4 of 4 EIA interface (Figure 7.5, bottom), four interface devices are connected to the switch. In this configuration, the switch selects one of two encoders to encode terminal data for transmission through an attached modem or DSU.

A second common switch derivation is a multiple fallback switch (Figure 7.6). Besides the EIA and telco versions, this switch is manufactured in a 1 of *n* version, with *n* being the number of possible selections. Figure 7.6 (top) shows two possible configurations for a 1 of 4 EIA fallback switch. At the left, the switch allows the

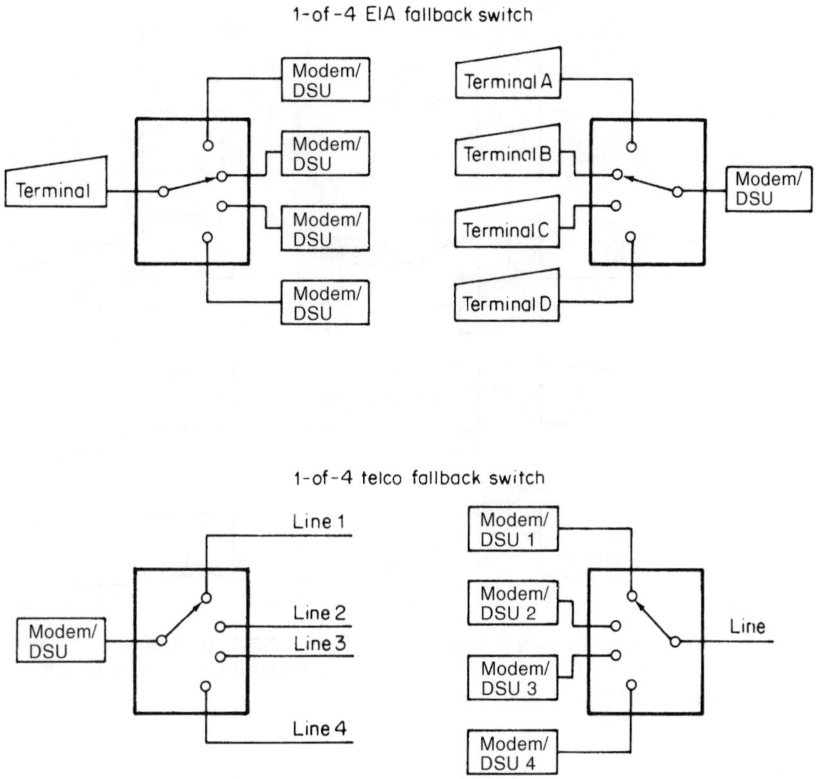

Figure 7.6 Multiple fallback switches. Multiple fallback switches, either EIA or telco, allow many terminals to share one modem (or DSU) or many modems (or DSUs) to share one telephone line

terminal to be connected to any one of four modems or DSUs, while at the right, any one of four terminals may be connected to a single modem or DSU. Similarly, Figure 7.6 (bottom) shows how the 1 of 4 telco fallback switch allows either four modems or DSUs to share a single line or four lines to share a single modem or DSU.

Chaining switches

No manufacturer produces a complete line of readymade switches, but it is often more convenient to deal with and install switches from a single maker. You can do so by developing the switching functions required from combinations of one vendor's switches. In Figure 7.7 (left), for instance, four fallback switches are chained together to perform the function of a bypass switch. A single backup terminal can be used to replace any one of four primary terminals. In Figure 7.7 (right), four fallback switches are chained so that a single backup modem or DSU may be used by any terminal if its primary modem or DSU fails.

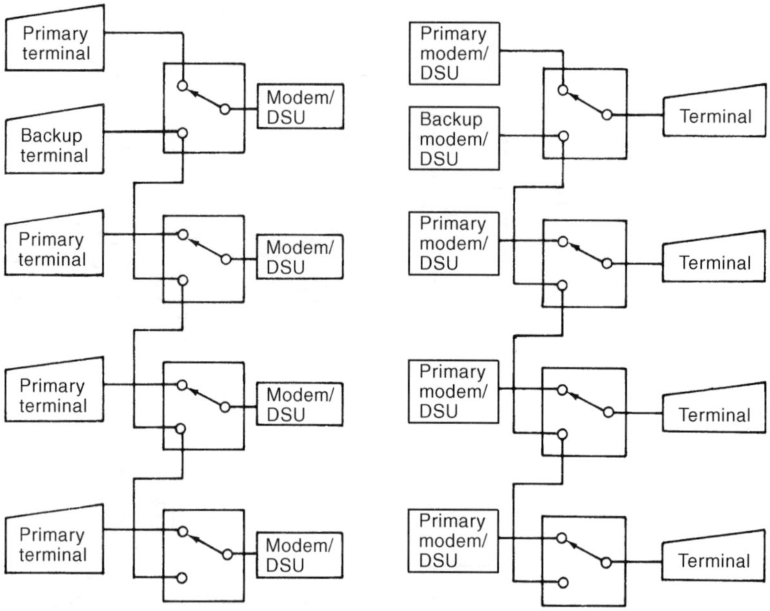

Figure 7.7 Chaining fallback switches. Chaining of simple switches yields a variety of functions, for instance permitting any of several terminal locations to select a single spare modem or DSU

Other switches can be similarly chained to develop additional switching functions or to increase the capacity of existing network devices. Even more usefully, different categories of switches and different types of switches within the same category can be chained. Figure 7.8 shows a 4 by 4 telco matrix switch chained to a 4 by 4 EIA matrix switch so that the user may interconnect any combination of lines, modems, DSUs, and front-end processor channels to arrange the information path desired. For this example, the number of possible configurations is increased to n^3 from the normal n^2 combinations available with a single n by n matrix switch.

Switch control

The four most common methods of transferring a switch are local and remote manual, American Standard Code for Information Interchange (ASCII) unattended remote, and via a business machine or central host computer. A local manual switch usually has a toggle or toggles, but many are also manufactured with pushbuttons and corresponding indicator lights. For a remote manual switch, one manufacturer produced a remote-control panel equipped with a pushbutton and a cable connecting it to the remote switch. This setup also has the advantage that shorter cable lengths can be run from communications components to the switch. Although toggle-operated units can be rack-mounted, they are normally available only in single-channel modules. Remote-control units, on the other hand, are

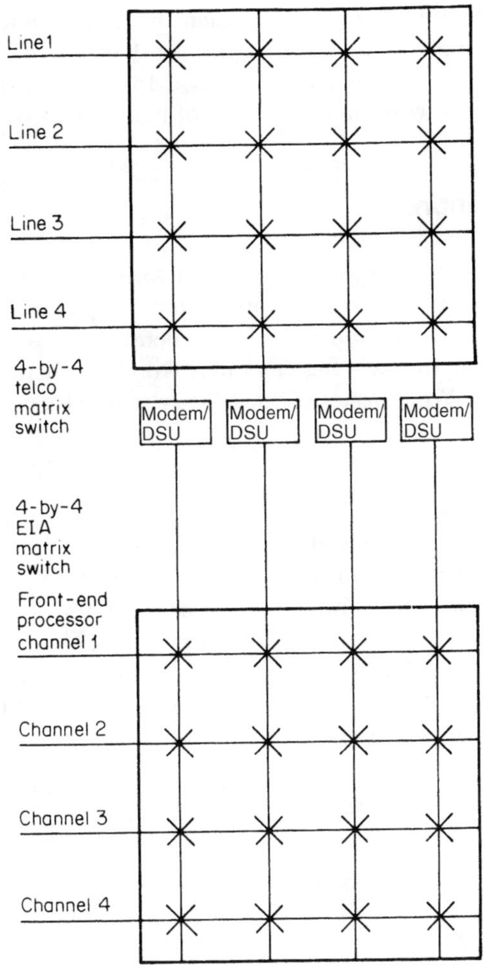

Figure 7.8 Chaining telco and EIA matrix switches. Here, an EIA matrix switch and a telco matrix switch are chained so the user may interconnect any combination of lines, modems, DSUs, and front-end processor channels

normally manufactured in 4, 8 or 16 unit configurations, and all units are master-switched from the remote control simultaneously, rather than one at a time.

The ASCII unattended remote control permits a switch to be controlled or monitored at any remote site at which a telephone line can be installed. An adapter interfaces the switch (or switches) to the telephone line and turns it on and off upon receiving a coded message consisting of the switch number and the state to which to transfer. The adapter then reports back the switch's new status. Also available is a query mode which allows the operator to check out a remote switch's position.

When a business machine (computer) is involved, switching is controlled directly by the machine, normally through a 5 V transistor–transistor logic (TTL) circuit. Let us now examine how such switches can be a cheaper alternative to obtaining overall network availability than the installation of typical redundant equipment. Furthermore, the cost of switches themselves can be kept low if the application allows a reasonable time interval for an operator to perform manual switching compared with the higher cost of master-controlled remotely operated switches.

Switching applications

When network equipment fails, a variety of communications switches can quickly get the network back into operation. They rapidly bring redundant equipment into place to meet established requirements for overall network availability. The cost of providing the switches can range from less than a few hundred dollars to well over \$50 000. What makes the price vary so much rests on the answers to such questions as:

- Which devices are most likely to fail?
- What tangible and intangible effects will a failed network device, such as a concentrator, have on the organization's operation?
- Would the operational loss be so great that it warrants the cost of including backup equipment and transmission lines?
- When a network component goes down, how much downtime, if any, is allowable to activate backup devices and get the network back into full-scale operation?
- To obtain speedy network recovery, what are the best types of switches for the application, and where should they be placed in the network?

The significance of these and similar questions, and their answers, will become apparent during the discussions of typical redundancy/switching configurations that follow.

The four basic types of communications switches—fallback, bypass, crossover, and matrix—come in two versions: EIA for switching digital signals and telco for switching two-wire and four-wire telephone lines. Chaining these switches provides a variety of extended switching functions. Furthermore, the switches can be activated or controlled in local or remote manual modes, in an unattended remote mode in which the switch is activated by a specified ASCII-character code, and in a computer-controlled remote mode. Switches becomes more expensive in going from local manual mode up to computer-controlled mode. Changing a network from primary to backup mode manually may, however, take 10 or 15 minutes, while a computer-controlled switch can activate all switch connections essentially instantaneously and automatically from a remote location.

In the first part of this section, the use of switches to substitute spares for such devices as modems (or DSUs) and terminals was discussed. In the second part, the discussion will center around the ramifications of switching between dual-collocated routers. Here, one router may be assigned completely to back up the other unit, or each router may be servicing its own networks during normal operation. In either case, on failure of one router, the other takes over all duties if it has enough capacity to do so. Although this discussion involves the use of routers, its points can be directly applied to most networking devices the user may wish to obtain a level of redundancy for, including statistical and intelligent multiplexers.

In the basic setup of the following applications, each router location services a number of local area networks and we desire to minimize the effect of the failure of a router or WAN transmission facility.

As will be seen, the applications tend to become more complex and more expensive. The actual choice depends to some extent on network application and to some extent on the severity of the consequences of a device failure.

Hot-start configuration

The two main methods of integrating collocated routers to service workstations on LANs are commonly called 'hot-start' and 'cold-start'. The hot-start approach shown in Figure 7.9 means that a backup router is energized, fully programmed with a duplicate of the software in the primary router and may be continuously tracking the traffic in and out of the primary router. When the primary router fails, a computer-controlled switch can put the backup router in control instantaneously.

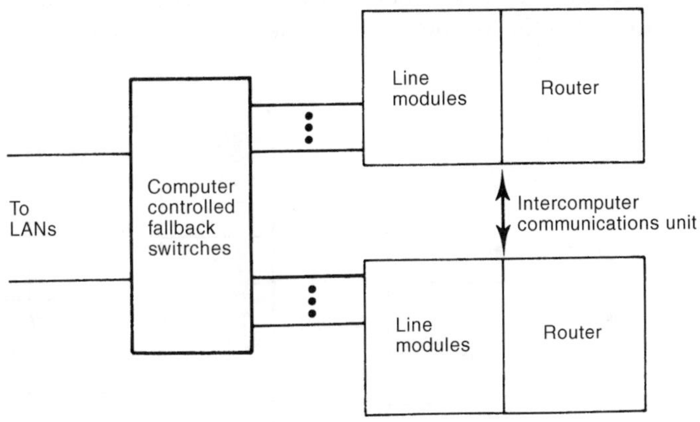

Figure 7.9 Hot-start configuration. When the primary router fails, a computer-controlled switch puts the backup router in control

Full effectiveness of such a hot-start arrangement requires the installation of an intercomputer (that is, interrouter) communications unit. When a failure such as memory-parity errors or power loss occurs, the router experiencing difficulty sends appropriate software commands through the communications unit. Additionally, an automatic command to a bank of computer-controlled fallback switches provides instantaneous transfer of the connection from each LAN to the operating router.

The near-instantaneous switching and the minimization of the loss of data are the important advantages of the hot-start configuration. There are, however, significant hardware costs associated with the computer-controlled switches and the intercomputer communications unit. In addition, the necessary software to permit the desired switching represents a complex development effort a vendor recovers via selling a hot-start configuration at an appropriate price. Overall, the cost for a hot-start configuration may well reach over $100 000, not counting the cost of the routers itself, but it may be well worth the money to ensure that a network remains continuously operational and available.

The remainder of this section will focus on variations of the cold-start redundancy configuration, with different methods of switchover available from manual (local toggles) and remote-control (pushbutton) switches into a communications network. However, computer-controlled switches can be used in any of these configurations with a consequent increase in complexity and cost combined with a salutary improvement in network uptime.

Cold-start configuration

Manual fallback switches represent one method of providing an alternate path between workstations located on remote LANs and the two routers (Figure 7.10). Here, the occurrence of a router failure will require manual intervention. When a failure occurs, it becomes necessary to switch the fallback units to insure that the remote LANs are connected to the operating router.

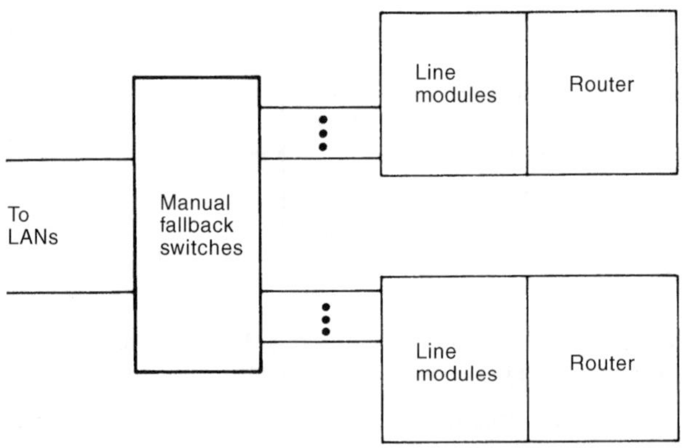

Figure 7.10 Cold-start configuration. Failure of a router requires manual intervention by the network operator

If the routers are initially sharing the LAN workload, the failure of one router may require the other router's software to be reconfigured to service the entire workload. This configuration can be completed in a few minutes by manually throwing the switches and reading the backup programs into the operational router's memory.

Some data being transmitted through the router may be lost during the reconfiguration time. The low cost of the cold-start configuration may, however, justify the extra time associated with satisfying retransmission requests for lost messages.

Sharing a backup router

The availability requirements of the network may be such that neither operating router has the reserve to serve as backup for the other, but it may be possible to service both devices with a single backup router, as shown in Figure 7.11. Here, fallback switches allow workstations on a LAN to be connected to a primary router or to a backup router. If the possibility of two routers failing at the same time is disregarded, the cost of the fallback switches is more than offset by the savings due to the ability to eliminate a second backup router.

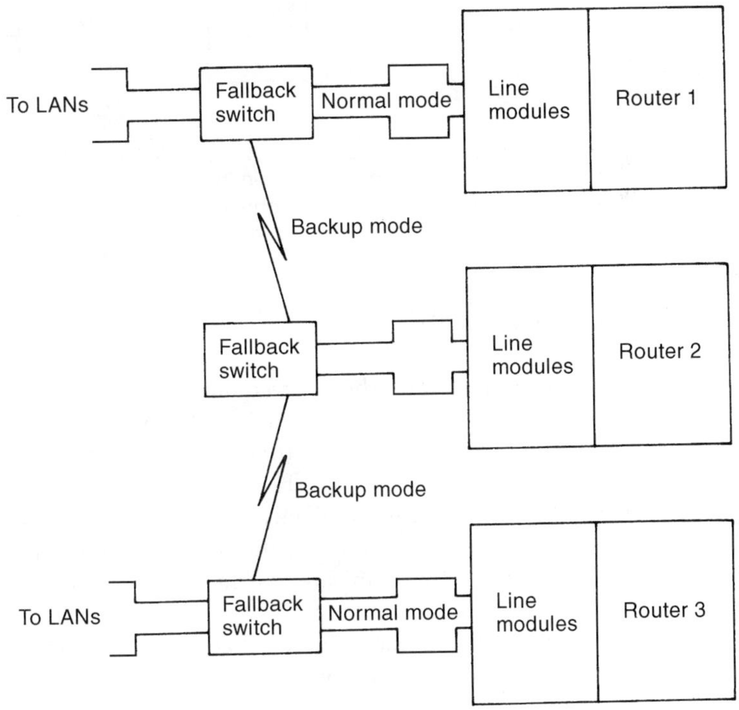

Figure 7.11 Sharing a backup router. Here, both primary routers, neither of which can completely take over for the other, are serviced by a third backup router

Router to router communications

If data transfer from each router takes place via a few high-speed lines, EIA fallback switches permit the transfer of DSUs or modems and lines between routers. In Figure 7.12, two switches permit each router to communicate over its own dedicated link. This type of configuration compensates for a router failure by permitting the remaining router to communicate with the distant location over its line and the line of the other device. However, the failure either of one of the dedicated lines or of a modem or DSU would require selection of one of the routers to use the remaining data communications link.

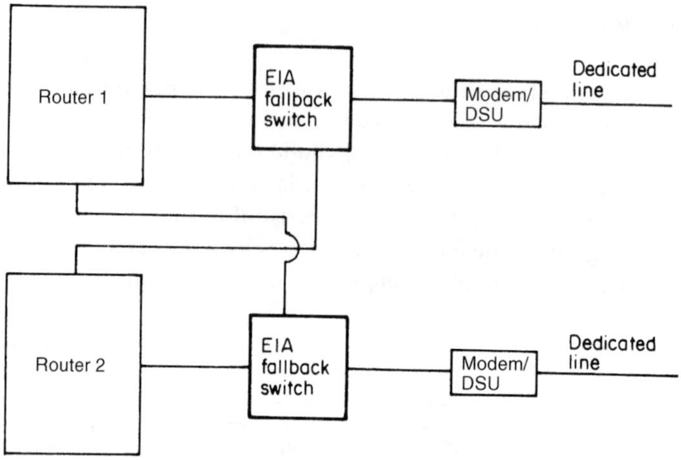

Figure 7.12 Router-to-router backup. Two EIA fallback switches permit each router to communicate over its own dedicated link in its primary mode of operation. When a fallback switch is activated, the router can be connected to the other router's modem or DSU and line

Adding a third EIA fallback switch

If you want to overcome the shortcomings of the preceding configuration, the inclusion of a third EIA fallback switch and another modem or DSU interfaced to the dual router configuration can either prevent or minimize the failure of a modem or DSU or a dedicated line, as shown in Figure 7.13. In the normal mode of operation, each router communicates via its own dedicated line. If a modem or DSU or dedicated line should fail, the proper positioning of two of the switches allows the router to communicate via the middle modem or DSU over an alternative path, either a dial-up network or another dedicated line. A disadvantage of this configuration is that each router has access to only one line at a time, unlike the configuration in the previous application.

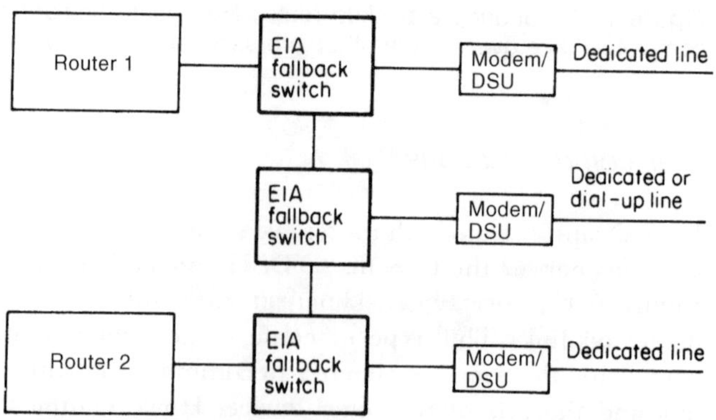

Figure 7.13 Adding a third EIA fallback switch. A third EIA fallback switch and another modem or DSU interfaced to the dual routers can prevent or minimize modem, DSU or line failure effects

Adding more switchable lines

Access to more than one dedicated line at a time may be obtainable by adding lines for each router and reconfiguring the EIA fallback switches, as shown in Figure 7.14. If one router should fail, the other can communicate over both dedicated lines, and it still has access to the backup line. In this manner, throughput degradation should be minimized.

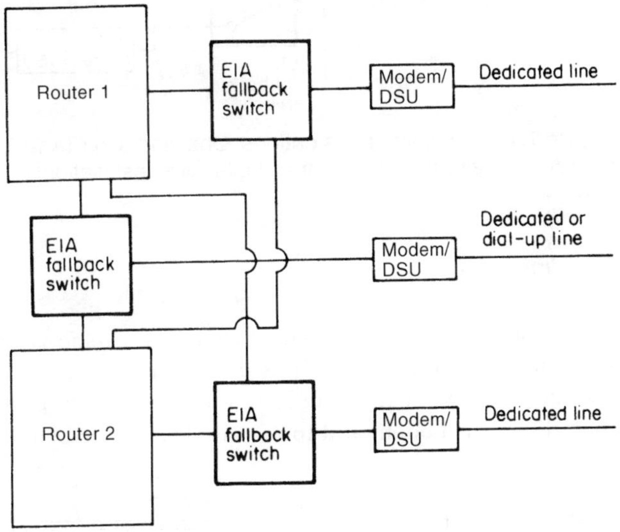

Figure 7.14 Adding more switchable lines. If one router should fail, the other can communicate over both dedicated lines and still have access to the backup line

Chaining adds options

Chaining two EIA fallback switches results in another way of providing an alternative link for a dual-router installation (Figure 7.15). Only one channel is required for each router. In normal operation, each switch interfaced to each router port remains in the primary modem or DSU position. If the dedicated line or the primary modem or DSU of either router should fail, the associated switch is positioned so that a path is provided to the backup DSU or modem. As in the other application, this backup DSU or modem can use a dial-up network or a dedicated line to communicate.

This configuration requires only one router port to provide a link in the event of DSU, modem or dedicated line failure. Should a router fail, however, the other one is not provided with access to the failing device's line. If workstations on LANs connected to the failing router are switched to the operational router, therefore, the operational link may not be sufficient to satisfy the increased traffic. Redundancy for this link through the use of EIA fallback switches can become rather complicated when more than a few lines require multiple access.

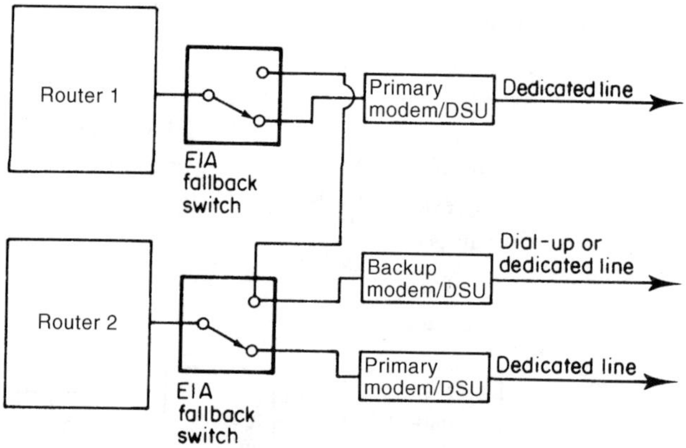

Figure 7.15 Chaining adds options. Chaining two EIA fallback switches is another way of providing an alternative link for a dual-router installation

Access to other lines

Use of one or more EIA matrix switches, as shown in Figure 7.16, can alleviate switching complexity as well as provide each router access to the other dedicated line. For example, with a single 4 by 4 switch, each router can have easy access to the spare DSU or modem and to any modem or DSU and line connected to the other

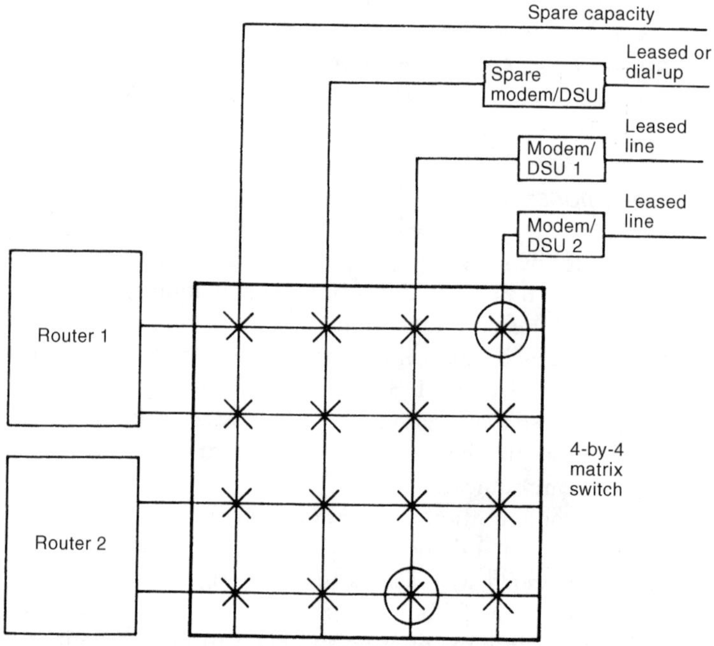

Figure 7.16 Access to other lines. One or more EIA matrix switches can alleviate switching complexity as well as provide each router access to the other dedicated line

router. Although only one spare modem or DSU is shown here, a second and its associated line facilities could be added, since the output side of the 4 by 4 switch can interface one additional device.

As shown by the circles, router 1 normally transmits data through modem 2 or DSU 2, and router 2 via modem 1 or DSU 1. Either router can be connected to the spare modem or DSU and associated line, should its primary modem, DSU or line fail. If a router fails, the other one can be connected to the failing device's primary modem or DSU and line, thus ensuring the continuation of full throughput. If each router communicates through more than one link, the use of an 8 by 8 or a 16 by 16 switch or the chaining of more than one matrix should be explored.

From the preceding examples, the utilization of switches provides the system designer with a low-cost option when designating a teleprocessing network that requires a level of redundancy which can range from a single redundant component requirement through multilayers of redundancy.

7.2 DATA COMPRESSION PERFORMING DEVICES

Although data compression is correctly considered as a technique used to reduce the duration of a transmission session, other benefits may be derived from the use of data compression performing devices. Since compression results in the transmission of a lessor number of characters in comparison to sending uncompressed data, the probability of a transmission error occurring is reduced. Similarly, a reduction in the number of characters can reduce the cost of transmission. Examples of cost reduction can range from the ability to connect additional terminals to a multidrop line to the use of a packet network where one of the cost elements associated with the use of the network is the number of characters transmitted and received by a terminal device.

From the preceding, it is obvious that the use of compression performing devices should be considered whenever possible since their use can boost network efficiency as well as potentially reduce the cost of operating a network. In this section, we will first review a few of the methods that can be employed to compress data. This will be followed by an examination of the benefits associated with the utilization of data compression performing devices. Using the previous information, we will then examine several compression performing data communication devices to investigate how they can be effectively used, focusing our effort upon the benefits that can be derived by their use.

For a detailed description of over 20 methods of data compression readers are referred to the author's book *Data and Image Compression*, 4th edition, also published by John Wiley & Sons Ltd.

Compression techniques

While the number of algorithms that have been developed over the years to compress data may exceed 100, in essence compression techniques can be subdivided into two classes—character and statistical encoding.

Character oriented

In character oriented compression techniques data is examined on a character by character basis, with special characters or a small group of characters employed to represent a larger string of characters. In this manner, redundancy can be reduced or eliminated prior to the transmission of data. At the receiver, a check of the incoming data occurs for the special characters or group of characters used to represent compressed data. Upon detection, the compressed data is decompressed according to a predefined algorithm, resulting in the duplication of the original data. Examples of character oriented compression techniques include null compression, run length encoding and pattern substitution.

Null compression

Null or space compression was probably the first data compression technique to be applied to data transmission. As illustrated in Figure 7.17, a special compression indicating character is employed to indicate the occurrence of null compression. This character is then followed by the count of the number of null or blank characters that occurred in the string. One of the earliest examples of the employment of null compression for data transmission is the IBM 3780 protocol which incorporates this technique.

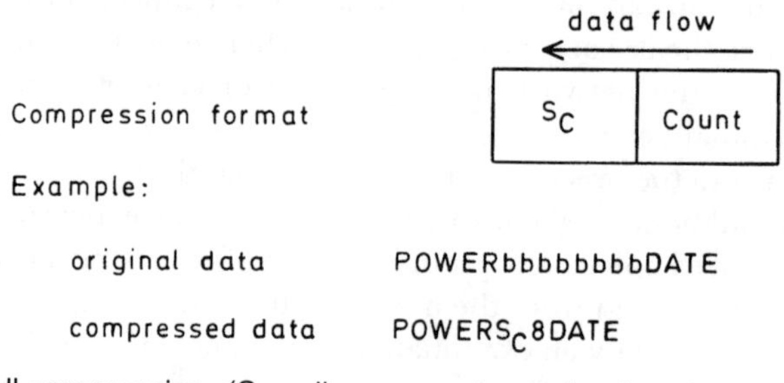

Figure 7.17 Null compression. (S_c null compression indicating character, b blank or null character)

Run length compression

Run length compression can be viewed as a superset of null compression. This technique is applicable to any repeating character string, whereas null compression is limited to a repeating sequence of null characters. Figure 7.18 illustrates the basic format and operation of run length compression.

As indicated in Figure 7.18, run length compression can be used to compress any string of repeating characters to include nulls. Since three characters in this compression format are substituted for the repeating string, one must encounter four or more repeating characters for this technique to compress data.

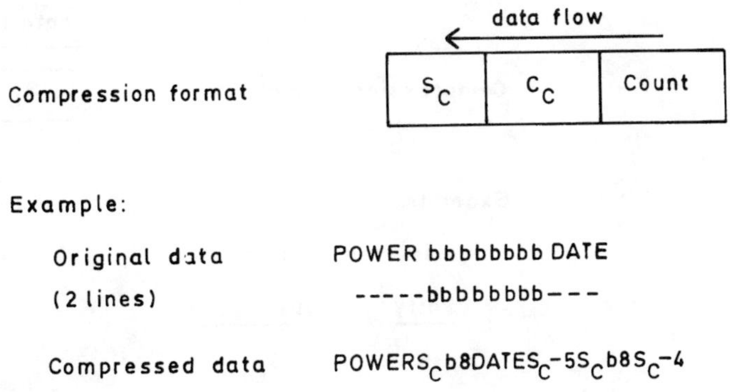

Figure 7.18 Run length compression. (S_c run length compression indicating character, C_c character repeated in string to be compressed, b blank or null character)

Pattern substitution

Pattern substitution requires data to be matched against entries in a pattern table. If the pattern is encountered in the table, the data that forms the pattern is replaced by a compression indicator which can range in scope from a single character to a group of characters. One version of pattern substitution that is frequently implemented in communications devices is diatomic encoding. In this compression technique, the most frequently encountered pairs of characters in the English language supplemented by common pairs of characters used in programming languages are used as entries in the pattern table.

An example of pattern substitution using diatomic compression is illustrated in Figure 7.19. Normally, the compression indicating substitution characters are obtained from undefined or unused characters in the character set being used. An example of the former might be the large number of undefined characters in the EBCDIC data code, while an example of the latter could be those characters from ASCII 127 through ASCII 255 which are known as extended ASCII.

Statistical encoding

In statistical encoding the probabilities of occurrence of single characters and patterns are employed to develop short codes to represent frequently occurring characters and patterns while longer codes are used for data that occurs less frequently. To better understand the utilization of statistical encoding assume the probability of occurrence of each character in a five-character set is as indicated in Table 7.1.

Based upon the data contained in Table 7.1, we can develop a coding scheme to minimize the average codeword length of each character by employing their probability of occurrence in our codeword construction scheme. For illustrative purposes, we will develop what is known as a Huffman code for the character set presented in Table 7.1

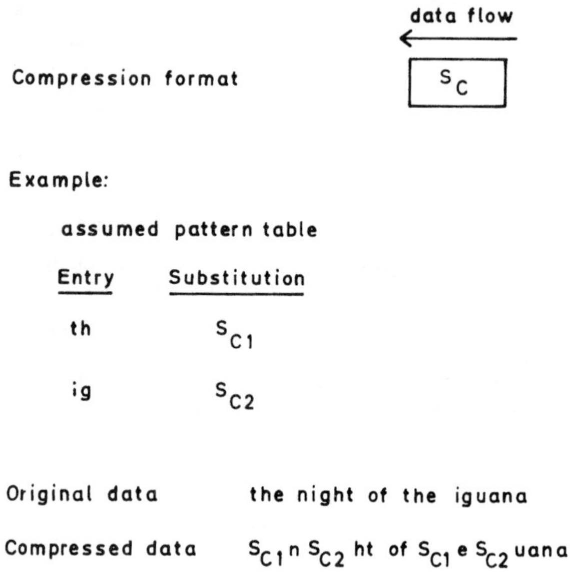

Figure 7.19 Pattern substitution using diatomic compression

Table 7.1 Five-character set probability of occurrence

Character	Probability of occurrence
A	0.250
B	0.250
C	0.250
D	0.125
E	0.125

Huffman coding

The Huffman code not only minimizes the average codeword length but, in addition, results in the generation of a code that is instantaneously decodable, that is, in the Huffman code every codeword is uniquely decipherable and no codeword is a prefix of another longer codeword. Figure 7.20 illustrates the development of a code set based upon the Huffman coding algorithm. First, the character set is arranged in a column in order of decreasing frequency of occurrence. Starting at the smallest probability of occurrence, the lines with the two smallest probabilities are merged, with their probabilities added to obtain a composite probability on a new line. This combining is continued until all lines have been merged. Then, for each pair of branches attached to a node we arbitrarily assign a 0 bit to one branch and a 1 bit to the other. The bit sequence assigned to each character is then determined by tracing the route from the 1.0 probability node backwards to the character, using the assigned bits in each branch that provides the route to the character.

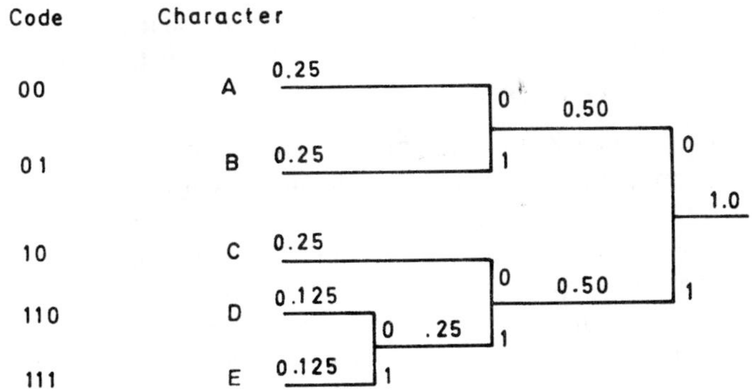

Figure 7.20 Huffman code set construction

Original data	ABADABD
Compressed data	00|01|00|110|00|01|110

Figure 7.21 Huffman coding

Figure 7.21 illustrates the compression of character data into a binary string based upon the utilization of the previously developed Huffman code set. Note that the original data consisting of the characters ABADABD would contain 56 bits if encoded with eight bits per character. In comparison, through the utilization of Huffman coding only 16 bits are required to encode the data.

LZW coding

Two professors at the Technion in Haifa, Israel's equivalent of the American Massachusetts Institute of Technology, can be considered as the founders of modern string-based statistical compression methods that were incorporated into a variety of hardware products ranging from modems to DSUs and remote bridges. The efforts of Professors Ziv and Lempel occurred during 1977 and 1978. While mostly theoretical, their papers resulted in the development of what are now referred to as LZ77 and LZ78 string compression algorithms.

The theoretical nature of the papers co-authored by Lempel and Ziv left the details of the implementation and operation of their dictionary-based string compression techniques to others. One person who made changes to the basic LZ77 and LZ78 algorithms was the Englishman Terry Welch. In fact, in recognition of his contribution, the resulting dictionary-based string compression technique he worked on during 1984 is referred to as LZW.

The LZW algorithm initially considers the character set as 256 individual string table entries whose codes range from 0 to 255. Then, the algorithm operates upon the next character in the input string of characters as follows:

1. If the character is in the table, get the next character.
2. If the character is not in the table, output the last known string's encoding and add the new string to the table.

Under the LZW algorithm, characters from the input data source are read and used to progressively form larger and larger strings until a string is formed that is not in the dictionary. When this occurs, the last known string's encoding is output and the new string is added to the table. The beauty about this technique is that the compressor and expander know that the initial string table consists of the 256 characters in the character set. Since the algorithm uses a numeric code as a token to indicate the position of a string in the dictionary's string table, this technique minimizes the token length as the dictionary begins to fill.

To illustrate the simplicity of the operation of the LZW algorithm, let's view an example of its use. Let's call the string previously read the prefix of the output string. Similarly, the last byte read becomes the suffix, where:

$$\text{prefix} + \text{suffix} = \text{new string}$$

Once a new string is formed, the suffix becomes the prefix (prefix = suffix).

Initially, each character in the character set is assigned a code value equivalent to its character code. Thus, in ASCII 'a' would have the code value 97, 'b' would have the code value 98, and so on. Let's assume the input string is 'ababc...'.

The first operation assumes that the prefix has a value of the null string which we will indicate as the symbol 'Δ'. Thus, the first operation results in the addition of the null string to 'a' which forms the new string 'a'. Since 'a' is in the dictionary we do not output anything. However, since the suffix becomes the prefix, 'a' now becomes the prefix for the next operation. This is illustrated in the top two lines in Figure 7.22.

Prefix	Suffix	New String	Output
Δ	a	a	–
a	b	ab	97
b	a	ba	98
a	b	ab	–
ab	c	abc	256
c	Δ	c	99

Figure 7.22 Encoding the string 'ababc'

Processing the next character in the input string (b) results in the addition of the prefix (a) and the suffix (b) to form the new string (ab). Since this represents a new string that is not in the dictionary, we follow rule 2. That is, we output the last known string's encoding which is 97 for the character 'a' and then add the new string to the dictionary or string table. Concerning the latter, since codes 0 through 255 represent the individual characters in an eight-bit character set, we can use the code 256 to represent the string 'ab'. However, doing so obviously requires the token to be extended. Most LZW implementations use between nine and 14 bits to represent a token whose length corresponds to the size of the dictionary. Next, 'b', which was the suffix when generating the string 'ab', becomes the prefix for the next operation as indicated in line 3 in Figure 7.22. Since the next character in the string to be encoded is 'a', it functions as the suffix in creating the new string 'ba'. As that string is not presently in the string table, the last known string 'b' is output using its ASCII code value of 98. Next, the string 'ba' is added to the string table using the next available code which is 257.

Benefits of compression

Table 7.2 lists some of the primary benefits that may result from the use of data compression performing devices. Since few of these benefits are self-explanatory, we will examine each in detail.

Table 7.2 Potential benefits of data compression

Reduces probability of transmission errors
Distorts clarity of transmitted data
Can reduce or eliminate RBT shifts
Provides ability to restructure a network
Provides capability to reduce costs

Transmitting fewer data equates to lowering the duration of a transmission session. Although we normally associate some cost per unit of transmission time leading to a reduction in communications costs when compressing data, there are other benefits that may result from compression. First, if we transmit fewer data while the probability of a transmitted bit being in error remains constant the overall probability of a transmission error occurring will decrease. Thus, data compression will reduce the probability of transmission errors occurring.

By converting text that is in a conventional code (such as standard ASCII) into a different code, compression algorithms may offer some security against illicit monitoring. As previously indicated in Figure 7.21, the original data stream 'ABADABD' was converted into the compressed binary string '0001001100001110.' Here, the 16-bit compressed string might be interpreted as two 8-bit characters by a person illicitly monitoring the transmission. Thus, compression can be considered as a technique that reduces the clarity of transmitted data to observers of the data traffic.

In many remote batch processing environments, data compression has been successfully used to reduce or eliminate workshifts. Typically, the primary function of second- and third-shift remote batch terminal (RBT) operations is the retrieval of output from deferred batch processing jobs run on a central computer in the evening. Since the limiting factors for this type of shift operation are the processing power of the central computer and the data transfer rate on the communications link, increasing the data transfer rate through compression may result in the reduction or elimination of an RBT shift.

One of the most promising features of data compression is its ability to provide a mechanism to alter the structure of a communications network. How such a change might occur can be seen from a brief examination of a typical network application.

Assume an existing multidrop line services eight terminal locations. Perhaps an expansion to add two additional terminal locations to the network is planned. Assume that the network planner estimated the activity on the multidrop line based upon its expansion to ten drops and determined that due to the additional data traffic the response time of the terminals would be unacceptable. As a result of this analysis, a separate line consisting of two drops might be installed. If data compression is implemented there would be less total traffic, reducing the time it takes a computer to poll and service each terminal on a multidrop line. Thus, by implementing data compression one might be able to service the two additional locations by extending the multidrop line without increasing the response time of the terminals.

Using compression performing devices

Often, you can obtain the benefits of data compression while avoiding the efforts required to analyze actual or potential data traffic and develop software to perform compression. This can be accomplished by leasing or purchasing data compression performing devices that are specifically designed to be used in a particular networking environment. We will thus examine several hardware products to obtain a better understanding of the use of compression performing devices. The products covered in this section were selected for illustrative purposes only and should not be construed as an endorsement of the device.

Compression DSUs

Codex Corporation, a subsidiary of Motorola Corporation, markets a compression-performing Data Service Unit (DSU) designed to enhance the transmission capability of 56 kbps digital lines. The Codex Compression DSU provides a variable compression ratio which depends upon the susceptibility of data to the compression algorithms built into the DSU. Under normal operations users can reasonably expect to achieve at least an average compression ratio of 2:1, resulting in an average throughput of 112 kbps over a 56 kbps transmission facility.

Figure 7.23 illustrates the use of a pair of compression-performing CSUs to boost router-to-router transmission over a 56 kbps digital transmission facility. Note that the serial ports of the routers are configured to transfer data to the DSU at 256 kbps, and the DSU will use flow control to temporarily terminate transmission from the routers when its buffers fill to the point where additional router transmission could result in the overflow of a buffer and loss of data. Once the DSU's buffer is emptied to a predefined level of occupancy, it will then disable flow control, allowing the attached router to resume transmission.

Through the use of compression-performing DSUs you can avoid the expense associated with obtaining a fractional T1 line instead of a 56 kbps line. With a retail

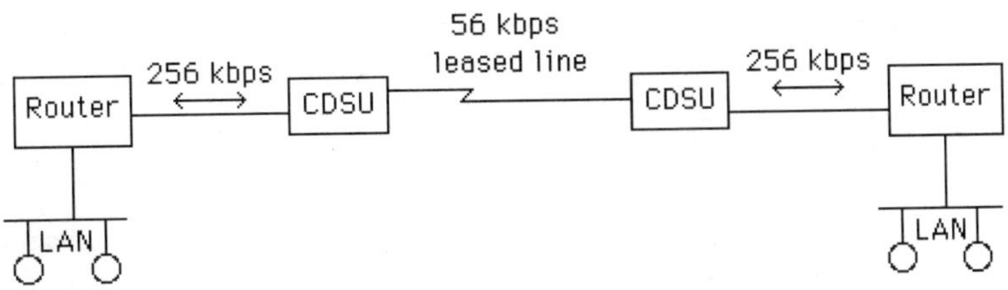

Figure 7.23 Using compression-performing DSUs to enhance router-to-router communications

cost under $1000 per unit, the ability to eliminate the need for a more expensive transmission facility allows a compression DSU to pay for itself in a short period of time.

Multifunctional compression

One of the more popular uses of compression performing devices is to reduce the quantity of data to be multiplexed, increasing the servicing capacity of the multiplexer. A natural evolution of the development of compression devices was to include both statistical multiplexing and data compression in one hardware device. One product representing this multifunctional capability is the Racal-Vadic Scotsman III.

Scotsman III

The Racal-Vadic Scotsman III is a stand-alone compression device that can be configured in a variety of ways to include the addition of an optional four-channel multiplexer. Figure 7.24 should be compared to the employment of a statistical multiplexer. Since some statistical multiplexers do not perform data compression, significant time delays would occur in attempting to multiplex four 9600 bps bisynchronous data sources onto one 9600 bps transmission line. In other situations, statistical multiplexers may service bisynchronous data through the use of bandpass channels. When this occurs, only asynchronous data is actually statistically multiplexed, while bisynchronous data is time division multiplexed onto a predefined time slot on the composite multiplexed link as illustrated in Figure 7.25. When bisynchronous data is multiplexed via a bandpass channel, the data rate of the bisynchronous input data reduces the portion of the composite channel's bandwidth available for multiplexing other data sources. As an example, one bisynchronous data source operating at 4800 bps would reduce the bandwidth of the composite channel for multiplexing other data sources by 4800 bps when the bisynchronous data is multiplexed via a bandpass channel. Thus, if the composite channel operates at 9600 bps, only 4800 bps would be available for the statistical multiplexing of other data sources.

Due to the limitations of servicing bisynchronous data with bandpass channels, most statistical multiplexers using this technique are limited to supporting only one

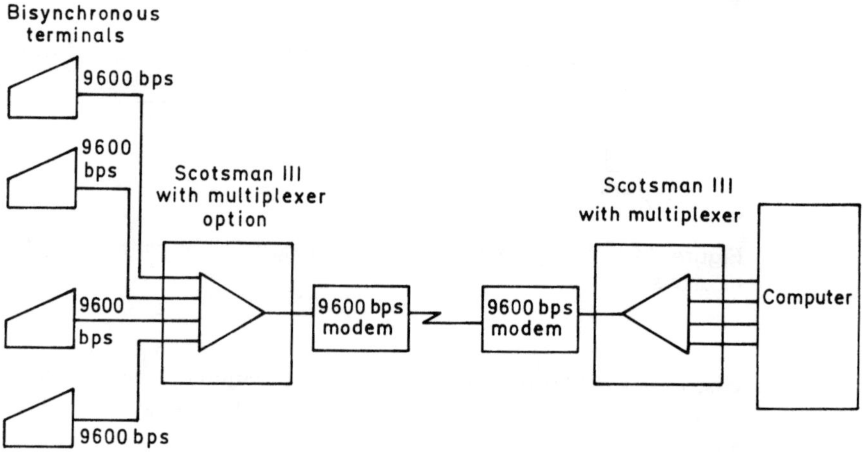

Figure 7.24 Using the Racal–Vadic Scotsman III as a multiplexer

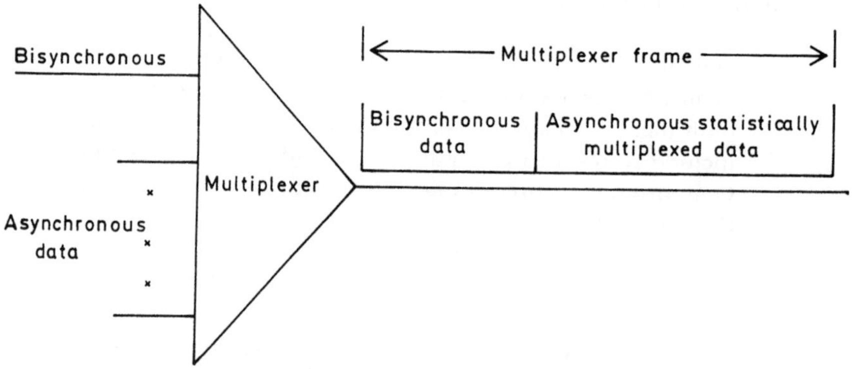

Figure 7.25 Bandpass multiplexing

bisynchronous data source. Thus, the Scotsman III can be effectively employed for situations where several high-speed bisynchronous data sources transmit to a common location.

7.3 FIBER OPTIC TRANSMISSION SYSTEMS

While the majority of attention focused upon fiber optic transmission relates to its use by communication carriers, this technology can be directly applicable to many corporate networks. Due to the properties of light transmission, fiber optic systems are well suited for many specialized applications to include the high-speed transmission of data between terminals and a computer or between computers located in the same building.

In this section we will first focus our attention upon the components of a fiber optic transmission system. After examining the advantages and limitations associated with such systems we will use the previous information presented as a foundation to analyze economics associated with cabling terminals to a computer, illustrating how this technology can be directly applicable to many computer installations. In doing so, we will examine the use of fiber optic modems and multiplexers to denote the advantages and limitations associated with their use in a communications network.

Fiber optics is a rapidly evolving technology that has moved from a laboratory and consumer-product curiosity into a viable mechanism for low-cost, high data rate communications. Permitting the transmission of information by light, fiber optics is most familiar to individuals by its incorporation into contemporary desk lamps that were first marketed in the 1970s. In such lamps, bundles of individual fibers were attached to a common light source at the base of the lamp's neck. Radiating outward in a geometric pattern, the fibers were shaped to form a contemporary design. By polishing the end of each fiber, the common light source was reflected at the tip of each fiber, resulting in a very impressive visual display. Since the early 1970s a significant amount of fiber research has resulted in such fibers becoming available for practical data communications applications.

System components

Similar to conventional transmission systems the major components of a fiber optic system include a transmitter, transmission medium and receiver. The transmitter employed in a fiber optic system is an electrical to optical (E/O) converter. The E/O converter receives electronic signals and converts those signals into a series of light pulses. The transmission medium is an optical fiber cable which can be constructed out of plastic or glass material. The receiver used in fiber optic systems is an optical to electric (O/E) converter. The O/E converter changes the received light signals into their equivalent electrical signals. The relationship of the three major fiber optic system components is shown in Figure 7.26.

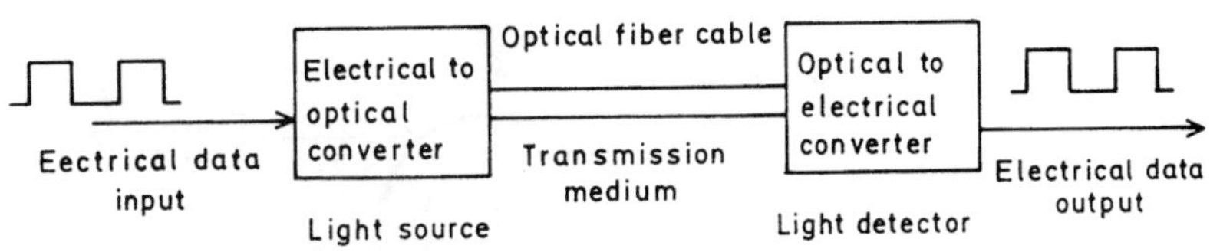

Figure 7.26 Fiber optic system component relationship. The electrical to optical converter produces a light source which is transmitted over the optical fiber cable. At the opposite end of the cable an optical to electrical converter changes the light signal back into an electrical signal. (Reprinted with permission from *Data Communications Management*, © 1986 Auerbach Publishers, New York)

The light source

Currently two types of light sources dominate the electronic to optical conversion market: the light emitting diode (LED) and the laser diode (LD). Although both devices provide a mechanism for the conversion of electrically encoded information into light encoded information, their utilization criteria depends upon many factors. These factors include response times, temperature sensitivity, power levels, system life, expected failure rate, and cost.

A LED is a PN junction semiconductor that emits light when biased in the forward direction. Typically a current between 25 to 100 mA is switched through the diode, with the wavelength of the emission a function of the material used in doping the diode.

The laser diode offers users a fast response time in converting electrically encoded information into light encoded information, thereby, they can be used for very high data transfer rate applications. The laser diode can couple a high level of optical power into an optical fiber, resulting in a greater transmission distance than obtainable with light emitting diodes. Although they can transmit further at higher data rates than LEDs, laser diodes are more susceptible to temperature changes and their complex circuitry makes such devices more costly.

Optical cables

Many types of optical cables are marketed, ranging from simple one-fiber cables to complex 18-fiber, commonly jacketed, cables. In addition, a large variety of specially constructed cables can be obtained on a manufactured-to-order basis from vendors.

In its most common form an optical fiber cable consists of a core area, cladding, and a protective coating. This is illustrated in the upper part of Figure 7.27. As a light beam travels through the core material the ratio of its speed in the core to the speed of light in free space is defined as the refractive index of the core.

A physical transmission property of light is that while traveling in a medium of a certain refractive index, if it should strike another material of a lower refractive index, the light will bend towards the material containing the higher index. Since

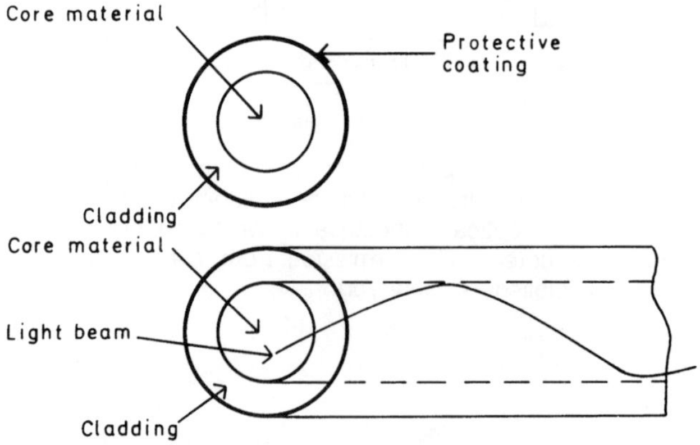

Figure 7.27 Optical fiber cable. (Reprinted with permission from *Data Communications Management*, © Auerbach Publishers, New York)

the core material of an optical fiber has a higher index of refraction than the cladding material, this index differential causes the transmitted light signal to reflect off the core-cladding junctions and propagate through the core. This is shown in the lower part of Figure 7.27.

The core material of fiber cables can be constructed with either plastic or glass. While plastic is more durable to bending, glass provides a lower attenuation of the transmitted signal. In addition, glass has a greater bandwidth, permitting higher data transfer rates when that material is used for fiber construction.

The capacity of a fiber optic data link of a given distance depends upon the numerical aperture (NA) of the cable as well as the core size, attenuation and pulse dispersion characteristics of the fiber. The NA value can be computed from the core and cladding refractive indices as follows:

$$NA = (m_1 - m_2)^{1/2}$$

where m is the core material refractive index, and m_2 the cladding material refractive index.

The numerical aperture value indicates the potential efficiency in the coupling of the light source to the fiber cable. Together with core material diameter, the numerical aperture indicates the level of optical power that can be transmitted into a fibre.

A function of both the fiber material and core/cladding imperfections, attenuation determines how much power can reach the far end of an unspliced link. When the numerical aperture, core diameter and attenuation are considered together one can determine the probable transmission power loss ratio.

Pulse dispersion is a measure of the widening of a light pulse as it travels down an optical fiber. Dispersion is a function of the cable's refractive index. Fibers with an appropriate refractive index permit an identifiable signal to reach the light detector at the far end of the cable.

Types of fibers

Two types of optical fibers are available for use in cables: step index and graded index.

In a step index fiber an abrupt refractive index change exists between the core material and the cladding. In a graded index fiber the refractive index varies from the center of the core to the core–cladding junction. The gradual variation of the refractive index in this type of fiber serves to minimize the optical signal dispersion as light travels along the fiber core. The minimization of dispersion results from the light rays near the core–cladding junction traveling faster than those near the center of the core. The minimization of signal dispersion permits greater bandwidth, allowing transmission to occur over greater distances at higher data rates than available with a step index fiber. Graded index fibers are thus commonly used for long haul, wide bandwidth communications applications. Conversely, step index fibers are used for communications applications requiring a narrow bandwidth and relatively low data rate requirement.

Common cable types

Five common types of optical cables are illustrated in Figure 7.28. Note that when a one-fiber cable has a transmitter connected to one cable end and a receiver to the opposite end, it functions as a simplex transmission medium. As such, transmission can only occur in one direction. The two-fiber cable can be considered a duplex transmission medium since each fiber permits transmission to occur in one direction. When transmission occurs on both fibers at the same time a full-duplex transmission sequence results. Thus, the 18-fiber cable is capable of providing nine duplex transmission paths.

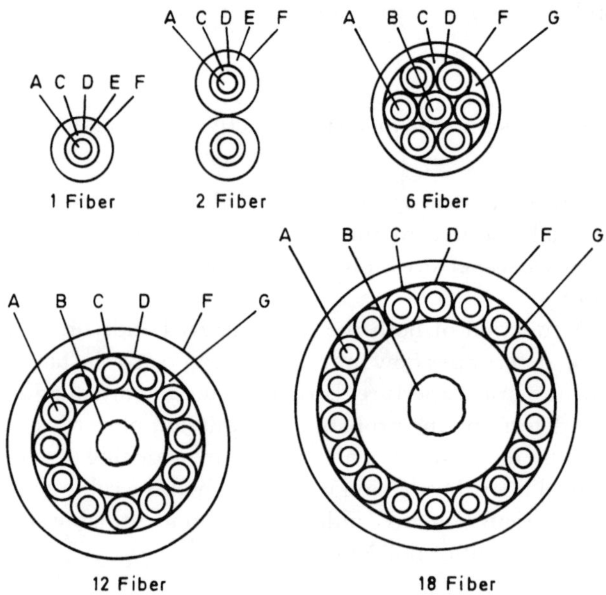

Figure 7.28 Common cable types. (Reprinted with permission from *Data Communications Management*, © 1986 Auerbach Publishers, New York.) (A Optical fiber, B Jacketed Kevlar strength member, C Engineering plastic tubes, D Plastic separator tape, E Braided Kevlar strength member, F PVC jacket, G Rip cord)

The light detector

To convert the received light signal back into a corresponding electrical signal a photodetector and associated electronics are required. Photodetection devices currently available include a PIN photodiode, an avalanche photodiode, a phototransistor and a photomultiplier. Due to their efficiency, cost, and light signal reception capabilities at red and near infrared (IR) wavelengths, PIN and avalanche photodiodes are most commonly employed as light detectors.

In comparison to PIN detectors, avalanche photodiode detectors offer greater receiver sensitivity. This increased light sensitivity results from their high signal-to-noise ratio, especially at high bit rates.

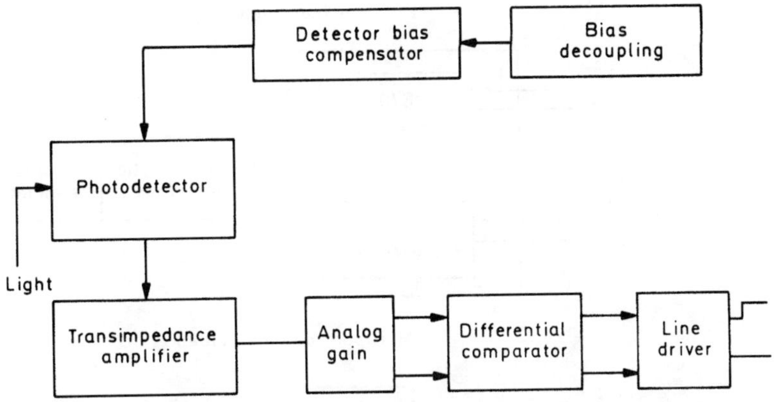

Figure 7.29 Light detector component. To filter dc input voltage, protect the photodiode and reduce the effect of electromagnetic interference, a light detector module has bias decoupling. Since an avalanche photodiode is a temperature sensitive device, a detector-bias compensator is used to compensate for temperature changes that could affect the diode.

The photodetector converts the received optical signal into a low-level electrical signal. This detector can be either an avalanche or PIN diode, depending upon the optical sensitivity requirements. The transimpedence amplifier is a low noise, current-to-voltage converter while the analog gain element increases the voltage gain from the preceding amplifier to the level required for the decision circuitry. The differential comparator converts the analog signal into digital form by interpreting signals below a certain threshold as a '0' and above that threshold as a '1'. The line driver regenerates and drives the squared signal from the comparator for transmission over metallic cable. (Reprinted with permission from *Data Communications Management*, © 1986 Auerbach Publishers, New York)

Since avalanche photodiode detectors require an auxiliary power supply which introduces noise, circuitry to limit such noise results in the device having a higher overall cost than a PIN photodiode. In addition, they are temperature sensitive and their installation environment requires careful examination. A block diagram of the major components of a light detector is given in Figure 7.29.

Other optical devices

Besides the optical devices previously discussed, two additional devices warrant coverage. The first device is an optical modem which performs 'optical' modulation and demodulation. The second device is an optical multiplexer which permits the transmission of many data sources over a single optical fiber.

Optical modem

An optical modem is a device housing both an optical transmitter and an optical detector as shown in the top portion of Figure 7.30. Similar to conventional modem development, a variety of optical modems have evolved. These variations range from single-channel stand-alone devices to multiport optical modems, the latter

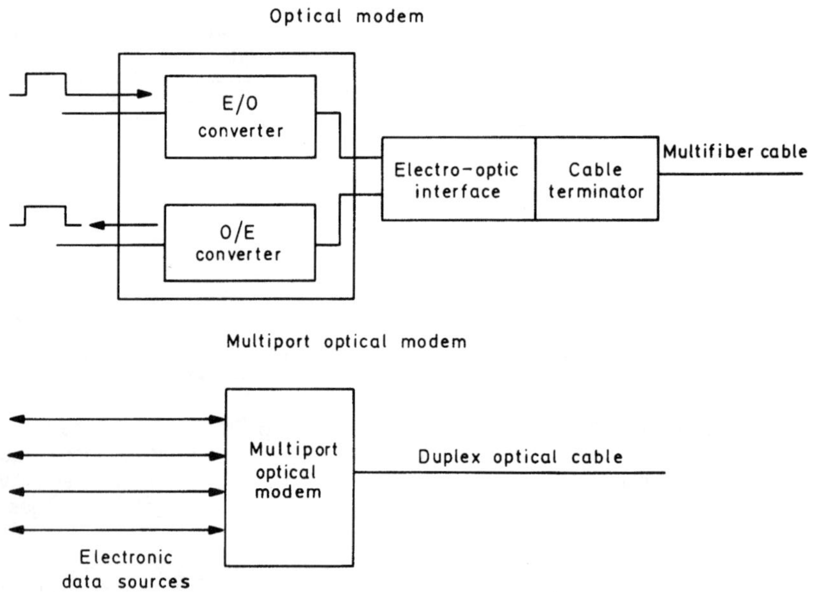

Figure 7.30 Optical modem. An optical modem can transmit and receive data over one multifiber cable, converting the electronic data source to light and the received light back into its corresponding electronic signal. Serving as a synchronous multiplexer, the multiport optical modem transmits data from up to four electronic sources as one optical signal. (Reprinted with permission from *Data Communcations Management,* © 1986 Auerbach Publishers, New York)

capable of functioning as a synchronous multiplexer and optical modem. The multiport optical modem permits conventional electronic bit streams from up to four data sources to be multiplexed and converted into one stream of light pulses for transmission on one optical fiber. In the lower part of Figure 7.30 a multiport optical modem servicing four data sources is shown.

Optical multiplexer

Functioning similar to a conventional time division multiplexer, an optical multiplexer accepts the RS-232 electrical input of many data sources and multiplexes such signals for transmission over a single optical fiber. Included in the optical modem are a light source generator and detector. Thus, such devices are equivalent to a multiplexer with a built-in modem since they convert electronic multiplexed data into an optical signal and detect and convert light signals to their equivalent electronic signals prior to demultiplexing.

Transmission advantages

When used for data transmission, fiber optic cables offer many advantages over cables with metallic electrical conductors. These advantages result from several

Table 7.3 Fiber optic system advantages

Large bandwidth	Mixed voice, video, and data on one line
No electromagnetic interference (EMI)	No specially shielded conduits required Cable routing simplified Bit error rate improved
Low attenuation	Permits extended cable distances
No shock, hazard, or short circuits	Can be used in dangerous atmosphers Common ground eliminated
High security	Transmission TEMPEST acceptable Tapping noticeable
Lightweight and small-size cable	Facilitates installation
Cable rugged and durable	Long cable life

distinct properties of the optical cables. The more common advantages associated with the utilization of a fiber optic transmission system are listed in Table 7.3.

Bandwidth

One of the advantages of fiber optic cable in comparison to metallic conductors is the wide bandwidth of optical fibers. With potential information capacity directly proportional to transmission frequency, light transmission on fiber cable provides a transmission potential for very high data rates. Currently, data rates of up to 10^{14} bps have been achieved on fiber optic links. When compared to the 9.6 to 19.2 kbps limitation of telephone wire pairs, fiber cable makes possible the merging of voice, video, and data transmission on one conductor. In addition, the wide bandwidth of optical fiber provides an opportunity for the multiplexing of many channels of lower speed, but which are still significantly higher data rate channels than are transmittable on telephone systems.

Electromagnetic non-susceptibility

Since optical energy is not affected by electromagnetic radiation, optical fiber cables can operate in a noise electrical environment. This means special conduits formerly required to shield metallic cables from radio interference produced by such sources as electronic motors and relays are not necessary.

Similarly, cable routing is easier since the rerouting necessary for metallic cables around fluorescent ballasts does not cause concern when routing fiber cables. Due to its electromagnetic interference immunity, fiber optic transmission systems can be expected to have a lower bit error rate than corresponding metallic cable systems. By not generating electromagnetic radiation, fiber optic cables do not generate crosstalk. This property permits multiple fibers to be routed in one common cable, simplifying the system design process.

Signal attenuation

The signal attenuation of optical fibers are relatively independent of frequency. In comparison, the signal attenuation of metallic cables increases with frequency. The lack of signal loss at frequencies up to 1 GHz permits fiber optic systems to be expanded as equipment is moved to new locations. In comparison, conventional metallic cable systems may require the insertion of line drivers or other equipment to regenerate signals at various locations along the cable.

Electrical hazard

On fiber optic systems light energy in place of electrical voltage or current is used for the transfer of information. The light energy alleviates the potential of a shock hazard or short circuit condition. The absence of a potential spark makes fiber optic transmission particularly well suited for such potentially dangerous industrial environment uses as petrochemical operations as well as refineries, chemical plants, and even grain elevators. A more practical benefit of optical fibers for most corporate networks is the complete electrical isolation they afford between the transmitter and receiver. This results in the elimination of a common ground which is a requirement of metallic conductors. In addition, since no electrical energy is transmitted over the fiber, most building codes permit this type of cable to be installed without running the cable through a conduit. This can result in considerable savings when compared to the cost of installing a conduit required for conventional cables, whose cost can exceed $2500 for a 300-foot metal pipe.

Security

Concerning security, the absence of radiated signals makes the optical fiber transmission TEMPEST acceptable. In comparison, metallic cables must often be shielded to obtain an acceptable TEMPEST level. Although fibers can be tapped like metallic cable, doing so would produce a light signal loss. Such a loss could be used to indicate to users a potential fiber tap condition.

Weight and size

Optical fibers are smaller and lighter than metallic cables of the same transmission capacity. As an example of the significant differences that can occur, consider an optical cable of 144 fibers with a capacity to carry approximately 100 000 telephone conversations. The cable would be approximately one inch in diameter and weigh about 6 ounces per foot. In comparison, the equivalent capacity copper coaxial cable would be about three inches in diameter and weigh about 10 pounds per foot. Thus, fiber optic cables are normally easier to install than their equivalent metallic conductor cables.

Durability

Although commonly perceived as being weak, glass fibers have the same tensile strength as steel wire of the same diameter. In addition, cables containing optical fibers are reinforced with both a strenghtening member inside and a protective jacket placed around the outside of the cable. This permits optical cables to be pulled through openings in walls, floors, and the like without fear of damage to the cable.

With better corrosion resistance than that of copper wire, transmission loss at splice locations has a low probability of occurring when optical fiber cable is used.

Limitations of use

As discussed, optical fiber cables offer many distinct advantages over conventional metallic cables. Unfortunately, they also have some distinct usage disadvantages. Two of the main limiting factors of fiber optic systems are cable splicing and system cost.

Cable splicing

When cable lengths of extended distances become necessary, optical fibers must be spliced together. To permit the transmission of a maximum amount of light between spliced fibers, precision alignment of each fiber end is required. This alignment is time consuming and depending upon the method used to splice the fibers, your installation cost can rapidly escalate.

Fibers may be spliced by welding, gluing, or through the utilization of mechanical connectors. All three methods result in some degree of signal loss between spliced fibers.

Welding or the fusion of fibers results in the least loss of transmission between splice elements. Due to the time required to clean each fiber end and then align and fuse the ends with an electric arc does not make splicing easily suitable for field operations. An epoxy, or gluing method of splicing requires the utilization of a bonding material that matches the refractive index of the core of the glass fiber. This method typically results in a higher loss than obtained with the welding process. While mechanical connectors have the highest data loss among the three methods, it is by far the easiest method to employ. Although mechanical connectors considerably reduce splicing time requirements, the cost of good quality connectors can be relatively expensive. Currently, typical connectors cost approximately $10.

System cost

Good quality, low-loss single-channel fiber optic cable costs between $1.50 and $2 per meter. A typical fiber optic modem having a transmission range of 1 km costs between $200 and $600. In comparison, conventional metallic cable for synchronous data transmission costs approximately 30 cents per foot while a line driver capable of regenerating digital pulses at data rates up to 19.2 kbps cost approximately $200.

Based upon the preceding, the cost of a limited-distance fiber optic system at data rates up to 19.2 kbps will generally exceed an equivalent conventional metallic based, transmission system. Only when high data rates are required or transmission distances expand beyond the capability of line drivers and to some extent limited distance modems are fiber optic systems economically viable.

Utilization economics

One of the most commonly used duplex fiber optic cables cost $2.50 per meter which is equivalent to 76 cents per foot. An optical modem containing an electric to optic and an optic to electric converter capable of transferring data at rates up to 10 Mbps costs approximately $600 per unit. The system cost of a pair of optical modems as a function of distance is illustrated in Figure 7.31. Note that as long as no splicing is required, costs are a linear function of distance.

Suppose a requirement materializes which calls for the communications linkage of four high-speed digital devices located in one building of an industrial complex to a computer center located in a different building 10 000 feet away. What fiber

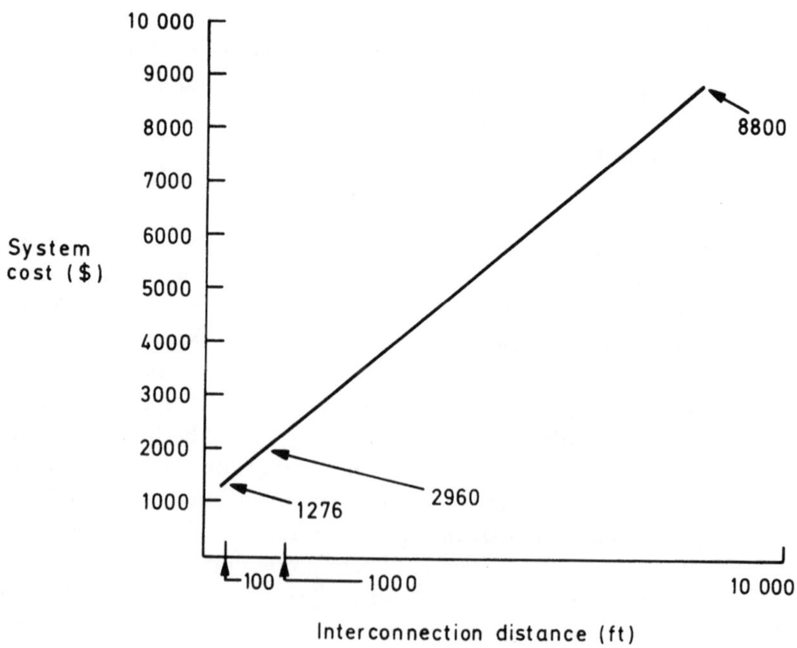

Figure 7.31 System cost varies with distance. Cost of a typical duplex optical fiber cable and a pair of optical modems capable of transmitting up to 10 Mbps. (Reprinted with permission from *Data Communications Management*, © 1986 Auerbach Publishers, New York)

optic systems can support this requirement and what are the economic ramifications of each configuration?

Dedicated cable system

Equivalent to individually connecting devices on metallic cables, four individual fiber optic systems and four cables could be installed. Here eight optical modems would be required, resulting in the cost of the modems being $4800. With four cables being required, 40 000 feet of cable would cost $30 400, resulting in a total cost of $35 200 for this type of network configuration. In addition, a substantial amount of personnel time may be required to install four individual 10 000 foot cables.

Multichannel cable

A second method that can be employed to link multiple devices at one location is by the employment of multichannel cable. We can examine the economics associated with multichannel cable by considering the cost of an eight-channel cable.

A typical eight-channel cable capable of supporting four duplex transmissions costs approximately $10 per meter or $3.05 per foot. On a duplex channel basis, this represents a cost of 76 cents per foot per channel. This is the same cost as an individual duplex cable. The use of most multichannel cable does not offer any appreciable savings over individual cable until ten or more channels are packaged together. Prior to excluding the use of a multichannel cable, one should carefully consider cable installation costs since the time required to install one multichannel cable can be significantly less than the time required to install individual cables.

Optical multiplexers

Similar to conventional metallic cables, the potential installation of parallel optical cables indicates that multiplexing should be considered. Prior to deciding upon the use of an optical multiport modem or optical multiplexer, one should examine the type of data to be transmitted as well as the data transfer rates required. If each data type is synchronous and no more than four data sources exist, the utilization of an optical multiport modem can be considered. If a mixture of asynchronous and synchronous data must be supported or, if more than four data sources exist, one should consider an optical multiplexer.

Currently, four-channel synchronous optical multiport modems cost approximately $1000 per unit. Returning to our system requirement, a pair of optical multiport modems and one cable could support the four data terminals. Here the total system cost would be reduced to $9600, of which $2000 would be for the pair of optical multiport modems and the remainder for the cable.

Suppose one or more of the data terminals was an asynchronous device, or more than four terminals required communications support. For such situations an

Table 7.4 Network configuration comparisons of four data sources and 10 000-foot transmission distance

Individual cable	Multichannel cable	Optical multiport moderm	Optical multiplexer
System cost			
$35 200	$35 200	$9600	$14 400
Expansion capability			
Extra cable and transmitter/receiver per data source	Requires more expensive cable and additional transmitters/receivers	Cannot support more than four channels	No cost to add support for up to four additional channels

optical multiplexer should be considered. One optical multiplexer currently marketed costs $3300 and supports up to eight data channels at data rates up to 64 kbps. Since an optical receiver/transmitter is included in the multiplexer, only the cost of one duplex cable must be added to the cost of a pair of multiplexers. Doing so, we will obtain the cost of an optical transmission system capable of supporting up to eight data sources at data rates up to 64 kbps. For such a system the cost of 10 000 feet of cable and the two optical multiplexers would be $14 400.

Table 7.4 provides a comparison of the four previously discussed network configurations. As indicated, both cost and expansion capacity varies widely between configurations. Although fiber optic systems represent new technology, the financial aspects and expansion capability must be considered similar to the process involved with conventional metallic cable based transmission systems.

7.4 SECURITY DEVICES

As networks expand and proliferate, more and more people have access to them, and it becomes harder to guard against unauthorized access to a network and entry into data files on computers which normally may be restricted to only a few users. A door that is locked and patrolled is, of course, some measure of security against sabotage or theft; but it offers little protection against the wily, white-collar evil-doer for whom entry is no serious problem. Hence, in the absence of adequate operational controls, someone may be billed for someone else's transactions; or, more seriously, a company with an account on a computer transmitting confidential bid information could have its message intercepted and read by a competitor.

To reduce the possibilities of these events occurring, a number of devices were introduced by manufacturers to encrypt both data files and transmitted data. Designed for use on wide area networks, such products were followed by security devices and software modules developed to protect LAN resources from both internal and external users. Concerning the latter, the growth in the use of the Internet resulted in the exposure of corporate networks connected to the Internet to a virtually unlimited number of hackers and other unscrupulous persons. This in turn resulted in the development of router access lists and firewalls as mechanisms to control access to local area networks. Prior to discussing how these devices operate and where they can be employed, a review of sign-on and database security features will be undertaken to provide you with an insight into the problems associated with unauthorized access as well as some of the methods that can reduce the probability of such access with computer software.

Password shortcomings

The most frequently employed methods of preventing unauthorized access to networks and databases, that of identification codes, has a number of shortcomings. The code can be glimpsed over a person's shoulder, found on a discarded printout in a trash basket, or on the ribbon of a printer terminal. A remedy, then, is to use characters that are non-printable. (The term 'character' in this context refers to any distinct electrical signal initiated at a terminal and does not necessarily imply a visible figure in the usual sense of the word.) This is the so-called password approach. It is a precaution that is fairly simple, being essentially an extension of the use of terminal identification numbers.

Whether printed or non-printed, however, careful consideration is required in putting together the elements of an adequate access control system. These include immunity to access by repeated random tries and the ability of the network to report repeated attempts at access.

Practically any element in a network can be the target of unauthorized access, but the most likely point of entry is a remote terminal device because it is out of sight of the office where a target computer will be under attack. The sign-on procedure, involving a code that must be keyed in to the terminal to gain clearance, is the standard method of assuring access. Figure 7.32(a) depicts the normal response of a computer to a sign-on request. If an illegal sign-on is attempted, the response might be as shown in Figure 7.32(b).

PLEASE SIGN-ON
ID $U_C U_C U_C V_C V_C V_C V_C$
GENERAL TIME SHARING COMPANY
ON AT DATE, TIME

(a)

PLEASE SIGN-ON
ID 127 ABCD
ILLEGAL SIGN-ON THRESHOLD EXCEEDED
CONTACT OPERATOR
OFF AT XX/YY

(b)

Figure 7.32 Identification messages. The format of the message requesting the user's identification and password (underlined) is similar to that shown in part(a). U_c is the user's identification number, while V_c represents a verification code, typically non-printable control characters. Notification that the maximum number of retries has been exceeded is shown in part (b). Operator intervention may follow

For example, in a time-sharing system, the procedure starts with the user establishing the telephone connection and identifying the type of terminal (code and transmission speed), usually by sending a carriage return. Typically, after getting a go-ahead from the computer, the user then transmits his or her own

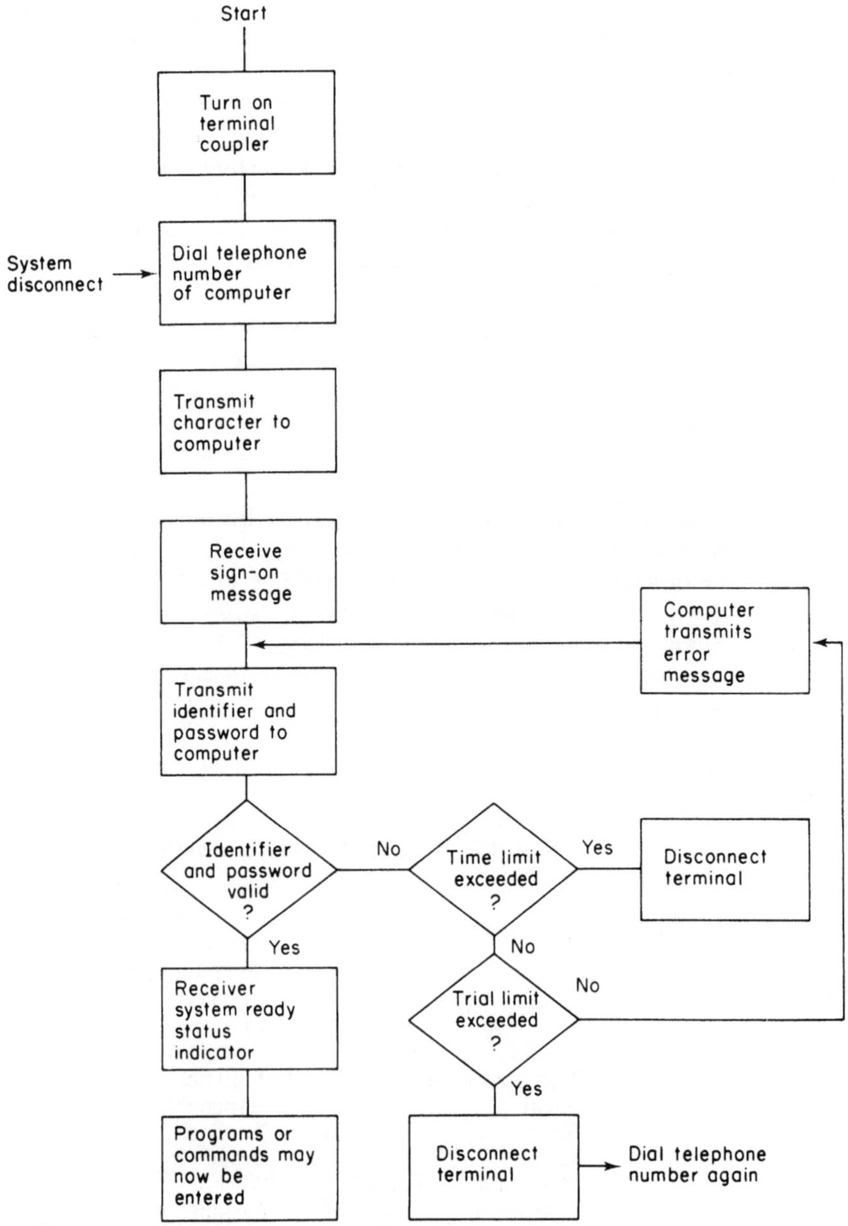

Figure 7.33 Sign-on procedure. In a typical sign-on procedure the response to a user attempting to gain access is a request for identification and password

personal identification number. If the number is invalid, as might happen if a wrong key is accidentally struck, the user receives a message, such as 'illegal sign-on'. The computer then allows the user a fixed amount of time, usually several seconds, to send another sign-on. If the time is up or the user exceeds a certain number of allowable retries, the computer automatically disconnects. In this case the user must redial and try again.

This procedure is susceptible to sophisticated attack, however, unless additionally protected; but more on that later. A flowchart outlining a typical sign-on procedure is shown in Figure 7.33.

Password combinations

A password approach incorporated into the sign-on procedure following the identification number has the obvious advantage of being less liable to illicit discovery. However, there are limitations. The most obvious limit is the number of characters on the terminal keyboard. For instance, a common terminal, the Teletype Corporation Model 33-ASR, limits the user to a subset of the American Standard Code for Information Interchange (ASCII). And not all control characters on a keyboard can be used in the password. Some systems reserve certain characters for terminal function control, system control, and communications control. Examples are a carriage return to indicate the end of a line and a backspace to cause deletion of the last character entered. A list of frequently unavailable non-printable characters is given in Table 7.5. For some terminals there may be few characters left. In such a case, a long password is needed to provide enough combinations. Since the list is extensive, it can easily be seen that not many characters in an average system may really be available for a password.

Experience has shown, however, that a three- to six-character password is the optimum length to provide enough characters in most systems to take care of the job. The optimum length, incidentally, is taken to mean one that does not defy an average user's ability to remember the password.

Table 7.5 Frequently unavailable characters

Character meaning	Possible reversed usage
Break	System interrupt
End of transmission	Communications
Bell	Terminal
Line feed	Terminal
Form feed	Terminal
Carriage return	Terminal, system
X-On	Terminal
Tape	Terminal
Cancel	System or communications
Space (field separation)	System
Delete line	System
Delete character	System
Rub-out	System or communications

If the code is three positions wide and there are two possible characters per position (A and B), the combinations are as follows:

$$\left\{\begin{matrix} AAA \\ AAB \\ ABA \\ ABB \\ BAA \\ BAB \\ BBA \\ BBB \end{matrix}\right\} \text{8 combinations} + \left\{\begin{matrix} AA \\ AB \\ BA \\ BB \end{matrix}\right\} \text{4 combinations} + \left\{\begin{matrix} A \\ B \end{matrix}\right\} \text{2 combinations} = \text{14 combinations}$$

Figure 7.34 Code combinations. By permitting the password to be smaller than the maximum number of chosen positions, extra combinations are possible. A and B denote non-printable characters, making up to 14 combinations

It may appear at first that the number of combinations possible in a password of a given length would simply be C, where C is the size of the available character repertoire and i the number of character positions to be filled. Actually, additional combinations are possible by using fewer than the maximum number of positions that one has decided upon. This is accomplished by typing a carriage return immediately following the shortened password to indicate that it is terminated.

Consider, for example, a three-position password in which each position can take either of two non-printable characters, which are designated A and B in Figure 7.34. By using all three positions there are eight possible combinations. But by using two positions and leaving the third position blank, four more combinations are possible. If only one position is filled, two more still are possible, for a total of 14 possibilities.

In this routine, the user's personal identification number, which is usually keyed in at the start of the request for access, tells the computer what to expect. Identification numbers and passwords must correlate. (From a practical standpoint, however, one- or two-position passwords might be inadvisable because it might make it easier for an intruder to come upon the correct password by simple random selection.)

In any event, the total number of realistic combinations, T, can be calculated as

$$T = \sum_{i=1}^{i=w} C^i$$

where i represents individual positions (first, second, third, etc.), w is the total number of positions in the password, and C is the number of characters available. For example, a four-position password with a character set of two provides 30 combinations. Increasing the character set to eight raises the combinations to 4680.

Illicit access

A potent tool available to the illicit network user is another computer. Here the human malefactor harnesses the machine, which need be little more than a microcomputer and intelligent modem, to repeatedly try a password, and retry if the password does not work.

If, for example, such a system were used to access a computer that permits five sign-on attempts before automatically disconnecting and the terminal transmits at 30 characters per second, then the five tries could be performed in less than 10 s. Upon sensing a disconnect, the microcomputer could redial and try five more code combinations. Over a three-day holiday weekend, with little traffic and the central computer unattended, this scheme could attempt over 40 000 passwords with nobody being the wiser. If instead of dial-in access, a hacker uses an Internet connection into a corporate network that operates at a much higher data rate, the ability to try different password combinations can dramatically increase. In fact, some of the more famous, or perhaps more infamous, hackers use electronic dictionaries as a mechanism to try hundreds of thousands of possible passwords over a long holiday weekend in an attempt to gain access to protected computers.

To counter this brute force approach, provision should be made to monitor and record the repetition of sign-on attempts. Once a predetermined number of tries has been made for a given identification number, the system could be programmed to inhibit additional sign-on efforts until an operator intervenes. The maximum number of tries can be established daily or at more frequent intervals. The threshold for a given identification number could be 10 attempts in a given hour or 20 attempts a day, depending on the individuality of the system. When the threshold is exceeded, the user is locked out of the network and receives a message requesting that he or she contact the system operator for return of service.

Automatic intervention does not eliminate the possible success of a concerted effort to gain access by numerous retries, but it does sharply reduce the probability of success. The only way in which access might be obtained is to try the maximum number of sign-on attempts short of intervention, disconnecting for the balance of the timing interval, and trying again. If this were to succeed at all, it could take months or years in a system with a small threshold, a long password, and a long time interval. Another technique that can be used to prevent the possible success of dictionary attacks is to create passwords using alphanumeric combinations. While CAT, DOG, and SCARF are in the dictionary, and a good hacker might add some custom programming to try 'catdog' and other combinations, passwords such as CAT4DOG and DOG4CAT will probably escape their method of attack. In addition, the lengthening of a password can have a significant effect upon brute force discovery methods.

As shown in Figure 7.35, with a three-position password drawing on 15 available characters, the probability of successful entry (a probability of 1) rises sharply if there is no intervention and is assured by the 3615th trial. With intervention after 20 repeated tries, however, the probability does not exceed 0.005. The diagram also shows that if the character repertoire is reduced to 10 and there is no intervention mechanism, the intruder is assured of access by the 1110th try, two-thirds sooner than with a 15-character repertoire. Another factor that can have a major bearing on access security is the number of characters reserved to perform communications functions. While a large number of such characters are standardized, installations that remove characters from the character set available for passwords run the risk of increasing the access vulnerability of the system.

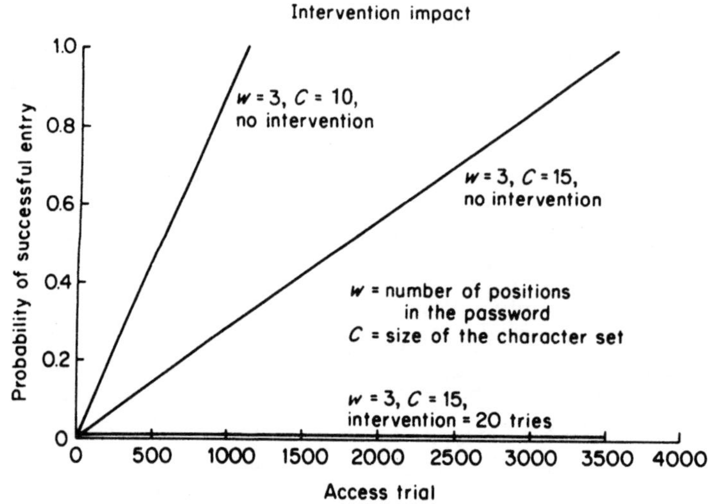

Figure 7.35 Intervention impact. With intervention software, the probability of successful entry by random password selection becomes extremely small

While the preceding discussion was primarily concerned with sign-on access, similar problems and solutions can be effected with respect to the passwords employed to protect data files.

Transmission security

Although a number of methods have been developed to promote access and database security further; unless the transmission medium utilized is secure, you can become vulnerable to having data transmission compromised by such means as line tapping or line monitoring. In addition, if transmission is over packet or message switching systems, or by courier or telegram, the message may be read by unauthorized personnel or obtained through active penetration by insiders as well as outsiders. In certain US Department of Defense installations, transmission is over secure transmission lines which makes access security their major concern. This is due to the fact that, although all transmission lines on the installation are secure, personnel with different security level clearances have access to the same computer and a method of differentiating who obtains what computer resources is primarily determined by a terminal's location and the identification code of the user.

For the situation where data transmission occurs over public, non-secure facilities, a method of making the transmitted data unintelligible to unauthorized parties becomes as important as having a good access security procedure. Fortunately, numerous techniques are available for users to make transmitted data unintelligible to unauthorised parties; the oldest and most widely used method is the various types of manual coding processes where, through the use of code books and pads, the original text is encoded before transmission and then decoded by the recipient of the message. The foundations for the various coding schemes go back thousands of years and have been used to protect a wide range of messages ranging from diplomatic and military communications from before the time of the Roman Empire through commercial messages of industrial companies today.

Manual techniques

In spite of the fact that most manual coding techniques can be broken by trained cryptanalysts, they do offer a measure of protection because of the time element: there may elapse a period of considerable length before the message is decoded by an unauthorized party. Thus, the information that company A will bid 2 million dollars on a contract whose bid is to be opened on 1 February is worthless to a competitor that decodes the message from the home office to company A's field agent after that date. In addition, manual coding schemes can also be used as a backup in the event of the failure of a coding device that you may have installed.

Caesar Cipher

One of the earliest coding techniques was the so-called Caesar Cipher, which was probably known and used long before Caesar was born. Using this technique, Julius Caesar enciphered his dispatches by displacing each letter by a fixed amount. If the displacement was two, then the message INVADE ENGLAND TONIGHT would be transformed into KPXCFG GPINCPF VQPKIJV and sent by messenger to the appropriate recipient. Upon receipt of the message, a reverse transformation would develop the plain text of the message. Although this scheme may appear primitive, some encryption devices use electronic circuits or software to perform a continuous and alternating displacement of the plain text to develop an encrypted message which could frustrate the best cryptanalyst.

Checkerboard technique

A group of encryption techniques were developed based upon what is known as the basic checkerboard. Here, the alphabet and numerals are written into a six by six block with coordinates for row and column used to specify the cells. This technique can involve the use of different indices as well as a rearrangement of cell data, as shown in Table 7.6. Again, with the advent of modern electronics and the use of

Table 7.6 Basic checkerboard

	Block 1							Block 2					
	A	E	I	O	U	V		G	H	I	J	K	L
A	A	B	C	D	E	F	A	Z	Y	X	W	V	0
E	G	H	I	J	K	L	B	U	T	S	R	Q	1
I	M	N	O	P	Q	R	C	P	O	N	M	L	2
O	S	T	U	V	W	X	D	K	J	I	H	G	3
U	Y	Z	0	1	2	3	E	F	E	D	C	B	4
V	4	5	6	7	8	9	F	A	9	8	7	6	5

Using the above blocks, the word CODE becomes: from Block 1: AI II AO AU; from Block 2: EJ CH EI EH microprocessors it is easy to construct a device to continuously change their indices after a certain period of transmission or to change the cells, or both.

A variation of the checkerboard technique is accomplished through the utilization of a keyword which is commonly referred to as a Vigenere cipher technique. In its simplest form the Vigenere cipher consists of a table of alphabets, as shown in Figure 7.36. For ease of remembrance, a meaningful phrase or mnemonic

Plain text	INVADE ENGLAND
Key	SECRET SECRETS
Cipher text	ARXRHXWRICEGV

Plain text

	A	B	C	D	E	F	G	H	I	J	K	L	M	N	O	P	Q	R	S	T	U	V	W	X	Y	Z
A	A	B	C	D	E	F	G	H	I	J	K	L	M	N	O	P	Q	R	S	T	U	V	W	X	Y	Z
B	B	C	D	E	F	G	H	I	J	K	L	M	N	O	P	Q	R	S	T	U	V	W	X	Y	Z	A
→ C	C	D	E	F	G	H	I	J	K	L	M	N	O	P	Q	R	S	T	U	V	W	X	Y	Z	A	B
D	D	E	F	G	H	I	J	K	L	M	N	O	P	Q	R	S	T	U	V	W	X	Y	Z	A	B	C
→ E	E	F	G	H	I	J	K	L	M	N	O	P	Q	R	S	T	U	V	W	X	Y	Z	A	B	C	D
F	F	G	H	I	J	K	L	M	N	O	P	Q	R	S	T	U	V	W	X	Y	Z	A	B	C	D	E
G	G	H	I	J	K	L	M	N	O	P	Q	R	S	T	U	V	W	X	Y	Z	A	B	C	D	E	F
H	H	I	J	K	L	M	N	O	P	Q	R	S	T	U	V	W	X	Y	Z	A	B	C	D	E	F	G
I	I	J	K	L	M	N	O	P	Q	R	S	T	U	V	W	X	Y	Z	A	B	C	D	E	F	G	H
J	J	K	L	M	N	O	P	Q	R	S	T	U	V	W	X	Y	Z	A	B	C	D	E	F	G	H	I
K	K	L	M	N	O	P	Q	R	S	T	U	V	W	X	Y	Z	A	B	C	D	E	F	G	H	I	J
L	L	M	N	O	P	Q	R	S	T	U	V	W	X	Y	Z	A	B	C	D	E	F	G	H	I	J	K
M	M	N	O	P	Q	R	S	T	U	V	W	X	Y	Z	A	B	C	D	E	F	G	H	I	J	K	L
N	N	O	P	Q	R	S	T	U	V	W	X	Y	Z	A	B	C	D	E	F	G	H	I	J	K	L	M
O	O	P	Q	R	S	T	U	V	W	X	Y	Z	A	B	C	D	E	F	G	H	I	J	K	L	M	N
P	P	Q	R	S	T	U	V	W	X	Y	Z	A	B	C	D	E	F	G	H	I	J	K	L	M	N	O
Q	Q	R	S	T	U	V	W	X	Y	Z	A	B	C	D	E	F	G	H	I	J	K	L	M	N	O	P
→ R	R	S	T	U	V	W	X	Y	Z	A	B	C	D	E	F	G	H	I	J	K	L	M	N	O	P	Q
→ S	S	T	U	V	W	X	Y	Z	A	B	C	D	E	F	G	H	I	J	K	L	M	N	O	P	Q	R
→ T	T	U	V	W	X	Y	Z	A	B	C	D	E	F	G	H	I	J	K	L	M	N	O	P	Q	R	S
U	U	V	W	X	Y	Z	A	B	C	D	E	F	G	H	I	J	K	L	M	N	O	P	Q	R	S	T
V	V	W	X	Y	Z	A	B	C	D	E	F	G	H	I	J	K	L	M	N	O	P	Q	R	S	T	U
W	W	X	Y	Z	A	B	C	D	E	F	G	H	I	J	K	L	M	N	O	P	Q	R	S	T	U	V
X	X	Y	Z	A	B	C	D	E	F	G	H	I	J	K	L	M	N	O	P	Q	R	S	T	U	V	W
Y	Y	Z	A	B	C	D	E	F	G	H	I	J	K	L	M	N	O	P	Q	R	S	T	U	V	W	X
Z	Z	A	B	C	D	E	F	G	H	I	J	K	L	M	N	O	P	Q	R	S	T	U	V	W	X	Y

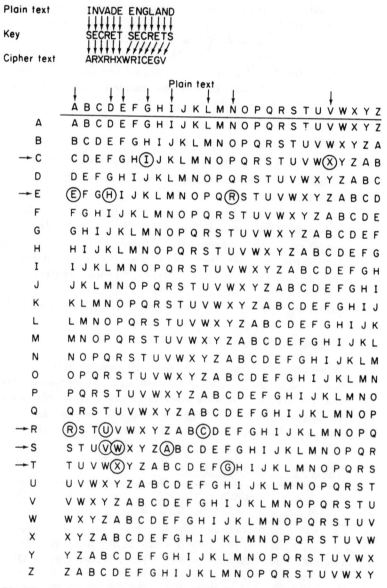

Figure 7.36 Vigenere cipher. A one-to-one correspondence between the plain text and the characters of a key is used to develop the cipher text. Here, the intersection of the plain text character I with the key character S produces a cipher text character A

is selected as the key, although this can be a major weakness, and incoherent keys which reduce the number of clues are preferable for use.

Here, encipherment begins by establishing a one-to-one correspondence between the characters of the plain text and the characters of the keys, with the key partially or completely repeated if shorter than the plain text. The cipher character is then obtained from the intersection of the appropriate keyletter row and the plain text column. For example, if SECRET SECRETS is used as the key to encipher the plain text message INVADE ENGLAND, the table provides a cipher text of ARXRMX WRICEGV. As shown, the cipher text is developed character by character, with the first character of the cipher text obtained from the intersection of the S character of the key SECRET with the first character (I) of the plain text, and so on. Since 676 memory locations are required for 26 characters of the alphabet or 1296 locations for the alphabet and digits, some devices encode data by using a fixed memory but generating pseudorandom numbers which are used to develop a pseudorandom key.

Automated techniques

Concurrent with the development of electromechanical devices, several methods of encoding information were developed. In 1917, Gilbert Vernam of the American Telephone and Telegraph Company (AT&T) developed a method for ensuring that the information contained in a punched paper tape would remain unintelligible to unauthorized users. In Vernam's technique, each text letter was enciphered with its own cipher letter. If the key tape was as long as the message, and its key perfectly random, the text was then theoretically unbreakable. The inconvenience of preparing thousands of feet for high-volume traffic as well as the security problems inherent in guarding tape supplies and accounting for both active and cancelled tape rolls prevented most users from considering this technique.

A practical compromise between convenience and security was the development of pseudorandom events which appear to be as unpredictable as those generated by white noise, sunspots, and other physical phenomena, but in reality are developed from a reproducible mathematical relationship. An example of this would be a program to manipulate two 18-bit registers where the product of each register's contents through a predetermined process is extracted into two numbers which return to the registers where the process is repeated again. Thus, the 36 bits of the two registers could produce a period length of $2^{36}-1$ before returning to a repeatable pattern. Owing to the continuing shrinkage of the size as well as a reduction in the costs of integrated circuits, pseudorandom keying devices became available to the commercial user at realistic prices. The key lengths of some of these devices were so long that the communications equipment whose transmission security they safeguarded could become obsolete before the end of the first key period.

In addition to a number of firms which manufacture encoding devices solely for use by intelligence agencies, the US Armed Forces cryptologic agencies, and other government users, several industrial firms actively entered the market and manufacture a family of security devices for commercial users. These manufacturers provide a family of devices designed to operate with a wide variety of terminal devices to include facsimile devices. Commercially available encryption devices utilize a built-in key generator which is similar to a multilevel register device. Using both standardized and proprietary encryption techniques, a user selects the code family by either turning appropriate rotary switches inside the device to a specific setting, entering the code through a built-in keypad, or entering a setting via a PC or terminal connected to the encrypter. Depending upon manufacturer, up to 32 trillion key settings or codes are available to the user. One device that offers 32 trillion code settings has those settings arranged as 16 million code families, with 2 million codes in each family. Using this arrangement the user's security officer can develop various code administration techniques in hierarchical arrangements.

Security devices available for the commercial market operate either off-line or on-line. The off-line devices preceded the development of the on-line devices and were primarily used with punched paper tape transmission over Telex and TWX networks. Off-line systems are now relegated to the preparation of enciphered message tapes prior to transmission over teleprinter circuits with the device connected to an auxiliary off-line teleprinter. This provides paper tape and keyboard/printer functions with the operator typing the clear message into the keyboard of the teleprinter and the encryption device simultaneously punching the encoded tape.

Some encryption devices also permit a two-pass operation to be used where a clear tape is punched by the operator and then read by the device at high speed to punch the encoded tape. The encoded tape is then transmitted by the subscriber over the network to the recipient as any punched paper tape normally would be transmitted. At the receiving end the punched paper tape of the receiving terminal is decoded off-line by reading it into the encryption device at that end and having the plain text printed at the connected teleprinter which is now turned to the off-line mode of operation. This is shown in Figure 7.37. Since the header of the message contains routing and destination information, it obviously cannot be encoded. The off-line device thus has either switches or buttons which allow the user to start and stop the encoding of information at the appropriate points within the message.

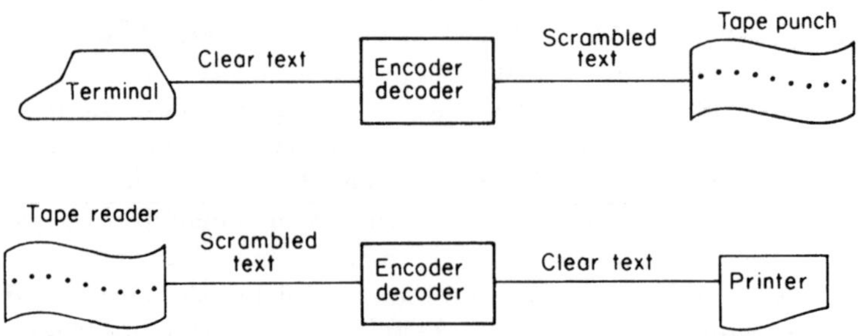

Figure 7.37 Off-line encoding (top) and decoding (bottom)

Modern developments

Advances in electronics to include large-scale integrated circuits and microprocessors produced a technology base for the development of a family of data security devices. These devices normally employ the National Bureau of Standards (NBS), renamed the National Institute of Standards and Technology (NIST), data encryption standard algorithm which provides a set of rules for performing the encryption and decryption of data which reduces the threat of code-breaking by illegal personnel to virtually an insignificant possibility. This is due to the fact that if the data is intercepted, the time required to decipher the encoded information would require many years of machine time, probably resulting in the information being of no value by the time it is decoded.

DES algorithm

The DES algorithm operates on 64 bits of plaintext to produce 64 bits of encrypted text. The actual encryption is based upon the use of a 56-bit binary number which serves as an encryption key.

DES involves 16 rounds of operation known as transformations that mix the data and key together in a prescribed manner. Each transformation defines how sequences of plaintext are changed in sequences of random bits.

New encryption devices use a feedback-shift register and associated circuits or a microprocessor to generate a pseudorandom bit sequence following the NBS algorithm. To make the transmitted data appear to be a random stream of ones and zeros, modulo 2 addition is employed to add the data to an apparently orderless bit stream developed by a pseudorandom number generator.

Modulo, or modulus, is a capacity or unit of measurement, and a modulo sum is a sum with respect to a modulus while the carry is ignored. When a two-digit decimal counter is used, it is a modulo 100 counter. Such a counter cannot distinguish the numbers 99 and 199. When 9 is added to 8, the sum is 17, but assuming that the modulus is 10 in this case, the modulo sum is 7. In modulo 2 addition, one and zero and zero and one result in one, and both one and one and zero and zero make zero. Modulo 2 addition and subtraction are illustrated in Table 7.7.

Table 7.7 Modulo 2 addition and subtraction

Modulo 2 addition				Modulo 2 subtraction			
0	0	1	1	0	0	1	1
+0	+1	+0	+1	−0	−1	−0	−1
–	–	–	–	–	–	–	–
0	1	1	0	0	1	1	0

When the bit sequence or key text generated by the pseudorandom number generator is added by modulo 2 addition to the original information or clear text, the result is an apparently random stream of ones and zeros referred to as the cipher

Table 7.8 Modulo 2 addition keeps data secure

Encoding	
Source (clear) text	1 0 1 1 1 1 0 1 1 1 1 0 0 0 1
Pseudorandom (key) generator	1 0 0 0 1 0 1 0 1 1 0 0 1 0 0
Modulo 2 addition (encoded) text	0 0 1 1 0 1 1 1 0 0 1 0 1 0 1
Decoding	
Encoding text	0 0 1 1 0 1 1 1 0 0 1 0 1 0 1
Pseudorandom (key) generator	1 0 0 0 1 0 1 0 1 1 0 0 1 0 0
Modulo 2 subtraction (clear) text	1 0 1 1 1 1 0 1 1 1 1 0 0 0 1

or encoded text. This encoded data is then transmitted to its destination where another encoder/decoder using an identical key text performs modulo 2 subtraction on the encoded data to develop the original clear text, as illustrated in Table 7.8.

The operation illustrated in Table 7.8 represents a simplistic example of the implementation of the DES algorithm. As previously mentioned, the DES algorithm involves a 16-transformation process in which shift registers are used to perform modulo 2 addition between plaintext and an encryption key. Thus, although the modulo 2 addition and subtraction process is critical to encrypting and decoding data, the 16-state transformation process is equally critical.

Another method employed to effect data security which eliminates the requirement of hardware devices is through the utilization of software packages that can be installed on many computers and which utilize the NBS algorithm. Some software packages can be utilized with most programming languages, including such languages as Fortran and Cobol through the use of call statements, and result in a data file being encrypted. Other packages and designed to work with specific applications, such as a database or electronic mail program. Once encrypted, the file can be transmitted to another location where another software package is available to enable the user at that location to decode the file. While this method eliminates hardware, users remotely located from the computer still face the threat of having their data intercepted if they are building a data file, since the file is not protected until it is stored and encrypted.

A problem associated with using software to encode and decode data according to the NBS data encryption algorithm is the overhead associated with the required processing. To alleviate this overhead most methods of implementing the algorithm have been through the utilization of stand-alone devices incorporating microprocessors to encode and decode data.

Public versus private keys

Encryption products based on the use of the DES algorithm are referred to as a private key system. That is, each recipient using an encryption device must have the same key as the originator to successfully decrypt a message. For this reason the key used by two or more persons must be kept private.

One of the problems associated with private key encryption is the distribution of keys. When transmitted they can be intercepted, and the logistics of maintaining separate keys for every user who requires communications with another user can quickly become a logistics nightmare. Recognizing these problems resulted in the development of a new category of encryption referred to as public key encryption.

Under public key encryption a mathematical relationship using prime numbers enables the creation of a public key that can be distributed to the public. That key can be used to encrypt data; however, due to the mathematical relationship between public and private keys only the holder of the private key can correctly decrypt a file encrypted with the public key linked to the private key. The use of public key encryption provides a far greater flexibility than private key encryption and is commonly used in World Wide Web browsers for electronic commerce.

On-line applications

With the commercial development of automatic cipher synchronization techniques, continuous on-line encryption devices are no longer the exclusive preserve of government agencies and the military and intelligence communities. These on-line encryption devices are capable of operating with both asynchronous and synchronous data streams at various data rates. In the on-line asynchronous mode of operation an encryption device uses the start–stop pulses of the individual characters to develop the synchronization between the sending and the receiving units.

Plain text, which can be typed on the terminal's keyboard or read from a storage device, is automatically encrypted on a character-by-character basis; the encrypted data is transmitted via the communications channel to the receiving unit where it is automatically deciphered and furnished to the receiving terminal. For synchronous transmission, encryption devices are automatically synchronized in time by a short character sequence typed by the user or by the Sync characters or Flag in the protocol, or through the depression of an initiate button on the front panel of one of the devices. Once synchronization has occurred, the two units step under the control of crystal-controlled clocks, which keep the units in synchronization. Figure 7.38 shows typical on-line encryption device applications.

As shown in Figure 7.38, on-line encryption for terminal-to-terminal transmission over common carrier facilities can be accomplished by the installation of an

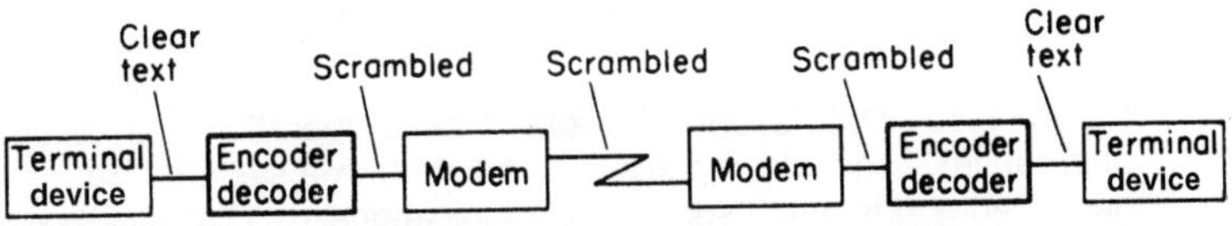

Figure 7.38 On-line encryption device applications. Terminal-to-terminal transmission via message-switching systems such as telex or TWX can be secured by interfacing security devices between the modem and each terminal device

encoder/decoder between the terminal and the terminal's associated modem or DSU. Clear text originating at the terminal device is then scrambled by the encoder and passed to the modem which transmits the data over the common carrier's facility. At the receiving end, scrambled data passes through the modem to a decoder which now produces clear text that is passed to the terminal at this end.

In Figure 7.39(a), a typical mainframe access application for an encryption device is shown. In this example, the terminal device which required the security measures associated with encryption devices is connected via a leased line to a time division multiplexer (TDM). Since this is a dedicated port connection, there is no problem

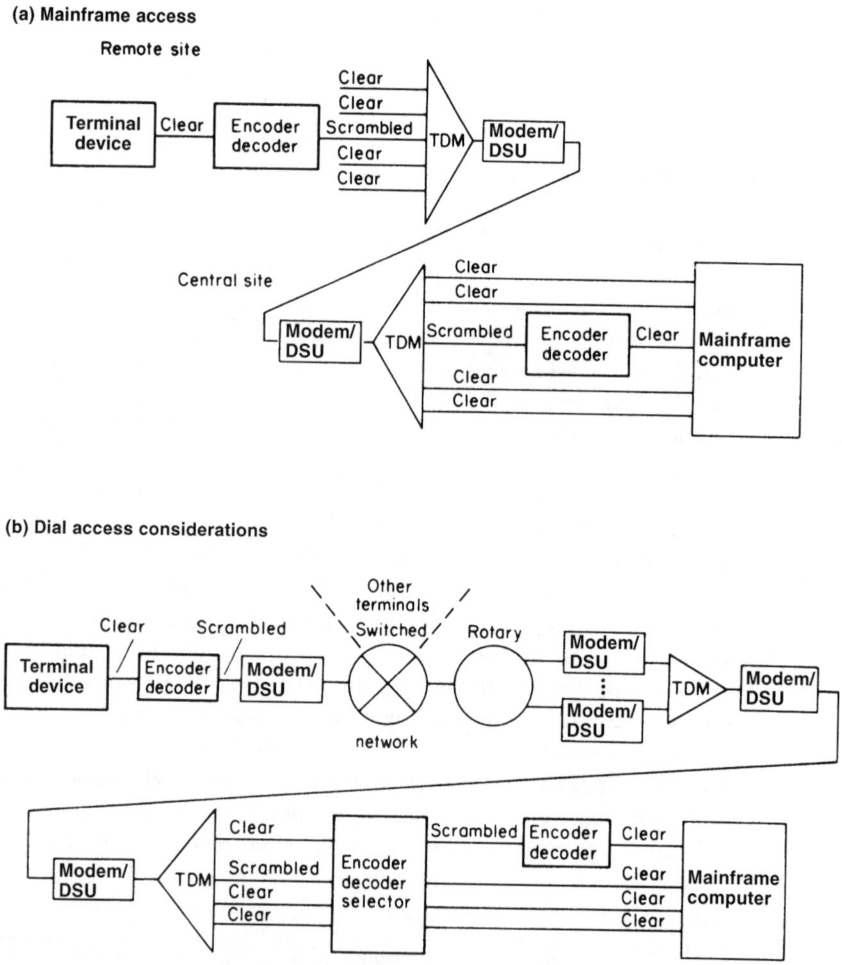

Figure 7.39 Data security for dial access operations. (a) Mainframe access encryption application. Terminals connected to a Mainframe system through multiplexers must have a dedicated port to effect secure communications. (b) Dial access considerations. To transmit encoded information via the switched network through a multiplexer, a selector must be employed at the computer site to switch the scrambled output of a TDM channel into a decoder to produce clear text

in determining what port of the demultiplexer to which a similar encoder/decoder should be interfaced.

In Figure 7.39(b), the terminal user now dials a rotary over the switched telephone network in order to connect to the multiplexer. For this configuration, the terminal device may be connected to any port on the multiplexer, which means that at the other end a method is necessary to determine which port of the demultiplexer has the scrambled data. The encoder/decoder selector performs this function by sampling the output of each of the demultiplexer's ports and switching the port that has the scrambled data to the encoder/decoder, which then decodes the scrambled text and passes it to the computer. This selector in effect performs limited switching and acts as a transparent device, passing the data from the other ports directly to the computer.

Before selecting an encryption device you should examine in detail the application requirements and develop a set of specifications which will aid in the selection process. Table 7.9 contains a checklist of some common encryption device specifications which should be considered. Although some devices are code transparent, many are not. Since some devices are manufactured for specific applications and their character sets to include the crypto and plain text character sets should be examined.

Table 7.9 Encryption device specification

Character set supported	Message key
Transmission rate supported	Key code change
Operational mode	Power supply
Crypto character set	Terminal interface
Plain text character set	Alarm circuitry
Carrier compatibility	

Similarly, the transmission speed supported should also be examined since some devices only support teletype operating speeds while other devices can operate at speeds up to 1 million bps.

The operational mode of encryption devices falls into three categories: off-line, on-line asynchronous and on-line synchronous, with the mode supported dependent upon your requirements and the capability of the device.

The crypto character set and plain text character set can vary from the Telex character set through the extended binary-coded decimal interchange code (EBCDIC) character set of 256 characters. The character set selected will depend upon the terminal character set, the computer's character set, and/or the character set supported by the communications facility used. Carrier compatibility is an important feature when the user wishes to transmit encrypted messages over a commercial network. Since certain control characters perform unique functions, carrier compatibility alleviates problems by the suppression of the generation of control characters during the encoding process as well as by sending the plain text control characters in the clear.

As mentioned previously, the message key and key code change determine the number of coding variations as well as the total codes available. Like most devices used in data communications, encryption devices have a wide range of interface options which the user must properly select from to match the terminal's requirements. To prevent transmission from being compromised if the device should fail, some encryption devices have built-in alarm circuitry which monitors the key generator output and will inhibit data transmission upon detection of encoding failure. Now that we have an appreciation for WAN security devices, let's turn our attention to LAN security.

LAN security

One of the major goals of any organization that connects an internal private network to a public network is to do so in a manner that minimizes the possibility of security threats to the internal network. Although we would obviously prefer to eliminate all risk, the complexity of hardware and software may result in loopholes hackers may be able to exploit. Thus, we should assume that there will always be some risk associated with the connection of a private internal network to a public network and consider the use of hardware and software to minimize that risk.

In this section we will examine the operation and utilization of two networking devices that can be used to minimize the threat associated with connecting a private local area network to a public network, such as the Internet. The first device we will examine is the router, which provides the ability to block data transfers based upon network and transport layer information. Although this level of protection may be sufficient for many networks, it precludes the ability to stop a variety of network attacks, as well as ignores such key security items as authentication and encryption. Thus, we will then turn our attention to the use of firewalls and their operational capability in the form of proxy services that can be used to plug the holes associated with reliance upon routers for network security.

Routers

Routers operate at layer 3 of the ISO Reference Model. This means that they can read network layer information. Recognizing that routers provide the basic mechanism by which private and public networks are interconnected, and also recognizing the need of organizations to control access to their internal network, router manufacturers added a packet filtering capability to their products. This capability, frequently implemented in the form of a list of access permissions, is also commonly referred to as a router's access list and is the focus of this section.

Access lists

An access list represents a sequential collection of permit and deny conditions that are applied to network addresses based upon a particular protocol. Since the most common internetworking protocol is TCP/IP, we will focus our attention upon router access lists designed to predefine the data flow between networks connected by routers transferring TCP/IP packets. Thus, the use of router access lists can be used to govern the data flow between segments on an internal corporate private local area network as well as between a private and a public network. Readers should note that although our examination of access lists will be focused upon the TCP/IP protocol, it is also applicable, if supported by the router manufacturer, to other network layer protocols to include NetWare IPX, AppleTalk, and DecNET.

The IP header contains both a source address and a destination address field. Under IPv4 those fields are 32 bits in length, while under the evolving IPv6 they are 128 bits in length. For both versions of IP the source address identifies the originator while the destination address identifies the recipient. Although IP represents the routable network layer of the TCP/IP protocol stack, end-to-end transmission is a layer 4 responsibility. At layer 4 the TCP/IP protocol suite supports two transport protocols, the Transport Control Protocol (TCP) and the User Datagram Protocol (UDP). Although TCP and UDP headers differ, both contain 16-bit source and destination ports that identify the port number of the sending and receiving process, respectively. Since each IP header is followed by either a TCP or a UDP header, a router can control the data flow of TCP/IP packets by operating upon up to four distinct fields that define the sender, recipient, and process. This method of control is commonly referred to as packet filtering and is controlled by the creation of an appropriate access list. Thus, let's turn our attention to the configuration of several access lists to meet different organizational requirements.

Configuring an access list

Although there are considerable differences between routers with respect to their use of access lists, most devices operate very similarly with respect to two key operations. First, they commonly compare packet fields sequentially against the contents of the access list. This means that the first match determines if a router will accept or reject a packet. Secondly, routers have a default of either permit or deny. This means that unless an access list has entries to override the default, the default condition will be applied to all non-listed situations. Since the best way to become familiar with access lists is by example, let's do so. In doing so we will use a generic command format for illustrative purposes that may or may not be applicable to specific routers.

For our first example we will assume that a router's access list is applicable to incoming packets received on a WAN connection and that the router using the list has only one LAN port. Thus, the access list does not include a port specification. Figure 7.40 illustrates an example of the network configuration associated with the router. In this example we will use an access list to control access to computers on the internal Ethernet LAN shown connected to the Internet via a router.

Based upon the preceding let's assume the format of the access list is as follows:

access-list {permit | deny} source address, port

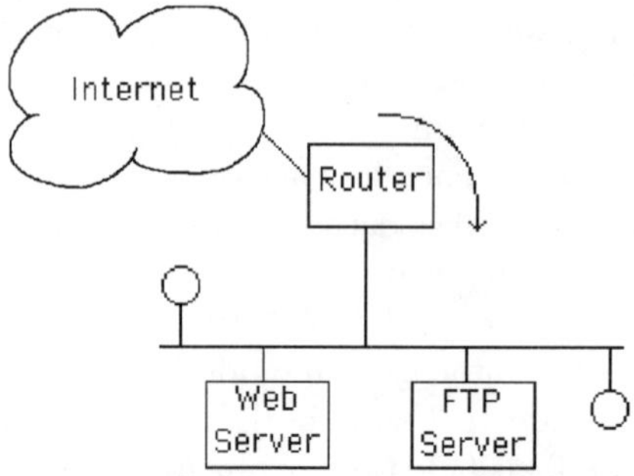

Figure 7.40 Using a router access list to control packet flow from the Internet onto a private Ethernet LAN. By programming access lists based upon source and destination address and well-known port, you can control access to computers on your internal private network

Table 7.10 Well-known TCP/UDP ports

Port number	Use
20	FTP (data)
21	FTP (control)
23	Telnet
25	SMTP
43	Whois
53	DNS
69	TFTP
70	Gopher
79	finger
80	HTTP

In this format, the port represents the numeric associated with the TCP or UDP process and is also referred to as the 'well-known port'. Table 7.10 lists 10 examples of well-known TCP/UDP ports.

If we assume the asterisk is used as a wildcard in IP address and port process value fields, two examples of access list entries follow:

access-list permit 198.*.*.*, *

access-list deny *.*.*.*, 80

In the first example any packet with a source address on a 198 network regardless of port value is permitted. In the second example, incoming HTTP (port 80) from any address is denied. Note that the two access list entries placed together in an access list would have the unintended effect of allowing all packets from network 198 to include HTTP packets to be permitted to flow through the router. Thus, if you wish to deny HTTP from network 198 you could either move the second access list entry to the top of the access list, or prefix the two access list entries with the following entry:

access-list deny 198.*.*.*, 80

Reversing access list entries would result in the following access list:

access-list deny *.*.*.*, 80

access-list permit 198.*.*.*, *

Now the first entry would deny all HTTP inbound traffic while the second entry would allow all traffic from network 198; however, since HTTP was previously barred, the second entry would only enable non-HTTP traffic from network 198.

The use of three entries in the access list would result in the following list:

access-list deny 198.*.*.*, 80

access-list permit 198.*.*.*, *

access-list deny *.*.*.*, 80

Now the first entry specifically denies HTTP access from network 198 while the second entry allows all traffic from network 198 other than HTTP that was previously barred. The third entry denies HTTP from all other network locations.

It should be noted that depending upon whether the general access default is permit or deny, you may at worst be required to modify your access list or at best have extraneous statements in your access list. For example, if your router's default is to deny all packets unless specifically permitted and you want to allow all packets from network 198 to include HTTP, you would only need one access list entry. That entry would be:

access-list permit 198.*.*.*, *

Extended access lists

Recognizing that routers can have multiple LAN interfaces, a logical extension of an access list is to apply list entries to specific ports. Thus, a second generic access list format is shown below:

access-list {permit | deny} port-out, source address, port

Here the port-out entry is a numeric which identifies a LAN interface.

To illustrate the application of this extended access list, let's assume a router is used to interconnect two LANs to the Internet as illustrated in Figure 7.41. Let's further assume that the Token-Ring network provides connectivity to an internal corporate network while the Ethernet LAN contains a public access Web server. Thus, one possible extended access list might be as follows:

access-list permit 1, *.*.*.*, 80

access-list permit 2, *.*.*.*, 25

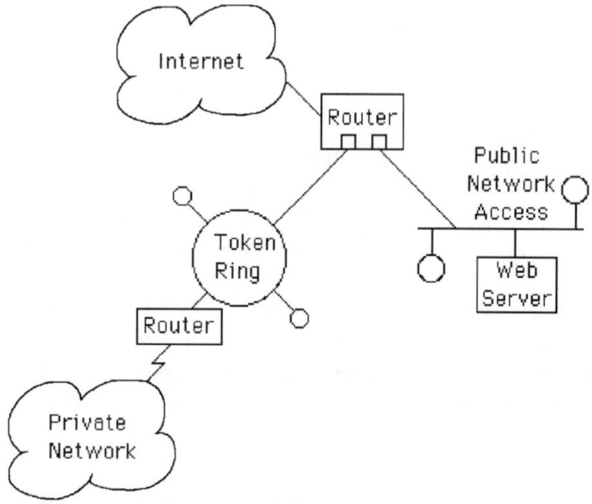

Figure 7.41 Connecting two LANs to the Internet. By applying router access-list entries to different router ports, you can control access to different networks

The first entry in the access list permits packets from any address containing HTTP (port 80) to be routed through the router and placed onto the Ethernet LAN. The second entry permits packets from any address containing SMTP (port 25) email to be routed through the router onto port 2 which is the Token-Ring network. If we presume a default of denial of service unless explicitly permitted, all other packets are sent to the great bit bucket in the sky. Thus, this short access list only permits email to flow onto the Token-Ring network and Web access to the Ethernet public access network.

Additional extensions

Some routers permit packet filtering based upon source and destination addresses as well as well-known ports for both inbound and outbound traffic. Thus, such routers may have an access list format similar to the following:

access-list {permit | deny} port-in, port-out, source, destination

When you obtain the ability to filter packets based upon inbound or outbound direction, you must usually include the direction of packet flow in the form of port-in to port-out port numbers. Other routers require only one physical port number to be specified; however, they then require a direction field in the access list to indicate if filtering should occur on inbound or outbound packets. Regardless of the method used, the ability to perform filtering based upon inbound and outbound packets provides additional security in the form of limiting access of organizational employees. For example, if your organization is using the Internet to connect several geographically dispersed corporate locations, you can use outbound filtering to limit employees to accessing only other corporate locations, preventing inadvertent downloading of viruses from an anonymous FTP site. Thus, the ability to perform filtering on both inbound and outbound traffic can enhance the ability of network managers to control communications.

Router access

Since access to the configuration capability of a router controls its filtering capability, it is important to control such access. All too often persons forget to change the default login password which can result in a hacker easily taking control of your router. Most routers provide access control not only via a console port, but, in addition, via Telnet and SNMP. This means that if a person knows or stumbles upon the IP address of an interface of the router via Telnet or SNMP, they can first try some common default passwords to gain access to the router's console capability. To prevent this situation from occurring, you should consider disabling such access. If you need the ability to remotely configure a router, you should change the default password. In doing so, you should use an alphanumeric string instead of a common name to prevent a dictionary attack. Many hackers purchase or otherwise acquire an electronic dictionary and write a program to try each entry in an attempt to break into routers and servers. Although such passwords as 'heather', 'administration', 'bozo', and 'georgia' might be easy to remember, they would all fail upon a dictionary attack. Thus, adding numerics to a name precludes the ability of a dictionary attack to be successful.

Now that we have an appreciation for the level of security afforded by the packet filtering capabilities of routers, let's discuss why you may wish to consider additional security in the form of a firewall by examining some common threats routers cannot prevent.

Threats not handled

There are numerous security threats that the packet filtering capability of routers cannot control. Table 7.11 lists six common threats presented in alphabetical order that can result in security-related problems that a router cannot detect. Although you could use a router's filtering capability to bar access to an FTP server, you cannot selectively control different FTP commands. Thus, the use of unauthorized commands would represent an all or nothing issue when working with router filters. To provide an additional level of security beyond packet level filtering, organizations commonly turn to the use of a firewall.

Firewalls

Unlike a router which simply passes packets from one interface to another, firewalls include a proxy service capability which results in IP packets being barred from directly passing from input to output destinations. Instead, the firewall obtains the

Table 7.11 Common security threats not controllable by routers

Threat	Description
Communications monitoring	A tap of your organization's Internet connection provides passwords and user account information as well as the ability to read data transferred
Dictionary attack	Hacker tries each entry in electronic dictionary to gain access to device
Password guessing	Hacker uses default installation password or common passwords such as 'cisco' for accounts
Terminal session monitoring	Hacker monitors active user, capturing keystrokes in an attempt to learn login to another system
Virus upload	A program containing a virus is placed onto a corporate FTP server or as an attachment to an email

capability to examine the contents of each packet at each layer in the ISO Reference Model up through the application layer. Depending upon the capabilities programmed into the firewall, the use of proxy services makes the device capable of detecting suspicious activity on a given connection, generating alerts in response to suspicious activity, differentiating between different file transfer modes, and making authentication and authorization decisions. Thus, a firewall can be considered to represent a much more sophisticated security device than that obtainable through the use of the packet filtering capability of a router.

Placement

When connecting an internal private network to the Internet or to a similar public network, a firewall is placed between the two, protecting inside users from outside users. Although you can place a firewall on an internal network and have all incoming access first directed to that device, if a person learns your internal network addressing scheme, it becomes possible to bypass the firewall and direct packets to recipients that may be better served by a packet examination process performed by the proxy service capability of most firewalls. Thus, the most common method used to install a firewall is to locate it on a separate network, a network commonly referred to as a demilitarized zone, or DMZ. The term DMZ or DMZ LAN obtains its name from the fact that LAN contains no directly connected organizational computers. Instead, a DMZ LAN has only two connections—a router connection

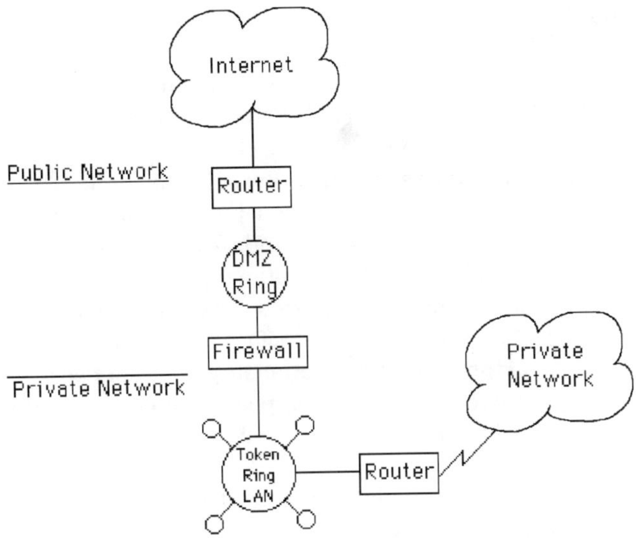

Figure 7.42 A firewall is normally connected to a DMZ LAN

and a firewall connection. Thus, all incoming packets must first be processed by the firewall prior to being transferred to the private network.

Figure 7.42 illustrates the placement of a firewall on a DMZ Token-Ring LAN which would consist of a MAU with two connections, one for the router and a second for the firewall. Note that the second firewall connection is to another Token-Ring; however, this ring represents part of the corporate network that is protected by the firewall.

Features

With over 50 vendors actively marketing firewalls, you would be quite correct in assuming that their features and operational capability can differ considerably between vendor products. Table 7.12 lists six basic firewall features you should, as a minimum, consider when acquiring this communications device. Due to the importance of these features, let us review each.

Table 7.12 Basic firewall features to consider

Proxy services
Address translation
Stateful inspection
Alerts
Authentication
Packet filtering

Proxy services

A proxy represents a code that performs handshaking for a specific application, such as FTP, Telnet, or HTTP. Through the proxy services capability of a firewall, specific users or groups of users can be allowed or denied access to a server or to a subset of a server's functionality. For example, through the use of FTP proxy services, you may be able to enable or disable the use of GET, MGET, PUT, MPUT, and other FTP commands for all addresses or selected IP addresses. Thus, it is important to examine both the type of proxy services supported by a firewall as well as the commands supported for each service.

Note that proxy services can vary considerably between different firewall products. However, this functionality does not exist in routers, nor in firewalls that simply provide an expanded packet filtering capability. Thus, proxy services represent a feature that can be used to differentiate a more capable firewall from less capable products.

To illustrate the configuration of proxy services and some additional firewall features, this author captured several configuration screens generated by the Technologic Interceptor firewall. Figure 7.43 illustrates the Interceptor's Advanced Policy Options screen in which the cursor is shown pointing to the HTTP Block Put and Post entries that were selected. In examining Figure 7.43 and subsequent

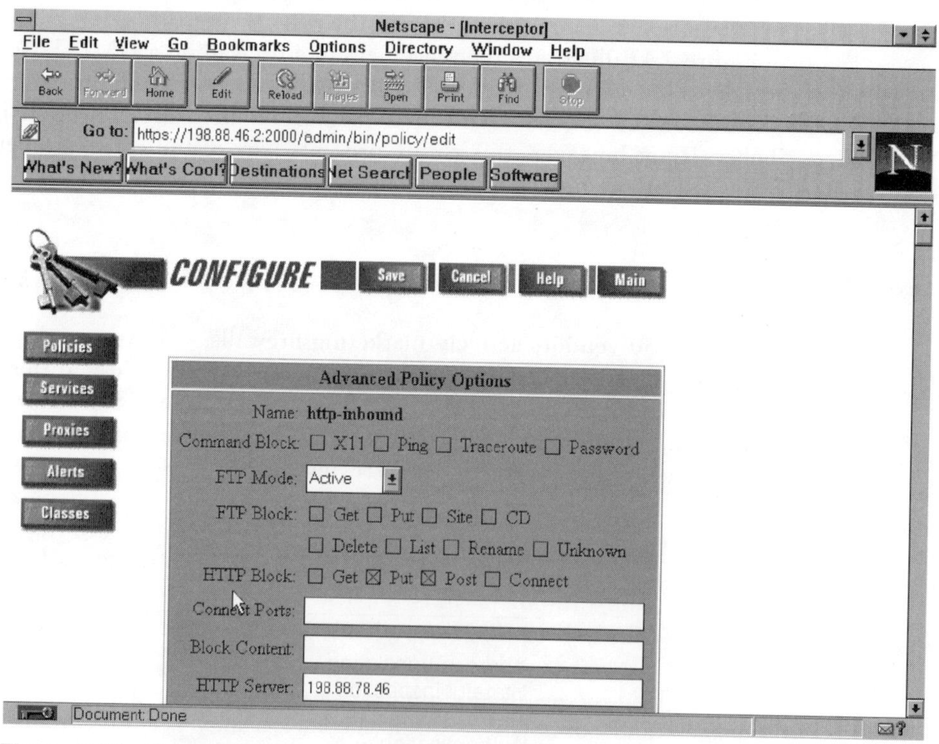

Figure 7.43 Using the Technologic Interceptor firewall configuration screen to block all FTP. Put commands

Interceptor screen displays, you will note they represent HTML screens displayed using a Netscape browser. The Technologic Interceptor firewall generates HTML forms to enable network managers to view, add, and modify firewall configuration data. To secure such operations the firewall uses encryption and authentication by supporting Netscape's Secure Sockets Layer (SSL) protocol for encrypting all traffic between the firewall and a Web browser used to configure the firewall while passwords are used for authentication. This means network managers can safely configure the firewall via the World Wide Web.

Using classes

The Technologic Interceptor firewall includes a class definition facility which provides users with a mechanism to replace address patterns, times of day, or URLs by symbolic names. Classes are initiated by selecting the Classes button at the left side of the configuration screen. By using an equal sign (=) as a prefix, they are distinguished from literal patterns.

Through the use of classes you can considerably facilitate the configuration of the firewall. For example, suppose you want to control access from users behind the firewall to Internet services. To do so you would first enter the IP addresses of computers that will be given permission to access common services you wish them to use. Then you would define a class name that would be associated with the group of IP addresses and create a policy that defines the services the members of the class are authorized to use. In addition to class names you can also use IP addresses as well as the asterisk (*) within an IP address as a global character.

Figure 7.44 illustrates the use of the Technologic Interceptor Edit Policy configuration screen to enable inbound traffic for HTTP. Note that this policy uses the specific address 198.88.46.7 in the box labeled 'From'. Thus, this policy would restrict inbound HTTP to a single IP address.

Address translation

One method commonly used to protect internal networks from unauthorized access attempts is obtained through address translation. When address translation is employed by a firewall, it hides IP addresses of the private network from outside users, functioning similar to the process used to obtain an unlisted telephone number. IP address translation can be accomplished in one of two ways—IP hiding or one-to-one mapping. Under IP hiding clients behind the firewall are restricted to initiating sessions and incoming packets are barred. When one-to-one mapping is employed, the firewall permits a bidirectional packet flow; however, addresses behind the firewall are hidden and substituted by the use of a different set of IP addresses.

Since IP addresses are becoming a scarce commodity until IPv6 arrives, network managers and administrators can consider using IP addresses specified in RFC 1918 which allocated three blocks of IP address space for private internets. Those address blocks allocated for private internets include:

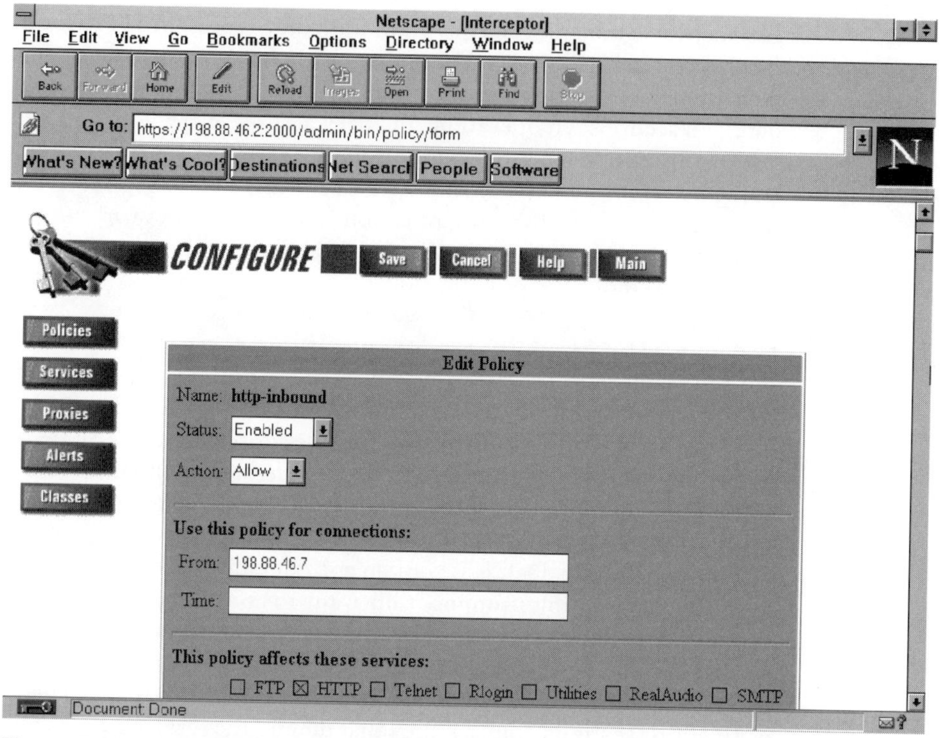

Figure 7.44 Using the Technologic Interceptor firewall to create a policy allowing inbound HTTP from a specific IP address

10.0.0.0 to 10.255.255.255
172.16.0.0 to 172.31.255.255
192.168.0.0 to 192.168.255.255

Any of the addresses in the preceding IP address blocks can be used behind a firewall, enabling the use of existing Class A, B, or C addresses to serve as translation addresses for use on the public side of the firewall.

Stateful inspection

The term stateful inspection was originally coined by Checkpoint Systems to reference the examination of packets at the network layer. Although this feature is similar to filtering, there are several key differences. First, the analysis is performed on each packet based upon the context of previous transmission. This is similar to tracking a series of telephone calls and permits a firewall to become aware of suspicious trends for which it may be configured either to bar further access attempts or to generate an alert to a designated person. Secondly, by tracking data from higher layers which are analyzed based upon the context of previous transmissions, the firewall becomes capable of providing a detailed audit trail of events.

Thus, a stateful inspection capability can provide both a higher level of protection than packet filtering as well as utilization information that may be useful in determining what services internal and external users have trouble accessing.

Alerts

Stateful inspection by itself may provide a limited protection capability unless a firewall automatically terminates suspicious activity or alerts a network manager or administrator to a possible attack on their network. Figure 7.45 illustrates the Technologic Interceptor Add Alert screen display with the IP-Spoof pattern shown selected. Note that the interceptor's patterns can be considered as equivalent to stateful inspection events.

In the example shown in Figure 7.45 the IP Spoof alert is used as a mechanism to denote a connection request occurring from a host claiming to have an IP address which does not belong to it. In actuality, it can be very difficult to note the occurrence of IP spoofing. This is because, unless the firewall previously obtained information about IP addresses, such as their locations on segments whose access is obtained via different firewall ports, or notes restrictions on service by IP address, it will assume an IP address is valid. In comparison, other patterns such as refused

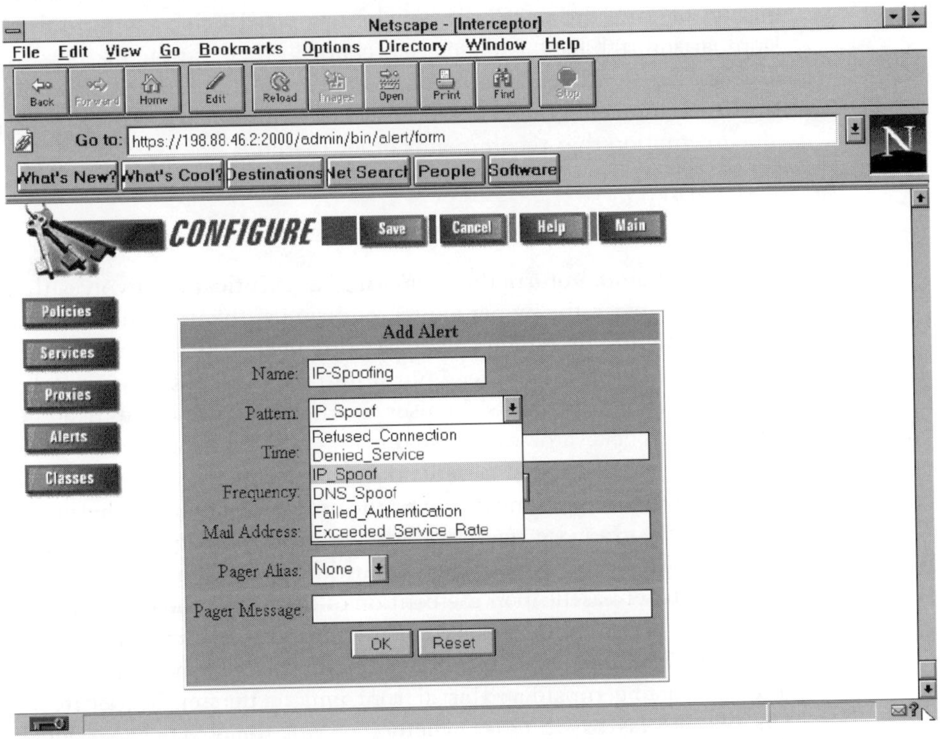

Figure 7.45 Using the Technologic Interceptor Add Alert configuration screen

connections or failed authentication are much easier to note. For each alert you would first specify a name for the alert definition, such as IP-Spoof for that pattern. After selecting the pattern, you can specify the days and times of day and the frequency of occurrence that, when matched, should generate an alert. The Interceptor supports two methods of alert generation—via either email or pager. If you select the use of a pager to transmit an alert you can include a message, such as a numeric alert code, to inform the recipient of the type of alert.

Authentication

Authentication is a mechanism used to verify the identity of a user requesting access to a specific service. A firewall can either directly perform authentication or operate in conjunction with one or more authentication servers located on the private side of the device.

There are a variety of methods used by firewalls or authentication servers to perform authentication. Those methods include static passwords, tokens, and one-time passwords.

Static passwords

Static passwords represent the least secure method of authentication as over a period of time the possibility of the password being compromised increases. In comparison, the use of tokens that change every minute or so as well as one-time passwords do a hacker no good if captured. Thus, most firewalls that perform authentication either do not support the use of static passwords or are provided with documentation that recommends other methods.

Tokens

The most common form of token-based authentication involves the use of a credit card token generator by personnel requiring authenticated access to a service. The token generator issues a new password every 60 seconds or so based upon a predefined algorithm and seed burnt into the card. If the firewall directly performs authentication, it prompts the user to enter a PIN and the token generated by the user's token generator card, usually a 6- or 8-digit random number. The firewall generates its own token based upon the user's PIN and compares the token it generated to the one transmitted by the user. If the two match, the requested service is allowed, otherwise access to the requested service is denied. If the actual authentication process runs on a separate server, such as FTP, Telnet, or a World Wide Web server, the firewall functions as a proxy, relaying authentication requests to the server providing the service that requires authentication.

Figure 7.46 illustrates the firewall authentication process. In this example, the firewall can be considered as a front end to the service, as it either enables or disables access based upon whether or not the token generated by the firewall matches the transmitted token.

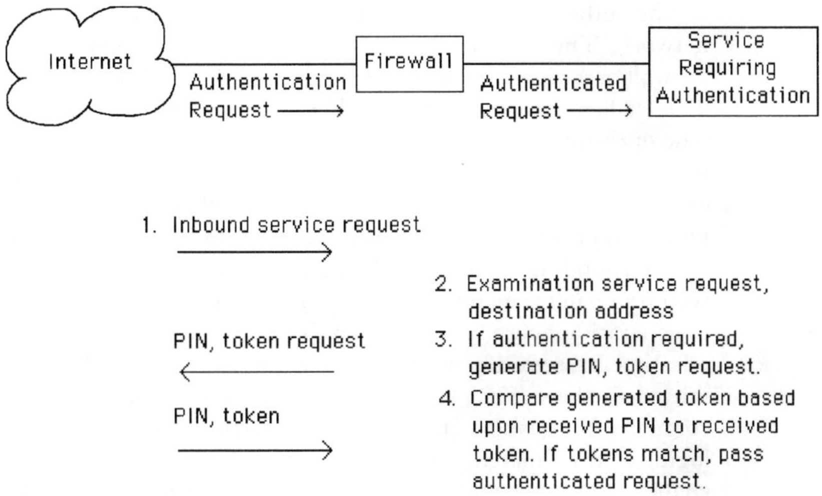
Internet
Firewall
Service Requiring Authentication
Authentication Request
Authenticated Request
1. Inbound service request
2. Examination service request, destination address
PIN, token request
3. If authentication required, generate PIN, token request.
PIN, token
4. Compare generated token based upon received PIN to received token. If tokens match, pass authenticated request.

Figure 7.46 The firewall authentication process

One-time passwords

Although one-time passwords can be considered similar to token-based authentication due to the fact that they generate passwords that are valid only for one-time use, there are two major differences between the two authentication methods. First, the token-based authentication method involves the use of hardware by the client in the form of a credit card-sized device that generates tokens. In comparison, one-time passwords are generated through the use of software. A second difference between the two is the fact that token-based authentication methods are commercial products whereas some one-time password generators are in the public domain and available for use without incurring additional expense.

Bellcore S/KEY

One example of a popular one-time password system is the Bellcore S/KEY. Currently there are two versions of S/KEY. An early version, referred to as the S/KEY reference implementation, is available from Bellcore via anonymous FTP; however, it has not been upgraded nor has the code been maintained for over the past five years. A second version of the product, simply referred to as S/KEY, is a commercial product from Bellcore that operates on a number of client and server platforms.

The S/KEY one-time password authentication system is based on the use of two software programs. One program, referred to as the key login program, operates on a server at the host site. The second program, which is called the S/KEY One-Time Password Generator, resides on the client computer. At the time this book was written, server software platforms supported included SunOS, IBM AIX, HP-UX, and Solaris. Client platforms supported include MS-Windows 3.1, Windows 95, Windows NT, SunOS, IBM AIX, HP-UX, Solaris, and Apple Macintosh.

The authentication process begins when a client attempts to access a protected network. The network server running S/KEY server software, which can be a firewall or a separate computer behind the firewall, issues a challenge to the client. The challenge consists of a number and a string of characters which forms a seed. The operator at the client workstation will enter the seed as well as a secret phrase into the S/KEY program running on the client workstation, resulting in the calculation of a response to the server's challenge. Thus, this method of one-time password entry is also commonly referred to as a challenge–response system.

In responding to the server's challenge, the client uses the challenge number and seed originally transmitted by the S/KEY server as well as the user's secret phrase to generate a secure hash function. This hash function is generated in the form of a one-time password consisting of six English words. This password is then transmitted to the S/KEY server. That server generates its own hash function and compares the result with the stored one-time password used for the most recent login. If they match, the user is allowed access and the server decrements the challenge number and stores the newly used one-time password for use with the next login attempt.

Since each access attempt commences with the generation and validation of a one-time password, this technique is similar to the use of a token generator in that it eliminates vulnerability from a 'sniffing' attack where a hacker learns and later uses the ID and password of a legitimate network user. Since unscrupulous persons can also reside on an internal private network, many organizations use token generators or one-time passwords either in conjunction with firewalls or with separate servers to protect their internal network from internal threats.

Packet filtering

Although the packet filtering capability of firewalls functions in a similar way to that router feature, the firewall is usually easier to configure and provides more flexibility in enabling or disabling access based upon the set of rules that can be developed. Since we previously examined the use of a router's basic access list to enable or disable network services based upon IP addresses, let's return to the use of the Technologic Interceptor firewall's configuration screen to examine how packet filtering is expanded upon by firewalls. Figure 7.47 illustrates the Technologic Interceptor Network Services display which lists the protocols for which this firewall accepts connections. Note that the HTTP protocol is shown selected as we will edit that service. Also note the columns labeled 'Max' and 'Rate'. The column labeled 'Max' indicates the maximum number of simultaneous connections allowed for each service while the column labeled 'Rate' indicates the maximum rate of new connections for each service on a per minute basis. By specifying entries for one or both columns you can significantly control access to the network services you provide as well as balance the loads on heavily used services.

Figure 7.48 illustrates the Interceptor Edit Service display configuration screen. In this example, HTTP service is enabled for up to 256 connections, and a queue size of 64 was entered, limiting TCP HTTP pending connections to that value. The Max Rate entry of 300 represents the maximum rate of new connections that will be

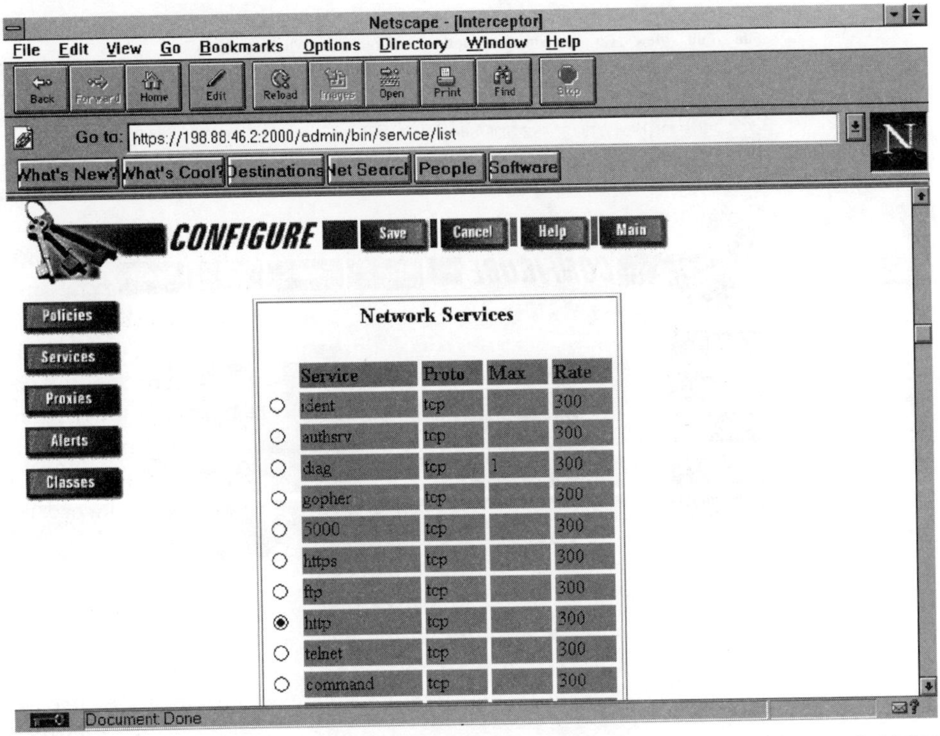

Figure 7.47 Using the Technologic Interceptor firewall configuration screen to edit HTTP network services

allowed to an HTTP service. Once this rate is exceeded, the firewall will temporarily disable access to that service for a period of one minute. If you allow both internal and external access to an internal Web server, the ability to control the maximum rate of incoming connections to a particular service can be an important weapon in the war against so-called denial of service attacks. Under this technique, a malicious person (or persons) programs a computer to issue bogus service initiation requests using random IP addresses that more than likely do not exist. Since each access request results in a server initiating a handshake response, the response is directed to a bogus IP address that does not respond. The server will typically keep the connection open for 60 to 120 seconds which represents a connection a valid user may not be able to use. Thus, if a hacker can issue 256 or more bogus connections and your Web server is configured to support that number of simultaneous connections, legitimate users could be barred from accessing the server. While there is no one uniform solution to this problem, you could use the Max Connects to limit inbound HTTP connections so you will always be able to let internal users access your Web server. In addition, if you specify a low Max Connects rate, you can negate some of the flooding of bogus connection attempts that will allow some legitimate users to reach your organization's Web server.

Since the firewall must examine each packet to determine whether or not to allow a connection, it is performing packet filtering. However, this is a much more

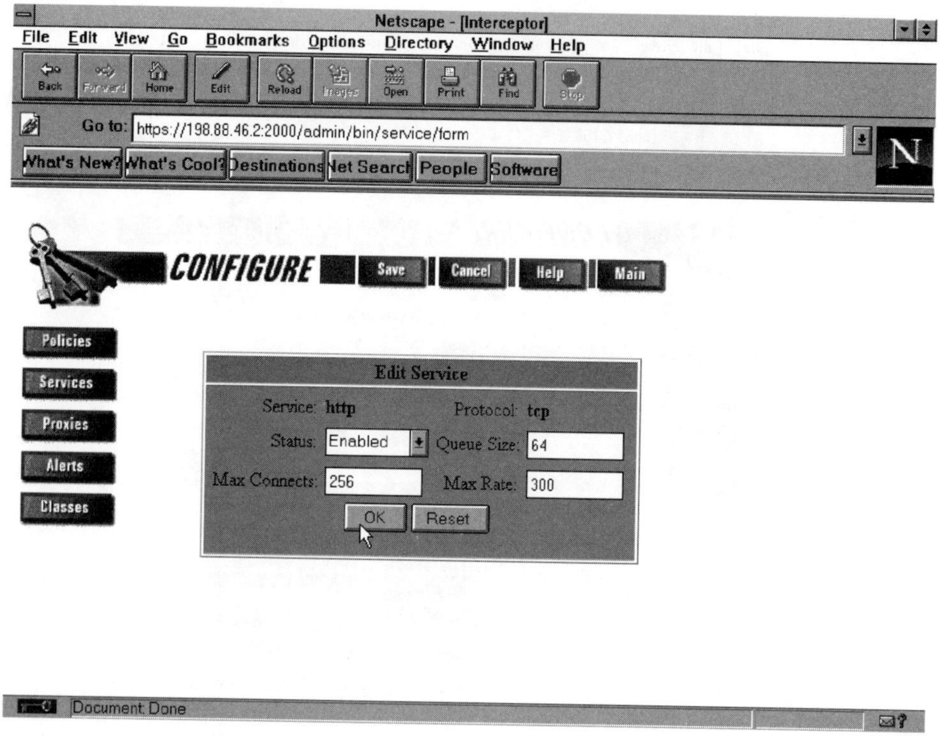

Figure 7.48 Using the Technologic Interceptor firewall Edit service configuration display to set a series of rules to govern access to HTTP

sophisticated type of packet filtering than that provided by most router manufacturers which is typically restricted to decisions being implemented based upon IP address and well-known port number. Thus, a firewall's packet filtering capability should also be carefully examined as it may provide you with the ability to control events that are beyond the capability of most routers.

The gap to consider

While routers and firewalls can be used to prevent unauthorized access to your network hosts, they do not guarantee the security of the communications connection between client and server, nor the security of the data being transferred. To obtain this security, you must use some type of encryption-based product. For example, when using Web browsers, you should consider the use of two related Internet protocols, Secure Sockets Layer (SSL) which was developed by Netscape, or the Secure Hypertext Transfer Protocol (S-HTTP) which was developed by Enterprise Integration Technologies. Both protocols support several cryptographic algorithms that use public key encryption methods. This allows a server to transmit upon request its public key to a client. The client uses that key to encrypt all communications for the session, enabling secure communications between client and server.

REVIEW QUESTIONS

7.1.1 What two types of switches are commonly used in networks?

7.1.2 Draw a network schematic illustrating the utilization of a fallback switch at a remote site and at a central computer site.

7.1.3 Discuss the utilization of a bypass switch at a central computer site.

7.1.4 What is the advantage of using a matrix switch over fallback or bypass switches?

7.1.5 How could you use a crossover switch at a remote site?

7.1.6 Draw a diagram illustrating how four fallback switches could be used to obtain a bypass switch capability linking m terminals to $m + 1$ modems or DSUs.

7.1.7 What are the major differences between manual and automatic switches and when might you consider using the latter device?

7.1.8 What are the major differences between the 'hot-start' and 'cold-start' methods of integrating collocated routers?

7.1.9 Assume that the terminals to be installed at a remote location perform critical applications and you desire to install dual multiplexer systems to provide access to the central computer site. Draw two network schematic diagrams illustrating the use of two different switches to provide the terminal's access to each multiplexer via a direct connect cable.

7.2.1 What is the primary difference between null suppression and run-length compression?

7.2.2 Discuss three benefits obtained from the compression of data prior to its transmission.

7.3.1 What are the two types of light sources used in a fiber optic system? What provides a higher data rate?

7.3.2 Where should you use a step index fiber cable? Where would you use a graded index fiber?

7.3.3 Describe how an optical modem operates.

7.3.4 Discuss five advantages associated with the use of fiber optic systems.

7.3.5 Discuss two limitations associated with the use of fiber optic systems.

7.4.1 What are some of the limitations of security access implemented through passwords?

7.4.2 What is the effect of reserving characters from a character set for additional system or communication functions upon access security implemented via passwords?

7.4.3 Assuming you required a measure of privacy in transmitting a message, discuss two techniques you could implement using a personal computer to scramble your message.

7.4.4 What is the effect of encoding both your header and text message for transmission on a Telex or TWX network?

7.4.5 If the data to be transmitted are 1 0 1 1 0 0 1 0 and the encoding key has the composition 0 1 1 0 1 0 1 0, what is the resulting encoded transmission if modulo 2 addition is used to develop the encoded text?

7.4.6 If the received data consists of the binary string 1 1 0 0 1 0 0 1 and your key is 0 1 0 1 0 1 1 0, what is the original encoded data if modulo 2 addition was used to form the encoded binary string?

7.4.7 What precaution should be taken when connecting encryption devices to a multiplexer?

7.4.8 What is the key advantage associated with the use of a public key system over a private key system?

7.4.9 What is a router access list?

7.4.10 Name four metrics that can be used for creating a router access list governing the flow of TCP/IP packets.

7.4.11 What does the following access list entry signify?

access-list permit 205.*.*.*, 23

7.4.12 Develop an access list for a three-port router where ports 1 and 2 are connected to LANs with network addresses 210 and 211, respectively, while port 3 is connected to the Internet. The access lists should only allow HTTP access to a server at address 210.21.31.14 and ftp access to a server at address 211.103.117.12.

7.4.13 How could you use an access list to prevent users on a corporate LAN connected to the Internet from downloading files with viruses?

7.4.14 How can you minimize the potential success of a dictionary attack aimed at gaining access to the control of a router?

7.4.15 Why is it recommended to locate a firewall on a DMZ LAN?

7.4.16 Define the term 'proxy service' with respect to a firewall.

7.4.17 What is the purpose of a firewall performing address translation?

7.4.18 Describe two differences between filtering and stateful inspection.

7.4.19 What is the purpose of authentication?

7.4.20 Describe three methods used to perform authentication.

APPENDIX A

SIZING DATA COMMUNICATIONS NETWORK DEVICES

A.1 DEVICE SIZING

Of many problems associated with the acquisition of data communications network devices, one item often requiring resolution is the configuration or sizing of the device. The process of ensuring that the configuration of the selected device will provide a desired level of service is the foundation upon which the availability level of a network is built.

The failure to provide a level of access acceptable to network users can result in a multitude of problems. First, a user encountering a busy signal might become discouraged, take a break or do something other than redial a telephone number of a network access port. Such action will obviously result in a loss of user productivity. If the network usage is in response to customer inquiries, a failure to certify a customer purchase, return, reservation or other action in a timely manner could result in the loss of that customer to a competitor. With a little imagination, it becomes easy to visualize that the lifeline of the modern organization is its data communications network. An unacceptable level of access to the network can be considered akin to a blockage in the human circulatory system-harm will result and additional analysis and testing may become necessary to alleviate the problem.

In this appendix we will examine the sizing process, including the application of Erlang and Poisson formulae. Readers are referred to the book *Practical Network Design* written by this author and published by John Wiley & Sons for detailed information concerning network design techniques, including the equipment sizing process. This book contains a large number of tables that can be used to simplify the network sizing process, of which extracts of those tables are contained in this appendix. Readers may also wish to consider the convenience diskette set which supplements the previously mentioned book. This diskette set contains a series of programs which can be used to automate various network design problems.

Sizing problem similarities

There are many devices that can be employed in a data communications network whose configuration or sizing problems are similar. Examples of such devices include the number of dial-in lines required to be connected to a telephone company rotary and the number of channels on such devices as multiplexers, data concentrators, and port selectors.

Basically, two methods can be used to configure the size of communications network devices. The first method, commonly known as experimental modeling, involves the selection of the device configuration based upon a mixture of previous experience and gut intuition. Normally, the configuration selected is less than the base capacity plus expansion capacity of the device. This enables the size of the device to be adjusted or upgraded without a major equipment modification if the initial sizing proves inaccurate. An example of experimental modeling is shown in Figure A.1.

A rack-mounted time division multiplexer is shown in Figure A.1(a). Initially, the multiplexer was obtained with five dual channel adapters to support ten channels of simultaneous operation. Since the base unit can support eight dual channel adapters, if the network manager's previous experience or gut intuition proves wrong, the multiplexer can be upgraded easily. This is shown in Figure

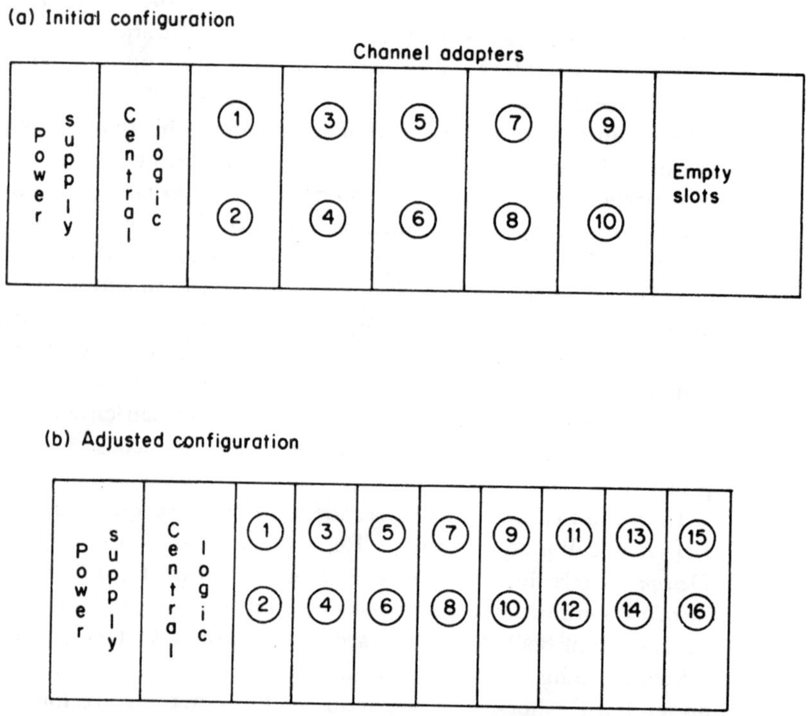

Figure A.1 Experimental modeling. Experimental modeling results in the adjustment of a network configuration based upon previous experience and gut intuition

A.1(b). Here, the addition of three dual channel adapters permits the multiplexer to support 16 channels in its adjusted configuration.

The second method which can be employed to size network components ignores experience and intuition. This method is based upon a knowledge of data traffic and the scientific application of mathematical formulae to traffic data, hence, it is known as the scientific approach or method of equipment sizing. Although some of the mathematics involved in determining equipment sizing can become quite complex, a series of tables can be employed to reduce many sizing problems to one of a simple table look-up process.

Although there are advantages and disadvantages to each method, the application of a scientific methodology to equipment sizing is a rigorously defined approach. This should provide a much higher degree of confidence and accuracy of the configuration selected when this method is used. On the negative side, the use of a scientific method requires a firm knowledge or an accurate estimate of the data traffic. Unfortunately, for some organizations, this may be difficult to obtain. In many cases a combination of two techniques will provide an optimum situation. For such situations sizing can be conducted using the scientific method with the understanding that the configuration selected may require adjustment under the experimental modeling concept. In the remainder of this appendix, we will focus our attention upon the application of the scientific methodology to equipment sizing problems.

Telephone terminology relationships

Most of the mathematics used for the sizing of data communications equipment evolved from work originally developed from the sizing problems of telephone networks. From a discussion of a few basic telephone network terms and concepts we will see the similarities between the sizing problems associated with data communications components and the structure of the telephone network. Building upon this foundation, we will learn how to apply the mathematical formulae developed for telephone network sizing to data communications network component configurations.

To study the relationships between the telephone network and communications component sizing problems let us examine a portion of the telephone network and study the structure and calling problems of a small segment formed by two cities, assuming that each city contains 1000 telephone subscribers.

The standard method of providing an interconnection between subscribers in a local area is to connect each subscriber's telephone to what is known as the local telephone company exchange. Other synonymous terms for the local telephone company exchange include the 'local exchange' and 'telephone company central office'. Whether the term central office or exchange is used, the subscriber's call is switched to the called party number at that location. If we assume each city has only one local exchange, then all calls originating in that city and with a destination located within that city will be routed through that exchange.

Since our network segment selected for analysis consists of two cities, we will have two telephone company exchanges, one located in each city. To provide a path

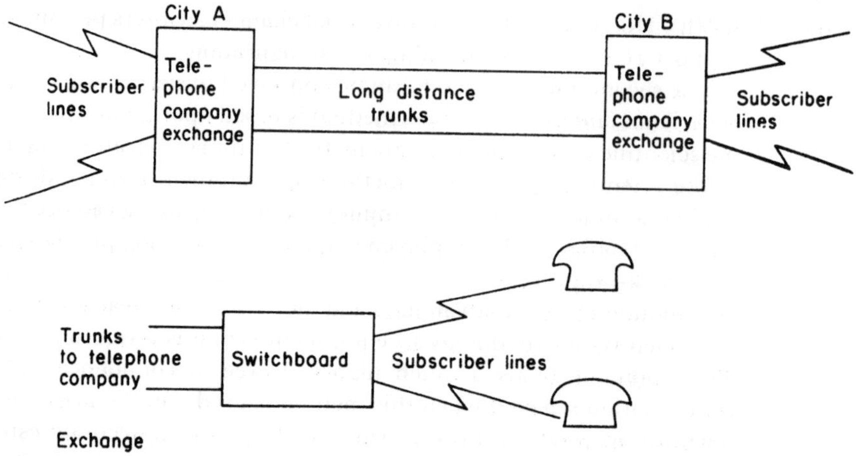

Figure A.2 Telephone traffic sizing problems. Although most subscriber calls are routed locally through the local telephone company exchange or local switchboard to parties in the immediate area, some calls require access to trunks. The determination of the number of trunks required to provide an acceptable grade of service is known as line dimensioning and is critical for the effective operation of the facility

between cities for intercity calling, a number of lines must be installed to link the exchanges in each city. The exchange in each city can then act as a switch, routing the local subscribers in each city desiring to contact parties in the other city.

As shown in the upper part of Figure A.2, a majority of the telephone traffic in the network segment consisting of the two cities will be among the subscribers of each city. Although there will be telephone traffic between the subscribers in each city, it will normally be considerably less than the amount of local traffic in each city. The path between the two cities connecting their telephone central offices is known as a trunk. One of the problems in designing the telephone network is determining how many trunks should be installed between company exchanges. A similar sizing problem occurs many times in each city at locations where private organizations desire to install switchboards. An example of the sizing problem with this type of equipment is illustrated in the lower portion of Figure A.2. In effect, the switchboard functions as a small telephone exchange, routing calls carried over a number of trunks installed between the switchboard and the telephone company exchange to a larger number of the subscriber lines connected to the switchboard. The determination of the number of trunks required to be installed between the telephone exchange and the switchboard is called dimensioning and is critical for the efficient operation of the facility. If insufficient trunks are available, company personnel will encounter an unacceptable number of busy signals when trying to place an outside telephone call. Once again, this will obviously affect productivity.

Returning to the intercity calling problem, consider some of the problems one faces in dimensioning the number of trunks between the two cities. Let us assume that, based upon a previously conducted study, it was determined that no more than 50 people would want to have simultaneous telephone conversations where the calling party was in one city and the called party in the other city. If 50 trunks were installed between cities and the number of intercity callers never exceeded 50, at any moment the probability of a subscriber completing a call to the distant city would always be unity, guaranteeing success. Although the service cost of providing 50 trunks is obviously more than providing a lesser number of trunks, no subscriber would encounter a busy signal.

Since some subscribers might postpone or choose not to place a long-distance call at a later time if a busy signal is encountered, a maximum level of service will produce a minimum level of lost revenue. If more than 50 subscribers tried to simultaneously call parties in the opposite city, some callers would encounter busy signals once all 50 trunks were in use. Under such circumstances, the level of service would be such that not all subscribers are guaranteed access to the long-distance trunks and the probability of making a long-distance call would be less than unity. Likewise, since the level of service is less than that required to provide all callers with access to the long-distance trunks, the service cost is less than the service cost associated with providing users with a probability of unity in accessing trunks. Similarly, as the probability of successfully accessing the long-distance trunks decrease, the amount of lost revenue or customer waiting costs will increase. Based upon the preceding, a decision model factoring into consideration the level of service versus expected cost can be constructed as shown in Figure A.3.

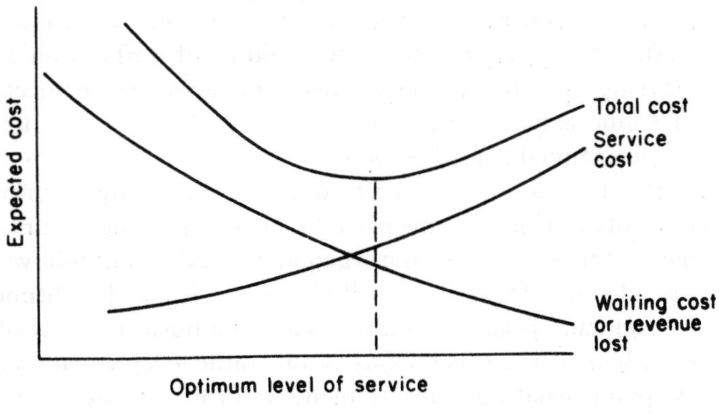

Figure A.3 Using a decision model to determine the optimum level of service. Where the total cost is minimal represents the optimum level of service one should provide

The decision model

For the decision model illustrated in Figure A.3, suppose the optimum number of trunks required to link the two cities is 40. The subscriber line to trunk ratio for this case would be 1000 lines to 40 trunks, for a 25:1 ratio.

To correctly dimension the optimum number of trunks linking the two cities requires both an understanding of economics as well as subscriber traffic. In dimensioning the number of trunks, a certain tradeoff will result that relates the number of trunks or level of service to the cost of providing that service and the revenue lost by not having enough trunks to satisfy the condition when a maximum number of subscribers dial subscribers in the other city. To determine the appropriate level of service, a decision model as illustrated in Figure A.3 is required. Here, the probability of a subscriber successfully accessing a trunk corresponds to the level of service provided. As more trunks are added, the probability of access increases as well as the cost of providing such access. Correspondingly, the waiting cost of the subscriber or the revenue loss to the telephone company decreases as the level of service increases. Where the total cost representing the combination of service cost and waiting cost is minimal represents the optimal number of trunks or level of service that should be provided to link the two cities.

Decision models can be easily adapted to data communications problems. As an example, consider a network designed to provide access to computational facilities for time-sharing customers. Here lost revenues could be equated to lost business at some fixed hourly rate while the service cost can be computed by determining the cost of extra equipment required to increase the level of service for customers.

Traffic measurements

Telephone activity can be defined by the calling rate and the holding time (duration of the call). The calling rate is the number of times a particular route or path is used per unit time period while the holding time is the duration of the call on the route or path. Two other terms that warrant attention are the offered traffic and the carried traffic. The offered traffic is the volume of traffic routed to a particular telephone exchange during a predetermined time period while the carried traffic is the volume of traffic actually transmitted through the exchange to its destination during a predetermined period of time.

The key factor required to dimension a traffic path is knowledge of the traffic intensity during the time period known as the busy hour. Although traffic varies by day and time of day, traffic is generally random but follows a certain consistency one can identify. In general, traffic peaks prior to lunchtime and then rebuilds to a second daily peak in the afternoon. The busiest period of the day is known as the busy hour (BH). It is the busy hour traffic level which is employed in dimensioning telephone exchanges and transmission routes since one wants to size the exchange or route with respect to its busiest period.

Telephone traffic can be defined as the product of the calling rate per hour and the average holding time per call. This measurement can be expressed mathematically as:

$$T = C \times D$$

where C is the calling rate per hour and D the average duration per call.

Using the above formula, traffic can be expressed in call-minutes (CM) or call-hours (CH) where a call-hour is the quantity represented by one or more calls having an aggregate duration of one hour.

If the calling rate during the busy hour of a particular day is 500 and the average duration of each call was 10 minutes, the traffic flow or intensity would be 500 x 10 or 5000 CM which would be equivalent to 5000/60 or approximately 83.3 CH.

Erlangs and call-seconds

The preferred unit of measurement in telephone traffic analysis is the erlang, named after A.K. Erlang, who was a Danish mathematician. The erlang is a dimensionless unit in comparison to the previously discussed call-minutes and call-hours. It represents the occupancy of a circuit where one erlang of traffic intensity on one traffic circuit represents a continuous occupancy of that circuit.

A second term often used to represent traffic intensity is the call-second (CS). The quantity represented by 100 call-seconds is known as 1 CCS. Here the first C represents the quantity 100 and comes from the French term cent. Assuming a one-hour unit interval, the previously discussed terms can be related to the erlang as follows:

$$1 \text{ erlang} = 60 \text{ call-minutes} = 36 \text{ CCS} = 3600 \text{ CS}$$

If a group of 20 trunks were measured and a call intensity of 10 erlangs determined over the group, then we would expect one-half of all trunks to be busy at the time of the measurement. Similarly, a traffic intensity of 600 CM or 360 CCS offered to the 20 trunks would warrant the same conclusion. A traffic conversion table is located in Table A.1 which will facilitate the conversion of erlangs to CCS and vice versa. Since the use of many dimensioning tables are based upon traffic

Table A.1 Traffic conversion table

Dimension		Erlangs (intensity) Call-hours	CCS
Minutes	Hours	(quantity)	(quantity)
12	0.2	0.2	6
24	0.4	0.4	12
36	0.6	0.6	18
48	0.8	0.8	24
60	1.0	1.0	36
120	2.0	2.0	72
180	3.0	3.0	108
240	4.0	4.0	144
300	5.0	5.0	180
360	6.0	6.0	210
420	7.0	7.0	252
480	8.0	8.0	288
540	9.0	9.0	324
600	10.0	10.0	360
900	15.0	15.0	540
1200	20.0	20.0	720
1500	25.0	25.0	900
1800	30.0	30.0	1080
2100	35.0	35.0	1260
2400	40.0	40.0	1440
2700	45.0	45.0	1620
3000	50.0	50.0	1800
6000	100.0	100.0	3600

intensity in erlangs or CCS, the conversion of such terms is frequently required in the process of sizing facilities.

Grade of service

One important concept in the dimensioning process is what is known as the grade of service. To understand this concept, let us return to our intercity calling example illustrated in example A.1, again assuming 50 trunks are used to connect the telephone exchanges in each city. If a subscriber attempts to originate a call from one city to the other when all trunks are in use, that call is said to be blocked. Based upon mathematical formulae, the probability of a call being blocked can be computed given the traffic intensity and number of available trunks. The concept of determining the probability of a call being blocked can be computed given the traffic intensity and number of available trunks. The concept of determining the probability of blockage can be easily adapted to the sizing of data communications equipment.

From a logical analysis of traffic intensity, it follows that if a call will be blocked, such blockage will occur during the busy hour since that is the period when the largest amount of activity occurs. Thus, telephone exchange capacity is engineered to service a portion of the busy hour traffic, the exact amount of service being dependent upon economics as well as the political process of determining the level of service one desires to provide to customers. One could over-dimension the route between cities and provide a trunk for every subscriber. This would ensure that a lost call could never occur and would be equivalent to connecting every terminal in a network directly to a front-end processor port. Since a 1:1 subscriber to trunk ratio is uneconomical and will result in most trunks being idle a large portion of the day, we can expect a lesser number of trunks between cities than subscribers. As the number of trunks decreases and the subscriber to trunk ratio correspondingly increases, we can intuitively expect some sizings to result in some call blockage. We can specify the number of calls we are willing to have blocked during the busy hour. This specification is known as the grade of service and represents the probability (P) of having a call blocked. If we specify a grade of service of 0.05 between the cities, we require a sufficient number of trunks so that only one call in every twenty or five calls in every one hundred will be blocked during the busy hour.

Route dimensioning parameters

To determine the number of trunks required to service a particular route one can consider the use of several formulae. Each formula's utilization depends upon the call arrival and holding time distribution, the number of traffic sources, and the handling of lost or blocked calls. Regardless of the formula employed, the resulting computation will provide one with the probability of call blockage or grade of service based upon a given number of trunks and level of traffic intensity.

Concerning the number of traffic sources, one can consider the calling population as infinite or finite. If calls occur from a large subscriber population and subscribers tend to redial if blockage is encountered, the calling population can be considered as infinite. The consideration of an infinite traffic source results in the probability of call arrival becoming constant and does not make the call dependent upon the state of traffic in the system. The two most commonly employed traffic dimensioning equations are both based upon an infinite calling population.

Concerning the handling of lost calls, such calls can be considered cleared, delayed, or held. When such calls are considered held, it is assumed that the telephone subscriber immediately redials the desired party upon encountering a busy signal. The lost call delayed concept assumes each subscriber is placed into a waiting mechanism for service and forms the basis for queuing analysis. Since we can assume a service or non-service condition, we can disregard the lost call delayed concept.

Traffic dimensioning formulae

The principal traffic dimensioning formula used in North America is based upon the lost call concept and is commonly known as the Poisson formula. In Europe, traffic formulae are based upon the assumption that a subscriber upon encountering a busy signal will hang up the telephone and wait a certain amount of time prior to redialing. The Erlang formula is based upon this lost call cleared concept.

A.2 THE ERLANG TRAFFIC FORMULA

The most commonly used telephone traffic dimensioning equation is the Erlang B formula. This formula is predominantly used outside the North American continent. In addition to assuming that data traffic originates from an infinite number of sources, this formula is based upon the lost calls cleared concept. This assumption is equivalent to stating that traffic offered to but not carried by one or more trunks vanishes and this is the key difference between this formula and the Poisson formula. The latter formula assumes that lost calls are held and it is used for telephone dimensioning mainly in North America. Since data communications system users can be characterized by either the lost calls cleared or lost calls held concept, both traffic formulae and their application to data networks will be covered in this appendix.

If E is used to denote the traffic intensity in erlangs and T represents the number of trunks designed to support the traffic, the probability $P(T, E)$ represents the probability that T trunks are busy when a traffic intensity of E erlangs is offered to those trunks. The probability is equivalent to specifying a grade of service and can be expressed by the erlang traffic formula as follows:

$$P(T,E) = \frac{E^T/T!}{1+E+(E^2/2!)+(E^3/3!)+\cdots+(E^T/T!)} = \frac{E^T/T!}{\sum_{m=0}^{T}(E^M/m!)}$$

where $T! = T(T-1)(T-2)\ldots 3\times 2\times 1.$

Table A.2 Factorial values

N	Factorial *N*
1	1
2	2
3	6
4	24
5	120
6	720
7	5040
8	40320
9	362880
10	3628800
11	3.99168E 07
12	4.79002E 08
13	6.22702E 09
14	3.71788E 10
15	1.30767E 12
16	2.09228E 13
17	3.55687E 14
18	6.40237E 16
19	1.21645E 17
20	2.48290E 18
21	5.10909E 19
22	1.12400E 21
23	2.58520E 22
24	6.20448E 23
25	1.55112E 25
26	4.03291E 26
27	1.03889E 28
28	3.04888E 29
29	8.84176E 30
30	2.65258E 32
31	8.22284E 33
32	2.63131E 35
33	8.68832E 36

In Table A.2 a list of factorials and their values are presented to assist the reader in computing specific grades of service based upon a given traffic intensity and trunk quantity.

To assist in the computation of a range of grades of service based upon specific traffic loads and channel or trunk capacity, a BASIC computer language program was developed. The BASIC program is listed in Appendix B and can be easily modified to compute the probability of all channels busy when a call is attempted for different levels of traffic loads by changing line 40 of the program. Currently, line 40 in conjunction with line 50 varies the traffic intensity from 0.5 through 40 erlangs in increments of 0.5 erlangs.

Extracts from the execution of the BASIC program are listed in Table A.3. While the use of the Erlang formula is normally employed for telephone dimensioning, it can be easily adapted to sizing data communications equipment. As an example of

Table A.3 Erlang B distribution extracts. Probability that all channels are busy when call attempted

	(Load in erlangs)				
Channels	5.500	6.000	6.500	7.000	7.500
1	0.846 154	0.857 143	0.866 667	0.875 000	0.882 353
2	0.699 422	0.720 000	0.737 991	0.753 846	0.767 918
3	0.561 840	0.590 164	0.615 234	0.637 546	0.657 510
4	0.435 835	0.469 565	0.499 939	0.527 345	0.552 138
5	0.324 059	0.360 400	0.393 910	0.424 719	0.453 016
6	0.229 022	0.264 922	0.299 099	0.331 330	0.361 541
7	0.152 503	0.185 055	0.217 365	0.248 871	0.279 209
8	0.094 897	0.121 876	0.150 100	0.178 822	0.207 455
9	0.054 814	0.075 145	0.097 803	0.122 101	0.147 397
10	0.029 265	0.043 142	0.059 772	0.078 741	0.099 544
11	0.014 422	0.022 991	0.034 115	0.047 717	0.063 557
12	0.006 566	0.011 365	0.018 144	0.027 081	0.038 206
13	0.002 770	0.005 218	0.008 990	0.014 373	0.021 566
14	0.001 087	0.002 231	0.004 157	0.007 135	0.011 421
15	0.000 398	0.000 892	0.001 798	0.003 319	0.005 678

the use of Table A.3, consider the following situation. Suppose one desires to provide customers with a grade of service of 0.1 when the specific traffic intensity is 7.5 erlangs. From Table A.3, 10 channels or trunks would be required since the use of the table requires one to interpolate and round to the highest channel. Thus, if it was desired to offer a 0.01 grade of service when the traffic intensity was 7 erlangs, one could read down the 7.0 erlang column and determine that between 13 and 14 channels are required. Since one cannot install a fraction of a trunk or channel, 14 channels would be required as we round to the highest channel number.

Multiplexer sizing

In applying the Erlang B formula to multiplexer sizing, an analogy can be made between telephone network trunks and multiplexer channel adapters. Let us assume that a survey of terminal users in a geographical area indicated that during the busy hour normally six terminals would be active. This would represent a traffic intensity of 6 erlangs. Suppose we wish to size the multiplexer to ensure that at most only one out of every 100 calls to the device encounters a busy signal. Then our desired grade of service becomes 0.01. From Table A.3, the 6 erlang column indicates that to obtain a 0.011 365 grade of service would require 12 channels while a 0.005 218 grade of service would result if the multiplexer had 13 channels. Based upon the preceding data, the multiplexer would be configured for 13 channels as illustrated in Figure A.4.

From a practical consideration, the Erlang B formula assumption that lost calls are cleared and that traffic not carried vanishes can be interpreted as traffic overflowing one dial-in port being switched to the next port on the telephone

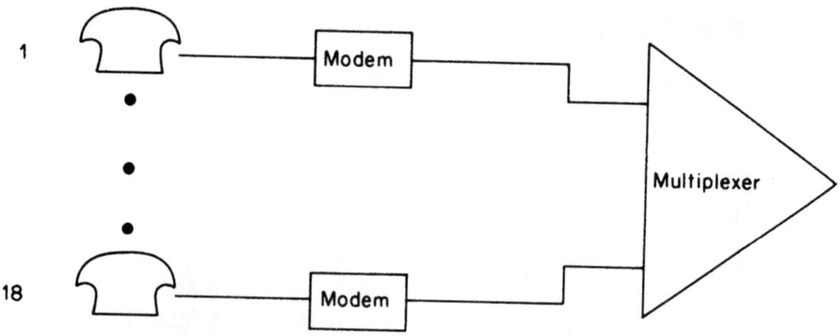

Figure A.4 Multiplexer channel sizing. Based upon a busy hour traffic intensity of 10 erlangs, 19 dial-in lines, modems and multiplexer ports would be required to provide a 0.01 grade of service

company rotary as each dial-in port becomes busy. Thus, traffic overflowing dial-in port m is offered to port $m + 1$ and the traffic lost by the mth dial-in port, Lm, is the total traffic offered to the entire group of dial-in ports multiplied by the probability that all dial-in ports are busy, thus

$$Lm = E \times P(m, E)$$

where E = traffic intensity in erlangs and m = number of ports or channels.

For the first dial-in port, when m is one, the proportion of traffic blocked becomes

$$P_1 = \frac{E}{E+1}$$

For the second dial-in port, the proportion of traffic lost by that port becomes

$$P_2 = \frac{E^2/2!}{1 + E + (E^2/2!)}$$

In general, the proportion of traffic lost by the mth port can be expressed as

$$P_m = \frac{E^m/m!}{\sum_{i=1}^{m}(E^i/i!)}$$

To reduce the complexity of calculation, let us analyze the data traffic carried by a group of four dial-in ports connected to a four-channel multiplexer when a traffic intensity of 3 erlangs is offered to the group.

For the first dial-in port, the proportion of lost traffic becomes

$$P_1 = \frac{3}{3+1} = 0.75$$

The proportion of lost traffic on the first port multiplied by the offered traffic provides the actual amount of lost traffic on port 1, thus

$$T_1 = P_1E = \left(\frac{E}{1+E}\right)E = \frac{E^2}{1+E} = 2.25 \text{ erlangs}$$

The total traffic carried on the first multiplexer port is the difference between the total traffic offered to that port and the traffic that overflows or is lost to the first port. Denoting the traffic carried by the first port as C_1 and the amount offered as A, we obtain

$$C_1 = A - T_1 = 3 - 2.25 = 0.75 \text{ erlangs}$$

Since we consider the rotary as a device that will pass traffic lost from port one to the remaining ports, we can compute the traffic lost by the second port in a similar manner. Substituting in the formula to determine the proportion of traffic lost, we obtain for the second port

$$P_2 = \frac{E^2/2!}{1 + E + (E^2/2!)} = 0.5294$$

The amount of traffic lost by the second port, T_2, becomes

$$T_2 = P_2E = 0.5294 \times 3 = 1.588 \text{ erlangs}$$

The traffic carried by the second port, C_2, is the difference between the traffic lost by the first port and the traffic lost by the second port, thus

$$C_2 = T_1 - T_2 = 2.25 - 1.588 = 0.662 \text{ erlangs}$$

In Table A.4, a summary of individual port traffic statistics are presented for the four-port multiplexer based upon a traffic intensity of 3 erlangs offered to the device. From Table A.4, the traffic carried by all four ports totaled 2.3817 erlangs. Since 3 erlangs were offered to the multiplexer ports, then 0.6183 erlangs were lost. The proportion of traffic lost to the group of four ports is T_4/E or $0.6183/3$ which is 0.2061. These calculations become extremely important from a financial standpoint

Table A.4 Individual port traffic statistics

Port	Proportion of lost traffic	Amount of lost traffic	Traffic carried
1	0.7500	2.25	0.75
2	0.5294	1.588	0.662
3	0.3462	1.038	0.550
4	0.2061	0.6183	0.4197

if a table lookup results in a device dimensioning which causes an expansion nest to be obtained to service one or only a few channel adaptors. Under such circumstances, one may wish to analyze a few of the individual high order channels to see what the effect of the omission of one or more of those channels will have upon the system.

If data tables are available, the previous individual port calculations are greatly simplified. From such tables the grade of service for channels 1 to 4 with a traffic intensity of 3 erlangs is the proportion of traffic lost to each port. Thus, if tables are available, one only has to multiply the grade of service by the traffic intensity to determine the traffic lost to each port.

A.3 THE POISSON FORMULA

The number of arrivals per unit time at a service location can vary randomly according to one of many probability distributions. The Poisson distribution is a discrete probability distribution since it relates to the number of arrivals per unit time. The general model or formula for this probability distribution is given by the following equation

$$\frac{P(r) = e^{-\lambda}(\lambda)^r}{r!}$$

where r is the number of arrivals, $P(r)$ the probability of arrivals, λ the mean of arrival rate, e the base of natural logarithms (2.718 28), and $r! = r$ factorial $= r(r-1)(r-2)\ldots 3 \times 2 \times 1$.

The Poisson distribution corresponds to the assumption of random arrivals since each arrival is assumed to be independent of other arrivals and also independent of the state of the system. One interesting characteristic of the Poisson distribution is that its mean is equal to its variance. This means that by specifying the mean of the distribution, the entire distribution is specified.

Multiplexer sizing

As an example of the application of the Poisson distribution let us consider a multiplexer location where user calls arrive at a rate of two per unit period of time. From the Poisson formula, we obtain

$$P(r) = \frac{2.71828^{-2}2^r}{r!}$$

Substituting the values 0, 1, 2, ..., 9 for r, we obtain the probability of arrivals listed in Table A.5, rounded to four decimal points. The probability of arrivals in excess of nine per unit period of time can be computed but is a very small value and was thus eliminated from consideration.

Table A.5 Poisson distribution arrival (rate of 2 per unit time)

Number of arrivals per period	Probability
0	0.1358
1	0.2707
2	0.2707
3	0.1805
4	0.0902
5	0.0361
6	0.0120
7	0.0034
8	0.0009
9	0.0002

The probability of the arrival rate being less than or equal to some specific number, m, is the sum of the probabilities of the arrival rate being 0, 1, 2, ... to m. This can be expressed mathematically as follows

$$P(r \leq m) = \sum_{r=0}^{m} \frac{e^{-\lambda}\lambda^r}{r!}$$

To determine the probability of four or fewer arrivals per unit period of time we obtain

$$P(r \leq 4) = \sum_{r=0}^{4} \frac{e^2 2^r}{r!}$$

$$P(r \leq 4) = 0.1358 + 0.2707 + 0.2707 + 0.1804 + 0.0902 = 0.9478$$

From the preceding, almost 95% of the time four or fewer calls will arrive at the multiplexer at the same time, given an arrival rate or traffic intensity of 2. The probability that a number of calls in excess of four arrives during the period is equal to one minus the probability of four or less calls arriving.

$$P(r \leq 4) = 1 - 0.9478 = 0.0522$$

If 4 calls arrive and are being processed, any additional calls are lost and cannot be handled by the multiplexer. The probability of this occurring is 0.0522 for a four-channel multiplexer, given a traffic intensity of 2 erlangs. In general, when E erlangs of traffic are offered to a service area containing m channels, the probability that the service area will fail to handle the traffic is given by the equation

$$P(r \geq m) = e^{-E} \sum_{r=m+1}^{\infty} \frac{E^r}{r!}$$

The key difference between the Poisson and Erlang formulae is that the Poisson formula assumes that lost calls are held or retired immediately after a busy signal is encountered. When the Erlang formula is used, it is assumed that lost calls are cleared.

In Appendix C the reader will find a BASIC program written to compute a table of grades of service for varying traffic intensities when lost calls are assumed to be held and thus follow the Poisson distribution. By changing the values of line 40 of the program in conjunction with line 50, different traffic intensities can be easily computed. Currently, the program computes grades of service as the traffic intensity varies from 0.5 to 40 erlangs in increments of 0.5 erlangs.

Formula comparison

In order to contrast the difference between Erlang B and Poisson formulae, let us return to the multiplexer examples previously considered. When 7 erlangs of traffic are offered and it is desired that the grade of service should be 0.01, 14 multiplexer channels are required when the Erlang B formula is employed. If the Poisson formula is used and the program in Appendix C is executed, one of the tables from running that program would appear as indicated in Table A.6. By using this table a grade of service of 0.01 for a traffic intensity of 7 erlangs results in a required channel capacity somewhere between 10 and 11. Rounding to the next highest number results in a requirement for 11 multiplexer channels. Now let us compare what happens at a higher traffic intensity. For a traffic intensity of 10 erlangs and the same 0.01 grade of service, it was determined that 19 channels were required

Table A.6 Poisson distribution program result extract B. Probability that all channels are busy when call attempted

	Traffic in erlangs (grade of service)				
Channel	5.500	6.000	6.500	7.000	7.500
1	0.995 913	0.997 521	0.998 497	0.999 088	0.999 447
2	0.973 436	0.982 649	0.988 724	0.992 705	0.995 299
3	0.911 623	0.938 031	0.956 964	0.970 364	0.979 743
4	0.798 300	0.848 796	0.888 150	0.918 234	0.940 854
5	0.642 481	0.714 942	0.776 327	0.827 008	0.867 937
6	0.471 079	0.554 319	0.630 957	0.699 290	0.758 562
7	0.313 961	0.393 695	0.473 474	0.550 287	0.621 843
8	0.190 511	0.256 017	0.327 239	0.401 283	0.475 358
9	0.105 640	0.152 759	0.208 423	0.270 905	0.338 029
10	0.053 774	0.083 920	0.122 611	0.169 500	0.223 587
11	0.025 247	0.042 617	0.066 834	0.098 516	0.137 756
12	0.010 984	0.020 088	0.033 874	0.053 345	0.079 235
13	0.004 447	0.008 824	0.016 021	0.026 995	0.042 660
14	0.001 681	0.003 625	0.007 095	0.012 807	0.021 558
15	0.000 594	0.001 396	0.002 950	0.005 712	0.010 254
16	0.000 196	0.000 505	0.001 154	0.002 402	0.004 602
17	0.000 059	0.000 171	0.000 425	0.000 953	0.001 953
18	0.000 015	0.000 053	0.000 146	0.000 357	0.000 784
19	0.000 001	0.000 014	0.000 045	0.000 125	0.000 297
20	0.000 001	0.000 001	0.000 011	0.000 039	0.000 104

when the Erlang formula was used. If the Poisson formula is used a 0.01 grade of service based upon 10 erlangs of traffic requires between 18 and 19 channels. Rounding to the next highest channel results in the Poisson formula providing the same value as computed previously with the Erlang formula.

In general, the Poisson formula produces a more conservative sizing at lower traffic intensities than the Erlang formula. At higher traffic intensities the results are reversed. The selection of the appropriate formula depends upon how one visualizes the calling pattern of users of the communications network.

Economic constraints

In the previous dimensioning exercises the number of trunks or channels selected was based upon a defined level of grade of service. Although we want to size equipment to have a high efficiency and keep network users happy, we must also consider the economics of dimensioning. One method that can be used for economic analysis is the assignment of a dollar value to each erlang-hour of traffic.

For a company such as a time-sharing service bureau, the assignment of a dollar value to each erlang-hour of traffic may be a simple matter. Here the average revenue per one hour time-sharing session could be computed and used as the dollar value assigned to each erlang-hour of traffic. For other organizations, the average hourly usage of employees waiting service could be employed.

As an example of the economics involved in sizing, let us assume lost calls are held, resulting in traffic following a Poisson distribution and that 7 erlangs of traffic can be expected during the busy hour. Let us suppose we initially desire to offer a 0.001 grade of service. From the extract of the execution of the Poisson distribution program presented in Table A.6, between 16 and 17 channels would be required. Rounding to the highest number, 17 channels would be selected to provide the desired 0.001 grade of service.

Multiplexers normally consist of a base unit of a number of channels or ports and an expansion chassis into which dual port adapter cards are normally inserted to expand the capacity of the multiplexer. Many times one may desire to compare the potential revenue loss in comparison to expanding the multiplexer beyond a certain capacity. As an example of this consider the data in Table A.7 which indicates that when the traffic intensity is 5 erlangs a 14-channel multiplexer would provide an equivalent grade of service. This means that during the busy hour 2 erlang hours of traffic would be lost and the network designer could then compare the cost of three additional ports on the multiplexer and modems and dial-in lines if access to the multiplexer is over the switched network to the loss of revenue by not being able to service the busy hour traffic.

A.4 APPLYING THE EQUIPMENT SIZING PROCESS

Many methods are available for end-users to obtain data traffic statistics required for sizing communications equipment. Two of the most commonly used methods are based upon user surveys and computer accounting information.

Table A.7 Poisson distribution program result extract A. Probability that all channels are busy when call attempted

	Traffic in erlangs (grade of service)				
Channel	3.000	3.500	4.000	4.500	5.000
1	0.950 213	0.969 802	0.981 684	0.988 891	0.993 262
2	0.800 851	0.864 111	0.908 422	0.938 900	0.959 572
3	0.576 809	0.679 152	0.761 896	0.826 421	0.875 347
4	0.352 767	0.463 366	0.566 529	0.657 703	0.734 973
5	0.184 735	0.274 553	0.371 161	0.467 895	0.559 505
6	0.083 916	0.142 384	0.214 867	0.297 068	0.384 037
7	0.033 507	0.065 286	0.110 672	0.168 947	0.237 814
8	0.011 902	0.026 736	0.051 131	0.086 584	0.133 369
9	0.003 801	0.009 871	0.021 361	0.040 255	0.068 090
10	0.001 100	0.003 312	0.008 129	0.017 090	0.031 825
11	0.000 290	0.001 017	0.002 837	0.006 666	0.013 692
12	0.000 069	0.000 286	0.000 912	0.002 402	0.005 450
13	0.000 014	0.000 073	0.000 271	0.000 802	0.002 015
14	0.000 001	0.000 016	0.000 074	0.000 249	0.000 694
15	0.000 001	0.000 002	0.000 017	0.000 071	0.000 223
16	0.000 001	0.000 002	0.000 002	0.000 017	0.000 065
17	0.000 001	0.000 002	0.000 002	0.000 002	0.000 016

Normally, end-user surveys require each terminal user to estimate the number of originated calls to the computer for average and peak traffic situations as well as the call duration in minutes or fractions of an hour, on a daily basis. By accumulating the terminal traffic data for a group of terminals in a particular geographical area one can then obtain the traffic that the multiplexer will be required to support.

Suppose a new application is under consideration at a geographical area currently not served by a firm's data communications network. For this application, 10 terminals with the anticipated data traffic denoted in Figure A.5 are to be installed at five small offices in the greater metropolitan area of a city. If each terminal will dial a centrally located multiplexer, how many dial-in lines, auto-answer modems, and multiplexer ports are required to provide users with a 98% probability of accessing the computer upon dialing the multiplexer? What would happen if a 90% probability of access was acceptable?

For the 10 terminals listed in Figure A.5 the average daily and peak daily traffic is easily computed. These figures can be obtained by multiplying the number of calls originated each day by the call duration and summing the values for the appropriate average and peak periods. Doing so, one obtains 480 minutes of average daily traffic and 1200 minutes of peak traffic. Dividing those numbers by 60 results in 8 erlangs average daily traffic and 20 erlangs of peak daily traffic.

Prior to sizing, some additional knowledge and assumptions concerning the terminal traffic will be necessary. First, from the data contained in most survey forms, information containing busy hour traffic is non-existent although such information is critical for equipment sizing. Although survey forms can be tailored to obtain the number of calls and call duration by specific time intervals, for most users the completion of such precise estimates is a guess at best.

Terminal	Calls originated per day Average	Peak	Call duration (minutes) Average	Peak
A	3	6	15	30
B	2	3	30	60
C	5	5	10	15
D	2	3	15	15
E	2	4	15	30
F	2	4	15	30
G	3	3	15	35
H	4	6	30	30
I	2	3	20	25
J	2	2	15	60

Figure A.5 Terminal traffic survey

Normally, busy hour traffic can be estimated fairly accurately from historical data or from computer billing and accounting tape data. Suppose that either source shows a busy hour traffic equal to twice the average daily traffic based upon an 8-hour normal operational shift. Then the traffic would be (8/8) x 2 or 2 erlangs while the busy hour peak traffic would be (20/8) x 2 or 5 erlangs.

The next process in the sizing procedure is to determine the appropriate sizing formula to apply to the problem. If we assume that users encountering a busy signal will tend to redial the telephone numbers associated with the multiplexer, the Poisson formula will be applicable. From Table A.8 the 2.0 erlang traffic column shows a 0.0165 probability (1.65%) of all channels busy for a device containing six channels, 0.0527 for five channels and 0.1428 for four channels, thus, to obtain a 98% probability of access based upon the daily average traffic would require six channels while a 90% probability of access would require five channels.

If we want to size the equipment based upon the daily peak traffic load, how would sizing differ? We would now use the 5 erlang traffic column of Table A.6 from the table, 11 channels would provide a 0.0137 probability (1.37%) of

Table A.8 Poisson distribution program result extract C. Probability that all channels are busy when call attempted

Channel	Traffic in erlangs (grade of service) 0.500	1.000	1.500	2.000	2.500
1	0.393 469	0.632 120	0.776 870	0.864 665	0.917 915
2	0.090 204	0.264 241	0.442 174	0.593 994	0.712 702
3	0.014 387	0.080 301	0.191 152	0.323 323	0.456 186
4	0.001 751	0.018 988	0.065 642	0.142 875	0.242 422
5	0.000 172	0.003 659	0.018 575	0.052 652	0.108 820
6	0.000 014	0.000 594	0.004 455	0.016 562	0.042 019
7	0.000 001	0.000 083	0.000 925	0.004 532	0.014 185
8	0.000 000	0.000 010	0.000 169	0.001 095	0.004 245
9	0.000 000	0.000 000	0.000 027	0.000 236	0.001 138
10	0.000 000	0.000 000	0.000 003	0.000 045	0.000 275
11	0.000 000	0.000 000	0.000 000	0.000 007	0.000 059
12	0.000 000	0.000 000	0.000 000	0.000 000	0.000 010

Probability of access (%)	Daily average	Traffic peak
90	5	9
98	6	10

Figure A.6 Channel requirements

encountering a busy signal while 10 channels would provide a 0.0318 probability. To obtain a 98% probability of access statistically would require 11 channels. Since there are only 10 terminals, logic would override statistics and 10 channels or one channel per terminal would suffice. It should be noted that the statistical approach is based upon a level of traffic which can be generated from an infinite number of terminals. Thus, one must also use logic and recognize the limits of the statistical approach when sizing equipment. Since a 0.0681 probability of encountering a busy signal is associated with nine channels and a 0.1334 probability with eight channels, nine channels would be required to obtain a 90% probability of access.

In Figure A.6 the sizing required for average peak daily traffic is listed for both 90 and 98% probability of obtaining access. Note that the difference between supporting the average and peak traffic loads is four channels for both the 90 and 98% probability of access scenarios, even though peak traffic is $2\frac{1}{2}$ times average traffic.

The last process in the sizing procedure is to determine the number of channels and associated equipment to install. Whether to support the average or peak load will depend upon the critical nature of the application, funds availability, how often peak daily traffic can be expected, and perhaps organizational politics. If peak traffic only occurs once per month, we could normally size equipment for the average daily traffic expected. If peak traffic was expected to occur twice each day, we would normally size equipment based upon peak traffic. Traffic between these extremes may require the final step in the sizing procedure to one of human judgement, incorporating knowledge of economics, and the application into the decision process.

APPENDIX B

ERLANG DISTRIBUTION PROGRAM

```
5 FOR Z=1 TO 5
6 PRINT
7 NEXT Z
10 REM A IS THE OFFERED LOAD IN ERLANGS
20 REM S IS THE NUMBER OF PORTS, DIAL IN LINES OR TRUNKS
30 DIM [[80], B[48,30]
35 C=0
40 FOR I=5 TO 400 STEP 5
45 C=C+1
50 A=I/10
60 A[C]=A
70 FOR S=1 TO 48
80 X=S
90 GOSUB 1000
100 N=(A**S)/F
110 D=1
120 FOR D1=1 TO S
130 X=D1
140 GOSUB 1000
150 D=D+(A**D1)/F
160 NEXT D1
170 B[S,C]=N/D
180 NEXT S
190 NEXT I
200 FOR I=1 TO 76 STEP 5
201 FOR Z=1 TO 5
202 PRINT
203 NEXT Z
```

```
204 PRINT "ERLANG B DISTRIBUTION"
205 PRINT & "PROBABILITY ALL CHANNELS BUSY WHEN CALL
    ATTEMPTED (GRADE OF SERVICE)"
206 PRINT "CHANNEL TRAFFIC IN ERLANGS"
210 PRINT USING 220;A[I],A[I+1],A[I+2],A[I+3],A[I+4]
220 IMAGE 5XDD.DDD
225 PRINT
230 FOR S=1 TO 48
235 IF B[S,I+4]<=1E-7 THEN 260
240 PRINT USING 250;S,B[S,I],B[S,I+1],B[S,I+2],B[S,I+3],B[S, I+4]
250 IMAGE DDD,5(2XD.DDDDDD)
260 NEXT S
270 NEXT I
800 STOP
990 REM SUBROUTINE TO COMPARE FACTORIAL S VALUES
1000 F=1
1005 IF X=0 THEN 1045
1010 FOR F1=X TO 1 STEP -1
1020 LET F=F*F1
1030 NEXT F1
1040 RETURN
1045 F=1
1050 RETURN
```

APPENDIX C

POISSON DISTRIBUTION PROGRAM

```
5 FOR Z= 1 TO 5
6 PRINT
7 NEXT Z
10 REM A IS THE OFFERED LOAD IN ERLANGS
20 REM S IS THE NUMBER OF PORTS, DIAL IN LINES OR TRUNKS
30 DIM [[80], B [48,80]
35 C=0
40 FOR I=5 TO 400 STEP 5
45 C=C+1
50 A=I/10
60 A[C]=A
65 K=0
70 FOR S=1 TO 47
75 K=K+1
80 X1=0
90 FOR X=0 TO S
100 GOSUB 1000
110 X1=X1+(A**X)/(F*2.71828**A)
120 NEXT X
130 B[K,C]=1-X1
140 B[K,C]=ABS9B[K,C]0
150 NEXT S
160 NEXT J
170 B[S,C]=N/D
180 NEXT S
190 NEXT I
200 FOR I=1 TO 76 STEP 5
201 FOR Z=1 TO 5
```

```
202 PRINT
203 NEXT Z
204 PRINT "POISSON DISTRIBUTION"
205 PRINT & "PROBABILITY ALL CHANNELS BUSY WHEN CALL
    ATTEMPTED (GRADE OF SERVICE)"
206 PRINT "CHANNEL TRAFFIC IN ERLANGS"
210 PRINT USING 220;A[I], A[I+1],A[I+2],A[I+3],A[I+4]
220 IMAGE 5XDD.DDD
225 PRINT
230 FOR S=1 TO 48
235 IF B[S,I+4] =1E-7 THEN 260
240 PRINT USING 250;S,B[S,I],B[S,I+1],B[S,I+2],B[S,I+3],B[S,I+4]
250 IMAGE DDD,5 (2XD.DDDDDD)
260 NEXT S
270 NEXT I
800 STOP
990 REM SUBROUTINE TO COMPARE FACTORIAL S VALUES
1000 F=1
1005 IF X=0 THEN 1045
1010 FOR F1=X TO 1 STEP-1
1020 LET F=F*F1
1030 NEXT F1
1040 RETURN
1045 F=1
1050 RETURN
```

APPENDIX D

MULTIDROP LINE ROUTING ANALYSIS

One of the most frequent problems encountered by organizations is to determine an economical route for the path of a multidrop circuit. This type of circuit is used to interconnect two or more locations that must be serviced by a common mainframe computer port. Although there are several commercial services that the reader can subscribe to as well as a free service offered by AT&T to obtain a routing analysis, in many situations this analysis can be conducted internally within the organization. Doing so not only saves time but may also eliminate some potential problems that can occur if one relies upon programs that do not consider whether the resulting number of drops on a circuit can support the data traffic while providing a desired level of performance.

In this appendix, the use of a simple algorithm that can be employed to minimize the routing distance and resulting cost of a multidrop circuit will be discussed. Since there is a finite limit to the number of drops a multidrop circuit can support, we will also investigate a method that will enable users to estimate the worst case and average terminal response times as the number of drops increase. Then if the response time exceeds the design goal of the organization, the networks manager can consider removing one or more drops and placing them on a different multidrop circuit.

The minimum-spanning-tree technique

When the total number of drops to be serviced does not exceed the capacity of the front-end processor software, the minimum-spanning technique can be used. This technique results in the most efficient routing of a multidrop line by the use of a tree architecture which is used to connect all nodes with as few branches as possible. When applied to a data communications network, the minimum-spanning-tree technique results in the selection of a multidrop line whose drops are interconnected by branches or line segments which minimize the total distance of the line connecting all drops. Since the distance of a circuit is normally proportional to its cost, this technique results in a multidrop line which is also cost optimized. To better understand the procedure used in applying the minimum-spanning-tree technique, let us examine an example of its use.

Figure D.1 illustrates the location of a mainframe computer with respect to four remote locations that require a data communications connection to the computer. Assuming that remote terminal usage requires a dedicated connection to the computer, such as busy travel agency offices might require to their corporate computer, an initial network configuration might require the direct connection of each location to the computer by separate leased lines. Figure D.2 illustrates this network approach.

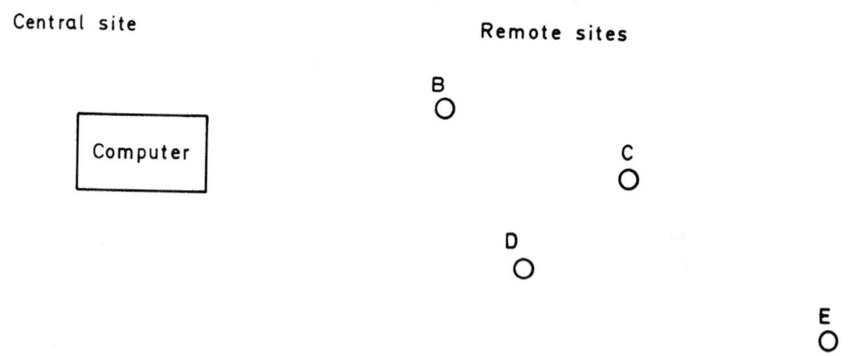

Figure D.1 Terminal locations to be services

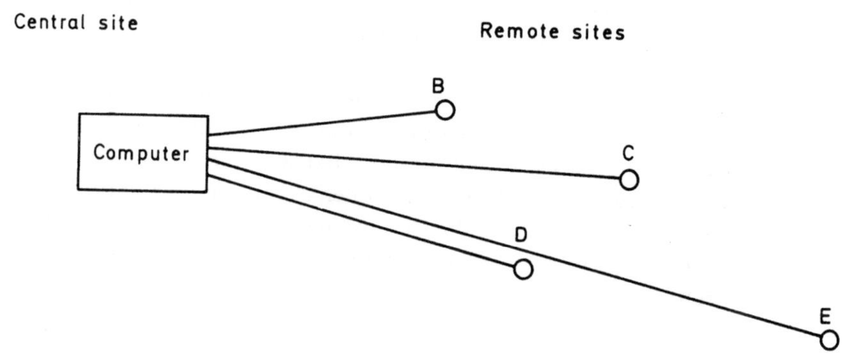

Figure D.2 Using separate leased lines

When separate leased lines are used to connect each terminal location to the mainframe, a portion of many line segments can be seen to run in parallel. Thus, from a visual perspective, it is apparent that the overall distance of one circuit linking all locations to the computer will be less than the total distance of individual circuits. Other factors that can reduce the cost of a composite multidrop circuit in comparison to separate leased lines include differences in the number of computer ports and modems required between the use of a multidrop line and individual leased lines.

A multidrop line requires the use of one computer port with a common modem servicing each of the drops connected to the port. Thus, a total of $n+1$ modems are required to service n drops on a multidrop circuit. In comparison, separate point-to-point leased lines would require one computer port per line as well as $n \times 2$ modems, where n equals the number of required point-to-point lines.

The minimum-spanning-tree algorithm

In the minimum-spanning-tree algorithm, the farthest point from the computer is first selected. In our example illustrated in Figure D.1 this would be location E. Next, the cost for connecting location E to each of the remaining locations is calculated. Since the cost of a line segment normally corresponds to the length of a line, users can simply consider the distance from location E to each of the remaining locations.

When a computer program is used to perform this operation, the user is normally required to enter the area code and telephone exchange of each location. The program uses this information to assign what is known as a (VH) coordinate pair to each entered location, where V and H refer to a vertical and horizontal coordinate system AT&T developed which subdivides the United States into grid squares. Then the computer uses the following equation to calculate the distance between pairs of (V,H) coordinate locations.

$$D = \text{Int}\left(\frac{[(V_1 - V_2)^2 + (H_1 - H_2)^2]^{1/2}}{10} + 0.5\right)$$

where D is the distance between locations, V_1 the V coordinate of first location, V_2 the V coordinate of second location, H_1 the H coordinate of first location, and H_2 the H coordinate of second location. In this equation, 0.5 is added to the result and the integer taken since the telephone company is permitted by tariff to round the answer to the next higher mile in performing its cost calculations.

Returning to the example in Figure D.1, location C is closest to location E, thus, these two locations would be connected to one another. Next, the distance from location C to all other points is computed and the closest point to location C is selected.

If the selected point was previously connected to the line, the line has looped back upon itself and the line segment will be defined as a cluster location to which more than one line segment can home upon. If the line did not loop back upon itself, as is the case in the example we are using, location C is linked to location B and B then becomes the next point of origination for calculating the distance to all other points. This procedure continues until all locations in the network are connected, with any resulting line clusters merged together by calculating the distance between the homing point of one cluster to the homing point of any other cluster using the minimum-distance path to connect clusters together. Figure D.3 illustrates the application of the minimum-spanning-tree algorithm to the four network locations previously illustrated in Figure D.1.

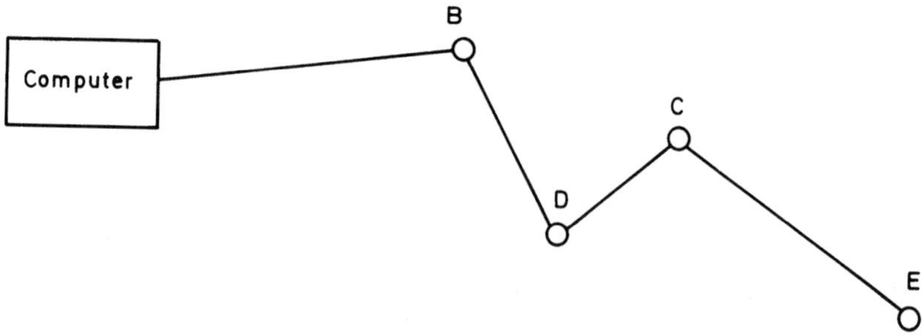

Figure D.3 Applying the minimum spanning-tree algorithm

Minimum-spanning-tree problems

The minimum spanning-tree algorithm, while economically accurate, does not consider two key variables which could make its implementation impractical—the terminal response time of the locations interconnected and the capacity of the front-end processor to service the total number of locations connected on one multidrop line.

Terminal response times

Normally, full-screen display terminals are used on a multidrop circuit. The terminal response time is defined as the time from the operator pressing the enter key to the first character appearing on the terminal's screen in response to the data sent to the computer. This response time depends upon a large set of factors, of which the major ones are listed in Table D.1.

The line speed refers to the transmission data rate which determines how fast data can be transported between the terminal and the computer once the terminal is polled or selected. The type of transmission line, full- or half-duplex, determines whether or not an extra delay will be incurred to turn the line around after each poll. If a half-duplex transmission protocol is used, then the modem turnaround time will effect the terminal response time.

The number of characters serviced per poll refers to how the communications software services a terminal capable of storing 1920 characters on a full screen of 25 lines by 80 characters per line. To prevent one terminal from hogging the line, most

Table D.1 Terminal response time factors

Line speed
Type of transmission line
Modem turnaround time
Number of characters serviced per polling used
Computer processing time
Polling service time

communications software divide the screen into segments and services a portion of the screen during each poll sequence.

This type of polling can occur 'round robin' where each terminal receives servicing in a defined order or it can occur based upon a predefined priority. Although the computer processing time can greatly affect the terminal response time, it is normally beyond the control of the communications staff. The polling service time is the time it takes to poll a terminal so the communications software can service another segment of the screen when the data to be read or written to the terminal exceeds one segment. Finally, the probability of a transmission error occurring will affect the probability of transmitting the same data again, since detected errors are corrected by the retransmission of data.

Probability of transmission errors

To estimate the average terminal response time requires an estimate of the average number of users that are using the terminals on a multidrop circuit. Next, the average number of characters to be transmitted in response to each depression of the enter key must be estimated. These data can then be used to estimate the average terminal response time.

Suppose there are 10 terminals on the multidrop circuit and at any one time four are active, with approximately 10 lines of 30 characters on the display when the enter key is pressed. If the communications software services segments of 240 characters, two polls will be required to service each terminal. Assuming a transmission rate of 4.8 kbps, which is equivalent to approximately 600 characters per second, it requires a minimum of 240/600 or 0.4 s to service the first segment on each terminal, excluding communications protocol overhead. Using a 25% overhead factor which is normally reasonable, the time to service the first segment becomes 0.5 s, resulting in the last terminal having its first segment serviced at a time of 2.0 s if all users requested servicing at the same time. Since 60 characters remain on each screen, the second poll requires 60/600 or 0.1 s per terminal plus 25% for overhead, or a total of 0.5 s until the fourth terminal is again serviced. Adding the time required to service each segment results in a total time of 2.5 s transpiring in the completion of the data transfer from the fourth terminal to the computer.

Assuming the average response is 300 characters, the transmission of two screen segments is also required in the opposite direction. The first segment would then require 2.0 s for displaying on each terminal while the second segment requires 0.3 s plus 25% overhead or 0.375 s until the first character starts to appear on the fourth terminal, for a total response time (inbound and outbound) of 2.5 plus 2.375 or 4.875 s.

If a 'round robin' polling sequence is used, the computer has an equal probability of polling any of the four terminals when the enter key is pressed. Thus, the computed 4.875 s response time is the worst case response time. The best case response time would be the response time required to service the first terminal sending data, which in the previous example would be 2.1 s inbound and 2.0 s outbound until the first character is received, or a total of 4.1 s. Thus, the average terminal response time would be $(4.1 + 4.875)/2$ or approximately 4.5 s.

Front-end processing limitations

A second limitation concerning the use of multidrop circuits is the capability of the front-end processor. In a large network that contains numerous terminal locations, the polling addressing capability of the front-end processor will limit the number of drops that can be serviced. Even if the processor could handle an infinite number of drops, polling delay times as well as the effect of a line segment impairment breaking access to many drops usually precludes most circuits to a maximum of 16 or 32 drops.

Large network design

When the number of drops in a network requires the use of multiple multidrop circuits, the network designer will normally consider the use of a more complex algorithm, such as the well known Esau and Williams formula. While such algorithms are best applied to network design problems by the use of computer programs, the reader can consider a practical alternative to these complex algorithms. This alternative is the subdivision of the network's terminal locations into servicing areas, based upon defining a servicing area to include a number of terminals that will permit an average response time that is acceptable to the end-user. Then each segment can be analyzed using the minimum-spanning-tree algorithm to develop a minimum cost multidrop line to service all terminals within the servicing area.

APPENDIX E

CSMA/CD NETWORK PERFORMANCE

Ethernet is a carrier sense multiple access with collision detection (CSMA/CD) network. Each station on the network listens for a carrier and attempts to transmit data when it senses the absence of that signal. Unfortunately, two stations may attempt to simultaneously transmit data, resulting in the occurrence of a collision. Even when one station thinks there is no carrier, it is quite possible that a carrier signal is propagating down the transmission path. Thus, a station transmitting data when its sampling of the line indicates the absence of a carrier may also result in a collision.

Because of the random nature of collisions, Ethernet bus performance is not deterministic and performance characteristics and message transmission delays are not predictable. However, over a period of time you can determine average and peak utilization, data elements which you may use to split one Ethernet LAN into two or more LANs via the use of bridges to increase individual network performance. In this appendix we will focus our attention upon determining the rate at which various length frames can flow on an Ethernet network. This information can be used by network managers and LAN administrators in conjunction with network monitoring tools to determine if a network should be segmented via the use of one or more bridges, upgraded to a higher operating rate network, or if LAN switches should be used to enhance network performance.

Determining the network frame rate

The top portion of Figure E.1 illustrates the IEEE 802.3 (Ethernet) frame format. In this illustration the seven-byte preamble field and the one-byte start of frame

Preamble	Destination Address	Source Address	Length or Type	Data	Frame Check Sequence
8	6	6	2	46≤ n≤ 1500	4

Frame size (bytes)

Field	Minimum size frame	Maximum size frame
Preamble	8	8
Destination address	6	6
Source address	6	6
Length or type	2	2
Data	46	1500
Frame check sequence	4	4
Total size	72	1526

Figure E.1 IEEE 802.3 (Ethernet) frame format

delimiter field were combined into a common eight-byte preamble field for simplicity. In actuality, the preamble field used by Ethernet is an eight-byte sequence of alternating 1's and 0's, while the IEEE 802.3 frame format uses a seven-byte preamble field of alternating 1's and 0's. The start of frame delimiter one-byte field used in IEEE 802.3 frames follows a seven-byte preamble field and the sequence of alternating 1's and 0's but ends with two set bits instead of the 1 and 0 used in the Ethernet preamble field. Since computations required to estimate Ethernet network performance are based upon frame length and not frame composition, the use of a common eight-byte preamble field, while not technically correct on a frame composition basis, does not affect our computations. As indicated by the tabulation of frame field lengths in the lower portion of Figure E.1, the frame size can vary from a minimum of 72 to a maximum of 1526 bytes.

Under Ethernet and 802.3 standards there is a dead time of 9.6 μs between frames. Using the frame size and dead time between frames you can compute the maximum number of frames per second that can flow on an Ethernet network. For our example, let us assume we have a 10 Mbps LAN, such as a 10BASE−2, 10BASE−5, or 10BASE−T network. Here the bit time then becomes $1/10^7$ seconds or 100 ns.

Now let us assume all frames are at the maximum length of 1526 bytes. Then, the time per frame becomes:

$$9.6\mu\text{s} + \left(1526 \text{ bytes} \times \frac{8 \text{ bits}}{\text{byte}} \times \frac{100 \text{ ns}}{\text{bit}}\right) = 1.23 \text{ ms}$$

Since one 1526-byte frame requires 1.23 ms then in one second (1000 ms) there can be 1000/1.23 or approximately 812 maximum-sized frames. Thus, the maximum transmission rate on an Ethernet network is 812 frames per second when information is transferred in 1500-byte units within a sequence of frames. One example of a situation in which data would be transferred in 1500-byte units is when a workstation downloads a file from a server or transfers a file to another workstation or to a server. When this type of data transfer occurs, the data field of a large number of sequential frames would be filled to their maximum size of 1500 bytes. If the last portion of the file being transferred is less than 1500 bytes, then the data field of the last frame used to transport the file would be less than 1500 bytes in length.

Now that the maximum number of frames that can traverse an Ethernet network when the data field is at its maximum size has been determined, let's compute the frame rate when the data field is at its minimum length. When that occurs the data field contains up to 46 characters of information, since PAD characters are required to fill the data field to a minimum length of 46 characters. This results in a minimum-size Ethernet frame being 72 characters in length.

For a minimum frame length of 72 bytes, the time per frame is:

$$9.6\,\mu\text{s} + \left(72 \text{ bytes} \times \frac{8 \text{ byte}}{\text{byte}} \times \frac{100\,\text{ns}}{\text{bit}}\right) = 67.2 \times 10^{-6} \text{ second}$$

Thus, in one second there can be a maximum of $1/67.2 \times 10^{-6}$, or 14 880 minimum-size 72-byte frames.

Table E.1 summarizes the frame processing requirements for a 10- and 100-Mbps Ethernet network under 50% and 100% load conditions based upon minimum and maximum frame sizes. Note that those frame processing requirements define the frame examination (filtering) operating rate of a bridge connected to an Ethernet. That rate indicates the number of frames per second a bridge connected to a 10- or 100-Mbps Ethernet local area network must be capable of examining under heavy (50% load) and full (100% load) traffic conditions. Those frame processing requirements also define the maximum frame forwarding rate for a bridge or router connected to a single network, since if all frames were routed off the network the forwarding rate would equal the filtering rate.

Table E1 Ethernet frame processing requirements (frames per second)

Average frame size (bytes)	Frames per second			
	10 Mbps		100 Mbps	
	50% load	100% load	50% load	100% load
1526	406	812	4060	8120
72	7440	14 880	74 400	148 800

In examining the entries in Table E.1, note that the frame processing capability of Fast Ethernet that operates at 100 Mbps is 10 times that of 10BASE-T. This results from the fact that Fast Ethernet uses the same access protocol and maintains a delay between frames in the form of idle characters one-tenth of the delay of

10BASE-T. However, emerging gigabit Ethernet does not provide a 10-fold processing increase above Fast Ethernet, as the former uses several techniques that alter the minimum length of frames and transmits a sequence of frames from a workstation (referred to as packet bursting) to extend its transmission distance. Thus, there is no direct correlation between the frame processing capability of gigabit Ethernet and Fast Ethernet.

INDEX